Oligodendroglioma

Clinical Presentation, Pathology, Molecular Biology, Imaging, and Treatment

Oligodendroglioma
Clinical Presentation, Pathology, Molecular Biology, Imaging, and Treatment

Edited by

Nina A. Paleologos
Director of Neuro-Oncology and
Medical Director of Neurology,
Advocate Medical Group, Advocate Healthcare; Professor (Adjunct) of Neurological Sciences,
Rush University Medical School, Chicago, IL, United States

Herbert B. Newton
Director, Neuro-Oncology Center
Medical Director, CNS Oncology Program
Advent Health Cancer Institute
Advent Health Medical Group, Orlando, FL, United States

Professor of Neurology & Neurosurgery (Retired), Division of Neuro-Oncology,
Esther Dardinger Endowed Chair in Neuro-Oncology, James Cancer Hospital & Solove Research Institute,
Wexner Medical Center at the Ohio State University, Columbus, OH, United States

ACADEMIC PRESS
An imprint of Elsevier

ELSEVIER

Academic Press is an imprint of Elsevier
125 London Wall, London EC2Y 5AS, United Kingdom
525 B Street, Suite 1650, San Diego, CA 92101, United States
50 Hampshire Street, 5th Floor, Cambridge, MA 02139, United States
The Boulevard, Langford Lane, Kidlington, Oxford OX5 1GB, United Kingdom

Notices
Knowledge and best practice in this field are constantly changing. As new research and experience broaden our understanding, changes in research methods, professional practices, or medical treatment may become necessary.

Practitioners and researchers must always rely on their own experience and knowledge in evaluating and using any information, methods, compounds, or experiments described herein. In using such information or methods they should be mindful of their own safety and the safety of others, including parties for whom they have a professional responsibility.

To the fullest extent of the law, neither the Publisher nor the authors, contributors, or editors, assume any liability for any injury and/or damage to persons or property as a matter of products liability, negligence or otherwise, or from any use or operation of any methods, products, instructions, or ideas contained in the material herein.

Library of Congress Cataloging-in-Publication Data
A catalog record for this book is available from the Library of Congress

British Library Cataloguing-in-Publication Data
A catalogue record for this book is available from the British Library

ISBN 978-0-12-813158-9

For information on all Academic Press publications
visit our website at https://www.elsevier.com/books-and-journals

Publisher: Nikki Levy
Acquisition Editor: Natalie Farra
Editorial Project Manager: Kristi Anderson
Production Project Manager: Paul Prasad Chandramohan
Cover Designer: Matthew Limbert

Cover images:
Upper middle: The photomicrograph shows a grade II oligodendroglioma, demonstrating classic low-grade histology with "fried egg" cells, used with permission from Dr. Seema Shroff, Advent Health Cancer Institute.

Lower left: Oligodendroglioma cells after in situ hybridization, demonstrating loss of chromosome 1p in the tumor cells, used with permission from Dr. Seema Shroff, Advent Health Cancer Institute.

Lower right: The MRI shows an oligodendroglioma in the left temporal region, showing up as a densely enhancing mass on the T1-GAD images, used with permission from Dr. Herbert B. Newton, Advent Health Cancer Institute.

Typeset by SPi Global, India

Dedication

I would like to thank my wife, Cindy, and all of the children in my wonderful mixed family—Alex and Ashley Newton and Sammi, Skylar, and Cameron Burrell—for their love, patience, and support while I was working on this book project.
I would also like to thank my Neuro-Oncology patients and their families for their amazing courage and strength in the face of adversity, which was a constant inspiration.

Herbert B. Newton

For Jeffrey and Nicole who light up my life, Stacy and Steve for their love and support, and for my patients and their families who are the bravest people I know

Nina A. Paleologos

Contents

Section A
Clinical presentation and quality of life

1. Clinical presentation in adults—Brain

Ugonma N. Chukwueke and David Reardon

2. Clinical presentation of spinal oligodendrogliomas

*Lily C. Pham, David Cachia,
Akash J. Patel and Jacob J. Mandel*

3. Oligodendrogliomas—Atypical clinical presentations

*P. Gage Gwyn, Sherif M. Makar and
Herbert B. Newton*

4. Clinical presentation of pediatric oligodendrogliomas

Scott L. Coven and Jonathan L. Finlay

5. Seizures, oligodendrogliomas, and brain-tumor-related epilepsy

Marta Maschio and Francesco Paladin

6. Living with oligodendroglioma

Ashlee R. Loughan, Deborah Hutchinson Allen and Sarah Ellen Braun

Section B
Pathology and molecular biology

7. Origin and development of oligodendroglioma

Josephine Volovetz, Defne Bayik and Justin D. Lathia

8. Histopathology and molecular biology of oligodendrogliomas

Josephine Volovetz, E. Yamamoto and R.A. Prayson

9. The role of biomarkers in the diagnosis and treatment of oligodendrogliomas

Laura E. Donovan and Andrew B. Lassman

10. Prospects of translational proteomics and protein microarrays in oligodendroglioma

Shabarni Gupta and Sanjeeva Srivastava

11. Pathology of pediatric oligodendroglioma

Hope T. Richard

Section C
Neuro-imaging

Section D
Surgical therapy

22. Standard external beam radiation therapy for oligodendroglioma

Gustavo Nader Marta, Fabio Y. Moraes, Erin S. Murphy and John H. Suh

23. Stereotactic radiosurgery in the management of oligodendroglioma

Fabio Y. Moraes, Gustavo Nader Marta, Erin S. Murphy and John H. Suh

24. Proton beam therapy for oligodendroglioma

Vonetta M. Williams, Simon S. Lo and Lia M. Halasz

25. Interstitial brachytherapy treatment for oligodendrogliomas

Ibrahim Abu-Gheida, Sarah Sittenfeld, Samuel T. Chao and John H. Suh

26. The role of radiation in pediatric oligodendrogliomas

Joelle P. Straehla and Karen J. Marcus

33. Pediatric oligodendroglioma

Lennox Byer, Cassie Kline-Nunnally,
Tarik Tihan and Sabine Mueller

34. Immune checkpoint blockade in glioma

Sherise D. Ferguson, Shiao-Pei Weathers
and Amy B. Heimberger

Contributors

Numbers in paraentheses indicate the pages on which the authors' contributions begin.

Ibrahim Abu-Gheida (287), Department of Radiation Oncology, Cleveland Clinic, Taussig Cancer Institute, Cleveland, OH, United States

Deborah Hutchinson Allen (55), Nursing Research & EBP, Duke University Health System & Duke Cancer Institute, Durham, NC, United States

John Anderson (157), University Hospitals, Cleveland Medical Center; Case Western Reserve University, Cleveland, OH, United States

Anthony Aquino (139), Department of Radiology, Ohio State Wexner Medical Center, Columbus, OH, United States

Defne Bayik (79), Cleveland Clinic Lerner College of Medicine, Lerner Research Institute, Cleveland, OH, United States

Eric C. Bourekas (139), Department of Radiology, Ohio State Wexner Medical Center, Columbus, OH, United States

Sarah Ellen Braun (55), Department of Psychology, Virginia Commonwealth University, Richmond, VA, United States

Jan C. Buckner (331), Department of Medical Oncology, Rochester, MN, United States

Lennox Byer (379), University of California, San Francisco (UCSF), School of Medicine, San Francisco, CA, United States

David Cachia (11), Department of Neurosurgery, Medical University of South Carolina, Charleston, SC, United States

Candice Carpenter (219), Department of Neurological Surgery, The Ohio State University Wexner Medical Center, The James Cancer Hospital and Solove Research Institute, Columbus, OH, United States

Marc C. Chamberlain (367), Seattle Genetics, Seattle, WA, United States

Samuel T. Chao (287), Department of Radiation Oncology, Cleveland Clinic, Taussig Cancer Institute, Cleveland, OH, United States

Ugonma N. Chukwueke (3), Center for Neuro-Oncology, Dana-Farber Cancer Institute; Department of Neurology, Harvard Medical School, Boston, MA, United States

Scott L. Coven (39), Department of Pediatrics, Division of Hematology/Oncology/BMT, Nationwide Children's Hospital and The Ohio State University, Columbus, OH, United States

Laura E. Donovan (109), Neuro-Oncology Division, Department of Neurology, Columbia University Irving Medical Center, New York, NY, United States

J. Bradley Elder (219), Department of Neurological Surgery, The Ohio State University Wexner Medical Center, The James Cancer Hospital and Solove Research Institute, Columbus, OH, United States

Sherise D. Ferguson (387), Department of Neurosurgery, The University of Texas MD Anderson Cancer Center, Houston TX, United States

Melvin Field (183,191), Orlando Neurosurgery, University of Central Florida College of Medicine, Orlando, FL, United States

Jonathan L. Finlay (39), Department of Pediatrics, Division of Hematology/Oncology/BMT, Nationwide Children's Hospital and The Ohio State University, Columbus, OH, United States

Liliana C. Goumnerova (229), Department of Neurosurgery, Boston Children's Hospital, Boston, MA, United States

Nandita Guha-Thakurta (147), Department of Diagnostic Radiology, Division of Diagnostic Imaging, The University of Texas MD Anderson Cancer Center, Houston, TX, United States

Shabarni Gupta (117), Department of Biosciences and Bioengineering, Indian Institute of Technology Bombay, Mumbai, India

P. Gage Gwyn (23), Neuro-Oncology Center, AdventHealth Cancer Institute, AdventHealth Medical Group, Orlando, FL, United States

Lia M. Halasz (279), Department of Radiation Oncology and Neurological Surgery, University of Washington School of Medicine, Seattle, WA, United States

Amy B. Heimberger (387), Department of Neurosurgery, The University of Texas MD Anderson Cancer Center, Houston TX, United States

John W. Henson (147), Ivy Center for Advanced Brain Tumor Treatment, Seattle, WA, United States

Arash Kardan (157), University Hospitals, Cleveland Medical Center; Case Western Reserve University, Cleveland, OH, United States

Cassie Kline-Nunnally (379), UCSF, Departments of Pediatrics and Neurology, Division of Hematology/Oncology, San Francisco, CA, United States

Friedrich-Wilhelm Kreth (209), Department of Neurosurgery, Hospital of the University of Munich, Campus Grosshadern, Munich, Germany

Michael A. Lach (173), Department of Diagnostic Radiology, Cleveland Clinic, Cleveland, OH, United States

Andrew B. Lassman (109), Neuro-Oncology Division, Department of Neurology, Herbert Irving Comprehensive Cancer Center, Columbia University Irving Medical Center, New York, NY, United States

Justin D. Lathia (79), Cleveland Clinic Lerner College of Medicine, Lerner Research Institute, Cleveland, OH, United States

Christine K. Lee (229), Department of Neurosurgery, Massachusetts General Hospital, Boston, MA, United States

Simon S. Lo (279), Department of Radiation Oncology and Neurological Surgery, University of Washington School of Medicine, Seattle, WA, United States

Ashlee R. Loughan (55), Department of Neurology, Virginia Commonwealth University & Massey Cancer Center, Richmond, VA, United States

Ihsan Mamoun (173), Department of Pediatric and Neuroimaging, Cleveland Clinic, Cleveland, OH, United States

Sherif M. Makar (23), Neuro-Oncology Center, AdventHealth Cancer Institute, AdventHealth Medical Group, Orlando, FL, United States

Jacob J. Mandel (11), Department of Neurology and Neurosurgery, Baylor College of Medicine, Houston, TX, United States

Karen J. Marcus (299), Department of Radiation Oncology, Dana-Farber/Brigham and Women's Hospital; Dana Farber/Boston Children's Cancer and Blood Disorders Center, Boston, MA, United States

Gustavo Nader Marta (263,271), Department of Radiation Oncology, Hospital Sírio-Libanês; Department of Radiology and Oncology, Division of Radiation Oncology, Instituto do Câncer do Estado de São Paulo (ICESP), Faculdade de Medicina da Universidade de São Paulo, Sao Paulo, Brazil

Marta Maschio (43), Center for Tumor-Related Epilepsy, Neuro-Oncology Unit, IRCCS - Regina Elena National Cancer Institute, Rome, Italy

Minesh P. Mehta (245), Department of Radiation Oncology, Florida International University; Miami Cancer Institute, Miami, FL, United States

Julie J. Miller (359), Department of Neurology, Pappas Center for Neuro-Oncology, Massachusetts General Hospital, Boston, MA, United States

Ahmed Mohyeldin (219), Department of Neurological Surgery, The Ohio State University Wexner Medical Center, The James Cancer Hospital and Solove Research Institute, Columbus, OH, United States

Fabio Y. Moraes (263,271), Department of Radiation Oncology, Hospital Sírio-Libanês, Sao Paulo, Brazil; Radiation Medicine Program, Princess Margaret Cancer Centre, University Health Network, Toronto; Division of Radiation Oncology, Department of Oncology, Kingston Health Sciences Centre, Queen's University, Kingston, ON, Canada

Sabine Mueller (379), UCSF, Departments of Neurology, Neurosurgery, Pediatrics, San Francisco, CA, United States

Erin S. Murphy (263,271), Department of Radiation Oncology, Cleveland Clinic, Taussig Cancer Institute, Cleveland, OH, United States

William C. Newman (183,191), Department of Neurological Surgery, University of Pittsburgh Medical Center, Pittsburgh, PA, United States

Herbert B. Newton (23,341,353), Neuro-Oncology Center; CNS Oncology Program, AdventHealth Cancer Institute, AdventHealth Medical Group, Orlando, FL; Division of Neuro-Oncology, Esther Dardinger Endowed Chair in Neuro-Oncology James Cancer Hospital & Solove Research Institute, Wexner Medical Center at the Ohio State University, Columbus, OH, United States

Francesco Paladin (43), Department of Neurology, Epilepsy Center, SS Giovanni and Paolo Hospital, Campo SS Giovanni e Paolo, Venice, Italy

Nina A. Paleologos (341,353), Advocate Medical Group, Advocate Healthcare; Rush University Medical School, Chicago, IL, United States

Akash J. Patel (11), Department of Neurology and Neurosurgery, Baylor College of Medicine, Houston, TX, United States

Katarina Petras (245), Department of Radiation Oncology, Robert H. Lurie Comprehensive Cancer Center, Northwestern University, Feinberg School of Medicine, Chicago, IL, United States

Lily C. Pham (11), Department of Neurology and Neurosurgery, Baylor College of Medicine, Houston, TX, United States

Kester A. Phillips (309), Inova Medical Group Hematology Oncology, Inova Fairfax Hospital, Falls Church, VA, United States

Raymond Poelstra (157), Charles F. Kettering Memorial Hospital, Dayton, OH, United States

R.A. Prayson (89), Department of Anatomic Pathology, Cleveland Clinic Lerner College of Medicine, Cleveland Clinic, Cleveland, OH, United States

David Reardon (3), Center for Neuro-Oncology, Dana-Farber Cancer Institute; Department of Medicine, Harvard Medical School, Boston, MA, United States

Marilyn Reed (157), Charles F. Kettering Memorial Hospital, Dayton, OH, United States

Tyler Richards (157), University Hospitals, Cleveland Medical Center; Case Western Reserve University, Cleveland, OH, United States

Hope T. Richard (129), Department of Pathology, Virginia Commonwealth University Health, Richmond, VA, United States

Michael W. Ruff (331), Department of Neurology, Department of Medical Oncology, Rochester, MN, United States

Sean Sachdev (245), Department of Radiation Oncology, Lou and Jean Malnati Brain Tumor Institute, Robert H. Lurie Comprehensive Cancer Center, Northwestern University, Feinberg School of Medicine, Chicago, IL, United States

Martin Satter (157), Charles F. Kettering Memorial Hospital; Wright State University, Dayton, OH, United States

David Schiff (309), Neuro-Oncology Center, University of Virginia Health System, Charlottesville, VA, United States

Samir Sejpal (245), University of Central Florida, Florida Hospital Cancer Institute, Orlando, FL, United States

Sarah Sittenfeld (287), Department of Radiation Oncology, Cleveland Clinic, Taussig Cancer Institute, Cleveland, OH, United States

H. Wayne Slone (139), Department of Radiology, Ohio State Wexner Medical Center, Columbus, OH, United States

Sanjeeva Srivastava (117), Department of Biosciences and Bioengineering, Indian Institute of Technology Bombay, Mumbai, India

Joelle P. Straehla, Department of Pediatric Hematology/Oncology, Dana-Farber/Boston Children's Cancer and Blood Disorders Center, Boston, MA, United States

John H. Suh (263,271,287), Department of Radiation Oncology, Cleveland Clinic, Taussig Cancer Institute, Cleveland, OH, United States

Sophie Taillibert (367), Pitié-Salpétrière University Hospital, Paris VI UPMC University, APHP, Paris, France

Niklas Thon (209), Department of Neurosurgery, Hospital of the University of Munich, Campus Grosshadern, Munich, Germany

Tarik Tihan (379), UCSF, Department of Pathology, San Francisco, CA, United States

Joerg-Christian Tonn (209), Department of Neurosurgery, Hospital of the University of Munich, Campus Grosshadern, Munich, Germany

Josephine Volovetz (79,89), Department of Anatomic Pathology, Cleveland Clinic Lerner College of Medicine, Lerner Research Institute, Cleveland Clinic, Cleveland, OH, United States

Joshua L. Wang (219), Department of Neurological Surgery, The Ohio State University Wexner Medical Center, The James Cancer Hospital and Solove Research Institute, Columbus, OH, United States

Shiao-Pei Weathers (387), Department of Neuro-Oncology, The University of Texas MD Anderson Cancer Center, Houston TX, United States

Patrick Y. Wen (359), Center for Neuro-Oncology, Dana-Farber Cancer Institute, Boston, MA, United States

Vonetta M. Williams (279), Department of Radiation Oncology, University of Washington School of Medicine, Seattle, WA, United States

E. Yamamoto (89), Department of Anatomic Pathology, Cleveland Clinic Lerner College of Medicine, Cleveland Clinic, Cleveland, OH, United States

Foreword

The modern era of neuro-oncology, which includes molecular analysis of tumor tissue for improved clinical diagnosis, prognostication, and treatment selection; molecular criteria for patient entry into clinical trials of promising investigational therapies for brain tumor; rigorous imaging-based criteria for the assessment of therapeutic responses in neuro-oncology; the recognition that gliomas with different molecular characteristics might be located in specific regions of the brain and display imaging features associated with their underlying molecular composition; and the incorporation of molecular testing into the official WHO classification of brain tumors, was ushered by the study of oligodendrogliomas, an uncommon brain cancer. A series of publications describing the often dramatic response of oligodendrogliomas to a chemotherapy regimen, called PCV [procarbazine, lomustine (CCNU), vincristine], and the subsequent observation that radiographic responses to PCV and long survival times could be predicted for patients with aggressive oligodendrogliomas based on a molecular signature involving chromosomes 1p and 19q (i.e., 1p/19q co-deletion), offered new hope to a field bereft of promising advances, and presented a new frame of reference for thinking about brain cancer and its treatment. Insights gleaned from the study of a relatively small number of carefully annotated patients introduced molecular medicine to neuro-oncology. Precisely, 30 years ago, I was in the right place at the right time with the right people, and fortunate to be able to contribute to this evolution in the understanding of cancers of the human brain.

Today, oligodendrogliomas are diagnosed with much greater precision than 30 years ago. Molecular characteristics (i.e., IDH1/2 mutation; 1p/19q co-deletion; TERT promoter mutation) complement a distinctive but nonuniform histological appearance to create cohorts of 'identical' cases. Not long ago, astrocytic mixed gliomas and small cell glioblastomas contaminated case series and clinical trials because they could be easily mistaken for oligodendrogliomas under the light microscope. Moreover, patients with oligodendrogliomas live much longer today because those at higher risk for early progression after surgery (i.e., anaplastic histopathology, incomplete surgical resection, or older age at diagnosis) are treated proactively and aggressively at initial presentation with radiotherapy and PCV chemotherapy (or increasingly the DNA-methylating agent, Temozolomide). But despite major progress in diagnosing and treating this type of brain cancer, the challenge of providing great care for patients with oligodendrogliomas is far from over, and much more research is needed. Today's approaches to treating these cancers have two serious shortcomings: (1) in spite of favorable molecular, histological, and other prognostic features, some patients with oligodendrogliomas relapse and die within a few years of surgery, radiotherapy, and chemotherapy, and (2) some patients who respond completely to these therapies slowly develop major cognitive, behavioral, and motor incapacities as a direct result of successful antitumor treatment. The solutions to these problems will not be easy to find, and are unlikely to emerge from the clinic, alone.

In my heyday in neuro-oncology, it was sufficient to have PCV at the ready and the opportunity to successfully treat a patient with a recurrent tumor that had textbook oligodendroglial histopathology in order to make the connection, and tease the problem apart. But future advances in the treatment of oligodendroglioma will likely require a much deeper understanding of the disease with new insights gleaned from the laboratory.

In this regard, significant progress in grasping the origins of oligodendroglioma has already occurred with the remarkable discovery 10 years ago that this cancer belongs to a family of brain tumors in which the earliest known, and likely initiating alteration, is a mutation in a Krebs cycle gene, known as isocitrate dehydrogenase (IDH). Mutations in either IDH1 or IDH2 lead to the excessive production of the 'onco' metabolite, 2-hydroxyglutarate (2-HG), which inhibits the function of enzymes that alter patterns of DNA methylation and the epigenome of the cell. In doing so, 2-HG production appears to enlarge the pool of undifferentiated cells in the maturing brain. Increasing the size of the brain's stem cell pool likely increases the risk of developing a cancer of the brain years later. How these progenitor cells change further to cause a slowly growing brain cancer with co-deletion of chromosomes 1p and 19q remains unknown, but promoter hyper-methylation with silencing of the DNA repair gene, MGMT, may explain the predictable sensitivity to DNA damaging chemotherapy that is typical of oligodendrogliomas. In addition, large sequencing projects are revealing other recurrent genomic alterations in oligodendrogliomas, including mutations of CIC on 19q, FUBP1 on 1p, and NOTCH1 on

9q that are providing us with a richer understanding of this cancer. These and other discoveries from the basic sciences offer great hope that oligodendrogliomas might be treated more effectively, more safely, and in an entirely new way a decade from now. Moreover, for oligodendrogliomas that affect younger adults and take years to develop the prospect of prevention or early diagnosis are other exciting ideas to contemplate.

This book is an important new resource in the brain tumor field and will remind us that analysis of unexpectedly good outcomes of patients with rare cancers can sometimes advance an entire field. I am indebted to Dr. Nina Paleologos and Dr. Herbert Newton for the opportunity to contribute a foreword to their textbook on oligodendrogliomas.

J. Gregory Cairncross

Clark H Smith Brain Tumour Centre and Charbonneau Cancer Institute, University of Calgary, Calgary, AB, Canada

Clinical presentation and quality of life

Chapter 1

Clinical presentation in adults—Brain

Ugonma N. Chukwueke*,† and David Reardon*,‡

*Center for Neuro-Oncology, Dana-Farber Cancer Institute, Boston, MA, United States; † Department of Neurology, Harvard Medical School, Boston, MA, United States; ‡ Department of Medicine, Harvard Medical School, Boston, MA, United States

Epidemiology

Oligodendroglioma and anaplastic oligodendrogliomas are rare primary brain tumors. Traditionally classified as World Health Organization (WHO) grade II and III tumors, among adults in the United States, oligodendroglioma and anaplastic oligodendroglioma account for 1% and 0.5% of brain and central nervous system (CNS) tumors annually.[1] Median age at the time of diagnosis of oligodendroglioma and anaplastic oligodendroglioma are 43 and 50 years, respectively (Fig. 1). Consistent with trends observed in malignant primary brain tumors, these tumors occur more commonly in men as compared to women. From the standpoint of race, the incidence of oligodendroglioma and anaplastic glioma is higher in non-Hispanic whites in comparison to other ethnic groups, although survival outcomes in blacks have been reported to be poorer relative to other races, even with similar treatment regimens.[1] To date, there have been few studies examining the impact of race on primary brain tumors, so the underpinnings of this disparity are poorly understood.[2]

In the United States, there are expected to be >1000 new cases of oligodendroglioma and >600 anaplastic oligodendroglioma over the next 2 years.[1] Additionally, it is expected that the incidence of primary brain tumors will continue to rise, notably in regions of higher economic development. Europe has the highest incidence of brain cancer with an annual age-standardize rate (ASR) of 5.5 per 100,000 persons. The lowest ASR is in sub-Saharan Africa at 0.8 per 100,000 persons. It is unclear whether the rising incidence and disparity between developed and developing nations are "true" or may be an observation effect, correlated with increasing availability of technology, allowing for timely evaluation and diagnosis. An additional consideration may be variability in data collection and surveillance.[3]

Despite investigation of multiple potential risk factors for the development of glioma, including oligodendroglial tumors specifically, more investigation is still warranted. Ionizing radiation has been studied frequently for its association with risk of intracranial malignancy. Longitudinal analyses of incidence of neoplasm in populations in Hiroshima and Nagasaki showed an elevated risk of glioma; however, without achieving statistical significance.[3] Similarly, work by Sadetzki et al. noted the incidence of glioma in individuals being treated with irradiation for tinea capitis has doubled, in comparison to population and sibling controls.[4] Additionally observed was a dose–response relationship.

In 2011, McCarthy et al. reviewed data pooled from seven case–control studies (five in the United States and two in Scandinavia) to identify possible risk factors for the development of oligodendroglial tumors (oligodendroglioma, anaplastic oligodendroglioma, and mixed glioma).[5] Among the factors identified included: history of asthma/allergy (decreased), family history (increased), and personal history of chicken pox (decreased). Ongoing work is required to further characterize whether these are reliable factors in reducing or enhancing the risk of tumor development, specifically, in oligodendroglial tumors.

This chapter will focus on the clinical presentation of oligodendroglioma in adults in the brain, with subsequent sections dedicated to discussion of these tumors in the pediatric population and other CNS structures.

Tumor classification

Until 2016, oligodendroglioma and anaplastic oligodendroglioma were defined exclusively by their histologic appearance. As they are derived from oligodendrocytes, the feature most consistent with a malignant process was the presence of a small, round nucleus surrounded by a halo of cytoplasm ("fried egg"), with calcifications and branching vessels. Anaplastic oligodendroglioma are further characterized by high cellularity and mitotic rate, with microvascular proliferation. The

Oligodendroglioma. https://doi.org/10.1016/B978-0-12-813158-9.00001-3

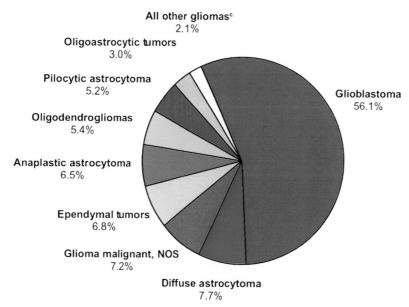

All other gliomas^c
2.1%

Oligoastrocytic tumors
3.0%

Pilocytic astrocytoma
5.2%

Oligodendrogliomas
5.4%

Anaplastic astrocytoma
6.5%

Ependymal tumors
6.8%

Glioma malignant, NOS
7.2%

Diffuse astrocytoma
7.7%

Glioblastoma
56.1%

(a) Percentages may not add up to 100% due to rounding; (b) ICD-O-3 codes = 9380-9384,9391-9460 (Table 2); (c) Includes histologies from unique astrocytoma variants, other neuroepithelial tumors, and neuronal and mixed neuronal-glial tumors (Table 2).

FIG. 1 Distribution of primary brain and other CNS gliomas by histology subtypes ($N = 100,619$), CBTRUS statistical report: NPCR and SEER, 2010–2014.

category of mixed oligoastrocytoma was previously included to capture tumors which were felt to morphologically harbor characteristics of both astrocytoma and oligodendroglioma.

In 2016, the WHO Classification of Brain Tumors was revised to reflect the emerging understanding and significance of the molecular characteristics of primary brain tumors. Inclusion of molecular-genetic alterations in the diagnosis of diffuse glial tumors is mandated by this scheme. In cases where testing is unavailable or incomplete, tumors may be assigned as "not otherwise specified" or NOS.[6] The historical paradigm of distinguishing gliomas solely on morphological features (mitotic rate, anaplasia, and vascular proliferation) has evolved into a system of classification based on both phenotypic and genotypic features. Additionally, the integration of histology and genetics allows for increased accuracy of diagnosis, as well as clarity in predicting the behavior of the tumors from a prognostic standpoint. For an integrated diagnosis of oligodendroglioma and anaplastic oligodendroglioma, in addition to the noted histological characteristics, co-occurrence of mutation in isocitrate dehydrogenase (IDH)-1 (cytoplasmic) or IDH-2 (mitochondrial) and deletion of the short arm of chromosome 1p and long arm of 19q (1p/19q co-deletion) are required. Over 90% of the IDH mutations result in a substitution of arginine (R) for histidine (H) at codon 132 or R132H.[7] Most WHO grade II and III tumors will harbor *IDH* mutations.[6]

While the prognostic difference between *IDH*-mutant and *IDH* wild-type gliomas has been established, whether this is also true in comparing histological grades II and III is under investigation.[8,9] In this revised system, the categories of oligoastrocytoma and glioblastoma with oligodendroglioma component (GBMO) are no longer included, as molecularly they are like astrocytoma or oligodendroglioma and it is recommended that they should be assigned to either group, whenever possible.[10] The inclusion of IDH mutation is critical in both understanding the molecular characteristics and natural history of oligodendroglioma and anaplastic tumors. This alteration is also informative in the collective understanding of clinical manifestations of disease, providing insight into management of tumor and associated neurologic symptoms such as seizure.

Clinical features

Although considered malignant primary brain tumors, there may be heterogeneity in how aggressive these tumors behave.[3] Symptoms and signs of either oligodendroglioma or anaplastic oligodendroglioma are largely contingent upon factors including: lesion size and location (supra- vs infratentorial), as well as the rate of growth. Consistent with other primary brain tumors in the adult population, there is a predilection for the supratentorial compartment with oligodendrogliomas. Most tumors are found in the cortex, with preference for the white matter, primarily frontal, followed by temporal, parietal, and occipital lobes, less frequently.[11] Infratentorial, spinal, and leptomeningeal spread have been reported to occur, however, rarely.[12] Although no longer considered in the revised 2016 WHO Classification of Brain Tumors, a gliomatosis or multifocal pattern may be seen; however, this has been more likely associated with tumors of astrocytic origin.

Unlike IDH wild-type gliomas, oligodendroglial tumors are indolent with respect to growth trajectory and behavior, thus accounting for associated clinical manifestations. In comparison to high-grade tumors, in which symptoms may progress over a period of days or weeks, the pattern observed in oligodendroglioma is over the course of months to years. Acute development of focal deficit, such as in association with ischemic or hemorrhagic stroke, is thought to be rare. Features of elevated intracranial pressure typically observed with high-grade glioma, such as nausea, syncope, or papilledema, are also less likely to occur but have been reported in single case reports.[13] Given the infiltrative nature of growth, these tumors may be identified during workup for other neurologic symptoms.

The most common presenting feature of oligodendroglioma is seizure, which is characteristic of low-grade glioma in general, occurring in up to 90% of patients.[14] This is a cause of tumor-related morbidity and reduction in quality of life. Seizures in patients with underlying tumors are localization-related or symptomatic, with semiology being determined by the involved cortical structure.[15] Like nontumor-related epilepsy, seizures may be focal in onset with evolution into a generalized event, or generalized from initial onset. Status epilepticus occurs in at least 10% of patients with glioma.[16] As they are a warning sign for the presence of pathology, initial presentation with seizure represents a positive prognostic factor.[17] Other factors may contribute to epileptogenesis in addition to tumor and location, such as its genetic profile and indolent growth pattern.[17,18] It is likely due to these and other underlying factors, which make treatment of tumor-related seizures challenging. Pharmacoresistant seizures have been associated with temporal or insular cortex location or partial-onset semiology.[19] Factors which may predict postoperative seizure control include: presence of generalized seizures, surgery <1 year after presentation, gross total resection, and preoperative seizure control with anti-seizure medication.[20]

Mutant IDH and seizure

One proposed mechanism for increased seizure activity is the presence of *IDH* mutations, which results in the formation of 2-hydroxyglutarate (2-HG), a substrate with structural characteristics like glutamate, an excitatory neurotransmitter. The presence of 2-HG is thought to activate N-methyl-D-aspartate (NMDA) receptors, thus contributing to epileptogenesis.[16] In an autosomal recessive disorder, L-2-hydroxyglutarate-acidurea (2HGA), 2-HG levels are elevated in serum, urine, and CSF with clinical manifestations including refractory seizures.[21]

The *IDH* mutations have been demonstrated to be associated with better survival, frontal lobe location, and seizures as initial presentation.[22] An alternate explanation for high likelihood of seizure in this population is the role of IDH in increasing glutamate concentrations in glioma cells, primarily in the extracellular space. This has been correlated with higher seizure frequency and tumor progression.[16] Recent work by Chen et al. has proposed an alternative mechanism for the effect of IDH and 2-HG, as not being limited to the tumor cells, but have an impact on cortical neurons.[23] The release of 2-HG by glioma cells leads to direct activation of neuronal NMDA receptors with downstream effects resulting in excitatory postsynaptic potentials and increased likelihood of action potentials.[23] There are ongoing clinical trials investigating the benefit of IDH inhibitors in the treatment of glioma; the impact of these agents on seizure control will also warrant additional investigation.

Prognosis and survival outcomes

Despite the indolent growth and behavior of oligodendroglial tumors, characterized by slow changes radiographically and clinically, it is expected that most patients will ultimately deteriorate. A 5-year survival for adult [age > 40 according to SEER (surveillance, epidemiology and end results) data] oligodendroglioma and anaplastic oligodendroglioma are 74.9% and 51.1%, respectively.[1] Prior to current understanding of the molecular features of oligodendroglioma, clinical criteria were established to predict prognosis and expected behavior of disease including age, tumor size and extent, histology, and the presence or absence of neurologic deficit.[20,24] From these factors, a prognostic aggregate score of 0–5 was calculated with each factor being worth 1 point with high-risk being classified as scores ≥3 and low-risk as ≤2. The following were considered "high-risk" features: age ≥ 40, tumor diameter ≥ 6 cm, tumor crossing midline, astrocytoma histology, and the presence of neurologic deficit.[20]

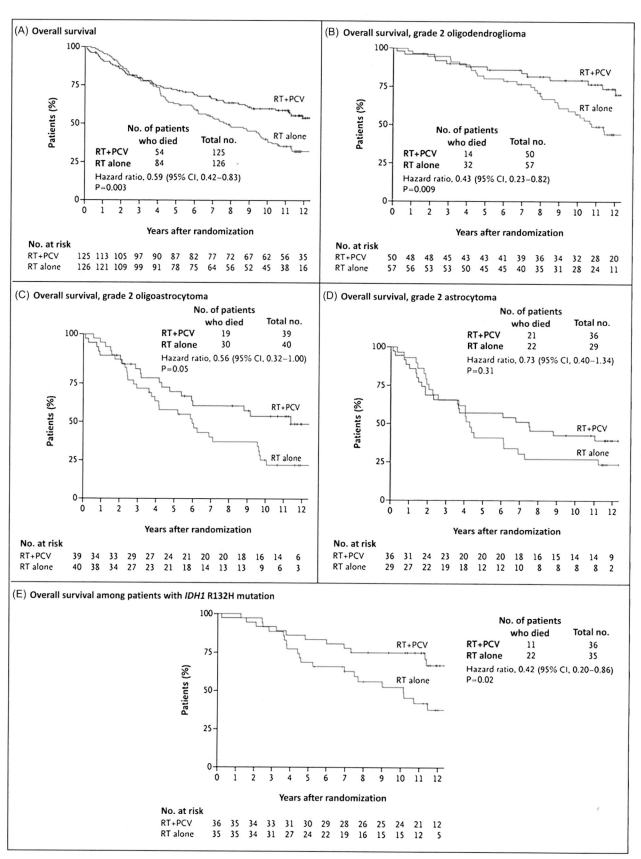

FIG. 2 Overall survival, according to treatment group.

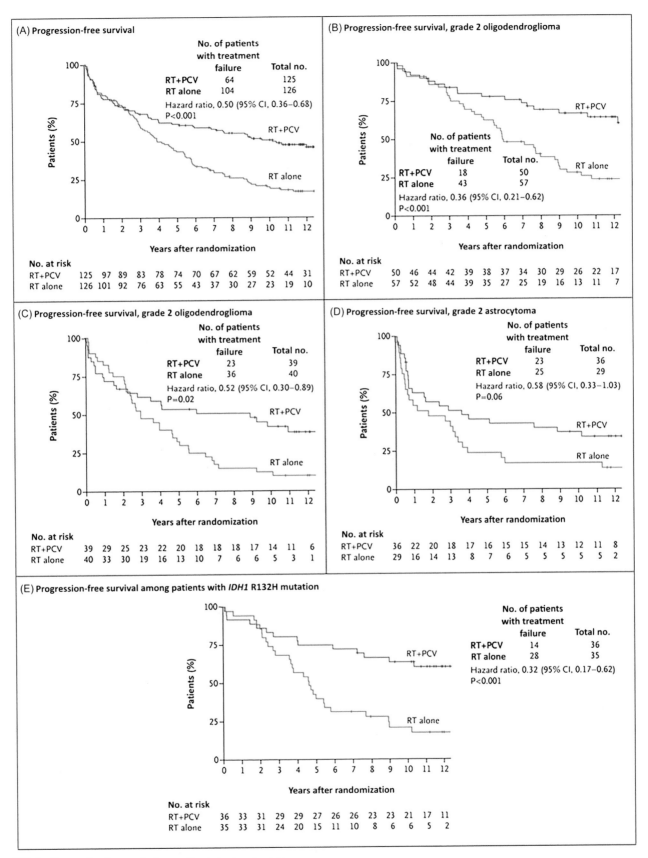

FIG. 3 Progression-free survival, according to treatment group.

RTOG 9802

In the RTOG 9802 trial, overall survival (OS) was evaluated in 251 patients with supratentorial WHO grade II gliomas who were randomized to either postoperative radiation or postoperative radiation followed by a chemotherapy regimen [procarbazine, lomustine (also called CCNU), and vincristine—PCV], for those who met high-risk criteria: age ≥ 40 or age 18–39 with either biopsy or subtotal resection. As this investigation preceded the revised CNS tumor classification scheme, histologies in this cohort included oligodendroglioma, astrocytoma, and mixed oligoastrocytoma. The recently published data were collected with median follow-up time of 11.9 years, in which the survival benefit was most apparent in the combined radiation and chemotherapy treatment cohort with median survival (OS) of 13.3 years vs 7.8 years.[25] Median progression-free survival (PFS) was also prolonged in the same cohort of 10.4 vs 4.0 years.[25] Within treatment groups, the survival benefit was enhanced in patients with oligodendroglial histology (pure or mixed) in comparison to astrocytoma (Figs. 2 and 3).

EORTC 26951 and RTOG 9402

In EORTC 26951, patients were randomized to either radiation alone or radiation and adjuvant PCV.[7] Of 368 patients, 80 had 1p/19q co-deleted tumors; within this subset, PFS was prolonged in the radiation and chemotherapy group with a median of 157 months, as compared to 50 months with radiation alone.[7] In this cohort, OS was also prolonged, having yet to be achieved in the co-deleted group treated with radiation and chemotherapy vs radiation alone, in which median OS was 112 months.

In RTOG 9402, 291 patients with anaplastic oligodendroglioma or anaplastic oligoastrocytoma were randomly assigned to either PCV followed by radiation ($n = 148$) or radiation alone without chemotherapy ($n = 143$). There was no difference in median survival between treatment groups: 4.6 years for PCV and radiation as compared to 4.7 years for radiation alone. However, between treatment groups, patients with co-deleted tumors were found to have longer OS than the retained population (14.7 years vs 2.6 years). This effect was magnified when comparing outcomes of the treatment regimens. Among those with co-deleted tumors, the combination of PCV and radiation increased median OS, 14.7 vs 7.3 years. There was no difference in OS in patients with retained tumors when comparing PCV and radiation vs radiation treatment alone. The result of these two cooperative group studies was the opening of a new trial in 2016, in which radiation with adjuvant chemotherapy (PCV) was compared to radiation with concurrent and subsequent adjuvant temozolomide in patients with anaplastic oligodendroglioma.

References

1. Ostrom QT, Gittleman H, Liao P, et al. CBTRUS statistical report: primary brain and other central nervous system tumors diagnosed in the United States in 2010–2014. *Neuro Oncol.* 2017;19(suppl_5):v1–v88.
2. Shin JY, Yoon JK, Diaz AZ. Racial disparities in anaplastic oligodendroglioma: an analysis on 1643 patients. *J Clin Neurosci.* 2017;37:34–39.
3. Ostrom QT, Gittleman H, Stetson L, Virk S, Barnholtz-Sloan JS. Epidemiology of intracranial gliomas. *Prog Neurol Surg.* 2018;30:1–11.
4. Sadetzki S, Chetrit A, Freedman L, Stovall M, Modan B, Novikov I. Long-term follow-up for brain tumor development after childhood exposure to ionizing radiation for tinea capitis. *Radiat Res.* 2005;163(4):424–432. Erratum in: Radiat Res. 2005; 164(2):234.
5. McCarthy BJ, Rankin KM, Aldape K, et al. Risk factors for oligodendroglial tumors: a pooled international study. *Neuro Oncol.* 2011;13(2):242–250.
6. Louis DN, Perry A, Reifenberger G, et al. The 2016 World Health Organization classification of tumors of the central nervous system: a summary. *Acta Neuropathol.* 2016;131(6):803–820.
7. van den Bent MJ, Brandes AA, Taphoorn MJ, et al. Adjuvant procarbazine, lomustine, and vincristine chemotherapy in newly diagnosed anaplastic oligodendroglioma: long-term follow-up of EORTC brain tumor group study 26951. *J Clin Oncol.* 2013;31(3):344–350.
8. Olar A, Wani KM, Alfaro-Munoz KD, et al. IDH mutation status and role of WHO grade and mitotic index in overall survival in grade II-III diffuse gliomas. *Acta Neuropathol.* 2015;129(4):585–596.
9. Killela PJ, Pirozzi CJ, Healy P, et al. Mutations in IDH1, IDH2, and in the TERT promoter define clinically distinct subgroups of adult malignant gliomas. *Oncotarget.* 2014;5(6):1515–1525.
10. Rogers TW, Toor G, Drummond K, et al. The 2016 revision of the WHO classification of central nervous system tumours: retrospective application to a cohort of diffuse gliomas. *J Neurooncol.* 2018;137:181–189.
11. Mulligan L, Ryan E, O'Brien M, et al. Genetic features of oligodendrogliomas and presence of seizures. The relationship of seizures and genetics in LGOs. *Clin Neuropathol.* 2014;33(4):292–298.
12. Strickland BA, Cachia D, Jalali A, et al. Spinal anaplastic oligodendroglioma with oligodendrogliomatosis: molecular markers and management: case report. *Neurosurgery.* 2016;78(3):E466–E473.
13. Roncone DP. Papilloedema secondary to oligodendroglioma. *Clin Exp Optom.* 2016;99(6):507–517.

14. van Breemen MS, Wilms EB, Vecht CJ. Epilepsy in patients with brain tumours: epidemiology, mechanisms, and management. *Lancet Neurol.* 2007; 6(5):421–430.

15. Rosati A, Tomassini A, Pollo B, et al. Epilepsy in cerebral glioma: timing of appearance and histological correlations. *J Neurooncol.* 2009; 93(3):395–400.

16. Kerkhof M, Vecht CJ. Seizure characteristics and prognostic factors of gliomas. *Epilepsia.* 2013;54(Suppl 9):12–17.

17. Vecht CJ, Kerkhof M, Duran-Pena A. Seizure prognosis in brain tumors: new insights and evidence-based management. *Oncologist.* 2014; 19(7):751–759.

18. Beaumont A, Whittle IR. The pathogenesis of tumour associated epilepsy. *Acta Neurochir.* 2000;142(1):1–15 Review.

19. Rudà R, Bello L, Duffau H, Soffietti R. Seizures in low-grade gliomas: natural history, pathogenesis, and outcome after treatments. *Neuro Oncol.* 2012;14(Suppl 4):iv55–64.

20. Pignatti F, van den Bent M, Curran D, et al. Prognostic factors for survival in adult patients with cerebral low-grade glioma. *J Clin Oncol.* 2002; 20(8):2076–2084.

21. Craigen WJ, Jakobs C, Sekul EA, et al. D-2-hydroxyglutaric aciduria in neonate with seizures and CNS dysfunction. *Pediatr Neurol.* 1994; 10(1):49–53.

22. Sanson M, Marie Y, Paris S, et al. Isocitrate dehydrogenase 1 codon 132 mutation is an important prognostic biomarker in gliomas. *J Clin Oncol.* 2009;27(25):4150–4154.

23. Chen H, Judkins J, Thomas C, et al. Mutant IDH1 and seizures in patients with glioma. *Neurology.* 2017;88(19):1805–1813.

24. Gorlia T, Wu W, Wang M, et al. New validated prognostic models and prognostic calculators in patients with low-grade gliomas diagnosed by central pathology review: a pooled analysis of EORTC/RTOG/NCCTG phase III clinical trials. *Neuro Oncol.* 2013;15(11):1568–1579.

25. Buckner JC, Shaw EG, Pugh SL, et al. Radiation plus procarbazine, CCNU, and vincristine in low-grade glioma. *N Engl J Med.* 2016; 374(14):1344–1355.

Chapter 2

Clinical presentation of spinal oligodendrogliomas

Lily C. Pham*, David Cachia†, Akash J. Patel* and Jacob J. Mandel*

*Department of Neurology and Neurosurgery, Baylor College of Medicine, Houston, TX, United States; †Department of Neurosurgery, Medical University of South Carolina, Charleston, SC, United States

Introduction

Initially defined in 1926 by Bailey and Cushing, oligodendrogliomas are exceedingly rare gliomas arising from oligodendrocytes.[1] Based on the most recent report from the Central Brain Tumor Registry of the United States 2010–2014, oligodendrogliomas make up 1.4% of all primary brain tumors, and represents 4.7% of all malignant primary brain tumors and other central nervous system tumors.[2] Within the already rare classification of oligodendrogliomas, spinal oligodendrogliomas are even rarer still representing 1.59% of all oligodendrogliomas. Primary intramedullary oligodendrogliomas represented 0.8%–4.7% of spinal cord and filum terminale tumors.[3] Historically, the majority of spinal oligodendrogliomas arise in the thoracic region (36.3%), followed by the lumbar (18.2%), cervical (18%), thoracolumbar (12.1%), cervicothoracic (9.1%) regions, and the whole spinal cord (8%).[4,5] The average size of a primary oligodendroglioma is 3.5 (±1.8) vertebral levels.[5]

The first description of a spinal oligodendroglioma was in 1931 by Kernohan et al.[6] Since then, approximately 60 reported cases of spinal oligodendrogliomas have been documented in the literature. The typical age of presentation is between 30 and 40 years of age, with the majority of patients presenting in adulthood.[3] There is a mild predilection toward males (1.33 times more likely to occur in males than in females), and these tumors are 2.5 times more prevalent in Caucasians than in African-American populations.[2] Rarely, spinal oligodendroglioma are also diagnosed in children. Since 1942, there have been five reported cases of spinal oligodendroglioma in children.[7] Usually, cases in the pediatric population have a poorer prognosis and a shorter duration of symptoms prior to presentation.[7,8]

Spinal oligodendrogliomas are often suspected based on history, clinical symptoms, physical examination, and subsequent imaging, then diagnosed following surgery based on histopathology. Only recently have cases with histopathologic diagnosis of oligodendrogliomas undergone tissue analysis and molecular characterization through gene deletions and by DNA analysis. Officially, oligodendrogliomas are now identified by an *IDH* (isocitrate dehydrogenase) gene family mutation and 1p/19q chromosomal co-deletion according to the 2016 World Health Organization (WHO) Classification of Tumors of the Nervous System.[9] Because of this fact, it is unclear how many of the 60 previously reported cases in the literature would still be classified as oligodendrogliomas based on current WHO diagnostic criteria.[9]

Clinical presentation

Clinical presentation of spinal oligodendroglioma is dependent on the age of presentation and the location of the tumor. The duration of symptoms reported prior to presentation has been variable, spanning from months to years. In the reported cases in the literature (Table 1), the most common presenting symptoms are weakness/paresis (69.5%), pain—including acute back pain, neck pain, and sciatic pain (50%), and sensory changes—including paresthesia and numbness (45%). While early presenting symptoms are related to the location of the tumor in the cord (cervical vs thoracic vs lumbar vs whole cord), symptomatology also differs depending on whether the tumor is intramedullary, intradural-extramedullary, or extramedullary (Fig. 1). Intramedullary tumors rarely produce pain and often occur with sensory dysesthesia and numbness with early loss of sphincter control, whereas pain, usually back or neck pain, is the presenting symptoms of extramedullary tumor and is only associated with loss of sphincter control if it is found in the lumbosacral region.[50,51]

TABLE 1 Reported cases of primary spinal cord oligodendrogliomas with presenting symptoms

Case	Age	Sex	Presenting symptoms	Imaging/cord level	Surgery	Radiation and/or chemo	Follow-up
1[10]	31	M	Spastic gait, generalized progressive weakness, atrophic distal low extremities, increased tone, decreased patellar reflex, increased ankle reflex, positive clonus, Babinski present, Romberg positive, spasms, spinal cord level, decreased temperature, decreased proprioception, decreased light touch	XR	1942—spinal fusion for scoliosis T10-L3 with tibial bone graft; 1948—laminectomy T8		
2[11]	68	M	Right lower extremity numbness, left lower extremity weakness, gait imbalance, falls, decreased pinprick	MRI with T2 hyperintense intramedullary lesion expanding conus at T9-T12, 1.2 cm extramedullary lesion at T11	Partial resection	RT, adjuvant TMZ × 8 cycles	1 year with no disease recurrence
3[4]	30	F	Nausea, vomiting, photophobia, anorexia, transient expressive aphasia; hydrocephalus, confusion, visual hallucinations	MRI with leptomeningeal enhancement of basal meninges, nonenhancing enlargement of C3 with hyperintensity on T2	Operation 1: R pterional craniotomy with biopsies; Operation 2: VPS and spinal cord biopsy; Operation 3: C3–4 laminectomy, partial C2 laminectomy, partial resection	RT	
4[12]	24	M	Left leg weakness and radiating pain in both legs, paresthesias, hypesthesias	MRI with intramedullary mass at T4-T8	Subtotal resection	RT	Imaging progression after 5 years but no new clinical symptoms
5[13]	18	F	Lower back pain, bilateral lower extremity weakness	MRI with intramedullary mass at T4-T8	Operation 1: T6-T12 laminectomy; Operation 2: T4-T8 laminectomy w/ subtotal resection	RT, TMZ after second subtotal resection	Recurrence of tumor after first operation with second operation 8 months after, no recurrence 1 year after completion of 18 months of TMZ

TABLE 1 Reported cases of primary spinal cord oligodendrogliomas with presenting symptoms—cont'd

Case	Age	Sex	Presenting symptoms	Imaging/cord level	Surgery	Radiation and/or chemo	Follow-up
6[8]	28	M	Leg weakness, leg numbness, constipation, headache	MRI with enhancing intramedullary mass at L3-S2	L3-S2 laminectomy		
7[3,14]	31	F	Left lower extremity paresthesia to left knee, bilateral lower extremity weakness, sensory deficit T7-S1	XR, T7-T11	Total resection		
8[15]	43	F	Back pain, bilateral sciatica, flacid paraplegia, hypoesthesia below L1	Conus medullaris, cauda equina	Partial resection	RT	Death in 4.5 months
9[16]	44	M	Back pain, decreased sensation below T9 dermatome, paraplegia		Laminectomy with partial removal, second partial removal 5 years later	RT	Death in 76 months
10[17]	27	F	Back pain, gait unsteadiness, lower extremity paresthesias, lower extremity weakness	XR	Laminectomy C4-T1, partial resection		Stable 54 months
11[3,18]	37	F	Back pain, lower extremity weakness, then lower extremity numbness, hyperesthesia	XR, T7	Partial resection		Stable for 16 months
12[3,19]	16	F	Back pain radiating to right hip, bilateral lower extremity weakness	T7			
13[3,19]	26	M	Left sciatica, cauda equina syndrome	XR, filum terminale	Total resection	RT	
14[20]	39	M	Abdominal pain, progressive weakness, sensory level at T8	XR T8–11			Death in 2 months
15[3,21]	36	F	Pain, spastic paraparesis, sensory level at T10	XR, T10-L1, extramedullary			
16[9,12]	36	M	Paresthesias, pain and fasciculations in left biceps	No imaging, C4–5	Total resection, C3–7 laminectomy		Survived 31 years
17[13]	37	M	Headache, vomiting, blurred vision, left sided weakness	C3-C7	Subtemporal decompression		Death in 9 months

Continued

TABLE 1 Reported cases of primary spinal cord oligodendrogliomas with presenting symptoms—cont'd

Case	Age	Sex	Presenting symptoms	Imaging/cord level	Surgery	Radiation and/or chemo	Follow-up
18[22]	45	F	Neck pain radiating to right arm, bilateral hand numbness, weakness of right sided extremities, patchy sensory loss	XR with widening of C3–4 foramen, location in C2–5	Partial resection		Death in 21 days
19[3,23]	40	M	Neck pain	XR, C3–7	Biopsy	RT	Death in 30 months
20[19,24]	52	M	Back ache, sensory-motor disturbances	XR, C4–5, destruction of vertebra	Biopsy		Death in 24 months
21[24]	29	F	Bilateral sciatic pain, sacral pain, bilateral lower extremity weakness, numbness of right lower extremity, saddle anesthesia	XR, T5–11	Biopsy		Death in 5 months
22[24]	18	F	Pain, decreased reflexes in lower extremities	XR, L1–3, conus, cauda	Partial resection	RT	
23[25]	40	M	Pain, flacid paresis	XR, L2–3	Partial resection	RT	
24[3,26]	25	F	Weakness bilateral lower extremities, pain, transverse section syndrome at T6	XR, T6–9	Partial resection		Death in 6 days
25[3,27]	39	M	Weakness and paresthesias in right lower extremity, anesthesia below L5	XR, filum	Total resection		Stable
26[3,28]	45	F	Back ache, bilateral sciatica	XR, myelogram with cysternal puncture, L1 extramedullar tumor, filum	Total resection, laminectomy T12-L4		
27[29]	8	M	Vomiting, headache, nuchal rigidity	Cervical	Autopsy		Death in 3 months
28[30]	16	M	Bilateral lower extremity weakness, then upper extremity weakness, sensory deficit C6-L1	XR spinal canal widening C2-L5, C2-L1	Biopsy, C2-L1 laminectomy	RT	
29[30]	8	F	Acute back ache, tetraparesis, sensory deficit below chest	Entire chord	Biopsy, autopsy		Death in 9 months

TABLE 1 Reported cases of primary spinal cord oligodendrogliomas with presenting symptoms—cont'd

Case	Age	Sex	Presenting symptoms	Imaging/cord level	Surgery	Radiation and/or chemo	Follow-up
30[31]	16	F	Weakness and paresthesias left upper extremity, flacid paresis of left upper extremity, touch hypesthesia L ulnar	XR cervical canal widening, C3–7	Partial resection, laminectomy C3-T3	RT	Progression in 9 months
31[3,32]	44	M	Buttock pain, numbness RLE, progressive paraparesis, bilateral sensory deficits	XR myelogram, lower thoracic	Autopsy, biopsy	RT	Death at 5 years
32[3,33]	13	M	Contracture of thoraco-lumbar muscles, meningeal syndrome	XR myelogram, T10	Autopsy		Death at 23 months
33[3,34]	28	F	Lower back pain, paresthesias, and weakness right lower extremity, spastic paraparesis, saddle anesthesia	XR myelogram, T10–11	Partial resection, autopsy		
34[35]	32	M	Headache, blurred vision, lower back pain, right lower extremity pain, right lower extremity paresis, papilledema	XR splaying of T12-L1 pedicles	Biopsy		Death
35[3]	26	F	Acute back ache, weakness bilateral lower extremity, transverse section syndrome at T12, paraparesis, sensory deficit below T10	XR, XR myelogram, T10–11	Laminectomy, medullary cyst resection, partial resection	RT	Death 8 years
36[36]		F	Lower back pain, paraparesis	T10-L2	Partial resection		Death
37[37]	13	M	Scoliosis, neck and shoulder pain, tetraparesis, hypaesthesia	MRI hyperintense T2 C5–7 with scattered lesions down to conus	C5-T9 laminotomy		No progression after 2 years
38[38]	3	M	Scoliosis	Thoracolumbar	Partial resection	RT	Ongoing treatment
39[39]	<16		Spinal deformity		Partial resection	RT	
40[40]	<3		Motor regression, gait abnormalities		Resection		

Continued

TABLE 1 Reported cases of primary spinal cord oligodendrogliomas with presenting symptoms—cont'd

Case	Age	Sex	Presenting symptoms	Imaging/cord level	Surgery	Radiation and/or chemo	Follow-up
41[41]	38 months	M	Scoliosis, spastic monoparesis right lower extremity	MRI T1 hyperintensity from T4 to conus medullaris and syringomyelia	Total laminotomy T7-L1, partial resection	RT	Stable 50 months after surgery
42[5]	12	M	Scoliosis, LE hyperreflexia and hyperesthesia, gait disturbance	MRI gad enhancing Cy-T12 with syringomyelia C7-T12	Laminectomy C7-T10, Partial resection × 2	RT	Recurrence after 10 months
43[42]	4	M	Hypotonic ataxia, regression of development	MRI	Partial resection × 3, oligodendrogliomatosis with primary spinal cord lesion		Recurrence in cerebellum, temporal lobe, occipital lobe, survived 7 years after diagnosis
44[43]	16	F	Neurogenic bladder, left lower extremity weakness (ankle), L S1 sensory deficit	MRI intrathecal extramedullary mass T11-L2 with ring-like enhancement	Total resection, T8-S1 laminectomy		No recurrence after 28 months
45[44]	56	M	Bilateral lower extremity pain	MRI intrathecal extramedullary mass L1-L2 with partial invasion of conus	Total resection		
46[45]	22	M	Left paresthesia, acute upper back pain, left Brown-Sequard syndrome	MRI intramedullary partially enhancing mass at C5–7	Partial resection	RT	Brain metastases after 2 years
47[46]	8	M	Neck pain and arm weakness	C3-T12	Partial resection, staged laminoplasty		Stable after 9 months
48[8,47]	44	M	Bilateral lower extremity weakness, urinary incontinence	T11-L2	Partial resection	RT	
49[17,48]	46	M	Neck pain, left upper extremity weakness, sensory loss C5-T1	MRI with intramedullary mass and synringomyelia at C3-T4	Partial resection	RT	Recurrence after 1 year
50[49]	3	F	neck pain, left leg weakness, asymmetric gait, and right head tilt	MRI intramedullary mass extending from C4 to T4	Subtotal resection × 2	18 cycles of TMZ, TMZ was reinitiated for nine additional cycles, Proton RT after second resection, ponatinib	Multiple recurrences, ongoing treatment

XR, X-ray; *MRI*, magnetic resonance imaging; *TMZ*, temozolomide; *RT*, radiation treatment; *VPS*, ventriculoperitoneal shunt.

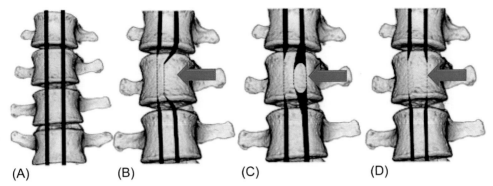

(A)   (B)   (C)   (D)

FIG. 1 Historic classification of spine tumors based on myelography: (A) normal, (B) extradural extramedullary, (C) intradural extramedullary, and (D) intradural intramedullary. *Adapted with permission from Mechtler LL, Nandigam K. Spinal cord tumors: new views and future directions.* Neurol Clin., *2013;31(1):241–268.*

In addition, primary intramedullary oligodendroglioma is more likely to be associated with meningeal spread and intracranial hypertension leading to fluctuating symptoms due to spontaneous hemorrhage.[41] Spinal oligodendrogliomas in the pediatric population tend to present as scoliosis, regression of motor developmental milestones, gait disturbance, and sensory deficits.[3,5,7,30,31,37–41,43,46]

Based on the available information from the 60 case reports in the literature, the distribution of spinal oligodendroglioma is: 25% cervical, 28.3% thoracic, 13.3% thoracolumbar, 13.3% lumbosacral, and 5% whole cord, with 15.7% unaccounted for due to poor documentation (Tables 1 and 2). In all, 66.6% of these cases were intramedullary tumors, 10% were intradural extramedullary, and 3.3% had both intramedullary and extramedullary components. The other 20% did not report location of the tumor in relation to the spinal cord and dura. These numbers are roughly similar to historically reported statistics in previous large case reports when taking into account the small number of overall cases in literature.[3,8,50,51]

Details regarding outcome is limited depending on the length of follow-up at the time of the case report. However, based on the available data, 13.3% showed progression/recurrence, 31.6% of patients had died by the time of follow-up, and 33.3% had no progression or recurrence at the time of follow-up. The earliest documented death following initial presentation was within 6 days, and the longest span of time without progression or recurrence was 31 years.[3,16] Of note, the survival rate of spinal oligodendrogliomas appears much more dismal in pediatric cases compared to adult cases. Most cases of primary oligodendroglioma in children result in death within 3–23 months after initial presentation, with the longest survival time being 7 years.[7] All of the reported cases without progression or recurrence were adults. Time to disease progression or death in the adult population ranges between months and years, without a clear pattern. Due to the infiltrative nature of spinal oligodendrogliomas, only a total of 6 cases out of 60 received a gross total resection.[3,16,43,44,51] There were no reported deaths in cases that received gross total resections. Most were either symptomatically stable or had oscillating symptoms without significant progression following surgery. Due to the rarity of the diagnosis of spinal oligodendrogliomas, there have been no randomized prospective studies to guide treatment. Recent studies of intracranial oligodendrogliomas have shown the benefit of using radiation and chemotherapy [procarbazine, CCNU, and vincristine (PCV); temozolomide] following diagnosis in patients with a 1p/19q co-deletion.[24,52] Yet despite similar histopathologic appearance it remains unclear whether spinal oligodendrogliomas are analogous molecularly to intracranial oligodendrogliomas. Treatment with radiation therapy and/or chemotherapy following surgical resection is therefore of unknown benefit, but should be considered in all patients unable to undergo a gross total resection.

Imaging

Magnetic resonance imaging (MRI) of the spine is the recommend diagnostic study for patients suspected of a spinal cord tumor. Spinal MRIs are capable of offering outstanding definition of the spinal cord and adjacent structures. The MRI is used to determine the exact location of the tumor in the spinal cord, whether the lesion is intramedullary, intradural-extramedullary, or extramedullary, and if it enhances following the injection of gadolinium contrast.[53]

Oligodendrogliomas typically appear on MRI as isointense to the spinal cord on T1-weighted images, hyperintense on T2-weighted and FLAIR images (Fig. 2), and demonstrate heterogeneous contrast enhancement.[54] Calcifications and hemorrhages may also be noted on imaging.[43] Syringomyelia (a process where a cyst or cavity develops within the spinal cord) has also been appreciated in several cases.[5,17,23,48]

TABLE 2 Primary spinal oligodendrogliomas with molecular testing

Case	Age	Sex	Surgery	Non-surgical treatment	Molecular mutation	Follow-up
1[11]	68	M	Partial resection	RT, adjuvant TMZ for 8 cycles	1p/19q co-deletion, IDH1 WT, IDH2 WT	1 year with no disease recurrence
2[4]	30	F	Operation 1: Right pterional craniotomy with biopsies; Operation 2: ventriculoperitoneal shunt and spinal cord biopsy; Operation3: C3–4 laminectomy, partial C2 laminectomy, partial resection	RT	1p19q negative, 1p36 deletion, 19q13 deletion	
3[12]	24	M	Subtotal resection	RT	1p/19q negative, MGMT positive	Radiographic progression after 5 years but asymptomatic
4[13]	18	F	Operation1: T6-T12 laminectomy; Operation 2: T4-T8 laminectomy with subtotal resection	RT, TMZ after second subtotal resection	1p/19q co-deletion, MGMT weakly positive	Recurrence after first operation with second operation 8 months later, no recurrence 1 year after completion of 18 months of TMZ, now stable
5[49]	3	F	Subtotal resection twice	18 cycles of TMZ, TMZ reinitiated for nine additional cycles, Proton RT after second resection, ponatinib	1p/19q negative, IDH WT, MGMT negative, ATRX preserved, p53 not overexpressed	Multiple recurrences, ongoing treatment

RT, radiation treatment; *TMZ*, temozolomide; *IDH*, isocitrate dehydrogenase; *WT*, wild type; *MGMT*, O^6-methylguanine–DNA methyltransferase.

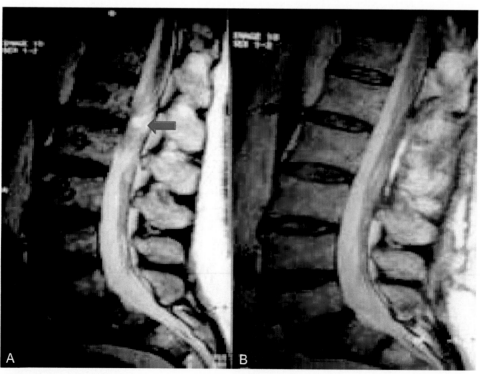

FIG. 2 Hyperintense intradural extramedullary mass (*red arrow*) on T2 weighted images with cystic components (A). Post-op MRI of the lumbosacral region revealing no residual tumor (B). *Adapted with permission from Gürkanlar D, et al., Primary spinal cord oligodendrioglioma. Case illustration. Neurocirugia (Astur). 2006;17(6):542–543.*

Alternatively, myelography or standard X-ray (XR) of the spine can often be used to diagnose spinal cord tumors. Myelography is an examination where a radiocontrast is injected into the cervical or lumbar spine followed by X-ray or computed tomography evaluation. Although it has largely fallen out of favor in lieu of the more detailed MRI, CT myelography can be performed in patients with contraindications to MRI such as a pacemaker. As with myelography, earlier case reports of spinal oligodendrogliomas utilized XR imaging of the affected spinal region. The most notable XR findings of spinal oligodendroglioma are scattered calcifications, which can be found in 28%–40% of the time within the tumor body.[3,8,12] However, XR imaging alone is no longer favored as the level of detail of the spinal cord and related structures on XR pale in comparison to that of a MRI.

Other studies
Pathology

Macroscopically, the classic gross appearance of spinal oligodendroglioma is described as a translucent, gelatinous solid tumor that appears to be gray, pink, and yellow in color.[43,51] These tumors are also easily friable and are often hemorrhagic on gross resection due to their close association with spinal vasculature.[3,43] The firm outward appearance is due to the high cellularity of oligodendroglial tumors, and the densely packed small cells at a microscopic level.[55]

Microscopically, the key features of spinal oligodendroglioma are similar to oligodendrogliomas found in the brain. These features were originally described by Bailey and Bucy in 1929 as uniform polyhedral cells with "honeycomb appearance," inside the cells are perinuclear clear halos surrounding a round and dark-staining/hyperchromatic nucleus, resembling a "fried egg".[51,55] Further, these cells have multifocal calcifications without any associated necrosis or significant presence of glial fibrils.[31,43,51]

Lumbar puncture

Prior to the advent of MRIs many patients presenting with oligodendrogliomas received lumbar punctures. While there are no pathognomonic CSF (cerebrospinal fluid) findings related to oligodendrogliomas, it has been noted that CSF protein levels are pathologically elevated. A study examining these levels found that the mean value of CSF protein was 1397 mg% (50–5000), with further elevation in lumbar tumors at 2850 mg%.[3] With improvement in imaging capabilities, lumbar punctures are no longer indicated as part of the initial workup for patients with spinal oligodendrogliomas.

Molecular signatures

Based on the latest WHO classification published in 2016, oligodendrogliomas are now identified by *IDH* gene family mutation and 1p/19q co-deletion.[9,55] The *IDH* gene mutations have been observed in patients with prolonged survival and give rise to changes in cellular metabolism affecting chromatin remodeling and transcription. The presence of the *IDH* gene mutation is a predictive factor in survival time, especially when considered in conjunction with a 1p/19q co-deletion.[56]

1p/19q co-deletion in oligodendroglial tumors was originally described in 1994 by Reifenberger et al.[57] Since then subsequent studies have shown that the presence of this translocation between 1p and 19q arms is associated with increased chemotherapy sensitivity, and ultimately improved outcome.[24,58] In the spinal oligodendroglioma literature, only five reported cases have utilized molecular signatures in addition to the histopathological diagnosis.[4,11–13] This low number is due to the rarity of the tumor, difficulty of obtaining substantial tissue due to eloquent location of the tumor, and the availability of genetic testing at the time of the report. As many mixed gliomas also exhibit a 1p/19q co-deletion and are *IDH* mutants, it is possible that there will now be an increase in the diagnosis of spinal oligodendrogliomas, simply based on molecular signatures.[55] Alternatively, it remains unclear how many cases of spinal oligodendrogliomas previously reported in the literature would still be classified as oligodendrogliomas because of their potential absence of IDH mutation or 1p/19q co-deletion.

Unlike intracranial gliomas, which have been reported to have an *IDH* mutation 75% of the time, the rate of *IDH* mutations in infratentorial and spinal gliomas has reported to be much lower.[59,60] A recent study of 44 patients with infratentorial or spinal cord grade 2 and 3 diffuse gliomas found only 7% were positive for *IDH*1 mutation, all of the cases with the mutation localizing to the brainstem.[60] None of the nine spinal cord gliomas (three diffuse astrocytomas, two WHO grade 2 oligodendrogliomas, and four anaplastic astrocytomas) were found to be *IDH*1 mutated by immunohistochemistry.[60] Therefore, it remains unclear how many of the cases previously reported in the literature would still be classified as oligodendrogliomas, if the diagnosis is dependent on the tumors having a molecular signature of a 1p/19q co-deletion and *IDH* mutation, in addition to oligodendroglial histopathology.

Due to the rarity of diagnosis and lack of molecular testing performed in the past for spinal oligodendrogliomas, an international registry and further molecular research is needed to help elucidate the nature of spinal oligodendrogliomas.

Conclusion

Spinal oligodendrogliomas are extremely rare, with only 60 reported cases. Clinical presentation of spinal oligodendroglioma is often dependent on the age of presentation and the location of the tumor, with the majority of spinal oligodendrogliomas arising in the thoracic region. The most common presenting symptoms are weakness/paresis, pain, and sensory changes. In addition to the location of the tumor in the spinal cord, presenting signs also depend on whether the tumor is intramedullary, intradural-extramedullary, or extramedullary. Patients presenting with a history of symptoms referable to the spinal cord and/or a concerning neurological examination should undergo prompt imaging with an MRI of the spine with and without contrast.

References

1. Strang RR, Nordenstam H. Intracerebral oligodendroglioma with metastatic involvement of the cauda equina. *J Neurosurg.* 1961;18(5):683–687.
2. Ostrom QT, et al. CBTRUS statistical report: primary brain and other central nervous system tumors diagnosed in the United States in 2010–2014. *Neuro Oncol.* 2017;19(Suppl 5):v1–v88.
3. Fortuna A, Celli P, Palma L. Oligodendrogliomas of the spinal cord. *Acta Neurochir.* 1980;52(3–4):305–329.
4. Guppy KH, et al. Spinal cord oligodendroglioma with 1p and 19q deletions presenting with cerebral oligodendrogliomatosis. *J Neurosurg Spine.* 2009;10(6):557–563.
5. Ushida T, et al. Oligodendroglioma of the "widespread" type in the spinal cord. *Childs Nerv Syst.* 1998;14(12):751–755.
6. Kernohan JW, Woltman HW, Adson AW. Intramedullary tumors of the spinal cord: a review of fifty-one cases, with an attempt at histologic classification. *Arch NeurPsych.* 1931;25(4):679–701.
7. Gilmer-Hill HS. Spinal oligodendroglioma with gliomatosis in a child. *J Neurosurg Spine.* 2000;92(1):109–113.
8. Hasturk AE, et al. A very rare spinal cord tumor primary spinal oligodendroglioma: a review of sixty cases in the literature. *J Craniovert Jun Spine.* 2017;8(3):253–262.
9. Louis DN, et al. The 2016 World Health Organization classification of tumors of the central nervous system: a summary. *Acta Neuropathol.* 2016;131(6):803–820.
10. Russell JR, Bucy PC. Oligodendroglioma of the spinal cord. *J Neurosurg.* 1949;6(5):433–437.
11. Strickland BA, et al. Spinal anaplastic oligodendroglioma with oligodendrogliomatosis: molecular markers and management: case report. *Neurosurgery.* 2016;78(3):E466–E473.
12. Yuh WT, Chung CK, Park SH. Primary spinal cord oligodendroglioma with postoperative adjuvant radiotherapy: a case report. *Korean J Spine.* 2015;12(3):160–164.
13. Wang F, Qiao G, Lou X. Spinal cord anaplastic oligodendroglioma with 1p deletion: report of a relapsing case treated with temozolomide. *J Neurooncol.* 2011;104(1):387–394.
14. Oljenick I. Intramedullaire gozwellen (diagnostick en heelknudige behande ling). *Ned Tijdschr Geneeskd.* 1936;80:1335–1342.
15. Woods W, Pimenta A. Intramedullary lesions of the spinal cord. Study of sixty-eight consecutive cases. *Arch Neurol Psychiatry.* 1944;52:383–399.
16. Love JG, Rivers MH. Thirty-one-year cure following removal of intramedullary glioma of cervical portion of spinal cord: report of case. *J Neurosurg.* 1962;19(10):906–908.
17. Padberg F, Davis L. Tumors of the spinal cord. I. Intramedullary tumors. *Q Bull Northwest Univ Med Sch.* 1952;26(3):204.
18. Enestrom S, Grontoft O. Oligodendroglioma of the spinal cord; report of one case. *Acta Pathol Microbiol Scand.* 1957;40(5):396–400.
19. Kornianskii GP. Tumors of the spinal cord in children. *Vopr Neirokhir.* 1959;23(1):39–47.
20. Coxe WS. Tumors of the spinal canal in children. *Am Surg.* 1961;27:62–73.
21. Klar E, Henn R. Experiences with 262 laminectomies. *Langenbecks Arch Klin Chir Ver Dtsch Z Chir.* 1961;296:614–659.
22. Chigasaki H, Pennybacker JB. A long follow-up study of 128 cases of intramedullary spinal cord tumours. *Neurol Med Chir (Tokyo).* 1968;10:25–66.
23. Slooff JL, Kernohan JW, CS MC. *Primary Intramedullary Tumors of the Spinal Cord and Filum Terminale.* Philadelphia: W.B. Saunders Co.; 1964. pp. 255.
24. van den Bent MJ, et al. Adjuvant procarbazine, lomustine, and vincristine chemotherapy in newly diagnosed anaplastic oligodendroglioma: long-term follow-up of EORTC brain tumor group study 26951. *J Clin Oncol.* 2013;31(3):344–350.
25. Love JG, Wagener HP, Woltman HW. Tumors of the spinal cord associated with choking of the optic disks. *AMA Arch Neurol Psychiatry.* 1951;66(2):171–177.
26. Backus ML. *Untersuchungen zur Statistik der Biologie und Pathologie Intrakranieller und Spinaler Raumfordernder Prozesse.* Köln: Inaug. Diss.; 1965.
27. Broder D. *Die Gliome des Rückenmarks—Klinik und Differentialdiagnose.* Köln: Inaug. Diss.; 1965

28. Ortiz Gonzales JM, Blázquez MG, Soto Cuenca M, Valenciano M. Tumores de la cauda equina. *Rev Esp Oncol.* 1965;12:117–126.
29. Toso V. Metastatic diffusion to the leptomeninges. *Acta Neurol (Napoli).* 1967;22(3):366–376.
30. O'Brien CP, et al. Extensive oligodendrogliomas of the spinal cord with associated bone changes. *Neurology.* 1968;18(9):887–890.
31. Garcia JH, Lemmi H. Ultrastructure of oligodendroglioma of the spinal cord. *Am J Clin Pathol.* 1970;54(5):757–765.
32. Maurice-Williams RS, Lucey JJ. Raised intracranial pressure due to spinal tumours: 3 rare cases with a probable common mechanism. *Br J Surg.* 1975;62(2):92–95.
33. Michel D, Lemercier G, Beau G, Tommasi M, Schott B. Gliomaseméningée et ventriculaire diffuse secondaire à un oligodendrogliome intramédullaire. A propos d'une observation. *Lyon Mßd.* 1975;234:37–41.
34. Wober G, Jellinger K. Intramedullary oligodendroglioma with meningocerebral dissemination (author's transl). *Acta Neurochir (Wien).* 1976;35(4): 261–269.
35. Ridsdale L, Moseley I. Thoracolumbar intraspinal tumours presenting features of raised intracranial pressure. *J Neurol Neurosurg Psychiatry.* 1978;41(8):737–745.
36. Guidetti B, Mercuri S, Vagnozzi R. Long-term results of the surgical treatment of 129 intramedullary spinal gliomas. *J Neurosurg.* 1981;54(3): 323–330.
37. Pagni CA, Canavero S, Gaidolfi E. Intramedullary "holocord" oligodendroglioma: case report. *Acta Neurochir.* 1991;113(1–2):96–99.
38. Wang KC, Chi JG, Cho BK. Oligodendroglioma in childhood. *J Korean Med Sci.* 1993;8(2):110–116.
39. Lunardi P, et al. Management of intramedullary tumours in children. *Acta Neurochir.* 1993;120(1–2):59–65.
40. Constantini S, et al. Intramedullary spinal cord tumors in children under the age of 3 years. *J Neurosurg.* 1996;85(6):1036–1043.
41. Nam D-H, et al. Intramedullary anaplastic oligodendroglioma in a child. *Childs Nerv Syst.* 1998;14(3):127–130.
42. Gilmer-Hill HS, et al. Spinal oligodendroglioma with gliomatosis in a child. Case report. *J Neurosurg.* 2000;92(1 Suppl):109–113.
43. Fountas KN, et al. Primary spinal cord oligodendroglioma: case report and review of the literature. *Childs Nerv Syst.* 2005;21(2):171–175.
44. Gurkanlar D, et al. Primary spinal cord oligodendroglioma. Case illustration. *Neurocirugia (Astur).* 2006;17(6):542–543.
45. Ramirez C, et al. Intracranial dissemination of primary spinal cord anaplastic oligodendroglioma. *Eur J Neurol.* 2007;14(5):578–580.
46. Tobias ME, et al. Surgical management of long intramedullary spinal cord tumors. *Childs Nerv Syst.* 2008;24(2):219–223.
47. Moorthy NL, Kondeti D, Chander M, Jadhav H, Ashok. Spinal cord oligodendroglioma: a case report. *IOSR J Dent Med Sci.* 2015;1:27–28.
48. Tunthanathip T, Oearsakul T. Primary spinal cord oligodendroglioma: a case report and review of the literature. *Chin Neurosurg J.* 2016;2:2.
49. Bruzek AK, et al. Molecular characterization reveals NF1 deletions and FGFR1-activating mutations in a pediatric spinal oligodendroglioma. *Pediatr Blood Cancer.* 2017;64(6). https://doi.org/10.1002/pbc.26346.
50. Rasmussen TB, Kernohan JW, Adson AW. Pathologic classification, with surgical consideration, of intraspinal tumors. *Ann Surg.* 1940;111(4): 513–530.
51. Nathoo AR, Halliday NP. Spinal cord oligodendroglioma. *Postgrad Med J.* 1967;43(506):789–791.
52. Cairncross G, et al. Phase III trial of chemoradiotherapy for anaplastic oligodendroglioma: long-term results of RTOG 9402. *J Clin Oncol.* 2013;31(3): 337–343.
53. Mechtler LL, Nandigam K. Spinal cord tumors: new views and future directions. *Neurol Clin.* 2013;31(1):241–268.
54. Traul DE, Shaffrey ME, Schiff D. Part I: spinal-cord neoplasms-intradural neoplasms. *Lancet Oncol.* 2007;8(1):35–45.
55. Wesseling P, van den Bent M, Perry A. Oligodendroglioma: pathology, molecular mechanisms and markers. *Acta Neuropathol.* 2015;129(6):809–827.
56. Horbinski C. What do we know about IDH1/2 mutations so far, and how do we use it? *Acta Neuropathol.* 2013;125(5):621–636.
57. Reifenberger J, et al. Molecular genetic analysis of oligodendroglial tumors shows preferential allelic deletions on 19q and 1p. *Am J Pathol.* 1994;145(5):1175–1190.
58. Cairncross JG, et al. Specific genetic predictors of chemotherapeutic response and survival in patients with anaplastic oligodendrogliomas. *J Natl Cancer Inst.* 1998;90(19):1473–1479.
59. Kloosterhof NK, et al. Isocitrate dehydrogenase-1 mutations: a fundamentally new understanding of diffuse glioma? *Lancet Oncol.* 2011;12(1):83–91.
60. Ellezam B, et al. Low rate of R132H IDH1 mutation in infratentorial and spinal cord grade II and III diffuse gliomas. *Acta Neuropathol.* 2012;124(3): 449–451.

Chapter 3

Oligodendrogliomas—Atypical clinical presentations

P. Gage Gwyn*, Sherif M. Makar* and Herbert B. Newton*,†,‡

*Neuro-Oncology Center, AdventHealth Cancer Institute, AdventHealth Medical Group, Orlando, FL, United States; †CNS Oncology Program, AdventHealth Cancer Institute, AdventHealth Medical Group, Orlando, FL, United States; ‡Division of Neuro-Oncology, Esther Dardinger Endowed Chair in Neuro-Oncology James Cancer Hospital & Solove Research Institute, Wexner Medical Center at the Ohio State University, Columbus, OH, United States

Introduction

Oligodendrogliomas (ODGs) are relatively rare neuroepithealial tumors accounting for approximately 5.6% of all intracranial tumors.[1] The relative frequency of histopathologically diagnosed oligodendroglial tumors is higher in adults at 5.9% vs 2.5%–3.5% in children in the spectrum of glial brain tumors.[1–5] Furthermore, it is highly unusual for ODG's to be diagnosed in the pediatric population, with the highest incidence between 6 and 12 years of age.[1–6] ODGs in the pediatric population appear to be genetically distinct from ODGs of adults, rarely exhibiting loss of heterozygosity (LOH) of 1p and 19q.[6,7] Additionally, the incidence of ODG's occurred 2.5% more frequently in whites than blacks.[1]

Primary ODGs occur more frequently in the cerebral hemispheres; in the frontal lobes at approximately 69% and in temporal lobes about 20%.[3,8,9] Some authors found a slight preponderance for lesions on the right side.[10] Rarely do ODGs occur infratentorially (only 2%–7%)[3,11,12] and several authors' research supports infratentorial ODGs may be more aggressive than hemispherical occurrences.[12,13] Seizure was the most common presenting symptom,[2,3,14] with HA, and mental status changes also frequently cited in the case presentations.[2,10] Additional symptoms correlated by anatomical site will be discussed sequentially by section in this chapter.

Optic pathway

Gliomas occurring in the optic pathway constitute approximately 1%–5% of all gliomas[15] and those that have been histologically confirmed to be ODGs are extremely rare. As of 2018 to our knowledge only four confirmed cases were reported in the literature. Two were pediatric[16,17] and two were adult.[15,18] Two were limited to the optic nerve[16,18] and two had more extensive disease occurring in optic nerve, optic chiasm, third ventricle, and hypothalamus.[15,17] Two cases received no adjuvant treatment after total resection,[16,18] the two cases with more extensive disease required radiation alone with 50 Gy/39 Fractions (Fx),[15] or combined treatment with temozolomide (75 mg/m²) with radiation.[17] Of note, in 1950 there was a publication reviewing 200 cases of ODGs and this included 10 tumors occurring in the "third ventricle, thalamus, lateral ventricle, or optic chiasm,"[11 (p. 968)] but the optic chiasm cases were not listed independently, so in actuality more than four cases to date may indeed exist.

Ciliary body/retina

There is a unique case report of an ODG occurring in the ciliary body reported in 2010.[8] These authors hypothesize that since gliomas arise from neural stem cells, ODGs could occur at any location neural stem cells exist. This case was a 52-year-old man who presented with signs of retinal detachment that included six months of deteriorating vision associated with metamorphopsia. Workup included physical examination, Doppler (demonstrating mass), and computed tomography (CT) (revealing semi-circular, nonhomogeneous, high-density area in the right eye extending from the vitreous cavity to the concave surface of the outer ring, and a crescent-shaped shadow with a clear border in the posterior vitreous).[8] The magnetic resonance imaging (MRI) demonstrated an abnormal T2 signal area in the lateral ciliary body (Fig. 1), and the patient

Oligodendroglioma. https://doi.org/10.1016/B978-0-12-813158-9.00003-7

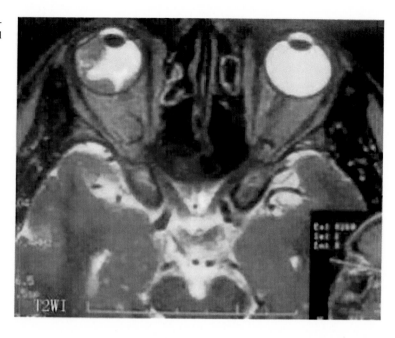

FIG. 1   Magnetic resonance imaging of ciliary body demonstrating oligodendroglioma tumor in the anterior lateral aspect of right orbit.

ultimately underwent enucleation with prosthetic implant. Immunohistochemistry results were as follows: OLIOG-2 (+), GFAP (−), NF (−), CD57 (focal +), CGA (−), CK (−), S-100 (scattered +), NSE (+/−) D34 (−), HMB45 (−), and KI-67 < 2%. The final diagnosis was ODG of the right eye, World Health Organization (WHO) grade II. The authors report that ciliary retinal stems cells have "the same properties of mobility and pluripotentiality as neural stem cells."[8 (p. 4)]

Another atypical ophthalmologic case of ODG that is likely the second case of retinal OGD was in 1969. A 3-year-old boy presented with new onset of headaches, fever, nausea, and vomiting with a longstanding history of right esotropia since the age of 6 months. Initially, he was thought to have heat stroke, but when the headaches reoccurred he was hospitalized and ophthalmology saw the patient. Examination revealed a large mass above and temporal to the right optic disc; ultimately, he required enucleation due to probable retinoblastoma. Soon after enucleation he became afebrile. Pathology's original impression was that this neoplasm most closely resembled an ODG due to "calcification without necrosis and cells with vacuolated cytoplasm and distinct cell borders."[19 (p. 286)] Slides were reviewed by several in house pathologists and sent for three additional opinions, all of whom agreed with the diagnosis of ODG. These authors report that prior to their 1969 published account, another case in 1947 was described in the literature and while they could not confirm if the 1947 case of retinal tumor was truly ODG, or some other glioma variant, they summarized it was a probable ODG of the retina, and that their case in 1969 was likely the *second* case of retinal ODG.[19]

Pineal region

Diverse tumor types can occur in the pineal region, including primary cancers such as germ cell tumors, pineal parenchymal tumors, gliomas, meningiomas, ependymomas, papillary tumors of the pineal region, and metastatic tumors, making it difficult to identify uncommon cancers (such as ODGs) arising in this area.[20–25] While exceedingly rare, case reports of ODG arising from the pineal region have been described.

In 2006 Das and colleagues published a case report of a 59-year-old woman with a 5-day history of HA, memory disturbance, and intermittent confusion.[22] Her medical history was positive for having had thyroidectomy and irradiation 14 years prior for thyroid cancer. Per report, neurological examination was essentially unremarkable except for nystagmus on lateral gaze. The MRI revealed a cystic irregularly enhancing mass in the pineal gland region, with compression of the cerebral aqueduct and mild ventriculomegaly. She underwent a frameless stereotactic approach to remove the tumor. The initial frozen section diagnosed the mass as a pineoblastoma. However, the permanent sections were more consistent with a high-grade glioneuronal tumor. Given the complexity of the case, the tissue was sent out for further pathological evaluation. It was then suggested that the diagnosis of an anaplastic ODG should be considered, and 1p/19Q deletion testing was conducted and found to be positive—consistent with anaplastic ODG. The postoperative MRI scan did not reveal any residual tumor, with resolution of hydrocephalus, and upward gaze paresis resolved after administration of high doses of dexamethasone. The patient underwent neuropsychiatric testing that revealed attention dysfunction and a delayed rate of information processing in the immediate postoperative phase, but the patient ultimately did return to neurological baseline.[22]

In the same year Kumar, Tatke, and Singh published their research results consisting of a retrospective review of 12 years of data and conducted histological analysis of these lesions of the pineal region.[25] Only 2 of 54 pineal tumors examined were anaplastic ODG, reinforcing how unusual this occurrence is.

Levidou et al. shared a case report in 2010 of a 37-year-old woman presenting with gaze impairment who denied nausea, vomiting, or other relevant past medical history (PMH).[23] The MRI revealed a cystic pineal/posterior third ventricular region mass. She underwent total excision of the tumor via midline infratentorial supracerebellar approach to the pineal region. Immunohistochemical (IHC) analysis revealed positivity for CD57, negative for EMA, Ker 18, Ker 7, synaptophysin, neurofilaments, CD99, and CD 10, as well as glial fibrillary acidic protein (GFAP). The Ki-67 proliferation index was low at approximately 1%. Given that the tissue was almost exhausted during immunohistochemistry an attempt was made to test for 1p and 19q, which did not show deletion. Therefore, on the basis of morphology and immunophenotype, a diagnosis of Grade II ODG was assigned. Immediate postoperative and 6-month MRI did not show any residual tumor.[23]

Lamis stated that by their estimate in 2014, only five cases of pineal region ODG existed in the literature, with the first low-grade case (i.e., grade II) being Levidou et al.,[23] reported in 2010; their report was only the second case of grade II ODG in the literature.[24] Their patient was a 22-year-old with PMH positive for congenital deafness secondary to rubella admitted with progressive HA, and the physical examination was positive for complete Parinaud's syndrome at presentation. The MRI revealed a heterogeneous, irregularly enhancing, solid lesion of the pineal gland, with compression of the cerebral aqueduct, and mild ventriculomegaly. A shunt was placed urgently with no improvement in her visual symptoms, and she then underwent a supracerebellar infratentorial approach for tumor resection. IHC analysis showed the neoplastic cells were positive for synaptophysin, neurofilaments, and GFAP. The Ki-67 proliferation index was low at <1% of neoplastic cells. FISH revealed 1p/19q deletion and therefore ODG grade II was confirmed.[24]

Naqvi et al. added a sixth case to the literature in 2017.[26] This was a 45-year-old man who presented with a 4-month HA associated with dizziness, and complaints of blurry vision. PMH was positive for diabetes and hypertension for 5 years. He underwent biopsy of the tumor; the neoplastic cells had round nuclei with a moderate degree of pleomorphism and were surrounded by a perinuclear halo. An increased mitotic rate was appreciated. IHC was positive for CD57, CD99, CD10, GFAP positive, weakly positive for synaptophysin, and high Ki-67 index at about 17%. Cytokeratin and extractable nuclear antigens (ENA) IHC stains were also negative. FISH was performed on a small number of cells and was positive for 1p/19q co-deletion.[26] The patient was offered chemotherapy and radiation therapy but declined both options, and was offered surgical management, but declined all treatment options.

Primary intraventricular (IV) ODG

While ODGs have a propensity to grow within the ventricles and often invade the ventricles, it is very unusual for them to truly arise from within the ventricles themselves.[11,27] It also can be a challenge for the clinician to separate them from a central neurocytoma due to similar features at the time of patient presentation.[28,29] Dickson described the first IV ODG case report in 1926,[30] and in actuality 1 of the 13 cases of ODG in Bailey and Bucy's 1929 report involved an ODG arising in the septum pellucidum.[31] By the year 1986 NG et al. reported that only 37 cases of primary intraventricular (IV) ODG were published and with their two cases the total was brought up to 39.[32] Through 2017 fewer than 80 cases were in the literature, demonstrating that while still not a common event, IV ODG well exceeds other rare sites of ODG occurrence.[27,33,34] Certainly, individual review of 80 cases is not feasible herein; therefore, key distinguishing characteristics of IV ODG will be presented.

While ODGs of primary intraventricular origin are uncommon in and of themselves, those arising in the septum pellucidum have been even more infrequently reported. In fact, by 1981 only 13 confirmed cases of ODG arising from the septum pellucidum had been reported in the literature, with the first being reported in 1939.[35] In 2009, a retrospective review was conducted of 70 cases in the years between 1926 and 2007, with a rate of occurrence of 20% in the septum pellucidum.[33] The cases are not discussed separately but are included with IV ODGs below.

Approximately 1.3%–10% of all ODGs arise in the ventricular system[3,36] and represent a distinct subgroup from intraparenchymal tumors.[37,38] IV ODG seems to present at a younger age than other ODG. Earnest and coworkers reported a mean age of 32.9 years with 14 of their 165 cases,[11] and other researchers have reported mean ages to be 28.0,[3] 30,[33] and 33.0 years.[36] There are also pediatric cases of primary IV ODG.[39] A slight female predominance for ODG occurring in this site has been noted.[33,37] Intraventricular ODGs tend to be low grade, with only three cases of anaplastic ODG reported between 1926 and 2007.[33,40]

On CT imaging IV ODG typically reveals a heterogeneous appearance, with moderate enhancement of the tumor after intravenous contrast.[27,37,38,41] Calcifications can be seen in some cases[27,37,38] and other researchers report that calcifications are frequently noted (i.e., wide ranges reported, 20%–91% of cases).[41] As seen on CT imaging, a majority of tumors caused widening of the ventricle(s) in which they arose with resulting hydrocephalus.[34,37,38,41] Peritumoral edema was not seen in the majority of IV ODG cases,[37,38] although in a report by Xiao et al., in which they studied 70 IV ODG patients, several cases did show peritumoral edema.[34] Additionally, Dokinskas and Simeone found a greater tendency for IV ODGs to have a tumor blush than peripheral ODGs and in their research (despite benign pathology), four of their cases demonstrated abnormal vessels angiographically.[37] Intraparenchymal ODGs often have ill-defined margins, but Dokinskas et al. found ODGs in the ventricles that were generally well defined.[37] However, in contrast, a more recent report by Xiao et al. also noted irregular, lobulated, and oval shaped IV tumors in their series of 70 patients.[34]

On MRI the tumor is typically hypointense in relation to the surrounding normal gray matter on T-1 weighted images and hyperintense on T-2 weighted images with heterogeneity of signal intensity being standard.[12,28,33,34,38,39,42,43] The Intraventricular ODGs tend to have a "chicken-wire" appearance indicative of a vascular pattern of dense capillaries that are branching and microcalcifications are often present. Angiography can be helpful in defining tumor extent, location, and vascularity.[27]

The IV ODGs are different from hemispheric ODGs in a few behaviors. Due to the location of intraventricular ODG's 76% patients present with signs of outlet obstruction including hydrocephalus and increased ICP.[33,37,39] Therefore, most patients will need CSF shunting.[27,33,37,39] Patients with IV ODG typically present with signs of increased ICP instead of the usual symptoms from hemispheric ODG due to the hydrocephalus from blocked CSF flow.[27,33,37,39] Patients presenting symptoms and physical examination findings are consequently consistent with increased ICP with HA and papilledema.[30,37,39,41]

Sixty-eight percent of IV ODG occur in the lateral ventricles and was more than twice as likely to occur in the right ventricle over the left,[33,39] rarely occurring in the third and fourth ventricles.[11,37,39] In contrast with these reports however, in a recent publication Xiao et al. reported three cases of IV ODG in the lateral ventricle, three cases in the third ventricle, and two cases in the fourth ventricle reflecting a slightly different rate of occurrence.[34]

While complete resection would be the treatment of choice, these tumors are often not amenable to complete resection due to microscopic infiltration of vital midline structures and growth along the ependymal lining of the ventricles,[32,38] as well as the often highly vascular nature of these tumors.[32]

Cerebellum

<10% of ODGs occur infratentorially,[44,45] with an overall rate of occurrence of ODGs located in the cerebellar hemispheres at approximately 3%.[11] When ODGs do occur infratentorially, approximately 66%–67% occur in the cerebellum, vs other infratentorial sites.[11,12,46] Krueger and Krupp in 1951 presented three cases of ODGs arising from the structures of the posterior fossa. Two of these were ODG tumors arising from the cerebellum.[47]

Patients with ODG arising from the cerebellum can present with many symptoms including: symptoms related to blockage of CSF outflow causing consequent hydrocephalus (e.g., headache, somnolence, and gait instability),[48] cerebellar syndrome,[5,45] seizures,[14] headache,[5,43,45,49] nystagmus,[46,47] papilledema,[5,45–47,50] visual issues/diplopia,[5] perioral numbness,[43] upper extremity tremor,[49] ataxia,[5,43,45–47,49,50] and nausea and/or vomiting.[45,46,49,50]

Posterior cranial fossa case reports are sporadic and include both low-grade[14,43,49–51] and high-grade[12,45,47,48] ODG dating back to 1928, when van Bogaert and Martin first reported an ODG arising from the cerebellum.[52] Recent case reports are beginning to clarify the anatomic locations more clearly than in historical publications.

In 1996, Alvarez et al. stated that only 56 cases of posterior fossa ODG could be identified out of 1593 cases published in the world literature,[53] providing further evidence that the posterior fossa and cerebellum are atypical sites for ODG to present. Furthermore, some authors argue that there are only two cases of *genuine* cerebellar ODG published in the literature that demonstrate the genetic hallmark of ODG with pathologically proven LOH and deletion of 1p/19q.[45,49] Regardless, pathologically confirmed cerebellar ODG cases will be presented herein, even if 1/19q is not reported in the publication, because the case predates the era when LOH 1p/19q became part of the consistent diagnostic workup for ODG's.

Wang et al. published the results in 1993 of 15 pediatric cases of ODG that were retrospectively examined and only two cases had tumors located in the cerebellum.[14] These two patients received subtotal resection plus ventriculoperitoneal shunting. One was lost after 1 month, and the other had been followed for a year at the time of publication. This study supported that reoccurrence of tumor was common for high-grade tumors and for those with positive CSF for tumor cells.[10,14]

Nadkarni and his fellows reported the case of an 18-year-old male presenting with headaches, vomiting, papilledema, nystagmus, and severe ataxia.[46] The CT scan showed a left cerebellar mass with calcifications that enhanced in a heterogeneous manner. The patient underwent craniotomy and had a histopathology confirmed ODG; he subsequently underwent craniospinal irradiation. The radiotherapy dosage was not provided in the publication and they reported that at one-year follow up he was tumor free.[46]

In 2007, Shah and his colleagues reported the case of an 11-year-old boy who had been in good health but presented after falling out of a tree and as part of the post-traumatic workup, underwent MRI revealing an incidental $2.0 \times 1.8 \times 1.3$ cm left cerebellar mass that was hypointense on T1- and hyperintense on T2-weighted images, respectively.[51] He had no focal neurological examination findings. He underwent craniotomy and immunohistochemistry revealed: GFAP (−); synaptophysin (−); Neu-N (−); S-100 (+); and Ki-67 labeling in <1% of cells. FISH was negative for chromosome 1p/19q deletion.[51] This publication states that only 10 cases of pediatric ODG had been reported to arise in the cerebellum prior to 2007.[51] Incidentally, their findings were consistent with other case reports in that deletion of 1p/19q were less likely to occur in pediatric cases.[7]

Lee and his colleagues examined 149 cases of pathologically proven ODG cases in Seoul, Korea between 1994 and 2008 and reported imaging findings for *only* infratentorial cases in their 2010 study, of which there were only six cases, and out of these six, four had cerebellar lesions (67%).[12] Prior studies suggested that infratentorial ODG's were relatively more common in younger age groups[11,48] but in their study all of the patients were over 50 years of age, which contrasts with prior results.[12] A limitation of their study was the small number of cases.

El Ouni et al. in 2010 published a case report of an 8-year-old female with a posterior fossa mixed density lesion that extended into the left cerebellar hemisphere, which showed calcifications and ring enhancement on CT.[5] The girl had vertigo, nausea, vomiting, and HA of two weeks duration, without seizures. On examination she was found to have ataxia, cerebellar syndrome, left sixth nerve palsy, and papilledema. An MRI was conducted and on the non-contrast T1-weighted images the tumor was hypointense, while on the contrast T1-weighted images there was peripheral enhancement of the tumor after gadolinium was administered. The fourth ventricle was compressed, with resultant hydrocephalus. The tumor was resected and histopathological analysis revealed a low-grade ODG, consisting of small round cells with nuclei, surrounded by a halo demonstrating the characteristic "fried egg" appearance."[5] The technical capacity to examine 1p/19Q deletion was not available. No metastasis was found outside the central nervous system. Treatment was local with prophylactic craniospinal radiation.[5]

In 2011, Furtado and his colleagues published two clinical and radiographical pediatric ODG case reports adding to the number of case reports.[45]

These cases highlight the fact that cerebellar ODGs are unusual but have some common clinical presenting features. While this chapter's focus is not on imaging specifically, these case report all address the fact that radiographically these cerebellar ODGs can lack consistent imaging findings that help differentiate ODG from other tumors. However, in these cases, ODGs tended to have moderate patchy contrast enhancement, to be hypointense on T1-weighted images and hyperintense on T2 and FLAIR images, and calcification in this area was not as common, with the tumor being more often solid than cystic.[12,43,45] Calcifications, enhancement, and edema are seen less frequently in infratentorial ODGs in children and adolescents than in adults.[54] Tumor enhancement is a great predictor of high-grade ODG and can be either patchy or homogeneous and when ring enhancement is present (albeit rare), it is purported to carry a poor prognosis.[5,9]

Brainstem

Among the atypical sites where ODG can occur, the brainstem appears to be among the most infrequent[14,47,53,54]; by 2012 Mohindra, Savaredekar, and Bal reported only six cases of intrinsic brainstem ODGs and ten cases of exophytic brainstem ODGs,[55] providing an indication of how rare this presentation is.

Primary brainstem ODG occurrence is as rare in pediatric patients as it is in adults.[55] Dohrmann, Farewell and Flannery published a synthesis of data spanning 41 years, reviewing all the cases of ODG that had occurred in children. Only one case of ODG was reported to have occurred in the brainstem from 1935 to 1975.

While having a primary ODG arising from the pons, medulla, or cerebellopontine angle is unusual, the presenting symptoms are as typical as other tumors in these critical locations, and include: ataxia,[9,44,47,48,53–57] upper motor-neuron signs,[9,13,54–56] headache,[14,54–57] visual issues/diplopia,[5,47,55,56] nystagmus,[13,44,47,53,56,57] papilledema,[5,47] nausea and/or vomiting,[47,53–57] dysphasia/impaired gag reflex,[13,53,55] seizures,[13,14,53] cranial nerve deficits and nerve palsies,[9,44,53–56,58] auditory deficit or disturbance,[9,53] hemiparesis,[9,54] and weight loss.[53] While case reports did not specifically mention classic brainstem syndromes with "crossed examination findings" (i.e., contralateral motor dysfunction with ipsilateral cranial nerve dysfunction), it was identifiable in patient presentations in several cases.[53,55]

Pons—In 2012, Mohindra, Savaredekar, and Baladded added their unique case to the literature of a 13-year-old girl with 6-month history of HA associated with projectile vomiting who presented with a 2-week history of visual changes, ataxic gait, and drooping eyelids.[55] On examination she had dysarthric speech, poor visual acuity in the left eye, bilateral ptosis, with the left eye deviated laterally and superiorly. The CT revealed a tumor in the brainstem, compressing and obscuring the fourth ventricle, leading to obstructive hydrocephalus. A shunt was placed, and then an MRI showed a large pontine mass, hypointense on T1-weighted, and isointense to hyperintense on T2-weighted images. The patient then underwent tumor debulking and did well with no change in deficits, except for minor exacerbation of right-sided motor weakness. The patient underwent craniospinal irradiation with 50 Gy. The histological features were consistent with ODG.[55] The patient was without tumor recurrence at 1 year. The authors state their case was distinctive because there was invasion of the midbrain from tumor in the pontine tegmentum.[55]

In 2015, Fukuoka et al. published two case reports of brainstem ODGs in pediatric patients stating that their findings were atypical for diffuse intrinsic pontine gliomas.[9] Finally, one case report was of an 8-year-old boy with MRI findings of "diffuse high-intensity changes in the pons, left middle cerebellar peduncle, and part of the left cerebellar hemisphere on T2-weighted images, with an enhanced spot lesion in the left cerebellar hemisphere.".[9] (p. 449) No calcifications were seen on CT. A biopsy was performed and on histological examination demonstrated highly cellular anaplastic oligodendroglial cells with perinuclear halos. IHC showed GFAP partially positive; oligodenrocyte transcription factor-2 (olig-2) positive in most neoplastic cells, negative for EMA and synaptophysin; p53 partially positive; IDH-1 wild type; FISH negative for 1p/19q deletion; and a MIB-1 labeling index of 35.7%.[9] Accordingly, the diagnosis was an anaplastic ODG. The tumor was screened for the presence of K27 M or G34 V mutations in histone H3.3 and was confirmed as having a K27 M mutation. Despite an aggressive course of radiotherapy (59.4 Gy in 33 fractions), the patient died 7 months from onset.[9]

While several authors have published case reports of ODGs located in the pons or medulla as presented above, bona fide brainstem ODGs are exceedingly rare in adults and some researchers feel that only two such cases have been published. One of these cases is that of a 42-year-old man with a non-enhancing mass in the right pons, published in 2015 by Hodges and colleagues, that demonstrated both deletion of 1p/19q and an IDH-2 mutation.[58] It was thought to be only the second case of a brainstem ODG with verified 1p/19q co-deletion, and was the first published report occurring simultaneously with an IDH-2 mutation.[58] IHC testing of the biopsied material revealed GFAP+; NeuN and chromogranin were negative, FISH was positive for 1p/19q co-deletion, and IDH-1/2 mutations were not present. However, subsequent sequencing of the IDH-1 and IDH-2 genes revealed an IDH-2 R172 M mutation.[58]

By 2018, only 6 ODG cases localized in the cerebellar peduncle had been published, with only two cases of true ODG originating from the cerebellar peduncle (i.e., not invasion into this structure from the cerebellum or deep cerebellar nuclei).[43] In 2018, Baran and coworkers reported the case of a 43-year-old woman who presented with complaints of headache, perioral numbness, and gait problems. On examination, she had findings of hemifacial palsy, right-sided dysmetria, dysdiadochokinesia, and wide-based gait. Imaging studies revealed a cyst-like, heterogeneous, hyperintense, well-demarcated lesion located in the right cerebellar peduncle. She had a near total excision of the tumor and pathology was consistent with a WHO grade II ODG (GFAP, olig2, IDH-1 mutation, synaptophysin, and EGFR were all positive; Ki-67 labeling index—1–6%). She was discharged 7 days after the craniotomy.[43]

Medulla—Although Bailey and Bucy's publication in 1929 mentions infratentorial ODG occurrences,[31] one of the earliest case reports specifically discussing a patient with ODG arising in the medulla was published by Krueger and Krupp in 1951.[47] In this publication, the authors presented three cases of ODGs arising from the structures of the posterior fossa, one of which was found in the medulla. The case was that of a 10-year-old boy who was admitted with persistent nausea and emesis. Earlier in the year he had developed headaches and double vision. On examination, he had an ataxic gait, papilledema, and nystagmus. A mass was found in the medulla, which was partially resected, and had histological features consistent with ODG. While the case itself was not unique in terms of presenting signs, symptoms, imaging, or pathological findings—as well as the typical grim clinical course—the fact that it was an ODG arising from the medulla did make it very unusual.[47]

Lee and his colleagues examined 149 cases of pathologically proven ODG cases in Seoul, Korea between 1994 and 2008 and reported imaging findings for only infratentorial cases in their 2010 study, of which there were only six, and only one of these was in the medulla.[12] Prior to this, only a few cases of ODG that had been reported in the literature had presented in the brainstem. Their findings were slightly different from other results, as their infratentorial tumors occurred in an older population (all of their cases occurred in patients that were 50 years of age or older); earlier brainstem ODG cases had primarily occurred in pediatric patients.[12]

In 2014, Reyes-Botero and his colleagues looked at 17 cases of brainstem gliomas in adults to examine their molecular analysis, and two of these were ODG grade II, occurring in the medulla oblongata (42.6 years old) and the pons (32.1 years old). These patients received radiation and chemotherapy with radiation, respectively, and survived 57.4 months and 57.7 months after treatment. Both were IDH-1 mutation negative and p53 negative, and one case that was tested for 1p/19q was non-co-deleted.[59]

Several authors have published case reports of ODGs located in the pons or medulla, but bonafide brainstem ODGs that are confirmed by IHC continue to be exceedingly rare in adults. One such confirmed case was published in 2014 by Hewer and colleagues, where a 55-year-old woman presented with dysphagia, generalized weakness, and hypesthesia of all four extremities. An MRI showed a mass centered in the medulla, with extension into the cerebellar hemispheres. After documentation of growth, a stereotactic biopsy was performed, which was consistent with WHO grade III anaplastic ODG. On LOH analysis, the tumor was co-deleted for 1p and 19q. She was treated with combined radiotherapy and Temozolomide chemotherapy, but developed an infection after 4 weeks and had to be placed into palliative care.[13]

Congenital presentation

Congenital ODG is rare, but a few cases have been reported.[6, 60–62] A brief summary of atypical cases in the literature of congenital ODG will be mentioned at this juncture as they typically have occurred in the cerebellum and brainstem.

In 1959, Svobada published a case of a 6-week old who died from congenital ODG and photos of autopsy in the case report show tumor attached to the peduncular region of mid-brain encasing the pituitary stalk and optic chiasm.[63]

Other cases reflecting this unusual occurrence of ODG occurring congenitally were in 1981 of a ODG occurring in the medulla oblongata in a neonate[61]; the second case reflective of congenital ODG was published in 1994, in a neonate delivered by emergency cesarean section who, two weeks after delivery had craniotomy for tumor in the area of the third ventricle blocking the foramen of Munro and causing obstructive hydrocephalus (this child was still alive at 2 years old at time of publication)[62]; and the third published case report of a tumor in a 34th gestational week fetus that covered the basal part of the brain and replacing the cerebellum.[60]

Kostadinvo and de la Monte published pathology findings in 2017 regarding an infant that died 3 days post-cesarean delivery, despite aggressive medical care, from a widely infiltrating brainstem ODG.[6] Histological staining demonstrated anaplastic ODG infiltrating the pons, fourth ventricle, midbrain, medulla, cerebellar white matter, posterior thalamus, and occipital white matter. IHC staining revealed Ki-67 positive, p53 negative, and myelin basic protein, scattered neoplastic cells with immunoreactivity for synaptopysin, NeuN, neuro-specific enolase, or GFAP.[6]

There are other case reports, but they are overall very infrequent; Chapter 4 expands on pediatric occurrence of ODGs in greater detail.

Leptomeningeal involvement

Leptomeningeal disease (LMD; also known as leptomeningeal metastasis or leptomeningeal gliomatosis) is typically a late complication of high-grade or malignant glial tumors, most commonly seen in Glioblastoma multiforme (GBM).[64] When LMD occurs, the primary tumor site sheds tumor cells into the CSF, where they can circulate and begin to grow on the surface of the brain, cranial nerves, brain stem, spinal cord, and spinal nerves, as well as within the ventricles. It has been reported to occur in 6%–7% of patients with GBM,[64,65] and has a median survival from diagnosis of roughly 4 months. It has also been described in patients with ODG, but with a lower rate of occurrence of approximately 4%–5%.[65,66] The signs and symptoms of LMD are quite variable, depending on where in the CNS the tumor is growing and affecting nerve function and can include cognitive deterioration, headache, neck pain, lower back pain, radicular pain (i.e., down an arm or leg), loss of balance, loss of vision and other cranial neuropathies, leg weakness, and bowel and bladder disturbance.[67–74]

It is quite rare for ODG patients to have LMD at presentation, and to our current knowledge this scenario has never been reported for a typical parenchymal oligodendroglial brain tumor. There have been several case reports of an ODG of the spinal cord that at presentation had cerebral symptoms and diffuse leptomeningeal spread involving the cerebral meninges.[75,76] The most recent case describes a 30-year-old woman without any significant PMH, who initially presented with a 3-week history of worsening headaches that were accompanied by nausea, emesis, photophobia, and anorexia.[75] An initial cranial CT scan and lumbar puncture were unremarkable. Six weeks later she had an episode of transient speech disturbance, followed by a tonic–clonic seizure. An MRI showed leptomeningeal enhancement of the basal meninges, as well as a high signal, non-enhancing enlarged region of the spinal cord at C3–4. An initial biopsy of the brain and involved meninges did not reveal any neoplastic cells or inflammation. She was placed on a course of steroids that improved her symptoms initially, but soon afterwards she continued to deteriorate and soon was unable to care for herself. The patient was eventually diagnosed with hydrocephalus, and so underwent a VP shunting procedure. However, several months later she continued to deteriorate, developing confusion and hallucinations; follow-up MRI scan showed increasing meningeal enhancement with improved hydrocephalus. In addition, there was progression of the cervical spinal cord lesion, with enlargement and prominent enhancement. She underwent a C3–4 laminectomy and partial inferior laminectomy of

C-2; the dorsum of the spinal cord at this level had an exophytic mass and the dura was thickened. The mass could only be 50% resected, due to being very infiltrative. The pathology was consistent with a WHO grade II ODG, with a Ki-67 labeling index of 5%, and co-deletion of 1p and 19q. She was treated with radiotherapy to the cervical region in combination with Temozolomide chemotherapy; whole brain RT was also suggested, but she declined. One year later she was stable, with improvement and regression of the spinal cord tumor, as well as in the amount of cerebral meningeal enhancement.

Primary diffuse leptomeningeal gliomatosis (PDLG) is a rare condition where the tumor arises and grows in the sub-arachnoid space and does not have any connection to a tumor in the parenchyma of the brain or spinal cord. The tumor is thought to be derived from heterotopic nests of neuroglial tissue within the leptomeninges.[77] PDLG has most commonly been described to occur with astrocytic tumors—often high-grade, including GBM—and carries a poor prognosis, with a median survival time of 4 months. However, it can also be of oligodendroglial origin in rare cases in both adults and children.[67–74,78] The reported cases of oligodendroglial PDLG have ranged in age from 2 to 60 years, with a slight male preponderance. The most common symptoms are headaches[65–67,69,70,79] and other signs of elevated intracranial pressure,[67,71,73,75,79–81] although other symptoms have been reported including seizures,[64,67] weakness/gait difficulty,[65,72,74,78,79] back pain,[65,72] and loss of cognitive function.[66,68,81] MRI usually demonstrates diffuse thickening and enhancement of the meninges, often involving the cerebrum, basilar regions, and spinal axis. In a few cases, there has been calcification noted in some areas of enhancement. Lumbar puncture for cerebrospinal fluid analysis and cytology has usually been non-diagnostic. In most of the cases, a biopsy for tissue evaluation was necessary to clarify and confirm the diagnosis; a few cases were diagnosed at autopsy. The histology has been a mix of grade II and grade III ODG, with the more benign cases being more common. Molecular phenotyping and testing for deletion of 1p and 19q has only been performed in three of the cases.[71,74,78] Two tumors had co-deletion of both 1p and 19q,[71,78] while another had only deletion of 1p.[74] Treatment has consisted of radiotherapy to the brain in most cases, along with radiotherapy to the spinal axis as well in several patients. Chemotherapy has also been attempted in several cases, with varying results. For example, in the report from Mathews and colleagues, a 50-year-old woman developed PDLG with ODG histology and was initially treated with a combination regimen consisting of cranial irradiation and daily oral Temozolomide for 6 weeks.[70] Unfortunately, the patient did not respond very long to this approach and had rapid progression of disease. Salvage chemotherapy with Etoposide was not successful and she quickly advanced to hospice care. In contrast, the patient reported by Michotte and coworkers had an excellent response to chemotherapy.[71] The patient was a 60-year-old man who developed PDLG with anaplastic ODG histology and co-deletion of 1p and 19q, and was felt to be too clinically unstable to undergo craniospinal radiotherapy. Instead, the patient received neoadjuvant Temozolomide 150 mg/m^2/day for 5 days every 4 weeks. After 6 cycles of Temozolomide, there was a dramatic improvement in the MRI scans, with resolution of the meningeal enhancement, as well as slow clinical improvement in his neurological condition. Radiotherapy had to be administered 1 year after diagnosis, when there was recurrence of diffuse enhancement on follow-up MRI.

Gliomatosis cerebri

Gliomatosis cerebri has been listed as a separate nosological entity in the WHO classification of CNS tumors for many years and is characterized by diffuse infiltration within the brain, affecting two or more cerebral lobes or regions (e.g., basal ganglia, thalamus)—yet maintaining the architecture of the involved neural tissue—and is most commonly described with astrocytic cell lineage.[82] More recently, with the 2016 update of the WHO Classification of Tumors of the Central Nervous System, gliomatosis cerebri has been dropped as a separate entity and instead has been relegated to a type of severely infiltrative growth pattern that can occur in any of the diffuse gliomas.[83] The main reason for dropping it as a separate type of glioma entity is that recent molecular studies have demonstrated that the phenotypes (e.g., IDH-1 mutation status, ATRX mutation status, 1p/19q deletion status, etc.) are similar between glial tumors that display a gliomatosis type growth pattern and others that have a more circumscribed growth pattern. Case reports of patients with gliomatosis cerebri in which the cell type is purely or predominantly oligodendroglial are very rare.[84–90] The majority of cases have been in adults,[84–88,90] although a few children have been reported as well.[89] The most common signs and symptoms are headaches,[85,86,89] seizures,[87,89,90] focal weakness,[84,88,89] papilledema,[84,89] and visual changes.[85,86] A biopsy procedure is always necessary to verify the neoplastic nature of the MRI findings, and to rule out other etiologies, such as vasculitis, viral infection, inflammatory disease, PML, and leukodystrophy. The oligodendroglial tissue obtained at biopsy can vary in grade, and be grade II or grade III.[82] Molecular testing for 1p and 19q deletion status has only been reported in one case thus far, by Farooq and colleagues—the tumor was non-deleted for 1p and deleted for 19q.[90] Treatment has consisted of a course of radiotherapy in most patients, with some responders. Chemotherapy has been attempted in several cases, with responses to PCV (procarbazine, CCNU, vincristine) and PAV (procarbazine, ACNU, vincristine).[84,87]

Spinal cord

While drop metastases are seen from ODG, primary spinal ODGs are a rare entity, with only about 50 cases of spinal ODG currently including all grades,[67,91–93] and of these ODGs only 10 cases to date are histologically confirmed high grade anaplastic ODGs.[92,93] The incidence of primary spinal cord ODG is reported to be higher in the adult population than in children.[94] Fountas et al.[91] recount the historical accrual of primary spinal ODG cases published through the years, and this will not be revisited herein.

This publication provides excellent review of anatomical location of reported cases of spinal ODGs and these authors state "there appears to be a predilection in descending order of frequency for the thoracic (30%), cervical (25%), and lumbar (5%) areas among 40 patients in the literature whose data were available."[91 (p. 173)] Additionally, in their research they note that "holocord" ODG is an entity in of itself that is an atypical presentation of ODG, and it should be mentioned that holocord ODG occurs in 7.5% of primary spinal ODG cases.[91] The patients with "holocord" ODG were all <16 years old.[91,94,95]

Spinal cord ODGs typically present with spinal cord symptoms such as weakness, sensory changes, pain, scoliosis, or genitourinary symptoms.[75,91–97]

Strickland et al. in 2016 published an illustrative case of anaplastic ODG presenting in the spinal cord as the primary site.[92] A 68-year-old man with a PMH of prostate cancer presented with a 1-month history of right leg numbness. He had a prior history of left leg numbness a year prior as well and having undergone decompression surgery of lumbar spine for stenosis. He had weight loss, weakness of lower extremities, numbness, and gait imbalance causing falls to the extent he needed to use a cane or wheelchair. Examination revealed no abnormalities with upper extremities. He maintained full motor strength in bilateral lower extremities. However, left lower extremity revealed left gastrocnemius muscle with minimal antigravity movement with atrophy. Patellar reflexes were absent bilaterally, decreased pinprick sensation bilaterally, and on the left he had non-dermatomal distribution sensory loss. MRI revealed nonenhancing T2 hyperintense intramedullary lesion at T-9 to T-12 expanding the conus (after diagnosis retrospective review revealed an extramedullary nodule at T-11). Surgery was performed in an attempt at resection and to gain tissue diagnosis; total resection was not possible. Histologically, the tissue showed 1p/19q co-deletion with no evidence of IDH1 or IDH1 mutation (IDH wild type). Postoperatively he received radiation with 39/2 Gy in 22 fractions of proton radiation from T1 to bottom of the thecal sac, followed by a boost to the T-9–T-12 area to a total dose for 43.2 Gy in 24 fractions to the primary disease followed by adjuvant TMZ 200 mg/m^2 on an every 28 day adjuvant basis for eight cycles, and at 1 year he was without radiographic recurrence and was clinically stable.[92]

Metastatic ODG to extracranial/extraneural sites

ODGs are the least likely CNS tumors to metastasize out of the CNS per Liwnicz and Rubinstein's 1979 study where they examined 116 cases of CNS tumors with extraneural metastasis.[98] Rates for metastasis were highest in patients with glioblastoma (41.4%), medulloblastoma (26.7%), ependymoma (16.4%), and astrocytoma (10.3%); patients with ODG were ranked fifth after these other tumor types in terms of metastatic disease (4.2%).[98]

Because systemic metastases are encountered so infrequently, clinicians often are working patients up for a second primary before considering the possibility of metastatic ODG.[99,100] It is imperative to heighten awareness that patients with prior history of a glial tumor should be evaluated for extraneural metastasis as a strong possibility, as well as for new primary diagnosis of non-CNS origin.[101]

Additionally, there are cases in which a decade or more has gone by before the patient has presented with metastatic disease after the original ODG diagnosis,[102–107] as well as cases in which metastasis transpires without intracranial reoccurrence.[99,108–113] This makes it all the more important for clinicians to be aware of the possibility of atypical metastatic presentations.

It is believed that several factors contribute to the low rates of glial tumor metastasis: (1) the blood–brain barrier, (2) lack of lymphatic pathways in the brain and spinal cord, (3) narrowing of the vessels with tumor growth, and (4) lack of a suitable environment for metastatic malignant clones outside of the cranial tissues.[98,101,114–118]

There are several well-accepted mechanisms responsible for spread of glial tumor cells outside of the CNS. Metastasis occurs via local invasion (subsequent access through the surgical defect into the extra-meningeal tissues), hematogenous spread, and seeding through the CSF,[119,120] via ventriculo-peritoneal shunting.[121]

Due to the emergence of new treatment options, ODG patients are living longer with their disease, resulting in more time for ODGs to spread. Therefore, patients who have extracranial sites of tumor should be scrutinized to rule out metastatic ODG as an etiology vs other primary tumors to avoid misdiagnosis, as has happened in previous cases.[100,101]

Another danger is that often a fine needle aspirate (FNA) is ordered instead of an excisional biopsy in the work-up of patients with pathologic lymphadenopathy. It is preferred that patients undergo an excisional lymph node biopsy, instead of an FNA, to avoid "crush" artifact. A total excisional lymph node biopsy will provide more tissue and a higher quality tissue sample for improved diagnostic outcomes. For example, Can et al. stated that they missed the opportunity to test for 1p/19q deletion status in a cell-block sample after FNA, because there was insufficient tissue.[101] Volavsek, Lamovec, and Popovic in 2009 had a case where a definitive diagnosis was elusive until excisional biopsy was performed; FNA was not definitive.[121]

Scalp/skin—Skin involvement of anaplastic ODG is rare and McLemore reports five cases existed in the literature prior to 2013[115,122–124] with the first case reported in 1951.[102] This first patient developed multiple tumor nodules of the scalp after craniotomy and surgical resection, and ultimately progressed to have widely metastatic disease and died.[122]

In 2013, two publications were released discussing case reports of patients with scalp involvement of ODG bringing the tally up to seven case reports.[102,125] McLemore et al. report a 37 year old female who presented with HA and tumor in the left parietal lobe and was diagnosed with WHO grade III ODG after having surgical excision of the tumor.[102] She was initially diagnosed at 10-years of age with ODG WHO grade II, treated with surgical excision and then followed. Seven months after the craniotomy for the recurrent tumor, she had thickening of the scalp at the site of the craniotomy incision. She underwent surgical reconstruction and excision of the area; pathology confirmed anaplastic ODG WHO grade III scalp metastasis.[102] The other case report in 2013 listed scalp among one of many sites of metastasis in a 49-year-old man admitted with widely metastatic ODG with lymph nodes, parotid gland, and bone involvement as part of the dissemination of the tumor.[125] He also had multiple masses over bilateral lower neck. Despite multiple courses of palliative chemotherapy his tumor grew progressively and MRI revealed right temporal extracranial subcutaneous and right frontal recurrent tumors.[125]

Breast—Alacaciouglu and colleagues published a case report in 2012 of a 58-year-old man with anaplastic ODG, who was in cycle #3 of adjuvant monthly TMZ, when he noted pathologic lymphadenopathy and a left breast nodule. The nodule was resected and pathology confirmed tissue infiltration of anaplastic ODG, that had features suggestive that the tumor had originated from the intracranial lesion,[120] confirming metastatic disease.

In 2013, two more case reports were published by Mazza et al. where ODG was found to have metastasized to the breast.[100] Their cases showed more widely disseminated metastasis of ODG along with the simultaneous breast metastasis. The second case actually had initially been misdiagnosed as breast cancer and the patient was not responding to typical chemotherapy targeted to breast cancer. After clinical conditions worsened and the patient had changed facilities the breast biopsy tissue was reexamined, confirming a misdiagnosis and was read out as an anaplastic ODG metastasis. IHC was positive for GFAP and S-100, and a bone lesion biopsy was also read out as ODG; her treatment regimen was changed. However, she ultimately died of widely metastatic disease.[100]

Multiple Solid Organs—Between the years 1951 to 2014 Li et al. conducted a literature review and in their publication present a table of 61 reported patients with ODG metastases outside the CNS.[126] Within this table they list approximately 14 cases that had organ involvement with metastatic disease, including the thymus gland,[127] lung,[122,128–130] (Fig. 2) chest wall or pleura,[103,109,110,127,129,131,132] liver,[109,119,123,130,131,133] (Figs. 3 and 4) spleen,[109] adrenal gland,[109,128] abdomen,[109] and iliopsoas muscle.[109]

FIG. 2 Enhanced CT scan of the chest shows multiple nodules and poorly defined nodular infiltration of both lung parenchyma.

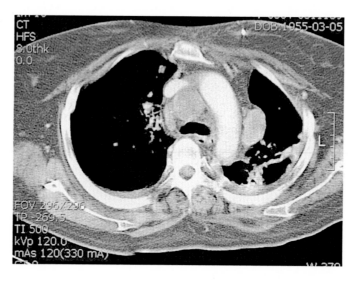

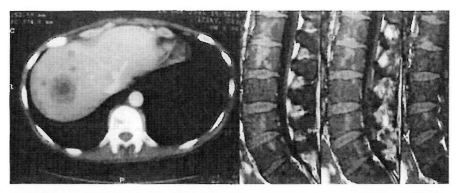

FIG. 3   CT imaging of multifocal metastatic OGD to liver and MR imaging of bone marrow involvement with ODG. Multifocal metastitic ODG (*left*) and metatatic ODG to bone marrow (*right*).

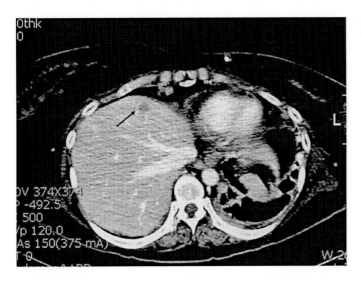

FIG. 4   Abdominal CT scan with contrast demonstrates a low attenuation lesion (*arrow*) in the liver.

Metastatic ODG to extraneural sites is an infrequent occurrence, as these numbers attest, and when it occurs it is usually a widely metastatic presentation.[111,116,120,121,123,124,127–134]

Lymph Nodes—As noted above, Li et al. conducted a literature review and in their publication have a table of 61 reported patients with ODG metastases outside the CNS between the years 1951 to 2014.[126] Within this table they list approximately 20 cases that had lymph node involvement with metastatic ODG. Many of these cases of metastatic ODG to the lymph nodes occurred simultaneously with widely metastatic disease, rarely occurring as solely a pathologic lymph node, and since this 2014 literature review, there have been additional published case reports with lymph node metastasis.[100,101,115,121,122,125,126]

Bone and Bone Marrow—While ODG is unlikely to metastasize to bone,[98] according to Zustovich and his colleagues who examined 32 cases of metastatic anaplastic ODG in the world literature between 1951 and 2008, when ODG does metastasize, the most frequent metastatic site was bone and bone marrow at 97%.[119] These authors[119] explained the affinity for ODG cells for bone based on the presence of neural cell adhesion molecule (NCAM), which is expressed by both gliomas and osteoblasts.[134,135]

Cases of bone metastasis are available,[99,100,104,108,110,114,119,121,125,126,133,136–138] and in the publication by Li et al.[126] referenced above where they synthesized a table of 61 reported patients with metastatic ODG, 31 patients were cited in which bone disease was part of their metastatic spectrum; 20 case reports included bone marrow involvement as a separate metastatic site.

Likewise, clinicians typically think of pancytopenia as the result of hematologic malignancies, rarely considering glial neoplasm invasion of the bone marrow as an etiology, which is in keeping with how infrequently this happens in clinical practice. However, there are several individual case reports in which bone marrow biopsy results demonstrate marrow invaded by ODG,[100,105,107,111,113,119,126,136,139] making it necessary to consider this as a possibility for patients with a past history of ODG and new cytopenias not easily explained by other causes.

In conclusion, Merrell and his colleagues stated that it is well established in the literature that deletions in 1p and 19q are associated with better response to chemotherapy and prolonged survival. However, they suggested that patients with 1p/19q co-deletions are prone to metastasis and that additional research would be necessary to substantiate or refute this hypothesis.[110]

References

1. Ostrom Q, Gittleman H, Xu J, et al. CBTRUS statistical report: primary brain and other central nervous system tumors diagnosed in the United States 2009–2013. *Neuro Oncol.* 2016;18(5):v1–v75.
2. Dohrmann GJ, Farmwell JR, Flanner JT. Oligodendrogliomas in children. *Surg Neurol.* 1978;10:21–25.
3. Chin HW, Hazel JJ, Kim TH, Webster JH. Oligodendrogliomas. I. A clinical study of cerebral oligodendrogliomas. *Cancer.* 1980;45(6):1458–1466.
4. Wesseling P, van de Bent M, Perry A. Oligodnedroglioma: pathology, molecular mechanism and markers. *Acta Neuropathol.* 2015;129:809–827.
5. El Ouni F, Gaha M, Moulahi H, Daadoucha A, Krifa H, Tlili K. Infratentorial oligodendroglioma in a child: case report and review of the literature. *Turk Neurosurg.* 2010;22(4):461–464.
6. Kostadinov S, de la Monte S. A case of congenital brainstem oligodendroglioma: pathology findings and review of the literature. *Case Rep Neurol Med.* 2017;2017. 1–4.
7. Kreiger PA, Okada Y, Simon S, Rorke LB, Lous DN, Golden JA. Losses of chromosomes 1p and 19q are rare in pediatric oligodendrogliomas. *Acta Neuropathol.* 2005;109:387–392.
8. Guo Q, Hao J, Sb S, et al. Oligodendroglioma of the ciliary body: a unique case report and the review of literature. *BMC Cancer.* 2010;10:579.
9. Fukuoka K, Yanagisawa T, Wantanabe Y, et al. Brainstem oligodendroglial tumors in children: two case reports and review of literatures. *Childs Nerv Syst.* 2015;31:449–455.
10. Mørk SJ, Lindegaard KF, Halvorsen TB, et al. Oligodendroglioma: incidence and biological behavior in a defined population. *J Neurosurg.* 1985;63(6):881–889.
11. Earnest E, Kernohan JW, Craig WM. Oligodendrogliomas. A review of 200 cases. *Arch Neurol Psychiatry.* 1950;63:964–976.
12. Lee IH, Kim ST, Suh YL, et al. Infratentorial oligodendrogliomas: imaging findings in six patients. *Acta Radiol.* 2010;51(2):213–217.
13. Hewer E, Beck J, Vassella E, Vajtai I. Anaplastic oligodendroglioma arising from the brain stem and featuring 1p/19q co-deletion. *Neuropathology.* 2014;34(1):32–38.
14. Wang KC, Chi JG, Cho BK. Oligodendroglioma in childhood. *J Korean Med Sci.* 1993;8(2):110–116.
15. Wong JYC, Uhl V, Wara W, Sheline G. Optic gliomas. A reanalysis of the University of California, San Francisco experience. *Cancer.* 1987;60:1847–1855.
16. Offret H, Gregoire-Casoux N, Frau E, Doyon D, Comoy J, Lacroi C. Solitary oligodendroglioma of the optic nerve. Apropos of a case. *J Fr Opthamol.* 1995;18(2):158–163.
17. Katayama K, Asano K, Ohkuma H, et al. Case of pediatric optic pathway oligodendroglioma presenting widespread invasion and dissemination in the cerebrospinal fluid. *Brain Tumor Pathol.* 2014;31(3):208–214.
18. Lucarini C, Tomei G, Gaini SM, Grimoldi N, Spagnoli D, Losa M. A case of optic nerve oligodendroglioma associated with an orbital non-Hodgkin's lymphoma in adult. Case report. *J Neurosurg Sci.* 1990;34(3–4):319–321.
19. Boniuk M, Bishop DW. Oligodendroglioma of the retina. *Surv Ophthalmol.* 1969;13(5):284–289.
20. Brastianos HC, Brastianos PK, Blakeley J. Pineal region tumors. In: Norden A, Reardon D, Wen P, eds. *Primary Central Nervous System Tumors.* New York, NY: Humana Press; 2011:435–455. Current Clinical Oncology. https://doi.org/10.1007/978-1-60761-166-0_19. Retrieved from.
21. Sonabend AM, Bruce JN. Pineal tumors. In: Winn HR, ed. *Youmans and Winn Neurological Surgery.* Philadelphia, PA: Elsevier; 2017:1048–1064.
22. Das S, Chandler JP, Pollack A, et al. Oligodendroglioma of the pineal region. Case report. *J Neurosurg.* 2006;105(3):461–464.
23. Levidou G, Korkolopoulou P, Agrogiannis G, Paidakakos N, Bouramas D, Patsouris E. Low-grade oligodendroglioma of the pineal gland: a case report and review of the literature. *Diagn Pathol.* 2010;5:59.
24. Lamis FC, de Paiva Neto MA, Stavale JN, Cavalheiro S. Low-grade oligodendroglioma of the pineal region: case report. *J Neurol Surg Rep.* 2015;76(1):e55–e58.
25. Kumar P, Tatke M, Sharma A, Daljit S. Histological analysis of lesions of the pineal region; a retrospective study of 12 years. *Pathol Res Pract.* 2006;202:85–92.
26. Naqvi S, Rupareliya C, Shams A, Hameed M, Mahuwala Z, Giyanwani PR. Pineal gland tumor but not pinealoma: a case report. *Cureus.* 2017;9(8).
27. Lee KS, Kelly Jr DL. Primary oligodendroglioma of the lateral ventricle. *South Med J.* 1990;83(2):254–255.
28. de Castro FD, Reis F, Guerra JG. Intraventricular mass lesions at magnetic resonance imaging: iconographic essay—part 2. *Radiol Bras.* 2014;47(4):245–250.
29. Okada M, Yano H, Hirose Y, et al. Olig2 is useful in the differential diagnosis of oligodendrogliomas and extraventricular neurocytomas. *Brain Tumor Pathol.* 2011;28(2):157–161.
30. Markwalder TM, Huber P, Markwalder RV, Seiler RW. Primary intraventricular oligodendrogliomas. *Surg Neurol.* 1979;11(1):25–28.
31. Bailey P, Busy PC. Oligodendrogliomas of the brain. *J Pathol Bacteriol.* 1929;32:735–751.

32. Ng S, Lui C, Wai Y, Liu Y, Tsai C, Chen L. Primary intraventricular oligoendroglioma—report of two cases. *Chang Gung Med J*. 1986;9:33–40.

33. Zada G, McNatt SA, Gonzalez-Gomez I, McComb JG. Anaplastic intraventricular oligodendroglioma: case report and review of the literature. *Surg Neurol*. 2009;71(6):693–700.

34. Xiao X, Zhou J, Wang J, et al. Clinical, radiological, pathological and prognostic aspects of intraventricular oligodendroglioma: comparison with central neurocytoma. *J Neurooncol*. 2017;135(1):57–65.

35. Geuna E, Regalia F, Pappadà G, Arrigoni M. Septum pellucidum oligodendroglioma. Case report and review of literature. *J Neurosurg Sci*. 1981;25(1):49–53.

36. Roberts M, German WJ. A long-term study of patients with oligodendrogliomas. *J Neurosurg*. 1966;24:697–700.

37. Dolinskas CA, Simeone FA. CT characteristics of intraventricular oligodendrogliomas. *AJNR Am J Neuroradiol*. 1987;8(6):1077–1082.

38. Atasoy C, Karagulle AT, Erden I, Akyar S. Primary oligodendroglioma of the lateral ventricle: computed tomography and magnetic imaging findings. *J Ankara Med School*. 2002;55(1):39–44.

39. Gai X, Li S, Wei Y, Yu S. A case report about oligodendrogliomas of the fourth ventricle. *Medicine (Baltimore)*. 2018;97(17).

40. Pomero E, Piana RL, del Pilar CM, Tampieri D. Intraventricularlocalization of an anaplastic oligodedendroglioma: a rare event. *Can J Neurol Sci*. 2012;39(5):649–651.

41. Hasuo K, Fukui M, Tamura S, et al. Oligodendrogliomas of the lateral ventricle: computed tomography and angiography. *J Comput Tomogr*. 1987;11(4):376–382.

42. Pant I, Chaturvedi S, Jha DK, Kumari R, Parteki S. Central nervous system tumors: radiologic pathologic correlation and diagnostic approach. *J Neurosci Rural Pract*. 2015;6(2):191–197.

43. Baran O, Kasimcan O, Oruckaptan H. Cerebellar peduncle localized oligodendroglioma: case report and review of the literature. *World Neurosurg*. 2018;113:62–66.

44. Ellenbogen JR, Perez S, Parks C, Crooks D, Malluci C. Cerebellopontine angle oliodendroglioma in a child; first case report. *Childs Nerv Syst*. 2014;30:185–187.

45. Furtado SV, Venkatesh PK, Ghosal N, Murthy GK, Hegde AS. Clinical and radiological features of pediatric cerebellar anaplastic oligodendrogliomas. *Indian J Pediatr*. 2011;78(7):880–883.

46. Nadkarni TD, Menon RK, Dsai KI, Goe AH. Cerebellar oligodendroglioma in a young adult. *J Clin Neurosci*. 2005;12:837–838.

47. Krueger EG, Krupp G. Oligodendrogliomas arising from structures of the posterior fossa. *Neurology*. 1952;2(6):461–470.

48. Packer RJ, Sutton LN, et al. Oligodendrogliomas of the posterior fossa in childhood. *Cancer*. 1985;56(10):195–199.

49. Gru AA, Fulling K, Perry A. A 39 year-old man with a cerebellar mass and pancytopenia. *Brain Pathol*. 2012;22:251–254.

50. Chitkara N, Chanda R, Thakur AK, Chanda S, Sharma NK. Posterior fossa oligodendroglioma. *Indian J Pediatr*. 2002;69:1099–1100.

51. Shah S, Schelper RL, Krishnamurthy S, Chang HT. An 11-year-old boy with an incidental mass in the left lateral cerebellum. *Neuropathology*. 2007;27(1):95–97.

52. Van Bogaert L, Martin P. Les tumeurs de quatrieme ventricule et le syndrome cerebellaux de la linge mediane: etiudes cliniques, histopathologiques et chirugicales. *Rev Neurol*. 1928;2:431–483.

53. Alvarez JA, Cohen ML, Hlavin ML. Primary intrinsic brainstem oligodendroglioma in an adult. Case report and review of the literature. *J Neurosurg*. 1996;86(6):1165–1169.

54. Abernathy CD, Camacho A, Kelly PJ. Stereotaxic suboccipital transcerebellar biopsy of pontine mass lesion. *J Neurosurg*. 1989;70:195–200.

55. Mohindra S, Savardekar A, Bal A. Pediatric brainstem oligodendroglioma. *J Neurosci Rural Pract*. 2012;3(1):52–54.

56. Guillamo JS, Monjour A, Taillandier L, et al. Brainstem gliomas in adults: prognostic factors and classification. *Brain*. 2001;124(Pt 12):2528–2539.

57. Mittelbronn M, Wolff M, Bültmann E, et al. Disseminating anaplastic brainstem oligodendroglioma associated with allelic loss in the tumor suppressor candidate region D19S246 of chromosome 19 mimicking an inflammatory central nervous system disease in a 9-year-old boy. *Hum Pathol*. 2005;36(7):854–857.

58. Hodges SD, Malafronte P, Gilhooly J, Skinner W, Carter C, Theeler BJ. Rare brainstem oligodendroglioma in an adult patient: presentation, molecular characteristics and treatment response. *J Neurol Sci*. 2015;355(1–2):209–210.

59. Reyes-Botero G, Giry M, Mokhtari K, et al. Molecular analysis of diffuse intrinsic brainstem gliomas in adults. *J Neurooncol*. 2014;116(2):405–411.

60. Narita T, Kurotaki H, Hashimoto T, Ogawa Y. Congenital oligodendroglioma: a case report of a 34th-gestational week fetus with immunohistochemical study and review of the literature. *Hum Pathol*. 1997;28(10):1213–1217.

61. Koeppen AH, Cassidy RJ. Oligodendroglioma of the medulla oblongata in a neonate. *Arch Neurol*. 1981;38(8):520–523.

62. Marwaha RK, Joss DV. A rare congenital intracranial tumour. *Med Pediatr Oncol*. 1994;22:348–349.

63. Svoboda DJ. Oligodendroglioma in a six-week old infant. *J Neuropathol Exp Neurol*. 1959;18:569–574.

64. Lawton CD, Nagasawa DT, Yang I, et al. Leptomeningeal spinal metastases from glioblastoma multiforme: treatment and management of an uncommon manifestation of disease. *J Neurosurg Spine*. 2012;17:438–448.

65. Roldan G, Scott J, George D, et al. Leptomeningeal disease from oligodendroglioma: clinical and molecular analysis. *Can J Neurol Sci*. 2008;35:204–209.

66. Roldan G, Chan J, Eliasziw M, Carincross JG, Forsyth P. Leptomeningeal disease in oligodendroglial tumors: a population-based study. *J Neurooncol*. 2011;104:811–815.

67. Rogers LR, Estes ML, Rosenbloom SA, et al. Primary leptomeningeal oligodendroglioma: case report. *Neurosurgery*. 1995;36:166–169.

68. Korein J, Feigin I, Shapiro MF. Oligodendrogliomatosis with intracranial hypertension. *Neurology*. 1957;7:589–594.

69. Chen R, Macdonald DR, Ramsay DA. Primary diffuse leptomeningeal oligodendroglioma. *J Neurosurg*. 1995;83:724–728.

70. Mathews MS, Pare LS, Kuo JV, et al. Primary leptomeningeal oligodendrogliomatosis. *J Neurooncol*. 2009;94:275–278.

71. Michotte A, Chaskis C, Sadones J, et al. Primary leptomeningeal anaplastic oligodendroglioma with a 1p36-19 q13 deletion: report of a unique case successfully treated with temozolomide. *J Neurol Sci.* 2009;287:267–270.

72. Ozkul A, Meteoglu I, Tataroglu C, et al. Primary diffuse leptomeningeal oligodendrogliomatosis causing sudden death. *J Neurooncol.* 2007;81:75–79.

73. Armao DM, Stone J, Castillo M, et al. Diffuse leptomeningeal oligodendrogliomatosis: radiologic/pathologic correlation. *AJNR Am J Neuroradiol.* 2000;21:1122–1126.

74. Bourne TD, Mandell JW, Matsumoto JA, et al. Primary disseminated leptomeningeal oligodendroglioma with 1p deletion. Case report. *J Neurosurg.* 2006;105:465–469.

75. Guppy KH, Akins PT, Moes GS, Prados MD. Spinal cord oligodendroglioma with 1p and 19q deletions presenting with cerebral oligodendrogliomatosis. Case report. *J Neurosurg Spine.* 2009;10:557–563.

76. Toso V. Metastatic diffusion to the leptomeninges. *Acta Neurol (Napoli).* 1967;22:366–376 (Italian).

77. Cooper IS, Kernohan JW. Heterotopic glial nests in the subarachnoid space: histopathologic characteristics, mode of origin and relation to meningeal gliomas. *J Neuropathol Exp Neurol.* 1951;10:16–29.

78. Rossi S, Rodriguez FJ, Mota RA, et al. Primary leptomeningeal oligodendroglioma with documented progression to anaplasia and t(1;19)(q10;p10) in a child. *Acta Neuropathol.* 2009;118:575–577.

79. Stödberg T, Deniz Y, Esteitie N, et al. A case of diffuse leptomeningeal oligodendrogliomatosis associated with HHV-6 variant A. *Neuropediatrics.* 2002;33(5):266–270.

80. Ng HK, Poon WS. Diffuse leptomeningeal gliomatosis with oligodendroglioma. *Pathology.* 1999;31(1):59–63.

81. Yomo S, Tada T, Hirayama S, et al. A case report and review of the literature. *J Neurooncol.* 2007;81(2):209–216.

82. Taillibert S, Chodkiewicz C, Laigle-Donadey F, et al. Gliomatosis cerebri: a review of 296 cases from the ANOCEF database and the literature. *J Neurooncol.* 2006;76:201–205.

83. Louis DN, Perry A, Reifenberger G, et al. The 2016 World Health Organization classification of tumors of the central nervous system: a summary. *Acta Neuropathol.* 2016;131:803–820.

84. Fukushima Y, Nakagawa H, Tamura M. Combined surgery, radiation, and chemotherapy for oligodendroglial gliomatosis cerebri. *Clin Neurol Neurosurg.* 1987;89:43–47.

85. Balko MG, Blisard KS, Samaha FJ. Oligodendroglial gliomatosis cerebri. *Hum Pathol.* 1992;23:706–707.

86. Gutowski NJ, Gómez-Ansón B, Torpey N, Revesz T, Miller D, Rudge P. Oligodendroglial gliomatosis cerebri: (1)H-MRS suggests elevated glycine/inositol levels. *Neuroradiology.* 1999;41(9):650–653.

87. Louis E, Keime-Guibert F, Delattre JY, et al. Dramatic response to chemotherapy in oligodendroglial gliomatosis cerebri. *Neurology.* 2003;60:151.

88. Di Ieva A, Gaetani P, Giannini M, et al. Oligodendroglial gliomatosis cerebri. *J Neurosurg Sci.* 2006;50:123–125.

89. Pal L, Behari S, Kumar S, et al. Gliomatosis cerebri—an uncommon neuroepithelial tumor in children with oligodendroglial phenotpye. *Pediatr Neurosurg.* 2008;44:212–215.

90. Farooq MU, Bhatt A, Chang HT. An uncommon cause of transient neurological dysfunction. *Neurohospitalist.* 2014;4:136–140.

91. Fountas KN, Karampelas I, Nikolakakos LG, Troup EC, Robinson JS. Primary spinal cord oligodendroglioma: case report and review of the literature. *Childs Nerv Syst.* 2005;21(2):171–175.

92. Strickland BA, Cachia D, Jalali A, et al. Spinal anaplastic oligodendroglioma with oligodendrogliomatosis: molecular markers and management: case report. *Neurosurgery.* 2015;78(3):E466–E473.

93. Nam DH, Cho BK, Kim YM, Chi JG, Wang KC. Intramedullary anaplastic ligodendroglioma in a child. *Childs Nerv Syst.* 1998;14(3):127–130.

94. Fortuna A, Celli P, Palma L. Oligodendrogliomas of the spinal cord. *Acta Neurochir (Wien).* 1980;52(3–4):305–329.

95. Pagni CA, Canavero S, Gaidolfi E. Intramedullary "holocord" oligodendroglioma: case report. *Acta Neurochir.* 1991;113:96–99.

96. Ramirez C, Delrieu O, Mineo JF, et al. Intracranial dissemination of primary spinal cord anaplastic oligodendroglioma. *Eur J Neurol.* 2007;14(5):578–580.

97. Aman RA, Padmosantjojo, Mahyuddin H, Atmadji LB, Soemitro D. Intramedullary oligodendroglioma: a case report. *Jpn J Canc Chemother.* 2000;27(suppl 2):571–573.

98. Liwnicz BH, Rubinstein LJ. The pathways of extraneural spread in metastasizing gliomas: a report of three cases and critical review of the literature. *Hum Pathol.* 1979;10(4):453–467.

99. Morrison T, Bilbao JM, Yang G, Perry JR. Bony metastases of anaplastic oligodendroglioma respond to temozolomide. *Can J Neurol Sci.* 2004;31(1):102–108.

100. Mazza E, Belli C, Terreni M, et al. Breast metastases from oligodendroglioma: an unusual extraneural spread in two young women and a review of the literature. *Crit Rev Oncol Hematol.* 2013;88(3):564–572.

101. Can B, Akpolat I, Meydan D, Üner A, Kandemir B, Söylemezoğlu F. Fine-needle aspiration cytology of metastatic oligodendroglioma: case report and literature review. *Acta Cytol.* 2012;56(1):97–103.

102. McLemore MS, Bruner JM, Curry JL, Prieto VG, Torres-Cabala CA. Anaplastic oligodendroglioma involving the subcutaneous tissue of the scalp: report of an exceptional case and review of the literature. *Am J Dermatopathol.* 2012;34(2):214–219.

103. Uzuka T, Kakita A, Inenaga C, Takahashi H, Tanaka R, Takahashi H. Frontal anaplastic oligodendroglioma showing multi-organ metastases after a long clinical course. Case report. *Neurol Med Chir (Tokyo).* 2007;47(4):174–177.

104. Maloney PR, Yamaki VN, Kumar R, et al. Osteosclerosis secondary to metastatic oligodendroglioma. *Rare Tumors.* 2017;9(1):6837.

105. Demeulenaere M, Duerinck J, DU Four S, Fostier K, Michotte A, Neyns B. Bone marrow metastases from a 1p/19q co-deleted oligodendroglioma—a case report. *Anticancer Res.* 2016;36(8):4145–4149.

106. Ng WH, Lim TC, Tan KK. Disseminated spread of recurrent oligodendroglioma (WHO grade II). *J Clin Neurosci.* 2006;13(5):602–607.

107. Al-Ali F, Hendon AJ, Liepman MK, Wisneiwski JL, Krinock MJ, Beckman K. Oligodendroglioma metastic to bone marrow. *AJNR Am J Neuroradiol.* 2005;26(9):2410–2414.

108. Pingi A, Trasimeni G, Di Biasi C, et al. Diffuse leptomeningeal gliomatosis with osteoblastic metastases and no evidence of intraaxial lesions. *AJNR Am J Neuroradiol.* 1995;16(5):1018–1020.

109. Bruggers C, White K, Zhou H, Chen Z. Extracranial relapse of an anaplastic oligodendroglioma in an adolescent: case report and review of the literature. *J Pediatr Hematol Oncol.* 2007;29(5):319–322.

110. Merrell R, Nabors LB, Perry A, Palmer CA. 1p/19q chromosome deletions in metastatic oligodendroglioma. *J Neurooncol.* 2006;80(2):203–207.

111. Tanaka Y, Nobusawa S, Ikota H, et al. Leukemia-like onset of bone marrow metastasis from anaplastic oligodendroglioma after 17 years of dormancy: an autopsy case report. *Brain Tumor Pathol.* 2014;31(2):131–136.

112. Jian Y, Gao W, Wu Y, et al. Oligodendroglioma metastasis to the bone marrow mimicking multiple myeloma: a case report. *Oncol Lett.* 2016;12(1):351–355.

113. Cordiano V, Miserocchi F, Storti M. Bone marrow metastases from anaplastic oligodendroglioma presenting with pancytopenia and hypogammaglobulinemia: a case report. *Tumori.* 2011;97(6):808–811.

114. Giordana MT, Ghimenti C, Leonardo E, Balteri I, Iudicello M, Duò D. Molecular genetic study of a metastatic oligodendroglioma. *J Neurooncol.* 2004;66(3):265–271.

115. Macdonald DR, OBrien RA, Gilbert JJ, et al. Metastatic anaplastic oligodendroglioma. *Neurology.* 1989;39:1593–1596.

116. Rubinstein LJ. Development of extracranial metastasis from a malignant astrocytoma in the absence of previous craniotomy. *J Neurosurg.* 1967;26:542–547.

117. Schweitzer T, Vince GH, Herbold C, Rosen K, Tonn JC. Extraneural metastases of primary brain tumors. *J Neurooncol.* 2001;53:107–114.

118. Subramanian A, Harris A, Piggott K, Shieff C, Bradford R. Metastasis to and from the central nervous stem: the relatively protected site. *Lancet Oncol.* 2002;3(8):2410–2414.

119. Zustovich F, Della Puppa A, Scienza R, Anselmi P, Furlan C, Cartei G. Metastatic oligodendrogliomas: a review of the literature and case report. *Acta Neurochir (Wien).* 2008;150(7):699–702. discussion 702-3.

120. Alacacioglu A, Unal S, Canapolat S, et al. Breast metastasis of anaplastic oligodendrogioma: a case report. *Tumori.* 2012;98:e162–e164.

121. Volavsek M, Lamovec J, Popović M. Extraneural metastases of anaplastic oligodendroglial tumors. *Pathol Res Pract.* 2009;205(7):502–507.

122. James TG, Pagel W. Oligodendroglioma and extracranial metastasis. *Br J Surg.* 1951;39(153):56–65.

123. Spataro J, Sacks O. Oligodendroglioma with remote metastasis. Case report. *J Neurosurg.* 1968;28(4):373–379.

124. Ordóñez NG, Ayala AG, Leavens ME. Extracranial metastases of oligodendroglioma: report of a case and review of the literature. *Neurosurgery.* 1981;8(3):391–396.

125. Sha SJ, Wu HP, Lu K, et al. Extraneural metastases of anaplastic oligodendroglioma. *APMIS.* 2014;122(7):660–662.

126. Li G, Zhang Z, Zhang J, et al. Occipital anaplastic oligodendroglioma with multiple organ metastases after a short clinical course: a case report and literature review. *Diagn Pathol.* 2014;9:17.

127. Noshita N, Mashiyama S, Fukawa O, Asano S, Watanabe M, Tominaga T. Extracranial metastasis of anaplastic oligodendroglioma with 1p19q loss of heterozygosity—case report. *Neurol Med Chir (Tokyo).* 2010;50(2):161–164.

128. Kernoharn JW. Oligodendroglioma. In: Minckler J, ed. *Pathology of the Nervous System.* New York: McGraw-Hill Co; 1971:1993–2007. 2nd ed. vol. 2.

129. Finsterer J, Breiteneder S, Mueller MR, et al. Pleural and bone marrow metastasis from supratentorial oligodendroglioma grade III. *Oncology.* 1998;55:345–348.

130. Han SR, Yoon SW, Yee GT, et al. Extraneural metastases of anaplastic oligodendroglioma. *J Clin Neurosci.* 2008;15(8):946–949.

131. Monzani V, Rovellini A, Masini B, Cappricci E, Miserocchi G. Metastatic oligodendroglioma. Case report. *J Neurol Sci.* 1996;40(3–4):239–241.

132. Lee CC, Jiang JS, Chen ET, Yokoo H, Pan YH, Tsai MD. Cytologic diagnosis of a metastatic oligodendroglioma in a pleural effusion. A case report. *Acta Cytol.* 2006;50:542–544.

133. Ambinder EB, Rowe SP. A case of anaplastic oligodendroglioma with extensive extraneural metastases imaged with FDG PET. *Clin Nucl Med.* 2017;42(12):968–970.

134. Garin-Chesa P, Fellinger EJ, Huvos AG, et al. Immunohistochemical analysis of neural cell adhesion in small round cell tumors of childhood and adolescence. *Am J Pathol.* 1991;139(2):275–286.

135. Wang X, Hisha H, Taketani S, et al. Neural cell adhesion molecule contributes to hemopoiesis-supporting capacity of stromal cell lines. *Stem Cells.* 2005;23:1389–1399.

136. Lavrador JP, Oliveira E, Tortosa F, Ortiz S, Pimentel J, Simas N. Metastatic grade III oligodendroglioma revealed by refractory thrombocytopenia. *J Neurosurg Sci.* 2017;61(5):547–550.

137. Kural C, Pusat S, Sentürk T, Seçer Hİ, Izci Y. Extracranial metastases of anaplastic oligodendroglioma. *J Clin Neurosci.* 2011;18(1):136–138.

138. Wu Y, Liu B, Qu L, Tao H. Extracranial metastasis in anaplastic oligodendroglioma. *J Clin Neurosci.* 2011;18:136–138.

139. Liu H, Lu X, Thakral B. A rare presentation of anaplastic oligodendroglioma metastatic to bone marrow, mimicking blast cells. *Br J Haematol.* 2016;174(3):344.

Chapter 4

Clinical presentation of pediatric oligodendrogliomas

Scott L. Coven and Jonathan L. Finlay

Department of Pediatrics, Division of Hematology/Oncology/BMT, Nationwide Children's Hospital and The Ohio State University, Columbus, OH, United States

Introduction

Central nervous system (CNS) tumors are the second most common cancers in children between 1 and 19 years of age. Additionally, CNS tumors are the leading disease-related cause of death in children under 19 years of age in the United States.[1,2] Over the past few decades, the outcome for patients with childhood CNS cancer has dramatically improved from 20% to almost 74%.[3,4] Despite the advances in diagnostic modalities and treatment methods, children with CNS tumors continue to experience a broad array of nonspecific clinical features.

Although gliomas are the most common histological CNS tumor type of childhood, pediatric oligodendroglioma remains relatively rare. Oligodendrogliomas were first described by Bailey and Cushing in 1926,[5,6] but have remained a diagnostic challenge in childhood and adolescence. In children aged 0–14 years, oligodendrogliomas comprise 0.8% of all primary CNS tumors. Additionally, for children 15–19 years, the rate of oligodendrogliomas increases to 1.7% for all primary CNS tumors.[2] However, oligodendroglioma remains more common in adults than pediatric patients (93.5% vs 6.5%).[7] Previous literature has documented the median age of diagnosis at 13.1 years, with a male preponderance.[8] Furthermore, the literature among pediatric oligodendroglioma is limited to few retrospective and/or institutional studies, which are summarized in Table 1.

The presenting features of pediatric oligodendroglioma have been well documented, with seizures the primary symptom in a majority of the retrospective studies. Tice et al. reviewed 39 patients, finding 85% of the patients presenting with seizures. Only eight patients included in the cohort had headaches, four patients had increased intracranial pressure (ICP), one patient presented with precocious puberty, and one patient had visual disturbances.[9] Bowers et al. documented among a 20-patient cohort, 17 patients with seizures, 5 patients with headache, and 3 patients with nausea/vomiting; 3 patients presented with a combination of less common symptoms of irritability, hydrocephalus, hemiparesis, tremors, decreased upward gaze, transient weakness, hypertension, and altered mental status.[10] Fausto et al. documented, among 50 pediatric cases of oligodendroglioma, seizure as the predominant clinical symptom, along with focal neurological symptoms.[11] Creach et al. documented among 37 patients with the diagnosis of pediatric oligodendroglioma, 23 patients presenting with seizure, 8 patients presenting with headache, and 4 presenting with other focal neurological signs.[12] Finally, Razack et al. described 19 patients presenting with seizures (10 patients), headache (3 patients), visual disturbances (2 patients), weakness (2 patients), decreased performance in school (1 patient), and cranial nerve palsy (1 patient).[13] These symptoms have remained well-documented within the pediatric oligodendroglioma literature over the past 20 years.

Several publications on pediatric oligodendroglioma have attempted to correlate the tumor type with its location. Tice et al. found the tumor location was primarily limited to the temporal and frontal regions in 82% of their cohort.[9] Pollack et al. found an almost uniform distribution in tumor location between the parietal ($n=27$), frontal ($n=25$), and temporal ($n=21$) lobes. Additionally, the occipital lobe also had significant involvement of lesions ($n=12$).[8] Razack et al., in a smaller cohort ($n=19$), determined the tumor locations to be almost evenly distributed among the cerebral hemispheres (temporal most common), and one located in the sellar region.[13] More recently, Lau et al. described similar results from the surveillance, epidemiology, and end result (SEER) database, revealing that the most common tumor locations were the temporal ($n=136$) and frontal ($n=114$) lobes among the 455-patient cohort. Interestingly, Lau et al. obtained adult oligodendroglioma data, and showed that 53.9% of the patients ($n=6546$) had involvement in the frontal lobe.[7]

Oligodendroglioma. https://doi.org/10.1016/B978-0-12-813158-9.00004-9

TABLE 1 Retrospective studies for pediatric oligodendroglioma

Author	Publication Year	Cohort Size	Average Age (Years)
Tice et al.	1993	39	9.82—mean
Razack	1998	19	13.1—mean
Bowers et al.	2002	20	7.2—median
Peters et al.	2004	32	10.3—median
Creach et al.	2011	37	11.1—median
Fausto et al.	2014	50	7—median
Lau et al.	2017	455	12—mean

This represents a significant change from pediatric oligodendroglioma, where children tend to have a more uniform distribution among several lobes within the brain.

Little is known about pediatric oligodendroglioma symptom interval with respect to diagnosis. However, since it is a low-grade lesion, delays in diagnosis are likely to exist. At our institution, we treated a 25-year-old male patient with an oligodendroglioma who presented with seizures after a delay of 730 days. Retrospectively, the patient underwent a CT and MRI evaluation at 13 years of age as part of a post-concussion evaluation. The MRI revealed a non-enhancing left temporal lobe lesion. The lesion was partially resected, rendering an inconclusive tissue diagnosis at that time. He was followed with surveillance MRIs over a 2-year period until he was lost to follow-up. The patient presented again at age 25 after the development of new seizures. An MRI revealed interval growth and infiltration of the previously demonstrated left temporal lobe lesion and after partial resection, pathology confirmed the diagnosis of an oligodendroglioma WHO grade II with 1p and 19q co-deletion and IDH mutation—the latter also demonstrated on the original biopsy of 12 years earlier.[14] This case emphasizes the diagnostic dilemma regarding low-grade tumors such as oligodendrogliomas. Our ability to determine when "progression" is initiated is essential to the future treatment of children with oligodendroglioma.

Oligodendrogliomas of childhood remains somewhat rare, but further studies are needed regarding clinical presentations to reduce potential delays in the diagnosis. Since the "time clock" is a moving target, improving awareness of pediatric oligodendroglioma may allow surgery alone to be the primary curative option, eliminating the need for adjuvant chemotherapy and/or radiation therapy or delay treatment until intervention is necessary.

References

1. Johnson KJ, Cullen J, Barnholtz-Sloan JS, et al. Childhood brain tumor epidemiology: A brain tumor epidemiology consortium review. *Cancer Epidemiol Biomarkers Prev.* 2014;23(12):2716–2736.
2. Ostrom QT, de Blank PM, Kruchko C, et al. Alex's lemonade stand foundation infant and childhood primary brain and central nervous system tumors diagnosed in the United States in 2007–2011. *Neuro Oncol.* 2015;16(Suppl 10):x1–x36.
3. Gatta G, Zigon G, Capocaccia R, et al. Survival of European children and young adults with cancer diagnosed 1995–2002. *Eur J Cancer.* 2009; 45(6):992–1005.
4. Ostrom QT, Gittleman H, Xu J, et al. CBTRUS statistical report: primary brain and other central nervous system tumors diagnosed in the United States in 2009–2013. *Neuro Oncol.* 2016;18(Suppl_5):v1–v75.
5. Bailey P, Cushing H. *A Classification of Tumors of the Glioma Group on a Histogenetic Basis with a Correlation Study of Prognosis.* Philadelphia: JB Lippincott; 1926.
6. Bailey P, Busy PC. Oligodendrogliomas of the bran. *J Pathol Bacteriol.* 1929;32:735–751.
7. Lau CS, Mahendraraj K, Chamberlain RS. Oligodendrogliomas in pediatric and adult patients: an outcome-based study from surveillance, epidemiology, and end results database. *Cancer Manag Res.* 2017;9:159–166.
8. Pollack IF, Claassen D, Al-Shboul Q, et al. Low-grade gliomas of the cerebral hemispheres in children: an analysis of 71 cases. *J Neurosurg.* 1995;82:536–547.
9. Tice H, Barnes PD, Goumnerova L, Scott RM, Tarbell NJ. Pediatric and adolescent oligodendrogliomas. *Am J Neuroradiol.* 1993;14:1293–1300.
10. Bowers DC, Mulne AF, Weprin B, et al. Prognostic factors in children and adolescents with low-grade oligodendrogliomas. *Pediatr Neurosurg.* 2002;37:57–63.

11. Rodriguez FJ, Tihan T, Lin D, et al. Clinicopathologic features of pediatric oligodendrogliomas. *Am J Surg Pathol*. 2014;38:1058–1070.

12. Creach KM, Rubin JB, Leonard JR, et al. Oligodendrogliomas in children. *J Neurooncol*. 2011;106:377–382.

13. Razack N, Baumgartner J, Bruner J. Pediatric oligodendrogliomas. *Pediatr Neurosurg*. 1998;28:121–129.

14. Navalkele P, Osorio DS, AbdelBaki MS, et al. Adolescent/young adult oligodendroglioma—When does the "time clock" to progression start? Poster Presentation: The Seventeenth International Symposium on Paeditric Neuro-Oncology (ISPNO), Liverpool, UK, June 2016. *Neuro-Oncology*. 2016;18:iii85.

Chapter 5

Seizures, oligodendrogliomas, and brain-tumor-related epilepsy

Marta Maschio* and Francesco Paladin[†]

*Center for Tumor-Related Epilepsy, Neuro-Oncology Unit, IRCCS - Regina Elena National Cancer Institute, Rome, Italy; [†] Department of Neurology, Epilepsy Center, SS Giovanni and Paolo Hospital, Campo SS Giovanni e Paolo, Venice, Italy

Introduction

Patients with brain tumor-related epilepsy (BTRE) present a complex therapeutic profile and require a unique and multidisciplinary approach.[1] BTRE results in a real challenge from many points of view—therapeutic, support, and psychosocial—due to the simultaneous presence of two pathologies: on one hand, the brain tumor, and on the other hand, epilepsy. A diagnosis of brain tumor as well as just the idea of cancer alone, in most patients, is enough to cause deep difficulties: behavioral, emotional, and intellectual. These problems can lead to a compromised quality of life (QoL) and a limited ability to live independently. Patient management requires careful consideration of many factors by clinicians. First, the impact that antiepileptic drugs (AEDs) can have on pharmacological therapies related to oncological disease. In fact, the concurrent use of AEDs with chemotherapy and supportive therapies can induce drug interactions and collateral effects that can differ from the non-oncological epileptic population.[2–4] Secondly, directly related to epilepsy, a relatively high rate of recurrence after a first seizure[5] and the impact of the disease progression on seizure frequency.[6] Finally, there is the matter of concern in maintaining a good QoL for these patients. Epilepsy still brings stigma with it and can cause the individual diagnosed with the disease to feel socially outcast and severely invalidated. In addition, seizures themselves can impact QoL. Considering all these factors, taking care of the "symptom" of epilepsy is of utmost importance for the patients if they are to resume their professional life, function successfully in a social context, and conduct a satisfying family life.[7] To accomplish this purpose, it is fundamental for healthcare professionals to see the patients as unique individuals with their particular needs. This requires a vision of patient management concerned not only about medical therapies (pharmacological, surgical, radiological, etc.) but also about emotional and psychological support for the individual, as well as their family, throughout all stages of the illness.

That's why epilepsy, though being a symptom of a rare tumor, is an illness that has tremendous weight, for patients, their families, and society at large.[1] All of these are key issues to understand the nature of this disease, about which we still know very little. The new insights gained from recent experimental studies, while leading to more questions than answers, continue to make this unique pathology a stimulating and dynamic area of research.

Definition of epilepsy

Epilepsy is one of the most common serious disorders of the brain, affecting at least 50 million people worldwide that knows no geographical, racial, or social boundaries. It accounts for 1% of the global burden of disease, determined by the number of productive life years lost due to disability or premature death.[8,9] Epilepsy represents more than just seizures for the affected individuals and their family: it leads to multiple interacting medical, psychological, economic, and social repercussions, all of which need to be considered. Fear, misunderstanding, and the resulting social stigma and discrimination surrounding the patients often force them "into the shadows."[10] The social effects may vary from country to country and culture to culture; all over the world, the social consequences of epilepsy are often more difficult to overcome than the seizures themselves. Significant problems are often experienced by people with epilepsy in the areas of personal relationships and sometimes, legislation. These problems may in turn undermine the treatment of epilepsy.[11] The best approach to take care of the epileptic patients must be a management by professionals from various sectors in every aspect of the lives of people with epilepsy. This should result in a multidisciplinary approach, coordinating health, education, social and

Oligodendroglioma. https://doi.org/10.1016/B978-0-12-813158-9.00005-0

professional activities, and psychology. Specialized medical professionals are important members of the team providing comprehensive care, especially at the tertiary level, for people with epilepsy. They are also essential for training and providing support and supervision to primary health-care providers in epilepsy care. Health professionals such as neurological nurses, psychologists, and social workers are important members of the multidisciplinary team providing comprehensive care, who play an important role in the diagnosis, treatment, and rehabilitation of people with epilepsy. Specialist training in epileptology is needed on multiple levels to reach all those concerned with epilepsy management.[10]

Epilepsy in oligodendroglioma: Epidemiology and incidence

Literature data indicate that the most common symptom in patients with brain tumors (BTs) is epilepsy. In patients with BT, seizures are the onset symptom in 20%–40% of cases, while a further 20%–45% of patients will develop them during the course of the disease. Overall, the incidence of epilepsy in BTs, regardless of histological type and anatomical site of the lesion varies from 35% to 70%[12–14] Epilepsy due to BTs constitutes 6%–10% of all cases of epilepsy as a whole and 12% of acquired epilepsy.[15,16] If seizures appear as the presenting symptom of BT, it seems to indicate a more favorable prognosis[17–20]: this could be due to early diagnosis resulting from following up on the manifestations of seizures, to a better position of the tumor for surgical intervention (a more superficial location), or to the presence of a more favorable histology (slow-growing tumors). Another crucial factor for the presence of seizures is the location of the tumor, with a higher seizure frequency being associated with supratentorial tumors (with respect to subtentorial tumors), superficially[21] located in the cortex.

Among the primary brain tumors, oligodendrogliomas (OD) are the third most common type of glioma, accounting for 2%–5% of all primary brain tumors. They commonly involve supratentorial regions of the brain, and most frequently manifest during the fourth or fifth decades of life, with a slight predominance in males.[22,23] Treatment options range from a conservative approach to more aggressive measures using a combination of antiepileptic drugs (AEDs), surgery, chemotherapy, and radiation therapy.[24] Studies on OD have attempted to identify prognostic factors that affect overall survival and tumor progression.[25,26] In these BTs, there is evidence of a particular relationship between seizures and survival: patients with refractory epilepsy seem to do worse than those with good control of seizures.[18] Furthermore, epilepsy in OD, among the symptoms, is the most relevant in this patient population[22] that imposes a severe burden on quality of life, psychosocial function, and economic independence.[18,27,28]

Seizures have been the most commonly reported presenting symptom, ranging in incidence from 35% to 85% of patients: they may be simple partial, complex partial, generalized, or a combination of these.[29] Although the precise nature of epileptogenesis in glioma is still unclear, Glutamate released by glioma cells has recently been reported as an important cause of seizures,[30] and tumors near the cortex and slowly progressive tumors are reported to be associated with a high incidence of seizures.[31] These data indicate that patients expected to survive longer have a higher risk of epilepsy. Therefore, control of seizures is without any doubt a major factor in maintaining quality of life in glioma patients.[32]

Detection, classification and documentation of seizures

Regarding the classification of seizures in epilepsy, generalized epileptic seizures are conceptualized as originating at some point within and rapidly engaging, bilaterally distributed networks that can include cortical and subcortical structures. Generalized seizures sometimes can be asymmetric. On the other hand, focal epileptic seizures are conceptualized as originating within networks limited to one hemisphere, but may also originate in subcortical structures. They may be discretely localized or more widely distributed. For each seizure type, ictal onset is consistent from one to another, with preferential propagation patterns that can involve the contralateral hemisphere. In some cases, there is more than one network, and more than one seizure type, but each individual seizure type has a consistent site of onset. The importance in the evaluation and management of individual patients is to recognize impairment of consciousness/awareness or other dyscognitive features, localization, and progression of ictal events. Therefore, focal seizures should be described according to their manifestations (e.g., dyscognitive, focal motor).[33,34]

If this is the way to classify the seizures in non-oncological epileptic patients, more difficulties can be encountered when seizures must be classified in BTRE. In fact, it can be extremely difficult for physicians and patients to recognize seizures, especially considering the range of clinical manifestations that can occur in the focal seizures, most often encountered by BTRE patients, such as epigastric aura described as pain, ictal fear-like panic attacks, dreamy state, experiential phenomena, mental or psychic symptoms, and hallucinations.[10,35] Other manifestations of these types of seizures could be confusional states, automatisms, or behavioral modifications, which can be confused with psychiatric disorders or other causes.[13] If medical professionals, family members, and the entire nonmedical support team that interact with the patient do not receive proper training to verify and quantify seizures, the number of seizures can be underestimated or missed altogether, which makes finding the right drug therapy difficult. Also, it is possible that some oncological centers which treat BTRE do not have the resources to properly train staff and patients' family members. Finally, evaluation of the efficacy of a therapeutic treatment in epileptic patients (i.e., non-oncological) is based on seizure count and therefore, an accurate quantification of seizures is fundamental, even more so in BTRE patients. For this reason, it is essential to consider routine to give a seizure diary to each BTRE patient and provide enough instruction to allow him/her or care givers to correctly document seizure frequency, and communicate the number during checkups.[11]

Epileptogenesis, biomarkers and seizures

The 2016 WHO classification[36] defined OD as IDH-mutant and 1p/19q- co-deleted tumor grade II or III (anaplastic).[37] This definition takes into account molecular and genetic factors that could play an important role in the etiology of seizures in brain tumors.[38]

In fact, recent studies of biomarkers associated with epilepsy in glioma have demonstrated the role of genetic biomarkers themselves in the etiology of seizures, such as isocitrate dehydrogenase 1 (IDH1) mutations and adenosine kinase (ADK) overexpression.[39,40] Among biomarkers, mutation of IDH1 seems to be the best predictor of preoperative seizure, but more robust data are necessary to confirm this observation.[39] Dysfunction of glutamate receptor signaling, or glutamate processing, and elevation of extracellular glutamate were suggested to be associated with the development of seizure.[41] Increasing evidence suggested that alterations of the GABA-ergic networks and also voltage-gated ion channels play an important role in the epileptogenesis of patients with glioma.[42,43] Also ADK expression in peritumoral areas may play a role in glioma-related seizures.[39,44] Finally, among the genetic biomarkers associated with regulation of cell proliferation,[39] microRNA (miRNA) are gaining increasing attention[45] for their possible role in seizure occurrence. In a recently published study, in which miRNAs expression profiling was analyzed in a cohort of patients affected by low grade-gliomas, miR-196b has been identified as a predictive marker of seizure occurrence.[46] Epidermal growth factor receptor (EGFR) amplification and low Ki-67 expression were indicated as independent risk factors for anaplastic glioma-related seizures, suggesting that Ki-67 is a glioma proliferation marker that participates in epileptogenesis by promoting proliferation of tumor cells.[47] A study performed by Kohling et al.[48] indicated that epileptiform activity is derived from peritumoral areas that were identified within 1–2 mm of the tumor border. Potentially, these observations reflect the contribution of cell proliferation to seizures, because tumor cell invasion into normal brain cells leads to alterations of neuronal signaling and epilepsy develops subsequently.

All these experimental reports have an important clinical practice: they suggest a possibility that drugs for epilepsy and on the other hand, drugs for brain tumors, can act together in this particular illness, possibly on common mechanisms. Identification of genetic biomarkers associated with seizures in patients with OD could provide new insights into targeted therapy for glioma-related seizures. New AEDs able to act on both targets should be developed.[39,49] Based on these observations, which indicate coincidence of seizure and tumor proliferation and invasion signaling pathways.

In fact, some AEDs per se may display an activity on brain cancer cells, as suggested by in vitro experiments,[3] as well as chemotherapy such as temozolomide (TMZ) can reduce the frequency of seizures.[50–52] Regarding the possible actions of AEDs on brain cancer cells including activity on cell proliferation, apoptosis, and migration,[53] particular attention has been devoted to two non-EIAED: valproic acid (VPA), known for a long time as a histone deacetylase inhibitor[54] and levetiracetam (LEV).[3] Recent studies evaluated the effect of VPA on survival of patients with newly diagnosed Glioblastoma (GBM) found that patients receiving VPA had a significantly longer survival than both those who did not receive an AED and those who received other AED.[55,56] As far as LEV is concerned, LEV was reported as the most potent MGMT inhibitor among several AEDs.[3] A recent study analyzed the benefit of LEV compared with other AED as a chemosensitizer to TMZ for patients with GBM: the median PFS and OS for patients who received LEV in combination with TMZ were significantly longer than those for patients who did not receive LEV.[51] Among drugs recently introduced in the management of epilepsy, both lacosamide (LCM) and brivaracetam (BRV) are devoid of enzyme-inducing activity on the cytochrome system, being good candidates for introduction in the management of brain tumor epilepsy. A recent study showed that BRV and LCM in vitro exert a dose-dependent cytotoxic effect on various glioma cell lines and this effect was concomitant with the modulation of a number of miRNAs.[45]

Pharmacoresistance

One of the characteristics of BTRE that have important implications in clinical practice is that it is often drug resistant. In patients with BTs, this remains unclear and appears to be multifactorial and related to mechanisms that bypass AEDs.[48]

The poor seizure control may result from the fact that the antiepileptic action of many AEDs is due to a modification of membrane excitability mediated by ion channels, but mechanisms of epileptogenesis in BTRE include also changes in pH, amino acids, proteins, etc. Seizures in BTRE may also be due to tumor progression (in this case, the first AED is often not sufficient) or to a low concentration of AEDs in serum or site of action that can result from drug interactions with cancer therapies.[14,57–59] In patients with drug-resistant epilepsy, a reduced intraparenchymal accumulation of AEDs has been demonstrated. This phenomenon can be due to overexpression of genes and proteins that mediate nonspecific resistance to treatment. These proteins have also been found in neurons and glia of the epileptogenic zone. In patients with BTRE, the growth of these intracellular proteins may be caused by the tumor. These transport proteins in cancer cells, discovered by Victor Ling, were resistant to CT and were named multidrug resistant proteins (MDR) or P-glycoprotein (P-gp). To date, there are several MDR—P-gp, MRP1, MRP2, among others. The primary function of these proteins is to pump lipophilic xenobiotics out of cells to prevent the accumulation of potentially toxic substances. In healthy brains, the multidrug-resistance gene *MDR1 (ABCB1*, P-glycoprotein) and multidrug-resistance-related protein (MRP, ABCC1) contribute to the function of the blood–brain barrier and blood–cerebrospinal fluid barrier.[60–64] Expression of multidrug-resistance proteins in tumor cells of patients with glioma is high, suggesting that they can diminish drug transport into the brain parenchyma: by doing so, they might diminish the access to intracerebral target tissue to effective drugs (i.e., antiepileptic drugs).[31,60] Insufficient concentration of AEDs in the blood can be the result of an active defense mechanism by MDR1, which restricts the penetration of lipophilic substances into the brain.[60] Carbamazepine (CBZ), phenytoin (PHT), phenobarbital (PB), lamotrigine (LTG), and felbamate (FBM) are substrates for this gene product; a nonspecific transporter can move gabapentin (GBP) out of the brain. LEV does not seem to be a substrate for MDR1 or other multidrug-resistance proteins and thus, this drug might be of potential use for patients with intractable epilepsy.[31]

For all these reasons, the integration of basic scientific knowledge about the pathophysiological mechanisms of BTRE with innovative clinical treatments based on individual genetic and proteomic profiling will lead to more effective and less costly therapies and most importantly, to a significant improvement in patients' QoL.

Impact of AEDs

The presence of epilepsy is considered the most important risk factor for long-term disability in BT patients.[65] For this reason, the problem of the proper drugs administration and their potential side effects (SEs) is a matter of great relevance. Good seizure control can significantly improve the patients' psychological and relational sphere (i.e., social, personal, and professional). Many studies (meta-analyzes) on non-oncological epileptic patients have been done, but it is difficult to transfer the results to clinical practice.[66] Prior to determining the efficacy of a given therapy, it is crucial to have documentation concerning whether the drug in question has been administered at the maximum possible dose for the patient and which type of add-on has been used. Often, these kinds of records have not been kept. It also must be taken into consideration that AEDs can induce many potentially serious SEs. In addition to these, there can also be drug interactions with the systemic treatments. Therefore, the evaluation of the efficacy of an AED must be based on the number and types of seizures as well as significant drug-related information that, as we have stated, is often not available. Elimination of seizures is the long-term goal, while improvement in seizure frequency is, of course, the initial objective at hand. However, the difficulty in standardizing methods of defining and measuring improvement needs to be considered.[11] Adverse effects of AEDs are more frequent in patients with tumor-related epilepsy than in the rest of the epileptic population;[17] in particular, many AEDs, in addition to the idiosyncratic, hematological, and systemic toxicity effects, also have effects on the central nervous system (CNS), which can strongly impact the patient's QoL, make it difficult to correctly assess the response to CT, and even mimic a progression of the tumor. Each AED can be associated with adverse effects, in both oncological and non-oncological patients. However, in brain cancer patients, the evaluation of AED SEs is crucial due to the fact that SE can affect the patients' perception of QoL more than seizure frequency.[67] Patients' priorities often have less to do with seizure freedom, than with the desire to have the least amount of SE induced by drugs which they perceive as being extremely limiting on their daily lives. With the older AEDs, there is a high incidence of serious SE (23.8%) and a mean of incidence of SE (20%–40%), higher than in the non-oncological epileptic population.[17] Only recently, studies have been published that have evaluated the percentage of SEs that appeared in BTRE patients. These data show that the new AEDs induce fewer SEs than the older ones.[1] Of all of the possible SE that can appear in BTRE patients undergoing AED therapy, rash is potentially very serious, especially during RT. This risk should not be underestimated, even with the use of new AEDs.[68] Other important SEs that must be taken into consideration regarding QoL of patients with BTRE are the possible effects on cognitive functions and in the sexual sphere.

To date, there have been no studies dedicated specifically to studying the impact of the older AEDs on cognitive function in patients with BTs. However, there have been studies on cognitive function in oncological patients in general;[69,70] while not examining the impact of AEDs in this specific area, these studies demonstrated that the older AEDs such as PHT, CBZ, VPA, and PB have the highest incidence of adverse effects on cognitive function. Finally, literature data indicate that some AEDs can induce negative effects on the sexual sphere,[71] a fundamental aspect of emotional well-being, and a significant contribution to a good QoL. For this reason, in patients with BTRE who may have a low life expectancy, and as a result, a possible fear of dying and/or sense of uncertainty due to the duplicity of their disease, the choice of the AED should take into account the possible effects on sexuality.

Pharmacological interaction

A pharmacological interaction occurs when one drug modifies the activity of another, increasing or reducing its effects.[72,73] Pharmacokinetic interactions occur when one drug interferes with the distribution in the organism of another drug, altering its concentration at the site of action. When a pharmacological interaction takes place between an AED and CT, as could be the case with BTRE, two contrasting effects might occur. In one case, there could be a rapid elimination of one or the other of the drugs (i.e., either the AED or the CT), which would result in either incomplete seizure control or a reduction of survival. In the other case, there could be a reduced elimination of one or the other of the drugs, resulting in an increased toxicity.[74] The hepatic isoenzymes catalyze the oxidation reactions both of drugs and of endogenously produced substances, and their function can be modified by the concomitant administration of other substances.[73] This means that P450-enzyme-inducing drugs provoke faster elimination of a second concomitantly administered drug and thus a reduction of its efficacy, or at least of its concentration at the site of action. On the other hand, CYP-enzyme inhibiting drugs decrease the metabolism of the second drug, increasing the risk of toxicity and of the appearance of adverse effects. Of the AEDs, PB, PRM, CBZ, and PHT act as enzyme inducers, and thus enhance the metabolism of corticosteroids and of many chemotherapeutic agents, including nitrosoureas, paclitaxel, cyclophosphamide, etoposide, topotecan, irinotecan, thiotepa, adriamycin, and methotrexate. This means that the administered doses of these drugs could be inadequate, and therefore less effective.[72] On the other hand, costicosteroids and many CT, too, are inducers of the CYP enzymes and can therefore, alter the efficacy or increase the toxicity of many AEDs administered concomitantly. There are, to date, no data in the literature suggesting that the new AEDs interfere with antitumor drugs.[72,74] In particular, among the new AEDs, topiramate (TPM) is metabolized by the CYP 3A4 pathway and LTG by glucuronidation. However, LEV, GBP, PGB, and LCM are not metabolized and are mostly excreted by the kidneys. Because LEV and GBP to date do not interact with other agents, they could represent a good option during CT.[32]

Impact of antitumor therapy on seizure control

The retrospective studies in low-grade gliomas indicate a 50% reduction in seizure frequency of 56%–77% and a seizure-freedom of 38%–80% during radiotherapy.[75,76] A randomized EORTC Phase III trial on external radiotherapy in LGG to a cumulative dose of 65 Gy showed that after early application of RT, 75% of patients became seizure-free and after late application of radiation therapy, 59%. The role played by RT in diminishing seizure activity could be due, aside from a direct antitumor effect, to damage of epileptogenic neurons or to a modification of the microenvironment around the brain tumor.[77] A number of studies have demonstrated that chemotherapy with alkylating agents (temozolomide, procarbazine + CCNU + vincristine) improve seizure control.[12,52,78,79] For example, administration of temozolomide results in a 50% or more reduction in seizures in 44%–59% of patients with low-grade gliomas, including OD. Overall, 20%–40% of patients become seizure-free by administration of systemic chemotherapy.[77,79] This evidence clearly indicates that patients with OD necessitate a "task force" (including at least epileptologists, neuro-oncologists, neurosurgeons, and neuro-radiologists) devoted to the decision-making, because cyto-reduction and seizure control are not two different goals.

Quality of life in BTRE patients

The diagnosis of epilepsy in a patient with no oncological disease already implicates an important change in one's concept of QoL that involves three main factors:

- Possible SEs from drugs
- The negative psychological impact caused by losing control of one's body and the surrounding environment during seizures
- The rejection and marginalization that still occurs today due to a societal view of individuals with epilepsy as "strange."

These three factors become even heavier to bear in patients that must confront both pathologies: epilepsy and BT. These patients are subjected to systemic treatments for the neoplastic disease as well as antiepileptic therapies, and therefore, are at even greater risk for SE and drug interactions. The loss of control of one's body during a seizure and the frustration that accompanies such an experience, represent for the patient a total lack of autonomy. The unpredictability of adverse events leads to an enormous sense of insecurity. In addition, seizures are a constant reminder to the patients of their illness and of being considered "different." Marginalization and rejection are especially felt by individuals who have a visible physical disability like hemiparesis or problems with speech (which may be due to the site of the tumor) and also by those whose physical aspect has been altered due to systemic therapies. All of these factors together with the label "epileptic" can cause the patient to feel extremely frustrated when attempting any type of social interaction. The QoL for patients with BT is affected by many factors, the most significant being the various therapies undertaken (e.g., CT, radiation, surgery, support therapies, and AEDs), possible physical disability due to the tumor location, and possible neurocognitive disturbances. Also, the appearance of cognitive disturbances among patients with BTRE can be due to tumor location and CT, as well as radiation. Therefore, the QoL in this particular patient population needs to be a primary objective, together with the knowledge that epilepsy can significantly affect the long-term disability.[11] To achieve such goals, periodic neurological and neuropsychological checkups are an important part of patient evaluation and of the patient-doctor feedback, which allow the monitoring of neurocognitive performances and possible collateral effects over time, and thus, enable the team of medical professionals to plan any necessary interventional strategies.[14] Health-related quality of life (HRQOL) expressed by the seizure type and frequency has been shown to affect patients with epilepsy.[80] Patients with gliomas taking AEDs had lower levels of cognitive functioning than those not taking AEDs. The older AEDs are known to produce decreased cognitive functioning, depression, and irritability,[81] effects that are amplified using AED polytherapy. Studies looking at the impact of primary brain tumors on HRQOL rarely documented seizures and antiepileptic medication use as important risk factors for impaired cognition and poor QoL, despite the fact that the incidence of seizures is high in patients with primary brain tumors. Epileptic seizures may cause long-term and acute complications, for example, fatal injuries, driving restrictions, social stigma, job limitations, and cognitive impairment, despite the affected persons being physically intact in between the episodes.[82] Patients with LGG perceived their overall health as being significantly poorer than their general population peers, in particular resulting in poorer role functioning due to physical and emotional problems, social functioning, mental health, and energy levels.[83,84] In addition, patients with LGG experienced a range of more specific HRQL deficits. Approximately one quarter of the patients reported serious problems with their neurocognitive functioning, particularly memory and concentration, correlated with a serious neurocognitive disability measured by objective neuropsychological tests.[85] With regard to epilepsy burden, Klein et al.[67] found that the inability to achieve complete seizure control, rather than the antiepileptic drugs themselves, negatively affected HRQOL. The neurocognitive deficits and epilepsy that are relatively prevalent among patients with LGG are associated with negative HRQL outcomes and thus, contribute additionally to the vulnerability of this population of cancer patients. [84]

Tumor-related epilepsy could negatively affect the QoL of patients with LGG, causing cognitive deterioration and profound morbidity,[86] along with psychological problems (intensified by the severity of epilepsy and by the epileptic drug treatment) that also could impact on cognitive functioning and HRQOL.[67,87] Given the high frequency of seizures, increasing survival with the use of chemotherapy and radiotherapy, and the risk of potential interaction of AEDs with CT, it might be necessary to incorporate a seizure-related questionnaire with the conventional HRQOL tools as an outcome measure in all PBT trials.[82]

Neurocognitive evaluation and possible rehabilitative programs

Only in recent years, interest has grown in the area of cognitive rehabilitation in patients with brain tumors and there is limited data regarding possible interventional strategies to improve cognitive dysfunctions in these patients. These data have demonstrated that patients with brain tumors who have cognitive impairment can participate meaningfully in a structured intervention of cognitive rehabilitation and experience increased independence and productivity with an improvement of their QoL.[88–91] In the case of brain cancer, which is characterized by progressive impairment of mental function, a beneficial treatment may be one that stabilizes or slows the progression of worsening symptoms, whether or not overall survival is extended.[92] Patients often have difficulty in performing more than one activity at the same time and can become easily overwhelmed when more than one thing is happening at one time.[92] They encounter problems in maintaining focused attention in daily life.

Even given the heterogeneity in methodology, all studies demonstrated a positive effect of rehabilitative training at different times: early,[91] at 3 months,[88] and at 6 months of follow-up,[90] with attention and verbal memory being the most improved cognitive functions. Future research is needed to explore variables that will help us identify patients most likely to benefit from cognitive training programs.[93]

Choice of AEDs

Whereas efficacy of AED is the point of reference for the choice of a given AED in patients with all types of epilepsy, to date, there is no evidence that the efficacy of AEDs differs for patients with BTRE compared to efficacy for patients with focal seizures from other aetiologies.[6] Therefore, for patients with BTRE, AED efficacy cannot be the reference point for initial AED choice. The choice of AEDs must take into account the following points. Recognizing interactions between AED and other agents used to treat brain tumors. Knowing that BTRE is a unique form of epilepsy in which the risk of seizures may alter according to tumor status: tumor recurrence and changes in surrounding brain tissue caused by treatment, such as necrosis after RT, may modify the risk of seizures. Considering that the older AEDs have the highest incidence of SE and of possible pharmacological interactions, and the new AEDs, such as LEV, LCM, GBP, and LTG, are reportedly safe in regard to cognitive function.[94] LEV reportedly improves memory in an animal model [95] and improves the QoL in patients with BTRE.[96] More recently, improvement in the verbal memory of patients with high-grade gliomas receiving LEV has been reported. [96]

Literature data indicated that most physicians prefer to initiate treatment with a non-enzyme-inducing AED, to minimize the risks of adverse interactions with anticancer agents.[6] AEDs that are frequently selected as first-line therapy include valproic acid and the second-generation non-enzyme-inducing AEDs LTG, LEV, oxcarbazepine, and TPM,[12] all of which have been approved in Europe as initial monotherapy for focal seizures. Zonisamide is also approved in Europe for initial monotherapy for focal seizures and represents an alternative option, although experience with this drug in patients with brain tumors is more limited. Experience with other second- generation AEDs such as lacosamide, pregabalin, and tiagabine is restricted to add-on therapy.[1,6] Nevertheless, at this moment treatment choices are largely determined by physicians' personal judgment and experience and by individual patient characteristics.[97] Treatment of epilepsy in patients with brain tumors needs a multidisciplinary approach that not only involves the use of AEDs but also gives consideration to surgery, cranial radiotherapy, chemotherapy, cognitive functions, and QoL.[76,98–101]

For all these reasons it should be preferred using new AEDs as a first choice and avoid the old AEDs.

Prophylaxis

Prophylactic use of AEDs has been assessed in patients who were undergoing surgery for a brain tumor. However, the results are conflicting, and therefore the need for anticonvulsive treatment before or after neurosurgery is uncertain.[31,102–104] A consensus statement from the Quality Standards Subcommittee of the American Academy of Neurology recommends not to use AEDs routinely as prophylaxis in patients with brain tumors and to withdraw these drugs in the first week after surgery if patients have never had a seizure.[17,31,105] In support of these data recently, researchers reported an extremely low incidence of seizures during craniotomy without AEDs[106,107] and if a prophylactic therapy during surgery should be chosen, recent data indicate that LEV is effective and superior to PHT.[32,108,109] Finally, if the use of prophylactic AEDs in patients with brain tumors is a complex and problematic issue, it can be equally, if not more difficult to establish if and when AED therapy should be suspended. Although it has been deemed appropriate to interrupt an AED 1 week following surgery,[17] clinicians often hesitate to suspend therapy for patients who must be treated with RT, worried about a possible increase of seizures during this procedure. Further studies are necessary to assess the value of continuing prophylactic antiepileptic therapy, at least until the end of RT sessions.

Health economics and BTRE

Health promotion as defined by the World Health Organization (WHO) is the process of enabling people to increase control over, and to improve, their health. It moves beyond a focus on individual behavior toward a wide range of social and environmental interventions. A population's health is the concern of many stakeholders who each represent a particular perspective and/or priority: from government institutions and policy makers, researchers and clinicians to health-care professionals and economists, to name only a few. Life is of highest value, and as health is fundamental to life, it leads to the creation of wealth.[110,111] The current policy agendas are faced with a complex health-care landscape that is changing due to people living longer, with costs that are escalating and resources that are diminishing. For this reason, today more than ever, professionals across disciplines are either preparing, participating in, or consulting studies pertaining to economic and societal costs related to health, health services, and illnesses themselves. Cost reduction will be a primary focus of all health-care systems regardless of political or geographical orientation. It needs to be structured by redefining health-care packages made available by public funds, which will mark the level of society's commitment to the well-being of its citizens. Patients will ultimately be made aware of these funds and the private sector will cover any additional costs not covered. For this reason, private interests will gain more importance. Within this context, it will be essential to create and foster sustainable clinical leadership: there can be no sustainable reform in the future without a solid core of medical professionals. While health-care costs have continued to be a primary concern of policy makers, there has also been a significant paradigm shift that sees the patient as the focal point. In fact, today's patient is increasingly well informed and motivated and is at the center of all efforts and all health-care provisions. There has been increased interest in monitoring the effectiveness of health-care provisions, quality control, and measuring the effectiveness of therapies by means of health technology assessment (HTA) and outcome indicators. Assessment can be done by a range of stakeholders: patients, patients groups, clinicians, health care organizations and providers of finance.[110,112] At present, there are no published data concerning the social and economic impact of BTRE. In terms of the health economic burden, disorders of the brain likely constitute the number one economic challenge for European health care.[113–117]

In a landmark study sponsored by the European Brain Council,[116] annual costs pertaining to neurological and mental disorders throughout Europe were examined and were found to have a high prevalence as well as short- and long-term impairments and disabilities, contributing to an emotional, financial, and social burden for patients and their families. That study was updated and expanded; the total cost of 19 disorders of the brain was estimated at €798 billion in 2010 for 30 European countries.[113] Direct costs constituted the majority of costs (37% direct health-care costs and 23% direct nonmedical costs) whereas the remaining 40% were indirect costs associated with patients' production losses. Within this study, the costs (in billion € Purchasing Power Parity—PPP for the year 2010) for brain tumors and epilepsy were as follows: brain tumor: €5.2; epilepsy: €13.8. The study declares that the cost model utilized clearly reveals that "brain disorders overall are much more costly than previously estimated constituting a major health economic challenge for Europe."[113]

Conclusions

In consideration of the possible long-term disability that epilepsy can provoke in patients with glioma, it is critical that when choosing an AED, the following must be taken into account: the frequency/seriousness of SEs, the drug's possible impact on both cognition and effectiveness of the systemic therapy, as well as the fact that interactions with other therapies may occur.[1] The therapeutic plan should take into consideration whether or not a specific antineoplastic therapy is needed (RT, CT, both) and whether or not a rapid titration could be necessary. Since epilepsy in these patients affects the QoL and creates a possible long-term disability, due to factors related to the epilepsy itself and to drug management (AEDs and systemic therapy), the choice of AED must also take into consideration possible effects of the AED on important aspects of patients' daily life, such as cognitive function and sexuality. Also the neuropsychological profile encompasses many challenges of living with both epilepsy and BT. Careful monitoring over time, using tests that adequately examine the full range of QoL issues, can lead to optimal patient management– and to positive reinforcement of each individual's capabilities that exist despite the pathology. Finally, a well-structured multidisciplinary team approach is essential in managing this group of patients with dual pathology.[82] For this reason it is necessary to develop a customized treatment plan for each individual patient with BTRE, the goals of which should be complete seizure control, minimal or no SEs, and elimination of cognitive impairment, and/or psychosocial problems.

Acknowledgments

The Author wishes to express her gratitude to Dr. Selvaggia Camilla Serini for reviewing the manuscript and for her precious support. The Author also thanks Dr. Serini for her precious and irreplaceable contribution, for her rare precision and attention in the performance of the work, and for her ability to work in synergy, enhancing and respecting the specificities of each part of the team.

References

1. Maschio M. Brain tumor-related epilepsy. *Curr Neuropharmacol.* 2012;10(2):124–133.
2. Hildebrand J. Management of epileptic seizures. *Curr Opin Oncol.* 2004;16(4):314–317.
3. Bobustuc GC, Baker CH, Limaye A, et al. Levetiracetam enhances p53-mediated MGMT inhibition and sensitizes glioblastoma cells to temozolomide. *Neuro Oncol.* 2010;12(9):917–927.
4. Bourg V, Lebrun C, Chichmanian RM, Thomas P, Frenay M. Nitroso-urea-cisplatin-based chemotherapy associated with valproate: increase of haematologic toxicity. *Ann Oncol.* 2001;12(2):217–219.

5. Anderson GD, Lin YX, Berge C, Ojemann GA. Absence of bleeding complications in patients undergoing cortical surgery while receiving valproate treatment. *J Neurosurg.* 1997;87(2):252–256.

6. Perucca E. Optimizing antiepileptic drug treatment in tumoral epilepsy. *Epilepsia.* 2013;54(Suppl 9):97–104.

7. Maschio M, Dinapoli L. Patients with brain tumor-related epilepsy. *J Neurooncol.* 2012;109(1):1–6.

8. Meinardi H, Scott RA, Reis R, Sander JW. ILAE Commission on the Developing World. The treatment gap in epilepsy: the current situation and ways forward. *Epilepsia.* 2001;42(1):136–149.

9. Reynolds EH. The ILAE/IBE/WHO epilepsy global campaign history. International league against epilepsy. International bureau for epilepsy. *Epilepsia.* 2002;43(Suppl 6):9–11.

10. WHO, International League Against Epilepsy, International Bureau for Epilepsy. *Atlas: Epilepsy Care in the World 2005;* 2005. Geneva.

11. Newton HB, Maschio M. Brain Tumor-related epilepsy: introduction and overview. In: Newton HB, Maschio M, eds. *Epilepsy and Brain tumors.* Amsterdam: Elsevier; 2015:1–9.

12. Rossetti AO, Stupp R. Epilepsy in brain tumor patients. *Curr Opin Neurol.* 2010;23(6):603–609.

13. Hildebrand J, Lecaille C, Perennes J, Delattre JY. Epileptic seizures during follow-up of patients treated for primary brain tumors. *Neurology.* 2005; 65(2):212–215.

14. Vecht CJ, Wilms EB. Seizures in low- and high-grade gliomas: current management and future outlook. *Expert Rev Anticancer Ther.* 2010 May;10(5): 663–669.

15. Bromfield EB. Epilepsy in patients with brain tumors and other cancers. *Rev Neurol Dis.* 2004;1(Suppl 1):S27–S33.

16. Forsgren I, Beghi E, Ekman M. Cost of epilepsy in Europe. *Eur J Neurol.* 2005;12(Suppl 1):54–58.

17. Glantz MJ, Cole BF, Forsyth PA, et al. Practice parameter: anticonvulsant prophylaxis in patients with newly diagnosed brain tumors. Report of the quality standards subcommittee of the american academy of neurology. *Neurology.* 2000;54(10):1886–1893.

18. Mirsattari SM, Chong JJ, Hammond RR, et al. Do epileptic seizures predict outcome in patients with oligodendroglioma? *Epilepsy Res.* 2011;94(1–2): 39–44.

19. Chang EF, Potts MB, Keles GE, et al. Seizure characteristics and control following resection in 332 patients with low-grade gliomas. *J Neurosurg.* 2008;108(2):227–235.

20. Kwan P, Brodie MJ. Early identification of refractory epilepsy. *N Engl J Med.* 2000;342(5):314–319.

21. Luyken C, Blümcke I, Fimmers R, et al. The spectrum of long-term epilepsy-associated tumors: long-term seizure and tumor outcome and neurosurgical aspects. *Epilepsia.* 2003;44(6):822–830.

22. Daumas-Duport C, Varlet P, Tucker ML, Beuvon F, Cervera P, Chodkiewicz JP. Oligodendrogliomas. Part I: patterns of growth, histological diagnosis, clinical and imaging correlations: a study of 153 cases. *J Neurooncol.* 1997;34(1):37–59.

23. Koeller KK, Rushing EJ. From the archives of the AFIP: oligodendroglioma and its variants: radiologic-pathologic correlation. *Radiographics.* 2005;25(6):1669–1688.

24. Allam A, Radwi A, El Weshi A, Hassounah M. Oligodendroglioma: an analysis of prognostic factors and treatment results. *Am J Clin Oncol.* 2000;23(2): 170–175.

25. Van den Bent MJ. Advances in the biology and treatment of oligodendrogliomas. *Curr Opin Neurol.* 2004;17(6):675–680.

26. Karim AB, Afra D, Cornu P, et al. Randomized trial on the efficacy of radiotherapy for cerebral low-grade glioma in the adult: European Organization for Research and Treatment of Cancer Study 22845 with the Medical Research Council study BRO4: an interim analysis. *Int J Radiat Oncol Biol Phys.* 2002;52(2):316–324.

27. Wiebe S, Bellhouse DR, Fallahay C, Eliasziw M. Burden of epilepsy: the ontario health survey. *Can J Neurol Sci.* 1999;26(4):263–270.

28. Berg AT, Kelly MM. Defining intractability: comparisons among published definitions. *Epilepsia.* 2006;47(2):431–436.

29. Engelhard HH. Current diagnosis and treatment of oligodendroglioma. *Neurosurg Focus.* 2002;12:1–7.

30. Buckingham SC, Campbell SL, Haas BR, et al. Glutamate release by primary brain tumors induces epileptic activity. *Nat Med.* 2011;17(10): 1269–1274.

31. Van Breemen MS, Wilms EB, Vecht CJ. Epilepsy in patients with brain tumors: epidemiology, mechanisms, and management. *Lancet Neurol.* 2007;6(5): 421–430.

32. Iuchi T, Hasegawa Y, Kawasaki K, Sakaida T. Epilepsy in patients with gliomas: incidence and control of seizures. *J Clin Neurosci.* 2015 Jan;22(1): 87–91.

33. Berg AT, Berkovic SF, Brodie MJ, et al. Revised terminology and concepts for organization of seizures and epilepsies: report of the ILAE Commission on classification and terminology, 2005–2009. *Epilepsia.* 2010;51(4):676–685.

34. Panayiotopoulos CP. The new ILAE report on terminology and concepts for the organization of epilepsies: critical review and contribution. *Epilepsia.* 2012;53(3):399–404.

35. Kargiotis O, Markoula S, Kyritsis AP. Epilepsy in the cancer patient. *Cancer Chemother Pharmacol.* 2011;67(3):489–501.

36. Louis DN, Perry A, Reifenberger G, et al. The 2016 world health organization classification of tumors of the central nervous system: a summary. *Acta Neuropathol.* 2016;131(6):803–820.

37. Weller M, van den Bent M, Tonn JC, et al. European association for neuro-oncology (EANO) task force on gliomas. European association for neuro-oncology (EANO) guideline on the diagnosis and treatment of adult astrocytic and oligodendroglial gliomas. *Lancet Oncol.* 2017;18(6): e315–e329.

38. Mulligan L, Ryan E, O'Brien M, Looby S, Heffernan J, O'Sullivan J, Clarke M, Buckley P, O'Brien D, Farrell M, Brett FM. Genetic features of oligodendrogliomas and presence of seizures. The relationship of seizures and genetics in LGOs. Clin Neuropathol. 2014;33(4):292-8.

39. Zhou XW, Wang X, Yang Y, et al. Biomarkers related with seizure risk in glioma patients: a systematic review. *Clin Neurol Neurosurg.* 2016;151:113–119.

40. Liang R, Fan Y, Wang X, Mao Q, Liu YH. The significance of IDH1 mutations in tumor-associated seizure in 60 chinese patients with low-grade gliomas. *ScientificWorld J.* 2013;2013:403942. Published online 2013 Nov 13.

41. Buckingham SC, Robel S. Glutamate and tumor-associated epilepsy: glial cell dysfunction in the peritumoral environment. *Neurochem Int.* 2013;63(7): 696–701.

42.. Avoli M, Louvel J, Pumain R, Köhling R. Cellular and molecular mechanisms of epilepsy in the human brain. *Prog Neurobiol.* 2005;77 (3):166–200.

43. Aronica E, Boer K, Becker A, et al. Gene expression profile analysis of epilepsy-associated gangliogliomas. *Neuroscience.* 2008;151(1):272–292.

44. de Groot M, Iyer A, Zurolo E, et al. Overexpression of ADK in human astrocytic tumors and peritumoral tissue is related to tumor-associated epilepsy. *Epilepsia.* 2012;53(1):58–66.

45. Rizzo A, Donzelli S, Girgenti V, et al. In vitro antineoplastic effects of brivaracetam and lacosamide on human glioma cells. *J Exp Clin Cancer Res.* 2017;36(1):76.

46. You G, Yan W, Zhang W, et al. Significance of miR-196b in tumor-related epilepsy of patients with gliomas. *PLoS One.* 2012;7(9).

47. Yang P, You G, Zhang W, et al. Correlation of preoperative seizures with clinicopathological factors and prognosis in anaplastic gliomas: a report of 198 patients from China. *Seizure.* 2014;23(10):844–851.

48. Köhling R, Senner V, Paulus W, Speckmann EJ. Epileptiform activity preferentially arises outside tumor invasion zone in glioma xenotransplants. *Neurobiol Dis.* 2006;22(1):64–75.

49. You G, Sha Z, Jiang T. The pathogenesis of tumor-related epilepsy and its implications for clinical treatment. *Seizure.* 2012;21(3):153–159.

50. Yuan Y, Xiang W, Qing M, Yanhui L, Jiewen L, Yunhe M. Survival analysis for valproic acid use in adult glioblastoma multiforme: a meta-analysis of individual patient data and a systematic review. *Seizure.* 2014;23(10):830–835.

51. Kim YH, Kim T, Joo JD, et al. Survival benefit of levetiracetam in patients treated with concomitant chemoradiotherapy and adjuvant chemotherapy with temozolomide for glioblastoma multiforme. *Cancer.* 2015;121(17):2926–2932.

52. Koekkoek JA, Dirven L, Heimans JJ, et al. Seizure reduction in a low-grade glioma: more than a beneficial side effect of temozolomide. *J Neurol Neurosurg Psychiatry.* 2015;86(4):366–373.

53. Eyal S, Yagen B, Sobol E, Altschuler Y, Shmuel M, Bialer M. The activity of antiepileptic drugs as histone deacetylase inhibitors. *Epilepsia.* 2004;45(7): 737–744.

54. Berendsen S, Broekman M, Seute T, et al. Valproic acid for the treatment of malignant gliomas: review of the preclinical rationale and published clinical results. *Expert Opin Investig Drugs.* 2012;21(9):1391–1415.

55. Guthrie GD, Eljamel S. Impact of particular antiepileptic drugs on the survival of patients with glioblastoma multiforme. *J Neurosurg.* 2013;118(4): 859–865.

56. Kerkhof M, Dielemans JC, van Breemen MS, et al. Effect of valproic acid on seizure control and on survival in patients with glioblastoma multiforme. *Neuro Oncol.* 2013;15(7):961–967.

57. Singh G, Rees JH, Sander JW. Seizures and epilepsy in oncological practice: causes, course, mechanisms and treatment. *J Neurol Neurosurg Psychiatry.* 2007;78(4):342–349.

58. Patsalos PN, Perucca E. Clinically important drug interactions in epilepsy: general features and interactions between antiepileptic drugs. *Lancet Neurol.* 2003;2(6):347–356.

59. Vecht CJ, Wagner GL, Wilms EB. Interactions between antiepileptic and chemotherapeutic drugs. *Lancet Neurol.* 2003;2(7):404–409.

60. Baltes S, Gastens AM, Fedrowitz M, Potschka H, Kaever V, Löscher W. Differences in the transport of the antiepileptic drugs phenytoin, levetiracetam and carbamazepine by human and mouse P-glycoprotein. *Neuropharmacology.* 2007;52(2):333–346.

61. Löscher W, Potschka H. Role of multidrug transporters in pharmacoresistance to antiepileptic drugs. *J Pharmacol Exp Ther.* 2002;301(1):7–14.

62. Hermann DM, Kilic E, Spudich A, Krämer SD, Wunderli-Allenspach H, Bassetti CL. Role of drug efflux carriers in the healthy and diseased brain. *Ann Neurol.* 2006;60(5):489–498.

63. Rogawski MA. Does P-glycoprotein play a role in pharmacoresistance to antiepileptic drugs? *Epilepsy Behav.* 2002;3(6):493–495.

64. Siddiqui A, Kerb R, Weale ME, et al. Association of multidrug resistance in epilepsy with a polymorphism in the drug-transporter gene ABCB1. *N Engl J Med.* 2003;348(15):1442–1448.

65. Taillibert S, Laigle-Donadey F, Sanson M. Palliative care in patients with primary brain tumors. *Curr Opin Oncol.* 2004;16(6):587–592.

66. Beghi E. Efficacy and tolerability of the new antiepileptic drugs: comparison of two recent guidelines. *Lancet Neurol.* 2004;3(10):618–621.

67. Klein M, Engelberts NH, van der Ploeg HM, et al. Epilepsy in low-grade gliomas: the impact on cognitive function and quality of life. *Ann Neurol.* 2003;54(4):514–520.

68. Maschio M, Dinapoli L, Vidiri A, Muti P. Rash in four patients with brain tumor-related epilepsy in monotherapy with oxcarbazepine, during radiotherapy. *J Neurol.* 2010;257(11):1939–1940.

69. Bosma I, Vos MJ, Heimans JJ, et al. The course of neurocognitive functioning in high-grade glioma patients. *Neuro Oncol.* 2007;9(1):53–62.

70. Taphoorn MJ, Klein M. Cognitive deficits in adult patients with brain tumors. *Lancet Neurol.* 2004;3(3):159–168. Review.

71. Maschio M, Saveriano F, Dinapoli L, Jandolo B. Reversible erectile dysfunction in a patient with brain tumor-related epilepsy in therapy with zonisamide in add-on. *J Sex Med.* 2011;8(12):3515–3517.

72. Patsalos PN, Fröscher W, Pisani F, van Rijn CM. The importance of drug interactions in epilepsy therapy. *Epilepsia.* 2002;43(4):365–385.

73. Spina E, Perucca E. Clinical significance of pharmacokinetic interactions between antiepileptic and psychotropic drugs. *Epilepsia.* 2002;43(Suppl 2):37–44.

74. Yap KY, Chui WK, Chan A. Drug interactions between chemotherapeutic regimens and antiepileptics. *Clin Ther.* 2008;30:1385–1407.

75. Warnke PC, Berlis A, Weyerbrock A, Ostertag CB. Significant reduction of seizure incidence and increase of benzodiazepine receptor density after interstitial radiosurgery in low-grade gliomas. *Acta Neurochir.* 1997;68(Suppl):90–92.

76. Rudà R, Magliola U, Bertero L, et al. Seizure control following radiotherapy in patients with diffuse gliomas: a retrospective study. *Neuro Oncol.* 2013;15(12):1739–1749.

77. Kerkhof M, Benit C, Duran-Pena A, Vecht CJ. Seizures in oligodendroglial tumors. *CNS Oncol.* 2015;4(5):347–356.

78. Van den Bent MJ, Afra D, de Witte O, et al. Long-term efficacy of early versus delayed radiotherapy for low-grade astrocytoma and oligodendroglioma in adults: the EORTC 22845 randomized trial. *Lancet.* 2005;366(9490):985–990.

79. Rudà R, Bello L, Duffau H, Soffietti R. Seizures in low-grade gliomas: natural history, pathogenesis, and outcome after treatments. *Neuro Oncol.* 2012;14(suppl 4):iv55–64.

80. Baker GA, Jacoby A, Buck D, Stalgis C, Monnet D. Quality of life of people with epilepsy: a European study. *Epilepsia.* 1997;38:353–362.

81. Meador KJ. Cognitive side effects of medications. *Neurol Clin.* 1998;16:141–155.

82. Rahman Z, et al. Epilepsy in patients with primary brain tumors: The impact on mood, cognition, and HRQOL. *Epilepsy Behav.* 2015;48:88–95.

83. Struik K, Klein M, Heimans JJ, et al. Fatigue in low-grade glioma. *J Neurooncol.* 2009;92:73–78.

84. Aaronson, et al. Compromised quality of life in patients with low-grade glioma. *J Clin Oncol.* 2011;29:4430–4435.

85. Klein M, Heimans JJ, Aaronson NK, et al. Effect of radiotherapy and other treatment-related factors on mid-term to long-term cognitive sequelae in low-grade gliomas: a comparative study. *Lancet.* 2002;360:1361–1368.

86. Englot DJ, Berger MS, Barbaro NM, Chang EF. Predictors of seizure freedom after resection of supratentorial low-grade gliomas. *A Rev J Neurosurg.* 2011;115:240–244.

87. Shields BE, Choucair AK, et al. *World Neurosurg.* 2014;82(1/2):e299–e309.

88. Locke DE, Cerhan JH, Wu W, et al. Cognitive rehabilitation and problem-solving to improve quality of life of patients with primary brain tumors: a pilot study. *J Support Oncol.* 2008;6:383–391.

89. Hassler MR, Elandt K, Preusser M, et al. Neurocognitive training in patients with high-grade glioma: a pilot study. *J Neurooncol.* 2010;97:109–115.

90. Gehring K, Sitskoorn MM, Gundy CM, et al. Cognitive rehabilitation in patients with gliomas: a randomized, controlled trial. *J Clin Oncol.* 2009;27:3712–3722.

91. Zucchella C, Capone A, Codella V, et al. Cognitive rehabilitation for early post-surgery inpatients affected by pri- mary brain tumor: a randomized, controlled trial. *J Neurooncol.* 2013;114:93–100.

92. Meyers CA, Hess KR. Multifaceted end points in brain tumor clinical trials: cognitive deterioration precedes MRI progression. *Neuro Oncol.* 2003;5(2):89–95.

93. Maschio M, Dinapoli L, Fabi A, Giannarelli D, Cantelmi T. Cognitive rehabilitation training in patients with brain tumor-related epilepsy and cognitive deficits: a pilot study. *J Neurooncol.* 2015 Nov;125(2):419–426. https://doi.org/10.1007/s11060-015-1933-8.

94. Eddy CM, Rickards HE, Cavanna AE, et al. The cognitive impact of antiepileptic drugs. *Ther Adv Neurol Disord.* 2011;4:385–407.

95. Celikyurt IK, Ulak G, Mutlu O, et al. Positive impact of levetiracetam on emotional learning and memory in naive mice. *Life Sci.* 2012;90:185–189.

96. Maschio M, Dinapoli L, Sperati F, et al. Levetiracetam monotherapy in patients with brain tumor-related epilepsy: seizure control, safety, and quality of life. *J Neurooncol.* 2011;104:205–214.

97. Perucca E, Tomson T. The pharmacological treatment of epilepsy in adults. *Lancet Neurol.* 2011;10(5):446–456.

98. Englot DJ, Chang EF, Vecht CJ. Epilepsy and brain tumors. *Handb Clin Neurol.* 2016;134:267–285.

99. Simonetti G, Gaviani P, Botturi A, Innocenti A, Lamperti E, Silvani A. Clinical management of grade III oligodendroglioma. *Cancer Manag Res.* 2015;7:213–223.

100. Vecht CJ, Kerkhof M, Duran-Pena A. Seizure prognosis in brain tumors: new insights and evidence-based management. *Oncologist.* 2014;19(7):751–759. Jul.

101. Chen DY, Chen CC, Crawford JR, Wang SG. Tumor-related epilepsy: epidemiology, pathogenesis and management. *J Neurooncol.* 2018;139(1):13–21.

102. Franceschetti S, Binelli S, Casazza M, et al. Influence of surgery and antiepileptic drugs on seizures symptomatic of cerebral tumors. *Acta Neurochir (Wien).* 1990;103:47–51.

103. Shaw MD. Post-operative epilepsy and the efficacy of anticonvulsant therapy. *Acta Neurochir Suppl (Wien).* 1990;50:55–57.

104. Cohen N, Strauss G, Lew R, Silver D, Recht L. Should prophylactic anticonvulsants be administered to patients with newly-diagnosed cerebral metastases? A retrospective analysis. *J Clin Oncol.* 1988;6:1621–1624.

105. Wen PY, Marks PW. Medical management of patients with brain tumors. *Curr Opin Oncol.* 2002;14:299–307.

106. Lwu S, Hamilton MG, Forsyth PA, et al. Use of peri-operative anti-epileptic drugs in patients with newly diagnosed high grade malignant glioma: a single center experience. *J Neurooncol.* 2010;96:403–408.

107. Sughrue ME, Rutkowski MJ, Chang EF, et al. Postoperative seizures following the resection of convexity meningiomas: are prophylactic anticonvulsants indicated? Clinical article. *J Neurosurg.* 2011;114:705–709.

108. Bähr O, Hermission M, Rona S, et al. Intravenous and oral levetiracetam in patients with a suspected primary brain tumor and symptomatic seizures undergoing neurosurgery: the HELLO trial. *Acta Neurochir (Wien).* 2012;154:229–235.

109. Kern K, Schebesch KM, Schlaier J, et al. Levetiracetam compared to phenytoin for the prevention of postoperative seizures after craniotomy for intracranial tumors in patients without epilepsy. *J Clin Neurosci.* 2012;19:99–100.

110. Unger F. Health is wealth: considerations to European healthcare. *Prilozi.* 2012;33:9–14.

111. Maschio M, Paladin F. Social cost of BTRE. In: Newton HB, Maschio M, eds. *Epilepsy and Brain Tumors.* Amsterdam: Elsevier; 2015.

112. Lukas CV, Holmes SK, Cohen AB, et al. Transformational change in health care systems: an organizational model. *Health Care Manage Rev.* 2007;2:309–320.

113. Gustavsson A, Svensson M, Jacobi F, et al. Cost of disorders of the brain in Europe 2010. *Eur Neuropsychopharmacol.* 2011;21:718–779.

114. Pugliatti M, Beghi E, Forsgren L, Ekman M, Sobocki P. Estimating the cost of epilepsy in Europe: a review with economic modeling. *Epilepsia.* 2007;48:2224–2233.

115. Pugliatti M, Sobocki P, Beghi E, et al. Cost of disorders of the brain in Italy. *Neurol Sci.* 2008;29:99–107.

116. Andlin-Sobocki P, J€onsson B, Wittchen HU, Olesen J. Cost of disorders of the brain in Europe. *Eur J Neurol.* 2005;12(suppl 1):1–27.

117. Olesen J, Gustavsson A, Svensson M, et al. The economic cost of brain disorders in Europe. *Eur J Neurol.* 2012;19:155–162.

Further Reading

118. De Groot M, Douw L, Sizoo EM, et al. Levetiracetam improves verbal memory in high-grade glioma patients. *Neuro Oncol.* 2013;15:216–223.

Chapter 6

Living with oligodendroglioma

Ashlee R. Loughan*, Deborah Hutchinson Allen[†] and Sarah Ellen Braun[‡]

*Department of Neurology, Virginia Commonwealth University & Massey Cancer Center, Richmond, VA, United States; [†]Nursing Research & EBP, Duke University Health System & Duke Cancer Institute, Durham, NC, United States; [‡]Department of Psychology, Virginia Commonwealth University, Richmond, VA, United States

---

"I have a good life.

I am happy, I have great friends and family, I do not live in pain, I am educated, I am comfortable.

I am healthy in the sense that I never get sick.

I am unhealthy in the reality that I have cancer in my brain."

-low grade glioma patient

In most brain tumor populations, current treatment options are not curative. Instead, medical teams focus on prolonging survival while trying to maintain or improve a patient's quality of life (QoL). QoL is a multidimensional construct covering life domains such as physical, psychological, and social, as well as symptoms induced by disease and its treatment (see Fig. 1).[1,2] Although domains can be separated, often changes in one domain can influence or affect perceptions in another. For example, changes in a patient's physical domain (e.g., inability to ambulate independently) can easily affect their psychological well-being or their social interactions with others. Up toward half of low-grade glioma patients report lessened overall QoL[3] with the most common complaints being cognitive decline[4,5] and/or physical ailments.[3] Assessing QoL can provide pivotal details in the pursuit of optimal patient care. With more targeted treatment approaches developing and the increased risk of side effects/neurotoxicity, there is growing support in considering a glioma patients QoL.[6] Currently, QoL has become an important complement to conventional medically assessed outcome parameters such as time to tumor progression, overall and progression free survival, and radiological response. This chapter discusses a variety of QoL components including physical, cognitive-behavioral, and psychosocial factors. The information draws from literature on primary brain tumors generally, as patients with oligodendrogliomas are appreciably similar to other primary brain tumor patients. In cases where there are pronounced differences or exceptions to discuss, percentage of sample or population clarification will be provided.

Physical factors in QoL

Patients with brain tumors, including oligodendrogliomas, experience many physical symptoms that often persist beyond diagnosis or treatment. These persistent symptoms can be distressing and negatively impact their QoL.[7–9] Although molecular and pathological features vary across tumor grades, most oligodendrogliomas exhibit a slow growth rate, therefore symptoms may not become apparent until the tumor displaces adjacent tissues.[10,11] Recent research involving alterations in genetic expression or identification of mutations are being associated with clinical outcomes or symptom burden[12–14] with tumor location, size, and tissue displacement factoring into the presentation of neurological symptoms.[10,15,16] To illustrate, it is predicted that tumors with HLA-G expression may display more aggressive tumor activity and contribute to a poor clinical outcome in patients with low-grade gliomas (LGG). Physical symptoms that are most common include headache and pain, seizure disorder, and a variety of focal neurological complaints (e.g., dizziness, weakness, gait disturbances, incoordination and imbalance, visual disturbances, hearing deficits, loss of smell, paresthesias), and mental status changes.[11,16] Seizures are common in 70%–90% of cases,[17] followed by headache and nausea

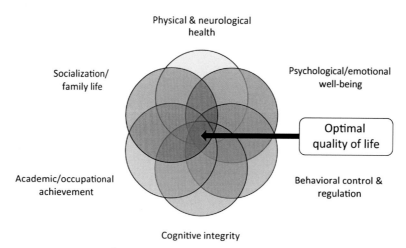

FIG. 1 Quality-of-life enmeshment of functionality.[1]

(10%–40% cases), mental status changes and focal symptoms (<5%–30% cases).[16] It appears that patients with right hemisphere LGG report less physical symptoms when compared to patients with left-hemisphere tumors. This suggests that right hemisphere tumors spare the dominant side, permitting a better QoL.[18] Mediating factors for symptom severity include preexisting conditions, psychosocial status, functional capacity, and tolerance of pain/discomfort. Despite patients experiencing long periods of time with a stable tumor status, nearly 40% of patients report a decline in QoL years after diagnosis.[7,8] The National Comprehensive Cancer Network (NCCN)[19,20] has established guidelines for assessing and managing symptoms across cancer treatment. These guidelines may be helpful to both providers and patients. Where appropriate, interventions for physical factors are discussed. Refer to Table 3 at the conclusion of this chapter for a complete discussion.

Treatment-related symptoms

In addition to the presence of focal symptoms from tumor location and central symptoms from cerebral edema, cancer treatment has been shown to produce systemic toxicities that may persist well beyond treatment cessation.[8,10,11,16,21] Symptoms resulting from treatment appear to vary in frequency, severity, and impact on QoL. Ediebah and colleagues[9] found that overall survival may be masked by the effect that treatment can have on QoL. The goal of surgery, often the first line of treatment for brain tumor patients, is to provide maximal resection while sparing eloquent brain/functional areas and preserving QoL.[10,11] Patients often experience improvement of symptoms after surgical resection, particularly a reduction in seizure activity[16,22]; however, motor, sensory, and cognitive deficits can persist.[8,23] Newer radiotherapy techniques have provided greater precision of therapy while sparing brain tissues and reducing symptoms.[10] Acute symptoms from radiotherapy (skin reactions, otitis media, headache, nausea and emesis,[21] fatigue and sleep disturbances)[24] tend to occur toward the end of the treatment and improve over the subsequent weeks. Acute symptoms from chemotherapy, particularly alkylating agents (temozolomide, PCV), include nausea and emesis, constipation, diarrhea, stomatitis, anorexia, bone marrow suppression, liver enzyme elevation, and fatigue.[11] As patient survival rates are increasing, late effects from radiotherapy and chemotherapy, such as neurocognitive effects are only recently being recognized and will be discussed in more detail later in this chapter.

Functional status

Assessing patient's functional status provides the clinician with meaningful trends of change in everyday function. Functional status instruments quantify one's ability to perform activities of daily living and serve as a proxy for QoL.[25–27] The Karnofsky Performance Status and Eastern Cooperative Oncology Group Performance Status scales are commonly used functional status instruments.[28] Physical functioning subscales are also commonly included in QoL instruments (e.g., FACT-Br or EORTC QLQ C30) to assess change over time.[11,23,29] Declining functional status is associated with poor QoL[11] and negatively impacts daily, social, and work activities.[23,26,29] Fox and colleagues[30] demonstrated the relationship between QoL and functional status by identifying two symptom clusters in 73 long-term survivors of primary brain tumor for all tumor types (see Fig. 2).

FIG. 2 Symptom clusters of QoL and functional status.[30]

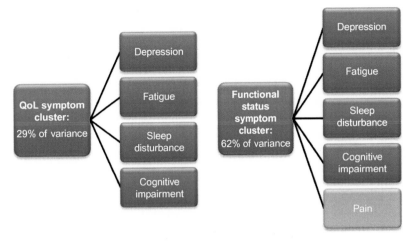

Seizures

Pathophysiological mechanisms from tumor burden increase intracranial pressure and focal inflammation to cause neuron hyper-excitability and potentiate seizure activity in susceptible tissues. It is estimated that >80% of patients with LGG experience seizures during their health trajectory; with seizures being the most common in oligodendrogliomas; followed by astrocytoma, meningioma, high-grade glioma, and primary CNS lymphoma.[17,31–33] Increased risk for seizure activity has been shown in patients with an IDH1/2 genetic tumor mutation,[14] those with anaplastic oligodendroglioma/astrocytoma, and those with tumors having lower MGMT and EGFR expression.[13] Seizure activity has also been shown to increase when nearing end-of-life.[34] Recent evidence demonstrates that a reduction in seizure activity during treatment may indicate tumor responsiveness to treatment; likewise, an increase in seizure activity during treatment may indicate tumor growth or recurrence.[35] While patients often experience less seizures after surgical debulking,[16,22] recommendations to continue antiepileptic treatment should be based on electroencephalography abnormalities, associated risk factors, and lesion localization.[11] Prophylactic antiepileptic use remains discouraged at this time.[32] Patients with seizure disorders tend to report worse QoL and greater distress[36,37]; some of these reports may be influenced by antiepileptic use as antiepileptic administration has been associated with adverse side effects including fatigue, drowsiness, and altered medication metabolism through the cytochrome P450 pathway.[11] Patients and their families require discussions regarding seizure risks, identification of potential triggers to avoid or minimize exposure, and incorporation of healthy behaviors into their daily activities as stress and extreme fatigue may exacerbate seizure activity.

Headaches

Headaches are another common symptom experienced by patients with gliomas. In fact, headaches are a presenting symptom at time of diagnosis in up to 40% of patients with LGG [16] and often present at the time of tumor progression.[15] Headaches may result from increased intracranial pressure, vascular permeability, inflammation, neuronal impairment, pain-sensitive structure distortion, and blood-brain barrier disruption.[15,37] Headache presentation varies according to tumor location and size, and may improve after tumor debulking surgery.[29] Most tumor-related headache symptoms tend to be unilaterally located; however, centrally located tumors tend to produce generalized, bilateral, or diffuse headache symptoms. Headache characteristics vary greatly, from tension-like characteristics to migraines. High-grade tumors with associated increased intracranial pressure tend to be very severe, worse in the morning, and induce nausea and vomiting.[15,37] Fox and colleagues[30] found that patients with headaches that persist after treatment completion reported poorer QoL. Headache symptoms from tumor growth are often controlled by corticosteroid therapy aimed to reduce vascular permeability and thereby decrease intracranial pressure from cerebral edema.[32] Care must be taken with steroid therapy as weight gain from steroid therapy may negatively impact body image and QoL. In most cases steroids may be slowly tapered off as causation of headache symptoms are treated (e.g., treatment for tumor progression).[23,29] Some patients require symptom control with pain medications (e.g., analgesics, opioids) and may benefit from consultation with a pain clinician specializing in headache management.

Nausea

Nausea, with or without emesis, can result from treatment, increased intracranial pressure, or tumor location.[11] Most chemotherapeutic agents are not highly emetogenic, but nausea has been reported as a side effect from PVC, temozolomide, and radiation therapy.[11] For example, two studies observed that nausea and vomiting were not significant in severity or frequency and therefore not included in brain tumor specific symptom clusters.[26,30] Patients experiencing treatment-related nausea benefit from pharmacologic agents (e.g., serotonin-3 receptor antagonists, steroids, and antianxiolytics), as well as nonpharmacologic interventions (e.g., relaxation, guided imagery).[36,38]

Fatigue

Fatigue is a common complaint reported by patients with gliomas, with 40%–70% endorsing an increase in fatigue throughout their illness trajectory.[32,39,40] Actually, fatigue is such a common symptom reported across all cancers that the NCCN[41] has produced guidelines for oncology providers and patients to consider when specifically assessing fatigue and its impact on QoL. Cancer-related fatigue is often described as feeling tired, weak, exhausted, sleepy, drowsy, and may preclude mental processing, especially thinking and concentration.[36,41] In a small cohort of patients with anaplastic oligodendrogliomas ($n = 32$), Habets and colleagues[8] found that right-sided tumor placement was associated with more fatigue, yet this has not been replicated. Twenty years of symptom research has found that patients with gliomas who report fatigue tend to report worse QoL and require increased support from caregivers for activities of daily living.[42–46] The implementation of psychostimulants as an intervention for fatigue has produced mixed results[24,40,47]; therefore, observation for trends in individual improvement should occur if therapy is initiated. Although stimulant use has not been associated with increased seizure activity risk in patients with glioma,[48] patients with LGG require close monitoring during initiation and titration to a therapeutic dose. Programs with moderate-intensity exercise or physical activity have recently demonstrated minimal improvement in fatigue.[49,50] When initiating a physical activity program, baseline assessment for balance, falls risk, and functional capacity should be performed.

Sleep

Sleep and wake disturbances are distressing to patients with gliomas and their caregivers.[26,51] Sleep disturbances range from difficulties falling asleep or returning to sleep, having frequent awakenings, excessive sleep, and daytime napping. Sleep and wake disturbances are associated with disruptions in everyday functioning, greater fatigue, cognitive dysfunction, depression, and QoL.[20,52] Postoperative sleep disturbances and incisional pain improve within a few weeks of surgery and are associated with greater improvements in physical function and QoL.[23] Age, tumor location (specifically right-sided tumors),[8] use of antiepileptics, steroid therapy, antipsychotics or stimulants, and antidepressants were found to be associated with insomnia.[53] For those nearing end of life, sleep disruptions are observed in up to 95% of patients as increasing intracranial pressure produce altered levels of consciousness and coma.[34] Interventions may include use of sleeping medication, antianxiolytics, adjusting steroid dose/frequency, and cognitive behavioral therapies. Mindfulness-based stress reduction has also been shown to improve sleep in oncology patients.[54] Given its strong association with fatigue and worse QoL, more research is warranted to ameliorate insomnia in neuro-oncology patients.

Nutrition

Cancer-related cachexia has been reported across many cancers; however, it is not commonly observed in patients with gliomas (<20% as compared to other cancer populations).[55] This may be due to the use of steroid therapy to control neurologic symptoms as steroids can increase appetite and increase body fat mass. Patients undergoing chemotherapy and radiotherapy may report more nausea or taste disturbances that may interfere with nutritional intake and place some patients with malignant gliomas at greater risk for malnourishment.[11] Patients presenting with ≥5% weight loss, persistent nausea and vomiting, presence of dysphagia and/or headache warrant nutritional screening for risk of developing malnutrition.[55]

Exercise

In cancer patients, moderate-intensity exercise reduces symptom presence, particularly fatigue, sleep disturbances, and cognitive dysfunction both during and after the treatment completion.[56] Jones and colleagues[57] found that patients with gliomas who exercised prior to treatment initiation tended to continue to exercise during treatment. Molassiotis and colleagues[52] found that glioma patients who added walking to their daily life improved their mood, coping skills, and QoL. Neuro-oncology patients who have neurological limitation with focal motor or sensory deficits should have physical therapy consultation prior to initiating exercise[47] and those with significant functional impairments may benefit from initiating a multidisciplinary rehabilitation program with structured exercise.[58]

Case Study 1

"I completely lost use of my left arm during my brain surgery. I did a lot of occupational therapy, physical therapy, and exercise and my brain responded, building new pathways, desperately wanting to reconnect with its lost limb. I am 5 years post-op and consensus among doctors is that I have regained all that I will regain. It is okay with me, I no longer have to pick up my arm to thread it into my shirt sleeve with my teeth. I can hold my cards while I play Gin Rummy. I gesture when I talk. No one can tell that I have a deficit, even though I can't type or text with two hands or tap my fingers to music. Cashiers sometimes give a puzzled look while I struggle with my wallet. Friends know that I can't carry my drink and a plate of food to the table in one trip. It takes a conscience effort to integrate my left arm and hand into some activities. When I'm tired, I sometimes realize that I'm sitting on it. Like the rest of me, my arm is neither fully healthy nor terribly sick. It is too good for occupational therapy, but too weak to use most gym equipment.

Two years after my surgery, my MRI showed some changes in my tumor, some growth. I got radiation. Radiation to your brain is exactly as terrifying as it sounds. Radiation was 3 years ago, as I write this. I have a burn scar on the side of my head with a radius similar to a grapefruit's. Some hair grows there, but my bald spot is mostly obscured by my long hair and the scarves I wear daily. People love my "bohemian" style, I'm grateful that I can hide my scars.

I also have moments of acute and profound weakness. Afternoons that are lost to deep sleep—the type of nap that you wake up from not knowing where you are and if it is day or night. The doctor asks if I am exercising, maybe I'm weak because I'm, you know, weak. I answer that I am active but not exactly athletic. He prescribes the gym. I get migraines and headaches now. My mom calls me to get the weather report, as I feel an overwhelming pressure as the outside pressure changes. I joke that it is like cows lying down in the fields before a storm. Is it funny because I am likening myself to a cow or because it is true? I can track all of my headaches and "wipe outs" as I call them, when I am flattened by fatigue without notice. Some are correlated to hormones, some to the weather, some to allergies, other episodes seem to pop up without cause. Maybe epilepsy (I have epilepsy now) medication side effects? I will probably never know."

—33-year-old female diagnosed with right frontal oligodendroglioma (grade II) treated with subtotal resection, radiation, and PCV chemotherapy.

Cognitive-behavioral factors in QoL

Although there has been a lack of consistency when defining cognitive impairment, there is no sparsity in the evidence demonstrating neurocognitive impairments in glioma patients throughout the disease course regardless of tumor grade or type when matched to noncancer peers.[59] Up to 91% of glioma patients complain of or demonstrate cognitive deficits during and after treatment,[60] with cognitive impairment consistently identified as patients' greatest concern following treatment.[61] As such, assessing and detecting cognitive decline is essential in this population. The International Cognitive and Cancer Task Force (ICCTF)[62] continues to stand behind their statement that objective [neuropsychological] tests remain the gold standard for measuring cognitive function when assessing for changes following cancer treatment. To determine impairment in this method, patient scores on neuropsychological measures are compared to either normative data (or a representative control group) and/or to an estimated prediagnosis capability.[63] Most comprehensive neurocognitive evaluations include extensive batteries measuring premorbid functioning, attention, verbal fluency, executive function, speed of processing, learning, recall, motor capabilities, and emotional adjustment—all the while being mindful of examinee burden and adjusting to the functional level of each patient.

Cognitive impairment

Cancer-related cognitive deficits are most consistent with subcortical abnormalities: slowed mental processing, reduced attention, reduced encoding and retrieval of information, and psychomotor slowing. Table 1 provides a detailed description of glioma-related cognitive impairments, prevalence, their associated etiologies, and impact on daily functioning within each domain. As patients with oligodendrogliomas are typically younger and have an extended expected survival rate, the long-term cognitive sequelae become especially salient. Although the "watch and wait" course of treatment has historically been accepted for this population, current literature is increasingly in support of active treatment which may include surgery, radiotherapy, and/or chemotherapy.

TABLE 1 Cancer-related cognitive impairments, associated etiologies, and commonly reported patient concerns[64]

Cognitive domain	Prevalence	Etiology	Impact on daily functioning
Frontal lobe functioning - Attention - Processing speed - Executive functioning	Common	Tumors located within the frontal lobes are the most obvious reasons for attention and executive deficits. However, tumors in other areas of the brain can cause frontal lobe deficits from mass effect, edema, or simply interfering with subcortical pathways associated with the frontal lobes. While deficits are most commonly reported during and shortly following a patient's chemotherapy regimen—some do complain of persisting impairments. Radiation therapy can cause necrosis, leukoencephalopathy, radiogenic tumors, vascular abnormalities, and/or endocrinopathies. All of these mechanisms damage white matter within the brain to some degree causing varying attention or executive dysfunction. When comparing, poorer outcomes for patients receiving radiation are most commonly noted.	- Distractibility - Lost train of thought - Difficulty managing conversations - Disorganization - Difficulty prioritizing tasks - Poor flexibility - Poor multitasking - Word finding - Daily forgetfulness
Memory - Learning - Recall of information	Common	Daily forgetfulness and difficulty retaining new information can be caused by treatment regimens (surgery, radiation therapy, chemotherapy), hormones, antiepileptic agents, corticosteroids, or psychological distress. In most cases a combination of these factors negatively impacts the patient's memory. A myth within the cancer population is that chemotherapy can lead to extensive memory loss in the form of dementia. Epidemiological studies have demonstrated that if there was no sign of cognitive impairment prior to cancer treatment, chemotherapy has no bearing on long-term risk of dementia.[65] Following radiation therapy, most brain tumor patients experience some degree of cerebral atrophy and hemispheric white matter changes which correlate to memory deficits.[66] Late-delayed (several months to years) encephalopathy is an irreversible and extremely serious complication that can result following radiation therapy. Although severity can vary, severe neurocognitive deterioration can lead to radiation-induced dementia. This tends to be a greater issue with low-grade gliomas given the time of initial diagnosis and extended length of overall survival.	- Forgetting conversations - Misplacing objects - Difficulty recalling names - Difficulty storing newly learned information - The need to write everything down - Missing appointments - Forgetting to take medication
Language	Limited	Tumors that are within or close to the language centers and found to be inoperable can cause detrimental cognitive effects. Language deficits most often result from tumors arising in the frontal or left temporal region (direct or mass effect) or neurosurgical intervention; not adjuvant treatment. When comparing treatment approaches, chemotherapy is viewed as less likely to affect language compared to radiation and even far less likely when compared to surgery or tumor location itself. Rarely does radiation or chemotherapy play an active role in language dysfunction.	- Word finding - Expressive aphasia - Receptive aphasia - Difficulty reading - Stuttering - Paucity in speech
Visuospatial	Limited	Tumors in the superior parietal lobe, intraparietal sulcus, or frontal eye fields may produce visuospatial attention concerns[67, 68] or tumors located in the putamen, pulivna, caudate nucleus, insula, superior-temporal gyrus, or paraventricular white matter can cause visuospatial neglect.[69] Findings are mixed when assessing visuospatial deficits following treatment with some reporting significant deficits[70, 71] and others noting no deficits or comparable scores to baseline testing.[72, 73] In agreement is the impact of other neurocognitive domains impeding visuospatial functioning indirectly including frontal lobe deficits leading to reduced planning, which impedes nonverbal, visual problem solving.	- Visual neglect - Reduced spatial awareness - Visual agnosia - Directional disorientation - Depth misperception - Flat affect/lack of expression

Surgery

There is mounting evidence in the literature that surgical resection is the recommended initial course of action for treating LGG including oligodendrogliomas. Klein and colleagues[74] reported that patients who underwent stereotactic needle biopsy had poorer cognitive outcomes when compared to those with a resection history. Santini et al.[75] investigated 22 glioma patients ($n = 14$ LGG) pre- and postoperatively. Prior to surgery, 59% demonstrated cognitive deficits (word fluency, attention, working and verbal memory, and executive functions), which highlights the mere presence of tumor on cognition. Interestingly, the subgroup of high-grade patients showed improvement following surgery, whereas patients with LGG fared worse.

Radiotherapy

Of all the potential etiologies of cognitive impairment in glioma treatment, none have been more criticized than radiotherapy. Although still a source of contention within the field, there appears to be enough evidence suggesting at least a moderate risk of cognitive impairment following radiotherapy treatment in LGG patients including oligodendroglioma, which is well known and more frequently studied within the high-grade glioma population. Klein et al.[74] demonstrated (via linear regression) that the use of radiotherapy was associated with poor scores on one of seven perception and psychomotor speed tasks and two of eight memory tasks. They suggest in the LGG sample (23.1% oligodendroglioma/6.7% oligoastocytoma) that cognitive functioning in irradiated patients, but not in nonirradiated patients, tends to decrease over time. In line with others,[76] there was also evidence of increased radiation dose (>2 Gy) producing increased long-term neurotoxicity. Correa and colleagues[77] demonstrated that treated LGG patients (44% oligodendroglioma/33% oligoastrocytoma) initially had slowed motor speed and significant executive dysfunction. Visual memory also demonstrated variance across time as treated patients initially performed worse, yet improved at the 6-month follow-up comparable to untreated patients. Unfortunately, both groups eventually declined slightly at the final 12-month follow-up. Douw et al.[66] revealed that 53% of their LGG sample (24% oligodendroglioma/5% oligoastocytoma) who had been treated with radiotherapy developed cognitive deficits at the 12-year neuropsychological follow-up in at least 5 of 18 test measures—with attention, executive functioning, and processing speed the most common. This was in comparison with only 27% of patients who were radiotherapy naïve.

Chemotherapy

Despite the well-documented sensitivity of chemotherapy in LGG, specifically in the setting of co-deletion of chromosomes 1p and 19q,[78,79] the role of chemotherapy as a single agent or in conjunction with surgery/radiotherapy has yet to be properly studied. Unfortunately, even less have correlated these treatment regimens with cognitive implications. For example, the Radiation Therapy Oncology Group (RTOG)[80] recently announced the results from a follow-up analysis in the doubling of overall survival in pure or mixed anaplastic oligodendroglioma patients with 1p/19q co-deletion who received procarbazine, lomustine, and vincristine (PCV) chemotherapy. This trial did not assess cognitive function to help determine the net clinical benefit of this survival advantage. Prabhu and colleagues[81] examined cognitive performance when utilizing the Mini Mental Status Exam (MMSE)[82] in 122 RT-only versus 116 RT+triple agent PCV treated LGG patients (34% oligodendroglioma/25% oligoastrocytoma). Results showed that regardless of treatment arm, there was an increase in MMSE scores over time. Reijneveld et al.[83] also utilized the MMSE when assessing QoL effects of radiotherapy versus temozolomide in high-risk LGG patients. Following treatment, no significant difference was found between groups (13% vs 14%). It should be noted that commonly used screening tools such as the MMSE or the Montreal Cognitive Assessment (MoCA)[84] have been shown to be rather insensitive in cancer populations, specifically in those undergoing brain tumor treatment, as mild impairments may go unnoticed. Lastly, Blonski et al.[85] invested a small sample of 10 LGG patients following treatment of temozolomide in facilitating the shrinkage and ultimately possible resection. Although there were no adverse effects on QoL, half the patients demonstrated impairment in verbal episodic memory and executive functioning. Future investigational research is warranted.

Seizures

What appears consistent within the literature is the negative effect seizure activity has on the cognitive functioning of glioma patients. Epilepsy affects approximately 70%–90% of LGG patients[17] with up to 71% prescribed antiepileptic agents.[86] Patient with tumor locations within the temporal lobe, parietal lobe, or cortex are at increased risk for seizures when compared against those with tumors in the infratentorial or white matter locations. Patients with higher seizure burden demonstrated greater cognitive impairment, with decreased information processing, attention, working memory capacity, executive functioning, and psychomotor slowing.[86] Although the control of seizures increased reported QoL, antiepileptic medicine has been associated with worse objective and self-reported cognitive functioning when compared to their nonmedicated peers.[74]

Often during and after glioma treatment, cognitive implications force patients to adapt to a new 'normal.' Cognitive impairments can range from subtle annoyances to severe challenges in functioning—mostly always impacting QoL. It is quite common for patients to experience daily challenges, even in tasks which once seemed routine or mindless processes. Frontal lobe (e.g., distractibility, slower thinking, difficulty prioritizing, reduced capability in multitasking, irritability, and lack of initiative) and memory deficits appear to be the most common causing day-to-day trials for patients and caregivers. As oligodendroglioma patients are often eager for normalcy, routine, and to connect to everyday experiences, these challenges can become quite frustrating. In some, hope may even be connected to fulfilling daily activities or life experiences[87] and not the typical medical or research questions such as overall survival or prospect of a cure.

Case Study 2

"Radiation has contributed to problems with focus and short-term memory. I try to cope by writing lists, and they help, but lists can't eradicate the problem entirely. Imagine standing at your front door, preparing to leave for work. You go through a quick list: is the stove off, is the toilet running, is the backdoor locked? Yes? Good. Do you have your work materials, your bag, your snack? Did you take your medicine? Do you have your keys? Your cellphone? Uh oh. Where is the phone? You quickly work your way through your small house looking for your phone. You somehow forget what you are looking for, but you know you will know it when you see it. Your phone is on the kitchen counter, you use it to check the time, and go back to the front door where you remember that you were supposed to pocket your phone so you can leave for work. I have described these moments to my doctors. These moments and others.

Sometimes I lose time, not in a worrisome type of way, I don't black out or go for hours without realizing time has passed. My episodes of lost time are smaller and more vague. Akin to setting your alarm clock really early so that you have time to shower, finish packing, and make it to the airport early. The alarm rings, you swing your feet off the bed, sit up and breathe for a minute to wake up. When you look back to the clock you are shocked to see that 45 min have passed. I think everyone has lost time in this way, it is human, it is relatable. Annoying but unconcerning. But it happens a lot for me. People will ask what I did on Saturday. My mind goes blank. What did I do? It comes to me slowly. I cooked a nice breakfast, took the dog to the park, and did laundry. Could that be it? Yes, that is a fairly typical free day for me.

I am tired. Many of these realities are physically taxing, some are mentally or emotionally taxing, sometimes coping takes a lot of work and a lot of time. When I fatigue, I have a difficult time finding words. Sometimes I have a lot to say but my mouth doesn't move fast enough and I slur. I pretend to be actively listening to people but I am so tired that nothing is heard. I ask the same question twice in a row.

I really hate admitting this, but I have a hard time reading like I once could, and like I want to. I can read short articles and stories but have a hard time delving into novels. Sometimes the issue is focus—the story doesn't grab me fast enough—but just as often it is this fatigue I have been trying to capture for you, lines blur and words invert and slip when I am tired (which is more often than not). As an exercise in understanding, try reading this passage from the bottom up. You can read the words and each word has meaning but it becomes hard to see the pattern. How do the words string together? What follows? You know that there is meaning, you can glimpse the meaning, but it is just beyond reach. You know that the problem is within you, that you are missing perspective or some key that you learned in Kindergarten. It should be right there. Plus, as you will see, it is physically difficult to read from the bottom up.

I teach 8 h a week. It is my career and I am really good at it. I don't often see my coworkers, but I was registering new students with a coworker the other night. She innocently asked me what I do for work. For her this is a side job! I can't imagine! This is my full time work. Do I tell her that I have cancer? That will make her feel bad. Do I tell her "this is my only job?" Then she will think I'm rich or married or a stay at home mom. The question is overwhelming. At the same time, it is flattering. I pass. She hasn't noticed my limp arm or how tough it was for me to stack those papers she handed me. She hasn't noticed my bald spot, or maybe she did and just thinks it is some thinning due to stress. Maybe she noticed those things and still thinks I am competent enough to work a day job! Being in this in between space isn't easy."

—33-year-old female diagnosed with right frontal oligodendroglioma (grade II) treated with subtotal resection, radiation, and PCV chemotherapy (continuation of Case Study 1).

Employment

Many glioma patients (at some point in their disease course) are met with the decision of maintaining employment. The literature claims that cancer survivors are more likely to be unemployed when compared to healthy controls (33.8% vs 15.2%), with the highest risk of unemployment attributable to CNS cancers.[88] Deficits in attention and recall can lead to decreased efficiency, increased frustration, and inability to complete required employment tasks across a wide range of professions. A report from the Finnish Cancer Registry documented that only 45% of brain tumor patients were employed 2–3 years after diagnosis compared with 69% of age- and sex-matched controls.[89] Being employed allows individuals to continue in the lifestyle they are accustomed to financially, retain employer-sponsored health insurance (in the United States), and find a sense of self-efficacy, as employment appears to be an important measure of self-worth and place in society.[90–92] Most patients who are successfully able to return to work do so within the first few months following treatment,[93] yet adaptations to duties or roles are often needed. Returning-to-work rates following diagnoses and treatment for brain tumors range from 27% to 80%[87,94–96] and frequently depend on diagnosis, tumor location, treatment regimen, time since treatment, and cognitive limitations. Work limitations are often multifaceted and associated with depressive symptoms, fatigue levels, cognitive limitations, sleep challenges, and negative problem-solving orientation.[97] Vocational rehabilitation has been shown to be beneficial,[98] though more research is needed.

Driving

For most people, particularly within the United States, driving is essential when trying to maintain autonomy in life. Unfortunately, when a glioma patient is managing side effects including seizure activity, motor limitations, medication contraindications, visual field cuts, or cognitive deficits, they may be forced to cease driving temporarily or even permanently. Practitioners report (81.7%) that seizure activity remains the primary symptom resulting in restriction of driving.[99] However, formalized laws on seizure activity and driving restrictions vary by state. For example, patients with documented seizure activity must be seizure-free anywhere from 3 to 24 months, depending on where the patient resides. Regardless of state laws, when polled, the majority of practitioners (56.9%) felt that 5–6 months was an adequate time to return to driving.[99] As mentioned previously, the most common primary brain tumor to experience seizure activity is oligodendrogliomas; followed by astrocytoma, meningioma, high-grade glioma, and primary CNS lymphoma.[33] The literature suggests that a thorough assessment be completed following brain injury (including glioma treatment) and prior to returning to driving; especially given the prevalence of cognitive deficits present following injury, yet as few as 25% of practitioners reported using formal, standardized testing to determine driving eligibility in their patients.[99] Assessment of visual perception, visual spatial abilities, attention, processing speed, self-awareness, executive functioning, mood, and sleep is imperative in determining an individual's capacity to drive safely.[100,101] In fact, the American Psychiatric Association[102] issued a position paper encouraging practitioners to advise their patients about the potential impact their disease and treatments could have on their driving capabilities and shortly following, the American Medical Associations Ethical and Judicial Council[103] did the same. Still, there remains no national recommendation or guideline regarding glioma patients and driving restrictions. Practitioners are encouraged to locate the closest vocational driving simulation program for patient referrals and evaluations.

Cognitive rehabilitation

Cognitive rehabilitation therapy (CRT) is a method used to improve and restore cognitive functioning. It is rooted in two clinical fields: stroke and traumatic brain injury (TBI) rehabilitation. CRT most commonly uses either a *restorative* or *compensatory* approach.[104] The restorative approach is based on the assumption that repetitive stimulation of an area of impairment will promote neurogenesis and ultimately restore lost function. In contrast, the compensatory approach holds the assumption that lost neurological function cannot be restored, rather teaching alternative strategies using residual strengths are necessary to accommodate areas of weakness and promote cognitive improvement. There have been direct cognitive benefits evident from CRT within cancer populations, including brain tumor patients. In fact, brain tumor patients have demonstrated CRT benefits at times better when compared to inpatient treatment for stroke and TBI.[105–107] In addition to the consultation between treatment team including oncology providers, rehabilitation therapists, and other medical providers treating issues such as sleep, mood, and appetite, the importance of including the caregiver is essential. The caregiver should learn the strategies alongside their loved one and help reinforce and generalize within the home environment. Adding a caregiver to the team has shown direct benefits to both the patient and the caregiver.[61]

Psychosocial factors in QoL

Psychological and social factors that impact to QoL in glioma patients, including those diagnosed with oligodendrogliomas, consist of emotional distress, illness-related adjustment and coping, and social support. In fact, emotional distress may be the most predictive factor of QoL in brain tumor patients.[108,109] Indeed, many of the physical symptoms of poor QoL (e.g., fatigue, decreased appetite, and trouble concentrating) are also symptoms of depression. Therefore, a strong relationship between QoL and emotional functioning is not surprising. There are also factors contributing to the psychosocial functioning in patients with oligodendrogliomas that may be less evident including future-related worries and existential issues.

Case Study 3

"When I was first informed of my diagnosis, it was hard to know where to focus my attention. At the forefront of my attention is always my children and family. However, a tumor diagnosis is not something that can simply be put to the side. Rather, the news of this diagnosis immediately brought with it anxiety and fear. The overwhelming amount of information to process and attend to was challenging. There were many new doctors to meet, a surgery, options to consider, and symptoms to manage and accept as the "new normal." There were opinions to filter, strength to muster, and questions left unanswered. There were schedules to coordinate, finances to figure out, and a list of medications that suddenly became ever growing. There were bald patches, an overwhelming feeling of fatigue, and medication side effects never considered. The focus was not straightforward, but rather multi-faceted, complicated by all the uncertainties that lie ahead. It is a new, unexpected life—a life in which seeking knowledge, finding support, and learning to manage now have new meaning. A oligodendroglioma diagnosis is not an easy thing to navigate. But, with positive support, appropriate strategies for coping, and a strong will to maintain a positive quality of life, I pray that I can keep my focus..." —34-year-old female diagnosed with right frontoparietal oligodendroglioma (grade II) treated with stereotactic biopsy, radiation, and TMZ chemotherapy.

Emotional distress

Emotional functioning comprises a large category of symptoms including but not limited to depression, anxiety, anger, fear, and other personality changes. Emotional distress may occur as a result of the tumor—that is, the tumor location may have an impact on mood and personality structures.[110] However, often, emotional distress is a response to the diagnosis and treatment of a primary brain tumor. A recent meta-analysis concluded that there were no significant differences in rates of depression based on tumor type[111] and given the prognosis, intensive treatment, and life changing nature of oligodendroglioma (low and high grade), increased emotional distress is common, especially in the later phases of the disease.[11] Major depressive disorder has been found to have a three times higher rate of incidence in primary brain tumor patients, higher than rates of other cancer diagnoses[112,113] and previous history of major depression is a well-established predictor of depression following diagnosis.[112,114] Notably, oligodendrogliomas, as well as other primary brain tumors, may initially present as psychiatric disorders prior to discovery of brain lesions and histopathology diagnosis.[110]

One study conducted in a sample of LGG (17% oligodendrogliomas/11% oligoastrocytoma) found that relative to a healthy control group, there were no differences in emotional functioning.[7] In a recently published meta-analysis, low tumor grade was shown to relate to lower prevalence of depression.[111] Furthermore, emotional distress has been shown to have a bearing on the prognosis in oligodendroglioma patients, such that heightened emotional distress predicts poorer prognosis.[115] Two systematic reviews in primary brain tumor patients have also reported on the relationship between depression and poor prognosis.[113,116] This, coupled with findings that emotional distress is correlated with disease progression[11] suggests that disease worsening may be related to emotional well-being. Suicidal ideation, a symptom of depression, was related to anxiety and overall QoL in a sample of primary brain tumors.[117] Emotional distress may also be affected by treatment type. One randomized-controlled trial, which included oligodendrogliomas, found that radiation therapy was related to worse emotional distress than temozolomide alone.[83]

Importantly, assessment of psychological functioning in primary brain tumor patients is worthy of continued attention. Given the conflation of disease symptoms and treatment side effects (appetite change, sleep change, fatigue, poor concentration, and psychomotor slowing) with depressive symptoms, the incidence rate of depression may be inflated.[111,118] This also makes the causal relationship between depression and overall QoL difficult to disentangle.[116] In one study, the somatic symptoms of depression captured by a validated self-report scale (Beck Depression Inventory) were correlated with functional status, but not the affective symptoms captured by the same depression scale.[119] To date, the Patient-Health Questionnaire-9 (PHQ-9) and the Hospital Anxiety and Depression (HAD) have both been partially validated for use as screening tools for depression in patients with gliomas.[120] Clinical interviews and QoL scales specific to chronic illness or even brain tumor patients are required to advance the study of psychiatric disorders in brain tumor patients.[113] In a recent systematic review and meta-analysis, clinician-administered depression measures showed significantly lower rates of depression than self-report scales.[111] Anxiety, too, may be poorly captured by traditional scales due to a lack of specificity to the anxiety symptoms experienced by patients with primary brain tumors.[114] Therefore, efforts to improve not only early detection and intervention, but also assessment of emotional functioning in oligodendroglioma patients are warranted.

Stress and coping

One aspect of an oligodendroglioma diagnosis, as well as other primary brain tumors, that may contribute to stress is future uncertainty and fear of death/dying. In one sample of oligodendroglioma patients, future uncertainty was found to significantly improve after the fourth cycle of chemotherapy.[121] Pain and role functioning were also found to improve after cycle 4 and 6, respectively, in this sample.[121] These findings suggest that coping with future uncertainty, which likely includes fear of death and dying given the terminal nature of such a diagnosis, may be an important aspect of coping for patients. Timely conversations about serious illness and its trajectory have been associated with more goal-concurrent care, better QoL, improved end-of-life care, and better patient and family coping.[122–125] Sadly, as patients with gliomas progress through their disease course, neurocognitive deficits can impair decision-making capacity. Gofton and colleagues[126] reported that 52% of brain tumor patients were found to be incompetent to assess their own situation during the last weeks of their life due to cognitive disturbances, aphasia, and/or delirium. As such, even with oligodendroglioma patients who tend to have increased survival rates, there is a push to integrate palliative care and discuss/offer an advanced care plan (ACP) to patients early in their disease course, as ACPs encourage patient participation and assist in matching end-of-life care with preferred care.[127] It has been found that a high rate of physicians (40%) were unaware of glioma patients end-of-life preferences, even though some of these patients had an ACP in place,[128]—which serves as a reminder that implementation of a plan might not be enough, and that communication between patient and provider is also essential. It may be important for specific training and preparation on end-of-life discussion to take place for healthcare professionals in order to gain confidence and comfort in leading these difficult conversations. Fig. 3 provides a useful ACP conversation model for practitioners. Regular evaluation of the documented or discussed ACP is advisable as wishes or plans may change depending on illness trajectory and patient capability. In addition, there needs to be a distinction within the field (and clearly communicated to the family) the difference between palliative care and hospice care services (see Table 2). Palliative care is appropriate at any stage of glioma treatment, regardless of grade or expected disease trajectory and it should be provided alongside treatment; it is not restricted to end-of-life care. Favorably for oligodendroglioma patients, the longer estimated survival allows family or caregivers to organize care within the home, which has repeatedly been noted to be preferred by the patient. Slower progression also allows for adjustment and planning or the development of necessary skills to manage a patient within the home during the final stages.

It should be noted that some patients find it difficult or are unwilling to have this conversation. Finding a balance between patient's willingness, cognitive capacity, and the need to maintain hope remains important. This also requires a discussion with the patient on the possibility of decline in their ability to communicate as progression of the disease

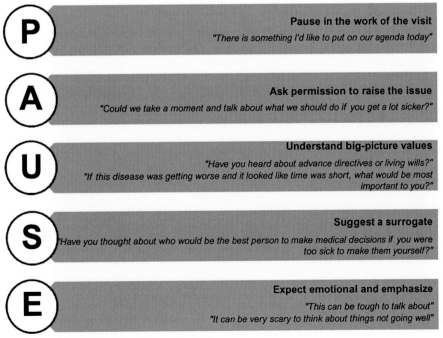

P Pause in the work of the visit
"There is something I'd like to put on our agenda today"

A Ask permission to raise the issue
"Could we take a moment and talk about what we should do if you get a lot sicker?"

U Understand big-picture values
"Have you heard about advance directives or living wills?"
"If this disease was getting worse and it looked like time was short, what would be most important to you?"

S Suggest a surrogate
"Have you thought about who would be the best person to make medical decisions if you were too sick to make them yourself?"

E Expect emotional and emphasize
"This can be tough to talk about"
"It can be very scary to think about things not going well"

FIG. 3 "PAUSE": an advance care planning conversation model.[129]

TABLE 2 Goals of palliative care[130] and hospice care[131]

Palliative Care Goals

- Provides relief from pain and other distressing symptoms
- Affirms life and regards dying as a normal process
- Intend neither to hasten or postpone death
- Integrates the psychological and spiritual aspects of patient care
- Offers support system to help patients live as actively as possible until death
- Offers a support system to help the family cope during the patient's illness and in their own bereavement
- Uses a team approach to address the needs of patients and their families, including bereavement counseling, if indicated
- Will enhance quality of life, and may also positively influence the course of illness
- Is applicable early during illness, in conjunction with other therapies that are intended to prolong life, such as chemotherapy or radiation therapy, and includes those investigations needed to better understand and manage distressing clinical complications

Hospice Care Goals

- Relieve the physical, mental, emotional, and spiritual suffering of patients and caregivers
- Promote the dignity and independence of our patients to the greatest extent possible
- Support patients and their families in finding personal fulfillment as they deal with end-of-life challenges

occurs, allowing a time for the patients to state their wishes for the dying process while they are still able to communicate fully with their family and medical team. When language presents as a barrier, alternative communication strategies should be implemented (writing, prompt cards, or pictures). Additionally, existential issues are well-documented in primary brain tumor patients, suggesting that meaning-making in disease trajectory is an important aspect of coping with brain tumor diagnosis.[108]

Case Study 4

"My doctors talk about my oligodendroglioma in the present tense—they are careful not to use the term "remission" as this type of cancer is characterized by long, microscopic tendrils. No one can be sure if it is living, waiting to return. My prognosis is very good. Doctors are optimistic about long term survival and a promising new study about my course of treatment has recently been published. Yet I still have to get an MRI every 3–4 months, just in case. Living with oligodendroglioma is like straddling two worlds—1 ft. in the hospital and 1 ft. in "normal" society. From my liminal perch I have learned this delicate dance, a precarious balance to find a place of comfort and joy. When people learn I have a brain tumor, their faces pale. I have mostly learned how to break the news to people, coworkers or old friends. When it becomes apparent that I am comfortable talking about it, everyone's first question is inevitably, "How did you know you had a brain tumor?" I can see them quickly figuring, their eyes flickering, as their minds tick, trying to calculate if last week's headache could be indicative of a tumor, as WebMD suggested. People have confirmed this. We can't help it, and it does not bother me. I have always been introverted and anxious, but brain surgery threw my panic disorder into overdrive. Lights, sounds, and smells were overpowering. It was hard to explain to people how exhausting a trip out for coffee was. I could hear a whispered conversation two tables over louder than I could hear the person across from me. The soft lights or the sounds of mugs and spoons clinking in the bus tray as patrons left were dizzying. I continue to struggle with anxiety but with medication and therapy, I am finally able to run more than one errand daily. I know that I am not going to go back to school and become a doctor, I know that I am not going to hike huge mountains on different continents, I know that I will never be rich. But I am so happy with what I have and the work I am doing and the health that I am working for.
My neuro oncologist sees me thriving—I am smart, active, good looking, well spoken, I haven't had a seizure in a year, I am working, I have healthy relationships, I am driving, caring for a pet. My problems don't register as problems. Yet when I take these problems to my primary care doctor, she won't touch me—I am too sick, I have brain cancer, I need a neurologist. I'm in between. I am well. I do take my dog to the park, I do my laundry, I cook, I write, I read. I don't drink, I don't stay out late, I am not the life of the party. Most of the time I don't have energy to make it to the party. My friends understand, they are helpful and giving. Having oligodendroglioma is uniquely isolating and freeing, once you find your balance."
—*33-year-old female diagnosed with right frontal oligodendroglioma (grade II) treated with subtotal resection, radiation, and PCV chemotherapy.*

Social support and caregiver burden

Another important contributing factor to QoL in oligodendroglioma patients is social support. Social support may include, but is not limited to, caregiver help with activities of daily living. One study found that up to 35% of primary brain tumor patients (17% oligodendroglioma) received help with activities of daily living and roughly half required help with instrumental activities of daily living.[132] Therefore, symptoms of diagnosis and treatment may influence family relationships making adjustment to this disease a family process rather than an individual process. Caregivers of medical patients report high levels of stress and poor physical and emotional health, as well as career sacrifices, monetary losses, and workplace discrimination.[133] This commonly results in depression, fatigue, and health issues themselves.[134] More specifically, caregivers of primary brain tumor patients use the term "shock" when explaining their feelings on hearing initial diagnosis[135] and although they have expectations of changes in the family dynamic, Schmer et al.[135] found that most were astonished at the quantity and depth of these changes. Family caregivers often feel inadequate and ill prepared for the daunting tasks and level of support required by their loved ones following diagnosis and throughout treatment.[136] There is a stress response to this level of care which may manifest as an emotional or physical response.[132] Schubart et al.[136] emphasized the need for a psychological component in the care of brain tumor patients to include the family at specific times through the disease trajectory: *diagnosis, chronic* (treatment) *phase, and terminal phase.* Although the psychosocial support for higher-grade glioma's are likely to be greater given the increased speed of disease trajectory, patients with oligodendroglioma whose tumor may be stable for some time, still have an intrusive illness that can affect virtually all aspects of life.

Case Study 5

"I should be eating lunch, but instead I'm looking up brain cancer statistics.

My wife was 14 weeks pregnant with our second child last fall when she started having bad headaches. We naturally assumed these were yet another cherry on the puke sundae of pregnancy symptoms, but the headaches persisted, and then got worse. Soon she was vomiting and shouting in pain while clutching her skull. On a Friday evening when we were supposed to be on the road for vacation to celebrate our wedding anniversary, we instead took a trip to the emergency department. As the ashen-faced doctor told us that there was a large mass in my wife's brain, two new avatars entered our lives: The Tumor, and what we came to nervous-laughingly call Google Neuroscience University.

Thankfully, my wife was diagnosed with an oligodendroglioma, a "good" form of brain cancer a term that gives extra poignancy to the phrase 'it's all relative.' Her prognosis is favorable, but carries with it an exquisitely torturous cone of uncertainty; my brother likens it to trying to forecast a tropical depression churning a thousand miles out in the Atlantic. It could peter out; it could devastate. My wife just turned 33. At any point she may begin to experience neurological deficits either caused by the cancer or its treatments. This radical uncertainty collides with an era in which the accumulated knowledge of the world's medical community is accessible with a few clicks.

Googling to predict the future is perhaps just the modern form of throwing sheep knuckles or consulting tea leaves, but it has the veneer of being more enlightened. So, I consume journal articles and message board anecdotes in a dizzying binge. If I can just place her specific circumstances on a spectrum somewhere, pin the tail on the actuarial donkey, perhaps I can settle my jittery mind. Except that it's impossible. These genetic markers are good. These markers of higher malignancy are bad. The extent of the resection is good. The preoperative size of the tumor is bad. Then a Google Alert hits my inbox with new trial results that warp the odds, and so I start over anew…

What I really want to ask Google is:

"…Will she see the girls graduate high school?"

"…Will I dance with her at their weddings?"

"…Will we hold a grandchild together?"

"…For how long will she be whole?"

I have periods where I can let it go, and periods where I'll let something slip and she'll look at me sharply and ask, "Have you been researching again?" As if neuroscience has become my equivalent of a drinking habit; which, symptomatically, maybe it has.

Someone on one of the online support groups commented that it took a long time before she was able to "stop wearing the cancer more than" her ill spouse did. I'm conflicted on that point. On the one hand, I do think my wife tends to have an attitude conducive to living fully in the moment. I'll always laugh at her relatively laissez-faire feelings toward the regular MRIs that show whether or not everything's stable in there. She jokes about "Schrödinger's Tumor": Either it has already grown, or it hasn't—the scan just opens the box. She doesn't let her cancer define her, and I admire and am gladdened by that.

At the same time, I feel that one of the roles I play for us is keeping a finger on the pulse, watching the radar to see how the storm's tracking. Because of the unique circumstances of my wife's case—a delay in postsurgery treatment due to pregnancy—she's an 'N' of one. It feels incumbent to have as much information as possible, be prepared to ask the right questions, weigh the values properly lest we unwittingly choose a path that cuts out years earlier than its crossroad.

That may all be bunk, by the way—a weak postfacto justification. Another narrative says that rubbing worry beads feels good. That I am letting myself wallow in an unhealthy bog, that my remembering self will look back and curse at missed opportunities to be completely present during these, what should probably be considered the halcyon days. After all, it's hard to know how long they'll last. One of brain cancer's particular brutalities is that everything can be going great until it's not.

Nowadays the main response to uncertainty is to try to fight it like a monster rather than come to terms with it like an unpleasant roommate. We have multiple weather models that can converge on a projection. We have Kaplan-Meier survival curves for diseases. We have data, our new rain god.

I wish I could say that I have reached some sort of equanimity, but the truth is more days than not feel heavy. I Google, in the end, because I want her in the world and by my side as long as possible."

—Spouse of a 33-year-old female diagnosed with right frontal anaplastic oligodendroglioma (grade III) treated with gross total resection and TMZ chemotherapy.

Interventions

To date there remains sparse evidence-based interventions available to cancer survivors in need. Over the past few decades, researchers have begun exploring both pharmacological and nonpharmacological interventions to address the well-known physical, cognitive-behavioral, and psychosocial challenges during and following cancer treatments, yet it is safe to say these explorations are in their infancy and as a field, we are in desperate need of many more. Additionally, the lack of randomized controlled trials investigating oligodendroglioma patients makes it difficult to gauge whether the same treatment strategies proven safe or effective in other cancers or neurological conditions should be pursued in patients with brain tumors. Table 3 provides a list of pharmacological and nonpharmacological intervention strategies that have been specifically investigated in brain tumor patients with a potential benefit identified.

TABLE 3 Potential interventions/strategies for physical, cognitive-behavioral, and psychosocial symptoms in brain tumor patients

Symptom	Intervention
Physical Factors	
Seizures	Pharmacologic • Antiepileptics: prioritize nonenzyme inducing agents Most commonly used: carbamazepine, lacosamide, lamotrigine, levetiracetam, oxacarbezepine, topiramate, valproic acid • Antianxiolytics: benzodiazepine Nonpharmacologic • Identify triggers, if appropriate • Stress management strategies • Sleep 6–8h every night
Fatigue	Pharmacologic • Psychostimulant: methylphenidate, dexamphetamine, modafinil, armodafinil • Corticosteroid: adjust dose, frequency, administer in morning Nonpharmacologic • Self-monitor fatigue levels • Assess contributing factors: anemia, pain, mood, sleep, nutrition, comorbidities, infection, functional status • Energy conserving strategies: delegate, set priorities, schedule rest, limit naps <1h, structure daily routine • Use assistive devices • Daily physical activity/exercise: walking, jogging, swimming, resistance training, yoga • Physical therapy consult if appropriate • Optimize sleep, good sleep hygiene • Psychosocial intervention: cognitive-behavioral therapy, mindfulness-based stress reduction • Bright white therapy
Headaches	Pharmacologic • Analgesics • Steroid therapy, adjust dose or frequency • Opioids, minimize to avoid dependence • Antianxiolytics, minimize to avoid dependence • Consider pain management consultation with headache specialist to develop plan of care Nonpharmacologic • Identify triggers, if appropriate • Stress management strategies • Sleep 6–8h every night

TABLE 3 Potential interventions/strategies for physical, cognitive-behavioral, and psychosocial symptoms in brain tumor patients—cont'd

Symptom	Intervention
Nausea	Pharmacologic • Assess emetogenic potential of chemotherapy agent • Neurokinin 1 antagonists: aprepitant, fosaprepitant, netupitant, rolapitant • Serotonin receptor antagonists: palonosetron, granisetron • Steroid: dexamethasone • Atypical antipsychotic: olanzapine • Other agents: benzodiazepine, phenothiazine, metoclopramide, haloperidol, scopolamine, cannabinoid Nonpharmacologic • Eat small frequent meals, 5–6 per day • Avoid high fat meals with strong aroma • Limit fluids to in-between meals instead of with meals • Elevate head for 30 min after meals • Use ginger: ale, tea, candy, capsules • Stress management strategies for anticipatory nausea
Sleep	Pharmacologic • Benzodiazepine • Nonbenzodiazepine: zolpidem, zaleplon, eszopiclone • CNS depressant: chloral hydrate Nonpharmacologic • Good sleep hygiene • Limiting naps <1 h • Psychosocial intervention: cognitive behavioral therapy, mindfulness-based stress reduction • Daily physical activity/exercise • Nutritional/herbal supplement: tryptophan, valerian, melatonin
Nutrition	• Maintain healthy normal weight: 18.45–24.9 kg/m^2 kg • Be physically active • Consume plant-based diet of 2.5 cups colorful vegetables and fruits daily, whole grain foods, and limit processed and red meats • Limit alcohol consumption to 1–2 drinks daily
Exercise	• 150 min of moderate-intensity exercise per week: optimal 30-min sessions 5 ×/week, 10–15 min sessions may be utilized initially • Moderate-intensity activities: walking as a pace
Cognitive-Behavioral Factors	
Cognitive impairment	Pharmacologic • Psychostimulant: methylphenidate, dexamphetamine, modafinil, armodafinil • NMDA receptor antagonist: memantine Nonpharmacologic • Cognitive training • Cognitive rehabilitation, speech therapy, occupational therapy • Exercise • Mindfulness-based stress reduction • Neurofeedback • Meditation • Good sleep hygiene
Employment	• Cognitive training • Cognitive rehabilitation: speech therapy, occupational therapy • Vocational rehabilitation • Reduced workload • Flexible scheduling • Quiet, distraction free environment • Social work consultation

Continued

TABLE 3 Potential interventions/strategies for physical, cognitive-behavioral, and psychosocial symptoms in brain tumor patients—cont'd

Symptom	Intervention
Driving	• Cognitive training • Cognitive rehabilitation • Driving evaluation
Psychosocial Factors	
Emotional distress	Depression pharmacologic • SSRIs • SNRIs Depression nonpharmacologic • Psychotherapy: mindfulness-based therapy, behavioral activation • Physical exercise • Pleasurable activities Anxiety pharmacologic • Benzodiazepine • SSRIs Anxiety nonpharmacologic • Psychotherapy: mindfulness-based therapy, behavioral activation, relaxation training
Stress and coping	Pharmacologic • SSRI • Benzodiazepine Nonpharmacologic • Psychotherapy: mindfulness-based therapy, behavioral activation, relaxation training • Advance care plan • Palliative care
Social support	Nonpharmacologic • Support groups • Individual therapy • Behavioral activation • Family system training • Social work support

References

1. Noggle CA, Dean RS. Neuropsychology and cancer: an emerging focus. In: *The Neuropsychology of Cancer and Oncology*: 2013:3–39. http://myaccess.library.utoronto.ca/login?url=http://search.proquest.com/docview/1518031536?accountid=14771%5Cnhttp://bf4dv7zn3u.search.serialssolutions.com/?ctx_ver=Z39.88-2004&ctx_enc=info:ofi/enc:UTF-8&rfr_id=info:sid/PsycINFO&rft_val_fmt=info:ofi/fmt.

2. Aaronson NK. Quality of life: what is it? How should it be measured? *Oncology (Williston Park)*. 1988;2(5):69–76. 64. https://doi.org/10.1159/000282146.

3. Gustafsson M, Edvardsson T, Ahlström G. The relationship between function, quality of life and coping in patients with low-grade gliomas. *Support Care Cancer*. 2006;14(12):1205–1212. https://doi.org/10.1007/s00520-006-0080-3.

4. Lehmann JF, DeLisa JA, Warren CG, DeLateur BJ, Bryant PL, Nicholson CG. Cancer rehabilitation: assessment of need, development, and evaluation of a model of care. *Arch Phys Med Rehabil*. 1978;59(9):410–419.

5. Lageman SK, Brown PD, Anderson SK, et al. Exploring primary brain tumor patient and caregiver needs and preferences in brief educational and support opportunities. *Support Care Cancer*. 2015;23(3):851–859. https://doi.org/10.1007/s00520-014-2413-y.

6. Taphoorn MJ, Stupp R, Coens C, et al. Health-related quality of life in patients with glioblastoma: a randomised controlled trial. *Lancet Oncol*. 2005;6 (12):937–944. https://doi.org/10.1016/s1470-2045(05)70432-0.

7. Boele FW, Douw L, Reijneveld JC, et al. Health-related quality of life in stable, long-term survivors of low-grade glioma. *J Clin Oncol*. 2015;33 (9):1023–1029. https://doi.org/10.1200/JCO.2014.56.9079.

8. Habets EJJ, Taphoorn MJB, Nederend S, et al. Health-related quality of life and cognitive functioning in long-term anaplastic oligodendroglioma and oligoastrocytoma survivors. *J Neurooncol*. 2014;116(1):161–168. https://doi.org/10.1007/s11060-013-1278-0.

9. Ediebah DE, Galindo-Garre F, Uitdehaag BMJ, et al. Joint modeling of longitudinal health-related quality of life data and survival. *Qual Life Res*. 2015;24(4):795–804. https://doi.org/10.1007/s11136-014-0821-6.

10. Forst D, Nahed B, Loeffler J, Batchelor T. Low-grade gliomas. *Oncologist*. 2014;19(4):403–413.

11. Simonetti G, Gaviani P, Botturi A, Innocenti A, Lamperti E, Silvani A. Clinical management of grade III oligodendroglioma. *Cancer Manag Res*. 2015;7:213–223. https://doi.org/10.2147/CMAR.S56975.

12. Fan X, Wang Y, Zhang C, Liu X, Qian Z, Jiang T. Human leukocyte antigen-G overexpression predicts poor clinical outcomes in low-grade gliomas. *J Neuroimmunol*. 2016;294:27–31. https://doi.org/10.1016/j.jneuroim.2016.03.015.

13. Yang P, Liang T, Zhang C, et al. Clinicopathological factors predictive of postoperative seizures in patients with gliomas. *Seizure*. 2016;35:93–99. https://doi.org/10.1016/j.seizure.2015.12.013.

14. Zhong Z, Wang Z, Wang Y, You G, Jiang T. IDH1/2 mutation is associated with seizure as an initial symptom in low-grade glioma: a report of 311 Chinese adult glioma patients. *Epilepsy Res*. 2015;109(1):100–105. https://doi.org/10.1016/j.eplepsyres.2014.09.012.

15. Alentorn A, Hoang-Xuan K, Mikkelsen T. Presenting signs and symptoms in brain tumors. In: *Handbook of Clinical Neurology*: 2016:19–26. Vol. 1342016. https://doi.org/10.1016/B978-0-12-802997-8.00002-5.

16. Mariş D, Nica D, Mohan D, Moisa H, Ciurea AV. Multidisciplinary management of adult low grade gliomas. *Chirurgia (Bucur)*. 2014; 109(5):590–599. http://www.embase.com/search/results?subaction=viewrecord&from=export&id=L603015330%250Ahttp://jq6am9xs3s.search. serialssolutions.com/?sid=EMBASE&issn=12219118&id=doi:&atitle=Multidisciplinary+management+of+adult+low+grade+gliomas&stitle= Chirurgia+%2528Bucur%25.

17. Smits A, Duffau H. Seizures and the natural history of world health organization grade II gliomas: a review. *Neurosurgery*. 2011;68(5):1326–1333. https://doi.org/10.1227/NEU.0b013e31820c3419.

18. Shields LBE, Choucair AK. Management of low-grade gliomas: a review of patient-perceived quality of life and neurocognitive outcome. *World Neurosurg*. 2014;82(1–2):e299–e309. https://doi.org/10.1016/j.wneu.2014.02.033.

19. National Comprehensive Cancer Network. NCCN Clinical Practice Guidelines in Oncology: Survivorship, version 2.2017.

20. National Comprehensive Cancer Network. NCCN Clinical Practice Guidelines in Oncology: Palliative care, version 1.2018.

21. Sarmiento JM, Venteicher AS, Patil CG. Early versus delayed postoperative radiotherapy for treatment of low-grade gliomas. *Cochrane Database Syst Rev*. 2015;6. https://doi.org/10.1002/14651858.CD009229.pub2.

22. Aghi MK, Nahed BV, Sloan AE, Ryken TC, Kalkanis SN, Olson JJ. The role of surgery in the management of patients with diffuse low grade glioma: a systematic review and evidence-based clinical practice guideline. *J Neurooncol*. 2015;125(3):503–530. https://doi.org/10.1007/s11060-015-1867-1.

23. Tankumpuan T, Utriyaprasit K, Chayaput P, Itthimathin P. Predictors of physical functioning in postoperative brain tumor patients. *J Neurosci Nurs*. 2015;47(1):E11–E21. https://doi.org/10.1097/JNN.0000000000000113.

24. Day J, Yust-Katz S, Cachia D, et al. Interventions for the management of fatigue in adults with a primary brain tumour. In: Day J, ed. *Cochrane Database of Systematic Reviews*. Chichester, UK: John Wiley & Sons, Ltd.; 2016:CD011376Vol. 4. https://doi.org/10.1002/14651858.CD011376. pub2.

25. Lovely MP. Quality of life of brain tumor patients. *Semin Oncol Nurs*. 1998;14(1):73–80.

26. Armstrong TS, Vera-Bolanos E, Acquaye AA, Gilbert MR, Ladha H, Mendoza T. The symptom burden of primary brain tumors: evidence for a core set of tumor- and treatment-related symptoms. *Neuro Oncol*. 2016;18(2):252–260. https://doi.org/10.1093/neuonc/nov166.

27. McClellan W, Klemp JR, Krebill H, et al. Understanding the functional late effects and informational needs of adult survivors of childhood cancer. *Oncol Nurs Forum*. 2013;40(3):254–262. https://doi.org/10.1188/13.ONF.254-262.

28. National Comprehensive Cancer Network. NCCN Clinical Practice Guidelines in Oncology: Central Nervous System Cancers, version 1.2017.

29. Kim C-W, Joo J-D, Kim Y-H, Han JH, Kim C-Y. Health-related quality of life in brain tumor patients treated with surgery: preliminary result of a single institution. *Brain Tumor Res Treat*. 2016;4(2):87–93. https://doi.org/10.14791/btrt.2016.4.2.87.

30. Fox SW, Lyon D, Farace E. Symptom clusters in patients with high-grade glioma. *J Nurs Scholarsh Off Publ Sigma Theta Tau Int Honor Soc Nurs*. 2007;39(1):61–67.

31. Kahlenberg CA, Fadul CE, Roberts DW, et al. Seizure prognosis of patients with low-grade tumors. *Seizure*. 2012;21(7):540–545. https://doi.org/ 10.1016/j.seizure.2012.05.014.

32. Schiff D, Lee EQ, Nayak L, Norden AD, Reardon DA, Wen PY. Medical management of brain tumors and the sequelae of treatment. *Neuro Oncol*. 2015;17(4):488–504. https://doi.org/10.1093/neuonc/nou304.

33. Pitkanen A, Schwartzkroin P, Moshe S. *Models of Seizures and Epilepsy*. https://doi.org/10.1016/B978-0-12-088554-1.X5000-2.

34. Thier K, Calabek B, Tinchon A, Grisold W, Oberndorfer S. The last 10 days of patients with glioblastoma: assessment of clinical signs and symptoms as well as treatment. *Am J Hosp Palliat Care*. 2016;33(10):985–988. https://doi.org/10.1177/1049909115609295.

35. Avila EK, Chamberlain M, Schiff D, et al. Seizure control as a new metric in assessing efficacy of tumor treatment in low-grade glioma trials. *Neuro Oncol*. 2017;19(1):12–21. https://doi.org/10.1093/neuonc/now190.

36. Lovely M, Stewart-Amidei C, Arzbaecher J, et al. *Care of the Adult Patient with a Brain Tumor*. Chicago: American Association of Neuroscience Nurses; 2014.

37. Perkins A, Liu G. Primary brain tumors in adults: diagnosis and treatment. *Am Fam Physician*. 2016;93(3):211–217.

38. National Comprehensive Cancer Network. NCCN Clinical Practice Guidelines in Oncology: Antiemesis, version 2.2017. https://www.nccn.org/ professionals/physician_gls/pdf/antiemesis.pdf; 2017.

39. Cahill J, LoBiondo-Wood G, Bergstrom N, Armstrong T. Brain tumor symptoms as antecedents to uncertainty: an integrative review. *J Nurs Scholarsh Off Publ Sigma Theta Tau Int Honor Soc Nurs*. 2012;44(2):145–155. https://doi.org/10.1111/j.1547-5069.2012.01445.x.

40. Boele FW, Heimans JJ, Aaronson NK, et al. Health-related quality of life of significant others of patients with malignant CNS versus non-CNS tumors: a comparative study. *J Neurooncol*. 2013;115(1):87–94. https://doi.org/10.1007/s11060-013-1198-z.

41. National Comprehensive Cancer Network. NCCN Clinical Practice Guidelines in Oncology: Cancer-Related Fatigue, Version 1.2008. https://www.nccn.org/professionals/physician_gls/pdf/fatigue.pdf; 2018.

42. Osoba D, Aaronson NK, Muller M, et al. Effect of neurological dysfunction on health-related quality of life in patients with high-grade glioma. *J Neurooncol*. 1997;34(3):263–278.

43. Lovely MP, Miaskowski C, Dodd M. Relationship between fatigue and quality of life in patients with glioblastoma multiformae. *Oncol Nurs Forum*. 1999;26(5):921–925.

44. Brown PD, Ballman KV, Rummans TA, et al. Prospective study of quality of life in adults with newly diagnosed high-grade gliomas. *J Neurooncol*. 2006;76(3):283–291. https://doi.org/10.1007/s11060-005-7020-9.

45. Janda M, Steginga S, Dunn J, Langbecker D, Walker D, Eakin E. Unmet supportive care needs and interest in services among patients with a brain tumour and their carers. *Patient Educ Couns*. 2008;71(2):251–258. https://doi.org/10.1016/j.pec.2008.01.020.

46. Hickmann A-K, Nadji-Ohl M, Haug M, et al. Suicidal ideation, depression, and health-related quality of life in patients with benign and malignant brain tumors: a prospective observational study in 83 patients. *Acta Neurochir*. 2016;158(9):1669–1682. https://doi.org/10.1007/s00701-016-2844-y.

47. Lovely MP. Symptom management of brain tumor patients. *Semin Oncol Nurs*. 2004;20(4):273–283.

48. Andrade C. A method for deciding about the possible safety of Modafinil and armodafinil in patients with seizure disorder. *J Clin Psychiatry*. 2016;77(1):e25–e28. https://doi.org/10.4088/JCP.15f10580.

49. Bigatão M dos R, Peria FM, Tirapelli DPC, Carlotti Junior CG. Educational program on fatigue for brain tumor patients: possibility strategy? *Arq Neuropsiquiatr*. 2016;74(2):155–160. https://doi.org/10.1590/0004-282X20160007.

50. Jones LW, Guill B, Keir ST, et al. Using the theory of planned behavior to understand the determinants of exercise intention in patients diagnosed with primary brain cancer. *Psychooncology*. 2007;16(3):232–240. https://doi.org/10.1002/pon.1077.

51. Pawl JD, Lee S-Y, Clark PC, Sherwood PR. Sleep loss and its effects on health of family caregivers of individuals with primary malignant brain tumors. *Res Nurs Health*. 2013;36(4):386–399. https://doi.org/10.1002/nur.21545.

52. Molassiotis A, Zheng Y, Denton-Cardew L, Swindell R, Brunton L. Symptoms experienced by cancer patients during the first year from diagnosis: patient and informal caregiver ratings and agreement. *Palliat Support Care*. 2010;8(3):313–324. https://doi.org/10.1017/S1478951510000118.

53. Robertson ME, McSherry F, Herndon JE, Peters KB. Insomnia and its associations in patients with recurrent glial neoplasms. *Springerplus*. 2016;5(1):823. https://doi.org/10.1186/s40064-016-2578-6.

54. Garland SN, Carlson LE, Stephens AJ, Antle MC, Samuels C, Campbell TS. Mindfulness-based stress reduction compared with cognitive behavioral therapy for the treatment of insomnia comorbid with cancer: a randomized, partially blinded, noninferiority trial. *J Clin Oncol*. 2014;32(5):449–457. https://doi.org/10.1200/JCO.2012.47.7265.

55. McCall M, Leone A, Cusimano MD. Nutritional status and body composition of adult patients with brain tumours awaiting surgical resection. *Can J Diet Pract Res*. 2014;75(3):148–151. https://doi.org/10.3148/cjdpr-2014-007.

56. Schmitz KH, Courneya KS, Matthews C, et al. American College of Sports Medicine roundtable on exercise guidelines for cancer survivors. *Med Sci Sport Exerc*. 2010;42(7):1409–1426. https://doi.org/10.1249/MSS.0b013e3181e0c112.

57. Jones LW, Guill B, Keir ST, et al. Exercise interest and preferences among patients diagnosed with primary brain cancer. *Support Care Cancer*. 2007;15(1):47–55. https://doi.org/10.1007/s00520-006-0096-8.

58. Levin GT, Greenwood KM, Singh F, Tsoi D, Newton RU. Exercise improves physical function and mental health of brain cancer survivors: two exploratory case studies. *Integr Cancer Ther*. 2015;15(2):190–196. https://doi.org/10.1177/1534735415600068.

59. Gehrke AK, Baisley MC, Sonck ALB, Wronski SL, Feuerstein M. Neurocognitive deficits following primary brain tumor treatment: systematic review of a decade of comparative studies. *J Neurooncol*. 2013;115(2):135–142. https://doi.org/10.1007/s11060-013-1215-2.

60. Tucha O, Smely C, Preier M, Lange KW. Cognitive deficits before treatment among patients with brain tumors. *Neurosurgery*. 2000;47:324–333-334. https://doi.org/10.1097/00006123-200008000-00011.

61. Locke DEC, Cerhan JH, Wu W, et al. Cognitive rehabilitation and problem-solving to improve quality of life of patients with primary brain tumors: a pilot study. *J Support Oncol*. 2008;6(8):383–391. http://www.ncbi.nlm.nih.gov/pubmed/19149323%5Cnhttp://www.scopus.com/inward/record.url?eid=2-s2.0-59449097658&partnerID=40&md5=f0f909f7f954f47216cec2758e57c203.

62. Wefel JS, Vardy J, Ahles T, Schagen SB. International cognition and cancer task force recommendations to harmonise studies of cognitive function in patients with cancer. *Lancet Oncol*. 2011;12(7):703–708. https://doi.org/10.1016/S1470-2045(10)70294-1.

63. Vardy J, Wefel JS, Ahles T, Tannock IF, Schagen SB. Cancer and cancer-therapy related cognitive dysfunction: an international perspective from the Venice cognitive workshop. *Ann Oncol*. 2008;19(4):623–629. https://doi.org/10.1093/annonc/mdm500.

64. Loughan A, Allen D, Baumstarck K, et al. Quality of life in neuro-oncology. In: Newton H, ed. *Handbook of Brain Tumor Chemotherapy, Molecular Therapeutics, and Immunotherapy*. 2nd ed. Academic Press: Elsevier; 2018.

65. Koppelmans V, Breteler MMB, Boogerd W, Seynaeve C, Schagen SB. Late effects of adjuvant chemotherapy for adult onset non-CNS cancer; cognitive impairment, brain structure and risk of dementia. *Crit Rev Oncol Hematol*. 2013;88(1):87–101. https://doi.org/10.1016/j.critrevonc.2013.04.002.

66. Douw L, Klein M, Fagel SS, et al. Cognitive and radiological effects of radiotherapy in patients with low-grade glioma: long-term follow-up. *Lancet Neurol*. 2009;8(9):810–818. https://doi.org/10.1016/S1474-4422(09)70204-2.

67. Corbetta M, Patel G, Review SGL. The reorienting system of the human brain: from environment to theory of mind. *Neuron*. 2008;306–324. https://doi.org/10.1016/j.neuron.2008.04.017.

68. Corbetta M, Kincade JM, Shulman GL. Neural systems for visual orienting and their relationships to spatial working memory. *J Cogn Neurosci.* 2002;14(3):508–523. https://doi.org/10.1162/089892902317362029.

69. Karnath H-O, Milner D, Vallar G. *The Cognitive and Neural Bases of Spatial Neglect.* Oxford Scholarship Online; 2002. https://doi.org/10.1093/acprof:oso/9780198508335.001.0001.

70. Wieneke MH. Neuropsychological assessment of cognitive functioning following chemotherapy for breast cancer. *Psychooncology.* 1995;4(1):61–66.

71. Freeman JR, Broshek DK. Assessing cognitive dysfunction in breast cancer: what are the tools? *Clin Breast Cancer.* 2002;3(Suppl. 3):S91–S99. http://www.embase.com/search/results?subaction=viewrecord&from=export&id=L36495688%5Cnhttp://sfx.umd.edu/hs?sid=EMBASE&issn=15268209&id=doi:&atitle=Assessing+cognitive+dysfunction+in+breast+cancer:+what+are+the+tools?&stitle=Clin.+Breast+Cancer&title=Cli.

72. Ahles TA, Saykin AJ, Furstenberg CT, et al. Neuropsychologic impact of standard-dose systemic chemotherapy in long-term survivors of breast cancer and lymphoma. *J Clin Oncol.* 2002;20(2):485–493. https://doi.org/10.1200/JCO.20.2.485.

73. Wefel JS, Lenzi R, Theriault RL, Davis RN, Meyers CA. The cognitive sequelae of standard-dose adjuvant chemotherapy in women with breast carcinoma: results of a prospective, randomized, longitudinal trial. *Cancer.* 2004;100(11):2292–2299. https://doi.org/10.1002/cncr.20272.

74. Klein M, Heimans JJ, Aaronson NK, et al. Effect of radiotherapy and other treatment-related factors on mid-term to long-term cognitive sequelae in low-grade gliomas: a comparative study. *Lancet.* 2002;360(9343):1361–1368. https://doi.org/10.1016/S0140-6736(02)11398-5.

75. Santini B, Talacchi A, Squintani G, Casagrande F, Capasso R, Miceli G. Cognitive outcome after awake surgery for tumors in language areas. *J Neurooncol.* 2012;108(2):319–326. https://doi.org/10.1007/s11060-012-0817-4.

76. DeAngelis LM, Delattre JY, Posner JB. Radiation-induced dementia in patients cured of brain metastases. *Neurology.* 1989;39(6):789–796. https://doi.org/10.1212/WNL.39.6.789.

77. Correa DD, Shi W, Thaler HT, Cheung AM, DeAngelis LM, Abrey LE. Longitudinal cognitive follow-up in low grade gliomas. *J Neurooncol.* 2008;86(3):321–327. https://doi.org/10.1007/s11060-007-9474-4.

78. Cairncross G, Berkey B, Shaw E, et al. Phase III trial of chemotherapy plus radiotherapy compared with radiotherapy alone for pure and mixed anaplastic oligodendroglioma: intergroup radiation therapy oncology group trial 9402. *J Clin Oncol.* 2006;24(18):2707–2714. https://doi.org/10.1200/JCO.2005.04.3414.

79. Van Den Bent MJ, Carpentier AF, Brandes AA, et al. Adjuvant procarbazine, lomustine, and vincristine improves progression-free survival but not overall survival in newly diagnosed anaplastic oligodendrogliomas and oligoastrocytomas: a randomized European Organisation for Research and Treatment of Cancer. *J Clin Oncol.* 2006;24(18):2715–2722. https://doi.org/10.1200/JCO.2005.04.6078.

80. Cairncross G, Wang M, Shaw E, et al. Phase III trial of chemoradiotherapy for anaplastic oligodendroglioma: long-term results of RTOG 9402. *J Clin Oncol.* 2013;31(3):337–343. https://doi.org/10.1200/JCO.2012.43.2674.

81. Prabhu RS, Won M, Shaw EG, et al. Effect of the addition of chemotherapy to radiotherapy on cognitive function in patients with low-grade glioma: secondary analysis of RTOG 98-02. *J Clin Oncol.* 2014;32(6):535–541. https://doi.org/10.1200/JCO.2013.53.1830.

82. Folstein MF, Folstein SE, McHugh PR. Mini-mental state. *J Psychiatr Res.* 1975;12(3):189–198. https://doi.org/10.1016/0022-3956(75)90026-6.

83. Reijneveld JC, Taphoorn MJB, Coens C, et al. Health-related quality of life in patients with high-risk low-grade glioma (EORTC 22033-26033): a randomised, open-label, phase 3 intergroup study. *Lancet Oncol.* 2016;17(11):1533–1542. https://doi.org/10.1016/S1470-2045(16)30305-9.

84. Nasreddine ZS, Phillips NA, Bédirian V, et al. The Montreal cognitive assessment, MoCA: a brief screening tool for mild cognitive impairment. *J Am Geriatr Soc.* 2005;53(4):695–699. https://doi.org/10.1111/j.1532-5415.2005.53221.x.

85. Blonski M, Taillandier L, Herbet G, et al. Combination of neoadjuvant chemotherapy followed by surgical resection as a new strategy for WHO grade II gliomas: a study of cognitive status and quality of life. *J Neurooncol.* 2012;106(2):353–366. https://doi.org/10.1007/s11060-011-0670-x.

86. Klein M, Engelberts NHJ, van der Ploeg HM, et al. Epilepsy in low-grade gliomas: the impact on cognitive function and quality of life. *Ann Neurol.* 2003;54(4):514–520. https://doi.org/10.1002/ana.10712.

87. Salander P, Bergenheim AT, Henriksson R. How was life after treatment of a malignant brain tumour? *Soc Sci Med.* 2000;51(4):589–598. http://ovidsp.ovid.com/ovidweb.cgi?T=JS&PAGE=reference&D=emed5&NEWS=N&AN=2000165064.

88. de Boer AGEM, Taskila T, Ojajärvi A, van Dijk FJH, Verbeek JHAM. Cancer survivors and unemployment. *JAMA.* 2009;301(7):753. https://doi.org/10.1001/jama.2009.187.

89. Taskila-Abrandt T, Martikainen R, Virtanen SV, Pukkala E, Hietanen P, Lindbohm ML. The impact of education and occupation on the employment status of cancer survivors. *Eur J Cancer.* 2004;40(16):2488–2493. https://doi.org/10.1016/j.ejca.2004.06.031.

90. Main DS, Nowels CT, Cavender TA, Etschmaier M, Steiner JF. A qualitative study of work and work return in cancer survivors. *Psychooncology.* 2005;14(11):992–1004. https://doi.org/10.1002/pon.913.

91. Verdonck-de Leeuw IM, van Bleek W-J, Leemans CR, de Bree R. Employment and return to work in head and neck cancer survivors. *Oral Oncol.* 2010;46(1):56–60. https://doi.org/10.1016/j.oraloncology.2009.11.001.

92. Feuerstein M, Todd BL, Moskowitz MC, et al. Work in cancer survivors: a model for practice and research. *J Cancer Surviv.* 2010;4(4):415–437. https://doi.org/10.1007/s11764-010-0154-6.

93. Kleinberg L, Wallner K, Malkin MG. Good performance status of long-term disease-free survivors of intracranial gliomas. *Int J Radiat Oncol Biol Phys.* 1993;26(1):129–133.

94. Gzell C, Wheeler H, Guo L, Kastelan M, Back M. Employment following chemoradiotherapy in glioblastoma: a prospective case series. *J Cancer Surviv.* 2014;8(1):108–113. https://doi.org/10.1007/s11764-013-0311-9.

95. Giovagnoli AR. Quality of life in patients with stable disease after surgery, radiotherapy, and chemotherapy for malignant brain tumour. *J Neurol Neurosurg Psychiatry*. 1999;67(3):358–363.

96. Kalkanis SN, Quiñones-Hinojosa A, Buzney E, Ribaudo HJ, Black PM. Quality of life following surgery for intracranial meningiomas at Brigham and Women's Hospital: a study of 164 patients using a modification of the functional assessment of cancer therapy-brain questionnaire. *J Neurooncol*. 2000;48(3):233–241.

97. Feuerstein M, Hansen JA, Calvio LC, Johnson L, Ronquillo JG. Work productivity in brain tumor survivors. *J Occup Environ Med*. 2007;49(7):803–811. https://doi.org/10.1097/JOM.0b013e318095a458.

98. Rusbridge SL, Walmsley NC, Griffiths SB, Wilford PA, Rees JH. Predicting outcomes of vocational rehabilitation in patients with brain tumours. *Psychooncology*. 2013;22(8):1907–1911. https://doi.org/10.1002/pon.3241.

99. Thomas S, Mehta MP, Kuo JS, Ian Robins H, Khuntia D. Current practices of driving restriction implementation for patients with brain tumors. *J Neurooncol*. 2011;103(3):641–647. https://doi.org/10.1007/s11060-010-0439-7.

100. Imhoff S, Lavallière M, Teasdale N, Fait P. Driving assessment and rehabilitation using a driving simulator in individuals with traumatic brain injury: a scoping review. *NeuroRehabilitation*. 2016;39(2):239–251. https://doi.org/10.3233/NRE-161354.

101. Wolfe PL, Lehockey KA. Neuropsychological assessment of driving capacity. *Arch Clin Neuropsychol*. 2016;31(6):517–529. https://doi.org/10.1093/arclin/acw050.

102. APA. Position statement on the role of psychiatrists in assessing driving ability. *Am J Psychiatry*. 1995;152:819.

103. American Medical Association. Physician's guide to assessing and counseling older drivers. *Ann Emerg Med*. 2004;43(6):746–747. https://doi.org/10.1016/j.annemergmed.2004.03.024.

104. Walsh S, Primeau M. Neuropsychological rehabilitation and habituation. In: Noggle C, Dean R, Barisa M, eds. *Neuropsychological Rehabilitation*. Springer; 2013.

105. Gehring K, Sitskoorn MM, Gundy CM, et al. Cognitive rehabilitation in patients with gliomas: a randomized, controlled trial. *J Clin Oncol*. 2009;27(22):3712–3722. https://doi.org/10.1200/JCO.2008.20.5765.

106. Huang ME, Cifu DX, Keyser-Marcus L. Functional outcome after brain tumor and acute stroke: a comparative analysis. *Arch Phys Med Rehabil*. 1998;79(11):1386–1390.

107. O'Dell MW, Barr K, Spanier D, Warnick RE. Functional outcome of inpatient rehabilitation in persons with brain tumors. *Arch Phys Med Rehabil*. 1998;79(12):1530–1534.

108. Pelletier G, Verhoef MJ, Khatri N, Hagen N. Quality of life in brain tumor patients: the relative contributions of depression, fatigue, emotional distress, and existential issues. *J Neurooncol*. 2002;57(1):41–49.

109. Bunevicius A, Tamasauskas S, Deltuva V, Tamasauskas A, Radziunas A, Bunevicius R. Predictors of health-related quality of life in neurosurgical brain tumor patients: focus on patient-centered perspective. *Acta Neurochir*. 2014;156(2):367–374. https://doi.org/10.1007/s00701-013-1930-7.

110. Bunevicius A, Deltuva VP, Deltuviene D, Tamasauskas A, Bunevicius R. Brain lesions manifesting as psychiatric disorders: eight cases. *CNS Spectr*. 2008;13(11):950–958.

111. Huang J, Zeng C, Xiao J, et al. Association between depression and brain tumor: a systematic review and meta-analysis. *Oncotarget*. 2017;8(55):94932–94943. https://doi.org/10.18632/oncotarget.19843.

112. Wellisch DK, Kaleita TA, Freeman D, Cloughesy T, Goldman J. Predicting major depression in brain tumor patients. *Psychooncology*. 2002;11(3):230–238. https://doi.org/10.1002/pon.562.

113. Rooney AG, Carson A, Grant R. Depression in cerebral glioma patients: a systematic review of observational studies. *J Natl Cancer Inst*. 2011;103(1):61–76. https://doi.org/10.1093/jnci/djq458.

114. Kilbride L, Smith G, Grant R. The frequency and cause of anxiety and depression amongst patients with malignant brain tumours between surgery and radiotherapy. *J Neurooncol*. 2007;84(3):297–304. https://doi.org/10.1007/s11060-007-9374-7.

115. Mauer MEL, Taphoorn MJB, Bottomley A, et al. Prognostic value of health-related quality-of-life data in predicting survival in patients with anaplastic oligodendrogliomas, from a phase III EORTC brain cancer group study. *J Clin Oncol*. 2007;25(36):5731–5737. https://doi.org/10.1200/JCO.2007.11.1476.

116. Ford E, Catt S, Chalmers A, Fallowfield L. Systematic review of supportive care needs in patients with primary malignant brain tumors. *Neuro Oncol*. 2012;14(4):392–404. https://doi.org/10.1093/neuonc/nor229.

117. Pranckeviciene A, Tamasauskas S, Deltuva VP, Bunevicius R, Tamasauskas A, Bunevicius A. Suicidal ideation in patients undergoing brain tumor surgery: prevalence and risk factors. *Support Care Cancer*. 2016;24(7):2963–2970. https://doi.org/10.1007/s00520-016-3117-2.

118. Rooney AG, Brown PD, Reijneveld JC, Grant R. Depression in glioma: a primer for clinicians and researchers. *J Neurol Neurosurg Psychiatry*. 2014;85(2):230–235. https://doi.org/10.1136/jnnp-2013-306497.

119. Richter A, Jenewein J, Krayenbühl N, Woernle C, Bellut D. Are preoperative sex-related differences of affective symptoms in primary brain tumor patients associated with postoperative histopathological grading? *J Neurooncol*. 2016;126(1):151–156. https://doi.org/10.1007/s11060-015-1950-7.

120. Rooney AG, McNamara S, MacKinnon M, et al. Screening for major depressive disorder in adults with cerebral glioma: an initial validation of 3 self-report instruments. *Neuro Oncol*. 2013;15(1):122–129. https://doi.org/10.1093/neuonc/nos282.

121. Ahluwalia MS, Xie H, Dahiya S, et al. Efficacy and patient-reported outcomes with dose-intense temozolomide in patients with newly diagnosed pure and mixed anaplastic oligodendroglioma: a phase II multicenter study. *J Neurooncol*. 2015;122(1):111–119. https://doi.org/10.1007/s11060-014-1684-y.

122. Mack JW, Cronin A, Taback N, et al. End-of-life care discussions among patients with advanced cancer: a cohort study. *Ann Intern Med*. 2012;156(3):204–210. https://doi.org/10.7326/0003-4819-156-3-201202070-00008.

123. Wright AA. Associations between end-of-life discussions, patient mental health, medical care near death, and caregiver bereavement adjustment. *JAMA.* 2008;300(14):1665. https://doi.org/10.1001/jama.300.14.1665.

124. Mack JW, Weeks JC, Wright AA, Block SD, Prigerson HG. End-of-life discussions, goal attainment, and distress at the end of life: predictors and outcomes of receipt of care consistent with preferences. *J Clin Oncol.* 2010;28(7):1203–1208. https://doi.org/10.1200/JCO.2009.25.4672.

125. Epstein AS, Prigerson HG, O'Reilly EM, Maciejewski PK. Discussions of life expectancy and changes in illness understanding in patients with advanced cancer. *J Clin Oncol.* 2016;34(20):2398–2403. https://doi.org/10.1200/JCO.2015.63.6696.

126. Gofton TE, Graber J, Carver A. Identifying the palliative care needs of patients living with cerebral tumors and metastases: a retrospective analysis. *J Neurooncol.* 2012;108(3):527–534. https://doi.org/10.1007/s11060-012-0855-y.

127. Houben CHM, Spruit MA, Groenen MTJ, Wouters EFM, Janssen DJA. Efficacy of advance care planning: a systematic review and meta-analysis. *J Am Med Dir Assoc.* 2014;15(7):477–489. https://doi.org/10.1016/j.jamda.2014.01.008.

128. Sizoo EM, Pasman HRW, Buttolo J, et al. Decision-making in the end-of-life phase of high-grade glioma patients. *Eur J Cancer.* 2012;48(2):226–232. https://doi.org/10.1016/j.ejca.2011.11.010.

129. Arnold R, Edwards K. www.vitaltalk.org; 2014. [accessed 01.01.18].

130. Walbert T, Chasteen K. Palliative and supportive care for glioma patients. *Cancer Treat Res.* 2015;163:171–184. https://doi.org/10.1007/978-3-319-12048-5_11.

131. Mission Hospice. Hospice.

132. Sherwood P, Given B, Given C, Schiffman R, Murman D, Lovely M. Caregivers of persons with a brain tumor: a conceptual model. *Nurs Inq.* 2004;11(1):43–53.

133. DesRoches C, Blendon R, Young J, Scoles K, Kim M. Caregiving in the post-hospitalization period: findings from a national survey. *Nurs Econ.* 2002;20(5):206–221.

134. Sherwood PR, Given BA, Given CW, et al. Predictors of distress in caregivers of persons with a primary malignant brain tumor. *Res Nurs Health.* 2006;29(2):105–120. https://doi.org/10.1002/nur.20116.

135. Schmer C, Ward-Smith P, Latham S, Salacz M. When a family member has a malignant brain tumor: the caregiver perspective. *J Neurosci Nurs.* 2008;40(2):78–84. https://doi.org/10.1097/01376517-200804000-00006.

136. Schubart JR, Kinzie MB, Farace E. Caring for the brain tumor patient: family caregiver burden and unmet needs. *Neuro Oncol.* 2008;10(1):61–72. https://doi.org/10.1215/15228517-2007-040.

Section B

Pathology and molecular biology

Chapter 7

Origin and development of oligodendroglioma

Josephine Volovetz, Defne Bayik and Justin D. Lathia
Cleveland Clinic Lerner College of Medicine, Lerner Research Institute, Cleveland, OH, United States

Abbreviations

CSF-1	colony stimulating factor 1
EGFR	epidermal growth factor receptor
MIF	macrophage inhibitory factor
NG2	neural/glial antigen 2
PDGFR	platelet-derived growth factor receptor
PD-L1	programmed death-ligand 1
RTK	receptor tyrosine kinase
SHH	sonic hedgehog
TGFβ	transforming growth factor beta
TLR	Toll-like receptor

Introduction

Oligodendrogliomas, defined in the 2016 World Health Organization Classification of Tumors of the Central Nervous System (CNS) as IDH-mutant and 1p/19q-codeleted, make up about 1.5% of all CNS tumors.[1] Low grade tumors are known as oligodendrogliomas and have an age-adjusted incidence of 0.24 per 100,000, while anaplastic oligodendrogliomas are higher grade and have an age-adjusted incidence of 0.11 per 100,000.[2] In patients with low-grade oligodendrogliomas who undergo radiation therapy as well as chemotherapy, median survival is 13 years.[3] To better understand these tumors and why they eventually become desensitized to the conventional therapies, it is important to understand how they initiate and evolve. The cell of origin can provide insight into the biology of oligodendrogliomas and inform new therapeutic targets. The purpose of this chapter is to explore possible cells of origin of oligodendrogliomas: cancer stem cells (CSCs) and oligodendrocyte precursor cells (OPCs). This chapter reviews asymmetric cell division, how it is utilized by neural stem cells (NSCs) to both self-renew and produce non-stem daughter cells, and how it may be altered in malignant transformation of cells. This chapter also reviews the role of the stem cell niche in regulating CSCs and the utility of preclinical models of oligodendrogliomas.

Cell of origin

Oligodendrogliomas are generally slow growing and known to have a better prognosis than other CNS tumors such as glioblastoma (GBM). Their cell of origin is unknown but studies have explored the possibility that oligodendrogliomas originate from CSCs versus progenitor cells. It is most likely that multiple cells contribute to the initiation of oligodendrogliomas, including both CSCs and oligodendrocyte progenitor cells.

Gliomas are known to be heterogeneous entities, composed of many different cell types. One explanation for the source of this heterogeneity lies in CSCs. The CSCs, also known as brain tumor-initiating cells, have the ability to self-renew, proliferate, and differentiate into cells that compose the tumor.[4, 5] This small population of cells in a tumor has the ability to initiate new tumors, while most of the other cells in the heterogeneous population are not tumorigenic.[5, 6] The cell surface marker CD133 has been found to be expressed on a minority of cells that otherwise lack neural differentiation markers and have the ability to differentiate into cells that are found in brain tumors.[4, 7] This marker is also found on embryonic NSCs.[8, 9]

Oligodendroglioma. https://doi.org/10.1016/B978-0-12-813158-9.00007-4

Other markers found to be enriched in CSCs include but are not limited to integrin α6, L1CAM, A2B5, CD15, and CD44.[2, 10–13] A fraction of brain tumor cells positive for the marker CD133 has been found to initiate brain tumors in nonobese diabetic, severe combined immunodeficient (NOD-SCID) mice.[14] Aggressive brain tumors, such as GBM, have been found to have a 19%–29% CD133-positive fraction.[14] When these CD133-positive cells are injected into mice, a tumor that recapitulates the phenotype of the original patient tumor results, whereas when CD133-negative cells are engrafted in these mice, no tumors form.[14]

High-grade oligodendrogliomas have been studied to determine whether CSCs are present and if they are associated with prognosis. The tumors with a favorable prognosis of progression-free survival of at least a year contain very few CD133-positive cells and have been found to have limited self-renewal capability when cells dissociated from those tumors are cultured in vitro.[15] On the other hand, high-grade oligodendrogliomas with poor prognosis have larger CD133-positive populations and contain more clonogenic cells, as evidenced by the ability of cells dissociated from the tumors to form neurospheres.[15] Although CSC populations, denoted by the presence of the CD133 marker, are small and rare, they correlate significantly with prognosis in high-grade oligodendroglioma.

Another marker of CSCs, Sox2, is a transcription factor that is responsible to help maintain NSCs' ability to self-renew.[16] Sox2 deletion in mouse oligodendroglioma cells leads to the loss of the ability of these cells to initiate tumors when transplanted into a mouse brain.[17]

The OPCs, which can be characterized by chondroitin sulfate proteoglycan NG2 expression as well as the presence of the α receptor of platelet-derived growth factor (PDGFRα), have been identified as the largest population of cycling cells in the brain, especially in the subventricular zone (SVZ) and white matter.[18] This indicates that they continue to divide, even in adult brains.[18]

Oligodendroglioma mouse models demonstrate high levels of cells positive for NG2.[19] Studies have shown that oligodendrogliomas can originate from OPCs in mouse models.[19, 20] Expression of NG2 and the PDGFRα has been identified in human oligodendrogliomas confirming that OPCs are present in these tumors.[21] In studies that have compared cells with the stem cell marker, CD15, to cells with OPC markers in murine tumor initiation, it has been found that CD15-positive cells did not form tumors, whereas those with an OPC marker but no stem cell marker did result in tumorigenesis.[19] When concurrent *p53* and *NF1* mutations are introduced into NSCs and their progeny are analyzed, OPCs are found to exhibit abnormal growth resulting in overexpansion.[22] This aberrant growth is specific to OPCs, supporting the idea that mutations in one cell type (NSCs) may lead to the development of the cell of origin for oligodendrogliomas.[22]

The data appear to support a hypothesis that OPCs are responsible for the initiation of oligodendrogliomas while CSCs are associated with a worse prognosis in high-grade tumors. However, whether oligodendrogliomas descend from OPCs or a combination of both OPCs and CSCs remains to be determined (Fig. 1).

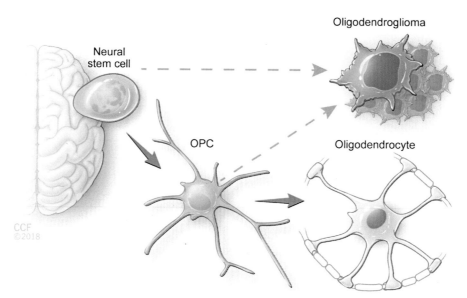

FIG. 1 NSCs are the precursor cells to OPCs and both may serve as the cellular origin of oligodendrogliomas. Mutations in NSCs or OPCs can lead to aberrant cell division and accumulation of dedifferentiated cells forming oligodendrogliomas.

Cell division and fate decision

Stem cells can maintain their population size through self-renewal while giving rise to differentiated daughter cells.[23] This process is tightly regulated by the balance between symmetric and asymmetric cell division. Unlike symmetric division in which two stem or differentiated daughter cells are generated, asymmetric division results in one stem and one non-stem cell.[24] This is achieved via cellular polarity through unequal distribution of intrinsic factors and/or differential exposure to extrinsic fate determinants.[23, 24] As with adult stem cells, NSCs are capable of dividing symmetrically or asymmetrically.[25, 26] During asymmetric cell division of NSCs, segregation of epidermal growth factor receptor (EGFR) marks stem-like cells, whereas Numb, TRIM32, and Cyclin D2 preferentially localize to non-stem daughter cells.[27–30] Differential accumulation of CD133 (prominin-1) and Numb determines the fate of the daughter cells derived from glioma CSCs[7] (Fig. 2). CD133 promotes stemness of daughter cells by activating the PI3K pathway, while Numb restricts cell proliferation in non-stem cells by limiting Notch activity and stabilizing p53.[7, 31–33]

The NSCs survive through adulthood and remain mostly quiescent except for maintaining a stable population in the SVZ of the forebrain and subgranular zone (SGZ) of the hippocampus.[34, 35] In a healthy state, around 8.6% of the NSCs localized to the SVZ (also referred to as type B cells or SVZ-B) divide asymmetrically to give rise to transit-amplifying cells (SVZ-C cells) and a small number of OPCs.[35–37] SVZ-C further generate neuroblasts (SVZ-A) and OPCs through asymmetric division.[37, 38] Thereafter, OPCs spread out from the SVZ to white and gray matter, where they divide actively to retain their population and generate oligodendrocytes.[39, 40] In this setting, co-segregation of NG2 and EGFR marks the OPCs, while NG2 negative cells differentiate into oligodendrogcytes.[41] Unlike NSCs that can proliferate indefinitely, OPCs undergo a limited number of cell divisions.[42]

Alterations in the OPC division pattern results in the accumulation of progenitor cells with high proliferative capacity and thus formation of oligodendroglioma.[41] Neoplastic OPCs mainly undergo symmetric rather than asymmetric cell division by equally distributing EGFR and NG2 into the daughter cells.[41] Genetic mutations in OPCs such as loss of Numb and TRIM32 are in part responsible for the shift in the mode of cell division.[41] Similar dysregulation in cell division profile of mammary stem cells has been linked to loss of p53.[21] Although p53 mutation provides proliferative advantage to NSCs, it is not sufficient for neoplastic transformation.[43, 44] Genetically engineered mouse models (GEMMs) have further suggested that additional oncogenic mutations (Kras, Nestin, PDGFR) and/or deletion of tumor-suppressor genes (pTEN, NF1) are required for the disruption of the self-renewal capacity of NSCs and formation of GBM.[45–48] In contrast, p53 mutated OPCs (Olig2-positive SVZ-C cells) potentially descend from CSCs with p53 mutation and accumulate cooperating oncogenic alterations throughout the cellular hierarchy. These are enriched in oligodendroglioma.[19, 44] However, it is noteworthy that this gene signature is not translated to human specimens, and p53 is frequently intact in human oligodendroglioma.[49, 50]

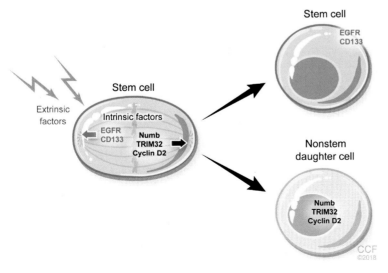

FIG. 2 Asymmetric distribution of intrinsic factors and exposure to environmental agents determine cell fate. Stem cells divide asymmetrically to give rise to non-stem cells, while maintaining their population. Unequal segregation of CD133 and EGFR during cell division establishes stem characteristics, while Numb, TRIM32, and Cyclin D2 are enriched in the nonstem daughter cell.

Mechanism of regulation: stem cell niches

A combination of intrinsic and extrinsic factors regulates NSC maintenance, differentiation, and repopulation. Environmental cues that are enriched in specific microenvironments, also referred to as stem cell niches, ensure maintenance of the stem cell state and regulate the fate decision process of immediate progenitors.[51] Stem cell niches consist of various types of cells, including immune cells, endothelial cells, and perivascular cells that regulate stem cell behavior through direct contact or by secreting extracellular matrix (ECM) components, extracellular vesicles, receptor ligands, and soluble mediators[52, 53] (Fig. 3). Similar to NSCs, survival of glioma CSCs depends on their microenvironment. These cells can reside in perivascular (PVN) and hypoxic niches or can even co-opt the SVZ niche as reported in high-grade gliomas.[54–56]

Unlike normal stem cell niches, CSC niches provide aberrant signaling that promotes tumorigenesis. Soluble factors and surface receptors in these niches dictate proliferation, differentiation, and survival of tumor-initiating cells by activating a combination of signaling pathways summarized in Table 1.

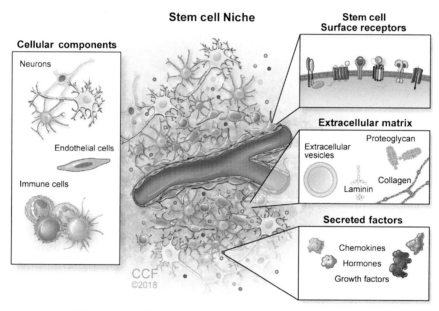

FIG. 3 The stem cell niche regulates cell fate decisions. Neurons, endothelial cells, and immune cells comprise the cellular compartment of the stem cell niche, while extracellular matrix provides scaffolding, and soluble factors and signaling through surface receptors drive stem cell properties. These niches are leveraged by brain tumors to ensure maintenance of CSCs.

TABLE 1 Signaling pathways involved in regulation of tumor-initiating cells

Receptor/soluble mediator	Intracellular signaling pathway	Cell type	Reference
EGFR	MAPK	NSCs, OPCs	19, 57, 58
PDGFR	MAPK, Akt	NSCs, OPCs	20, 59, 60
CD133	PI3K/Akt	NSCs	7, 33
Notch	Nestin	NSCs, OPCs	61, 62
NG2	RTK	OPCs	41, 63
L1CAM	Olig2	NSCs	10
Wnt	β-catenin	NSCs, OPCs	64, 65
SHH	Sox, Gli1, Nanog	NSCs, OPCs	64, 66–68

Preclinical models

The biology and origin of oligodendrogliomas, as well as responses to possible therapeutics, are studied in a variety of preclinical models, a majority of which are murine. Because the cell of origin for oligodendrogliomas is not known, mouse models of oligodendroglioma have been created with a variety of mutations. The EGFR is often found to be dysregulated in oligodendrogliomas and has provided a target for mutation in order to generate preclinical models.[69, 70] One model that aims to recapitulate oligodendroglioma via its EGFR expression has been created using transgenic mice expressing *v-erbB*, a homolog of *EGFR*.[71] This gene is under the control of a promoter found in oligodendrogliomas and NSCs: S100β.[71] This mutation has a tumor-initiating penetrance of about 20%, and the tumors formed had classical low-grade oligodendroglioma features such as monotonous cells with perinuclear halos.[71] The INK4a/ARF locus is a known tumor suppressor found to be deleted in high-grade tumors; it has been mutated in transgenic mice in order to create a model of oligodendroglioma.[71–73] Another mouse model has been created from the overexpression of platelet-derived growth factor B (PDGF-B), which is often found to be altered in oligodendrogliomas.[74] This model starts with low-grade tumors that have the ability to transform into high-grade tumors with new mutations.[55]

In an effort to define disease progression and response in preclinical models that are more similar to humans, especially with regards to the immune system, canine models have been explored. Spontaneous canine gliomas have been evaluated and found to have features of human oligodendrogliomas, such as positive staining for the marker Olig2.[75] These canine models of oligodendroglioma have an immunologic environment that is similar to that seen in humans.[75] Comparison of gene expression between these canines and humans, however, does show differentially expressed genes; therefore, despite similarities in immune environment, it is important to take differences between canine and human biology into account when interpreting canine oligodendroglioma studies in the future.[75]

Conclusion

Classification of oligodendroglioma tumors as grade II (low-grade) and grade III (anaplastic) determines the type of the treatment. A "watch and wait" approach is employed for grade II oligodendrogliomas in many cases, whereas standard therapy for grade III oligodendrogliomas involves combination of chemotherapy with radiotherapy that may or may not be preceded by surgical resection.[76] Although this treatment strategy prolongs survival, oligodendrogliomas, especially high-grade tumors, remain an incurable disease, and neurotoxicity associated with the conventional treatment negatively impacts patients' quality of life.[77] Alternative therapeutic approaches addressing tumor heterogeneity, particularly the presence of chemo/radio-resistant CSCs and immunosuppressive immune infiltrates, can potentially improve disease management.

Despite recent advancements in the field of oligodendroglioma research, cellular origin of these tumors remains ambiguous. This issue is particularly important for the success of the treatment strategies considering the distinct properties of CSCs and OPCs including their response to temozolomide and anti-EGFR therapy.[19, 78] Dissecting the cellular heterogeneity of oligodendrogliomas and identifying genetic/epigenetic/metabolic signatures associated with each tumor-initiating cell-type promises improved therapeutic efficacy. A prominent example is the isocitrate dehydrogenase (*IDH*) mutation status, which is closely linked to cellular architecture of oligodendrogliomas, more specifically with the maintenance of progenitor cells. *IDH* mutation results in an increase in oncometabolite 2-hydroxyglutarate (2-HG) production which in return blocks histone demethylase activity.[79–81] This altered methylation pattern, also known as CpG island methylator phenotype (CIMP), prevents differentiation of progenitor cells into terminally differentiated cells leading to the accumulation of dedifferentiated cells.[79] The discovery of *IDH* as a driver mutation was accompanied by the development of IDH inhibitors that are currently in clinical trials for treatment of oligodendroglioma[82] [NCT 02977689]. It is noteworthy that *IDH* mutation frequently coexists with other genetic alterations such as deletion of the 1p/19q locus, far-upstream binding protein 1 (*FUBP1*), and homolog of *Drosophila capicua* (*CIC*) as well as mutations in telomere reverse transcriptase (*TERT*) and *PIK3CA*.[83–86] Loss of the 1p/19q region and *FUBP1* gene has been linked to enhanced *MYC* activation, while mutations in *CIC* and *PIK3CA* caused deregulation of receptor tyrosine kinase (RTK) signaling.[83–86] Collectively, these studies suggest that genetic composition of tumor-initiating cells control their survival and proliferation capacity. Recent characterization of oligodendroglioma CSC and OPC gene signatures with single-cell RNA sequencing expand our understanding of development programs altered in tumor-initiating cells and provide insight into how to therapeutically target them effectively.[49]

Another potential therapeutic approach is to target the cross-talk between oligodendroglioma-initiating cells and immune cells. The IDH mutated gliomas are characterized by limited infiltration of pro-inflammatory cells, particularly cytotoxic T lymphocytes.[87–89] Thus, phase I/II clinical trials are underway to improve recruitment and activation of antitumoral cells by combining dendritic cell vaccines with T cell-activating strategies [NCT02924038, NCT03014804]. However, CSCs employ additional mechanisms to escape immune recognition, which form the basis of their resistance to immunotherapy. Studies with high-grade gliomas demonstrated that glioma CSCs drive immunosuppressive regulatory T cells (T_{regs}), tumor-associated macrophages (TAMs), and myeloid-derived suppressor cells (MDSCs) by secreting immunomodulatory mediators such as TGFβ, CSF-1, periostin, and MIF.[90–93] In return, these immunosuppressive cells promote disease progression not only by enabling immune evasion but also by promoting CSC maintenance, cancer cell invasiveness and proliferation.[94, 95] Furthermore, CSCs escape immunosurveillance by downregulating inflammatory receptors such as TLR4 or NKG2D while upregulating immune checkpoint inhibitor PD-L1.[96–98] Collectively, these findings suggest that combinational targeting of CSC-immune axis can further improve the efficacy of oligodendroglioma treatment.

Acknowledgment

We thank Amanda Mendelsohn (Center of Medical Art and Photography, Cleveland Clinic) for assistance with figure preparation.

References

1. Louis DN, Perry A, Reifenberger G, et al. The 2016 World Health Organization classification of tumors of the central nervous system: a summary. *Acta Neuropathol.* 2016;131(6):803–820.
2. Ogden AT, Waziri AE, Lochhead RA, et al. Identification of A2B5+CD133- tumor-initiating cells in adult human gliomas. *Neurosurgery.* 2008;62 (2):505–514 [discussion 514–505].
3. Buckner JC, Shaw EG, Pugh SL, et al. Radiation plus procarbazine, CCNU, and vincristine in low-grade glioma. *N Engl J Med.* 2016;374 (14):1344–1355.
4. Singh SK, Clarke ID, Terasaki M, et al. Identification of a cancer stem cell in human brain tumors. *Cancer Res.* 2003;63(18):5821–5828.
5. Reya T, Morrison SJ, Clarke MF, Weissman IL. Stem cells, cancer, and cancer stem cells. *Nature.* 2001;414(6859):105–111.
6. Hamburger AW, Salmon SE. Primary bioassay of human tumor stem cells. *Science.* 1977;197(4302):461–463.
7. Lathia JD, Hitomi M, Gallagher J, et al. Distribution of CD133 reveals glioma stem cells self-renew through symmetric and asymmetric cell divisions. *Cell Death Dis.* 2011;2.
8. Pfenninger CV, Roschupkina T, Hertwig F, et al. CD133 is not present on neurogenic astrocytes in the adult subventricular zone, but on embryonic neural stem cells, ependymal cells, and glioblastoma cells. *Cancer Res.* 2007;67(12):5727–5736.
9. Uchida N, Buck DW, He D, et al. Direct isolation of human central nervous system stem cells. *Proc Natl Acad Sci U S A.* 2000;97(26):14720–14725.
10. Bao S, Wu Q, Li Z, et al. Targeting cancer stem cells through L1CAM suppresses glioma growth. *Cancer Res.* 2008;68(15):6043–6048.
11. Lathia JD, Gallagher J, Heddleston JM, et al. Integrin alpha 6 regulates glioblastoma stem cells. *Cell Stem Cell.* 2010;6(5):421–432.
12. Read TA, Fogarty MP, Markant SL, et al. Identification of CD15 as a marker for tumor-propagating cells in a mouse model of medulloblastoma. *Cancer Cell.* 2009;15(2):135–147.
13. Yan Y, Zuo X, Wei D. Concise review: emerging role of CD44 in cancer stem cells: a promising biomarker and therapeutic target. *Stem Cells Transl Med.* 2015;4(9):1033–1043.
14. Singh SK, Hawkins C, Clarke ID, et al. Identification of human brain tumour initiating cells. *Nature.* 2004;432(7015):396–401.
15. Beier D, Wischhusen J, Dietmaier W, et al. CD133 expression and cancer stem cells predict prognosis in high-grade oligodendroglial tumors. *Brain Pathol.* 2008;18(3):370–377.
16. Favaro R, Valotta M, Ferri AL, et al. Hippocampal development and neural stem cell maintenance require Sox2-dependent regulation of Shh. *Nat Neurosci.* 2009;12(10):1248–1256.
17. Favaro R, Appolloni I, Pellegatta S, et al. Sox2 is required to maintain cancer stem cells in a mouse model of high-grade oligodendroglioma. *Cancer Res.* 2014;74(6):1833–1844.
18. Geha S, Pallud J, Junier MP, et al. NG2+/Olig2+ cells are the major cycle-related cell population of the adult human normal brain. *Brain Pathol.* 2010;20(2):399–411.
19. Persson AI, Petritsch C, Swartling FJ, et al. Non-stem cell origin for oligodendroglioma. *Cancer Cell.* 2010;18(6):669–682.
20. Lindberg N, Kastemar M, Olofsson T, Smits A, Uhrbom L. Oligodendrocyte progenitor cells can act as cell of origin for experimental glioma. *Oncogene.* 2009;28(23):2266–2275.
21. Shoshan Y, Nishiyama A, Chang A, et al. Expression of oligodendrocyte progenitor cell antigens by gliomas: implications for the histogenesis of brain tumors. *Proc Natl Acad Sci U S A.* 1999;96(18):10361–10366.
22. Liu C, Sage JC, Miller MR, et al. Mosaic analysis with double markers reveals tumor cell of origin in glioma. *Cell.* 2011;146(2):209–221.
23. Yamashita YM, Yuan H, Cheng J, Hunt AJ. Polarity in stem cell division: asymmetric stem cell division in tissue homeostasis. *Cold Spring Harb Perspect Biol.* 2010;2(1):a001313.
24. Doe CQ. Neural stem cells: balancing self-renewal with differentiation. *Development.* 2008;135(9):1575–1587.
25. Gomez-Lopez S, Lerner RG, Petritsch C. Asymmetric cell division of stem and progenitor cells during homeostasis and cancer. *Cell Mol Life Sci.* 2014;71(4):575–597.

26. Noctor SC, Martinez-Cerdeno V, Ivic L, Kriegstein AR. Cortical neurons arise in symmetric and asymmetric division zones and migrate through specific phases. *Nat Neurosci.* 2004;7(2):136–144.

27. Sun Y, Goderie SK, Temple S. Asymmetric distribution of EGFR receptor during mitosis generates diverse CNS progenitor cells. *Neuron.* 2005;45 (6):873–886.

28. Shen Q, Zhong W, Jan YN, Temple S. Asymmetric Numb distribution is critical for asymmetric cell division of mouse cerebral cortical stem cells and neuroblasts. *Development.* 2002;129(20):4843–4853.

29. Schwamborn JC, Berezikov E, Knoblich JA. The TRIM-NHL protein TRIM32 activates microRNAs and prevents self-renewal in mouse neural progenitors. *Cell.* 2009;136(5):913–925.

30. Tsunekawa Y, Osumi N. How to keep proliferative neural stem/progenitor cells: a critical role of asymmetric inheritance of cyclin D2. *Cell Cycle.* 2012;11(19):3550–3554.

31. Pece S, Serresi M, Santolini E, et al. Loss of negative regulation by Numb over Notch is relevant to human breast carcinogenesis. *J Cell Biol.* 2004;167 (2):215–221.

32. Colaluca IN, Tosoni D, Nuciforo P, et al. NUMB controls p53 tumour suppressor activity. *Nature.* 2008;451(7174):76–80.

33. Wei Y, Jiang Y, Zou F, et al. Activation of PI3K/Akt pathway by CD133-p85 interaction promotes tumorigenic capacity of glioma stem cells. *Proc Natl Acad Sci U S A.* 2013;110(17):6829–6834.

34. Merkle FT, Alvarez-Buylla A. Neural stem cells in mammalian development. *Curr Opin Cell Biol.* 2006;18(6):704–709.

35. Doetsch F, Garcia-Verdugo JM, Alvarez-Buylla A. Cellular composition and three-dimensional organization of the subventricular germinal zone in the adult mammalian brain. *J Neurosci.* 1997;17(13):5046–5061.

36. Ponti G, Obernier K, Guinto C, Jose L, Bonfanti L, Alvarez-Buylla A. Cell cycle and lineage progression of neural progenitors in the ventricular-subventricular zones of adult mice. *Proc Natl Acad Sci U S A.* 2013;110(11):E1045–E1054.

37. Menn B, Garcia-Verdugo JM, Yaschine C, Gonzalez-Perez O, Rowitch D, Alvarez-Buylla A. Origin of oligodendrocytes in the subventricular zone of the adult brain. *J Neurosci.* 2006;26(30):7907–7918.

38. Hack MA, Saghatelyan A, de Chevigny A, et al. Neuronal fate determinants of adult olfactory bulb neurogenesis. *Nat Neurosci.* 2005;8(7):865–872.

39. Zhu X, Hill RA, Dietrich D, Komitova M, Suzuki R, Nishiyama A. Age-dependent fate and lineage restriction of single NG2 cells. *Development.* 2011;138(4):745–753.

40. Dawson MR, Polito A, Levine JM, Reynolds R. NG2-expressing glial progenitor cells: an abundant and widespread population of cycling cells in the adult rat CNS. *Mol Cell Neurosci.* 2003;24(2):476–488.

41. Sugiarto S, Persson AI, Munoz EG, et al. Asymmetry-defective oligodendrocyte progenitors are glioma precursors. *Cancer Cell.* 2011;20(3):328–340.

42. Durand B, Gao FB, Raff M. Accumulation of the cyclin-dependent kinase inhibitor p27/Kip1 and the timing of oligodendrocyte differentiation. *EMBO J.* 1997;16(2):306–317.

43. Gil-Perotin S, Marin-Husstege M, Li J, et al. Loss of p53 induces changes in the behavior of subventricular zone cells: implication for the genesis of glial tumors. *J Neurosci.* 2006;26(4):1107–1116.

44. Wang Y, Yang J, Zheng H, et al. Expression of mutant p53 proteins implicates a lineage relationship between neural stem cells and malignant astrocytic glioma in a murine model. *Cancer Cell.* 2009;15(6):514–526.

45. Munoz DM, Guha A. Mouse models to interrogate the implications of the differentiation status in the ontogeny of gliomas. *Oncotarget.* 2011;2 (8):590–598.

46. Zhu Y, Guignard F, Zhao D, et al. Early inactivation of p53 tumor suppressor gene cooperating with NF1 loss induces malignant astrocytoma. *Cancer Cell.* 2005;8(2):119–130.

47. Zheng H, Ying H, Yan H, et al. p53 and Pten control neural and glioma stem/progenitor cell renewal and differentiation. *Nature.* 2008;455 (7216):1129–1133.

48. Friedmann-Morvinski D, Bushong EA, Ke E, et al. Dedifferentiation of neurons and astrocytes by oncogenes can induce gliomas in mice. *Science.* 2012;338(6110):1080–1084.

49. Tirosh I, Venteicher AS, Hebert C, et al. Single-cell RNA-seq supports a developmental hierarchy in human oligodendroglioma. *Nature.* 2016;539 (7628):309–313.

50. Wesseling P, van den Bent M, Perry A. Oligodendroglioma: pathology, molecular mechanisms and markers. *Acta Neuropathol.* 2015;129(6):809–827.

51. Walker MR, Patel KK, Stappenbeck TS. The stem cell niche. *J Pathol.* 2009;217(2):169–180.

52. Quail DF, Joyce JA. The microenvironmental landscape of brain tumors. *Cancer Cell.* 2017;31(3):326–341.

53. Fuentealba LC, Obernier K, Alvarez-Buylla A. Adult neural stem cells bridge their niche. *Cell Stem Cell.* 2012;10(6):698–708.

54. Lim DA, Cha S, Mayo MC, et al. Relationship of glioblastoma multiforme to neural stem cell regions predicts invasive and multifocal tumor phenotype. *Neuro Oncol.* 2007;9(4):424–429.

55. Calabrese C, Poppleton H, Kocak M, et al. A perivascular niche for brain tumor stem cells. *Cancer Cell.* 2007;11(1):69–82.

56. Seidel S, Garvalov BK, Wirta V, et al. A hypoxic niche regulates glioblastoma stem cells through hypoxia inducible factor 2 alpha. *Brain.* 2010;133(Pt 4):983–995.

57. Gonzalez-Perez O, Romero-Rodriguez R, Soriano-Navarro M, Garcia-Verdugo JM, Alvarez-Buylla A. Epidermal growth factor induces the progeny of subventricular zone type B cells to migrate and differentiate into oligodendrocytes. *Stem Cells.* 2009;27(8):2032–2043.

58. Ivkovic S, Canoll P, Goldman JE. Constitutive EGFR signaling in oligodendrocyte progenitors leads to diffuse hyperplasia in postnatal white matter. *J Neurosci.* 2008;28(4):914–922.

59. Jackson EL, Garcia-Verdugo JM, Gil-Perotin S, et al. PDGFR alpha-positive B cells are neural stem cells in the adult SVZ that form glioma-like growths in response to increased PDGF signaling. *Neuron.* 2006;51(2):187–199.

60. Kim Y, Kim E, Wu Q, et al. Platelet-derived growth factor receptors differentially inform intertumoral and intratumoral heterogeneity. *Genes Dev.* 2012;26(11):1247–1262.

61. Park HC, Appel B. Delta-Notch signaling regulates oligodendrocyte specification. *Development.* 2003;130(16):3747–3755.

62. Jeon HM, Jin X, Lee JS, et al. Inhibitor of differentiation 4 drives brain tumor-initiating cell genesis through cyclin E and notch signaling. *Genes Dev.* 2008;22(15):2028–2033.

63. Wade A, Robinson AE, Engler JR, Petritsch C, James CD, Phillips JJ. Proteoglycans and their roles in brain cancer. *FEBS J.* 2013;280(10):2399–2417.

64. Chen HL, Chew LJ, Packer RJ, Gallo V. Modulation of the Wnt/beta-catenin pathway in human oligodendroglioma cells by Sox17 regulates proliferation and differentiation. *Cancer Lett.* 2013;335(2):361–371.

65. Zhang N, Wei P, Gong A, et al. FoxM1 promotes beta-catenin nuclear localization and controls Wnt target-gene expression and glioma tumorigenesis. *Cancer Cell.* 2011;20(4):427–442.

66. Tong CK, Fuentealba LC, Shah JK, et al. A dorsal SHH-dependent domain in the V-SVZ produces large numbers of oligodendroglial lineage cells in the postnatal brain. *Stem Cell Reports.* 2015;5(4):461–470.

67. Clement V, Sanchez P, de Tribolet N, Radovanovic I, Ruiz i Altaba A. HEDGEHOG-GLI1 signaling regulates human glioma growth, cancer stem cell self-renewal, and tumorigenicity. *Curr Biol.* 2007;17(2):165–172.

68. Zbinden M, Duquet A, Lorente-Trigos A, Ngwabyt SN, Borges I, Ruiz i Altaba A. NANOG regulates glioma stem cells and is essential in vivo acting in a cross-functional network with GLI1 and p53. *EMBO J.* 2010;29(15):2659–2674.

69. Reifenberger J, Reifenberger G, Liu L, James CD, Wechsler W, Collins VP. Molecular genetic analysis of oligodendroglial tumors shows preferential allelic deletions on 19q and 1p. *Am J Pathol.* 1994;145(5):1175–1190.

70. Libermann TA, Nusbaum HR, Razon N, et al. Amplification, enhanced expression and possible rearrangement of EGF receptor gene in primary human brain tumours of glial origin. *Nature.* 1985;313(5998):144–147.

71. Weiss WA, Burns MJ, Hackett C, et al. Genetic determinants of malignancy in a mouse model for oligodendroglioma. *Cancer Res.* 2003;63 (7):1589–1595.

72. Sharpless NE. INK4a/ARF: a multifunctional tumor suppressor locus. *Mutat Res.* 2005;576(1–2):22–38.

73. Ivanchuk SM, Mondal S, Dirks PB, Rutka JT. The INK4A/ARF locus: role in cell cycle control and apoptosis and implications for glioma growth. *J Neurooncol.* 2001;51(3):219–229.

74. Di Rocco F, Carroll RS, Zhang J, Black PM. Platelet-derived growth factor and its receptor expression in human oligodendrogliomas. *Neurosurgery.* 1998;42(2):341–346.

75. Filley A, Henriquez M, Bhowmik T, et al. Immunologic and gene expression profiles of spontaneous canine oligodendrogliomas. *J Neurooncol.* 2018;.

76. Stupp R, Brada M, van den Bent MJ, Tonn JC, Pentheroudakis G, Group EGW. High-grade glioma: ESMO clinical practice guidelines for diagnosis, treatment and follow-up. *Ann Oncol.* 2014;25(Suppl 3):iii93–iii101.

77. Simonetti G, Gaviani P, Botturi A, Innocenti A, Lamperti E, Silvani A. Clinical management of grade III oligodendroglioma. *Cancer Manag Res.* 2015;7:213–223.

78. Hide T, Takezaki T, Nakatani Y, Nakamura H, Kuratsu J, Kondo T. Combination of a ptgs2 inhibitor and an epidermal growth factor receptor-signaling inhibitor prevents tumorigenesis of oligodendrocyte lineage-derived glioma-initiating cells. *Stem Cells.* 2011;29(4):590–599.

79. Lu C, Ward PS, Kapoor GS, et al. IDH mutation impairs histone demethylation and results in a block to cell differentiation. *Nature.* 2012;483 (7390):474–478.

80. Turcan S, Rohle D, Goenka A, et al. IDH1 mutation is sufficient to establish the glioma hypermethylator phenotype. *Nature.* 2012;483(7390):479–483.

81. Turcan S, Makarov V, Taranda J, et al. Mutant-IDH1-dependent chromatin state reprogramming, reversibility, and persistence. *Nat Genet.* 2018;50 (1):62–72.

82. Fujii T, Khawaja MR, DiNardo CD, Atkins JT, Janku F. Targeting isocitrate dehydrogenase (IDH) in cancer. *Discov Med.* 2016;21(117):373–380.

83. Bettegowda C, Agrawal N, Jiao Y, et al. Mutations in CIC and FUBP1 contribute to human oligodendroglioma. *Science.* 2011;333(6048):1453–1455.

84. Arita H, Narita Y, Fukushima S, et al. Upregulating mutations in the TERT promoter commonly occur in adult malignant gliomas and are strongly associated with total 1p19q loss. *Acta Neuropathol.* 2013;126(2):267–276.

85. Broderick DK, Di C, Parrett TJ, et al. Mutations of PIK3CA in anaplastic oligodendrogliomas, high-grade astrocytomas, and medulloblastomas. *Cancer Res.* 2004;64(15):5048–5050.

86. Kamoun A, Idbaih A, Dehais C, et al. Integrated multi-omics analysis of oligodendroglial tumours identifies three subgroups of 1p/19q co-deleted gliomas. *Nat Commun.* 2016;7:11263.

87. Venteicher AS, Tirosh I, Hebert C, et al. Decoupling genetics, lineages, and microenvironment in IDH-mutant gliomas by single-cell RNA-seq. *Science.* 2017;355(6332).

88. Amankulor NM, Kim Y, Arora S, et al. Mutant IDH1 regulates the tumor-associated immune system in gliomas. *Genes Dev.* 2017;31(8):774–786.

89. Kohanbash G, Carrera DA, Shrivastav S, et al. Isocitrate dehydrogenase mutations suppress STAT1 and CD8 + T cell accumulation in gliomas. *J Clin Invest.* 2017;127(4):1425–1437.

90. Lottaz C, Beier D, Meyer K, et al. Transcriptional profiles of CD133 + and CD133- glioblastoma-derived cancer stem cell lines suggest different cells of origin. *Cancer Res.* 2010;70(5):2030–2040.

91. Wu A, Wei J, Kong LY, et al. Glioma cancer stem cells induce immunosuppressive macrophages/microglia. *Neuro Oncol.* 2010;12(11):1113–1125.

92. Zhou W, Ke SQ, Huang Z, et al. Periostin secreted by glioblastoma stem cells recruits M2 tumour-associated macrophages and promotes malignant growth. *Nat Cell Biol.* 2015;17(2):170–182.

93. Otvos B, Silver DJ, Mulkearns-Hubert EE, et al. Cancer stem cell-secreted macrophage migration inhibitory factor stimulates myeloid derived suppressor cell function and facilitates glioblastoma immune evasion. *Stem Cells.* 2016;34(8):2026–2039.

94. Ouzounova M, Lee E, Piranlioglu R, et al. Monocytic and granulocytic myeloid derived suppressor cells differentially regulate spatiotemporal tumour plasticity during metastatic cascade. *Nat Commun.* 2017;8:14979.

95. Ye XZ, Xu SL, Xin YH, et al. Tumor-associated microglia/macrophages enhance the invasion of glioma stem-like cells via TGF-beta1 signaling pathway. *J Immunol.* 2012;189(1):444–453.

96. Alvarado AG, Thiagarajan PS, Mulkearns-Hubert EE, et al. Glioblastoma cancer stem cells evade innate immune suppression of self-renewal through reduced TLR4 expression. *Cell Stem Cell.* 2017;20(4):450–461 [e454].

97. Parsa AT, Waldron JS, Panner A, et al. Loss of tumor suppressor PTEN function increases B7-H1 expression and immunoresistance in glioma. *Nat Med.* 2007;13(1):84–88.

98. Zhang X, Rao A, Sette P, et al. IDH mutant gliomas escape natural killer cell immune surveillance by downregulation of NKG2D ligand expression. *Neuro Oncol.* 2016;18(10):1402–1412.

Chapter 8

Histopathology and molecular biology of oligodendrogliomas

Josephine Volovetz , Yamamoto E. and Prayson R.A.
Department of Anatomic Pathology, Cleveland Clinic Lerner College of Medicine, Cleveland Clinic, Cleveland, OH, United States

Introduction

Oligodendrogliomas have long been recognized as gliomas that have a distinctive morphologic appearance. The precise histogenesis of these tumors is still a matter of some debate. It has also been recognized that some tumors demonstrate overlapping morphologic features between oligodendroglioma and astrocytoma, the so-called mixed glioma or oligoastrocytoma. More recent molecular and genetic work has more clearly defined the oligodendroglioma and has allowed for more ready distinction of oligodendroglioma from astrocytoma. Tumors formerly designated as mixed glioma can usually be classified as either oligodendroglioma or astrocytoma based on molecular phenotype. The culmination of all this work was an updating of the World Health Organization (WHO) Classification of Tumors of the Central Nervous System and recommendation that integrated diagnoses, employing both conventional histologic phenotype and molecular diagnostics, be rendered.[1] The goal of this chapter is to review the typical histologic features of oligodendroglioma, review the salient molecular findings employed in the clinical diagnoses of these tumors, and to review histopathologic differential diagnostic considerations.

Gross appearance

Most oligodendrogliomas appear to arise in white matter regions of the brain, most commonly involving the frontal or parietal lobes. Tumors arising in the cerebellum and spinal cord are more unusual. Similar to their diffuse astrocytoma counterparts, these tumors are infiltrative by nature, which renders them not amenable to surgical resection. Infiltration of tumor into the overlying cortex or adjacent gray matter areas is common, grossly resulting in a blurring or obscuring of the gray-white interface. Bilateral involvement can occur, as tumor cells can infiltrate commissural structures. Tumors can occasionally extend into the leptomeninges. Tumors are typically solid (although occasionally cystic areas may be present), soft, and gray-pink in coloration. Areas of hemorrhage and calcification are fairly common findings.

Histologic appearance

Oligodendrogliomas have distinct histologic features that help to discern them from other tumors of the central nervous system, as well as features they may share with other types of tumors, which will be discussed later. Oligodendrogliomas characteristically are generally composed of rounded cells which have very few cellular processes and scant cytoplasm (Fig. 1).[1,2] The nucleus typically has a small distinct nucleolus. Nuclear pleomorphism is mild in lower grade tumors and can become more pronounced in higher grade neoplasms. Although the cells appear uniformly distributed overall, some tumors contain hyperdense nodules focally. More cellularity can be tolerated in low-grade oligodendrogliomas (WHO grade II) versus their low-grade diffuse astrocytoma counterparts (Fig. 2). An often noticeable but artefactual trait of oligodendroglioma cells is the presence of perinuclear clearing (Fig. 1). Autolytic absorption of water after paraffin embedding can lead to a clearing around the nucleus, producing a perinuclear halo. These halos are not visible in specimens that have been immediately fixed or examined in the context of frozen section consultation. Although common in oligodendrogliomas, this artifact is not a specific finding; it can also be seen in other neoplasms such as astrocytomas, clear cell ependymomas, and neurocytomas (Fig. 3). Frozen section may also introduce some degree of nuclear pleomorphism and nuclear hyperchromasia, which can make distinction of oligodendrogliomas from diffuse astrocytomas difficult in this context (Fig. 4); in most cases, distinction at the time of surgery is not critical to intraoperative patient management.

Oligodendroglioma. https://doi.org/10.1016/B978-0-12-813158-9.00008-6

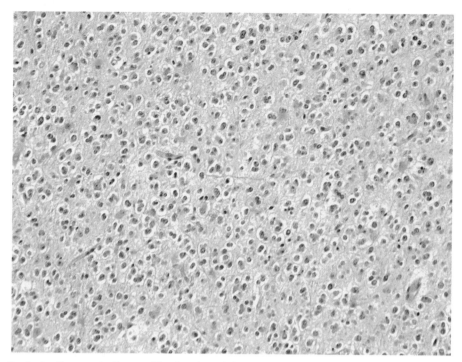

FIG. 1   Low-grade oligodendroglioma marked by a proliferation of rounded cells with scant cytoplasm and pericellular clearing (an artifact of delayed formalin fixation) (hematoxylin and eosin, original magnification 200×).

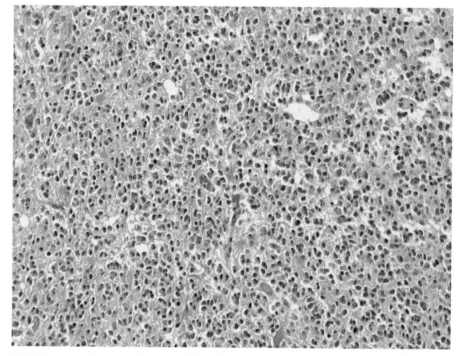

FIG. 2   Low-grade oligodendroglioma with an area of moderate hypercellularity. Areas of increased cellularity may be focally seen in a low-grade tumor (hematoxylin and eosin, original magnification 200×).

In general, oligodendrogliomas are diffusely infiltrative tumors. Most arise in the white matter and then infiltrate into adjacent gray matter structures (Fig. 5). Oligodendroglioma cells tend to localize closer to or satellite around cells (neurons) in the cortical gray matter. Along with their perineuronal affinity, cells can be found aggregated along the vasculature and in the subpial regions (Fig. 6).

Some oligodendrogliomas are associated with increased numbers of cells with increased eosinophilic cytoplasm. These cells resemble gemistocytes that are encountered in some astrocytomas and are sometimes referred to as minigemistocytes,

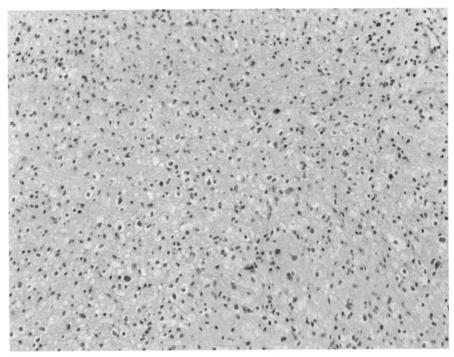

FIG. 3 Although a common feature of oligodendrogliomas, not everything with pericellular clearing is an oligodendroglioma. This tumor is marked by scattered, more pleomorphic cells and represents a diffuse astrocytoma, WHO grade II (hematoxylin and eosin, original magnification 200×).

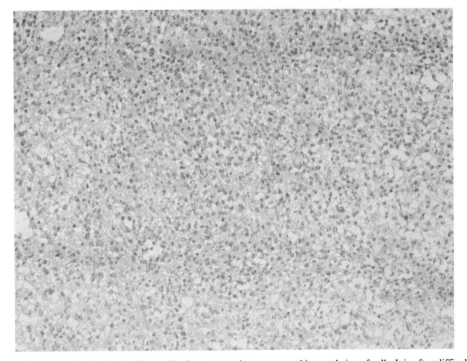

FIG. 4 Frozen section of a low-grade oligodendroglioma showing a somewhat monomorphic population of cells. It is often difficult to definitively distinguish between oligodendroglioma and astrocytoma at the time of frozen section; fortunately, this is not necessary (hematoxylin and eosin, original magnification 200×).

because they are generally a bit smaller in size than true gemistocytes (Fig. 7). Other cells in these lesions appear more polygonal than round with eccentric, eosinophilic cytoplasm; these cells are referred to as gliofibrillary oligodendrocytes.[3] The cytoplasm of these cells appears to contain fibrils that surround the nucleus. Both types of cells are more likely to be found near or around vasculature. It is not unusual to find some scattered reactive astrocytes associated with the tumor, particularly at the infiltrating edge of the lesion. On glial fibrillary acidic protein (GFAP) staining, reactive astrocytes

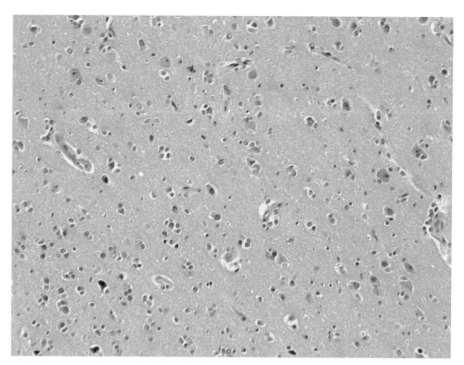

FIG. 5 Low-grade oligodendroglioma with tumor cells infiltrating into the overlying cortex (hematoxylin and eosin, original magnification 200 ×).

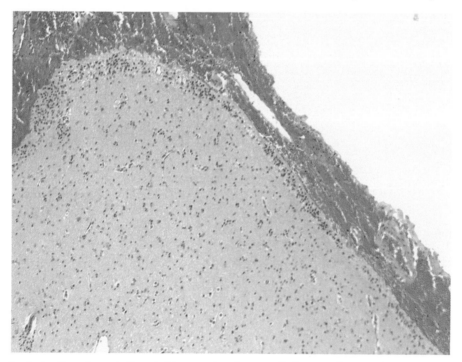

FIG. 6 Low-grade oligodendroglioma demonstrating subpial aggregation of tumor cells (hematoxylin and eosin, original magnification 100 ×).

are marked by a prominent starburst-like pattern of long cytoplasmic processes in contrast to minigemistocytes, which contain fewer and shorter cytoplasmic processes.

Tumoral calcifications can be found in the majority of oligodendrogliomas[4] (Fig. 8). These calcifications are also known as calcospherites and are larger than mineralization that can also be found within blood vessel walls in the tumor. Focal areas of "microcystic" change can also be seen in oligodendrogliomas (Fig. 9). Microcystic degenerative changes are not specific to oligodendrogliomas; they can be found in a variety of other tumors including infiltrating astrocytomas, clear cell ependymomas, and neurocytomas. Microcystic changes are generally not a feature of gliosis.

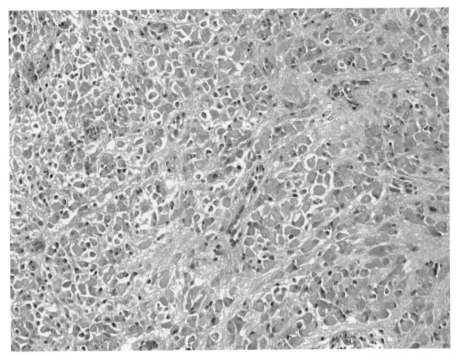

FIG. 7 Prominent minigemistocytes with abundant eosinophilic cytoplasm mark some oligodendrogliomas and should not be confused with gemistocytic astrocytomas (hematoxylin and eosin, original magnification 200×).

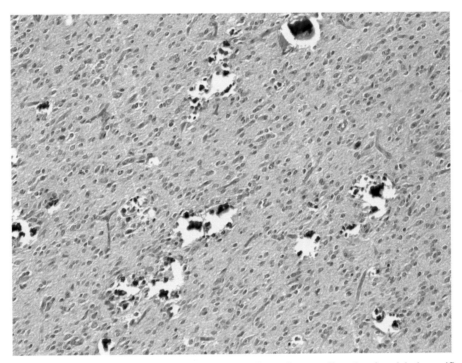

FIG. 8 Microcalcifications are a frequent concomitant finding in oligodendrogliomas (hematoxylin and eosin, original magnification 200×).

In part because of the paucity of cytoplasmic processes in oligodendroglioma tumor cells, the tumor's vascularity is readily discernible and manifests as a prominent arcuate capillary pattern with shorter angled vascular segments crossing each other, forming a branching network (Fig. 10). The pattern has also been referred to as a "chicken wire" pattern. Intra-tumoral hemorrhages are visualized at times. In higher grade oligodendrogliomas, the vascular proliferative changes, similar to those seen in glioblastomas, may be encountered. These changes are marked by a hyperplasia of cellular elements that normally comprise blood vessel walls (i.e., endothelial cells, smooth muscle cells, pericytes, and fibroblasts).

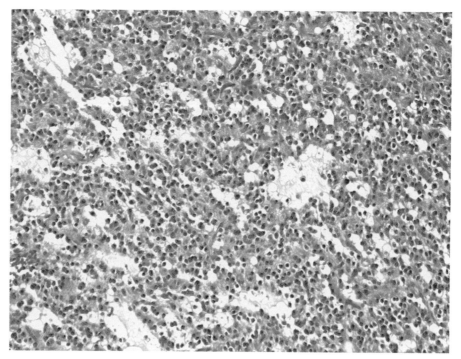

FIG. 9 Focal areas of microcystic change in a low-grade oligodendroglioma (hematoxylin and eosin, original magnification 200×).

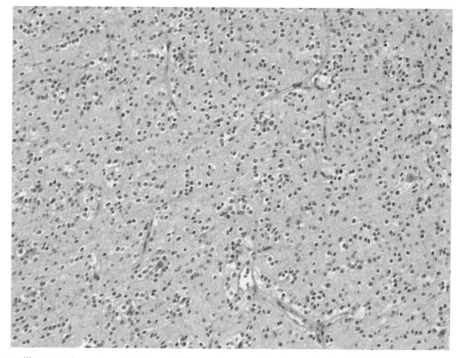

FIG. 10 An arcuate capillary vascular pattern or "chicken wire" vascular pattern in a low-grade oligodendroglioma (hematoxylin and eosin, original magnification 200×).

Grading

The WHO schema for grading brain tumors is the approach that is universally employed. The 2016 WHO guidelines stratify oligodendrogliomas into two grades: grade II oligodendroglioma, IDH-mutant and 1p/19q co-deleted, and grade III anaplastic oligodendroglioma, IDH-mutant and 1p/19q co-deleted.[1,5] On rare occasions, molecular testing for IDH status and 1p/19q deletion may be inconclusive, but the histologic oligodendroglioma features are present; in these cases, a diagnosis of oligodendroglioma, not otherwise specified (NOS), can be made. This subset of patients represents mostly pediatric oligodendrogliomas, of which a majority are low-grade tumors.[6] The new classification system requires integration of both morphologic and molecular findings in defining the entity, a major shift in brain tumor classification from the 2007 WHO guidelines.[7] Low-grade tumors are morphologically marked by mild to moderate hypercellularity and occasional mitotic figures (up to 4 mitotic figures per 10 high-power fields is acceptable). Grade III tumors are generally more uniformly hypercellular with more nuclear pleomorphism (Fig. 11). Higher grade tumors frequently contain vascular proliferative changes and/or necrotic areas (Figs. 12 and 13). Mitotic counts in anaplastic tumors (WHO grade III) are often 5 or more per 10 high-power fields (Fig. 14).

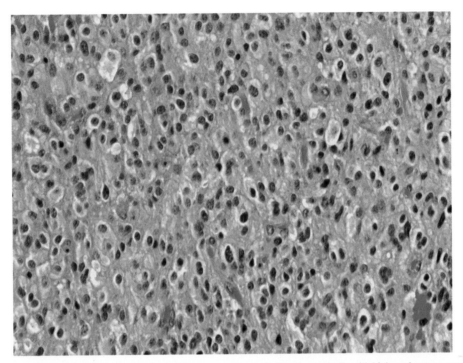

FIG. 11 Nuclear pleomorphism in an anaplastic oligodendroglioma can cause confusion diagnostically with high-grade astrocytomas (hematoxylin and eosin, original magnification 400 ×).

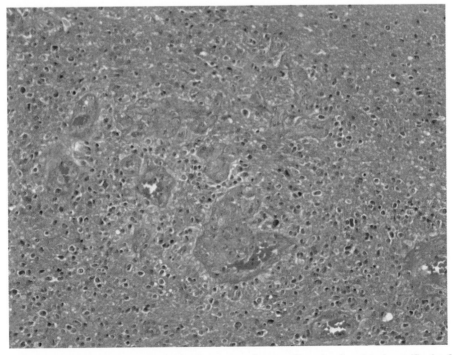

FIG. 12 An anaplastic oligodendroglioma with vascular proliferative changes (hematoxylin and eosin, original magnification 200 ×).

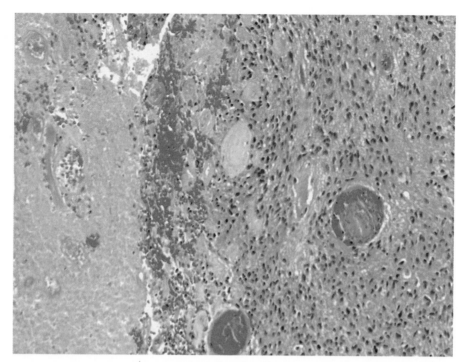

FIG. 13   A geographic area of necrosis in an anaplastic oligodendroglioma (hematoxylin and eosin, original magnification 200×).

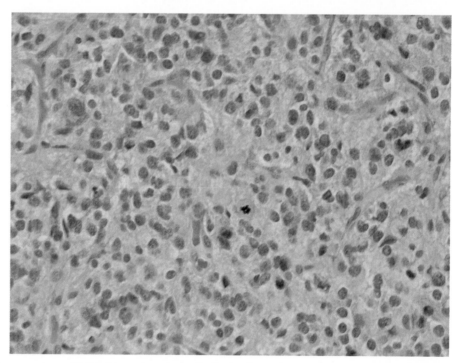

FIG. 14   Mitotic figures can be readily found in most anaplastic oligodendrogliomas (hematoxylin and eosin, original magnification 400×).

Ki-67 (MIB-1), a protein associated with cellular proliferation, can be visualized through staining and has been found to correlate with overall survival and disease-free survival.[8] Grade III tumors generally have a higher Ki-67 labeling index (percent of positive staining tumor cells) versus low-grade tumors. However, there can be considerable tumor heterogeneity within an individual neoplasm with respect to cell proliferation as well as morphology, and so extensive sampling of the tumor both surgically and histologically is important in arriving at an accurate assessment of tumor grade. A precise cutoff with respect to Ki-67 labeling index for distinguishing between grade II and III tumors does not exist. Although grade III lesions are associated with a higher mitotic index, studies investigating the role of the mitotic index in prognosis found that it was associated with outcomes only in tumors that are IDH wild-type but not in IDH-mutant tumors.[9] Now that oligodendrogliomas are defined as IDH-mutant, the prognostic benefit from mitotic rate is less clear.

Clinically relevant molecular markers

The focus of this discussion is on the utilization of molecular markers in clinical practice with respect to (1) establishing a diagnosis of oligodendroglioma given the WHO 2016 guidelines and (2) helping distinguish between histopathologic differential diagnostic considerations. A more detailed discussion of genetics and the molecular aspects of these tumors is addressed elsewhere in this text.

As previously mentioned, IDH-mutant status and 1p/19q co-deletion consisting of whole arm losses are required for a diagnosis of oligodendroglioma or anaplastic oligodendroglioma.[10] Most tumors stain uniformly with antibody against the R132H-mutation of IDH1. If the immunostaining is negative, then DNA genotyping is indicated to test for less common types of IDH1 and IDH2 mutation which are not as clinically significant. The major differential diagnostic consideration usually involves ruling out a diffuse astrocytoma. A subset of astrocytomas may also demonstrate IDH1 or IDH2 mutations; they do not demonstrate chromosome 1p/19q co-deletions. Immunostaining with IDH1-R132H antibody has been found to help distinguish adult oligodendrogliomas from other central nervous system tumors such as neurocytomas, dysembryoplastic neuroepithelial tumors (DNETs), clear cell ependymomas, clear cell meningiomas, primary glioblastomas with oligodendroglioma-like differentiation, and pilocytic astrocytomas with oligodendroglioma-like differentiation.[11]

The MGMT promoter methylation is found in about half of oligodendrogliomas.[12] MGMT methylation is associated with IDH mutations due to increases in methylation by 2-hydroxyglutarate, which is produced as a result of the mutation.[13] Another common mutation seen in oligodendrogliomas is a mutation of the TERT (telomerase reverse transcriptase) promoter.[14,15]

Oligodendrogliomas generally demonstrate retention of ATRX (α-thalassemia/mental retardation syndrome X-linked) and a lack of nuclear p53 accumulation, which is the opposite of what is normally seen with astrocytomas and supports the idea that the 1p/19q co-deletion is mutually exclusive with the TP53 mutation.[16] ATRX and p53 status can be evaluated by immunohistochemistry and can provide quick and useful information in sorting out a differential diagnosis of oligodendroglioma from diffuse astrocytoma.

1p/19q co-deletion has been associated with overexpression of Olig2, an oligodendrocyte lineage gene.[17] Olig2 is a marker of oligodendrogliomas and stains oligodendroglial cell nuclei, but alone, it is not specific to oligodendrogliomas, as it can also stain the nuclei of infiltrating astrocytoma cells.[18,19] Other oligodendrocyte lineage markers that stain positively in oligodendrogliomas include Olig1 and Sox10, although they also lack specificity to oligodendrogliomas, as they are also expressed in other gliomas.[20–22]

MAP2 (microtubule-associated protein 2) is another immunomarker that fairly consistently stains the perinuclear cytoplasm of oligodendrogliomas.[23] Other markers that have been found to stain oligodendrogliomas include S-100 protein, Leu 7, and synaptophysin, but they are all nonspecific.[24–26] Vimentin positivity may be encountered in some anaplastic oligodendrogliomas, but it is not often visualized in lower grade tumors and is nonspecific.[27] GFAP staining is nonspecific and is often seen in higher grade lesions, staining the cytoplasm.[26,28] It is also a marker for minigemistocytes and gliofibrillary oligodendrocytes.

Differential diagnosis

Diffuse astrocytoma

In 2016, WHO guidelines stated that diffuse infiltrating gliomas of either astrocytic or oligodendroglial origin are grouped together due to similarities in growth pattern, behavior, and shared genetic driving mutations.[1,5] However, distinction of the two tumor types is important given the different prognosis and therapeutic approaches to tumor management.

Diffuse astrocytomas are the most common low-grade glioma. Age of presentation overlaps with that of oligodendroglioma and typically occurs in the late 30s. These tumors present throughout the central nervous system, are most frequently found in the cerebral hemispheres. Macroscopically, diffuse astrocytomas and oligodendrogliomas may appear similar and both have nondescript borders due to the infiltrating nature of the tumor. Microscopically, diffuse astrocytoma tumor cells are generally more pleomorphic, with greater nuclear irregularity in comparison to the uniformity seen in oligodendrogliomas (Fig. 15). Diffuse astrocytomas have more angular/elongated and hyperchromatic nuclei and eosinophilic, fibrillary cytoplasm. Diffuse astrocytomas are graded using similar histologic parameters to oligodendrogliomas; however,

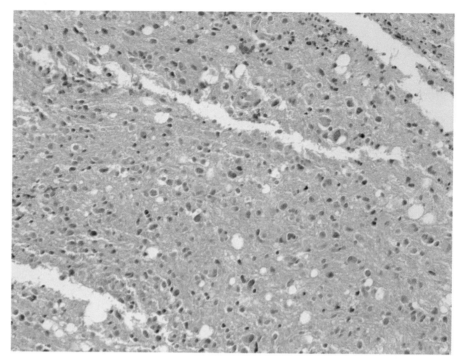

FIG. 15   A diffuse astrocytoma marked by scattered pleomorphic nuclei (hematoxylin and eosin, original magnification 200×).

thresholds for determining grades are a bit different. Diffuse astrocytomas are stratified into grade II, III, and IV lesions.[1] Grade III tumors (anaplastic astrocytomas) are generally more cellular than grade II lesions and mitotic figures are more readily discernible. Necrosis and vascular proliferative changes are only seen in grade IV astrocytoma (glioblastoma) and are the histologic hallmarks of that neoplasm. Grade IV oligodendroglioma does not exist in the current WHO construct.[1]

Mutations in IDH1 and IDH2 are present in a majority of gliomas.[1,29,30] The IDH mutations frequently occur with glioma-associated mutations in TP53 and ATRX, and together are useful in positively identifying astrocytomas.[31] In contrast, the absence of 1p/19q co-deletions and TERT promoter mutations are also suggestive of a tumor of nonoligodendroglial origin.[32] Other molecular findings more commonly encountered in diffuse astrocytomas, particularly higher grade tumors, include loss of heterozygosity on chromosome 19q, epidermal growth factor (EGFR) overexpression or amplification, and O(6)-methylguanine-DNA-methyltransferase (MGMT) methylation abnormalities.[33–35]

Gemistocytic astrocytoma is a diffuse astrocytoma variant characterized by the presence of increased numbers of gemistocytes, astrocytes with abundant eosinophilic cytoplasm, short cytoplasmic processes, and eccentrically placed nuclei.[1,36] While cellular density may vary, densely populated gemistocytes may be confused with the minigemistocytes that are sometimes encountered in oligodendrogliomas. Gemistocytes express glial-type intermediate filaments (GFAP positivity), and aberrant diffuse immunopositivity for p53 is common.

A variant of astrocytoma that presents a particular differential diagnostic challenge is the small cell variant.[1,37] Tumor cells in this variant are marked by a somewhat more homogenous appearance of cells with scant cytoplasm and generally rounded nuclei. Distinction from anaplastic oligodendrogliomas may be difficult based solely on morphology. Nucleoli tend to be less distant in small cell astrocytomas than in anaplastic oligodendrogliomas. In many cases, use of previously outlined molecular markers may be needed to tell the two lesions apart; small cell astrocytomas are frequently not IDH1 or IDH2 mutated, almost always demonstrate EGFR amplification or overexpression, and do not demonstrate co-deletion of chromosome 1p/19q.[1,37]

Oligoastrocytoma/mixed glioma

Oligoastrocytoma was defined as a diffusely infiltrative glial neoplasm exhibiting histologic features of both oligodendroglioma and astrocytoma. It has been long recognized that astrocytomas and oligodendrogliomas may contain small areas or intermixed cells morphologically resembling the other tumor type. As such, determining a diagnosis of oligodendroglioma, astrocytoma, or oligoastrocytoma was subject to high inter-rater variability. The identification of unique genetic signatures for oligodendrogliomas and astrocytomas led to the realization that nearly all neoplasms then fit into either an astrocytoma or oligodendroglioma molecular phenotype.[1] With that said, there are few reports of neoplasms with true mixed histological and genetic features.[38] In the new classification system, the term oligoastrocytoma/mixed glioma was removed in favor of genetically identified oligodendrogliomas and astrocytomas, and rare dual genotype oligoastrocytomas are designated as NOS.[1]

Pilocytic astrocytoma

Pilocytic astrocytomas are an astrocytoma variant with specific clinical, genetic, and histologic characteristics. It is the most frequent primary brain tumor in children and adolescents, although it may also present in adulthood. Pilocytic astrocytomas predominantly occur with the highest frequency in the cerebellum; other common sites of origin include supratentorial areas, optic nerve, hypothalamus, and brainstem in children.[39] Pilocytic astrocytomas are typically well-demarcated tumors; on imaging, they classically appear as cystic tumors with an enhancing mural nodule. The tumor is slow-growing and has a favorable prognosis, with a 10-year survival of over 90% with surgical resection.[40]

Most pilocytic astrocytomas are composed of elongated cells, resembling atypical astrocytes. Tumors demonstrate low to moderate cellularity with areas of dense fibrillation accompanied frequently by brightly eosinophilic Rosenthal fibers juxtaposed with areas in which cells are more loosely arranged with microcystic spaces. Often in these looser areas, eosinophilic granular bodies may be evident (Fig. 16). Microvascular proliferative changes, vascular sclerosis, and infarct-like necrosis may be seen. Mitotic figures are rare. Occasionally, tumors may have areas in which cell nuclei are more rounded, resembling oligodendroglioma (Fig. 17).

Fusion genes between KIAA1549 and BRAF leading to constitutively active BRAF is the most frequent genetic alteration in the majority of pilocytic astrocytomas, especially those in the cerebellum.[41] In adults, lack of IDH1/2 mutations, absence of 1p/19q co-deletion, and/or presence of BRAF fusion mutation help to identify pilocytic astrocytoma. However, it should be noted that childhood oligodendrogliomas do not present with IDH1 and IDH2 mutations or 1p/19q co-deletions.[40]

Papillary glioneuronal tumor

Papillary glioneuronal tumors are rare, low-grade tumors primarily affecting young adults.[42] Typically, they are present in the cerebral hemispheres, often in the temporal lobes. As the name suggests, this neoplasm has a biphasic phenotype with components of neuronal and glial cells.[42] GFAP positive glial cells are arranged in a papillary architecture with hyalinized blood vessels at the center of the papillae. Between the glial cells lining the papillae is the second population of neurocytic cells, morphologically resembling oligodendroglial cells that stain strongly for synaptophysin, neuron specific enolase,

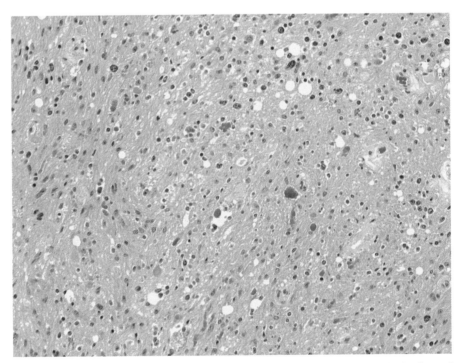

FIG. 16 Pilocytic astrocytoma with Rosenthal fibers and eosinophilic granular bodies, features not seen in oligodendrogliomas (hematoxylin and eosin, original magnification 200×).

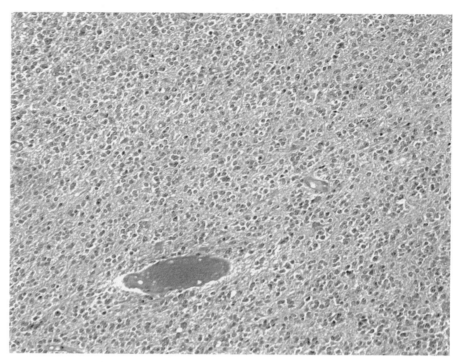

FIG. 17 Pilocytic astrocytomas may sometimes be composed of cells with rounded nuclei, and focally resemble oligodendroglioma (hematoxylin and eosin, original magnification 200 ×).

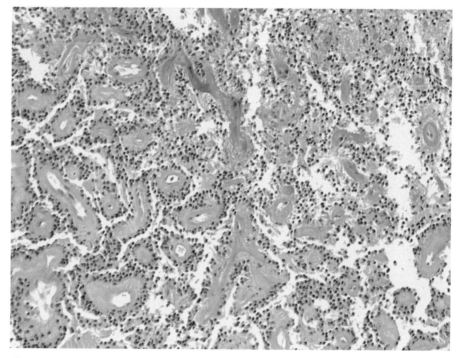

FIG. 18 A papillary glioneuronal tumor characterized by a papillary lined by glial cells and intervening rounded cells, resembling oligodendroglioma, that are neurocytic in nature (hematoxylin and eosin, original magnification 200 ×).

NeuN, and class III beta-tubulin (Fig. 18). It should be noted that some oligodendrogliomas do express these neuronal markers, and definitive diagnosis should not be based on the presence or absence of these neuronal markers. Papillary glio-neuronal tumors do not have the 1p/19q co-deletion as seen in oligodendroglioma.[43] In addition, recent studies identified a SLC44A1-PRKCA fusion in these tumors, resulting in aberrant MAPK signaling as a specific marker of these tumors. This has been evaluated in a low number of cases due to the rarity of the tumor; future studies may reveal this as a promising marker.[44]

FIG. 19 A cortical based, multinodular architectural pattern is characteristic of the dysembryoplastic neuroepithelial tumor (hematoxylin and eosin, original magnification 100×).

Dysembryoplastic neuroepithelial tumor

The DNET are rare glioneuronal neoplasms that are typically identified in children and adolescents suffering from medically intractable partial complex seizures.[45] Although they can arise essentially anywhere in the brain, the most common site of origin is the temporal lobe.[45] The DNETs are well-circumscribed lesions. The tumors typically have a multinodular architectural pattern with most of the nodules being situated in the cortex (Fig. 19). This is in contrast to oligodendrogliomas, which typically arise in the white matter and are not multinodular. Microscopically, nodules are composed of rounded cells resembling oligodendrocytes intermixed with normal appearing neurons, often arranged against a microcystic background (Fig. 20). Areas of focal cortical dysplasia may be identified in the cortex adjacent to the neoplasm in nearly two-thirds of cases.[46,47] The histological similarity may make it difficult on small biopsies or subtotal resections to distinguish an oligodendroglioma from a DNET microscopically due to inability to identify more readily associated macroscopic features of DNET. In these cases, molecular testing for IDH-1 immunoreactivity and 1p/19q co-deletion can be helpful, if positive, to confirm oligodendroglioma.[48,49]

Ganglioglioma

Gangliogliomas, similar to DNETs, are glioneuronal tumors encountered predominantly in patients with medically intractable epilepsy and most commonly in the temporal lobe.[50] Macroscopically, tumors are typically well circumscribed and may be solid, cystic, or have combined solid and cystic components. Microscopically, they are composed of dysplastic neuronal cells intermixed with a component resembling a low-grade glioma. Although the glioma component in the majority of cases resembles an astrocytoma, focal area may have the appearance of an oligodendroglioma. Glial cells are often positive for GFAP, S-100, vimentin, and neurons are immunoreactive for synaptophysin, class III tubulin, neurofilament protein, and chromogranin A.[50] Calcifications, similar to oligodendrogliomas, are common. Other features include Rosenthal fibers, eosinophilic granular bodies, and perivascular chronic inflammation. Adjacent cortex is often marked by focal cortical dysplasia. Approximately 20% of gangliogliomas contain a BRAF V600E mutation, which has been shown to predominantly localize to the neuronal rather than glial tumor component.[51] The identification of an IDH mutation suggests an alternate diagnosis of oligodendroglioma or diffuse astrocytoma.[52]

FIG. 20   Dysembryoplastic neuroepithelial tumors are classically composed of oligodendroglial-like cells with intermixed normal appearing neurons arranged against a microcystic background (hematoxylin and eosin, original magnification 200×).

FIG. 21   A lateral ventricular central neurocytoma marked by rounded cells with a salt and pepper nuclear chromatin pattern and scant cytoplasm (hematoxylin and eosin, original magnification 200×).

Central neurocytoma

Central neurocytomas are low-grade (WHO grade II) neuronal tumors located in the ventricular system. Nearly 50% of central neurocytomas arise in the lateral ventricles.[53] Similar tumors encountered outside of the ventricles are referred to as extraventricular neurocytomas. At first glance, neurocytomas mimic oligodendrogliomas with small uniform cells with rounded nuclei and scant cytoplasm demonstrating the "fried egg" appearance of the pericellular halo due to delayed formalin fixation (Fig. 21).

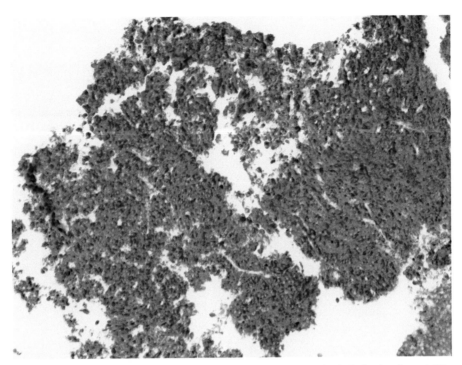

FIG. 22 A central neurocytoma demonstrating diffuse positive staining with antibody to synaptophysin, indicative of neural differentiation (hematoxylin and eosin, original magnification 200×).

There is a subtle "salt and pepper" nuclear chromatin pattern, which is not typically a feature of oligodendrogliomas.[54] Fibrillary regions within cellular areas and presence of ill-defined rosettes argue in favor of neurocytoma; whereas, parenchymal infiltration with perineuronal satellitosis suggests oligodendroglioma.[55] In addition, neurocytomas appropriately demonstrate immunoreactivity with neuronal markers including neuron-specific enolase and synaptophysin (Fig. 22); however, central neurocytomas can also focally stain for GFAP.[53] Neurocytomas have not been shown to harbor IDH mutations or 1p/19q co-deletions.[49,56,57]

Clear cell ependymoma

Ependymomas are slow growing tumors that occur in both children and adults. They arise throughout the neuroaxis, but primarily in the ventricular system and spinal canal. Ependymomas have a sharply delineated border and are noninfiltrating. Ependymal cells lining the ventricular system are of variable morphology, and different morphologic patterns have given rise to several subtypes of ependymoma.[58] The rare clear cell ependymoma displays crowded, uniform cells with rounded nuclei (Fig. 23). Cells contain a central nucleolus and cytoplasmic clearing that may be suggestive of oligodendroglioma. Ependymomas characteristically have specialized structures including perivascular pseudorosettes and true rosettes.[59] When such structures cannot be identified by light microscopy, electron microscopy may be helpful to identify features of ependymal cells including tight junction complexes, microvilli and cilia at the apical surfaces, and ciliary body attachments (blepharoplasts).[59] Clear cell ependymomas have not been identified as having IDH1 mutations.[48]

Rosette-forming glioneuronal tumor of the fourth ventricle

Rosette-forming glioneuronal tumors of the fourth ventricle (RGNT) are rare, slow growing tumors containing both neurocytic and glial components.[60] These tumors typically present in the third to fourth decades of life. The majority of RGNT arise in the midline posterior fossa or near the fourth ventricle. However, anecdotal cases have also been reported in the brainstem, cerebellum, pineal gland, optic chiasm, and septum pellucidum.[61] These tumors are biphasic, demonstrating both neurocytic and glial regions. The neurocytic component is composed of cords of small round cells with eosinophilic cytoplasm which are organized into pseudorosettes around an eosinophilic center (Fig. 24). Rosettes are strongly immunoreactive for synaptophysin. The glial component stains for GFAP and may resemble pilocytic astrocytoma.

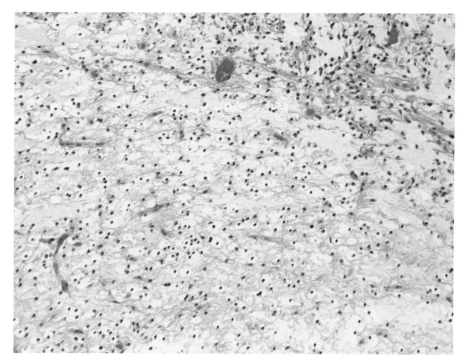

FIG. 23 An intraventricular clear cell ependymoma marked by areas resembling oligodendroglioma (hematoxylin and eosin, original magnification 200 ×).

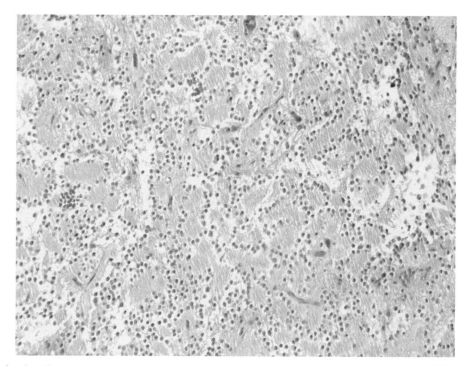

FIG. 24 A rosette forming glioneuronal tumor of the fourth ventricle marked by rounded cells arranged around an eosinophilic core forming pseudoro-settes (hematoxylin and eosin, original magnification 200 ×).

Inflammatory related lesions

Several inflammatory lesions may mimic oligodendrogliomas. These lesions are often rich with macrophages that resemble dysplastic oligodendrocytes. In contrast to the clear perinuclear halo seen in oligodendrogliomas, macrophages have a foamy or granular appearance (Fig. 25). The identification of macrophages with CD68 or KiM1P may aid in the proper

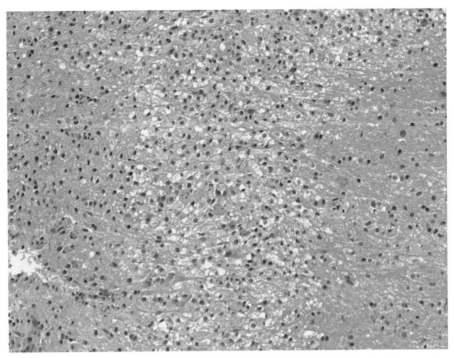

FIG. 25 Macrophages in a demyelinating lesion, seen here, may occasionally, particularly on frozen section, be misinterpreted as oligodendroglial cells (hematoxylin and eosin, original magnification 200 ×).

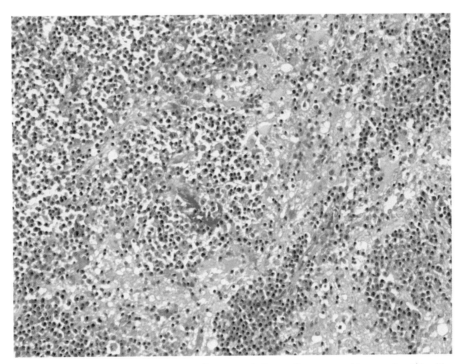

FIG. 26 A primary lymphoma involving the central nervous system is composed of generally monomorphic appearing cells with scant cytoplasm, resembling oligodendroglioma (hematoxylin and eosin, original magnification 200 ×).

diagnosis.[62] Primary lymphomas of the central nervous system can also resemble oligodendrogliomas on low-power visualization due to the generally monomorphic appearance of the lymphomatous cells (Fig. 26). The vast majority of primary central nervous system lymphomas are B cell lymphomas (diffuse large B cell type) and stain with immunomarkers which target B cells, such as CD19 and CD20, which definitively distinguishes lymphoma from oligodendroglioma.[63]

References

1. Louis DN, Ohgak IH, Wiestler OD, et al. *WHO Classification of Tumours of the Central Nervous System, Revised.* 4th ed. Lyon: IARC Press; 2016.
2. Burger PC. Use of cytological preparations in the frozen section diagnosis of central nervous system neoplasia. *Am J Surg Pathol.* 1985;9(5):344–354.
3. Herpers MJHM, Budka H. Glial fibrillary acidic protein (GFAP) in oligodendroglial tumors: gliofibrillary oligodendroglioma and transitional oligoastrocytoma as subtypes of oligodendroglioma. *Acta Neuropathol.* 1984;64(4):265–272.
4. Chen WY, Liu HC, Lam PC. Calcification in oligodendroglioma. *Zhonghua Yi Xue Za Zhi (Taipei).* 1990;45(3):143–146.
5. Louis DN, Perry A, Reifenberger G, et al. The 2016 World Health Organization classification of tumors of the central nervous system: a summary. *Acta Neuropathol.* 2016;131(6):803–820.
6. Rodriguez FJ, Tihan T, Lin D, et al. Clinicopathologic features of pediatric oligodendrogliomas: a series of 50 patients. *Am J Surg Pathol.* 2014;38(8): 1058–1070.
7. Louis DN, Ohgaki H, Wiestler OD, et al. The 2007 WHO classification of tumours of the central nervous system. *Acta Neuropathol.* 2007;114(2): 97–109.
8. Coleman KE, Brat DJ, Cotsonis GA, Lawson D, Cohen C. Proliferation (MIB-1 expression) in oligodendrogliomas. *Appl Immunohistochem Mol Morphol.* 2006;14(1):109–114.
9. Olar A, Wani KM, Alfaro-Munoz KD, et al. IDH mutation status and role of WHO grade and mitotic index in overall survival in grade II-III diffuse gliomas. *Acta Neuropathol.* 2015;129(4):585–596.
10. Hu N, Richards R, Jensen R. Role of chromosomal 1p/19q co-deletion on the prognosis of oligodendrogliomas: a systematic review and meta-analysis. *Interdiscip Neurosurg.* 2016;5:58–63.
11. Boots-Sprenger SHE, Sijben A, Rijntjes J, et al. Significance of complete 1p/19q co-deletion, IDH1 mutation and MGMT promoter methylation in gliomas: use with caution. *Mod Pathol.* 2013;26(7):922–929.
12. Leu S, von Felten S, Frank S, et al. IDH/MGMT-driven molecular classification of low-grade glioma is a strong predictor for long-term survival. *Neuro-Oncology.* 2013;15(4):469–479.
13. Cahill DP, Louis DN, Cairncross JG. Molecular background of oligodendroglioma: 1p/19q, IDH, TERT, CIC and FUBP1. *CNS Oncol.* 2015;4(5): 287–294.
14. Arita H, Narita Y, Fukushima S, et al. Upregulating mutations in the TERT promoter commonly occur in adult malignant gliomas and are strongly associated with total 1p19q loss. *Acta Neuropathol.* 2013;126(2):267–276.
15. Sahm F, Reuss D, Koelsche C, et al. Farewell to oligoastrocytoma: in situ molecular genetics favor classification as either oligodendroglioma or astrocytoma. *Acta Neuropathol.* 2014;128(4):551–559.
16. Capper D, Reuss D, Schittenhelm J, et al. Mutation-specific IDH1 antibody differentiates oligodendrogliomas and oligoastrocytomas from other brain tumors with oligodendroglioma-like morphology. *Acta Neuropathol.* 2011;121(2):241–252.
17. Durand KS, Guillaudeau A, Weinbreck N, et al. 1p19q LOH patterns and expression of p53 and Olig2 in gliomas: relation with histological types and prognosis. *Mod Pathol.* 2010;23(4):619–628.
18. Marie Y, Sanson M, Mokhtari K, et al. OLIG2 as a specific marker of oligodendroglial tumour cells. *Lancet.* 2001;358(9278):298–300.
19. Bouvier C, Bartoli C, Aguirre-Cruz L, et al. Shared *oligodendrocyte lineage* gene expression in gliomas and oligodendrocyte progenitor cells. *J Neurosurg.* 2003;99(2):344–350.
20. Bannykh SI, Stolt CC, Kim J, Perry A, Wegner M. Oligodendroglial-specific transcriptional factor SOX10 is ubiquitously expressed in human gliomas. *J Neuro-Oncol.* 2006;76(2):115–127.
21. Azzarelli B, Miravalle L, Vidal R. Immunolocalization of the oligodendrocyte transcription factor 1 (Olig1) in brain tumors. *J Neuropathol Exp Neurol.* 2004;63(2):170–179.
22. Aguirre-Cruz L, Mokhtari K, Hoang-Xuan K, et al. Analysis of the bHLH transcription factors Olig1 and Olig2 in brain tumors. *J Neuro-Oncol.* 2004;67(3):265–271.
23. Blümcke I, Becker AJ, Normann S, et al. Distinct expression pattern of microtubule-associated protein-2 in human oligodendrogliomas and glial precursor cells. *J Neuropathol Exp Neurol.* 2001;60(10):984–993.
24. Motoi M, Yoshino T, Hayashi K, Nose S, Horie Y, Ogawa K. Immunohistochemical studies on human brain tumors using anti-Leu 7 monoclonal antibody in paraffin-embedded specimens. *Acta Neuropathol.* 1985;66(1):75–77.
25. Nakopoulou L, Kerezoudi E, Thomaides T, Litsios B. An immunocytochemical comparison of glial fibrillary acidic protein, S-100p and vimentin in human glial tumors. *J Neuro-Oncol.* 1990;8(1):33–40.
26. Dehghani F, Schachenmayr W, Laun A, Korf HW. Prognostic implication of histopathological, immunohistochemical and clinical features of oligodendrogliomas: a study of 89 cases. *Acta Neuropathol.* 1998;95(5):493–504.
27. Kubo O, Tajika Y, Toyama T, et al. Clinicopathological study of oligodendroglioma with special reference to immunohistochemical investigation. *No Shinkei Geka.* 1988;16(9):1029–1035.
28. Jagadha V, Halliday WC, Becker LE. Glial fibrillary acidic protein (GFAP) in oligodendrogliomas: a reflection of transient GFAP expression by immature oligodendroglia. *Can J Neurol Sci.* 1986;13(4):307–317.
29. Pisapia DJ. The updated World Health Organization glioma classification. Cellular and molecular origins of adult infiltrating gliomas. *Arch Pathol Lab Med.* 2017;141:1633–1645.
30. Parsons DW, Jones S, Zhang X, et al. An integrated genomic analysis of human glioblastoma multiforme. *Science.* 2008;321(5897):1807–1812.
31. Liu X-Y, Gerges N, Korshunov A, et al. Frequent ATRX mutations and loss of expression in adult diffuse astrocytic tumors carrying IDH1/IDH2 and TP53 mutations. *Acta Neuropathol.* 2012;124(5):615–625.

32. Ichimura K, Yoshitaka Narita B, et al. Diffusely infiltrating astrocytomas: pathology, molecular mechanisms and markers. *Acta Neuropathol.* 2015;129:789–808.

33. Malley DS, Hamoudi RA, Kocialkowski S, et al. A distinct region of the MGMT CpG island critical for transcriptional regulation is preferentially methylated in glioblastoma cells and xenografts. *Acta Neuropathol.* 2011;121(5):651–661.

34. Bady P, Sciuscio D, Diserens A-C, et al. MGMT methylation analysis of glioblastoma on the Infinium methylation BeadChip identifies two distinct CpG regions associated with gene silencing and outcome, yielding a prediction model for comparisons across datasets, tumor grades, and CIMP-status. *Acta Neuropathol.* 2012;124(4):547–560.

35. Wesseling P, van den Bent M, Perry A. Oligodendroglioma: pathology, molecular mechanisms and markers. *Acta Neuropathol.* 2015;129(6):809–827.

36. Krouwer HGJ, Davis RL, Silver P, et al. Gemistocytic astrocytomas: a reappraisal. *J Neurosurg.* 1991;74(3):399–406.

37. Perry A, Aldape KD, George DH, et al. Small cell astrocytoma: an aggressive variant that is clinicopathologically and genetically distinct from anaplastic oligodendroglioma. *Cancer.* 2004;101(10):2318–2326.

38. Huse JT, Diamond EL, Wang L, et al. Mixed glioma with molecular features of composite oligodendroglioma and astrocytoma: a true "oligoastrocytoma"? *Acta Neuropathol.* 2015;129(1):151–153.

39. Burkhard C, Di Patre P-L, Schüler D, et al. A population-based study of the incidence and survival rates in patients with pilocytic astrocytoma. *J Neurosurg.* 2003;98(6):1170–1174.

40. Collins VP, Jones DTW, Giannini C. Pilocytic astrocytoma: pathology, molecular mechanisms and markers. *Acta Neuropathol.* 2015;129:775–788.

41. Jacob K, Albrecht S, Sollier C, et al. Duplication of 7q34 is specific to juvenile pilocytic astrocytomas and a hallmark of cerebellar and optic pathway tumours. *Br J Cancer.* 2009;101(4):722–733.

42. Demetriades AK, Al Hyassat S, Al-Sarraj S, et al. Papillary glioneuronal tumour: a review of the literature with two illustrative cases. *Br J Neurosurg.* 2013;27(3):401–404.

43. Myung JK, Byeon S, Kim B, et al. Papillary glioneuronal tumors. *Am J Surg Pathol.* 2011;35(12):1794–1805.

44. Pages M, Lacroix L, Tauziede-Espariat A, et al. Papillary glioneuronal tumors: histological and molecular characteristics and diagnostic value of SLC44A1-PRKCA fusion. *Acta Neuropathol Commun.* 2015;3:85.

45. Daumas-Duport C, Scheithauer BW, Chodkiewicz JP, et al. Dysembryoplastic neuroepithelial tumor: a surgically curable tumor of young patients with intractable partial seizures. Report of thirty-nine cases. *Neurosurgery.* 1988;23(5):545–556.

46. Zhang J-G, Hu W-Z, Zhao R-J, et al. Dysembryoplastic neuroepithelial tumor: a clinical, neuroradiological, and pathological study of 15 cases. *J Child Neurol.* 2014;29(11):1441–1447.

47. Suh Y-L. Dysembryoplastic neuroepithelial tumors. *J Pathol Transl Med.* 2015;49(6):438–449.

48. Capper D, Reuss D, Schittenhelm J, et al. Mutation-specific IDH1 antibody differentiates oligodendrogliomas and oligoastrocytomas from other brain tumors with oligodendroglioma-like morphology. *Acta Neuropathol.* 2011;121(2):241–252.

49. Fujisawa H, Marukawa K, Hasegawa M, et al. Genetic differences between neurocytoma and dysembryoplastic neuroepithelial tumor and oligodendroglial tumors. *J Neurosurg.* 2002;97(6):1350–1355.

50. Hirose T, Scheithauer BW, Lopes MB, et al. Ganglioglioma: an ultrastructural and immunohistochemical study. *Cancer.* 1997;79(5):989–1003.

51. Koelsche C, Wöhrer A, Jeibmann A, et al. Mutant BRAF V600E protein in ganglioglioma is predominantly expressed by neuronal tumor cells. *Acta Neuropathol.* 2013;125(6):891–900.

52. Horbinski C, Kofler J, Yeaney G, et al. Isocitrate dehydrogenase 1 analysis differentiates gangliogliomas from infiltrative gliomas. *Brain Pathol.* 2011;21(5):564–574.

53. Sharma MC, Deb P, Sharma S, et al. Neurocytoma: a comprehensive review. *Neurosurg Rev.* 2006;29(4):270–285.

54. Chen C-L, Shen C-C, Wang J, et al. Central neurocytoma: a clinical, radiological and pathological study of nine cases. *Clin Neurol Neurosurg.* 2008;110(2):129–136.

55. Sharma S, Deb P. Intraoperative neurocytology of primary central nervous system neoplasia: a simplified and practical diagnostic approach. *J Cytol.* 2011;28(4):147–158.

56. Myung JK, Cho HJ, Park C-K, et al. Clinicopathological and genetic characteristics of extraventricular neurocytomas. *Neuropathology.* 2013;33(2):111–121.

57. Tong CY, Ng HK, Pang JC, et al. Central neurocytomas are genetically distinct from oligodendrogliomas and neuroblastomas. *Histopathology.* 2000;37(2):160–165.

58. Pajtler KW, Witt H, Sill M, et al. Molecular classification of ependymal tumors across all CNS compartments, histopathological grades, and age groups. *Cancer Cell.* 2015;27(5):728–743.

59. Min K-W, Scheithauer BW. Clear cell ependymoma: a mimic of oligodendroglioma: clinicopathologic and ultrastructural considerations. *Am J Surg Pathol.* 1997;820–826.

60. Zhang J, Babu R, Mclendon RE, et al. A comprehensive analysis of 41 patients with rosette-forming glioneuronal tumors of the fourth ventricle. *J Clin Neurosci.* 2013;20:335–341.

61. Xiong J, Liu Y, Chu S-G, et al. Rosette-forming glioneuronal tumor of the septum pellucidum with extension to the supratentorial ventricles: rare case with genetic analysis. *Neuropathology.* 2012;32:301–305.

62. Kuhlmann T, Lassmann H, Brück W. Diagnosis of inflammatory demyelination in biopsy specimens: a practical approach. *Acta Neuropathol.* 2008;115(3):275–287.

63. Giannini C, Dogan A, Salomão DR. CNS lymphoma: a practical diagnostic approach. *J Neuropathol Exp Neurol.* 2014;73(6):478–494.

Chapter 9

The role of biomarkers in the diagnosis and treatment of oligodendrogliomas

Laura E. Donovan* and Andrew B. Lassman[†]

*Neuro-Oncology Division, Department of Neurology, Columbia University Irving Medical Center, New York, NY, United States; [†]Neuro-Oncology Division, Department of Neurology, Herbert Irving Comprehensive Cancer Center, Columbia University Irving Medical Center, New York, NY, United States

Introduction

Oligodendrogliomas are rare primary brain tumors, accounting for less than 10% of glioma diagnoses annually.[1] They are classified as low grade (grade II) or anaplastic (grade III) based on the number of mitotic figures, degree of nuclear atypia, and the presence of other malignant features. Regardless of grade, oligodendrogliomas tend to be less aggressive, more sensitive to chemotherapy, and are associated with nearly double the median overall survival compared to their astrocytic counterparts.[2]

Biology

Prior to 2016, oligodendrogliomas were defined histologically as diffusely infiltrating tumors composed of monomorphic cells similar to oligodendrocytes with perinuclear halos and thin arborizing capillaries, giving them the classic "fried egg" appearance with the so-called "chicken wire vasculature" on microscopy. The majority are located in the frontal lobes and microcalcifications are common. A third category of gliomas, oligoastrocytomas, was reserved for tumors with mixed oligodendrocytic and astrocytic histology. This approach to diagnosis resulted in considerable heterogeneity within glioma subgroups, with studies demonstrating substantial interobserver variability regarding the typing and grading of gliomas.[3]

In an effort to refine the diagnostic criteria, the most recent World Health Organization classification guidelines for brain tumors moves away from pure histopathology based on light microscopy to an integrated molecular approach to the classification of brain tumors. The diagnosis of oligodendroglioma is now reserved for tumors with mutations in *isocitrate dehydrogenase (IDH) 1* or *IDH2* genes with concurrent losses in chromosome arms 1p and 19q (1p/19q co-deletion) (Fig. 1). Under the new classification scheme, the diagnosis of oligoastrocytoma is strongly discouraged; similarly, glioblastoma with oligodendroglial differentiation has been removed.[4]

This new classification scheme emerged from over a decade of research into the biology of gliomagenesis. In 2008, Parsons, et al. identified mutations in *IDH* in a small subset of human glioblastomas.[5] These mutations were associated with significantly improved overall survival compared to *IDH* wild-type tumors and appeared to be the most common in secondary glioblastomas arising from lower grade tumors.[5] Subsequent analysis of grade I–IV gliomas demonstrated *IDH 1* or *2* mutations were present in over 80% of grade II gliomas (all subtypes) and over 80% of grade III oligodendroglial tumors.[6] Again, *IDH* mutations correlated with longer overall survival. Based on analysis of co-occurring mutations, it was postulated that *IDH* mutations are an early event in gliomagenesis, and some tumors also then lose 1p/19q and become oligodendrogliomas, whereas others lose the *alpha thalassemia/mental retardation—X linked (ATRX)* gene and become astrocytomas.[7]

Molecular studies in the early 1990s first reported that loss of heterozygosity (LOH) of chromosome arms 1p and 19q occurred with high frequency in oligodendrogliomas compared with mixed gliomas.[8] Follow-up studies confirmed this finding with 1p/19q co-deletion reported in 74% of well-differentiated oligodendrogliomas and 83% of anaplastic oligodendrogliomas, compared with only 38% of tumors then classified as "oligoastrocytomas." Conversely, mutations in TP53 or LOH on chromosome 17p rarely occurred in oligodendrogliomas and were much more commonly associated with astrocytic histology, suggesting a clear genetic distinction between these tumors.[9]

Oligodendroglioma. https://doi.org/10.1016/B978-0-12-813158-9.00009-8

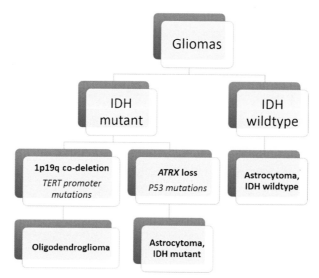

FIG. 1 Current molecular classification of gliomas.

Co-deletion of 1p and 19q results from an unbalanced translocation leaving a single arm of chromosome 1p and 19q.[10] Sequencing of the coding regions of 1p and 19q identified mutations in two potential tumor suppressor genes: Far upstream element (FUE) binding protein 1 (*FUBP1*) on chromosome arm 1p and *CIC* (homolog of the *Drosophila capicua* gene) on chromosome arm 19p. In Drosophila, *CIC* encodes a transcription factor (cic) regulated by the receptor tyrosine kinase (RTK) pathway. The activation of RTK signaling through mitogen-associated protein kinases (MAPK) leads to the degradation of cic protein and activation of previously suppressed genes. *CIC* is highly conserved and the human correlate is thought to function similarly. *FUBP1* regulates expression of *MYC,* and inactivating mutations in *FUBP1* are believed to result in the activation of *MYC,* a known oncogene.[11,12]

Retrospective analyses and subsequent prospective clinical trials confirmed that the presence of 1p/19q co-deletion in gliomas was both prognostic for longer survival regardless of treatment, and specifically predictive of response to alkylator chemotherapy.[13–17] Less commonly, 1p and 19q co-deletion is seen in other tumor types including glioblastoma, where it does not always carry a favorable prognosis.[18, 19] However, in combination with *IDH* mutations, this genetic signature is most consistent both histologically and clinically with oligodendrogliomas, prompting the shift in WHO classification criteria.[20–22] The vast majority of these tumors also harbor mutations in *FUBP1, CIC,* and the *Telomerase reverse transcriptase (TERT)* promoter gene (discussed below), but these molecular alterations are not required for diagnosis.[12, 23, 24]

Maintenance of telomeres is critical for cancer cell survival by preventing cellular senescence or death. This can occur by increasing levels of telomerase, an enzyme that directly maintains telomeres, or through a telomerase-independent mechanism, known as alternative telomerase lengthening (ALT). The ALT is seen in cancers harboring mutations in *ATRX* or *death-domain associated protein (DAXX)*, while mutations in the *TERT* promoter result in increased telomerase expression.[25, 26] *ATRX* and *TERT* mutations are mutually exclusive.[7]

Both *ATRX* and *TERT* promoter mutations are found in gliomas. In *IDH* mutant tumors specifically, mutations in *ATRX* result in the development of astrocytic tumors while *TERT* promoter mutations are almost exclusively seen in combination with 1p/19q co-deletion, resulting in oligodendroglial tumors. *TERT* promoter mutations are also commonly seen in IDH wild-type glioblastomas.[25–27]

Observational studies suggest that TERT promoter mutations may hold some prognostic significance in gliomas.[27] In a 2017 meta-analysis of 28 studies including over 11,500 glioma cases, *TERT* promoter mutations were associated with worse overall and progression-free survival; however, there was substantial heterogeneity within the analyses associated with both tumor grade and *IDH* mutation status. *TERT* promoter mutations only held a negative prognostic value for patients with glioblastoma compared to those with lower grade (II/III) tumors. In patients with grade II or III tumors, *TERT* promoter mutations conferred a poorer prognosis in *IDH* wild-type tumors, but were associated with improved outcomes in patients with *IDH* mutant tumors.[28] Within the subset of patients with molecularly defined oligodendrogliomas, the impact of *TERT* promoter mutations remains unclear, with some studies showing no impact on survival and others reporting improved overall survival in this group as well.[27, 29]

Treatment

There remains no universally accepted standard of care in the treatment of oligodendrogliomas. Trials are limited by long follow-up times, often over a decade or more, during which knowledge of tumor biology shifts and treatment paradigms can change. For anaplastic oligodendrogliomas, a combination of radiation and/or chemotherapy is generally recommended; however, the most effective chemotherapy regimen and the timing of radiation remain debated.

Radiotherapy was first established as an effective treatment for anaplastic gliomas in the 1970s, more than doubling the median overall survival compared to surgery alone; however, studies during that time repeatedly failed to demonstrate a clear benefit of adjuvant chemotherapy or identify the subgroup of patients predicted to benefit most.[30, 31] In 1988, a case series published in the Annals of Neurology first suggested that anaplastic oligodendrogliomas represented a subgroup of high-grade gliomas that were chemosensitive. In a series of eight patients with multiply recurrent, radiation refractory, anaplastic oligodendrogliomas, 100% had a sustained response to combination chemotherapy. The likelihood that this was due to chance alone was considered statistically impossible with a *P* value <.00002–00397.[32]

The majority of patients in this series received an alkylator-based regimen containing procarbazine, lomustine (CCNU), and vincristine—also known as PCV—first described by Levin et al. in 1980.[33] Moving forward, a prospective phase II study of PCV for recurrent anaplastic oligodendrogliomas in the early 1990s appeared to confirm the benefit with a reported overall response rate of 75%, half of which were complete responses, including some lasting more than 16 months.[34]

Based on these results, two phase III, multicenter, randomized trials opened in 1994 to formally assess the benefit of PCV chemotherapy in this population. Radiation Therapy Oncology Group (RTOG) 9402, conducted mainly in North America, randomized patients to receive radiotherapy alone or with four cycles of an intensified PCV regimen followed by radiotherapy.[35] Simultaneously, across the Atlantic, the European Organization for the Research and Treatment of Cancer (EORTC) 26,951 trial randomized patients to radiation alone or radiation followed by six cycles of traditional PCV.[36] Both were powered for overall survival and progression-free survival as primary end points to determine whether PCV improved the outcome in newly diagnosed high-grade oligodendroglial tumors.

In 2006, interim results of both RTOG 9402 and EORTC 26951 demonstrated statistically significant improved PFS by adding PCV to radiotherapy for anaplastic oligodendroglial tumors. At that time, neither study demonstrated an improvement in overall survival. However, in both trials, neither median overall nor progression-free survival had been reached in patients with 1p/19q co-deleted tumors, suggesting that these data were immature. Regardless, this lack of overall survival benefit further called into question whether or not the toxicity of PCV was justified for an improvement in progression-free survival without a perceived improvement in survival.[35, 36]

Furthermore, in 2005, prior to the release of the interim analyses, Stupp et al. published results of the landmark EORTC 26981/22981—NCIC CE3 Phase III trial demonstrating an overall survival benefit in patients with glioblastoma with the combination of temozolomide (TMZ), an oral alkylating agent designed to penetrate the blood-brain barrier, and radiotherapy given concurrently for 6 weeks followed by six cycles of adjuvant TMZ. The chemotherapy regimen, single agent TMZ, was well tolerated, with only 16% experiencing grade III/IV hematologic toxicities and less than 5% discontinuing for toxicity.[37] While the "Stupp" trial for GBM did not compare TMZ against PCV, it became clear that TMZ was a simpler and more tolerable regimen, and combined with the absence of a clear survival benefit from PCV in RTOG 9402 and EORTC 26951, TMZ became the chemotherapy of choice for all gliomas, both on and off-label, including oligodendrogliomas.[38]

The efficacy of TMZ in all gliomas was further reinforced by the interim analyses of NOA-04, a phase III, double-randomized trial in patients with anaplastic gliomas with a primary end point of time to treatment failure that randomized patients to receive either radiation followed by chemotherapy or vice versa. The chemotherapy arm was further randomized to receive either PCV or TMZ. At the time of disease progression, patients in the chemotherapy arm crossed over to radiotherapy while those in the radiation arm were randomized to receive PCV or TMZ at progression. Preliminary results, published in 2009, demonstrated no apparent difference between any of the treatment arms, reinforcing the belief that TMZ was equally effective as PCV but with less toxicity.[39] However, at the time of publication, the data were immature. Furthermore, the primary outcome was defined as progression after receiving both chemotherapy and radiotherapy, meaning second or in some cases, third progression, raising concerns that the trial was not designed to detect a difference between treatment arms.[40]

As a result of these combined experiences, starting in the mid-2000s, TMZ became widely used as first-line chemotherapy for the treatment of high-grade gliomas including anaplastic oligodendroglioma despite a lack of prospective studies directly comparing PCV to TMZ in these patients. Upfront chemotherapy, predominantly with TMZ, also became increasingly common for 1p/19q co-deleted tumors, deferring radiotherapy until progression, as the field moved away from radiation out of concern for long-term cognitive side effects.[41, 42] Subsequently, the CODEL (1p/19q CO-DELeted anaplastic gliomas) trial was initiated and designed to test the efficacy of TMZ in co-deleted tumors with three arms: radiation alone, TMZ alone, and radiation with concurrent TMZ followed by adjuvant TMZ.

While prospective studies continued to accrue and mature, Lassman et al. published a large, retrospective analysis of outcomes in over 1000 patients with histologically (by local pathologist) defined anaplastic oligodendroglial tumors treated with various front-line approaches. Of 631 patients with complete information regarding 1p/19q status (determined locally), 301 (48%) were 1p/19q co-deleted. Median time to progression (mTTP) among co-deleted cases was longer in patients treated with combination of chemoradiotherapy (PCV or TMZ, $N=133$, mTTP 7.2 years) than chemotherapy alone (PCV or TMZ, $N=93$, mTTP 3.8 years, $P=.003$) or radiation alone ($N=54$, TTP 2.5 years, $P < -.001$), suggesting that the combination may be more effective in this population.

Furthermore, among 89 patients receiving upfront chemotherapy alone with either PCV ($N=21$) or TMZ ($N=68$), the median TTP was significantly longer in patients treated with PCV (7.6 years vs 3.3 years, $P=.186$). Although there was a survival trend favoring PCV (10.5 years vs 7.2 years for TMZ), it did not reach statistical significance ($P=.16$); however, the follow-up for the TMZ group was shorter (3.6 years compared to 7 years in patients receiving PCV) and with longer follow-up survival curves may separate further.[43]

In 2013, the long-term analyses of ETORC 26951 and RTOG 9402 were published with surprising results. With a median follow-up of over 11 years in both trials, PCV profoundly prolonged survival in co-deleted cases when added to RT in both studies, with a Hazard Ratio for death of approximately 0.6 in both studies, representing a 40% reduction in the risk of death.[15, 16]

Following the publication of these updated results, and the clear demonstration that radiotherapy alone without chemotherapy as an initial treatment strategy for co-deleted tumors was associated with shortened survival, the CODEL trial was modified to eliminate the arm that randomized patients to radiotherapy alone. In addition, early results from the TMZ monotherapy arm, presented at the 2015 Society for Neuro-Oncology meeting, reported disappointing efficacy. A total of 36 patients—12 in each arm—were analyzed. Progression during treatment occurred in 5/12 patients treated with TMZ alone vs 0/24 patients treated with RT. With median follow-up of 3.5 years, a total of 7/12 patients in the TMZ alone arm progressed compared to 3/24 patients treated with RT. A total of 4/12 patients in the TMZ alone group died (HR 9.2, $P=.048$). Thus, the TMZ monotherapy arm of CODEL was closed for inferiority and the trial became a randomization to RT with either TMZ or PCV.[44] However, with these results not expected for many years, the debate surrounding the role of PCV in anaplastic oligodendrogliomas resurfaced.[41, 45, 46]

Support for TMZ was buttressed from long-term results of RTOG BR0131, a phase II single arm study designed to evaluate the safety and efficacy of upfront TMZ in 40 patients with pure anaplastic or mixed oligodendrogliomas. All patients received dose-dense TMZ (150 mg/m^2, 7 days on/7 days off) for six cycles. Patients with stable or progressive disease went on to receive RT with concurrent TMZ while those with a complete response to TMZ deferred RT. The primary end point was 6-month progression-free survival (PFS6). The study reached its primary end point with 93% of participants progression free at 6 months. In 18 patients with co-deleted tumors, all were free from progression following completion of six cycles of TMZ at 6 months; however, ultimately 16 of these 18 patients went on to receive chemoradiation. Overall survival was not reached in the study after a median follow-up of 8.7 years. While median progression-free survival for the entire cohort was 5.8 years, this was not reached in the subset of patients with 1p/19q co-deleted tumors. However, this study is limited by small sample size, open label design, and a high withdrawal rate.[47, 48]

Not surprisingly, the final analysis of NOA-04 in 2016 demonstrated significantly longer time to treatment failure in patients with molecularly defined oligodendrogliomas (CIMP^{CODEL}: IDH mutant, 1p/19q co-deleted). Further exploration of treatment outcomes in this population suggested that PCV may have some benefit over TMZ with prolonged progression-free survival of 9.4 years in the PCV arm compared to 4.5 years in TMZ arm. Based on extrapolation with the survival data previously published in the EORTC and RTOG trials, the authors concluded that chemotherapy alone was probably inferior to RT with PCV in this population.[49]

Summary recommendations

There remains no clear standard of care for the treatment of anaplastic oligodendrogliomas and significant debate over the best course of treatment remains within the Neuro-Oncology community.[41] TMZ is by far easier to administer and better tolerated. However, the toxicity profile of PCV is not trivial. In the initial phase II study of PCV in anaplastic oligodendrogliomas, some degree of neutropenia or thrombocytopenia was reported in upwards of 80% of cycles, with 40% experiencing grade III or IV neutropenia and 15% experiencing grade III or IV thrombocytopenia. Mild to moderate nausea occurred in 70%–80% of cycles and nearly 50% of participants lost >5% of their body weight. Up to one-third of patients had debilitating fatigue and the majority developed some degree of neuropathy including 8% of patients with paralytic ileus.[34]

In the randomized phase III studies, up to 38% of patients were required to discontinue PCV due to toxicity and an additional 9% chose not to continue.[15, 16] In comparison, only 8% of patients were required to stop treatment due to toxicity and 4% voluntarily withdrew for unacceptable side effects in prospective phase III trials with TMZ.[37]

The PCV is also a much more complicated regimen administered over 6–8 weeks, depending on the intensity.[32–36] A typical regimen includes CCNU dosed at $110 \, mg/m^2$ on day 1 followed by procarbazine $60 \, mg/m^2$/day on days 8–21, both oral agents are taken at home. Procarbazine is an old drug that comes only in 50 mg tablets, which requires the total dose to be rounded and divided over 14 days. Patients must come to an infusion center on days 8 and 29 to receive IV vincristine $1.3 \, mg/m^2$ with a 2-mg cap. In many cases, myelosuppression prevents administration of the second vincristine dose.[33] TMZ, on the other hand, is administered orally at a dose of $150–200 \, mg/m^2$ on days 1–5 of a 28-day cycle.[37] This often does not require a calendar and the dose is easily calculated with available capsule sizes.

While TMZ has not been directly compared to PCV (without RT) in a fully powered head to head trial for co-deleted tumors, existing data suggest PCV may be more effective. The magnitude of benefit in RTOG 9402 and EORTC 26951 with the addition of PCV to radiotherapy was surprising. While RTOG 0131 suggests that TMZ may not be inferior to PCV, this was a single arm study with only 23 patients harboring tumors with 1p/19q co-deletion. Several studies demonstrated a continued reduction in tumor size after PCV, with Payne et al. reporting a nadir on average 2.7 years after stopping treatment.[50–52] This same pattern was not seen with TMZ where tumor regrowth typically occurred within a year of stopping the treatment.[53]

Based on the existing data, the Spanish Medical Oncology Society and Neuro-Oncology investigative group (SEOM/ GEINO) published guidelines in 2018 recommending the combination of RT and PCV for patients with anaplastic oligodendroglioma. In patients with poor performance status or in the elderly, RT followed by TMZ is recommended instead.[54]

Ongoing challenges and future directions

While CODEL is designed to evaluate if PCV or TMZ is superior in conjunction with RT, these results are not expected for many years. Furthermore, although RT appears to be important, given patients with anaplastic oligodendrogliomas are expected to survive up to a decade or longer, late effects of radiation are of real concern. Retrospective data evaluating patients with oligodendrogliomas (grade II and III) treated with upfront chemotherapy using a regimen similar to PCV (procarbazine, nimustine, and vincristine) demonstrated good long-term outcomes with 75% overall survival being 112 months (median overall survival not reached).[55] However, whether RT can be deferred in favor of upfront chemotherapy until the time of progression remains unclear and an ongoing prospective trial in France is designed to address this question.

Given the toxicity associated with PCV, some have questioned whether vincristine is necessary or whether it simply adds additional neurotoxicity with little benefit. Vincristine is a vinca alkaloid with limited CNS penetration in animal models.[56, 57] A retrospective analyses of 98 patients treated at MD Anderson Cancer Center between 1993 and 2009 demonstrated no difference in progression-free or overall survival in patients with histologic anaplastic oligodendrogliomas treated with procarbazine and CCNU (PC) alone compared to PCV. Patients treated with PC reported no neurotoxicity compared to 14% in the PCV arm.[58] PC has not been prospectively studied and in that cohort, data on co-deletion of 1p/19q was not available for most patients, so while plausible, it remains to be seen if the regimens would be equivalent in a cohort of molecularly defined anaplastic oligodendrogliomas.

In addition to identifying the optimal treatment, challenges include accurately identifying a subgroup of patients with anaplastic oligodendrogliomas that behave much more aggressively. The characteristics of this group are incompletely understood. An analysis of short-term (<7.3 years) (STS) and long-term (>7.3 years) (LTS) survivors from RTOG 9402 found *IDH1* mutations, *CIC*, *FUBP1*, and *TERT* promoter mutations were more common in LTS vs STS, but this was not statistically significant.[59] Mutations in *TCF12*, a transcription factor highly expressed in cells of oligodendrocyte lineage and neural progenitor cells, have been identified in anaplastic oligodendrogliomas and are linked to a more aggressive phenotype with shorter overall survival; however, more work is needed in this area.[60]

Conclusions

Biomarkers have revolutionized the modern approach to diagnosis and treatment of primary brain tumors, specifically gliomas.[61–63] Oligodendrogliomas are molecularly distinct from their astrocytic counterparts, which is reflected in the different biology and behavior of these tumors. As such, the ideal treatment strategy for these tumors may be different. We await the results of ongoing prospective trials to better clarify the optimal approach to maximize both overall survival and quality of life for patients with these rare tumors.

References

1. Ostrom QT, Gittleman H, Liao P, et al. CBTRUS statistical report: primary brain and other central nervous system tumors diagnosed in the United States in 2010–2014. *Neuro Oncol.* 2017;19(suppl. 5):v1–v88.
2. Van Den Bent MJ, Bromberg JE, Buckner J. Low-grade and anaplastic oligodendroglioma. *Handb Clin Neurol.* 2016;134:361–380.
3. van den Bent MJ. Interobserver variation of the histopathological diagnosis in clinical trials on glioma: a clinician's perspective. *Acta Neuropathol.* 2010;120(3):297–304.
4. Louis DN, Perry A, Reifenberger G, et al. The 2016 World Health Organization classification of tumors of the central nervous system: a summary. *Acta Neuropathol.* 2016;131(6):803–820.
5. Parsons DW, Jones S, Zhang X, et al. An integrated genomic analysis of human glioblastoma multiforme. *Science.* 2008;321(5897):1807–1812.
6. Yan H, Parsons DW, Jin G, et al. IDH1 and IDH2 mutations in gliomas. *N Engl J Med.* 2009;360(8):765–773.
7. Cancer Genome Atlas Research N, Brat DJ, Verhaak RG, et al. Comprehensive, integrative genomic analysis of diffuse lower-grade gliomas. *N Engl J Med.* 2015;372(26):2481–2498.
8. Reifenberger J, Reifenberger G, Liu L, James CD, Wechsler W, Collins VP. Molecular genetic analysis of oligodendroglial tumors shows preferential allelic deletions on 19q and 1p. *Am J Pathol.* 1994;145(5):1175–1190.
9. Bigner SH, Rasheed BK, Wiltshire R, McLendon RE. Morphologic and molecular genetic aspects of oligodendroglial neoplasms. *Neuro Oncol.* 1999;1(1):52–60.
10. Jenkins RB, Blair H, Ballman KV, et al. A t(1;19)(q10;p10) mediates the combined deletions of 1p and 19q and predicts a better prognosis of patients with oligodendroglioma. *Cancer Res.* 2006;66(20):9852–9861.
11. Bettegowda C, Agrawal N, Jiao Y, et al. Mutations in CIC and FUBP1 contribute to human oligodendroglioma. *Science.* 2011;333(6048):1453–1455.
12. Wesseling P, van den Bent M, Perry A. Oligodendroglioma: pathology, molecular mechanisms and markers. *Acta Neuropathol.* 2015;129(6):809–827.
13. Cairncross JG, Ueki K, Zlatescu MC, et al. Specific genetic predictors of chemotherapeutic response and survival in patients with anaplastic oligo-dendrogliomas. *J Natl Cancer Inst.* 1998;90(19):1473–1479.
14. Ino Y, Betensky RA, Zlatescu MC, et al. Molecular subtypes of anaplastic oligodendroglioma: implications for patient management at diagnosis. *Clin Cancer Res.* 2001;7(4):839–845.
15. Cairncross G, Wang M, Shaw E, et al. Phase III trial of chemoradiotherapy for anaplastic oligodendroglioma: long-term results of RTOG 9402. *J Clin Oncol.* 2013;31(3):337–343.
16. van den Bent MJ, Brandes AA, Taphoorn MJ, et al. Adjuvant procarbazine, lomustine, and vincristine chemotherapy in newly diagnosed anaplastic oligodendroglioma: long-term follow-up of EORTC brain tumor group study 26951. *J Clin Oncol.* 2013;31(3):344–350.
17. McNamara MG, Jiang H, Lim-Fat MJ, et al. Treatment outcomes in 1p19q co-deleted/partially deleted gliomas. *Can J Neurol Sci.* 2017;44(3):288–294.
18. Boots-Sprenger SH, Sijben A, Rijntjes J, et al. Significance of complete 1p/19q co-deletion, IDH1 mutation and MGMT promoter methylation in gliomas: use with caution. *Mod Pathol.* 2013;26(7):922–929.
19. Appin CL, Gao J, Chisolm C, et al. Glioblastoma with oligodendroglioma component (GBM-O): molecular genetic and clinical characteristics. *Brain Pathol.* 2013;23(4):454–461.
20. Chan AK, Yao Y, Zhang Z, et al. Combination genetic signature stratifies lower-grade gliomas better than histological grade. *Oncotarget.* 2015;6(25):20885–20901.
21. Weller M, Weber RG, Willscher E, et al. Molecular classification of diffuse cerebral WHO grade II/III gliomas using genome- and transcriptome-wide profiling improves stratification of prognostically distinct patient groups. *Acta Neuropathol.* 2015;129(5):679–693.
22. Mellai M, Annovazzi L, Senetta R, et al. Diagnostic revision of 206 adult gliomas (including 40 oligoastrocytomas) based on ATRX, IDH1/2 and 1p/19q status. *J Neurooncol.* 2017;131(2):213–222.
23. Masui K, Cloughesy TF, Mischel PS. Review: molecular pathology in adult high-grade gliomas: from molecular diagnostics to target therapies. *Neuropathol Appl Neurobiol.* 2012;38(3):271–291.
24. Cahill DP, Louis DN, Cairncross JG. Molecular background of oligodendroglioma: 1p/19q, IDH, TERT, CIC and FUBP1. *CNS Oncol.* 2015;4(5):287–294.
25. Killela PJ, Reitman ZJ, Jiao Y, et al. TERT promoter mutations occur frequently in gliomas and a subset of tumors derived from cells with low rates of self-renewal. *Proc Natl Acad Sci U S A.* 2013;110(15):6021–6026.
26. Arita H, Narita Y, Fukushima S, et al. Upregulating mutations in the TERT promoter commonly occur in adult malignant gliomas and are strongly associated with total 1p19q loss. *Acta Neuropathol.* 2013;126(2):267–276.
27. Lee Y, Koh J, Kim SI, et al. The frequency and prognostic effect of TERT promoter mutation in diffuse gliomas. *Acta Neuropathol Commun.* 2017;5(1):62.
28. Vuong HG, Altibi AMA, Duong UNP, et al. TERT promoter mutation and its interaction with IDH mutations in glioma: combined TERT promoter and IDH mutations stratifies lower-grade glioma into distinct survival subgroups-A meta-analysis of aggregate data. *Crit Rev Oncol Hematol.* 2017;120:1–9.
29. Pekmezci M, Rice T, Molinaro AM, et al. Adult infiltrating gliomas with WHO 2016 integrated diagnosis: additional prognostic roles of ATRX and TERT. *Acta Neuropathol.* 2017;133(6):1001–1016.
30. Walker MD, Green SB, Byar DP, et al. Randomized comparisons of radiotherapy and nitrosoureas for the treatment of malignant glioma after surgery. *N Engl J Med.* 1980;303(23):1323–1329.
31. Walker MD, Alexander Jr E, Hunt WE, et al. Evaluation of BCNU and/or radiotherapy in the treatment of anaplastic gliomas. A cooperative clinical trial. *J Neurosurg.* 1978;49(3):333–343.

32. Cairncross JG, Macdonald DR. Successful chemotherapy for recurrent malignant oligodendroglioma. *Ann Neurol.* 1988;23(4):360–364.

33. Levin VA, Edwards MS, Wright DC, et al. Modified procarbazine, CCNU, and vincristine (PCV 3) combination chemotherapy in the treatment of malignant brain tumors. *Cancer Treat Rep.* 1980;64(2–3):237–244.

34. Cairncross G, Macdonald D, Ludwin S, et al. Chemotherapy for anaplastic oligodendroglioma. National Cancer Institute of Canada clinical trials group. *J Clin Oncol.* 1994;12(10):2013–2021.

35. Intergroup Radiation Therapy Oncology Group T, Cairncross G, Berkey B, et al. Phase III trial of chemotherapy plus radiotherapy compared with radiotherapy alone for pure and mixed anaplastic oligodendroglioma: Intergroup Radiation Therapy Oncology Group Trial 9402. *J Clin Oncol.* 2006;24(18):2707–2714.

36. van den Bent MJ, Carpentier AF, Brandes AA, et al. Adjuvant procarbazine, lomustine, and vincristine improves progression-free survival but not overall survival in newly diagnosed anaplastic oligodendrogliomas and oligoastrocytomas: a randomized European Organisation for Research and Treatment of Cancer phase III trial. *J Clin Oncol.* 2006;24(18):2715–2722.

37. Stupp R, Mason WP, van den Bent MJ, et al. Radiotherapy plus concomitant and adjuvant temozolomide for glioblastoma. *N Engl J Med.* 2005;352(10):987–996.

38. Panageas KS, Iwamoto FM, Cloughesy TF, et al. Initial treatment patterns over time for anaplastic oligodendroglial tumors. *Neuro Oncol.* 2012;14(6):761–767.

39. Wick W, Hartmann C, Engel C, et al. NOA-04 randomized phase III trial of sequential radiochemotherapy of anaplastic glioma with procarbazine, lomustine, and vincristine or temozolomide. *J Clin Oncol.* 2009;27(35):5874–5880.

40. DeAngelis LM. Anaplastic glioma: how to prognosticate outcome and choose a treatment strategy. [corrected]. *J Clin Oncol.* 2009;27(35):5861–5862.

41. Lassman AB. Procarbazine, lomustine and vincristine or temozolomide: which is the better regimen? *CNS Oncol.* 2015;4(5):341–346.

42. Panageas KS, Iwamoto FM, Cloughesy TF, et al. Initial treatment patterns over time for anaplastic oligodendroglial tumors. *Neuro Oncol.* 2012;14(6):761–767.

43. Lassman AB, Iwamoto FM, Cloughesy TF, et al. International retrospective study of over 1000 adults with anaplastic oligodendroglial tumors. *Neuro Oncol.* 2011;13(6):649–659.

44. Jaeckle KV, Vogelbaum M, Ballman K, et al. ATCT-16: codel (alliance-N0577; EORTC-26081/2208; NRG-1071; NCIC-CEC-2): phase III randomized study of RT VS. RT + TMZ vs. TMZ for newly diagnosed 1p/19q-codeleted anaplastic glioma. Analysis of patients treated on the original protocol design. *Neuro Oncol.* 2015;17(Suppl 5):v4–v5.

45. Lassman AB. Oligodendrogliomas: questions answered, answers questioned. *Oncology (Williston Park).* 2013;27(4):326–328.

46. Rinne ML, Wen PY. Treating anaplastic oligodendrogliomas and WHO grade 2 gliomas: PCV or temozolomide? The case for temozolomide. *Oncology (Williston Park).* 2015;29(4):265–275.

47. Vogelbaum MA, Hu C, Peereboom DM, et al. Phase II trial of pre-irradiation and concurrent temozolomide in patients with newly diagnosed anaplastic oligodendrogliomas and mixed anaplastic oligoastrocytomas: long term results of RTOG BR0131. *J Neurooncol.* 2015;124(3):413–420.

48. Vogelbaum MA, Berkey B, Peereboom D, et al. Phase II trial of preirradiation and concurrent temozolomide in patients with newly diagnosed anaplastic oligodendrogliomas and mixed anaplastic oligoastrocytomas: RTOG BR0131. *Neuro Oncol.* 2009;11(2):167–175.

49. Wick W, Roth P, Hartmann C, et al. Long-term analysis of the NOA-04 randomized phase III trial of sequential radiochemotherapy of anaplastic glioma with PCV or temozolomide. *Neuro Oncol.* 2016;18(11):1529–1537.

50. Peyre M, Cartalat-Carel S, Meyronet D, et al. Prolonged response without prolonged chemotherapy: a lesson from PCV chemotherapy in low-grade gliomas. *Neuro Oncol.* 2010;12(10):1078–1082.

51. Stege EM, Kros JM, de Bruin HG, et al. Successful treatment of low-grade oligodendroglial tumors with a chemotherapy regimen of procarbazine, lomustine, and vincristine. *Cancer.* 2005;103(4):802–809.

52. Mason WP, Krol GS, DeAngelis LM. Low-grade oligodendroglioma responds to chemotherapy. *Neurology.* 1996;46(1):203–207.

53. Ricard D, Kaloshi G, Amiel-Benouaich A, et al. Dynamic history of low-grade gliomas before and after temozolomide treatment. *Ann Neurol.* 2007;61(5):484–490.

54. Balana C, Alonso M, Hernandez A, et al. SEOM clinical guidelines for anaplastic gliomas (2017). *Clin Transl Oncol.* 2018;20(1):16–21.

55. Hata N, Yoshimoto K, Hatae R, et al. Deferred radiotherapy and upfront procarbazine-ACNU-vincristine administration for 1p19q codeleted oligodendroglial tumors are associated with favorable outcome without compromising patient performance, regardless of WHO grade. *Onco Targets Ther.* 2016;9:7123–7131.

56. Greig NH, Soncrant TT, Shetty HU, Momma S, Smith QR, Rapoport SI. Brain uptake and anticancer activities of vincristine and vinblastine are restricted by their low cerebrovascular permeability and binding to plasma constituents in rat. *Cancer Chemother Pharmacol.* 1990;26(4):263–268.

57. Boyle FM, Eller SL, Grossman SA. Penetration of intra-arterially administered vincristine in experimental brain tumor. *Neuro Oncol.* 2004;6(4):300–305.

58. Webre C, Shonka N, Smith L, Liu D, De Groot J. PC or PCV, that is the question: primary anaplastic oligodendroglial tumors treated with procarbazine and CCNU with and without vincristine. *Anticancer Res.* 2015;35(10):5467–5472.

59. Holdhoff M, Cairncross GJ, Kollmeyer TM, et al. Genetic landscape of extreme responders with anaplastic oligodendroglioma. *Oncotarget.* 2017;8(22):35523–35531.

60. Labreche K, Simeonova I, Kamoun A, et al. TCF12 is mutated in anaplastic oligodendroglioma. *Nat Commun.* 2015;6:7207.

61. Lassman AB, Cloughesy TF. Biomarkers in NOA-04: another piece to the puzzle. *Neuro Oncol.* 2016;18(11):1467–1469.

62. Lassman AB. Success at last: a molecular factor that informs treatment. *Curr Oncol Rep.* 2013;15(1):47–55.

63. Weller M, Pfister SM, Wick W, Hegi ME, Reifenberger G, Stupp R. Molecular neuro-oncology in clinical practice: a new horizon. *Lancet Oncol.* 2013;14(9):e370–e379.

Chapter 10

Prospects of translational proteomics and protein microarrays in oligodendroglioma

Shabarni Gupta and Sanjeeva Srivastava

Department of Biosciences and Bioengineering, Indian Institute of Technology Bombay, Mumbai, India

Introduction

Gliomas are the most commonly occurring and heterogeneous type of tumors accounting for 80% of all brain malignancies.[1] Broadly, gliomas have been classified into four histological grades—the slow growing and potentially curable grade I gliomas; diffuse low-grade (grade II); more aggressive intermediate grade gliomas (grade III) constituting the lower-grade gliomas; followed by the highly aggressive glioblastoma multiforme (grade IV).[2] Accurate classification of glioma is undeniably one of the most important facets dictating the clinical discourse of a patient. However, the heterogeneous nature of gliomas often poses a problem in accurate diagnosis and therapy.

Over the years, the WHO has made several amendments to its classification system to address this issue. The limitations of inter- and intraobserver variability and poor correlation with clinical outcomes in purely histopathology-based diagnosis have led the scientific community to integrate molecular approaches to classify and treat gliomas.[3] This change has been deemed particularly important for diffuse lower grade gliomas, which tend to be highly invasive in nature and difficult to completely resect.[4] With a small percentage of these tumors progressing into grade IV glioma, molecular markers have been essential to classify and predict the outcomes of these tumors, empowering clinicians to better diagnose and treat them.[4]

The most recent and impacting reforms in glioma classification came through in 2015 with three studies providing a road map dissecting the molecular heterogeneity observed across various grades of tumor using genomic approaches.[4–6] In regard to low-grade gliomas, these studies have established three types of molecular signatures (Fig. 1). Type I harbors mutations in the IDH1/2 genes and co-deletion of 1p and 19q chromosome arms, an indicator of good clinical prognosis.[4,6] They often also acquire mutations in the TERT, CIC, FUBP1, and NOTCH genes.[4,6] This type histologically presents themselves as oligodendrogliomas and show the proneural TCGA glioblastoma expression subtype.[6] Type II harbors mutation in IDH1/2 genes, however, has an intact 1p/19q chromosome arms.[4,6] This type acquires mutations in the TP53 and ATRX gene and is also associated with the proneural TCGA glioblastoma expression subtype.[6] Type II low-grade tumors show intermediate clinical prognosis and are histologically similar to astrocytomas.[4] Type III tumors do not harbor mutations in the IDH genes; however, they do show mutations in the promoter of the TERT gene.[4,6] By virtue of this, they follow an aggressive clinical course similar to that of glioblastomas.[4,6] They also acquire mutations in EGFR, CDKN2, MDM4. PTEN, and NF1 genes, and harbor a poor clinical prognosis.[4,6] It is thus important to acknowledge the power of genomics in this era, as NGS is now an accessible modality enabling clinicians to not only accurately diagnose the tumor, but also establishing the appropriate line of therapeutic action, which is considerably different among these three types. However, genomics and transcriptomics by themselves are not sufficient to understand the complex biochemical processes resulting from the aberration of one or a network of proteins that underlie oncogenesis. It has been established through comparative analysis that the mRNA levels do not share a good correlation with protein abundances, especially in cancer tissues,[8] emphasizing the need to understand the dynamic interplay of proteins which lead to disease. Furthermore, the dearth in glioma diagnosis in today's age is perhaps a rapid, accessible, easily interpretable diagnostic modality, which could capture these genomic signatures, provide with similar efficacies and provide an in-depth understanding of an individual's pathophysiology. This is where proteomics comes into play.

Proteins are the functional units driving all the biochemical processes in a cell. The expression profile of proteins, its posttranslational modifications and interaction partners reflects a cell's phenotype. It has been well established that aberrations in key biological signaling pathways have a cascading effect in cells. Proteomics provides the expression profile of

Oligodendroglioma. https://doi.org/10.1016/B978-0-12-813158-9.00010-4

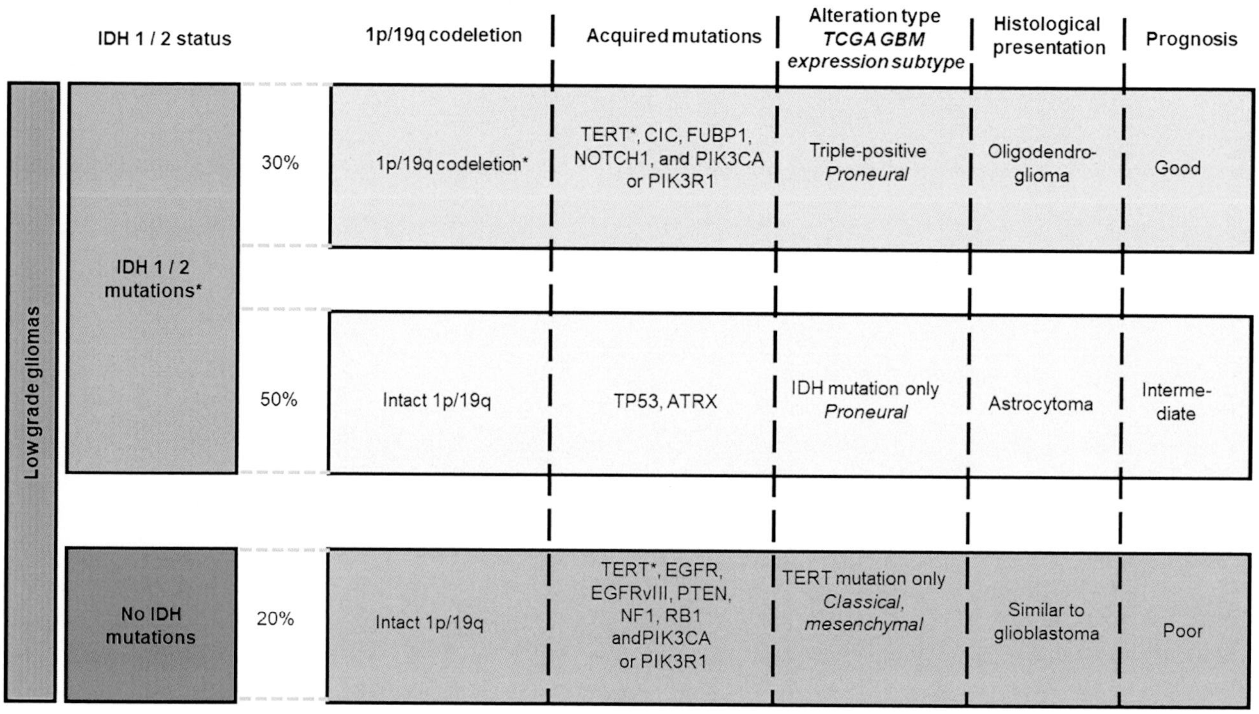

IDH 1 / 2 status			1p/19q codeletion		Acquired mutations	Alteration type *TCGA GBM* *expression subtype*	Histological presentation	Prognosis
Low grade gliomas	IDH 1 / 2 mutations*	30%	1p/19q codeletion*		TERT*, CIC, FUBP1, NOTCH1, and PIK3CA or PIK3R1	Triple-positive *Proneural*	Oligodendro-glioma	Good
		50%	Intact 1p/19q		TP53, ATRX	IDH mutation only *Proneural*	Astrocytoma	Interme-diate
	No IDH mutations	20%	Intact 1p/19q		TERT*, EGFR, EGFRvIII, PTEN, NF1, RB1 andPIK3CA or PIK3R1	TERT mutation only *Classical, mesenchymal*	Similar to glioblastoma	Poor

FIG. 1 Schematic representation of molecular alterations in low-grade gliomas as reported in Refs. 4, 5, 7. "*" denotes that alterations in these genes are determinants for subtype grouping.

all the proteins in a given cell, tissue, or organism at a given time under specific conditions. This makes it a powerful tool to compare the protein expression profiles of several subtypes of tumors to understand molecular aberrations that may be specific to each type.[7] Furthermore, they can be easily translated into a clinical diagnostic modality like western blots, ELISA, RIA, etc., thus catering to many health-care institutions. In addition to diagnosis, incorporating a panel of protein biomarkers could individualize therapy and allows clinicians to target a specific signaling pathway involved in the malignant phenotype. In the following sections, we summarize the insights that proteomics has and can offer in glioma pathology with special focus on oligodendrogliomas.

Source of biospecimen understates proteome discovery

Unlike the genome, the proteome of an organism is highly dynamic and is constantly changing, depending on its condition. Proteomics emerged as a field to complement the other "-omic" technologies like genomics and transcriptomics to capture the nebulous phenotype which alters under specific conditions. Proteomics uses an arsenal of technologies to capture the dynamic snap shot of a cell depending on the biological problem under question. Therefore, in the context of tumor proteomics, especially brain malignancies, the first aspect that a researcher must consider is the biospecimen under study, as it would vastly affect its downstream processing and choice of technology employed to understand disease proteomics.

Over the years, glioma proteomics has employed biospecimens from tissue specimens from patient biopsies, serum, plasma, proximal fluids like CSF, and cell lines and animal models, each having their pros and cons[9] (Fig. 2). Tissue-derived markers capture the cellular processes providing valuable insights into disease mechanism. Protein markers identified using tissue proteomics can be used in translational aspects like potential IHC markers for tissue biopsies, identifying signaling pathways for therapeutics, and predicting prognosis. Serum/plasma is an attractive source of biomarkers for any disease, especially in the case of brain malignancies, as whole blood collection is minimally invasive with high patient compliance as compared to any other approach. Plasma or serum can capture tissue leakage proteins, or nonsecretory proteins, which could show aberrant abundance in circulation due to tissue damage and apoptosis, like in the case of cancer. Serum also contains immunoglobulins and therefore is the only source for detecting autoantibodies, a valuable early diagnosis marker. They are also a source for exosomes, which are being explored as potential biomarkers for several malignancies. Masking of potential biomarkers by highly abundant proteins in serum, in addition to their high dilution in the peripheral system, makes it a challenging biospecimen to work with. Few of these hurdles are overcome by CSF, which

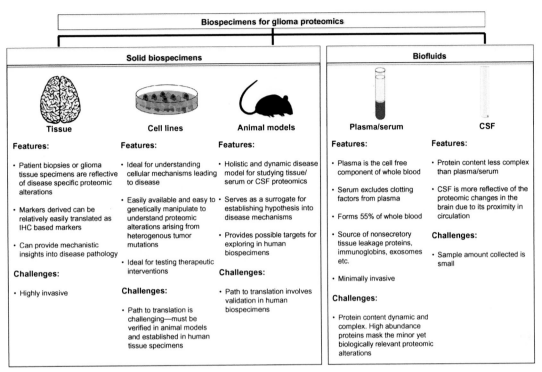

FIG. 2 Schematic representation of the biospecimens commonly used for glioma proteomics studies alongside features and challenges involved with their use in biomarker and therapeutic research.

is an extracellular fluid circulating around the spinal cord and the subarachnoid cavity in the brain. Although similar to serum, its protein content is far less complex than the later. Moreover, its protein profile is largely a derivative of the physiology of the CNS, making it an excellent choice for biomarker discovery in glioma. Markers from serum, plasma, or CSF can be translated into ELISA or western blot-based tests for diagnosis. Cell lines and animal models provide a platform not only to discover proteomic alterations, but also to serve as in vivo models for time bound assays monitoring progression or testing therapeutics. Findings from these in vivo models must however be validated on tissue or serum/plasma/CSF for its translation.

Sample preparation from any of the above sources is key to the success of any proteomics-based experiment. Generally, for sample acquisition of proteins, its labile nature must be taken into consideration. Proteins being prone to degradation due to freeze-thaw processes and proteolytic enzymes must be processed in lysis buffers containing chaotropes, protease and phosphatase inhibitors, and detergents and/or reducing agents. Biofluids like serum/plasma/CSF, and other biospecimens must be aliquoted to avoid freeze-thaw cycles. Depletion strategies may be employed to reduce the abundance of commonly occurring proteins like immunoglobulins in these specimens, to enrich the biologically relevant pool of biomarkers.

Proteomic technologies for data acquisition
Gel-based proteomics

Historically, gel-based technologies dominated the field of proteomics and continue to be a classical approach to achieve separation of proteins in biospecimens. Fundamentally, biospecimens are processed and separated in a polyacrylamide gel matrix electrophoretically, typically on the basis of their mass-to-charge ratio (m/z). In complex biosamples like tissue lysates and serum samples, merely molecular weight differences among proteins do not allow a large depth of resolution. For this reason, proteins are separated in two dimensions. Proteins in biospecimens are first separated on the basis of their isoelectric points (in a process called isoelectric focusing) and next they are separated in the second dimension on the basis of their m/z ratios.[10, 11] This technique is called two-dimensional gel electrophoresis (2-DE). The proteins in these gels are stained using total protein stains like Coomassie blue and each protein in the sample can be distinctly visualized separately from the other as a protein spot. Protein spots can be visualized on two gel matrices containing samples from two conditions (e.g., healthy vs diseased) and differences between samples can be identified as presence, absence, or altered levels of certain protein spots. Software algorithms help in identifying and recording coordinates of a protein spot that may be of interest. These spots are then excised from gels and proteins from these gels are digested in a process called in-gel digestion and further analyzed to reveal its identity. This is typically done using a mass spectrometer.

In order to avoid technical and artefactual errors arising from gel to gel, a technique called difference in gel electrophoresis (2D-DIGE) was developed.[12] In this technique the samples from the two conditions under study are labeled using fluorescent cyanine dyes, like Cy3 and Cy5. A third sample or an internal standard required for normalizing the two samples can be labeled using Cy2. These three differentially labeled samples are then pooled and can be simultaneously run in 2-DE. The gel is then scanned at different wavelengths to excite each dye. This enables the simultaneous visualization of protein expression profiles of each sample in a single gel, thus reducing technical variation. The differentially expressed protein spots can then be identified using mass spectrometry (MS).

Mass spectrometry

Instrumentation

The MS is the workhorse that has catapulted proteomics to its success. This analytical tool ionizes compounds and helps identify them based on their m/z ratio.[13] It is composed of three principal components—the ionization source, analyzer, and detector.

The sample under consideration first encounters the ionization source. Considering the labile nature of proteins, in proteomics-based experiments, soft ionization sources like matrix-assisted laser desorption/ionization (MALDI) or electrospray ionization (ESI) are employed.[14] In MALDI the sample is mixed with an energy absorbent organic material, also termed the matrix. When the matrix containing the sample crystallizes,[15] it is ionized using a laser beam generating single protonated ions. The ESI mass spectrometer on the other hand deals with samples in the liquid state. It generates ions by spraying a dilute solution of analyte (peptide/protein) molecules through a fine metal capillary tip (i.e., Taylor cone), which leads to the formation of very fine droplets.[16] The droplets, when subjected to an electrical field, become highly charged, and as the solvent evaporates, the peptide molecules in the droplet get ionized by proton exchange. The ESI is often coupled with a separation method like liquid chromatography (LC) prior to ionization. Overall, these ionization techniques help in imparting a charge to the peptides and are then separated as per their m/z ratio in the mass analyzer.

There are several types of mass analyzers which are constantly evolving to enhance the sensitivity of mass spectrometers.[14] Time of flight (TOF) analyzers identify molecules based on the amount of time taken by its ions to reach the detector in vacuum. These analyzers now employ delay in extraction approach for the release of ions and incorporate multiple reflectron lenses to enhance the linear path taken by these ions to reach a second detector; this improves its resolution. The TOF analyzers are commonly used along with MALDI. Quadrupole mass filters employ a combination of RF (radio frequency) and DC (direct current) voltages across the four parallel rods, which allow the user to filter and detect peptides of specific m/z ratios. These mass filters are immensely useful for the emerging targeted proteomics approach. Ion traps are another kind of mass filter which also use RF and DC voltage; however, its design involves three electrodes containing two rod-shaped (entrance and exit) electrodes and a central ring-shaped electrode used to filter ions in this path. Fourier transform MS (FT-MS) is a type of MS offering the most mass resolution. In Fourier transform ion cyclotron resonance (FT-ICR), ions generated from an external source are trapped and are subjected to cyclotron motion in a magnetic field. These ions are then excited by applying an RF sweep. When the cyclotron frequency of ions is in resonance with the applied RF, the detector records an image current which is corelated to its m/z ratio.

The new age detectors in mass spectrometers are essentially electron multipliers which provide amplification of the ion signals by emitting electrons from its surface when an ion hits it.[14, 17]

Proteomic approaches

The two commonly used strategies for proteome discovery are the top-down and the bottom-up approaches. The top-down strategy, as the name suggests, subjects its analysis to intact proteins,[18] whereas the bottom-up strategy employs proteolyzed or fragmented proteins in the form of peptides for analysis.[19] Techniques like 2-DE have been traditionally used for top-down approaches where gel pieces containing intact proteins are excised and then in-gel digested for mass analysis using MS. Separation of intact proteins can also be achieved using liquid chromatography, reverse-phase liquid chromatography, and ion-exchange chromatography which is coupled to MS/MS.[18] Alternatively, techniques like MALDI, surface-enhanced laser desorption/ionization (SELDI), or ESI coupled to FT-MS mass analyzers like FT-ICR or Orbitraps, allow a breadth of proteome coverage in top-down proteomics.[18]

Bottom-up proteomics, also termed "shot-gun proteomics," incorporates complex specimens which are enzymatically fragmented.[19] The fragmented peptides are fractionized using strong cation exchange (SCX) chromatography or multidimensional protein identification technology (MudPIT).[20, 21] Ionized precursor fragments are selected in the first stage of MS and the fragmented ions are subsequently analyzed in the second stage of MS (MS/MS). The spectra thus observed is compared with existing databases and protein sequences, which help overlap the fragmented peptides to reveal the identity of the proteins and its abundance in the sample.

While top-down proteomics has the advantage of analyzing intact proteins, which reflects in its ability to completely characterize all proteoforms, spliced variants, and posttranslational modifications, its processing requires mass spectrometers with extremely high resolution. This is because intact proteins are difficult to ionize, which affects its sensitivity and detection limits.[22]

A third kind of strategy in proteomics is a targeted approach. Targeted proteomics utilizes the technology of selected reaction monitoring (SRM; also termed as multiple reaction monitoring, MRM) where specifically fragmented peptide ions from a protein of interest can be detected in a complex sample using a triple quadrupole mass spectrometer.[23] This emerging technology is deemed to be extremely promising for sensitive and high-throughput next-generation diagnosis.

Quantitation strategies in proteomics

Proteomics is broadly quantitative in its essence. Quantitative proteomics in the MS arena revolves around two broad strategies: label-based and label-free approaches. Label-free quantitation of proteins using MS employs spectral counting which measures the number of peptides from a protein detected in an MS/MS spectra and is reflective of the abundance of that protein in a given sample.[24] Measuring the intensity of the peptide peak at both chromatography and MS levels are also useful for peptide/protein quantitation in label-free quantitation.[24]

Label-based detection employs a wide range of labeling techniques, which allow multiplexing of samples, increasing sensitivity, and throughput. In this approach, samples under different conditions are differentially labeled and subjected to liquid chromatography-based separation followed by MS/MS. Labeling techniques like stable isotope labeling of amino acids in cell culture (SILAC), isotope coded affinity tag (ICAT), isobaric tag for relative and absolute quantification (ITRAQ) of peptides, tandem mass tags (TMT), etc., offer multiplexing; up to 10-plexing in some cases, thus reducing technical variation and helping in accurately quantitating proteins with high sensitivity and dynamic range.[25, 26]

Protein microarrays

Types of protein arrays

Protein microarrays are chips harboring an array of proteins, peptides, antibodies, aptamers, or lysates embedded on its surface allowing researchers to study protein interactions, functions, or detect the presence of proteins in samples in a high-throughput manner.[27] Protein arrays are a powerful proteomics and interactomics tool and can be broadly classified into three types.[28]

The first type is termed an analytical protein array, which consists of a chip with an array of antibodies imprinted on its surface. It allows a researcher to screen for the presence of target proteins in samples by probing them with a number of reporter antibodies in a high-throughput manner. Clinically, analytical microarrays can, in principle, be used as a diagnostic tool wherein a chip could contain antibodies against a panel of biomarkers and is subjected to a tissue lysate or serum and probed for the presence of the marker proteins, increasing the throughput of traditional ELISA.[29]

Functional microarrays are the second type of microarray, which aid in detecting protein-protein interactions, thereby providing functional insights of a systems cellular biology. Functional arrays have proteins or peptides imprinted on their surface and can be probed with query proteins, peptides, small molecules, or nucleotides to deduce novel interactions between them.[28] Functional biomarkers are powerful tools in basic research to understand protein function; they also have great clinical applicability as a platform to screen and test for the presence of autoantibodies. Autoantibodies are antibodies generated against aberrant self-proteins (autoantigens). Autoantigens are known to be produced at the onset of diseases like cancers and are termed tumor-associated antigens (TAA). Functional microarrays with a multitude of proteins can be printed and screened with patient sera for the presence of autoantigens against each of the proteins printed on the chip.[30] The cell-free expression systems (CFES) containing a milieu of protein expression machinery has also been used to generate functional arrays by imprinting DNA cloned with a gene of interest. When CFES is applied onto such a DNA chip, proteins are expressed and get bound to the chip to form a functional protein array. Nucleic acid programmable protein array (NAPPA) is a popular example of such surface chemistry with great clinical applicability.[28, 31]

Analytical and functional microarrays are types of forward phase arrays. Reverse phase arrays are the third type of arrays in which cell or tissue lysates from a number of samples can be printed on a chip in a high-throughput manner, and can be probed for the presence of a protein of interest using a target antibody against it.[27] Tissue microarrays are a validation platform where numerous tissue sections are imprinted on a chip and can be probed, in principle, similar to reverse phase protein arrays.[30]

Microarray analysis

Protein microarrays allow high-throughput screening of protein presence, interaction, and function. Its high-throughput nature makes data acquisition lucrative; however, difficult to analyze. To understand the various constrains of microarray data analysis, one must understand the various steps involved in a protein microarray experiment. Here, we shall summarize a hypothetical experimental setup describing autoantibody screening in an oligodendroglioma cohort using human proteome arrays (HuProt v3.1) harboring over 16,152 full-length human proteins covering ∼81% of the human proteome (Fig. 3).

A. Experimental setup

    (i) Serum samples from healthy controls and oligodendroglioma patients must be appropriately collected and stored to avoid degradation or contamination from whole blood constituents.

    (ii) HuProt slides containing 16,152 full-length GST (glutathione S-transferase) tagged human proteins (one for each serum sample) must be subjected to blocking using a blocking solution like 2% BSA in SuperBlock, and incubated for 2h on an orbital shaker at room temperature (RT). Slides must be handled with tweezers at all time with printed proteins faced side up. Care must be taken to hold the slide off the edges using tweezers and not on the protein printed interface.

    (iii) The slides must be washed for 4–5min using TBST in specialized Wheaton glass chambers with a magnetic bead circulating the wash solution. Care must be taken that the magnetic bead does not scratch the printed interface.

    (iv) The slides must then be rinsed using distilled water to remove any residual TBST by centrifuging the slides at 900rpm for 2min.

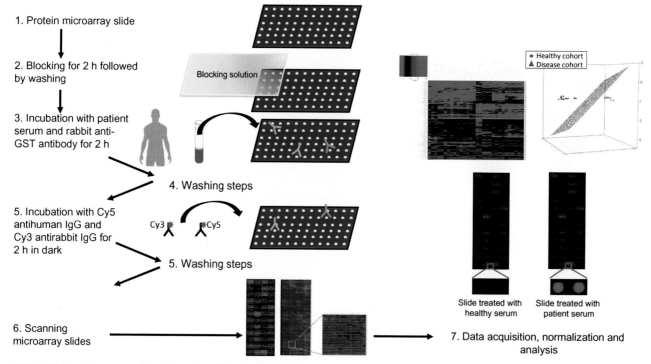

FIG. 3 Schematic representation of overall workflow involved in autoantibody screening using protein microarrays. Step 1 represents a cartoon of a HuProt chip imprinted with ∼16,000 full-length proteins. Step 2 shows blocking of the protein microarray slide with blocking reagent like SuperBlock. Step 3 involves usage of patient serum as source of primary antibody for probing on protein microarray slide. Anti-GST antibody is used to map all the GST tagged proteins on the slide. The slides are washed in step 4 as described in section "Microarray analysis". Step 5 involves incubation with respective secondary antibodies. Step 6 involves scanning of microarray slides followed by step 7 involving data acquisition, normalization, and analysis.

   (v) Diluted serum serves as the source of primary antibodies. As a quality control check and normalization measure, anti-GST antibody is also added to this solution, the purpose of which will be described later. The chips are incubated for 2 h at RT.

   (vi) The primary antibody solution is discarded and the chips are subjected to washing with TBST 3 × 5 min. Steps (iii) and (iv) are repeated.

   (vii) The chips are then incubated with Cy labeled secondary antibodies, for example, Cy5 labeled anti-human IgG and Cy3 labeled secondary antibody against the host in which the anti-GST primary antibody was raised.

   (viii) The chips are incubated in the dark for 2 h with gentle shaking.

   (ix) The secondary antibody is then discarded and steps (vi), (iii), and (iv) are repeated in that order. These slides can then be scanned and analyzed.

B. Data acquisition and analysis

   (i) The slides are scanned using microarray readers or scanners, which contain lasers for excitation and a PMT (photomultiplier tube) detector. Scanners could also contain LED (light-emitting diode) white light sources with multiple filters in combination to CCD (charge-coupled device) cameras. The laser power and PMT gains are parameters that must be optimized for each protocol during such experiments.[28]

   (ii) The scanner in this experiment would excite the chip at two wavelengths to acquire images for the protein features (spots) illuminated by the Cy3 and Cy5 dyes conjugated in the secondary antibodies used during the experiment.

   (iii) One can export this image in ".tiff" format. Each protein in this chip is GST tagged and hence, the anti-GST tagged antibody illuminates all the protein features in a given chip in the Cy3 channel. This is an extremely crucial aspect of quality control of an experiment. The Cy3 image of the slide is superimposed with a ".gal" template. This grid-based template contains a map of all the features and its purpose is to align each feature to its correct identity. Often during the protein printing process, certain features tend to get misaligned, not get printed at all, or be merged to form artifacts. Using the ".gal" file, the user can manually align the misaligned spots or flag spots that do not pass the requisite quality controls, so that the readouts taken in the Cy5 channel (spots illuminated due to autoantibodies in the serum) are accurate. The Cy3 channel readouts also help to normalize readings in the Cy5 channel by considering noise from the local background in the chip, in cases where the washing is not uniform.

   (iv) Once the grid in the form of the gal file is laid, the data is exported as a ".gpr," ".CEL," or ".txt" file, which varies according to the scanner used. This process assigns numeric values to each protein feature on the basis of the pixel intensities in the scanned image. Scanning parameters can affect these values and hence must be maintained post optimization across all the samples in a given experiment. A step-wise schematic of these steps is detailed in Fig. 4.

   (v) Normalization is performed on the basis of the intensities of spots obtained in the Cy3 channel and also control spots in the Cy3 and Cy5 channels. Once normalized, the data is statistically accessed for its significance and spots in the Cy5 channel showing considerable difference from the healthy cohorts are further accessed for their validity.

   (vi) Depending on the scope, various normalization strategies like quantile normalization, cyclic loess, and variance stabilizing normalization can be applied. Robust linear model normalization must be used in consultation with a statistician. Various statistical tools can be applied further to access the most significant differentially expressed proteins in isolation or combination. These include the Student's t-test, Wilcoxon rank sum test, rank product, significance analysis of microarrays, etc. There are a number of open source platforms like BRBarray tools and Bioconductor, which allow for such high-throughput analysis.

   (vii) In the clinical setting, where there is often more than a single gold standard marker, a panel of biomarkers can be used to capture the heterogeneity of cancers. This also increases the over all sensitivity and specificity of a finding. Hence, biologists look to deduce a panel of classifier proteins, which can be obtained by correspondence analysis. Using the recursive feature elimination of Support Vector Machine platforms, a panel of significant classifier proteins can be deduced. This can be tested on a validation cohort of diseased and healthy samples, and the separation between the cohorts using the panel can be visualized using multidimensional scaling (MDS) plots.

   (viii) To obtain further insight into the biology of the tumor, all the significantly differentially regulated proteins can be subject to pathway analysis tools. Several open source and commercial online platforms map a set of input proteins to cellular biochemical pathways. These platforms aid in the identification of which pathways are dysregulated and provide insight into the pathobiology of the disease. Hypotheses based on the findings can be further assessed using in vivo models for devising novel therapeutic strategies.

1. Scanning (laser settings)

2. Scanning (hardware settings)

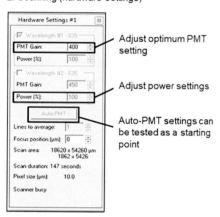

Adjust optimum PMT setting

Adjust power settings

Auto-PMT settings can be tested as a starting point

3. Scanning (adjust to optimal PMT)

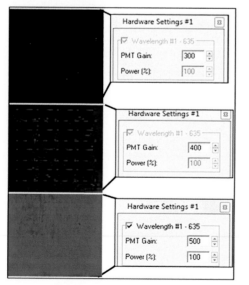

4. Slide scanning

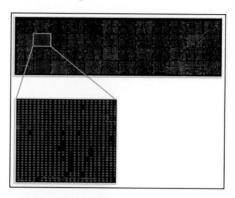

5. Data acquisition (over laying the gal file onto scanned image file)

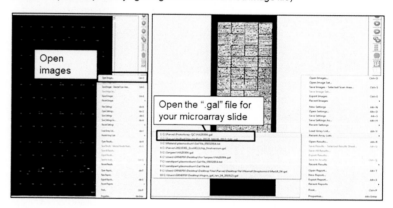

Open images

Open the ".gal" file for your microarray slide

6. Adjust individual feature parameters

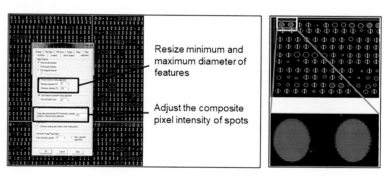

Resize minimum and maximum diameter of features

Adjust the composite pixel intensity of spots

7. Export data as ".gpr" file and analyse

FIG. 4 Layout of the software settings involved in microarray slide scanning and data acquisition: (1) laser setting panel in GenePix pro software; (2) hardware settings where PMT gain and laser power can be adjusted; (3) outcomes when PMT gains are increased or decreased for user to determine optimal PMT; (4) scanned image of a protein microarray slide with optimal settings; (5) the steps to overlay a.gal file over a scanned microarray image file; and (6) zoomed in section to show how one must manually adjust features to the correct assigned spot so that the intensity corresponding to that protein feature is correctly and carefully assigned.

Proteomics in oligodendroglioma research

While a lot of scientific literature focuses on glioma proteomics, only a handful discuss oligodendrogliomas. Until date, researchers have primarily explored gel- and MS-based techniques to understand the oligodendroglioma proteome. Bouamrani et al. used SELDI-TOF on minimally invasive biopsies to discriminate glioblastomas from oligodendrogliomas, which led to the identification of three potential markers.[32] Park et al. performed a single case study to investigate the molecular factors associated with transformation of oligodendroglioma to anaplastic oligodendroglioma using 2-DE followed by MALDI-TOF.[33] Of 23 markers, 2 were validated using western blot and immunocytochemistry. Okamoto et al. investigated the chemosensitivity of oligodendrogliomas with and without the short arm of chromosome 1 (1pLOH) using 2-DE and LC-MS/MS and found seven candidate proteins associated with chemoresistance in tumors with 1pLOH.[34] These were proposed therapeutic targets and prognostic markers. Rostomily et al. performed a similar study to understand the proteomic alterations in oligodendrogliomas with and without the 1p/19q co-deletion using quantitative proteomics and ICAT labels in conjunction with LC-MS/MS.[35] Grzendowski et al. used 2D-DIGE followed by mass spectrometric analysis to identify differentially expressed proteins in oligodendrogliomas with and without 1p/19q co-deletion.[36] They validated 4 of the 22 markers identified. Along with sequencing of sodium bisulfite-treated tumor DNA in the same study, they established epigenetic downregulation due to methylation of certain genes. Yang et al. used iTRAQ-based quantitative proteomics to identify biomarkers associated with the deletion of 1pLOH. Of 13 markers found in this study 4 were validated in a validation set of 39 samples using western blot and immunohistochemistry.[37] Thirant et al. used 2D-DIGE followed by MALDI-TOF to understand the proteomic alterations in oligodendrogliomas with IDH1 mutations.[38] A study by Khaghani-Razi-Abad et al. used a gel-based approach to identify differentially expressed proteins in oligodendroglial tumors followed by their identification using MALDI-TOF-TOF. Zhuang et al. identified glutamine synthetase as a marker differentiating oligodendrogliomas and astrocytomas using 2-DE followed by LC-MS/MS.[39] Lin et al. performed a similar analysis of oligodendrogliomas and astrocytomas to identify differentially expressed proteins using iTRAQ-based quantitation.[40] Three of the identified markers were validated using western blotting. Hong et al. performed a global proteomics analysis of anaplastic oligodendrogliomas by separating proteins from tissue lysates on an SDS PAGE gel, in-gel digesting them, and performing LC-MS/MS analysis.[41] It must be noted that most of these studies published until 2016 have specimens characterized as per the 2007 WHO classification system and essentially utilize tissue-based proteomics. Anaplastic oligodendroglioma cell lines like UPN933 have been used in a study by Redzic et al. to study unfolded protein response when compared to the U87MG cell line derived from glioblastomas using MS.[42] The CSF, serum, and plasma-based proteomics have been used in glioma proteomics studies, but there is certainly a paucity of studies dedicated to oligodendroglial tumors.[43, 44] Similarly, protein microarrays have been used to screen autoantibodies across various grades of gliomas.[45] Protein arrays have also been used for subtyping and performing expression analysis in glioblastomas.[46, 47] Perhaps autoantibody signatures and expression profiles distinguishing molecular subtypes of gliomas could be a promising diagnostic approach.

Conclusions

Clinical translation is an intrinsic rationale of most proteomics research across the world. In biological problems as pertinent and complex as Neuro-Oncology, proteomics integrates powerfully with the other -omic technologies, and provides valuable insights into not just disease biology but also aid in diagnosis, therapeutics, and predicting disease severity. In the recent past, several independent studies in glioma using proteomics have shed light on its biology and highlighted several potential biomarkers.[17, 44] However, it has been a challenging path for these markers to reach clinical translation.

The reason for this could be multifaceted. It is only recently (i.e., 2015) that studies recognized five molecular subtypes of glioma that behave distinctly from each other. When speaking of low-grade gliomas, there are three molecular subgroups. In proteomic studies to date, most studies were using the 2007 WHO grading of gliomas, which largely overlap these subgroups. Thus, even while there may have been promising biomarkers emerging from these studies, perhaps they do not apply to a larger validation cohort which may have a blend of these molecular subtypes. Furthermore, the interobserver variability with histopathology could also play a role in misclassification. It would therefore be crucial for proteomics researchers to follow the model established in the Eckel-Passow et al. study,[6] by dissecting glioma subgroups based on proteomic signatures and devising a model around their clinical behavior and response to therapy on a large cohort. This is pertinent because most existing studies in oligodendroglioma proteomics use a considerably small sample size. With data-independent acquisition methods like SWATH MS complementing shot-gun and targeted approaches like SRM, proteomics can contribute immensely to clinically translatable diagnostic and therapeutic modalities.[48, 49] With the power of resourceful bioinformatic databases, coupled with technological advances in instrumentation and global collaborations to expand sample size, there is much hope that proteomics could reduce the gap between discovery, validation, and clinical translation of diagnostic and prognostic biomarkers in Neuro-Oncology.

Acknowledgment

The authors acknowledge Ministry of Human Resources and Development (MHRD) for funding this work through the Uchhatar Avishkar Yojana (UAY) – Ministry of Human Resource Development project F. No.21-105/2015-TS.II/TC_#34 granted to Sanjeeva Srivastava.

Competing interests

The authors declare no competing financial interests.

References

1. Ostrom QT, Gittleman H, Liao P, et al. CBTRUS Statistical Report: primary brain and other central nervous system tumors diagnosed in the United States in 2010–2014. *Neuro-Oncol.* 2017;19(suppl_5):v1–v88. https://doi.org/10.1093/neuonc/nox158.

2. Louis DN, Ohgaki H, Wiestler OD, et al. The 2007 WHO classification of tumours of the central nervous system. *Acta Neuropathol (Berl).* 2007;114 (2):97–109. https://doi.org/10.1007/s00401-007-0243-4.

3. Louis DN, Perry A, Reifenberger G, et al. The 2016 World Health Organization classification of tumors of the central nervous ystem: a summary. *Acta Neuropathol (Berl).* 2016;131(6):803–820. https://doi.org/10.1007/s00401-016-1545-1.

4. Cancer Genome Atlas Research Network, Brat DJ, Verhaak RGW, et al. Comprehensive, integrative genomic analysis of diffuse lower-grade gliomas. *N Engl J Med.* 2015;372(26):2481–2498. https://doi.org/10.1056/NEJMoa1402121.

5. Suzuki H, Aoki K, Chiba K, et al. Mutational landscape and clonal architecture in grade II and III gliomas. *Nat Genet.* 2015;47(5):458–468. https://doi.org/10.1038/ng.3273.

6. Eckel-Passow JE, Lachance DH, Molinaro AM, et al. Glioma groups based on 1p/19q, IDH, and TERT promoter mutations in tumors. *N Engl J Med.* 2015;372(26):2499–2508. https://doi.org/10.1056/NEJMoa1407279.

7. Wen PY, Reardon DA. Neuro-oncology in 2015: progress in glioma diagnosis, classification and treatment. *Nat Rev Neurol.* 2016;12:69–70. https://doi.org/10.1038/nrneurol.2015.242.

8. Kosti I, Jain N, Aran D, Butte AJ, Sirota M. Cross-tissue analysis of gene and protein expression in normal and cancer tissues. *Sci Rep.* 2016;6. https://doi.org/10.1038/srep24799.

9. Kalinina J, Peng J, Ritchie JC, Van Meir EG. Proteomics of gliomas: initial biomarker discovery and evolution of technology. *Neuro-Oncol.* 2011;13 (9):926–942. https://doi.org/10.1093/neuonc/nor078.

10. Klose J. Protein mapping by combined isoelectric focusing and electrophoresis of mouse tissues. A novel approach to testing for induced point mutations in mammals. *Humangenetik.* 1975;26(3):231–243.

11. O'Farrell PH. High resolution two-dimensional electrophoresis of proteins. *J Biol Chem.* 1975;250(10):4007–4021.

12. Unlü M, Morgan ME, Minden JS. Difference gel electrophoresis: a single gel method for detecting changes in protein extracts. *Electrophoresis.* 1997;18(11):2071–2077. https://doi.org/10.1002/elps.1150181133.

13. Aebersold R, Mann M. Mass spectrometry-based proteomics. *Nature.* 2003;422:198–207. https://doi.org/10.1038/nature01511.

14. Parker CE, Warren MR, Mocanu V. Mass spectrometry for proteomics. In: Alzate O, ed. *Neuroproteomics. Frontiers in Neuroscience.* Boca Raton (FL): CRC Press/Taylor & Francis; 2010. *http://www.ncbi.nlm.nih.gov/books/NBK56011/.*

15. Singhal N, Kumar M, Kanaujia PK, Virdi JS. MALDI-TOF mass spectrometry: an emerging technology for microbial identification and diagnosis. *Front Microbiol.* 2015;6:791. https://doi.org/10.3389/fmicb.2015.00791.

16. Banerjee S, Mazumdar S. Electrospray ionization mass spectrometry: a technique to access the information beyond the molecular weight of the analyte. *Int J Anal Chem.* 2012;2012. https://doi.org/10.1155/2012/282574.

17. Somasundaram K, Nijaguna MB, Kumar DM. Glioma proteomics: methods and current perspective. *Brain Tumors—Curr Emerg Ther Strategy.* 2011; https://doi.org/10.5772/24710.

18. Catherman AD, Skinner OS, Kelleher NL. Top down proteomics: facts and perspectives. *Biochem Biophys Res Commun.* 2014;445(4):683–693. https://doi.org/10.1016/j.bbrc.2014.02.041.

19. Zhang Y, Fonslow BR, Shan B, Baek M-C, Yates JR. Protein analysis by shotgun/bottom-up proteomics. *Chem Rev.* 2013;113(4):2343–2394. https://doi.org/10.1021/cr3003533.

20. Ball CH, Roulhac PL. Multidimensional techniques in protein separations for neuroproteomics. In: Alzate O, ed. Neuroproteomics. *Frontiers in Neuroscience.* Boca Raton (FL): CRC Press/Taylor & Francis; 2010. *http://www.ncbi.nlm.nih.gov/books/NBK56015/.*

21. Zhang X, Fang A, Riley CP, Wang M, Regnier FE, Buck C. Multi-dimensional liquid chromatography in proteomics. *Anal Chim Acta.* 2010;664 (2):101–113. https://doi.org/10.1016/j.aca.2010.02.001.

22. Zhang H, Ge Y. Comprehensive analysis of protein modifications by top-down mass spectrometry. *Circ Cardiovasc Genet.* 2011;4(6):711. https://doi.org/10.1161/CIRCGENETICS.110.957829.

23. Doerr A. Targeted proteomics. *Nat Methods.* 2011;8:43. https://doi.org/10.1038/nmeth.f.329.

24. Zhu W, Smith JW, Huang C-M. Mass spectrometry-based label-free quantitative proteomics. In: *Biomed Res Int:* 2010. https://doi.org/10.1155/2010/840518.

25. Ong S-E, Mann M. Mass spectrometry–based proteomics turns quantitative. *Nat Chem Biol.* 2005;1(5):252–262. https://doi.org/10.1038/nchembio736.

26. Chahrour O, Cobice D, Malone J. Stable isotope labelling methods in mass spectrometry-based quantitative proteomics. *J Pharm Biomed Anal.* 2015;113:2–20. https://doi.org/10.1016/j.jpba.2015.04.013.

27. Reymond Sutandy F, Qian J, Chen C-S, Zhu H. Overview of protein microarrays. *Curr Protoc Protein Sci.* 2013;27. https://doi.org/10.1002/0471140864.ps2701s72.

28. Gupta S, Manubhai KP, Kulkarni V, Srivastava S. An overview of innovations and industrial solutions in protein microarray technology. *Proteomics.* 2016;16(8):1297–1308. https://doi.org/10.1002/pmic.201500429.

29. Sauer U. Analytical protein microarrays: advancements towards clinical applications. *Sensors.* 2017;17(2). https://doi.org/10.3390/s17020256.

30. Atak A, Mukherjee S, Jain R, et al. Protein microarray applications: autoantibody detection and posttranslational modification. *Proteomics.* 2016;16(19):2557–2569. https://doi.org/10.1002/pmic.201600104.

31. Wang J, Figueroa JD, Wallstrom G, et al. Plasma autoantibodies associated with basal-like breast cancers. *Cancer Epidemiol Biomark Prev Publ.* 2015;24(9):1332–1340. https://doi.org/10.1158/1055-9965.EPI-15-0047.

32. Bouamrani A, Ternier J, Ratel D, et al. Direct-tissue SELDI-TOF mass spectrometry analysis: a new application for clinical proteomics. *Clin Chem.* 2006;52(11):2103–2106. https://doi.org/10.1373/clinchem.2006.070979.

33. Park C-K, Kim JH, Moon MJ, et al. Investigation of molecular factors associated with malignant transformation of oligodendroglioma by proteomic study of a single case of rapid tumor progression. *J Cancer Res Clin Oncol.* 2008;134(2):255–262. https://doi.org/10.1007/s00432-007-0282-1.

34. Okamoto H, Li J, Gläsker S, et al. Proteomic comparison of oligodendrogliomas with and without 1pLOH. *Cancer Biol Ther.* 2007;6(3):391–396.

35. Rostomily RC, Born DE, Beyer RP, et al. Quantitative proteomic analysis of oligodendrogliomas with and without 1p/19q deletion. *J Proteome Res.* 2010;9(5):2610–2618. https://doi.org/10.1021/pr100054v.

36. Grzendowski M, Wolter M, Riemenschneider MJ, et al. Differential proteome analysis of human gliomas stratified for loss of heterozygosity on chromosomal arms 1p and 19q. *Neuro-Oncol.* 2010;12(3):243–256. https://doi.org/10.1093/neuonc/nop025.

37. Yang L, Xu X, Liu X, et al. iTRAQ-based quantitative proteomic analysis for identification of oligodendroglioma biomarkers related with loss of heterozygosity on chromosomal arm 1p. *J Proteomics.* 2012;77:480–491. https://doi.org/10.1016/j.jprot.2012.09.028.

38. Thirant C, Varlet P, Lipecka J, et al. Proteomic analysis of oligodendrogliomas expressing a mutant isocitrate dehydrogenase-1. *Proteomics.* 2011;11(21):4139–4154. https://doi.org/10.1002/pmic.201000646.

39. Zhuang Z, Qi M, Li J, et al. Proteomic identification of glutamine synthetase as a differential marker for oligodendrogliomas and astrocytomas. *J Neurosurg.* 2011;115(4):789–795. https://doi.org/10.3171/2011.5.JNS11451.

40. Lin S, Li J, Wang P, Zang Y, Zhao T. Differential proteomics analysis of oligodendrogliomas and astrocytomas using iTRAQ quantification. *J Proteomics Bioinform.* 2017;10(4):128–134. https://doi.org/10.4172/jpb.1000433.

41. Hong Y, Park EC, Shin E-Y, et al. Proteomic and bioinformatic analysis of recurrent anaplastic oligodendroglioma. *J Anal Sci Technol.* 2013;4(1):6. https://doi.org/10.1186/2093-3371-4-6.

42. Redzic JS, Gomez JD, Hellwinkel JE, Anchordoquy TJ, Graner MW. Proteomic analyses of brain tumor cell lines amidst the unfolded protein response. *Oncotarget.* 2016;7(30):47831–47847. https://doi.org/10.18632/oncotarget.10032.

43. Somasundaram K, Nijaguna MB, Kumar DM. Serum proteomics of glioma: methods and applications. *Expert Rev Mol Diagn.* 2009;9(7):695–707. https://doi.org/10.1586/erm.09.52.

44. Niclou SP, Fack F, Rajcevic U. Glioma proteomics: status and perspectives. *J Proteomics.* 2010;73(10):1823–1838. https://doi.org/10.1016/j.jprot.2010.03.007.

45. Syed P, Gupta S, Choudhary S, et al. Autoantibody profiling of glioma serum samples to identify biomarkers using human proteome arrays. *Sci Rep.* 2015;5. https://doi.org/10.1038/srep13895.

46. Hutter G, Sailer M, Azad TD, et al. Reverse phase protein arrays enable glioblastoma molecular subtyping. *J Neurooncol.* 2017;131(3):437–448. https://doi.org/10.1007/s11060-016-2316-5.

47. Ross JL, Cooper LAD, Kong J, et al. 5-Aminolevulinic acid guided sampling of glioblastoma microenvironments identifies pro-survival signaling at infiltrative margins. *Sci Rep.* 2017;7(1):15593. https://doi.org/10.1038/s41598-017-15849-w.

48. Aebersold R, Bensimon A, Collins BC, Ludwig C, Sabido E. Applications and developments in targeted proteomics: from SRM to DIA/SWATH. *Proteomics.* 2016;16(15–16):2065–2067. https://doi.org/10.1002/pmic.201600203.

49. Faria SS, Morris CFM, Silva AR, et al. A timely shift from shotgun to targeted proteomics and how it can be groundbreaking for cancer research. *Front Oncol.* 2017;7. https://doi.org/10.3389/fonc.2017.00013.

Chapter 11

Pathology of pediatric oligodendroglioma

Hope T. Richard

Department of Pathology, Virginia Commonwealth University Health, Richmond, VA, United States

Introduction

Oligodendroglial tumors represent approximately 6% of infiltrative gliomas in adults and <1% of all brain tumors in the pediatric population (<15 years old).[1] This group of tumors includes oligodendroglioma (WHO grade II) and anaplastic oligodendroglioma (WHO grade III) as well as the newly described disseminated oligodendroglial-like leptomeningeal neoplasm (DOLN). The diagnosis of oligodendroglioma portends a good prognosis as compared to other gliomas in both patient populations, due to their responsiveness to medical therapy; however, the pediatric population shows longer overall survival than their adult counterparts, particularly following complete surgical resection.[2] The most common clinical signs and symptoms include seizure and headache in both adults and children, which also show similar imaging findings.[3,4] In adults, the diagnosis of oligodendroglioma is based on specific histopathological and molecular features (IDH1 mutation and 1p/19q co-deletion). Although the histological features are similar in the pediatric population, the common molecular features noted in adult tumors are absent, making a definitive diagnosis challenging.[5] These diagnostic challenges will be discussed in detail in this chapter.

Clinical signs and symptoms

Gliomas in general show a wide variety of affected patient age groups and patient populations, with similar clinical presentations. Oligodendrogliomas are far more common in the adult population, occurring predominately in the fourth decade of life. These lesions are uncommon in children, accounting for less than 1% of primary central nervous system (CNS) tumors and 5%–18% of all gliomas that occur in children.[1,6,7] There is no significant gender predilection reported in either patient population. Due to the infiltrative nature of these tumors, many patients will present with seizures and headache, regardless of age. Focal neurologic deficits, and signs/symptoms related to increased intracranial pressure may also occur, depending on the exact location of the tumor within the cortex. A difference in tumor location has been reported for oligodendrogliomas in adults and children, the frontal lobe being a more common location in adults, as opposed to the temporal lobe in the pediatric population.[2]

Neuroimaging

Oligodendrogliomas typically arise within the cerebral hemispheres, involving the cortex and subcortical white matter. The frontal and temporal lobes are the most frequent sites of tumor origin, with parietal and occipital origination less common. Additionally, the temporal lobe has been reported to be more common in the pediatric population.[2] Rarely, multilobe involvement can occur, as well as localization within the deep gray structures, posterior fossa, and spinal cord.[7–10] Those cases previously reported as showing leptomeningeal involvement likely represent the now-recognized entity of disseminated oligodendroglial-like leptomeningeal neoplasms (DOLNs).[11,12]

On imaging, oligodendrogliomas have certain characteristics regardless of the age of the patient. Computerized tomography (CT) scans typically show a well-demarcated hypo- to isodense lesion within the cerebral cortex and subcortical white matter. Calcifications may be present and are more commonly encountered than in their astrocytic counterparts. Magnetic resonance imaging (MRI) shows oligodendroglial tumors to be hypointense on T1, and hyperintense on T2 and fluid-attenuated inversion recovery (FLAIR) images.[7,13–15] The lesions are well demarcated with generally minimal surrounding vasogenic edema, similar to that seen on CT scan. Additionally, cystic change, hemorrhage, or calcification may be seen on occasion with MRI. In higher-grade tumors (WHO grade III), lesions may show contrast enhancement with a heterogeneous appearance on most image sequences. When ring enhancement is encountered, there is a correlation with particularly poor prognosis.

Advanced MRI sequences have recently been shown to provide additional biologic, physiologic, and metabolic features of pediatric oligodendroglioma. Diffusion-weighted imaging (DWI) and DTI can help differentiate low-grade from high-grade tumors with high-grade tumors showing a low apparent diffusion coefficient (ADC) value.[7,16–18] The recently developed three-dimensional (3D) gradient-echo susceptibility weighted imaging (SWI) sequences are useful in detecting blood, blood products, and calcifications, which may also be useful in differentiating high-grade from low-grade tumors because hemorrhagic foci are typically only associated with high-grade lesions, whereas calcifications are more likely to be associated with low-grade lesions.[7,19–21] Similarly, perfusion weighted imaging (PWI) is useful in determining the vascularity of the tumor, which then may correlate with grade. High vascularity is reflective of high tumor angiogenesis and higher tumor grade, whereas the opposite is seen in low-grade lesions.[22] Finally, [1]H-magnetic resonance spectroscopy (MRS) allows for measurement of metabolites in brain tumors in a noninvasive fashion. Typically, high-grade tumors show a high choline-to-creatinine ratio, low N-acetylaspartate-to-creatinine ratio, as well as lipid and lactate peaks. MRS of high-grade oligodendrogliomas may reveal a high choline peak with associated lactate and lipid peaks.[23]

DOLNs show unique imaging characteristics as compared to intraparenchymal oligodendrogliomas. These lesions typically show leptomeningeal enhancement with cystic or nodular subpial T2 hyperintense lesions.[24,25] A discrete intraparenchymal lesion is also commonly identified in addition to the leptomeningeal lesions.

Pathology (gross and microscopic)

Oligodendroglial tumors are typically soft and gray-tan, sometimes with mucoid or hemorrhagic areas or flecks of calcification on biopsy grossly. Necrosis is commonly limited to anaplastic forms. At autopsy, these lesions appear as well-defined, gray soft masses involving the cortex and underlying white matter with expansion of the involved gyri and blurring of the gray-white junction. These findings are similar among all low-grade glial neoplasms.

Intraoperative examination of biopsy and/or resection material is important in the overall treatment of the patient. On intraoperative cytologic imprints/smears, oligodendroglial tumors contain cells with small, round nuclei typically lacking fine fibrillary processes, and a fine capillary network (Fig. 1A and B). Minigemistocytes may be present in these lesions (Fig. 2). High-grade lesions tend to be more hypercellular, with substantial nuclear atypia and endothelial proliferation, with or without necrosis.

Examination of WHO grade II oligodendrogliomas on permanent sections shows a glial neoplasm with a monomorphous population of infiltrative tumor cells having uniformly round nuclei and surrounding perinuclear clearing (Fig. 3A). This feature is artifactual, secondary to formalin fixation and unfortunately lacking in frozen section, smears, and rapidly fixed specimens. A fine capillary network is also typically identified as well as microcysts and microcalcifications (Fig. 3B). Additionally, tumor cells have a tendency to accumulate in subpial zones and form secondary structures such as perineuronal and perivascular satellitosis (Fig. 4A and B). Occasional mitotic figures and scattered atypical nuclei may be present and are still compatible with a WHO grade II designation.[26,27] Uncommon findings in otherwise typical oligodendrogliomas include cells with granular eosinophilic cytoplasm or signet ring morphology, and rhythmic cellular palisading in a polar spongioblastoma-like pattern.[28,29]

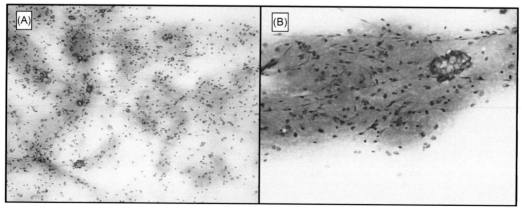

FIG. 1   A (20×) and B (40×): Intraoperative cytologic imprints/smears (Diff Quick). Neoplastic cells with small, round nuclei typically lacking fine fibrillary processes, and a fine capillary network. Focal calcifications are also noted.

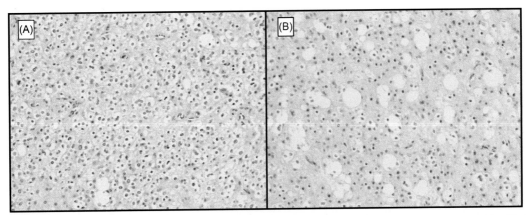

FIG. 2 Menigemistocytes showing abundant eosinophilic cytoplasm and eccentrically placed oval nuclei are also commonly identified in oligodendroglial tumors.

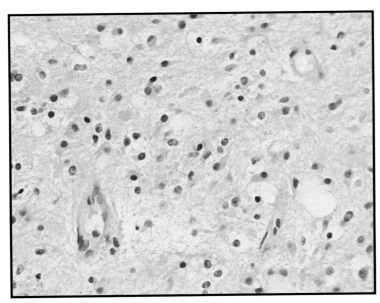

FIG. 3 Oligodendroglioma (WHO grade II). (A) Monomorphous population of infiltrative tumor cells with uniformly round nuclei and surrounding perinuclear clearing (B); a fine capillary network is also typically identified as well as microcystic changes.

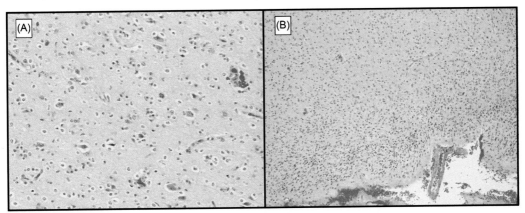

FIG. 4 Tumor cells have a tendency to form (A) secondary structures such as perineuronal and perivascular satellitosis and (B) may accumulate in subpial zones.

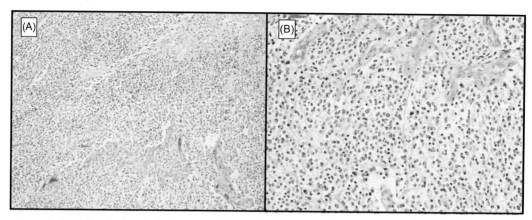

FIG. 5   A (10×) and B (20×) Anaplastic oligodendroglioma (WHO grade III). The classic oligodendroglial histological features are noted as in low-grade tumors with additional hypercellularity, numerous mitotic figures, nuclear atypia, and endothelial proliferation.

In higher-grade lesions, anaplastic oligodendroglioma (WHO grade III), the above histological features are noted in addition to hypercellularity, numerous mitotic figures, nuclear atypia, and endothelial proliferation (Fig. 5A and B). Necrosis may also be present. Hypercellular nodules with nuclear atypia and increased proliferation may be encountered in otherwise typical grade II oligodendroglioma; however, a grade III designation should not be used in the absence of diffuse features of anaplasia. Because of this, the classification and grading of oligodendroglial tumors may be less reproducible than it is for astrocytic tumors.

The more recently described entity DOLN shows the typical histological features of low-grade (WHO grade II) oligodendrogliomas, with the addition of desmoplasia in the involved leptomeninges. Rare cases may show anaplastic features.

Immunohistochemical analysis has not proven helpful in differentiating oligodendroglial and astrocytic tumors. Both tumors are positive for S100, whereas GFAP highlights minigemistocytes and other cells of astrocytic origin, with the neoplastic oligodendroglial cells typically being negative. MAP2, Olig1, and Olig2 are positive in oligodendroglial tumors; however, they may also show positivity in astrocytic lesions. Other nonspecific markers, pancytokeratin and vimentin, may be positive with no notable staining for epithelial membrane antigen (EMA).

Ultrastructural examination by electron microscopy can be performed; however, the findings are often nonspecific with concentric arrays of membranes (membrane lamination or whorls). Minigemistocytes often encountered in oligodendrogliomas contain tight bundles of intermediate filaments in their cytoplasm. Neural features have also been documented, including occasional synapse-like structures and neurosecretory granules.[30]

Molecular

The molecular alterations associated with adult oligodendrogliomas have been well studied and represent an important clinicopathologic subset of gliomas. These tumors are characterized by an unbalanced translocation of chromosomes 1 and 19, which results in the characteristic whole arm co-deletion of 1p and 19q.[5,31,32] Similar to other low-grade gliomas, oligodendrogliomas often show mutations in IDH1 or IDH2.[33] These are diagnostic features of oligodendrogliomas, which can clarify a histological diagnosis. In children, these molecular features are rare, limited mainly to oligodendrogliomas arising in children >10 years old.[5,20,26,34]

No defining genetic signature has been assigned to pediatric oligodendrogliomas[5]; however, some tumors arising in children may show overexpression of p53 by immunohistochemistry or deletion of the phosphatase and tensin homolog (PTEN) tumor suppressor gene. Deletion of p16, an alteration also shared with adult oligodendroglial tumors, is similarly more frequent in older children. Rarely, epidermal growth factor receptor (EGFR) amplifications may be encountered in high-grade examples.[34,35] For DOLN, isolated 1p deletions with concurrent BRAF-KIAA1549 gene fusion are common, which are not seen in classic cortical oligodendrogliomas.[12]

Differential diagnosis

Oligodendroglial tumors in the pediatric population are diagnostically more challenging than their adult counterparts due to a number of histologically similar tumors common to this age group. Differentiation between oligodendrogliomas and infiltrative astrocytomas requires careful histological inspection for classic oligodendroglial morphologies, including minigemistocytes and gliofibrillary oligodendrocytes.[36] Clear cell ependymoma can mimic oligodendroglial tumors, although ependymomas should show at least focal evidence of perivascular pseudorosettes, and dot-like EMA positivity on immunohistochemistry. Pilocytic astrocytomas may have areas resembling oligodendroglioma; however, there should also be the classic features of biphasic architecture, Rosenthal fibers, and eosinophilic granular bodies. The genetic alteration of a BFAF:KIAA fusion may also be associated with pilocytic astrocytomas[37]; however a single study has also associated these mutations with pediatric oligodendrogliomas.[35] Dysembryoplastic neuroepithelial tumor (DNET) typically contains oligodendroglial-like cells and may even have areas of classic oligodendroglioma. These tumors should also show mucin-rich nodules and a specific glioneuronal unit with floating neurons, which are absent in oligodendrogliomas. Additionally, molecular analysis may show a BRAF V600E mutation not seen in oligodendrogliomas.[38] Radiographic features of oligodendroglioma and DNT are also dissimilar and can be useful in differentiating these lesions. Neurocytomas may also simulate an oligodendroglial tumor by virtue of round regular nuclei; however, the diagnosis of neurocytoma can be confirmed via identification of diffuse cellular staining for neuronal markers. Clear cell meningioma potentially enters the differential diagnosis; however, these lesions should be dural-based with an immunohistochemical profile that is GFAP negative and EMA positive with PAS-positive glycogen-rich cytoplasm. Additionally, there are often areas of easily identifiable meningioma morphology present.

Prognosis and treatment

In both adult and pediatric patients, oligodendroglial tumors tend to be slower growing with a longer survival relative to their grade-matched astrocytoma counterparts. Time to recurrence and malignant progression are also prolonged in comparison. Specifically in pediatric cases, gross total resection and low tumor grade are strongly associated with improved progression-free survival (PFS) and overall survival (OS).[2,39–41] Overall, 5-year PFS and OS rates of 81 and 84%, respectively, have been reported for these tumors.[39,42] The presence of p53 overexpression, *PTEN* deletion, and enhancement on imaging have been found to be associated with shorter OS in pediatric patients.[42–44] Deletions of 1p/19q in pediatric oligodendroglial tumors, unlike similar alterations in their adult counterparts, do not afford a favorable outcome. Additionally, deep/central tumors that cannot be resected show a significantly shorter median survival of less than 2 years, as compared to their cerebral-centered counterparts.[44]

Following incomplete resection or in cases of high-grade tumors having undergone any level of resection, radiation and chemotherapy may increase overall survival (OS) time. Several clinical trials have studied the effects of radiation and chemotherapy alone or in conjunction for these lesions. These studies have found increased progression-free survival with early adjuvant radiation with a similar overall survival in low-grade gliomas following gross total resection. For high-grade tumors and low-grade tumors with marginal resection, a combination of chemotherapy and radiation has proven to be more effective.[2,3,45–50] The typical chemotherapeutic regimen included procarbazine, lomustine, and vincristine (PCV), although temozolamide (TMZ) has more recently shown promise as a chemotherapeutic alternative.[5] DOLNs have a variable clinical course, with most of which being indolent despite the often extensive tumor burden.

References

1. Ostrom QT, Gittleman H, Liao P, et al. CBTRUS statistical report: primary brain and central nervous system tumors diagnosed in the United States in 2007–2011. *Neuro Oncol.* 2014;16(Suppl 4):iv1–63.
2. Lau CS, Mahendraraj K, Chamberlain RS. Oligodendrogliomas in pediatric and adult patients: an outcome-based study from the Surveillance, Epidemiology, and End Result database. *Cancer Manag Res.* 2017;9:159–166.
3. Koeller KK, Rushing EJ. From the archives of the AFIP: oligodendroglioma and its variants: radiologic-pathologic correlation. *Radiographics.* 2005;25(6):1669–1688.
4. Lee YY, Van Tassel P. Intracranial oligodendrogliomas: imaging findings in 35 untreated cases. *AJR Am J Roentgenol.* 1989;152(2):361–369.
5. Nauen D, Haley L, Lin MT, et al. Molecular analysis of pediatric oligodendrogliomas highlights genetic differences with adult counterparts and other pediatric gliomas. *Brain Pathol.* 2016;26(2):206–214.
6. Chawla S, Krejza J, Vossough A, et al. Differentiation between oligodendroglioma genotypes using dynamic susceptibility contrast perfusion-weighted imaging and proton MR spectroscopy. *AJNR Am J Neuroradiol.* 2013;34(8):1542–1549.
7. Wagner MW, Poretti A, Huisman TA, et al. Conventional and advanced (DTI/SWI) neuroimaging findings in pediatric oligodendroglioma. *Childs Nerv Syst.* 2015;31(6):885–891.
8. Fukuoka K, Yanagisawa T, Watanabe Y, et al. Brainstem oligodendroglial tumors in children: two case reports and review of literatures. *Childs Nerv Syst.* 2015;31(3):449–455.
9. Ellenbogen JR, Perez S, Parks C, et al. Cerebellopontine angle oligodendroglioma in a child: first case report. *Childs Nerv Syst.* 2014;30(1):185–187.
10. Nam DH, Cho BK, Kim YM, et al. Intramedullary anaplastic oligodendroglioma in a child. *Childs Nerv Syst.* 1998;14(3):127–130.
11. Schniederjan MJ, Alghamdi S, Castellano-Sanchez A, et al. Diffuse leptomeningeal neuroepithelial tumor: 9 pediatric cases with chromosome 1p/19q deletion status and IDH1 (R132H) immunohistochemistry. *Am J Surg Pathol.* 2013;37(5):763–771.

12. Rodriguez FJ, Schniederjan MJ, Nicolaides T, et al. High rate of concurrent BRAF-KIAA1549 gene fusion and 1p deletion in disseminated oligodendroglioma-like leptomeningeal neoplasms (DOLN). *Acta Neuropathol.* 2015;129(4):609–610.

13. Connor SE, Gunny R, Hampton T, et al. Magnetic resonance image registration and subtraction in the assessment of minor changes in low grade glioma volume. *Eur Radiol.* 2004;14(11):2061–2066.

14. Pallud J, Capelle L, Taillandier L, et al. Prognostic significance of imaging contrast enhancement for WHO grade II gliomas. *Neuro Oncol.* 2009;11(2):176–182.

15. Gerin C, Pallud J, Deroulers C, et al. Quantitative characterization of the imaging limits of diffuse low-grade oligodendrogliomas. *Neuro Oncol.* 2013;15(10):1379–1388.

16. Kono K, Inoue Y, Nakayama K, et al. The role of diffusion-weighted imaging in patients with brain tumors. *AJNR Am J Neuroradiol.* 2001;22(6):1081–1088.

17. Jäger HR, Waldman AD, Benton C, et al. Differential chemosensitivity of tumor components in a malignant oligodendroglioma: assessment with diffusion-weighted, perfusion-weighted, and serial volumetric MR imaging. *AJNR Am J Neuroradiol.* 2005;26(2):274–278.

18. Jenkinson MD, Smith TS, Brodbelt AR, et al. Apparent diffusion coefficients in oligodendroglial tumors characterized by genotype. *J Magn Reson Imaging.* 2007;26(6):1405–1412.

19. Louis D, Ohgaki H, Wiestler O, et al. *WHO Classification of Tumours of the Central Nervous System.* 4th ed. Lyon: IARC; 2007.

20. Kreiger PA, Okada Y, Simon S, et al. Losses of chromosomes 1p and 19q are rare in pediatric oligodendrogliomas. *Acta Neuropathol.* 2005;109(4):387–392.

21. Engelhard HH, Stelea A, Mundt A. Oligodendroglioma and anaplastic oligodendroglioma: clinical features, treatment, and prognosis. *Surg Neurol.* 2003;60(5):443–456.

22. Cha S, Tihan T, Crawford F, et al. Differentiation of low-grade oligodendrogliomas from low-grade astrocytomas by using quantitative blood-volume measurements derived from dynamic susceptibility contrast-enhanced MR imaging. *AJNR Am J Neuroradiol.* 2005;26(2):266–273.

23. Xu M, See SJ, Ng WH, et al. Comparison of magnetic resonance spectroscopy and perfusion-weighted imaging in presurgical grading of oligodendroglial tumors. *Neurosurgery.* 2005;56(5):919–926. discussion 919–926.

24. Preuss M, Christiansen H, Merkenschlager A, et al. Disseminated oligodendroglial-like leptomeningeal tumors: preliminary diagnostic and therapeutic results for a novel tumor entity [corrected]. *J Neurooncol.* 2015;124(1):65–74.

25. Rodriguez FJ, Perry A, Rosenblum MK, et al. Disseminated oligodendroglial-like leptomeningeal tumor of childhood: a distinctive clinicopathologic entity. *Acta Neuropathol.* 2012;124(5):627–641.

26. Rodriguez FJ, Tihan T, Lin D, et al. Clinicopathologic features of pediatric oligodendrogliomas: a series of 50 patients. *Am J Surg Pathol.* 2014;38(8):1058–1070.

27. Razack N, Baumgartner J, Bruner J. Pediatric oligodendrogliomas. *Pediatr Neurosurg.* 1998;28(3):121–129.

28. Takei Y, Mirra SS, Miles ML. Eosinophilic granular ceels in oligodendrogliomas. An ultrastructural study. *Cancer.* 1976;38(5):1968–1976.

29. Kros JM, van den Brink WA, van Loon-van Luyt JJ, et al. Signet-ring cell oligodendroglioma—report of two cases and discussion of the differential diagnosis. *Acta Neuropathol.* 1997;93(6):638–643.

30. Liberski PP. The ultrastructure of oligodendroglioma: personal experience and the review of the literature. *Folia Neuropathol.* 1996;34(4):206–211.

31. Griffin CA, Burger P, Morsberger L, et al. Identification of der(1;19)(q10;p10) in five oligodendrogliomas suggests mechanism of concurrent 1p and 19q loss. *J Neuropathol Exp Neurol.* 2006;65(10):988–994.

32. Jenkins RB, Xiao Y, Sicotte H, et al. A low-frequency variant at 8q24.21 is strongly associated with risk of oligodendroglial tumors and astrocytomas with IDH1 or IDH2 mutation. *Nat Genet.* 2012;44(10):1122–1125.

33. Yan H, Parsons DW, Jin G, et al. IDH1 and IDH2 mutations in gliomas. *N Engl J Med.* 2009;360(8):765–773.

34. Raghavan R, Balani J, Perry A, et al. Pediatric oligodendrogliomas: a study of molecular alterations on 1p and 19q using fluorescence in situ hybridization. *J Neuropathol Exp Neurol.* 2003;62(5):530–537.

35. Kumar A, Pathak P, Purkait S, et al. Oncogenic KIAA1549-BRAF fusion with activation of the MAPK/ERK pathway in pediatric oligodendrogliomas. *Cancer Genet.* 2015;208(3):91–95.

36. Inagawa H, Ishizawa K, Hirose T. Qualitative and quantitative analysis of cytologic assessment of astrocytoma, oligodendroglioma and oligoastrocytoma. *Acta Cytol.* 2007;51(6):900–906.

37. Jones DT, Hutter B, Jäger N, et al. Recurrent somatic alterations of FGFR1 and NTRK2 in pilocytic astrocytoma. *Nat Genet.* 2013;45(8):927–932.

38. Chappé C, Padovani L, Scavarda D, et al. Dysembryoplastic neuroepithelial tumors share with pleomorphic xanthoastrocytomas and gangliogliomas BRAF(V600E) mutation and expression. *Brain Pathol.* 2013;23(5):574–583.

39. Creach KM, Rubin JB, Leonard JR, et al. Oligodendrogliomas in children. *J Neurooncol.* 2012;106(2):377–382.

40. Shaw EG, Scheithauer BW, O'Fallon JR, et al. Oligodendrogliomas: the Mayo Clinic experience. *J Neurosurg.* 1992;76(3):428–434.

41. Snyder LA, Wolf AB, Oppenlander ME, et al. The impact of extent of resection on malignant transformation of pure oligodendrogliomas. *J Neurosurg.* 2014;120(2):309–314.

42. Pollack IF, Finkelstein SD, Burnham J, et al. Association between chromosome 1p and 19q loss and outcome in pediatric malignant gliomas: results from the CCG-945 cohort. *Pediatr Neurosurg.* 2003;39(3):114–121.

43. Tice H, Barnes PD, Goumnerova L, et al. Pediatric and adolescent oligodendrogliomas. *AJNR Am J Neuroradiol.* 1993;14(6):1293–1300.

44. Peters O, Gnekow AK, Rating D, et al. Impact of location on outcome in children with low-grade oligodendroglioma. *Pediatr Blood Cancer.* 2004;43(3):250–256.

45. Rusthoven CG, Carlson JA, Waxweiler TV, et al. The impact of adjuvant radiation therapy for high-grade gliomas by histology in the United States population. *Int J Radiat Oncol Biol Phys.* 2014;90(4):894–902.

46. van den Bent MJ, Afra D, de Witte O, et al. Long-term efficacy of early versus delayed radiotherapy for low-grade astrocytoma and oligodendroglioma in adults: the EORTC 22845 randomised trial. *Lancet*. 2005;366(9490):985–990.

47. van den Bent MJ, Carpentier AF, Brandes AA, et al. Adjuvant procarbazine, lomustine, and vincristine improves progression-free survival but not overall survival in newly diagnosed anaplastic oligodendrogliomas and oligoastrocytomas: a randomized European Organisation for Research and Treatment of Cancer phase III trial. *J Clin Oncol*. 2006;24(18):2715–2722.

48. van den Bent MJ, Brandes AA, Taphoorn MJ, et al. Adjuvant procarbazine, lomustine, and vincristine chemotherapy in newly diagnosed anaplastic oligodendroglioma: long-term follow-up of EORTC brain tumor group study 26951. *J Clin Oncol*. 2013;31(3):344–350.

49. Cairncross JG, Macdonald DR. Successful chemotherapy for recurrent malignant oligodendroglioma. *Ann Neurol*. 1988;23(4):360–364.

50. Cairncross G, Berkey B, Shaw E, et al. Phase III trial of chemotherapy plus radiotherapy compared with radiotherapy alone for pure and mixed anaplastic oligodendroglioma: Intergroup Radiation Therapy Oncology Group Trial 9402. *J Clin Oncol*. 2006;24(18):2707–2714.

Section C

Neuro-imaging

Chapter 12

CT imaging of oligodendroglioma

Anthony Aquino, H. Wayne Slone and Eric C. Bourekas

Department of Radiology, Ohio State Wexner Medical Center, Columbus, OH, United States

Introduction

In the practice of imaging of neoplasms of the brain, magnetic resonance imaging (MRI) is superior to computed tomography (CT) in fully characterizing the features and extent of a given neoplastic process. However, certain brain lesions, particularly oligodendroglioma, have particular imaging characteristics that can allow interpreting physicians to suggest this diagnosis, and to some degree, assess the extent of the disease process. CT features of oligodendroglioma will be reviewed, in order to demonstrate its usefulness in assessing this disease entity and in determining how CT can be complimentary to MRI.

On initially assessing a CT scan in a patient with oligodendroglioma, particularly in a patient who has not yet been diagnosed, the initial findings that need to be recognized are those that are less specific to this neoplasm, but that guide physicians that the patient's pathology may indeed be neoplastic. For example, several lower-grade primary brain lesions may present as a solitary focal area of low attenuation in the brain on CT. Considering these neoplastic etiologies as a differential consideration at presentation when initial CT is obtained is critical. For example, confusing this entity for pathology related to trauma or ischemia can drastically alter management (see Fig. 1). As expected, oligodendroglioma typically presents as a hypodense lesion on CT, not necessarily a distinguishing CT feature for this lesion. It has been estimated that approximately 60% of these tumors are hypoattenuating.[1] While internal hemorrhage (see Fig. 2) or cystic changes are occasionally visualized, these are not distinguishing features. There are however more specific features of oligodendroglioma that can be noted on CT that can suggest this as a possible diagnosis, when MRI has not yet been obtained.

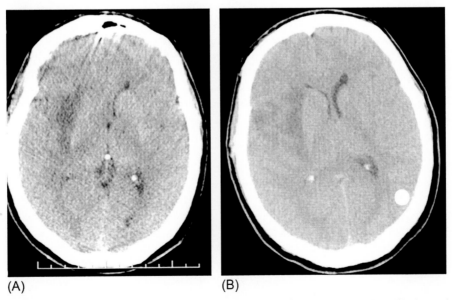

(A)   (B)

FIG. 1 (A) A noncontrast head CT image from a 45-year-old male who presented after a motor vehicle accident. The image demonstrated confluent hypoattenuation involving the cortical and subcortical region in the right insula as well as in the right frontal region (not shown), reported as a nonhemorrhagic contusion on the initial CT. The radiologist interpreting the follow-up head CT the next day (not shown, essentially unchanged in the appearance) raised concern for a right MCA territory acute infarct. (B) A noncontrast head CT image from a later date, after the patient presented with altered mental status, with the findings essentially unchanged. MRI was subsequently performed, and pathology would reveal a WHO Grade II oligodendroglioma.

Oligodendroglioma. https://doi.org/10.1016/B978-0-12-813158-9.00012-8

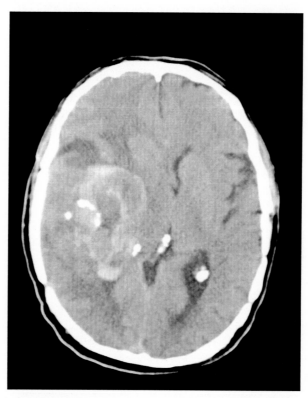

FIG. 2 Noncontrast head CT in this 59-year-old male with a history of previously resected anaplastic variant oligodendroglioma demonstrates a mass with pronounced hemorrhagic component, as well as coarse calcifications. Pathology proved recurrent anaplastic (WHO Grade III) oligodendroglioma.

A feature of oligodendroglioma that can be assessed on CT, as well as on MRI, is the typical location of the lesion. First, like most primary brain neoplasms in adults, this neoplasm is usually in the supratentorial brain, most commonly in the frontal lobes. These lesions occur less commonly in the occipital lobe than in the other portions of the cerebral hemisphere.[2] Of note, those lesions containing 1p and 19q deletions characteristically occur in the frontal lobe, and often cross midline, while their counterparts lacking those deletions may be more likely to occur in the insular or temporal region.[3] Additionally, while several types of neoplasm can cause hypodensity in the subcortical white matter, oligodendroglioma often presents as a round or ovoid mass-like lesion involving the immediate subcortical white matter and the overlying cortical mantle. This location preference often leads to loss of the interface between gray and white matter as a dominant finding on CT in these patients (see Figs. 3–6). Like all neoplastic processes, these lesions can sometimes occur in atypical locations. For example, 3%–8% of these tumors have been reported to occur within the ventricular system.[4] While rare, intraventricular oligodendrogliomas are worthy of mention, as they can have notable differences in their imaging appearance on CT. In particular, these lesions typically demonstrate increased attenuation compared to surrounding parenchyma, and enhancement is essentially always observed.[4]

The aforementioned preferential location of this lesion within the cortex and underlying peripheral white matter, leads us to an additional potential imaging feature that can often be better appreciated on CT than on MRI, which is changes of the adjacent calvarium. Ultimately, if these tumors demonstrate sufficient size, they can cause thinning and scalloping of the adjacent calvarium (see Fig. 7). Those tumors that demonstrate deletions of the 1p and 19q alleles have been particularly noted to be less likely to cause changes to the adjacent bone.[5] As one would expect, given the superior resolution of bone on CT, and in particular cortex of the calvarium relative to MRI, assessment for these findings would be a particular area where CT can add value to imaging of oligodendroglioma patients, even in lesions that have already been characterized with MRI.

Perhaps the most characteristic CT feature of oligodendroglioma is calcification. Evaluation of calcification is one area where CT is superior to MRI. In this finding CT can be a complement to MRI, as this feature is often much less conspicuous on MRI. Oligodendroglioma is the most likely primary brain tumor to develop calcifications, with reported rates ranging from 40% to 80%.[5] Tumors demonstrating 1p and 9q deletions are considered more likely to demonstrate calcifications. In general, these tumors are noted to contain calcifications with a coarse configuration, which may differentiate this lesion from other calcifying brain tumors on CT. Also, a band of calcification in a gyriform pattern along the cortex is relatively

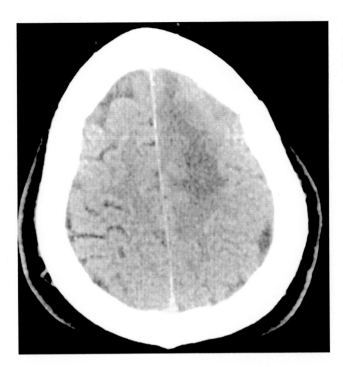

FIG. 3 Noncontrast head CT was performed in a 57-year-old male presenting with seizure. This lesion, which proved to be an oligodendroglioma (WHO Grade II), demonstrates characteristically common frontal lobe location, with involvement of the cortex and subcortical white matter. Postcontrast imaging was not performed.

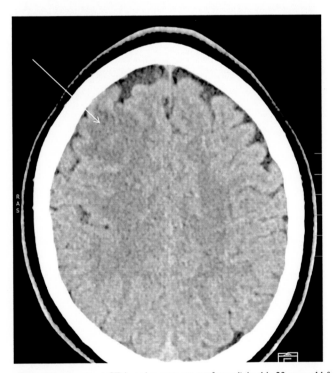

FIG. 4 Noncontrast head CT was performed (postcontrast CT imaging was not performed) in this 23-year-old female presenting with headache and seizure. Pathology of this infiltrative right frontal lobe hypodense lesion (arrow) demonstrated WHO Grade II oligodendroglioma.

specific for this tumor (see Fig. 8). As anticipated, appearance becomes less specific with smaller lesions, which may not contain these characteristic calcifications. Nonetheless, this CT imaging feature of these tumors is a definite area where the CT imaging modality can truly aid and be an adjunct to MRI in suggesting this diagnosis when the differential considerations for a neoplasm may include other possibilities based on MRI (see Figs. 9–11).

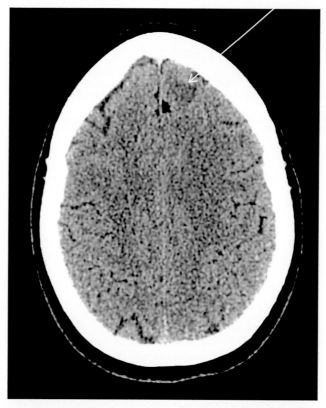

FIG. 5 Noncontrast head CT image in this 34-year old female demonstrates a subtle hypoattenuating left medial frontal lesion in the cortical-subcortical region. Pathology on resection confirmed WHO Grade II oligodendroglioma.

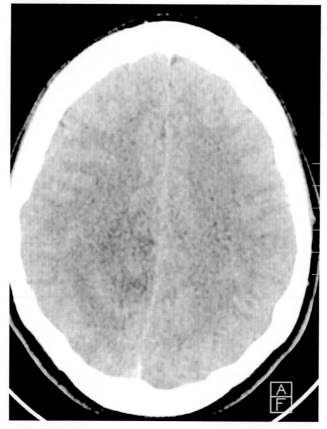

FIG. 6 Noncontrast head CT images in this 30-year-old male (postcontrast CT imaging was not performed) show a medial right parietooccipital cortical-subcortical hypodense lesion. Note the absence of calcification in this case. Pathology on resection confirmed WHO Grade II oligodendroglioma.

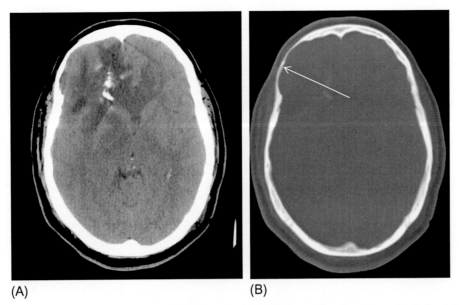

(A)   (B)

FIG. 7   (A) A noncontrast head CT image from a 53-year-old male patient demonstrating confluent hypoattenuation in the right frontal cortical and subcortical region, crossing midline, and containing internal calcifications. (B) From the same head CT examination, when viewed with bone window, clearly demonstrates scalloping and thinning of the adjacent right frontal calvarium (arrow). Pathology demonstrated WHO Grade II Oligodendroglioma.

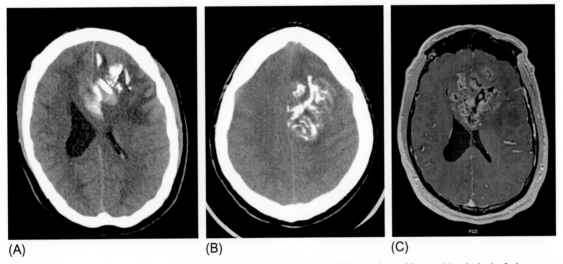

(A)   (B)   (C)

FIG. 8   (A) A noncontrast head CT image from a 57-year-old male and (B) a noncontrast CT image from a 33-year-old male, both of whom presented with seizure. The left frontal lobe mass in each of these images demonstrate significant coarse calcification, including characteristic "gyriform" calcification along the cortex. Pathology in both cases demonstrated anaplastic (WHO Grade III) oligodendroglioma. (C) A postcontrast T1 MRI image is from the same patient as in B, at essentially the same level. Note the extensive heterogeneous enhancement (this area demonstrated only minimal intrinsic T1 signal prior to contrast administration). The previously demonstrated CT image in B would be needed to strongly consider the diagnosis of oligodendroglioma in this case.

Although CT imaging of the head, especially when performed in the acute setting or as a compliment to MRI, is more often performed without contrast, the potential contrast enhancement observed on CT in patients with oligodendroglioma warrants discussion. While a lack of contrast enhancement is considered typical, contrast enhancement can sometimes be observed on CT. In fact, the presence of enhancement can be a distinguishing feature that separates oligodendroglioma from other low-grade gliomas. With reported rates of enhancement ranging from 20% to 50%,[4, 6] the enhancement of these lesions is attributed to a "chicken-wire" network of capillaries that is pathologically observed in these tumors.[5] When these tumors are found in children, contrast enhancement as well as the previously discussed characteristic calcifications, are found to be less common.[7] The WHO Grade III anaplastic variant of oligodendroglioma tends to demonstrate enhancement more often, but ultimately this is not considered a reliable feature in grading these tumors (Fig. 12).

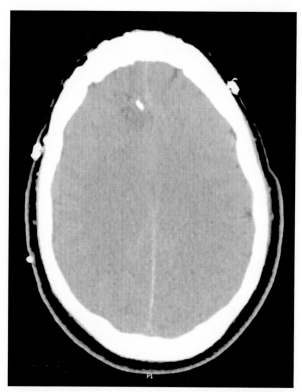

FIG. 9   Noncontrast head CT image in a 45-year-old male with history of right frontal approach resection of oligodendroglioma (WHO Grade II) demonstrates focal cortical hypodensity with internal coarse calcification. Postcontrast CT imaging was not performed. On subsequent resection, pathology confirmed recurrent WHO Grade II oligodendroglioma.

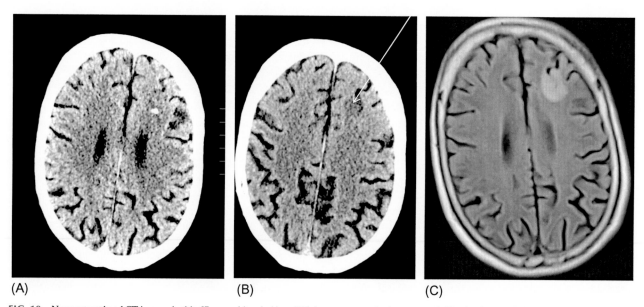

(A)   (B)   (C)

FIG. 10   Noncontrast head CT images in this 67-year-old male (A and B) demonstrate a single-coarse calcification in the left frontal subcortical region as the dominant CT finding, with some mild adjacent superior focal hypoattenuation (arrow). Axial FLAIR MRI image (C) in the same patient demonstrates a more striking area of abnormality, with increased FLAIR signal involving the corresponding left frontal cortical-subcortical parenchyma. Pathology demonstrated WHO Grade II oligodendroglioma.

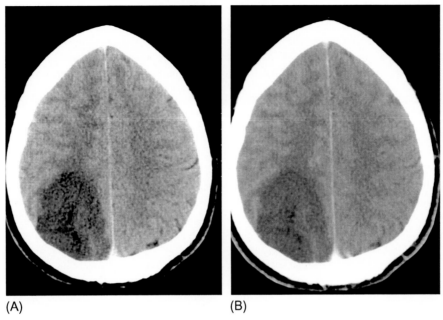

(A) (B)

FIG. 11 Precontrast (A) and postcontrast (B) head CT images in this 47-year-old male patient demonstrate a large hypodense mass in the right parietal cortical-subcortical region, with no significant enhancement appreciated. Pathology demonstrated WHO Grade II oligodendroglioma.

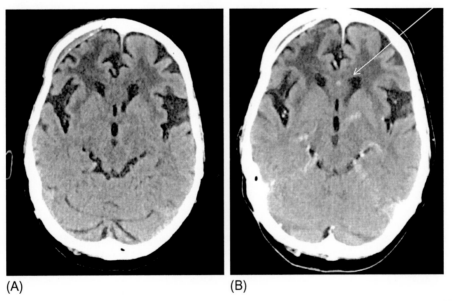

(A) (B)

FIG. 12 Noncontrast (A) and post-contrast (B) head CT images from a 68-year-old female with a history of oligodendroglioma (WHO Grade II), status post-radiation therapy (apparent postsurgical changes were related to overlying scalp infection). A focus of enhancement is now clearly seen in the genu of the corpus callosum (arrow), as well as confluent hypodensity in the surrounding bifrontal region. The enhancement, as well as the "butterfly" pattern spanning the anterior corpus callosum, is more often observed with the anaplastic variant of oligodendroglioma. Biopsy confirmed WHO Grade III anaplastic oligodendroglioma.

Another potential CT technique in which brain neoplasms can be characterized is CT cerebral perfusion. This technique capitalizes on the direct proportional relationship between concentration of iodinated contrast in a given tissue and attenuation. This study is essentially performed with cinematic CT imaging of the passage of administered contrast through the intracranial vasculature, with subsequent generation of time-contrast concentration curves in pixels of parenchyma, as well as in an arterial and venous regions of interest. Parameters such as mean transit time (MMT), cerebral blood flow (CBF), and cerebral blood volume (CBV) can then be generated. The CBV is ascertained utilizing the area under the time-concentration curve for a given parenchymal tissue divided by the area under an arterial curve.[8] It is this parameter that is typically used to further characterize brain neoplasms. As one would expect, higher-grade brain neoplasms, due to presumed increased vascularity, should demonstrate increased CBV on perfusion imaging. However, this relationship does not appear to hold true in the case of evaluating oligodendroglioma. On assessment, these tumors appear to demonstrate areas of greater CBV values than tumors of similar histological grade, which is attributed to the rich capillary network supplying these tumors.[9] While perfusion imaging of these tumors will likely more often be encountered on MRI imaging than CT, it can be helpful for providers to be aware of this technique if and when it is performed on patients with this neoplasm.

An additional technique in CT imaging that can have implications in patients with oligodendroglioma is Dual Energy CT (DECT). This technique applies two different energy settings simultaneously in order to aid in separating specific materials based on attenuation characteristics at both high-energy and low-energy settings.[10] Very common utilities of DECT in Neuroradiology include differentiation of acute hemorrhage in an acute stroke from hyperdense staining from iodinated contrast, as well as attempting to identify enhancing tumor in the setting of superimposed hemorrhage. While both of these examples capitalize on ability to separate iodine from hemorrhage, the characteristic calcifications that often are present in oligodendroglioma present another use for DECT. If this neoplasm is being initially discovered on CT, with no prior known diagnosis, it can be useful for an interpreting physician to be certain that hyperdense foci within this lesion are due to calcification instead of acute blood. Furthermore, even in those patients with a known diagnosis, it can be essential to evaluate for hemorrhagic changes in the tumor if the patient is presenting with acute clinical changes. Essentially, the dual-energy technique can create "virtual non-calcium" images that will subtract any hyperdensity due to calcification, and therefore any hyperdensity on these images on a noncontrast head CT can be presumed to be due to hemorrhage. Most significantly, the ability of DECT to differentiate material composition has been shown to clearly increase accuracy of differentiating intracranial calcification from hemorrhage in the emergency setting.[11]

Ultimately, CT is overall inferior to MRI in the imaging assessment of oligodendroglioma, as is the case with other brain neoplasms. However, a goal of this chapter was to illuminate how CT can act as an adjunct to MRI in assessment of these tumors, particularly in visualization of intratumoral calcifications and in assessment for changes of the calvarium adjacent to these lesions. An additional goal was to introduce ancillary techniques such as dual energy CT and CT perfusion and how they can be applied in patients with these brain neoplasms when clinically necessary. Also, the information and figures provided here can hopefully help in the evaluation of oligodendroglioma when CT is performed in lieu of MRI, either in the acute setting, as the initial evaluation, or in patients with conditions that preclude MRI.

REFERENCES

1. Lee YY, Van Tassel P. Intracranial oligodendrogliomas: imaging findings in 35 untreated cases. *Am J Roentgenol*. 1989;152:361–369.
2. Reni M, Mazza E, Zanon S, et al. Central nervous system gliomas. *Crit Rev Oncol Hematol*. 2017;113:213–234.
3. Zlatescu M, Tehrani Y, et al. Tumor location and growth pattern correlate with genetic signature in oligodendroglial neoplasms. *Cancer Res*. 2001;61:6713–6715.
4. Koeller K, Rushing E. Oligodendroglioma and its variants: radiologic-pathologic correlation. *RadioGraphics*. 2005;25(6):1669–1688.
5. Nadgir R, Yousem D. *Neuroradiology: The Requisites*. 4th ed. Elsevier; 2017.
6. Smits M. Imaging of oligodendroglioma. *Br J Radiol*. 2016;89(1060).
7. Tice H, Barnes P, et al. Pediatric and adolescent oligodendrogliomas. *Am J Neuroradiol*. 1993;14(6):1293–1300.
8. Hoeffner E, Case I, Jain R, et al. Cerebral perfusion CT: technique and clinical applications. *Radiology*. 2004;231(3):632–644.
9. Ding B, Ling H, Chen K, et al. Comparison of cerebral blood volume and permeability in preoperative grading of intracranial glioma using CT perfusion imaging. *Neuroradiology*. 2006;48:773–781.
10. De Cecco C, Darnell A, Rengo M, et al. Dual energy CT: oncologic applications. *AJR*. 2012;S98–S105.
11. Hu R, Basheli L, Young J, et al. Dual-energy head CT enables accurate distinction of intraparenchymal hemorrhage from calcification in emergency department patients. *Radiology*. 2016;280(1):177–183.

Chapter 13

Routine and advanced magnetic resonance imaging of oligodendrogliomas

Nandita Guha-Thakurta* and John W. Henson[†]

*Department of Diagnostic Radiology, Division of Diagnostic Imaging, The University of Texas MD Anderson Cancer Center, Houston, TX, United States;
[†]Ivy Center for Advanced Brain Tumor Treatment, Seattle, WA, United States

Introduction

The diagnosis, management, and experimental therapeutics of diffuse gliomas depend heavily on neuroimaging. This chapter will cover these subjects with a focus on imaging features that are specific to oligodendrogliomas. We will discuss routine diagnostic imaging techniques as well as advanced imaging features. The recent redefinition of oligodendroglioma based on the presence of 1/p19q deletion provides a new opportunity to examine whether molecularly defined subsets of gliomas have unique correlates with anatomic and physiologic features as defined by imaging.[1]

Diagnostic imaging of diffuse gliomas
Basic principles

The diagnosis of a diffuse glioma of the adult cerebral hemispheres can be suspected with a reasonable degree of confidence on neuroimaging prior to histological examination of pathological tissue. The details of the clinical history, such as duration of symptoms and presence of any known illnesses, can be used to refine the possible causes of a brain lesion.

Gliomas occur on a spectrum ranging from a well-defined mass lesion with little evidence of infiltration into the surrounding brain parenchyma, to diffuse, highly infiltrative lesions. Most lesions have a combination of these findings, with a focal mass and surrounding T2 changes consistent with edema or infiltration.

The signal change in the lesion and adjacent brain is almost always hyperintense compared to normal brain on T2-weighted images [i.e., fast spin echo T2, and fluid attenuated inversion recovery (FLAIR) T2], although areas of densely cellular tumor in high-grade tumors may demonstrate minimal hyperintensity, or rarely hypointensity in comparison to normal brain. Infiltrating tumor (as compared to a mass lesion) typically shows mild T2 hyperintensity whereas vasogenic edema is very bright on T2-weighted images with a pattern of "finger-like" extension into the white matter. These two appearances are sometimes easily distinguished, but the very bright signal intensity of edema can obscure the underlying tumor infiltration. Notably, the presence of vasogenic edema implies leaky vasculature and so should be considered a worrisome sign of anaplasia. The T2-weighted hyperintensity of infiltrating tumor can have varying degrees of increased signal. Areas of heterogeneous signal may reflect cystic degeneration, hemorrhage, and/or calcification, with the latter being an important consideration in the subsequent discussion of signal heterogeneity of oligodendrogliomas.

Diffusion weighted imaging (DWI) is based on the Brownian random movement of water molecules in brain and tumor tissue. The measured water diffusion coefficient is the apparent diffusion coefficient (ADC) map, a quantifiable parameter (value in units of mm^2/s), which can be obtained on routine MRI (magnetic resonance imaging) studies using the region of interest tool. Densely cellular tumors show restricted, or decreased, diffusion of water compared to normal brain, with areas of restricted diffusion appearing as hypointensity on ADC maps. In comparison to tumor, areas of brain edema have increased rates of diffusion of water and are hyperintense on ADC maps.

Susceptibility weighted imaging (SWI) depends on the distortion of the local magnetic field by substances which have paramagnetic, diamagnetic, or ferromagnetic properties resulting in signal loss called susceptibility effect, for instance, as with the iron in ferritin following hemorrhage. Calcification and old hemorrhage both show signal loss on SWI.

Areas of susceptibility effect with hemosiderin may be markedly hypointense on SWI and produce the so-called "blooming" effects that can overstate the size of the lesion.

Following intravenous administration of the contrast agent gadolinium, tumors show hyperintensity on T1-weighted images compared to the precontrast image when large numbers of vessels or newly developed tumor-related vessels (neovascularity) that lack a blood-brain barrier are present. While the presence of enhancement in infiltrative astrocytomas is an indication of anaplastic changes, it is not necessary so in the case of oligodendrogliomas.

Technical considerations

Serial comparison of imaging studies allows the neuroradiologist and brain tumor specialist to determine changes within the tumor during conventional therapy, periods of surveillance, or quantitatively for purposes of a clinical trial.[2] For this reason, patients benefit from a standardized brain tumor imaging protocol to eliminate technical differences that could confound the interpretation of changes between studies.

The MRI with gadolinium is the preferred imaging modality, but contrast-enhanced CT (computed tomography) can be used for tumor assessment in patients who cannot undergo MRI due to large body size, presence of metallic foreign bodies, an allergy to gadolinium, and in certain patients with a pacemaker. Additionally the same imaging modality, viz. MRI or CT, should be maintained in the serial evaluation of every patient. Patients who cannot be administered intravenous gadolinium (e.g. during pregnancy, allergy to gadolinium), noncontrast MRI is preferred over noncontrast CT due to its better tissue resolution. Previously contrast was withheld in patients with renal insufficiency due to risk of nephrogenic systemic fibrosis (NSF). NSF a serious systemic disease, although rare, has been associated with some gadolinium based contrast agents (GBCAs) when used in renally impaired patients. However, recently the American College of Radiology (ACR) guidelines state that group II GBCAs, viz. gadobenate dimeglumine, gadobutrol, gadoterate acid, gadoteridol, are associated with few, if any, cases of NSF. Therefore currently, standard or lower than standard doses of group II GBCA's can be administered in patients with renal insufficiency if a contrast MRI examination has to be performed.

Imaging parameters for the T1-weighted imaging (T1WI) spin echo should be standardized in protocols. Thin section (e.g., 1.5 mm), T1-weighted images without FLAIR should be used for the postcontrast axial sequence employed for tumor measurements. These high-resolution postcontrast axial images contain a complete three-dimensional volumetric set of intensity values for each voxel in the entire field of view, and can be reformatted into multiplanar coronal and sagittal images, saving significant scanning time. However, if the patient moves during the acquisition of the volumetric sequence, the motion artifact will be translated into the reformatted planes, limiting optimal evaluation.

The axial angle should be prescribed using a line parallel to the anterior and posterior commissures as determined from the sagittal T1W image. If these reference points are not easily discernable, the axial plane should be placed parallel to the body of the corpus callosum, as defined on the sagittal T1W image.

Optimal slice thickness is based on the principle that the smallest enhancing lesion, for measurement purposes, should be no less than two times the slice thickness. Thus, a slice thickness of 3 mm on postcontrast axial T1-weighted images permits reproducible measurement of 6-mm diameter enhancing tumor nodules.

Gadolinium enhancement is a time- and dose-dependent phenomenon, and variation in these two factors can result in significant apparent changes in lesion size. The postcontrast T1W images should be acquired at a standardized time interval after intravenous administration of gadolinium. Sequences not affected by gadolinium (e.g., FSE T2W images and diffusion images, over a total of 6 min) can also be acquired after contrast injection. The same dose of gadolinium should be used for every study unless there is a marked change in the patient's body weight.

Unique imaging features of oligodendrogliomas

Oligodendrogliomas have sufficiently unique biological features in their formation and growth rate that their imaging findings allow the neuroradiologist to suggest a preoperative histologic diagnosis and grade. However, MRI does not replace the need for biopsy, histopathology, and molecular studies in the diagnosis of diffuse gliomas other than for some inaccessible brainstem tumors.

One situation in which a preoperative diagnosis of probable oligodendroglioma can serve the patient and clinician, however, might be in the case of a large lesion located in an eloquent or functional brain region where an aggressive resection presents significant risk of neurologic injury. In this situation a needle biopsy diagnosis of anaplastic oligodendroglioma, which is a treatment-sensitive tumor, could lead to neoadjuvant treatment that might reduce the size of the tumor, following which surgical decision-making might be affected.

Currently, there is no reason to believe that imaging can suggest a particular therapy based on the unique findings seen with any diffuse glioma, since available treatment options are limited.

Imaging can be used to predict the clinical behavior of a newly diagnosed lesion, which can be helpful in directing the brain tumor specialist decision-making regarding treatment approach. As discussed below, the nature of contrast enhancement, proton magnetic resonance spectroscopy (MRS), and relative cerebral blood volume (rCBV) maps can all predict higher grade. Contrary to low-grade astrocytomas, enhancement may be seen in low-grade oligodendrogliomas but the pattern of enhancement differs from that of anaplastic astrocytomas.

Researchers have sought to demonstrate a unique set of imaging features for oligodendrogliomas. The recent redefinition of 1p/19q deletion as the diagnostic feature of oligodendrogliomas has made this subject of even greater interest. Smits recently assessed the literature regarding comparison of imaging findings between diffuse gliomas with 1p/19 co-deletion (oligodendrogliomas) and those without 1p/19 deletion (astrocytomas).[3] Most of these studies were performed before the genetic definition of oligodendrogliomas and thus compared oligodendrogliomas and oligoastrocytomas with and without 1p/19q co-deletion, assessing differences based on their genetics.

The conclusions from the numerous papers that have sought to distinguish oligodendrogliomas from astrocytomas suggest that there are indeed important imaging differences that reflect the distinct genetic background of these tumors.[4–10]

The conventional imaging features that distinguish oligodendrogliomas from astrocytomas are as follows. 1p/19q co-deletion correlates with frontal and noninsular/nontemporal location, prominent cortical involvement, a diffuse margin of the tumor edge on T1- and T2-weighted images, heterogeneous signal on T2-weighted images, and the presence of calcification on CT. A heterogeneous MRI signal in this tumor could be due to calcification, hemorrhage, or cystic changes.

Lobar distribution

Although oligodendrogliomas occur preferentially in the frontal lobes, and may be seen in any lobe, their occurrence in the insular region and temporal lobes is uncommon (Fig. 1). Bilateral hemispheric involvement also may be more common than in astrocytomas.

Cortical involvement

Involvement of the cortex and adjacent white matter is a common feature of oligodendrogliomas (Fig. 1), although not all studies have found this correlation. It is interesting to note that cortical involvement of oligodendrogliomas may account for their extension along the pia or into the subarachnoid space.

Margin

Numerous studies have shown that a diffuse margin on T1- and T2-weighted images is statistically more common in oligodendrogliomas, whereas astrocytomas are more likely to have a sharp border with the adjacent brain parenchyma (Fig. 1). Authors have hypothesized that this relates to the greater infiltrative nature of oligodendrogliomas, although at least one study did not find correlation with the histopathologic findings of a discreet border.

Enhancement

WHO grade II oligodendrogliomas frequently show contrast enhancement, whereas this is not a common feature of grade II astrocytomas. While contrast enhancement places patients with histopathological low-grade astrocytomas into a high-risk category, the presence of enhancement does not predict high grade with oligodendrogliomas. Several studies have suggested that the presence of enhancement in grade II tumors (Fig. 2) may be predominantly due to increased vascularity (identified on pathological specimens as the chicken-wire pattern), rather than abnormal neovascularity with a leaky blood-brain barrier and therefore increased permeability. However, the pattern of enhancement can help predict the grade of oligodendrogliomas. Low-grade oligodendrogliomas can show a punctate, or stippled pattern of enhancement, whereas anaplastic oligodendrogliomas are more likely to show areas of patchy or heterogeneous enhancement (Figs. 3 and 4). Atypical robust enhancement in grade II oligodendrogliomas can be seen (Fig. 2), and a lack of enhancement can occur in anaplastic oligodendrogliomas (Fig. 5).

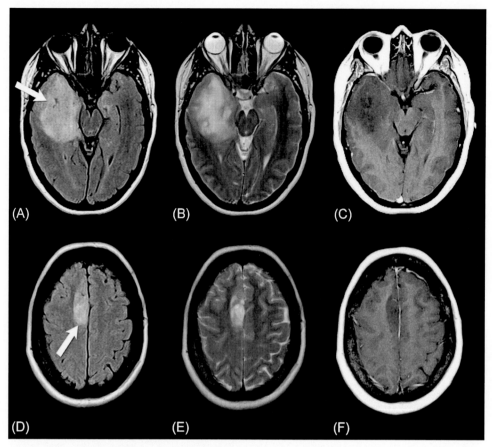

FIG. 1 Multicentric low-grade oligodendroglioma. Axial FLAIR (A, D), and T2-weighted MR images (B, E), demonstrate expansile, hyperintense masses involving the white matter and cortex in an uncommon right temporal lobe (A–C) location, and the more typical right frontal (D–F) lobe, without any enhancement on the postcontrast T1-weighted images (C, F).

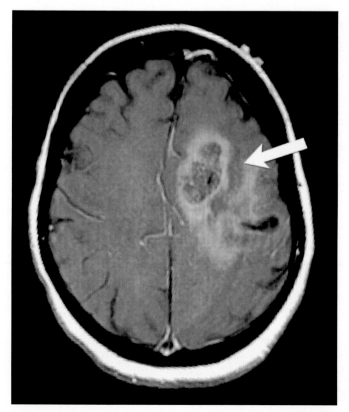

FIG. 2 Enhancing low-grade oligodendroglioma. Axial T1-weighted post-contrast MR image at the level of the centrum semi-ovale demonstrates a heterogeneously enhancing infiltrative lesion.

FIG. 3 Enhancing anaplastic oligodendrogliomas. (A) Axial T1-weighted postcontrast MR image at the level of the lateral ventricle demonstrates a patchily enhancing lesion. (B) A heterogeneously enhancing infiltrative mass centered in the right anterior corpus callosum extending to the left of the midline, involving the bilateral periventricular regions and inseparable from the ventricular wall.

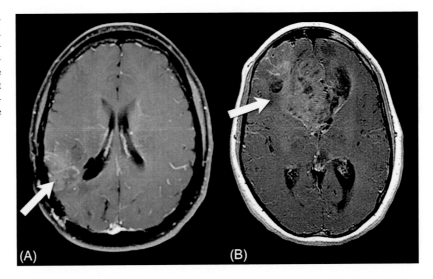

FIG. 4 Anaplastic oligodendroglioma with calcification. (A) Sagittal T1 postcontrast T1-weighted image demonstrates an enhancing mass with cystic areas, and calcification on the noncontrast CT image (arrow in B).

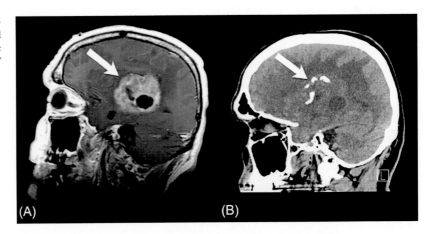

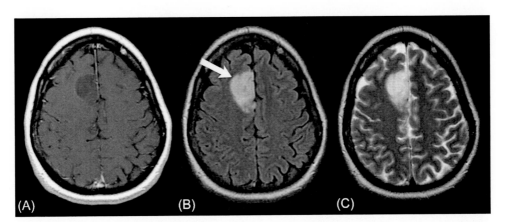

FIG. 5 Nonenhancing anaplastic oligodendroglioma. (A) Axial T1-weighted postcontrast image demonstrates a mass without significant enhancement in the right paramedian frontal lobe with cortical involvement, an ill-defined posterior border and hyperintensity on the FLAIR (B) as well as on the T2-weighted (C) images.

Multifocality

Multiple areas of tumor within the brain are common in astrocytomas, often with one area showing abnormal enhancement and another demonstrating nonenhancing T2 signal hyperintensity. According to one study, multifocality is less common in patients with oligodendrogliomas. Fig. 1 shows an oligodendroglioma with this reportedly uncommon feature.

Calcification

For unknown reasons, calcification is a common finding in oligodendrogliomas, occurring in about a quarter of cases. On CT, the calcification is easily visualized as coarse hyperdense areas within the tumor, as shown in Fig. 6, and can account for the heterogeneous signal on MRI discussed next.

Heterogeneous appearance

This is another feature that is much more common in oligodendrogliomas than in grade II-III astrocytomas. The focal areas of T2 signal loss on MRI are likely a result of calcification and possibly of hemosiderin, although the latter was not found as an explanatory factor in at least one study that correlated pathology and imaging findings. Hemorrhage is uncommon but can present acutely and on noncontrast CT is identified as hyperdensity (Fig. 7).

Growth rate

The classic study of Mandonnet et al. showed that untreated grade II oligodendrogliomas show a pattern of steady slow growth over time that is remarkably consistent between tumors.[11] The rate is about 3 mm increase in diameter each year. Implications from this study are low level of variation in the rate of enlargement and the importance of examining serial images starting with comparison to the earliest time point possible.

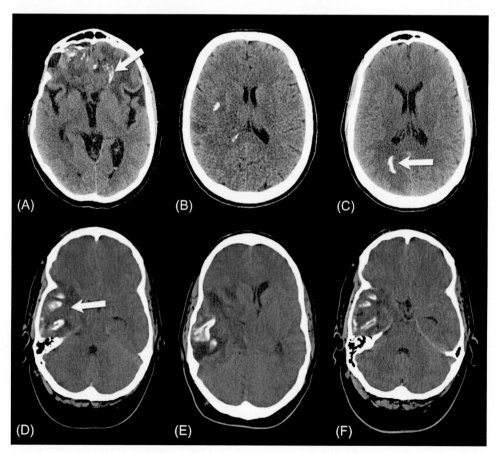

FIG. 6   Axial noncontrast CT images demonstrate hypoattenuating oligodendroglial lesions with hyperdensities (arrows) reflective of calcification in varying patterns; stippled (A), nodular (B), curvilinear (C), gyriform (D–F).

FIG. 7   Anaplastic oligodendroglioma with hemorrhage. Axial noncontrast CT image at the level of the centrum semiovale demonstrates a mass with areas of hyperdensity (arrow) reflective of acute hemorrhage resulting in local mass effect and mild leftward midline shift.

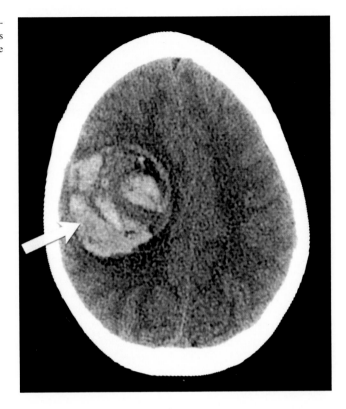

Response to treatment

Anaplastic oligodendrogliomas are well known to show marked shrinkage during the administration of radiation or chemotherapy (Fig. 8). This pattern is uncommon with low-grade oligodendrogliomas and with astrocytomas of any grade.

Many years after irradiation of the brain for any lesion, multiple foci of susceptibility effect may appear in the brain (Fig. 9). These represent microhemorrhages in radiation-associated cavernous angiomas. The development of clinically significant hemorrhage within one of these lesions is a well-known event.

Advanced imaging of diffuse gliomas

Perfusion images assess the degree of vascularity in brain and tumor by measuring the signal change produced as a rapidly injected bolus of gadolinium passes through the tumor while very rapid sequential images are obtained. Dynamic susceptibility is the MR technique used to acquire the most commonly assessed perfusion measure called rCBV, or relative

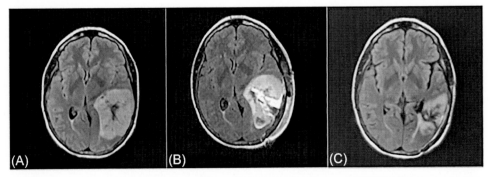

FIG. 8   Marked shrinkage can occur in anaplastic oligodendrogliomas during treatment. Preoperative T2 FLAIR images prior to surgery (A), postoperatively after biopsy (B), (C).

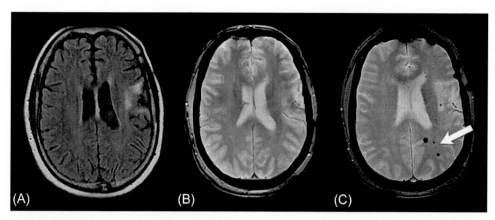

FIG. 9 An anaplastic oligodendroglioma case status post resection followed by radiation and chemotherapy completed in 1997 demonstrates a stable left frontotemporal resection bed with FLAIR hyperintensity on the axial FLAIR images (A). Axial T2*-weighted image (B) is unremarkable in 2006 (B), with the development of susceptibility effect on surveillance MR in 2015 (arrow in C) reflective of radiation-associated cavernous angiomas.

cerebral blood volume. Protocols for gadolinium injections for perfusion are widely available, and typically include the use of an 18-gauge intravenous needle, power injector, and 1 cc prebolus gadolinium contrast injection to decrease any effect of leaky vessels. Spin echo or gradient echo imaging technique can be employed. The gradient echo technique is more sensitive to the presence of large, tumor or brain, vessels in the region of interest and is typically used to evaluate blood flow and volume. The first pass bolus of gadolinium produces sudden loss of signal, and voxels are compared to each other with regard to the degree of signal loss and converted into contrast medium concentration. To analyze perfusion data small regions of interest are compared to unaffected contralateral brain, usually white matter, and a relative value is obtained.

Proton magnetic resonance spectroscopy (H1-MRS) is an MRI technique that permits semiquantitative measurement of a small number of metabolites within brain tissue. The most useful of these measurements is the ratio of choline (measures the rate of cell membrane turnover) and creatine (measures the more static cellular energy). Densely cellular tumors show an elevated choline to creatine ratio of greater than 3:1. Multivoxel H1-MRS allows comparison of numerous adjacent areas within the tumor and can be used to identify areas with the highest grade, which can be targeted for biopsy.

Positron emission tomography (PET) images assess the concentration of a radioactive compound following injection into the patient. The most common is radioactive fluorodeoxyglucose (FDG-PET), although multiple other compounds have been studied. Because FDG-PET measures glucose metabolism, it provides a semiquantitative measure of the metabolic activity in the tumor.

Diffusion tensor imaging (DTI) is a special case of DWI that exploits the directional (nonisotropic) nature of water movement parallel to the myelin fibers, and tractography is the method that identifies the white matter tracts in the brain. The most common use of DTI is in surgical planning so that the major functional pathways, such as the corticospinal tract, can be avoided during resection.

Blood-oxygen level-dependent (BOLD) imaging detects small changes in the oxygenated state of hemoglobin that occur during the activation of a brain region executing a certain task, such as during finger tapping. This permits the neurosurgeon to identify functionally important brain areas adjacent to the tumor as part of presurgical planning, thereby avoiding resection and injury to these areas.

Features of advanced imaging in oligodendrogliomas

Advanced imaging studies have sought to characterize differences between tumors with and without loss of 1p/19q and between low- and high-grade tumors. While some differences have been described, the current data can be summarized with the statement that the imaging features are not sufficiently distinctive to allow a confident preoperative histopathologic diagnosis. Preoperative planning is an area where tractography and motor mapping studies are of increasing value to achieve maximally safe resection.

Studies of PWI have shown that low-grade oligodendrogliomas have a higher rCBV than low-grade astrocytomas or histopathologically apparent low-grade oligodendrogliomas lacking 1p/19q deletion (the latter are now classified as astrocytomas). In one study a correlation between EGFR and VEGF expression and rCBV could not be demonstrated.[12] Studies have generally found that anaplastic oligodendrogliomas have higher rCBV than low-grade tumors. It seems likely that the relatively high rCBV seen in low-grade oligodendrogliomas is due to known hypervascularity rather than due to endothelial proliferation seen in high-grade gliomas.[5, 13–17]

H1-MRS has not shown clear value in distinguishing gliomas with and without loss of 1p/19q. Most of the studies that purported a difference between oligodendrogliomas and astrocytomas did not include analysis of 1p/19q and thus the oligodendroglial group also included astrocytomas. The choline to creatine ratio is higher in anaplastic than in low-grade oligodendrogliomas and this can be useful in deciding on a biopsy site in a tumor with features that are otherwise suggestive of a low-grade tumor.

Some authors have sought to combine rCBV and H1-MRS in the differentiation of various tumor categories, but the results do not point to a significant clinical value.[16, 18–20]

Imaging in clinical trials

The response assessment in neuro-oncology (RANO) working group has published criteria for imaging response evaluation in clinical trials.[21] As with the previous Macdonald criteria, lesions are measured as the product of two-dimensional, perpendicular diameters of the enhancing component of the tumor. Lesions must have diameters of at least 10 mm to be considered measureable. When there are multiple lesions, the products of two-dimensional diameters are summed together as a measure of overall disease burden for serial measurement. Lesion measurements must not include the resection cavity or cystic components. Response categories include complete response, partial response, stable disease, and progressive disease (Table 1). Considerations must be made for the possibility of pseudo-progression if the timing after chemoradiation therapy is appropriate. The introduction of bevacizumab can complicate the measurement of enhancing lesions. Separate criteria now have been devised for immunotherapy (iRANO) because of the possibility of worsening enhancement due to the immune response,[22] and for nonenhancing low-grade tumors.[23]

TABLE 1 Response criteria in clinical trials

	Macdonald	RECIST	RANO
Complete response (CR)	Complete disappearance of enhancing lesion No new lesions No corticosteroids Clinically stable or improved	Complete disappearance of enhancing lesions No new lesions	Complete disappearance of enhancing lesions No new lesions Stable or improved nonenhancing FLAIR lesion No corticosteroids Clinically stable or improved
Partial response (PR)	≥50% decrease in size of enhancing lesion No new lesions Stable or reduced corticosteroids Clinically stable or improved	≥30% decrease in size of enhancing lesions No new lesions	≥50% decrease in size of enhancing lesions No new lesions Stable or improved nonenhancing FLAIR lesion Stable or reduced corticosteroids Clinically stable or improved
Stable disease (SD)	Not meeting other criteria	Not meeting other criteria	Not meeting other criteria
Progressive disease (PD)	≥25% increase in size of enhancing lesion New lesions Clinical deterioration	≥20% increase in size of enhancing lesions New lesions	≥25% increase in size of enhancing lesions New lesions Significantly increased nonenhancing FLAIR lesion Clinical deterioration

Conclusion

Oligodendrogliomas are infiltrative glial neoplasms, typically hyperintense on T2-weighted images, and can demonstrate varying degrees of enhancement irrespective of the grade. Heterogeneous appearance is secondary to cystic degeneration, calcification, and/or hemorrhage. Redefinition of this tumor based on the molecular characteristic of 1p19q deletion will allow future studies to better characterize features unique to this tumor, and assess response to treatment, as well as develop novel therapies.

References

1. Louis DN, Ohgaki H, Wiestler OD, Cavenee WK, World Health Organization, International Agency for Research on Cancer. *WHO classification of tumours of the central nervous system.* revised 4th ed. Lyon: International Agency for Research on Cancer; 2016.
2. Henson JW, Ulmer S, Harris GJ. Brain tumor imaging in clinical trials. *AJNR Am J Neuroradiol.* 2008;29(3):419–424.
3. Smits M. Imaging of oligodendroglioma. *Br J Radiol.* 2016;89(1060).
4. White ML, Zhang Y, Kirby P, Ryken TC. Can tumor contrast enhancement be used as a criterion for differentiating tumor grades of oligodendrogliomas? *AJNR Am J Neuroradiol.* 2005;26(4):784–790.
5. Cha S, Tihan T, Crawford F, et al. Differentiation of low-grade oligodendrogliomas from low-grade astrocytomas by using quantitative blood-volume measurements derived from dynamic susceptibility contrast-enhanced MR imaging. *AJNR Am J Neuroradiol.* 2005;26(2):266–273.
6. Jenkinson MD, du Plessis DG, Smith TS, Joyce KA, Warnke PC, Walker C. Histological growth patterns and genotype in oligodendroglial tumours: correlation with MRI features. *Brain.* 2006;129(Pt 7):1884–1891.
7. Megyesi JF, Kachur E, Lee DH, et al. Imaging correlates of molecular signatures in oligodendrogliomas. *Clin Cancer Res.* 2004;10(13):4303–4306.
8. Zlatescu MC, Tehrani Yazdi A, Sasaki H, et al. Tumor location and growth pattern correlate with genetic signature in oligodendroglial neoplasms. *Cancer Res.* 2001;61(18):6713–6715.
9. Kim JW, Park CK, Park SH, et al. Relationship between radiological characteristics and combined 1p and 19q deletion in World Health Organization grade III oligodendroglial tumours. *J Neurol Neurosurg Psychiatry.* 2011;82(2):224–227.
10. Sherman JH, Prevedello DM, Shah L, et al. MR imaging characteristics of oligodendroglial tumors with assessment of 1p/19q deletion status. *Acta Neurochir.* 2010;152(11):1827–1834.
11. Mandonnet E, Delattre JY, Tanguy ML, et al. Continuous growth of mean tumor diameter in a subset of grade II gliomas. *Ann Neurol.* 2003;53(4):524–528.
12. Spampinato MV, Smith JK, Kwock L, et al. Cerebral blood volume measurements and proton MR spectroscopy in grading of oligodendroglial tumors. *AJR Am J Roentgenol.* 2007;188(1):204–212.
13. Kapoor GS, Gocke TA, Chawla S, et al. Magnetic resonance perfusion-weighted imaging defines angiogenic subtypes of oligodendroglioma according to 1p19q and EGFR status. *J Neuro-Oncol.* 2009;92(3):373–386.
14. Lev ML, Ozsunar Y, Henson JW, et al. Glial tumor grading and outcome prediction using dynamic spin-echo MR susceptibility mapping compared to conventional contrast enhanced MRI: confounding effect of elevated relative cerebral blood volume of oligodendrogliomas. *Am J Neuroradiol.* 2004;25:214–221.
15. Fellah S, Caudal D, De Paula AM, et al. Multimodal MR imaging (diffusion, perfusion, and spectroscopy): is it possible to distinguish oligodendroglial tumor grade and 1p/19q codeletion in the pretherapeutic diagnosis? *AJNR Am J Neuroradiol.* 2013;34(7):1326–1333.
16. Xu M, See SJ, Ng WH, et al. Comparison of magnetic resonance spectroscopy and perfusion-weighted imaging in presurgical grading of oligodendroglial tumors. *Neurosurgery.* 2005;56(5):919–926 [discussion 919–926].
17. Whitmore RG, Krejza J, Kapoor GS, et al. Prediction of oligodendroglial tumor subtype and grade using perfusion weighted magnetic resonance imaging. *J Neurosurg.* 2007;107(3):600–609.
18. Chawla S, Krejza J, Vossough A, et al. Differentiation between oligodendroglioma genotypes using dynamic susceptibility contrast perfusion-weighted imaging and proton MR spectroscopy. *AJNR Am J Neuroradiol.* 2013;34(8):1542–1549.
19. Chawla S, Wang S, Wolf RL, et al. Arterial spin-labeling and MR spectroscopy in the differentiation of gliomas. *AJNR Am J Neuroradiol.* 2007;28(9):1683–1689.
20. Rijpkema M, Schuuring J, van der Meulen Y, et al. Characterization of oligodendrogliomas using short echo time 1H MR spectroscopic imaging. *NMR Biomed.* 2003;16(1):12–18.
21. Wen PY, Macdonald DR, Reardon DA, et al. Updated response assessment criteria for high-grade gliomas: response assessment in neuro-oncology working group. *J Clin Oncol.* 2010;28(11):1963–1972.
22. Okada H, Weller M, Huang R, et al. Immunotherapy response assessment in neuro-oncology: a report of the RANO working group. *Lancet Oncol.* 2015;16(15):e534–e542.
23. van den Bent MJ, Wefel JS, Schiff D, et al. Response assessment in neuro-oncology (a report of the RANO group): assessment of outcome in trials of diffuse low-grade gliomas. *Lancet Oncol.* 2011;12(6):583–593.

Chapter 14

Advanced [11C]methionine and [18F]FDG positron emission tomography for diagnosis, treatment, and follow-up of oligodendrogliomas

Tyler Richards*,†, John Anderson*,†, Marilyn Reed‡, Raymond Poelstra‡, Martin Satter‡,§ and Arash Kardan*,†

*University Hospitals, Cleveland Medical Center, Cleveland, OH, United States; †Case Western Reserve University, Cleveland, OH, United States
‡Charles F. Kettering Memorial Hospital, Dayton, OH, United States; §Wright State University, Dayton, OH, United States

Updated diagnostic criteria for oligodendrogliomas

The updated 2016 WHO brain tumor classification transformed the diagnosis of gliomas from a schema based on histologic morphology to one based on a tumor's molecular and genetic characteristics.[1] Oligodendrogliomas are now defined by the presence of an isocitrate dehydrogenase 1 (IDH1) or isocitrate dehydrogenase 2 (IDH2) gene mutation and by a coexistent deletion of chromosomal arms 1p and 19q—the so-called 1p/19q co-deletion. This new molecular classification has made the diagnosis of oligodendrogliomas more straightforward, entirely doing away with the problematic hybrid entity, oligoastrocytoma, a tumor demonstrating histologic features of both oligodendroglioma and astrocytoma. This diagnostic paradigm shift has also allowed diagnostic classification to more closely resemble and inform treatment strategies and outcomes; an improvement from the prior morphologic classification which was fraught with discrepancies in interobserver reliability and varying therapeutic outcomes between otherwise histologically similar tumors.[2]

Despite the diagnostic clarity provided by molecular and genetic characteristics, oligodendroglial tumor grading remains a histologic determination. Like other glial neoplasms, oligodendrogliomas demonstrate a wide range of behavior. On one end of the spectrum, the WHO grade II well-differentiated oligodendroglioma is a relatively slow growing and indolent tumor. At the other end of the spectrum, the WHO grade III anaplastic oligodendroglioma is a rapidly growing aggressive tumor. Approximately, two-thirds of oligodendroglial tumors are well-differentiated WHO II tumors, while the remaining third are of the anaplastic variant.[3] Histologic grading between the low-grade and anaplastic variants of oligodendroglioma is based on the criteria including cellular density, nuclear and cellular atypia, microvascular proliferation, mitotic activity, and the presence of necrosis.

Clinical management of oligodendrogliomas

Multiple trials have shown oligodendroglial tumors, with their entity defining IDH mutation and 1p/19q co-deletion, are generally more sensitive to chemotherapy and radiation therapy when compared to equivalent WHO grade II and III astrocytomas, and demonstrate greater median overall survival and progression-free survival times.[4] As also may be expected, survival rates for patients with low-grade oligodendrogliomas are better than those of their age-matched counterparts with anaplastic oligodendrogliomas. Yang et al. showed survival rates for patients with WHO grade III gliomas, anaplastic astrocytomas, and oligodendrogliomas of 60%–80% at 1 year and 26%–46% at 5 years, while WHO grade II astrocytomas and oligodendrogliomas had survival rates of 94% at 1 year and 67% at 5 years.[5] Therefore, treatment strategies and prognostic projections are dependent on the biochemical identification of IDH mutation and 1p/19q co-deletion, as well as histologic grading. Evaluation of these diagnostic features is in turn reliant on accurate tissue sampling.

Multiple studies have shown a correlation with a more extensive initial resection and improved outcomes when compared to partial resection or biopsy, even for low-grade gliomas such as WHO II oligodendrogliomas. These outcomes include, among others, improved overall and progression-free survival and a lower degree of disability.[6–8] Even a small (≥1 cm) residual tumor, as determined by postoperative magnetic resonance imaging (MRI), is associated with a recurrence rate of 69% for low-grade gliomas.[9] Unfortunately, oligodendrogliomas are classically diffuse infiltrative tumors with a preponderance to involve cortical structures in the frontal lobes. Adequate surgical margins are, therefore, often limited due to adjacent eloquent structures. During surgery, microscopic tumor infiltration, which is often invisible to the surgeon, further limits intraoperative differentiation of normal tissue from that involved by tumor. While additional intraoperative tools, such as frozen section microscopy and intraoperative MRI, can help guide surgical resection, they are not perfect. These methods still rely upon "blind" sampling by the surgeon and may not meet the physical or timing restraints of the surgery. Further complicating accurate tissue sampling is the frequent intratumoral cellular heterogeneity, and often the resected or biopsied tissue may under-sample the most biologically aggressive part of the tumor which determines treatment and prognosis.[10] Thus, noninvasive preoperative guidance is of utmost importance.

Limitations of anatomic imaging

Conventional anatomic imaging modalities like magnetic resonance (MR) and computed tomography (CT) reveal morphological information but are of comparatively limited value for the assessment of more specific and reproducible information about the biology and metabolic activity of a tumor. Indeed, differentiating between WHO grade II and III tumors, or even between an astrocytoma and an oligodendroglioma, is often precluded due to indistinguishable morphological imaging characteristics. Current criteria often used in clinical practice and by the Response Assessment in Neuro-Oncology (RANO) working group for assessing gliomas in clinical trials are heavily reliant on MRI appearance of postcontrast T1-weighted enhancement and fluid-attenuated inversion recovery (FLAIR) signal abnormalities.[11] While somewhat useful in avidly enhancing high-grade gliomas, these criteria lack sensitivity and specificity for low-grade gliomas, such as oligodendrogliomas, many of which fail to demonstrate enhancement on postcontrast imaging. Sequelae of treatment have varying effects on enhancement and FLAIR signal which only further complicates imaging of oligodendrogliomas following therapy (see Fig. 1). Not surprisingly, MRI has been shown to be limited in its ability to delineate the full extent of glial tumors.[12,13] Furthermore, when it comes to evaluating the molecular markers and their biochemical consequences, which now define the classification of oligodendrogliomas and gliomas, MRI and CT are figuratively in the dark. However, functional imaging with positron emission tomography (PET) has brought these once concealed molecular processes into the light.

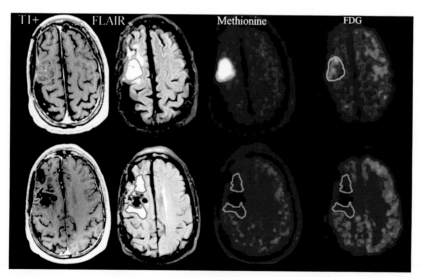

FIG. 1   Row 1, MR T1 postgadolinium, MR FLAIR, [11C]methionine PET, [18F]FDG PET; row 2, MR T1 postgadolinium, MR FLAIR, [11C]methionine PET, [18F]FDG PET; 59-year-old male presented with a seizure after being diagnosed 16 years prior with an anaplastic astrocytoma treated initially with PCV chemotherapy and radiation followed by surgical resection. Imaging demonstrated an area of patchy enhancement and increased FLAIR signal in the right frontal lobe near the vertex (row 1) with associated increased [11C]methionine and [18F]FDG uptake, which was surgically resected and found to be consistent with anaplastic oligodendroglioma (given 1p/19q co-deletion present). More inferiorly within the right frontal lobe (row 2) was an additional area of increased FLAIR signal which demonstrated decreased [11C]methionine and [18F]FDG uptake consistent with radiation necrosis. This demonstrates the nonspecificity of MRI in surveillance and added value of PET imaging in these cases. *Courtesy of the Charles F. Kettering Memorial Hospital Department of Nuclear Medicine/PET.*

PET in neuro-oncology

PET imaging allows highly sensitive measurements (in the picomolar range) of biochemically active molecules labeled with positron-emitting radionuclides (radiotracers). Positrons emitted from the nucleus of PET radiopharmaceuticals subsequently undergo annihilation events upon contact with electrons after having traveled a short distance from their source radiotracer molecule into the tissue (mean range, 0.2–1.5 mm, depending on the mean energy of the positron). Each annihilation event produces a pair of high-energy (511 keV) gamma photons which are emitted at 180° orientation to one another. These annihilation photon pairs are emitted out of the patient and then detected by surrounding tomographic detectors arranged in a ring around the axial plane of the patient. If both annihilation photons hit opposing detectors in the same time window (500–1000 ps), an annihilation event is registered and assigned to a single line of response. The tomographic reconstruction software then utilizes the information about each detected coincident photon pair and maps the location of the emission. It can then produce a three-dimensional (3D) image of the radiopharmaceutical distribution with spatial resolution in the 4–5 mm range.[14] The PET data can then be fused with anatomic cross-sectional imaging, most commonly with MRI in the setting of neuro-oncology, to provide additional anatomic correlation and attenuation correction.

Molecular imaging with PET is being increasingly implemented in neuro-oncology, since it provides additional metabolic and biochemical information about the state of the tumor, both for patient management and evaluation of therapy. Various molecular processes have been designated as key to tracking tumor biology and pathogenesis. Such biochemical pathways include those related to glucose metabolism, amino acid transport, protein expression, cellular proliferation, membrane biosynthesis, and ischemia. As greater understanding of these molecular mechanisms underlying the biology of brain tumors has grown, so too have the number of PET radiopharmaceuticals designed to investigate these pathways.

The ideal PET radiopharmaceutical should be biologically indistinguishable from its nonradiolabeled molecule, should be inseparable from its radionuclide, and demonstrate minimal nonspecific uptake that would reduce contrast between target tissue and background, that is, healthy brain tissue in the setting of neuroimaging. In practice, to achieve higher signal to background PET images, several hundred million cells in relatively close proximity must accumulate the radiotracer to visualize it against the background uptake.[6–8] PET imaging can help visualize the abnormal metabolism or function of a tumor resulting in an earlier and more accurate diagnosis. In short, changes in biochemistry precede anatomic changes enabling PET to potentially detect brain tumors before morphologic changes are detectable on conventional anatomic imaging. And once detected, PET can be used to assess the metabolic/functional status of that tumor in response to therapy.

[¹⁸F]FDG PET in neuro-oncology

The glucose analog, 2-[¹⁸F]fluoro-2-deoxy-D-glucose ([¹⁸F]FDG), is the most frequently used radiotracer to measure the local metabolic rate of glucose that represents a common pathway of neurochemical activity in the brain.[15,16] Imaging of brain tumors with [¹⁸F]FDG was the first oncologic application of PET (see Fig. 2).[17–19] [¹⁸F]FDG is actively transported across the blood-brain barrier via a glucose cell membrane transporter, in proportion to the glucose metabolic rate, where it is phosphorylated and thus sequestered intracellularly.

For oligodendrogliomas, and gliomas as a whole, a higher histologic grade is correlated with higher [¹⁸F]FDG uptake. The increased uptake has been suggested to be related to changes in transport rate via modulation of glucose transport proteins.[20] [¹⁸F]FDG PET was proposed to be useful for imaging of gliomas because of increased glucose metabolism in high-grade gliomas as well as a positive correlation between the glycolysis rate and malignancy. The amount of [¹⁸F]FDG accumulation in a primary brain tumor correlates with the histologic tumor grade, cell density, and

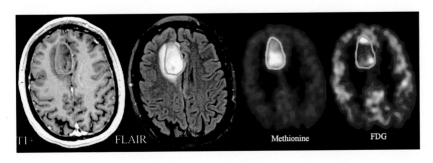

FIG. 2 MR postgadolinium T1, MR FLAIR, [¹¹C] methionine PET, [¹⁸F]FDG PET; 50-year-old female at the time of initial diagnosis demonstrated a parasagittal right frontal mass with minimal patchy enhancement and increased FLAIR signal. PET imaging demonstrated both increased [¹¹C]methionine and [¹⁸F]FDG uptake. Final pathology was consistent with a WHO grade II oligodendroglioma (1p/19q co-deletion present). *Courtesy of the Charles F. Kettering Memorial Hospital Department of Nuclear Medicine/PET; images were processed utilizing Mirada XD3 software.*

survival.[21–25] Furthermore, high uptake of [^{18}F]FDG in a previously known low-grade tumor establishes the diagnosis of anaplastic transformation.[26,27]

Some studies have demonstrated the diagnostic limitations of [^{18}F]FDG PET in the evaluation of brain tumors.[28,29] Glucose is the primary energy substrate of the brain, providing approximately 95% of the required adenosine triphosphate (ATP) and is also tightly connected to neuronal activity. High physiologic brain glucose metabolism will then correlate with high uptake of [^{18}F]FDG in the brain especially within areas containing high amounts of gray matter such as the cerebral cortex, basal ganglia, and the thalamus. This relatively high gray matter background metabolism poses a particular problem for oligodendrogliomas which classically demonstrate an affinity for the cerebral cortex.[30]

Because of the high rate of physiologic glucose metabolism in normal brain tissue, the detectability of tumors with only modest increases in glucose metabolism, such as low-grade oligodendrogliomas, can be difficult. The clear demarcation between high-grade glioma and normal brain tissue, such as that seen with aggressive glioblastoma, can be blurred in the setting of low-grade well-differentiated oligodendrogliomas, which often demonstrate [^{18}F]FDG uptake similar to normal white matter. As a whole, oligodendrogliomas demonstrate widely variable glucose metabolism when compared to background parenchymal uptake. In fact, more often than not, no measurable difference in [^{18}F]FDG uptake between normal tissue and low-grade oligodendroglioma can be demonstrated.[30]

Further limiting the clinical utility of [^{18}F]FDG, the coupling of [^{18}F]FDG uptake and glucose metabolism in many tumors can be radically different from that in normal tissue. It was demonstrated that [^{18}F]FDG uptake in tumors may be increased, while glucose metabolism is not.[31] Posttreatment changes in tumor can also change a tumor's metabolic characteristics and [^{18}F]FDG uptake. For instance, high-grade tumors, which demonstrated relatively increased [^{18}F]FDG uptake on pretreatment imaging, may have uptake that is only similar to or slightly above that in white matter following treatment.[19] Finally, [^{18}F]FDG is known to accumulate in macrophages and inflammatory tissue, potentially making the distinction between anaplastic oligodendroglioma, or other high-grade glioma, from an acute or subacute inflammatory process challenging.

Amino acid PET in neuro-oncology

Stemming from the limitations of [^{18}F]FDG, an additional class of PET imaging agents, amino acid, and amino acid analog radiotracers, have been developed. The essential amino acids have been of particular use in the imaging of brain tumors. Essential amino acids are not synthesized in the body and therefore, require an exogenous source, usually in the diet. Among many other processes, they are necessary proteins involved in mitosis, cellular metabolism, and neurotransmitter synthesis. These amino acids require specific cellular membrane transporters in order to cross the blood-brain barrier. A number of these transport systems have been identified, and they have been an important area of research for both functional imaging and drug delivery. Amino acid imaging is based on the observation that amino acid transport and utilization is generally increased in malignant transformation for cellular processes requiring these essential amino acid building blocks.[32,33]

These agents are useful for imaging of brain tumors, because of relatively high uptake in tumor tissue and low uptake in normal brain tissue, providing a high tumor-to-normal contrast ratio. In gliomas, increased transport via the L-type amino acid transport system for large neutral amino acids, and chiefly its LAT1 and LAT2 subtypes, has been shown to be the predominant cause for the increased uptake of the most commonly used amino acid PET agents.[34] It should be noted that the upregulation of L-type amino acid transporter systems occurs in the absence of increased vascular permeability and occurs in all phases of the cell cycle.[35] While transport does not depend on breakdown of the blood-brain barrier, studies have suggested that uptake of the amino acid PET radiotracers may be enhanced by its breakdown.[36]

The most studied PET amino acid tracer is [^{11}C]methionine. It is composed of the essential amino acid, methionine, chemically labeled to the positron-emitting isotope of carbon-11 (^{11}C). Because of the short half-life of ^{11}C (20 min), an onsite cyclotron is a prerequisite for [^{11}C]methionine use in brain tumor imaging and has limited a more widespread implementation into clinical practice. For instance, the longer half-life of fluorine-18 (^{18}F) of 110 min allows for off-site production of ^{18}F-labeled amino acid PET radiotracers and transport to a clinical site. Due to these logistical advantages, several ^{18}F-labeled aromatic amino acid analogs have been developed for tumor imaging.

For instance, 3-O-methyl-6-[^{18}F]fluoro-L-DOPA ([^{18}F]FDOPA) has also been investigated for brain tumor imaging with PET.[37] As the fluorinated form of the amino acid L-3,4-dihydroxyphenylalanine (L-DOPA), the famous precursor to such neurotransmitters as dopamine, norepinephrine, and epinephrine, it was initially utilized to measure dopamine synthesis in the basal ganglia. The diagnostic accuracy of [^{18}F]FDOPA has been reported to be superior to that of [^{18}F]FDG and similar to that of [^{11}C]methionine in evaluating recurrent low- and high-grade gliomas, despite the mild limitation of high physiologic uptake of [^{18}F]FDOPA in the basal ganglia.[38,39]

Another ^{18}F-labeled amino acid agent whose use has become increasingly popular in Japan and Europe is O-(2-[18F] fluoroethyl)-L-tyrosine ([^{18}F]FET). It is a tyrosine analog which is taken up by tumor cells via L-type amino acid transporters. It has been shown to have a similar distribution of uptake as [^{11}C]methionine, but unlike other amino acid agents, such as [^{11}C]methionine and [^{18}F]FDOPA, it is not incorporated in proteins or metabolized.[40,41] This metabolic stability has fostered the use of dynamic PET imaging. Limited reports utilizing dynamic [^{18}F]FET imaging have demonstrated unique activity curves which correlate with increased accuracy of glioma grading and better differentiation between glioma recurrence from postradiation changes.[42–44]

Other prospective amino acid agents are also actively being studied including α-[^{11}C]-methyl-L-tryptophan and [^{18}F]Fluciclovine. [^{18}F]Fluciclovine is a synthetic nonnaturally occurring amino acid analog of L-leucine which was initially used for suspected cases of prostate cancer recurrence.[45–47] Each has shown promising, although limited, data supporting potential use in delineation of glioma tumor extent, differentiation of recurrence from radiation changes, and response to therapy. Further validation of these agents with more robust prospective studies and larger patient samples is, however, still needed before they see the same popularity as with more extensively tested agents.

Despite its logistical challenges, [^{11}C]methionine has the longest and most validated clinical history and has been the most commonly used amino acid agent in studies.[48] Like the other amino acid tracers, its uptake is mainly determined by a specific carrier-mediated transport, predominately via L-type amino acid membrane transporters. As well as protein synthesis, methionine is important for its role as a precursor to S-adenosylmethionine, the most predominant biologic methyl group donor, which among other processes is a key component in genetic regulation. Methionine is also well known for its role in the formation of tetrahydrofolate, the active form of folate, an important factor for DNA synthesis. Given its close association with DNA regulation and synthesis, it is not surprising that increased transport of methionine has been demonstrated in malignant cell lines, including gliomas.[49] Uptake of [^{11}C]methionine has similarly been correlated with proliferative activity, as well as tumor cell and microvessel density.[50–53] The uptake of [^{11}C]methionine in gliomas is also influenced by its specific activity in plasma, transfer across the blood-brain barrier, intracellular metabolism, and incorporation into proteins.[54,55] As with other amino acid tracers, transport across the blood-brain barrier is the rate-limiting step for [^{11}C]methionine uptake, and while disruption of the blood-brain barrier is not a prerequisite for increased uptake, a damaged blood-brain barrier may actually enhance leakage of the tracer to the extracellular space and contribute to the increased uptake seen in malignant gliomas.[56] In contrast, transport across the blood-brain barrier is not the rate-limiting step for [^{18}F]FDG. The significance of this contrasting relationship is illustrated in the setting of low-grade meningiomas and high-grade glioblastomas. [^{11}C]methionine uptake is often higher in meningioma (absent blood-brain barrier) than in glioblastoma (leaky, but not absent blood-brain barrier). In contrast, [^{18}F]FDG uptake is higher in glioblastoma than in meningioma.

[^{11}C]methionine and other amino acid PET radiotracers offer distinct advantages over [^{18}F]FDG due to its better tumor to normal brain contrast, and they provide metabolic and functional information that is unable to be obtained by conventional anatomic imaging. Even with these advantages, amino acid radiotracers play a limited role in the current management of patient with oligodendrogliomas as well as other gliomas. Conventional MRI remains the most commonly utilized modality for initial diagnosis, treatment planning, and follow-up of these patients. Despite its limited utilization, [^{11}C]methionine PET has shown utility that often outperforms MRI and [^{18}F]FDG PET in these settings and may play a larger role in the care of these patients in the future.

Distinguishing tumor from nonneoplastic etiologies

[^{11}C]methionine PET is effective in differentiating both low- and high-grade gliomas from nonneoplastic pathology with a reported sensitivity of 76%–100% and specificity of 75%–100%.[57] These reported sensitivities and specificities, while good, likely underestimate the performance of [^{11}C]methionine PET for the diagnosis of gliomas. Most studies include a higher proportion of low-grade gliomas than is seen in clinical practice, which is likely related to the frequent nonspecific findings of low-grade gliomas on conventional MRI sequences. In these cases, [^{11}C]methionine PET is often used as a problem solving tool to add valuable metabolic data that is not available on anatomic imaging. Given the lower [^{11}C]methionine uptake in low-grade compared to high-grade gliomas, the reported sensitivities and specificities likely underestimate the effectiveness of [^{11}C]methionine PET to diagnose gliomas. In oligodendrogliomas, [^{11}C]methionine PET studies typically show diffuse, high uptake of the radiotracer even in WHO grade II tumors.[58] If there is no [^{11}C]methionine PET uptake in a lesion, an oligodendroglioma is very unlikely. False positives in [^{11}C]methionine PET do occasionally occur, and these are most often due to infectious etiologies such as brain abscesses, inflammatory etiologies such as leukoencephalitis and radiation necrosis, ischemia, intraparenchymal hemorrhage, and demyelinating diseases.[57]

Differentiating oligodendrogliomas from astrocytomas

The ability to differentiate oligodendrogliomas from astrocytomas preoperatively is important given that oligodendrogliomas have a better prognosis and are more chemosensitive.[59] If a tumor involves eloquent cortex, less aggressive surgery is often pursued in oligodendrogliomas given that any residual tumor after surgery will likely be responsive to adjuvant therapy. Currently, preoperative diagnosis is obtained by stereotactic biopsy; however, noninvasive imaging diagnosis may be possible in the near future. When attempting to differentiate oligodendrogliomas from other tumors, characteristic conventional MRI and CT findings include coarse calcifications (present in up to 90% of patients),[60] cortical/subcortical location, indistinct tumor borders, and patchy enhancement (approximately 50%),[61,62] which are all uncommon in low-grade astrocytomas. The majority of studies examining the use of MR and PET imaging to diagnose oligodendrogliomas were done prior to the current molecular diagnosis of oligodendrogliomas implemented in the WHO 2016 diagnostic criteria. Prior to the WHO classification using molecular markers, multiple studies showed that oligodendrogliomas defined by histology (histologic oligodendrogliomas) demonstrated higher amino acid PET uptake compared to astrocytomas of the same grade.[63] When incorporating 1p/19q co-deletion status, the amino acid PET uptake becomes more controversial. In a study including 102 patients with WHO grade II or III oligodendroglial tumors by histology, Saito et al. found that the presence of the 1p/19q co-deletion was associated with a significantly higher [11C]methionine uptake both when combining both tumor grades and when limited to grade II oligodendroglial tumors.[64] Iwadate et al. also found that tumors with oligodendroglial histology had significantly higher [11C]methionine uptake than astrocytomas of the same grade; however, oligodendroglial tumors with the 1p/19q co-deletion demonstrated significantly lower uptake than histologic oligodendroglial tumors without the co-deletion.[65] An additional problem that arises when attempting to utilize [11C]methionine PET for noninvasive diagnosis of oligodendrogliomas is that although oligodendrogliomas demonstrate uptake higher than astrocytomas of the same grade, they demonstrate overlapping [11C]methionine uptake with higher-grade astrocytomas.[66] These findings illustrate the pitfalls of attempting to use PET as a single modality and [11C]methionine as a single radiotracer when evaluating brain tumors. A multimodality diagnostic approach is better suited to evaluate patients at initial diagnosis of their brain tumors including co-registration of [11C]methionine PET to anatomic MR imaging in addition to utilization of other imaging modalities such as diffusion-weighted imaging, perfusion MRI, MR spectroscopy, and additional radiotracers including [18F]FDG. A multimodality imaging approach may prove to be reliable for noninvasive diagnosis of gliomas in the future.

Differentiation of low grade from anaplastic oligodendrogliomas

Noninvasive differentiation of WHO grade II and III oligodendrogliomas by imaging has potential value given that the anaplastic subtype generally necessitates more aggressive treatment. While the presence of enhancement on MRI and increased cerebral blood volume on MR perfusion studies can be helpful in differentiating high-grade from low-grade astrocytomas, these parameters are not as useful in differentiating anaplastic from low-grade oligodendrogliomas given the frequent contrast enhancement and high cerebral blood volumes of low-grade oligodendrogliomas. This provides a potential useful role for [11C]methionine PET.

Prior to the molecular criteria for diagnosis of oligodendrogliomas, studies showed that histologic anaplastic oligodendrogliomas demonstrated significantly higher [11C]methionine uptake compared to low-grade oligodendrogliomas.[65,67] Iwadate et al. further stratified the tumors using 1p/19q co-deletion and found that grade II oligodendrogliomas with the co-deletion (molecular oligodendrogliomas) had a significantly lower mean [11C]methionine uptake compared to molecular anaplastic oligodendrogliomas (mean tumor to normal brain uptake ratio of 1.94 vs 3.49, respectively).[65] [18F]FDG has also shown to be significantly higher in high-grade compared to low-grade gliomas; however, in oligodendrogliomas, it has been demonstrated that [11C]methionine PET may outperform [18F]FDG PET for differentiating grade II from grade III tumors.[67] Although grade III oligodendrogliomas demonstrate significantly higher [11C]methionine uptake than grade II oligodendrogliomas, there is still some overlap between the range of uptake between the two grades. Because of the overlap, [11C]methionine PET cannot replace histologic diagnosis from tissue biopsy at this point. In order to adequately differentiate low-grade from anaplastic oligodendrogliomas by imaging, using a combined radiotracer approach of [11C]methionine and [18F]FDG, as well as co-registration with functional and anatomic MR imaging, may prove to be the optimal method to differentiate these tumors (see Fig. 3).

Prognostic value of [11C]methionine PET

Increased [11C]methionine PET uptake in gliomas has been shown to be an independent prognostic factor for decreased survival. Riboms et al. subdivided gliomas into astrocytomas and oligodendrogliomas based on histology and found that [11C]methionine PET remained an independent prognosticator in only oligodendrogliomas.[68] All studies that evaluated the prognostic utility of [11C]methionine PET in oligodendrogliomas were done prior to the new WHO 2016 molecular criteria for these tumors. While it has not been studied using [11C]methionine PET, a study utilizing [18F]FET found that amino acid

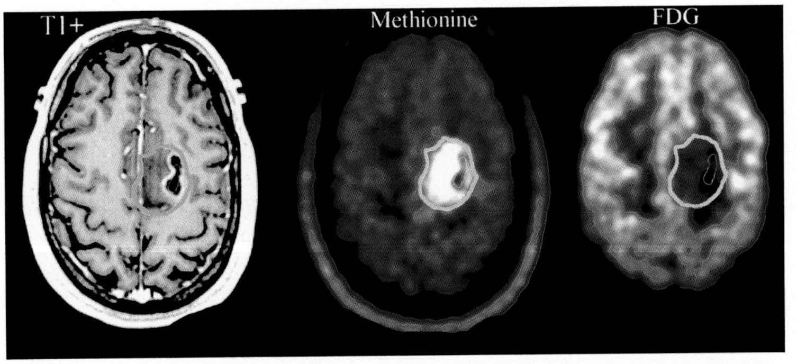

FIG. 3 MR T1 postgadolinium, [11C]methionine PET, [18F]FDG PET; 32-year-old male at the time of initial diagnosis demonstrates a left frontal partially cystic mass with rim enhancement around the cystic portion of the tumor. PET imaging demonstrated increased [11C]methionine uptake and decreased [18F]FDG uptake suggestive of low-grade histology; however, final pathology was consistent with anaplastic oligodendroglioma (1p/19q co-deletion present). Given that the MRI demonstrated the only findings of high-grade histology, this study illustrates the importance of incorporating all imaging modalities to assist in making the diagnosis by imaging. *Courtesy of the Charles F. Kettering Memorial Hospital Department of Nuclear Medicine/PET; images were processed utilizing Mirada XD3 software.*

uptake remained an independent prognostic indicator even when adjusting for many glioma tumor markers including the 1p/19q co-deletion.[69] Kaschten et al. evaluated the use of a dual-radiotracer PET study including both [18F]FDG and [11C]methionine to predict survival.[70] They found that both [18F]FDG and [11C]methionine contributed valuable prognostic information and that the prognostic value of both radiotracers when used together was better than either when used alone.[70] As expected, patients that had both high [18F]FDG and high [11C]methionine had the worst Kaplan Meier survival curves and patients with low uptake of both radiotracers demonstrated the best survival curves.[70] In addition, there were a significant amount of tumors that demonstrated high [11C]methionine and low [18F]FDG uptake which followed a separate survival curve slightly worse than that of high uptake of both radiotracers.[70] A separate group of gliomas with high [18F]FDG uptake and low [11C]methionine uptake also demonstrated a distinct survival curve which was closer to the curve of tumors with low uptake of both radiotracers.[70] These data support that the dual radiotracer technique using [18F]FDG and amino acid PET for prognosis is superior to using either radiotracer alone. Incorporation of all imaging and clinical data including anatomic imaging, MR or CT perfusion and permeability data, and MR spectroscopy if available will possibly lead to even more refined prognostic information and may even help guide treatment in the current era of precision medicine.

Guiding stereotactic biopsy

Although CT and MR image-guided stereotactic biopsy is the current standard of care for diagnosis of brain tumors, there are significant limitations to this approach. Brain tumors often show significant intratumoral heterogeneity, the highest-grade portion of the tumor is often not sampled, and nondiagnostic biopsies due to sampling of gliosis or necrosis are not uncommon. It has been reported that in low- to intermediate-grade brain tumors, the initial grade of the tumor from stereotactic biopsy is increased in over half of the tumors after review of the surgically resected specimen.[71] [11C]methionine PET has shown to be an excellent tool for stereotactic biopsy planning. In a study that included 32 patients with brain tumors (some of which were oligodendrogliomas), Pirotte et al. found that all stereotactic biopsy trajectories that had an associated area of increased [11C]methionine uptake were diagnostic for tumor.[72] Furthermore, all stereotactic biopsies that utilized CT or MRI for planning that were outside the area of [11C]methionine uptake on PET were nondiagnostic. In a similar study on pediatric patients, PET-guided stereotactic biopsies were performed in 35 patients with suspected brain tumors ([11C]methionine PET was used in 24 of the cases).[73] They found all PET-guided trajectories yielded tumor, but seven of the MRI-guided trajectories were nondiagnostic.[73] In addition, the PET-guided trajectories which were targeted at the area of highest PET uptake yielded a higher-grade tumor diagnosis compared to the MRI-guided biopsies in 16 cases.[73] Focal areas of highest [11C]methionine uptake within a tumor, also known as "hot spots," have been used for stereotactic biopsy planning in attempt to sample the highest-grade areas of brain tumors (see Fig. 4). In a study on 12 presumed low-grade gliomas by MRI which were found to have at least one [11C]methionine "hot spot," 8 of these tumors (67%) were found to be high-grade gliomas.[58] This [11C] methionine "hot spot"-guided stereotactic biopsy approach has strong potential in oligodendrogliomas given the typical high diffuse uptake in low-grade oligodendrogliomas and significantly higher uptake in anaplastic oligodendrogliomas. Other PET imaging methods such as dynamic 18F-FET PET, which measures radiotracer uptake over time, have shown excellent results for identifying and targeting areas of anaplasia in a background of more low-grade surrounding tumor.[69,74]

FIG. 4 Row 1, MR T1 postgadolinium, [^{11}C]
methionine PET, [^{18}F]FDG PET; row 2, MR T1
postgadolinium, [^{11}C]methionine PET, [^{18}F]FDG
PET; 29-year-old female that presented with a
seizure at the time of initial diagnosis (row 1) demonstrated a right frontal nonenhancing mass with
increased [^{11}C]methionine uptake and decreased
[^{18}F]FDG uptake consistent with low-grade histology. Final pathology was consistent with low-
grade oligoastrocytoma with positive 1p/19q
co-deletion (later reclassified as a low-grade oligodendroglioma by current WHO diagnostic criteria).
Repeat imaging 2 months later (row 2) demonstrated no intraaxial enhancement; however, there
was residual increased [^{11}C]methionine uptake with
a new small focus of increased [^{18}F]FDG uptake
which was concerning for recurrent tumor with
high-grade transformation. These areas of focal
increased PET uptake or "hot spots" can be targeted
by stereotactic biopsies and may represent areas
of higher-grade malignancy. *Courtesy of the
Charles F. Kettering Memorial Hospital
Department of Nuclear Medicine/PET; images
were processed utilizing Mirada XD3 software.*

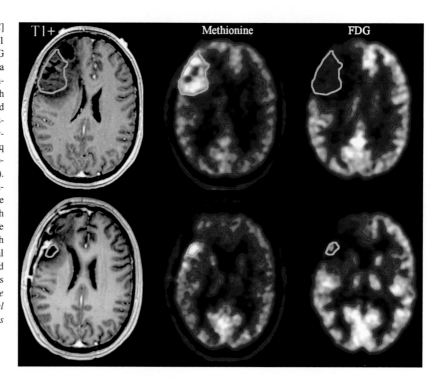

Defining tumor margins and surgical planning

Conventional MRI has limited sensitivity and specificity for defining tumor borders, especially in nonenhancing low-grade gliomas. In these cases, surgical planning is typically based on the area of abnormal hyperintense FLAIR signal, however, this often does not clearly differentiate between tumor tissue and peritumoral edema, gliosis, and other benign processes. Due to the high incidence of diffuse uptake in both low-grade and anaplastic oligodendrogliomas, [^{11}C]methionine PET has great potential as a surgical planning tool to define the margins of these tumors. While no studies have been done limited strictly to oligodendrogliomas, Pirotte et al. studied the use of both [^{18}F]FDG and [^{11}C]methionine for guidance in surgical resection of both low- and high-grade gliomas. Of these, [^{11}C]methionine ($n=31$) and [^{18}F]FDG ($n=6$) were used for preoperative planning in 28 histologic low-grade oligodendrogliomas and 5 histologic anaplastic oligodendrogliomas.[75] They found that [^{11}C]methionine was superior to [^{18}F]FDG PET for surgical planning and ended up using primarily [^{11}C]methionine in their study.[75] Also, they found that the target volumes derived from MRI compared to PET were dif-ferent in almost all cases. In 76 out of 82 cases which had a [^{11}C]methionine PET examination, the uptake on the scan was either smaller ($n=27$), larger ($n=34$), or nonoverlapping ($n=15$) than the planned MRI volume.[75] [^{11}C]methionine PET influenced their final target volume in 88% of low-grade and 78% of high-grade glioma cases.[75] In the 11 patients that had the entire PET tumor volume resected, as confirmed by postoperative PET showing no residual uptake, 8 of these patients had no evidence for viable tumor in biopsies of the surgical resection margins.[75] When limited to histologic oligoden-droglial tumors, [^{11}C]methionine PET provided useful information that influenced final surgical target volume in 35 out of the 37 patients. The only two times that it was not useful was when the [^{11}C]methionine PET-based contour volume was contained within the MRI contour volume and the MRI contour volume was used alone for the final surgical target volume.[75] They concluded that the margin of infiltrative low-grade gliomas may be better visualized using [^{11}C]methionine PET and neurosurgeons should consider integrating these studies into their neuronavigational surgeries.[75]

In a study which featured 10 oligoastrocytomas and 5 oligodendrogliomas by histology, patients that underwent com-plete surgical resection of the PET avid portion of their tumor (either [^{11}C]methionine ($n=43$) or [^{18}F]FDG ($n=23$)) had a statistically significant increase in mean overall survival of 32.5 months compared to 17.6 months in those with subtotal PET resection.[76] There was a trend toward increased survival in complete resection of the MRI signal abnormality; however, this was not as substantial and not statistically significant.[76] In a study on pediatric brain tumors, 37 out of the 50 cases that underwent PET-guided surgical resection had no residual PET uptake on the postoperative scans.[73]

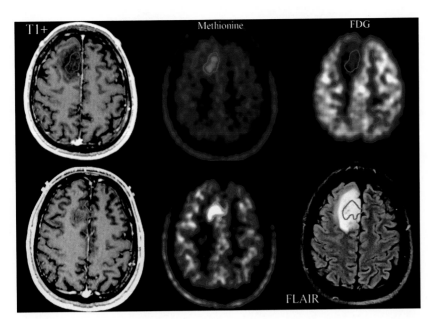

FIG. 5 Row 1, MR T1 postgadolinium, [¹¹C]methionine PET, [¹⁸F]FDG PET; row 2, MR T1 postgadolinium, [¹¹C]methionine PET, MR FLAIR; 45-year-old male at the time of initial diagnosis (row 1) demonstrated a nonenhancing parasagittal right frontal mass with increased [¹¹C]methionine uptake and decreased [¹⁸F]FDG uptake suggestive of low-grade histology. Intraoperatively, the resection was presumed to be a gross total resection, however, postoperative imaging (row 2) demonstrated an area of residual increased FLAIR signal with corresponding increased [¹¹C]methionine uptake consistent with residual viable neoplasm. Reresection confirmed residual WHO grade II oligodendroglioma (1p/19q co-deletion present). *Courtesy of the Charles F. Kettering Memorial Hospital Department of Nuclear Medicine/PET; images were processed utilizing Mirada XD3 software.*

All of these patients were found to have negative surgical margins.[73] PET imaging was also found to be useful for determining which cases should undergo a second resection as well as guiding surgery in these patients (see Fig. 5).[73]

While [¹¹C]methionine PET alone performs well to delineate tumor borders in gliomas, Kinoshita et al. developed a method using both [¹¹C]methionine and [¹⁸F]FDG which had an even higher sensitivity and specificity for detecting infiltrative tumor cells at the borders of the tumor core.[77] At the edges of the tumor in areas which demonstrated increased T2 signal but no enhancement, they calculated a decoupling score, which when elevated, reflects areas of "decoupling" of the normal linear relationship between [¹¹C]methionine and [¹⁸F]FDG uptake in normal brain tissue.[77] In the areas which demonstrated increased [¹¹C]methionine uptake out of proportion to the [¹⁸F]FDG uptake, these areas were found to have infiltrative tumor.[77] When the cutoff for the decoupling score was set at 3.0, there was an increase in the specificity for infiltrative glioma cells in the biopsy specimen to 93.5% compared to 87% using [¹¹C]methionine alone without any change in the sensitivity at 87.5%.[77]

While the WHO current recommendations for anaplastic oligodendrogliomas is maximal safe surgical resection,[78] recent studies have brought the survival benefit of complete surgical resection of low-grade oligodendrogliomas into question. At a minimum, the benefit of complete surgical resection is significantly attenuated compared to astrocytomas.[79,80] This is likely related to the high sensitivity of oligodendrogliomas to chemotherapy. Even taking this into account, more accurate delineation of the tumor borders by PET imaging is still very helpful for surgical planning. For example, when abnormal MRI signal from the tumor involves eloquent areas of the brain, [¹¹C]methionine PET may better delineate the borders of the tumor, so that surgical morbidity can be reduced as much as possible. Furthermore, if the plan is to undergo subtotal resection based on the size or location of the tumor, the more metabolically active portions of the tumor as defined by either increased [¹¹C]methionine or [¹⁸F]FDG uptake can be targeted for resection. By resecting these areas of higher metabolic uptake that likely reflect more aggressive areas of the tumor, the tumor may follow a more benign clinical course typical of a low-grade oligodendroglioma. Furthermore, in patients foregoing initial surgical resection and instead undergoing chemotherapy, radiotherapy, laser interstitial thermal therapy, or radiosurgery, [¹¹C]methionine and [¹⁸F]FDG PET may be helpful to evaluate new areas of enhancement that may be related to malignant progression versus treatment-related necrosis (see Fig. 6).

Role of [¹¹C]methionine in radiation therapy and chemotherapy

Patients with both low-grade and anaplastic oligodendrogliomas are typically treated with maximal safe surgical resection followed by chemotherapy and radiation. Conventional MRI is used most commonly to identify areas of residual tumor after surgery and plan radiotherapy. Conventional MRI has limited specificity after surgery to identify residual tumor, and the signal abnormalities on FLAIR and postcontrast T1-weighted sequences are often unable to differentiate gliosis,

FIG. 6 Row 1, MR T1 postgadolinium, [11C]methionine PET, [18F]FDG PET; row 2, MR T1 postgadolinium, [11C]methionine PET, [18F]FDG PET; row 3, MR T1 postgadolinium, [11C]methionine PET, [18F]FDG PET; 35-year-old male at the time of initial diagnosis (row 1) demonstrated a parasagittal frontal nonenhancing mass with increased [11C]methionine uptake and decreased [18F]FDG uptake consistent with low-grade histology. Final pathology was consistent with WHO grade II oligodendroglioma (1p/19q co-deletion present). Follow-up imaging 6 and 7 years later (rows 2 and 3, respectively) after chemotherapy and laser interstitial thermal therapy (LITT), showed increased size, enhancement, and volume of [11C]methionine uptake overtime; however, there remains decreased [18F]FDG uptake suggestive of low-grade histology. Biopsies during LITT and surgical resection after these imaging studies demonstrated persistent WHO grade II oligodendroglioma histology. The new enhancement was likely due to treatment-related necrosis. *Courtesy of the Charles F. Kettering Memorial Hospital Department of Nuclear Medicine/PET; images were processed utilizing Mirada XD3 software.*

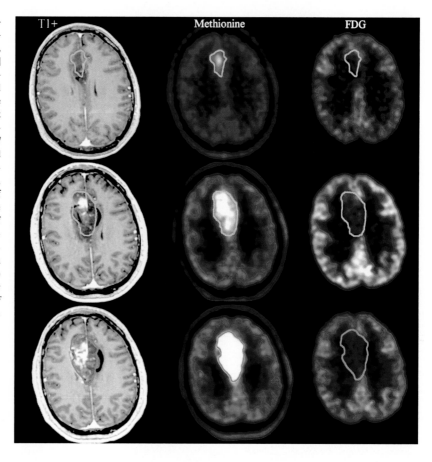

ischemia, and necrosis from viable tumor tissue. In addition, it has been demonstrated that gliomas, including oligodendrogliomas, have tumor cells that extend beyond the area of abnormal MRI signal.[81] Studies have shown that [11C]methionine PET has better sensitivity and specificity for glioma tumor tissue compared to conventional MRI, so [11C]methionine has great potential for identifying residual tumor and defining gross tumor volumes for radiotherapy. It has been demonstrated that the area of [11C]methionine uptake, which corresponds to residual viable tumor, does not match the contrast enhancement and abnormal T2-weighted signal on conventional MRI.[82] Often, the area of [11C]methionine uptake extends outside both the area of enhancement and T2 signal abnormality.[82] Also, in almost all cases, the area of abnormal T2 signal has areas without corresponding [11C]methionine uptake.[82] Having an accurate gross tumor volume is crucial for both treating the entirety of the viable tumor and sparing as much normal brain tissue as possible from radiation-induced toxicity. Grosu et al. found that integrating [11C]methionine or ^{123}I-alpha-methyl-tyrosine single-photon emission (SPECT) CT into their radiation therapy planning to define tumor volumes increased overall survival in patients with recurrent high-grade gliomas (one of which was an anaplastic oligodendroglioma) treated with hypofractionated reirradation.[83] Although defining gross tumor volume by [11C]methionine PET has been shown to be feasible in low-grade gliomas including oligodendrogliomas,[84] the impact of [11C]methionine PET for radiation therapy on survival has not been studied in these low-grade tumors. Given the high [11C]methionine uptake in oligodendrogliomas, using [11C]methionine PET for radiation therapy planning has potential to improve patient outcomes.

Oligodendrogliomas typically show an excellent response to chemotherapy and are typically treated with either temozolomide or the procarbazine, CCNU, and vincristine (PCV) regimen. Response rates for the PCV regimen in addition to radiotherapy have been reported to achieve a 5-year progression free survival around 75%.[59] Given the previously described limitations of MRI to delineate these tumors, especially after surgery, [11C]methionine PET has been shown to be an excellent means of follow-up in oligodendrogliomas after treatment. [11C]methionine PET not only allows for measurement of change in the volume of uptake but also change in the intensity of uptake compared to normal brain tissue. This adds an additional dimension compared to the current standard of measuring only changes in the volume of abnormal FLAIR signal and enhancement. In patients with oligodendrogliomas, it has been demonstrated that measuring both the mean uptake and volume of increased uptake of [11C]methionine before and after chemotherapy yields a larger decrease compared to using FLAIR volumes alone.[85] Nariai et al. also found [^{11}C]methionine PET to be an effect means of monitoring response to radiotherapy.[86] After adjuvant radiotherapy with or without chemotherapy, they found a significant decrease in the [^{11}C]methionine uptake in high-grade astrocytomas and both low- and high-grade oligodendroglial tumors.[86] Using [^{11}C]methionine PET to monitor treatment response needs to be further studied on a larger scale to determine if it provides more accurate assessment of response to chemotherapy and radiotherapy compared to MRI.

Tumor surveillance

In patients with oligodendrogliomas that have undergone treatment, imaging surveillance is essential to monitor for changes that suggest tumor recurrence. On conventional MRI, tumor recurrence typically presents as a mass with hyperintense FLAIR signal with or without enhancement, often in the radiation field. However, conventional MR imaging is often not definitive, and recurrence can be difficult to differentiate from radiation necrosis or treatment-related changes. Furthermore, contrast enhancement and abnormal increased FLAIR signal are often diminished by corticosteroids which are often necessary to reduce edema in progressing tumors. This confounds contrast enhancement and the area of FLAIR signal abnormality as a means of evaluating for residual tumor.[87,88] Since anatomic imaging is limited in its assessment for underlying tumor, there is incentive to develop alternate methods to provide better assessment for recurrent disease. [^{18}F]FDG was the first PET agent to be described to be helpful in distinguishing tumor recurrence from radiation necrosis.[17] Nevertheless, [^{18}F]FDG has limited specificity for tumor recurrence, because it is known to also accumulate in macrophages and inflammatory tissue which are often present in cases of radiation necrosis. Also, [^{18}F]FDG has limited sensitivity due to limited contrast between normal brain tissue and lower-grade tumors, which increases the potential for false-negative [^{18}F]FDG PET studies.[70] [^{11}C]methionine PET co-registered to MRI has been shown to be significantly better at differentiating tumor recurrence from radiation necrosis with an approximate sensitivity and specificity of 75% in gliomas.[89–92] The somewhat limited specificity has been postulated to relate to accumulation of [^{11}C]methionine in microglial cells and astrocytes involved in treatment-induced gliosis. Alternatively, it may be related to treatment-induced vascular proliferation and breakdown of the blood-brain barrier.[93] The limited sensitivity is likely related to the lower uptake of [^{11}C]methionine in tumors after treatment and inherent lower level of uptake in some low-grade astrocytomas. Although the sensitivity of [^{11}C]methionine PET for tumor recurrence has not been specifically studied in oligodendrogliomas, it may be even higher than gliomas as a whole given the higher uptake of [^{11}C]methionine in oligodendrogliomas compared to astrocytomas.

Patients with low-grade oligodendrogliomas will occasionally forego initial treatment and instead choose to undergo surveillance for progression of their tumor. The clinical course of these tumors typically involves an initial period of indolent tumor growth followed by malignant progression and more rapid growth. Once there are signs of malignant progression, the patient typically must begin more aggressive treatment. Again, [^{18}F]FDG was the initial PET agent to show utility for detection of anaplastic transformation, as illustrated by high [^{18}F]FDG uptake in a previously known low-grade tumor.[26,27] In extreme, exceptionally rare cases, anaplastic oligodendrogliomas can metastasize,[94] and [^{18}F]FDG can be useful for delineating the sites of metastatic disease (see Fig. 7). Interindividual variability in [^{11}C]methionine uptake limits the ability to grade tumors on initial imaging.[95] On the contrary, [^{11}C]methionine has shown to be very effective in detecting anaplastic transformation in patients with a previous baseline [^{11}C]methionine PET study (see Fig. 8). If subsequent [^{11}C]methionine PET studies demonstrate new foci of higher [^{11}C]methionine that was not present on the baseline study, this was found to have a sensitivity of 90% and specificity of 92.3% for anaplastic transformation.[96] Co-registration of conventional MRI and [^{11}C]methionine PET with as much additional data as possible from MR perfusion, spectroscopy, and additional radiotracers will likely provide be the most sensitive method for detecting early tumor recurrence or high-grade transformation.

Conclusion

Oligodendrogliomas have undergone a substantial shift in their diagnosis based on the new molecular diagnosis in the WHO 2016 criteria, which uses the presence of the 1p/19q co-deletion and IDH mutation to define these tumors. They are more radiosensitive and chemosensitive and thus have a better prognosis compared to astrocytomas of the same grade. Therefore, accurate diagnosis is essential for managing therapy. Noninvasive differentiation of these tumors by imaging may be possible in the future, and utilization of [^{18}F]FDG, [^{11}C]methionine, and other amino acid PET radiotracers have shown utility in helping to differentiate these tumors from astrocytomas in vivo. [^{11}C]methionine PET has also shown utility in prognostication and grading oligodendrogliomas. Both low-grade and anaplastic oligodendrogliomas typically demonstrate avid, diffuse radiotracer uptake on [^{11}C]methionine PET examinations. Given that oligodendrogliomas demonstrate variable patchy enhancement and ill-defined borders on conventional MRI, [^{11}C]methionine PET may prove to be superior to MRI for defining tumor volumes for both surgical planning and postoperative radiotherapy, as well as assessing response to adjuvant chemoradiotherapy. In patients that undergo complete surgical resection of

FIG. 7 Noncontrast coronal CT, coronal [^{18}F]FDG PET, coronal [^{18}F]FDG PET 3D MIP; this patient was initially diagnosed with a WHO grade II astrocytoma, which was treated with surgical resection and adjuvant radiotherapy. Nine years later the patient developed a recurrence in the brain which was resected and pathology was consistent with anaplastic oligodendroglioma (1p/19q co-deletion present). Seven months later the patient presented with a neck mass which was biopsied, and the final pathology was consistent with metastatic anaplastic oligodendroglioma. [^{18}F]FDG PET at that time demonstrates metastatic disease to the right cervical region as well as the left side of the sacrum, left femur, and L2 vertebral body. *Courtesy of the Charles F. Kettering Memorial Hospital Department of Nuclear Medicine/ PET; images were processed utilizing Mirada XD3 software.*

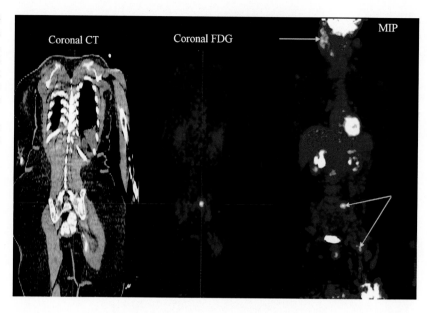

FIG. 8 Row 1, MR T1 postgadolinium, [^{11}C] methionine PET, [^{18}F]FDG PET; row 2, MR T1 postgadolinium, [^{11}C]methionine PET, [^{18}F]FDG PET; 63-year-old female at the time of initial diagnosis (row 1) demonstrates a right parietal nonenhancing cortical/subcortical mass with increased [^{11}C]methionine uptake and decreased [^{18}F]FDG uptake. Final pathology was consistent with anaplastic oligodendroglioma (1p/19q co-deletion present). Follow-up imaging 7 years later (row 2) demonstrated patchy areas of enhancement in the right temporal and right occipital white matter with increased [^{11}C]methionine uptake and now patchy areas of increased [^{18}F]FDG uptake. Final pathology confirmed malignant degeneration into an IDH wild-type anaplastic glioma (at least WHO grade III; 1p/19q co-deletion absent). *Courtesy of the Charles F. Kettering Memorial Hospital Department of Nuclear Medicine/PET; images were processed utilizing Mirada XD3 software.*

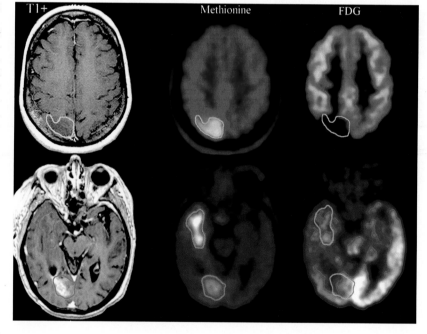

oligodendrogliomas, [^{11}C]methionine PET has shown to be useful for detection of tumor recurrence and differentiation of recurrent tumor from radiation necrosis. Although [^{11}C]methionine PET has shown good results for diagnosis, treatment, and surveillance in oligodendrogliomas, the optimal strategy for imaging these patients likely includes co-registration of [^{11}C]methionine PET with anatomic CT or MR imaging, perfusion MRI, MR spectroscopy, and possibly additional radiotracers such as [^{18}F]FDG which have shown to provide complimentary information. In medical centers without a cyclotron, other ^{18}F amino acid analogs with longer half-lives such as [^{18}F]FET have shown potential to be an adequate substitute for [^{11}C]methionine for similar uses. Additional research is needed to determine the optimal yet cost-effective imaging strategy for these patients and to determine if the molecular and MR imaging characteristics can help establish a personalized treatment approach to achieve the most favorable outcomes for these patients.

References

1. Louis DN, et al. The 2016 World Health Organization Classification of Tumors of the Central Nervous System: a summary. *Acta Neuropathol (Berl)*. 2016;131:803–820.
2. van den Bent MJ. Interobserver variation of the histopathological diagnosis in clinical trials on glioma: a clinician's perspective. *Acta Neuropathol (Berl)*. 2010;120:297–304.
3. Ostrom QT, et al. CBTRUS statistical report: primary brain and other central nervous system tumors diagnosed in the United States in 2011–2015. *Neuro-Oncology*. 2018;20:iv1–iv86.
4. van den Bent MJ, Smits M, Kros JM, Chang SM. Diffuse infiltrating oligodendroglioma and astrocytoma. *J Clin Oncol*. 2017;35:2394–2401.
5. Yang P, et al. Management and survival rates in patients with glioma in China (2004–2010): a retrospective study from a single-institution. *J Neuro-Oncol*. 2013;113:259–266.
6. Aghi MK, et al. The role of surgery in the management of patients with diffuse low grade glioma: a systematic review and evidence-based clinical practice guideline. *J Neuro-Oncol*. 2015;125:503–530.
7. Smith JS, et al. Role of extent of resection in the long-term outcome of low-grade hemispheric gliomas. *J Clin Oncol Off J Am Soc Clin Oncol*. 2008;26:1338–1345.
8. McGirt MJ, et al. Extent of surgical resection is independently associated with survival in patients with hemispheric infiltrating low-grade gliomas. *Neurosurgery*. 2008;63:700–707. author reply 707–708.
9. Shaw EG, et al. Recurrence following neurosurgeon-determined gross-total resection of adult supratentorial low-grade glioma: results of a prospective clinical trial. *J Neurosurg*. 2008;109:835–841.
10. Paulus W, Peiffer J. Intratumoral histologic heterogeneity of gliomas. A quantitative study. *Cancer*. 1989;.
11. Wen PY, et al. Updated response assessment criteria for high-grade gliomas: response assessment in neuro-oncology working group. *J Clin Oncol Off J Am Soc Clin Oncol*. 2010;28:1963–1972.
12. Watanabe M, Tanaka R, Takeda N. Magnetic resonance imaging and histopathology of cerebral gliomas. *Neuroradiology*. 1992;34:463–469.
13. Johnson PC, Hunt SJ, Drayer BP. Human cerebral gliomas: correlation of postmortem MR imaging and neuropathologic findings. *Radiology*. 1989;170:211–217.
14. la Fougere C. Molecular imaging of gliomas with PET: opportunities and limitations. *Neuro-Oncology*. 2011;13:806–819.
15. Di Chiro G, et al. Glucose utilization of cerebral gliomas measured by [^{18}F] fluorodeoxyglucose and positron emission tomography. *Neurology*. 1982;32:1323–1329.
16. Di Chiro G. Positron emission tomography using [^{18}F] fluorodeoxyglucose in brain tumors. A powerful diagnostic and prognostic tool. *Investig Radiol*. 1987;22:360–371.
17. Patronas NJ. Work in progress: [^{18}F] fluorodeoxyglucose and positron emission tomography in the evaluation of radiation necrosis of the brain. *Radiology*. 1982;.
18. Di Chiro G, et al. Cerebral necrosis after radiotherapy and/or intraarterial chemotherapy for brain tumors: PET and neuropathologic studies. *AJR Am J Roentgenol*. 1988;150:189–197.
19. Wong TZ, van der Westhuizen GJ, Coleman RE. Positron emission tomography imaging of brain tumors. *Neuroimaging Clin N Am*. 2002;12:615–626.
20. Nagamatsu S, Sawa H, Wakizaka A, Hoshino T. Expression of facilitative glucose transporter isoforms in human brain tumors. *J Neurochem*. 1993;61:2048–2053.
21. Di Chiro G, et al. Glucose utilization of cerebral gliomas measured by [^{18}F] fluorodeoxyglucose and positron emission tomography. *Neurology*. 1982;32:1323–1329.
22. Alavi JB, et al. Positron emission tomography in patients with glioma. A predictor of prognosis. *Cancer*. 1988;62:1074–1078.
23. Herholz K, et al. Correlation of glucose consumption and tumor cell density in astrocytomas. A stereotactic PET study. *J Neurosurg*. 1993;79:853–858.
24. Patronas NJ, et al. Prediction of survival in glioma patients by means of positron emission tomography. *J Neurosurg*. 1985;62.
25. Barker FG, et al. 18-Fluorodeoxyglucose uptake and survival of patients with suspected recurrent malignant glioma. *Cancer*. 1997;79:115–126.
26. De Witte O, et al. Prognostic value positron emission tomography with [^{18}F]fluoro-2-deoxy-D-glucose in the low-grade glioma. *Neurosurgery*. 1996;39:470–476.
27. Padma MV, et al. Prediction of pathology and survival by FDG PET in gliomas. *J Neuro-Oncol*. 2003;64(3):227–237.
28. Ricci PE. Differentiating recurrent tumor from radiation necrosis: time for re-evaluation of positron emission tomography? *AJNR Am J Neuroradiol*. 1998;19(3):407–413.
29. Olivero WC, Dulebohn SC, Lister JR. The use of PET in evaluating patients with primary brain tumours: is it useful? *J Neurol Neurosurg Psychiatry*. 1995;58:250–252.
30. Giammarile F, et al. High and low grade oligodendrogliomas (ODG): correlation of amino-acid and glucose uptakes using PET and histological classifications. *J Neuro-Oncol*. 2004;68:263–274.
31. Krohn KA, Mankoff DA, Muzi M, Link JM, Spence AM. True tracers: comparing FDG with glucose and FLT with thymidine. *Nucl Med Biol*. 2005;32:663–671.
32. Isselbacher KJ. Sugar and amino acid transport by cells in culture–differences between normal and malignant cells. *N Engl J Med*. 1972;286:929–933.
33. Busch H. The uptake of a variety of amino acids into nuclear proteins of tumors and other tissues. *Cancer Res*. 1959;.
34. Galldiks N, Law I, Pope WB, Arbizu J, Langen K-J. The use of amino acid PET and conventional MRI for monitoring of brain tumor therapy. *NeuroImage Clin*. 2017;13:386–394.

35. Sasajima T, et al. Proliferation-dependent changes in amino acid transport and glucose metabolism in glioma cell lines. *Eur J Nucl Med Mol Imaging*. 2004;31:1244–1256.

36. Roelcke U. Alteration of blood-brain barrier in human brain tumors: comparison of [18F]fluorodeoxyglucose, [11C]methionine and rubidium-82 using PET. *J Neurol Sci*. 1995;132(1).

37. Beuthien-Baumann B, et al. 3-O-methyl-6-[18F]fluoro-L-DOPA and its evaluation in brain tumour imaging. *Eur J Nucl Med Mol Imaging*. 2003;30:1004–1008.

38. Chen W, et al. 18F-FDOPA PET imaging of brain tumors: comparison study with 18F-FDG PET and evaluation of diagnostic accuracy. *J Nucl Med Off Publ Soc Nucl Med*. 2006;47:904–911.

39. Becherer A, et al. Brain tumour imaging with PET: a comparison between [18F]fluorodopa and [11C]methionine. *Eur J Nucl Med Mol Imaging*. 2003;30:1561–1567.

40. Weber WA, et al. O-(2-[18F]fluoroethyl)-L-tyrosine and L-[methyl-11C]methionine uptake in brain tumours: initial results of a comparative study. *Eur J Nucl Med*. 2000;27:542–549.

41. Langen K-J, et al. O-(2-[18F]fluoroethyl)-L-tyrosine: uptake mechanisms and clinical applications. *Nucl Med Biol*. 2006;33:287–294.

42. Albert NL, et al. Early static (18)F-FET-PET scans have a higher accuracy for glioma grading than the standard 20–40 min scans. *Eur J Nucl Med Mol Imaging*. 2016;43:1105–1114.

43. Calcagni ML, et al. Dynamic O-(2-[18F]fluoroethyl)-L-tyrosine (F-18 FET) PET for glioma grading: assessment of individual probability of malignancy. *Clin Nucl Med*. 2011;36:841–847.

44. Ceccon G, et al. Dynamic O-(2-18F-fluoroethyl)-L-tyrosine positron emission tomography differentiates brain metastasis recurrence from radiation injury after radiotherapy. *Neuro-Oncology*. 2017;19:281–288.

45. Kamson DO, et al. Tryptophan PET in pretreatment delineation of newly-diagnosed gliomas: MRI and histopathologic correlates. *J Neuro-Oncol*. 2013;112:121–132.

46. Kondo A, et al. Phase IIa clinical study of [18F]fluciclovine: efficacy and safety of a new PET tracer for brain tumors. *Ann Nucl Med*. 2016;30:608–618.

47. Parent EE, Schuster DM. Update on 18F-Fluciclovine PET for prostate cancer imaging. *J Nucl Med Off Publ Soc Nucl Med*. 2018;59:733–739.

48. Singhal T, Narayanan TK, Jain V, Mukherjee J, Mantil J. 11C-L-methionine positron emission tomography in the clinical management of cerebral gliomas. *Mol Imaging Biol MIB Off Publ Acad Mol Imaging*. 2008;10:1–18.

49. Stern PH, Wallace CD, Hoffman RM. Altered methionine metabolism occurs in all members of a set of diverse human tumor cell lines. *J Cell Physiol*. 1984;119:29–34.

50. Bergstrom M, et al. Comparison of the accumulation kinetics of L-(methyl-11C)-methionine and D-(methyl-11C)-methionine in brain tumors studied with positron emission tomography. *Acta Radiol*. 1987;28:225–229.

51. Sato N. Evaluation of the malignancy of glioma using 11C-methionine positron emission tomography and proliferating cell nuclear antigen staining. *Neurosurg Rev*. 1999;22:210–214.

52. Kracht L, et al. Methyl-[11C]-L-methionine uptake as measured by positron emission tomography correlates to microvessel density in patients with glioma. *Eur J Nucl Med Mol Imaging*. 2003;20:868–873.

53. Okita Y. (11)C-methionine uptake correlates with tumor cell density rather than with microvessel density in glioma: a stereotactic image-histology comparison. *NeuroImage*. 2010;49:2977–2982.

54. Derlon JM, et al. [11C]L-methionine uptake in gliomas. *Neurosurgery*. 1989;25:720–728.

55. Goldman S. Regional methionine and glucose uptake in high-grade gliomas: a comparative study on PET-guided stereotactic biopsy. *J Nucl Med*. 1997;38(9).

56. Sasaki M, et al. A comparative study of thallium-201 SPET, carbon-11 methionine PET and fluorine-18 fluorodeoxyglucose PET for the differentiation of astrocytic tumours. *Eur J Nucl Med*. 1998;25:1261–1269.

57. Glaudemans AWJM, et al. Value of 11C-methionine PET in imaging brain tumours and metastases. *Eur J Nucl Med Mol Imaging*. 2013;40:615–635.

58. Roessler K, et al. Surgical target selection in cerebral glioma surgery: linking methionine (MET) PET image fusion and neuronavigation. *Minim Invasive Neurosurg*. 2007;50:273–280.

59. Buckner JC, et al. Radiation plus procarbazine, CCNU, and vincristine in low-grade glioma. *N Engl J Med*. 2016;374:1344–1355.

60. Smits M. Imaging of oligodendroglioma. *Br J Radiol*. 2016;89.

61. Megyesi JF, et al. Imaging correlates of molecular signatures in oligodendrogliomas. *Clin Cancer Res Off J Am Assoc Cancer Res*. 2004;10:4303–4306.

62. Jenkinson MD, et al. Histological growth patterns and genotype in oligodendroglial tumours: correlation with MRI features. *Brain J Neurol*. 2006;129:1884–1891.

63. Shinozaki N, et al. Discrimination between low-grade oligodendrogliomas and diffuse astrocytoma with the aid of 11C-methionine positron emission tomography. *J Neurosurg*. 2011;114:1640–1647.

64. Saito T, et al. 11C-methionine uptake correlates with combined 1p and 19q loss of heterozygosity in oligodendroglial tumors. *AJNR Am J Neuroradiol*. 2013;34:85–91.

65. Iwadate Y, et al. Molecular imaging of 1p/19q deletion in oligodendroglial tumours with 11C-methionine positron emission tomography. *J Neurol Neurosurg Psychiatry*. 2016;87:1016–1021.

66. Jansen NL, et al. Prediction of oligodendroglial histology and LOH 1p/19q using dynamic [(18)F]FET-PET imaging in intracranial WHO grade II and III gliomas. *Neuro-Oncology*. 2012;14:1473–1480.

67. Derlon JM, et al. Non-invasive grading of oligodendrogliomas: correlation between in vivo metabolic pattern and histopathology. *Eur J Nucl Med*. 2000;27:778–787.

68. Ribom D, Smits A. Baseline 11C-methionine PET reflects the natural course of grade 2 oligodendrogliomas. *Neurol Res.* 2005;27:516–521.

69. Thon N, et al. Dynamic 18F-FET PET in suspected WHO grade II gliomas defines distinct biological subgroups with different clinical courses. *Int J Cancer.* 2015;136:2132–2145.

70. Kaschten B. Preoperative evaluation of 54 gliomas by PET with fluorine-18-fluorodeoxyglucose and/or carbon-11-methionine. *J Nucl Med.* 1998;39(5).

71. Jackson RJ, et al. Limitations of stereotactic biopsy in the initial management of gliomas. *Neuro-Oncology.* 2001;3:193–200.

72. Pirotte B. Combined use of 18F-fluorodeoxyglucose and 11C-methionine in 45 positron emission tomography-guided stereotactic brain biopsies. *J Neurosurg.* 2004;101:476–483.

73. Pirotte BJ, et al. Clinical impact of integrating positron emission tomography during surgery in 85 children with brain tumors. *J Neurosurg Pediatr.* 2010;5:486–499.

74. Kunz M, et al. Hot spots in dynamic (18)FET-PET delineate malignant tumor parts within suspected WHO grade II gliomas. *Neuro-Oncology.* 2011;13:307–316.

75. Pirotte B, et al. Integrated positron emission tomography and magnetic resonance imaging-guided resection of brain tumors: a report of 103 consecutive procedures. *J Neurosurg.* 2006;104:238–253.

76. Pirotte BJM, et al. Positron emission tomography-guided volumetric resection of supratentorial high-grade gliomas: a survival analysis in 66 consecutive patients. *Neurosurgery.* 2009;64:471–481. discussion 481.

77. Kinoshita M, et al. A novel PET index, 18F-FDG-11C-methionine uptake decoupling score, reflects glioma cell infiltration. *J Nucl Med.* 2012;53:1701–1708.

78. Im JH, et al. Recurrence patterns after maximal surgical resection and postoperative radiotherapy in anaplastic gliomas according to the new 2016 WHO classification. *Sci Rep.* 2018;8.

79. Ding X, et al. The prognostic value of maximal surgical resection is attenuated in oligodendroglioma subgroups of adult diffuse glioma: a multicenter retrospective study. *J Neuro-Oncol.* 2018;140:591–603.

80. Wijnenga MMJ, et al. The impact of surgery in molecularly defined low-grade glioma: an integrated clinical, radiological, and molecular analysis. *Neuro-Oncology.* 2018;20:103–112.

81. Pallud J, et al. Diffuse low-grade oligodendrogliomas extend beyond MRI-defined abnormalities. *Neurology.* 2010;74:1724–1731.

82. Grosu A-L, et al. L-(Methyl-11C) methionine positron emission tomography for target delineation in resected high-grade gliomas before radiotherapy. *Int J Radiat Oncol Biol Phys.* 2005;63:64–74.

83. Grosu AL. Reirradiation of recurrent high-grade gliomas using amino acid PET (SPECT)/CT/MRI image fusion to determine gross tumor volume for stereotactic fractionated radiotherapy. *Int J Radiat Oncol Biol Phys.* 2005;63:511–519.

84. Nuutinen J, et al. Radiotherapy treatment planning and long-term follow-up with [(11)C]methionine PET in patients with low-grade astrocytoma. *Int J Radiat Oncol Biol Phys.* 2000;48:43–52.

85. Tang BN, et al. Semi-quantification of methionine uptake and flair signal for the evaluation of chemotherapy in low-grade oligodendroglioma. *J Neuro-Oncol.* 2005;71:161–168.

86. Nariai T, et al. Usefulness of L-[methyl-^{11}C] methionine-positron emission tomography as a biological monitoring tool in the treatment of glioma. *J Neurosurg.* 2005;103:498–507.

87. Cairncross JG, et al. Steroid-induced CT changes in patients with recurrent malignant glioma. *Neurology.* 1988;38:724–726.

88. Ostergaard L. Early changes measured by magnetic resonance imaging in cerebral blood flow, blood volume, and blood-brain barrier permeability following dexamethasone treatment in patients with brain tumors. *J Neurosurg.* 1999;90(2).

89. Thiel A, et al. Enhanced accuracy in differential diagnosis of radiation necrosis by positron emission tomography-magnetic resonance imaging co-registration: technical case report. *Neurosurgery.* 2000;46:232–234.

90. Tsuyuguchi N, et al. Methionine positron emission tomography for differentiation of recurrent brain tumor and radiation necrosis after stereotactic radiosurgery—in malignant glioma. *Ann Nucl Med.* 2004;18:291–296.

91. Terakawa Y, et al. Diagnostic accuracy of 11C-methionine PET for differentiation of recurrent brain tumors from radiation necrosis after radiotherapy. *J Nucl Med.* 2008;49:694–699.

92. Van Laere K, et al. Direct comparison of 18F-FDG and 11C-methionine PET in suspected recurrence of glioma: sensitivity, inter-observer variability and prognostic value. *Eur J Nucl Med Mol Imaging.* 2004;32:39–51.

93. Kim YH, et al. Differentiating radiation necrosis from tumor recurrence in high-grade gliomas: assessing the efficacy of 18F-FDG PET, 11C-methionine PET and perfusion MRI. *Clin Neurol Neurosurg.* 2010;112:758–765.

94. Zustovich F, et al. Metastatic oligodendrogliomas: a review of the literature and case report. *Acta Neurochir.* 2008;150:699–702 discussion 702–703.

95. Ceyssens S, et al. [^{11}C]methionine PET, histopathology, and survival in primary brain tumors and recurrence. *Am J Neuroradiol.* 2006;27:1432–1437.

96. Ullrich RT, et al. Methyl-L-11C-methionine PET as a diagnostic marker for malignant progression in patients with glioma. *J Nucl Med Off Publ Soc Nucl Med.* 2009;50:1962–1968.

Chapter 15

Neuroimaging of pediatric oligodendrogliomas

Michael A. Lach* and Ihsan Mamoun†

**Department of Diagnostic Radiology, Cleveland Clinic, Cleveland, OH, United States;* †*Department of Pediatric and Neuroimaging, Cleveland Clinic, Cleveland, OH, United States*

Epidemiology

Being typically a tumor of adulthood, oligodendrogliomas are rare in the pediatric population. The true incidence of which is difficult to determine and is documented as accounting for <1%–2% of all pediatric brain tumors and 5%–18% of pediatric gliomas.[1–5] Of all the oligodendrogliomas that occur, about 6% of them occur in children,[6] with a peak age of 6–12 years and with a slight male Caucasian predominance.[3,4] Few cases have been reported in infancy and in the rarest of circumstances, congenital brainstem/infratentorial,[7–9] and congenital supratentorial oligodendrogliomas have been described.[9,10]

Location

Pediatric oligodendrogliomas are diffusely infiltrating tumors most often found in the cerebral hemispheres, usually involving either cortical or subcortical/white matter.[11] While pediatric gliomas can technically arise anywhere in the central nervous system,[12,13] pediatric oligodendrogliomas have classically been described as most frequently occurring in the frontal region mimicking their adult counterparts[2,5,12,14–17]; however, were seen most frequently in the temporal lobe in a retrospective study performed by Tice et al., in 1993,[11] whose results were mirrored in a large cohort study of pediatric patients by Lau et al., in 2017,[4] where pediatric oligodendrogliomas were exceedingly more common in the temporal lobe. These tumors more infrequently occur in the parietal and occipital lobes, as well as within the posterior fossa/cerebello-pontine angle, optic nerve, basal ganglia, and brain stem, with spinal cord and intraventricular oligodendrogliomas being even more rare.[2,11,13,15,16,18,19]

Clinical presentation

The location and grade of the tumor will ultimately dictate the clinical presentation and can be grouped into generalized and localizing symptoms.[12] The most common presentation is seizure activity,[1,2,20] possibly due to its tendency to involve the cortical gray matter.[3] Additional symptoms can manifest as signs of increased intracranial pressure, such as headache, often seen more frequently in the higher grade tumors given more rapid growth,[13] including nausea, vomiting, lethargy, visual field defects, or limb weakness.[12,13] Back pain as well as increased head circumference are less frequent, though have also been described.[2]

Prognostic indicators

The WHO grading system recognizes two distinct grades of oligodendrogliomas: the more common, well-differentiated grade II and grade III anaplastic type. These distinctions, as with other pediatric and glial tumors in general, imply important prognostic factors with the higher grades demonstrating decreased overall survival.[2,13] Supratentorial location, aside from histologic grade, implies a favorable prognostic factor overall, given the improved total gross resection rate/greater extent of resection, with less favorable outcomes demonstrated for centrally located oligodendrogliomas, for example, within the thalamus, basal ganglia, or mesencephalon—which are more often associated with incomplete resection and are less suitable for radical surgery.[2,5,12,20,21]

Molecular analysis

In the adult population, molecular markers such as the 1p/19q co-deletion and isocitrate dehydrogenase (IDH) mutation are common; however, these are notoriously absent in the pediatric population. When found in the pediatric population, these patients tend to be the older children and adolescents.[16,22,23] In adult anaplastic oligodendrogliomas, the 1p/19q co-deletion has been correlated with increased chemosensitivity and overall improved survival;[13,15,24,25] however, this has been shown to confer a poorer prognosis in the pediatric population.[22] Additional reported markers in oligodendrogliomas include deletions of the cyclin-dependent kinase inhibitor 2A (CDKN2A) gene on chromosome 9p, mutations in PTEN, and amplification of the EGFR, which are indeterminate factors in the pediatric population.[13,16,26]

Neuroimaging

Neuroimaging plays a key role in the presurgical evaluation of pediatric oligodendrogliomas, being able to provide such essential information as location and extent, signal characteristics, enhancement pattern, as well as diffusion and susceptibility characteristics.[5]

In a case report demonstrating a rare congenital supratentorial oligodendroglioma, Wu-Shiun Hsieh et al., in 2002, described initial imaging evaluation with head ultrasonography in the setting of slightly tense anterior fontanel in a 2-day-old male infant.[10] Ultrasound appearance is interpreted as a discrete mass demonstrating mixed echogenicity within the left temporoparietal region in this particular case study.

While the cross-sectional imaging appearance can vary, pediatric oligodendrogliomas are typically described as slow growing, infiltrative yet well circumscribed and sharply marginated, round or oval masses that tend to expand the cerebral cortex with variable degrees of white matter involvement, and otherwise typically without significant mass effect.[6] These tumors are most commonly hypodense (approximately 60%) or isoattenuating (approximately 20%) on computed tomography (CT), and less than 40% contain calcifications,[6] a characteristic seen far less frequently than in adult tumors. The tumor may appear cystic as well as hemorrhagic.[1,4,16] The CT findings for a high-grade lesion include mass effect, surrounding vasogenic edema, and enhancement.

Findings on magnetic resonance imaging (MRI) are rather nonspecific,[5] though the modality allows for better soft tissue contrast and therefore superior delineation of tumor involvement compared to CT.[4] Oligodendrogliomas are typically hypointense to gray matter on T1-weighted images and hyperintense to gray matter on T2-weighted images.[6] Internally, the signal intensity on fluid-attenuation inversion recovery (FLAIR) images is most commonly hyperintense.[5] Partial enhancement following gadolinium administration is less common than in adult oligodendrogliomas, and when present is more commonly seen in aggressive grade III oligodendrogliomas, as well as in tumors that grow more as solid masses and less commonly in purely infiltrative tumors.[13] In a small pediatric cohort by Wagner et al., in 2015, the only tumor that showed contrast enhancement was the anaplastic oligodendroglioma.[5] However, in a retrospective study including both adult and pediatric patients performed by White et al., in 2005, there was no statistically significant difference in tumoral contrast enhancement between low- and high-grade oligodendrogliomas (Fig. 1A–C).[27] Though not always seen, varying degrees of adjacent T2/FLAIR hyperintensity typically represents vasogenic edema surrounding the tumor cells,[5] which in addition to the degree of mass effect, are findings that can suggest a more aggressive pathology.[3,27]

Advanced neuroimaging

Low-grade oligodendrogliomas are typically without restricted diffusion on diffusion weighted imaging (DWI), while high-grade tumors may show restricted diffusion.[5] Susceptibility-weighted imaging (SWI) can be utilized to detect intratumoral hemorrhage as well as calcifications[28]; again, a feature less commonly seen in pediatric oligodendrogliomas (Fig. 2A–E).[15] Generally, calcifications are typical of low-grade tumors, while hemorrhagic foci are more commonly seen in high-grade lesions.[5]

Perfusion-weighted imaging (PWI) is used to further characterize these tumors and can be performed with either CT or MRI. In children, MR perfusion is often favored to limit exposure to ionizing radiation.[6] The PWI most frequently involves injection of a gadolinium-based contrast agent and then measuring the local magnetic susceptibility differences (T2*) between the contrast within the vessels and surrounding tissue.[6,29] When viewed dynamically, the amount of signal dropout through the tissue can be calculated to provide the transit time. With this information, relative cerebral blood volume

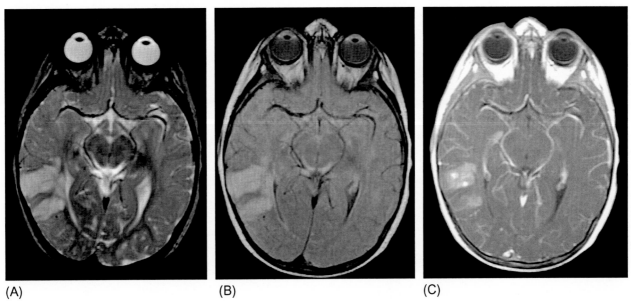

(A) (B) (C)

FIG. 1 A 11-month-old infant presented with seizures. (A and B) Axial T2 and T2/FLAIR images demonstrate a T2 hyperintense infiltrative lesion within the right temporal lobe without significant mass effect. (C) Axial contrast-enhanced T1 image shows intense infiltrative and nodular enhancement corresponding with the T2 abnormality; however, final pathology revealed low-grade oligodendroglioma.

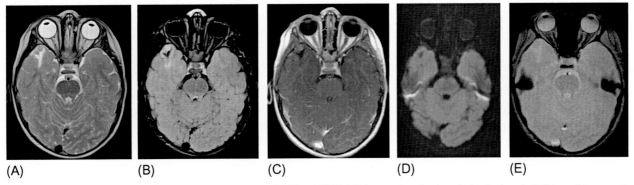

(A) (B) (C) (D) (E)

FIG. 2 A 3-year-old male presented with seizures. (A and B) Axial T2 and T2/FLAIR images show focal cortical and subcortical T2 hyperintense cystic changes within the right anterior temporal lobe. (C) Axial T1 postcontrast image demonstrates minimal cortical nodular enhancement at the right temporal pole. (D) High B-value axial DWI shows no restricted diffusion associated with the lesion. (E) Gradient echo (GRE) images are without susceptibility to suggest either calcification or hemorrhage. Final pathology revealed low-grade oligodendroglioma.

(rCBV) can be estimated by measuring the area under the tracer concentration-time curve.[29] In most gliomas, such as astroglial tumors, rCBV can be used to differentiate between high- and low-grade lesions, as rCBV is often elevated in the former and absent in the latter (thought to be due to increased angiogenesis seen in high-grade tumors). However, in oligodendrogliomas, the rCBV is often elevated in both low- and high-grade tumors.[15,30] This lack of correlation between rCBV and grade is consistent with the increased microvessel density frequently seen in both low- and high-grade oligodendrogliomas.[31] Therefore, it has been suggested that the rate of change of rCBV rather than the absolute value may be helpful in predicting high-grade morphology and malignant transformation while overall increased rCBV can be, but not always, helpful in differentiating low-grade oligodendrogliomas from low-grade astrocytomas (Fig. 3A–C).[6,30,31]

Spectroscopy is nonspecific, usually showing elevated choline and decreased NAA levels while the presence of lipid and lactate peaks are correlated with higher grade tumors.[5]

By measuring changes in blood oxygenation based on regional hemodynamic differences in tissue perfusion associated with various stimuli, functional MRI (fMRI) can be used to map specific territories activated during various forms of motor, sensory, or cognitive stimuli.[29] fMRI can be used for presurgical planning to identify eloquent brain/cortex in an effort to increase maximal resection while decreasing residual deficits.[6] Utilizing anisotropic water movement along white matter tracts, diffusion tensor imaging (DTI) can also be used in the planning stage to identify and localize vital neural tracts in order to anticipate the best surgical approach and degree of resection.[6,29]

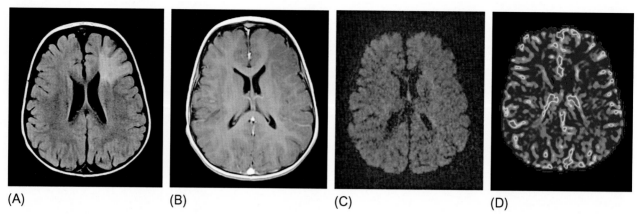

FIG. 3   A 17-month-old female presented with fever and seizures. (A) Axial T2/FLAIR sequence shows homogeneous hyperintensity involving the sub-cortical left frontal lobe without significant mass effect. (B) There is no associate enhancement on the axial T1 postcontrast image. (C) The lesion does not restrict diffusion on the axial DWI image. (D) There is no elevated CBV on the perfusion-weighted image to suggest angiogenesis. Final pathology revealed low-grade (WHO grade II) oligodendroglioma.

Role of positron emission tomography (PET)

18F-Fluorodeoxyglucose (FDG) uptake is increased in high-grade and anaplastic tumors and thus can be used for grading and prognostic purposes if the initial imaging is unclear. While this concept is not unique to pediatric oligodendrogliomas, it can be useful given the lack of specificity of contrast enhancement and rCBV as described. Co-registration with MRI is ideal to combine the physiologic data with the excellent anatomic detail in order to improve the tumor characterization.[32] In the study of pediatric gliomas in general, amino acid tracers, especially C-methionine, have also been utilized given their high affinity for tumor tissue with relatively low background signal in normal brain.[32]

Differential diagnosis

Given the rather nonspecific imaging and clinical findings, several other masses should be considered in the differential including ganglioglioma, dysembryoplastic neuroepithelial tumor, pleomorphic xanthoastrocytoma, pilocytic astrocytoma, and desmoplastic infantile ganglioglioma.

Ganglioglioma

Most commonly presenting with seizures, these slow growing, low-grade (WHO I) tumors are cortically based supratentorial lesions, most commonly located within the temporal lobes (particularly mesial regions), followed by the frontal lobes and are most commonly cystic or mixed solid/cystic in appearance.[1,15,33] On CT, gangliogliomas are most commonly hypodense or mixed density and contain calcifications up to 50% of the time.[33] On MRI, the solid component may be iso- or hypointense to gray matter on T1 and hyperintense on T2, while the degree of contrast enhancement can differ (enhancement seen in approximately 60%) with variable patterns.[13,33]

Dysembryoplastic neuroepithelial tumor (DNET)

Benign, WHO grade I glioneuronal tumors most commonly presenting with intractable complex partial seizures in children and young adults less than 20 years old.[15,34] These tumors are typically well defined and cortically based, most commonly occurring in the temporal lobes, though can occur in the frontal lobes as well.[33,34] On CT, DNETs are typically hypodense and cystic with minimal mass effect and calcifications are seen in 20%–40% of the time.[15,33,34] On MR, these lesions are usually iso- to hypointense on T1 and hyperintense "bubbly" internal structure on T2 images with minimal vasogenic edema. Approximately 30% are associated with focal cortical dysplasia.[15] While typically nonenhancing, occasional nodular, ring-like or faint patchy enhancement can be seen.[15,33] Bony remodeling of the adjacent skull as well as internal hemorrhagic changes can also be seen.[33]

Pleomorphic xanthoastrocytoma (PXA)

Most commonly presenting with seizures, this rare tumor accounts for <1% of all astrocytic neoplasms with a median age documented from 14 to 20 years old at the time of diagnosis.[1,15] These tumors most commonly occur in the temporal lobes and are typically large and superficial lesions adjacent to the cortex.[15] Most commonly described as a well-defined mixed solid/cystic mass which is predominantly hypodense on CT, with a possible enhancing solid mural nodule.[33] On MRI, the cystic portion is isointense to CSF while the solid component is hypo- to isointense to gray matter on the T1 and hyperintense on the T2-weighted images, typically without restriction on DWI.[6,33] Given the peripheral location, scalloping of the inner table can be seen as well as involvement of the adjacent meninges, and can also be associated with focal cortical dysplasia. While most of these tumors are WHO grade II, there is an anaplastic WHO grade III PXA which confers shorter survival.[15,33]

Pilocytic astrocytoma

The most common primary CNS tumor in children, this lesion is classified as WHO grade I and carries an excellent prognosis, typically seen in the first two decades without a clear gender predilection, though there is a higher incidence in patients with neurofibromatosis type 1.[1,15] These tumors can arise anywhere in the CNS, though most commonly occur in the cerebellum, followed by the optic pathway and hypothalamus, and the presenting symptoms are dictated by tumor location.[12,15] These masses are well circumscribed and typically possess the characteristic appearance of a cystic lesion with a robustly enhancing solid mural nodule with minimal if any associated vasogenic edema, and calcifications are possible, though rare (Fig. 4A–C).[1,12,15,35,36]

Desmoplastic infantile ganglioglioma

Desmoplastic infantile astrocytomas are rare, WHO grade I lesions typically occurring in children less than 2 years of age, presenting with seizures, increasing head circumference, and/or bulging fontanelles, or other focal deficits.[1,15,34] These are typically large cystic lesions, often involving the entire cerebral hemisphere with an enhancing mural nodule.[34] There is frequent leptomeningeal involvement and the solid components can show restricted diffusion, despite the WHO grade I classification.[1,15]

Management

As with other pediatric glial tumors, treatment is centered on primary aggressive resection and good progression-free survival and overall survival is seen following maximal surgical resection,[2,37] especially among children and adolescents.[1,4] It follows that hemispheric location and therefore amenability to radical surgery usually indicates a good/better prognosis.[21]

In the adult population, it has been proposed that sequential treatment with radiation therapy and chemotherapy, taking into account the expected side effects and efficacy, is the optimal approach given relative sensitivity of oligodendroglial

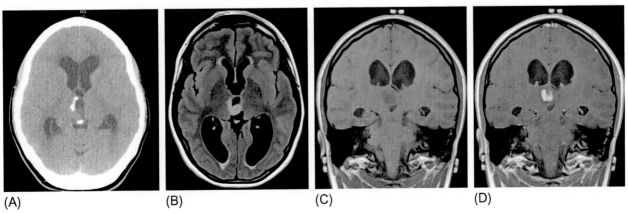

(A)   (B)   (C)   (D)

FIG. 4 (A) Axial noncontrast CT image shows an isodense mass within and expanding the third ventricle containing multiple calcifications. (B, C, and D) The solid mass is homogeneously hyperintense on the axial T2/FLAIR image and demonstrates robust, somewhat nodular enhancement following contrast administration on the coronal T1 images. While the imaging features are more typical for oligodendroglioma given the calcifications pathology revealed pilocytic astrocytoma.

tumors to chemoradiation.[38,39] The role of adjuvant chemotherapy and radiation in pediatric patients appears more contingent on histologic grade and tumor location. In peripherally located low-grade pediatric oligodendrogliomas, Peters et al., 2004, demonstrated an overall excellent prognosis, even with minor resection and without postoperative adjunctive treatment.[20] Additionally, Bowers et al., in 2002, suggested that based on the lack of evidence for subsequent tumor progression in their study of low-grade pediatric oligodendrogliomas following only gross total resection, additional adjuvant chemoradiation therapy is unnecessary.[37] These results were echoed by Mishra et al., in 2006, whose group demonstrated no survival benefit of administering radiation as part of the initial management of low-grade gliomas and should be avoided based on the well-described late effects of radiation therapy.[13,40]

In children with high-grade tumors and anaplastic oligodendrogliomas, a population in which there is more likely to be disease progression despite resection, or in patients with central tumors, there appears to be a clearer role for adjuvant therapy. Peters et al., in 2004, described central tumor location as an independent negative prognostic factor (even in the setting of low-grade tumors) and endorses immediate adjuvant postoperative therapy.[20] Allison et al., in 1997, suggested pre- and postoperative chemotherapy in a subset of pediatric patients with advanced tumors, in an effort to avoid the consequences of radiation therapy.[38] This sentiment is echoed by Packer et al., in 1997, who describes the efficacy of carboplatin and vincristine as chemotherapy options for progressive gliomas in an effort to avoid radiotherapy or further, possibly damaging aggressive surgical resection.[41] Procarbazine, CCNU, and vincristine (PCV) chemotherapy regimen has been shown to be efficacious in high-grade pediatric oligodendrogliomas.[42] Initial treatment with temozolomide has been shown to be safe with improved survival outcome in pediatric high-grade gliomas,[43] while complete response has been achieved in a pediatric patient with anaplastic oligodendroglioma following surgical resection and temozolomide alone.[44]

As it stands, while a trend toward longer survival with greater extent of resection appears evident,[2] the exact role of radiation and chemotherapy in the treatment of pediatric oligodendrogliomas has yet to be fully elucidated, as there is limited data given relative lack of prospective studies in such low-incidence tumors.

Follow-up imaging

MRI is the gold standard for follow-up imaging in the postprocedure context given high structural detail. Tumor progression is typically defined as a change in any number of imaging characteristics including increased tumor size, surrounding T2/FLAIR changes, or new/increased contrast enhancement[45]; though can be nonspecific in identifying viable or recurrent tumor from treatment-induced changes. While contrast enhancement can be seen in both residual or recurrent tumor as well as posttreatment changes, PWI has been found to be helpful for differentiating recurrent tumor from posttreatment changes including radiation necrosis, as new or increased perfusion is typically absent in posttreatment change.[6]

Unfortunately, FDG-PET is generally not sensitive for identifying the recurrence of low-grade tumors, and in some cases anaplastic tumors, given the high physiologic metabolism of the normal background brain parenchyma and the wide range of metabolic activity that can be seen in the treatment bed.[46] That being said, there is a potential role for FDG-PET to identify tumor recurrence as the posttreatment bed typically demonstrates FDG uptake equal to or lower than the neighboring normal brain.[32] Thus, when interpreting the posttreatment scan, it is important to analyze FDG activity not by the absolute value, but whether the uptake is higher than expected by referencing the background activity of the adjacent normal brain tissue, and if it is, recurrence may be suspected.[32] Co-registration with MRI can increase the sensitivity and specificity for evaluating recurrent tumor and the benefit of delayed FDG-PET to differentiate between tumor and normal gray matter that has been recognized.[32,46]

References

1. Karajannis M, Gardner S, Allen J. Primary nervous system tumors in infants and children. In: Daroff R, Jankovic J, Mazziotta J, Pomeroy S, eds. *Bradley's Neurology in Clinical Practice*. 7th ed. Elsevier Limited; 2016:1065–1083.
2. Wu C-T, Tsay P-K, Jaing T-H, Chen S-H, Tseng C-K, Jung S-M. Oligodendrogliomas in children: clinical experiences with 20 patients. *J Pediatr Hematol Oncol*. 2016;38(7):555–558.
3. Koeller KK, Rushing EJ. Oligodendroglioma and its variants: radiologic-pathologic correlation. *Radiographics*. 2005;25(6):1669–1688.
4. Lau CS, Mahendraraj K, Chamberlain RS. Oligodendrogliomas in pediatric and adult patients: an outcome-based study from the surveillance, epidemiology, and end result database. *Cancer Manag Res*. 2017;9:159.
5. Wagner MW, Poretti A, Huisman TA, Bosemani T. Conventional and advanced (DTI/SWI) neuroimaging findings in pediatric oligodendroglioma. *Childs Nerv Syst*. 2015;31(6):885–891.
6. Borja MJ, Plaza MJ, Altman N, Saigal G. Conventional and advanced MRI features of pediatric intracranial tumors: supratentorial tumors. *Am J Roentgenol*. 2013;200(5):W483–W503.

7. Kostadinov S, de la Monte S. A case of congenital brainstem oligodendroglioma: pathology findings and review of the literature. *Case Rep Neurol Med.* 2017;2017.

8. Narita T, Kurotaki H, Hashimoto T, Ogawa Y. Congenital oligodendroglioma: a case report of a 34th-gestational week fetus with immunohistochemical study and review of the literature. *Hum Pathol.* 1997;28(10):1213–1217.

9. Richard H, Stogner-Underwood K, Fuller C. Congenital oligodendroglioma: clinicopathologic and molecular assessment with review of the literature. *Case Rep Pathol.* 2015;2015.

10. Hsieh W-S, Lien R-I, Lui T-N, Wang C-R, Jung S-M. Congenital oligodendroglioma with initial manifestation of jaundice. *Pediatr Neurol.* 2002;27 (3):230–233.

11. Tice H, Barnes PD, Goumnerova L, Scott RM, Tarbell NJ. Pediatric and adolescent oligodendrogliomas. *Am J Neuroradiol.* 1993;14(6):1293–1300.

12. Sievert AJ, Fisher MJ. Pediatric low-grade gliomas. *J Child Neurol.* 2009;24(11):1397–1408.

13. Kieran MW, Chi S, Manley P. Tumors of the brain and spinal cord. In: Orkin S, Nathan D, Ginsburg D, Look A, Fisher D, Lux S, eds. *Nathan and Oski's Hematology and Oncology of Infancy and Childhood.* 8th ed. Philadelphia: Elsevier Limited; 2015:1779–1885.

14. Wang KC, Chi JG, Cho BK. Oligodendroglioma in childhood. *J Korean Med Sci.* 1993;8(2):110–116.

15. Zamora C, Huisman TA, Izbudak I. Supratentorial Tumors in Pediatric Patients. *Neuroimag Clin.* 2017;27(1):39–67.

16. Rodriguez FJ, Tihan T, Lin D, et al. Clinicopathologic features of pediatric oligodendrogliomas: a series of 50 patients. *Am J Surg Pathol.* 2014;38 (8):1058.

17. Razack N, Baumgartner J, Bruner J. Pediatric oligodendrogliomas. *Pediatr Neurosurg.* 1998;28(3):121–129.

18. Bruzek AK, Zureick AH, McKeever PE, et al. Molecular characterization reveals NF1 deletions and FGFR1-activating mutations in a pediatric spinal oligodendroglioma. *Pediatr Blood Cancer.* 2017;64(6).

19. Matsumoto H, Yoshida Y. Primary intraventricular oligodendroglioma: a case report of the usefulness of Olig2 immunohistochemistry for diagnosis. *Neuropathology.* 2015;35(6):553–560.

20. Peters O, Gnekow A, Rating D, Wolff J. Impact of location on outcome in children with low-grade oligodendroglioma. *Pediatr Blood Cancer.* 2004;43 (3):250–256.

21. Lundar T, Due-Tønnessen BJ, Egge A, Scheie D, Stensvold E, Brandal P. Neurosurgical treatment of oligodendroglial tumors in children and adolescents: a single-institution series of 35 consecutive patients. *J Neurosurg Pediatr.* 2013;12(3):241–246.

22. Raghavan R, Balani J, Perry A, et al. Pediatric oligodendrogliomas: a study of molecular alterations on 1p and 19q using fluorescence in situ hybridization. *J Neuropathol Exp Neurol.* 2003;62(5):530–537.

23. Wesseling P, van den Bent M, Perry A. Oligodendroglioma: pathology, molecular mechanisms and markers. *Acta Neuropathol.* 2015;129(6):809–827.

24. Kreiger PA, Okada Y, Simon S, Rorke LB, Louis DN, Golden JA. Losses of chromosomes 1p and 19q are rare in pediatric oligodendrogliomas. *Acta Neuropathol.* 2005;109(4):387–392.

25. Capper D, Reuss D, Schittenhelm J, et al. Mutation-specific IDH1 antibody differentiates oligodendrogliomas and oligoastrocytomas from other brain tumors with oligodendroglioma-like morphology. *Acta Neuropathol.* 2011;121(2):241–252.

26. Nutt CL. Molecular genetics of oligodendrogliomas: a model for improved clinical management in the field of neurooncology. *Neurosurg Focus.* 2005;19(5):1–9.

27. White ML, Zhang Y, Kirby P, Ryken TC. Can tumor contrast enhancement be used as a criterion for differentiating tumor grades of oligodendrogliomas? *Am J Neuroradiol.* 2005;26(4):784–790.

28. Berberat J, Grobholz R, Boxheimer L, Rogers S, Remonda L, Roelcke U. Differentiation between calcification and hemorrhage in brain tumors using susceptibility-weighted imaging: a pilot study. *Am J Roentgenol.* 2014;202(4):847–850.

29. Wycliffe N, Holshouser B, Bartnik-Olson B. Pediatric neuroimaging. In: Swaiman K, Ashwal S, Ferriero D, et al., eds. *Swaiman's Pediatric Neurology.* 6th ed. Elsevier Limited; 2017:173–211.

30. Cha S, Tihan T, Crawford F, et al. Differentiation of low-grade oligodendrogliomas from low-grade astrocytomas by using quantitative blood-volume measurements derived from dynamic susceptibility contrast-enhanced MR imaging. *Am J Neuroradiol.* 2005;26(2):266–273.

31. Jenkinson MD, Smith TS, Joyce KA, et al. Cerebral blood volume, genotype and chemosensitivity in oligodendroglial tumours. *Neuroradiology.* 2006;48(10):703–713.

32. Chen W. Clinical application of PET in pediatric brain tumors. *PET Clinics.* 2008;3(4):517–529.

33. Miller E, Widjala E. Neuroimaging in pediatric epilepsy. In: Coley B, ed. *Caffey's Pediatric Diagnostic Imaging.* 12th ed. Philadelphia: Elsevier Limited; 2013:408–416.

34. Argyropoulou MIRA, Gunny RS, et al. Paediatric neuroradiology. In: Adams ADA, Gilliard JH, Schaefer-Prokop CM, eds. *Grainger and Allison's Diagnostic Radiology.* 6th ed. Elsevier Limited; 2015:1978–2041.

35. Koeller KK, Rushing EJ. From the archives of the AFIP: pilocytic astrocytoma: radiologic-pathologic correlation. *Radiographics.* 2004; 24(6):1693–1708.

36. Lee Y-Y, Van Tassel P, Bruner JM, Moser RP, Share JC. Juvenile pilocytic astrocytomas: CT and MR characteristics. *Am J Roentgenol.* 1989;152 (6):1263–1270.

37. Bowers DC, Mulne AF, Weprin B, Bruce DA, Shapiro K, Margraf LR. Prognostic factors in children and adolescents with low-grade oligodendrogliomas. *Pediatr Neurosurg.* 2002;37(2):57–63.

38. Allison RR, Schulsinger A, Vongtama V, Barry T, Shin KH. Radiation and chemotherapy improve outcome in oligodendroglioma. *Int J Radiat Oncol Biol Phys.* 1997;37(2):399–403.

39. Bromberg JE, Van Den Bent MJ. Oligodendrogliomas: molecular biology and treatment. *Oncologist.* 2009;14(2):155–163.

40. Mishra KK, Puri DR, Missett BT, et al. The role of up-front radiation therapy for incompletely resected pediatric WHO grade II low-grade gliomas. *Neuro-Oncology.* 2006;8(2):166–174.

41. Packer RJ, Ater J, Allen J, et al. Carboplatin and vincristine chemotherapy for children with newly diagnosed progressive low-grade gliomas. *J Neurosurg.* 1997;86(5):747–754.

42. Rizk T, Mottolese C, Bouffet E, et al. Cerebral oligodendrogliomas in children: an analysis of 15 cases. *Childs Nerv Syst.* 1996;12(9):527–529.

43. Jung T-Y, Kim C-Y, Kim D-S, et al. Prognosis of pediatric high-grade gliomas with temozolomide treatment: a retrospective, multicenter study. *Childs Nerv Syst.* 2012;28(7):1033–1039.

44. Sorge C, Li R, Singh S, et al. Complete durable response of a pediatric anaplastic oligodendroglioma to temozolomide alone: case report and review of literature. *Pediatr Blood Cancer.* 2017;64(12).

45. Olson JD, Riedel E, DeAngelis LM. Long-term outcome of low-grade oligodendroglioma and mixed glioma. *Neurology.* 2000;54(7):1442–1448.

46. Chen W. Clinical applications of PET in brain tumors. *J Nucl Med.* 2007;48(9):1468–1481.

Section D

Surgical therapy

Chapter 16

Oligodendrogliomas: Basic techniques in surgery

William C. Newman* and Melvin Field†

*Department of Neurological Surgery, University of Pittsburgh Medical Center, Pittsburgh, PA, United States; †Orlando Neurosurgery, University of Central Florida College of Medicine, Orlando, FL, United States

Introduction

Oligodendrogliomas are a group of glial tumors defined by the 1p/19q chromosomal co-deletion and whose management involves maximal safe resection when possible or, at a minimum, biopsy for definitive diagnosis with subsequent radiation and/or chemotherapy. Basic preoperative work-up and evaluation includes a detailed neurologic examination, diagnostic imaging and high-resolution imaging for both operative planning and neuronavigation. In this chapter, we will discuss the basics of preoperative work-up and operative planning. A more detailed description of pre- and postoperative surgical adjuncts will be provided in the chapter on Advanced Techniques in Surgery.

Preoperative work-up
Preoperative clinical testing

The most common presenting symptom for oligodendrogliomas is a seizure,[1] and as a result, many of these patients will present to the emergency department with a first-time seizure. Clinical evaluation should consist of a detailed neurologic examination to determine any neurologic deficits. If the patient presented with focal seizures, a detailed history should inquire as to difficulties with speech (i.e., word finding difficulty, difficulties with comprehension, etc.) or motor weakness (clumsiness, loss of coordination, etc.). More detailed neuro-psychological evaluation for the purposes of elucidating subtle neurologic deficits can be deferred for a later time and performed by a trained neuropsychologist. This testing and its importance will be discussed in the *Advanced Techniques* chapter.

Diagnostic imaging

Standard radiologic work-up of these lesions often begins with a computed tomography (CT) scan and should include magnetic resonance imaging (MRI). CT scans demonstrate mixed-density (iso- to hypodense) lesions with areas of significant hyperdensity correlating with intratumoral calcification. Calcifications can occur anywhere within the tumor. Occasionally, areas of hyperdensity may represent hemorrhage, however, this is less common than calcification (Fig. 1). Cystic degeneration within the lesion may also be observed. Given the typically slow-growing nature of these lesions, changes in the overlying calvarium may also be seen as a result of pressure remodeling. On postcontrast CT scans, oligodendrogliomas will show variable enhancement patterns that range from nonenhancing to avidly enhancing.

On T1-weighted imaging, MRI demonstrates a hemispheric mass that is hypo- to isointense to white matter. Postcontrast T1-weighted sequences demonstrate heterogeneous contrast enhancement, with no identified correlation between extent of enhancement and histologic grading. T2-weighted sequences often demonstrate hyperintensity with areas of blooming artifact secondary to intratumoral calcification. On diffusion-weighted imaging, oligodendrogliomas typically do not restrict (Fig. 1).

MR spectrometry can be a useful adjunct for radiographically distinguishing grade II from grade III lesions. Grade II lesions will demonstrate decreased *N*-acetylaspartate (NAA, marker of viable neuronal density) levels with increased choline (marker of membrane synthesis and cellular density) and no lactate peak (measure of anaerobic metabolism).

Oligodendroglioma. https://doi.org/10.1016/B978-0-12-813158-9.00016-5

183

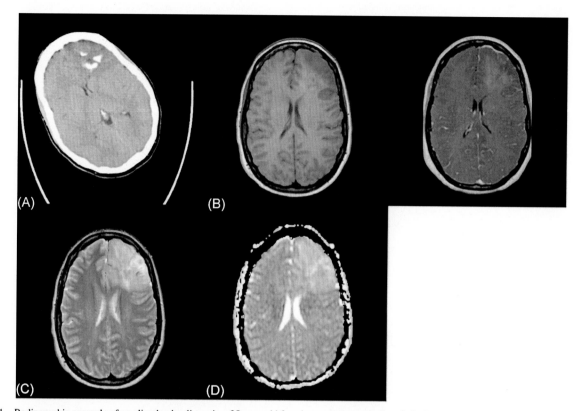

FIG. 1 Radiographic example of an oligodendroglioma in a 35-year-old female presenting to the hospital with new onset generalized seizure. (A) Non-contrasted CT axial image showing a hypodense left frontal mass with associated calcifications. (B) T1-weighted axial MR image showing a hypodense mass relative to surrounding normal brain parenchyma. (C) T1-weighted Gadolinium enhanced axial MR image showing minimal to no enhancement of the hypodense mass. (D) T2-weighted axial MR image showing hyperintensity of the left frontal mass relative to the normal surrounding brain parenchyma as well as areas of hypointensity within the mass due to calcifications often referred to as blooming artifact. (E) ADC (diffusion weighted) axial MR image showing no restricted diffusion within the mass.

In contrast, grade III lesions, will demonstrate a lactate peak.[2] In addition, a choline/creatine ratio threshold of 2.33 was found to distinguish high- from low-grade oligodendrogliomas, with 100% sensitivity and 83% specificity.[3]

Positron emission tomography (PET) using ^{11}C-methionine is a very infrequently used diagnostic test for differentiating oligodendrogliomas from anaplastic oligodendrogliomas. In this test, oligodendrogliomas will show an uptake pattern similar to that of normal white matter whereas anaplastic oligodendrogliomas will show an uptake pattern similar to that of normal gray matter.[4]

MR perfusion studies looking at relative cerebral blood volume (rCBV) have been unreliable for predicting initial tumor grade, but may be useful for tracking tumor progression.[2] After initial biopsy or resection of a grade II lesion, serial MR perfusion studies that show increased rCBV overtime have correlated with progression to a grade III lesion. These changes may occur as early as 12 months after diagnosis, before other signs of malignant transformation become apparent.[5]

Image-guided stereotactic craniotomy and neuronavigation

Image-guided stereotactic craniotomy is a term used routinely when discussing almost any form of surgery for a primary brain tumor in the developed world. Such surgery involves making an opening in the bone and biopsying, debulking, or removing an intracranial mass while utilizing imaging data that helps the surgeon to localize or "navigate" the region often with millimeter precision. Computerized and robotic systems that aid in this technique are known as neuronavigation systems. These systems use fiducial-based coordinates to fuse imaging data—CT, MR, angio, fMR, and diffusion tensor imaging (DTI) to real-time fiducials that are fixed to the patient via a head holder, mask, or some other cranial fixation device that moves as the head moves. Cameras and sensors in the operating room that are integrated to the navigated systems then provide real-time feedback regarding the patient positioning related to active and passive probes that tell the surgeon where in the head he is located using the fused imaging. Many people refer to this as "GPS for the brain" in that these systems show the surgeon a roadmap of where they are using preoperative and sometimes intraoperative imaging as a guide. For oligodendroglioma, such technology is integral to ensuring a maximally safe resection. Intraoperatively, such tumors may look indistinguishable from edematous surrounding brain parenchyma to the naked eye or under a microscope, and yet look markedly different on MRI. This makes it very difficult for the surgeon to know where the tumor starts or ends. In such cases, the surgeon can use the navigation imaging tools to guide the resection of the mass while avoiding critical functional structures also seen on the navigation imaging (Fig. 2).

Prior to an image-guided stereotactic craniotomy or biopsy using neuronavigation, the patient's preoperative imaging is uploaded to the navigation system's workstation in the operating room. In addition, many systems also allow the surgeon to create preoperative plans and approaches for a tumor's resection or biopsy that are then fused to these data sets and imported to the workstation on the day of surgery. After the surface calibration and image quality is reviewed to ensure the data set will provide accurate representations of the region of interest, multiple image modalities can then be fused and overlayed onto a base sequence. This allows the surgeon to utilize a patient's standard MRI, CT, functional MRI (fMRI), magnetoencephalogram (MEG), DTI, and other imaging data modalities seamlessly to guide a tumor removal during surgery (Fig. 3). Advantages and limitations of these imaging modalities are outlined in Table 1. After the patient is positioned on the operating room table, cranial fixation is routinely applied, which in our practice is achieved by using the three-pin Mayfield head holder (Integra, New Jersey). This allows for stability of the operative field and creates a defined position for registering the patient's surface anatomy to the preoperative imaging. Once fixated, surface landmarks are selected on the software and then a registration probe (a fine-tipped device of known length linked to the navigation system) is used to make contact with surface landmarks. Alternatively, some systems allow for auto-registration with intraoperative imaging

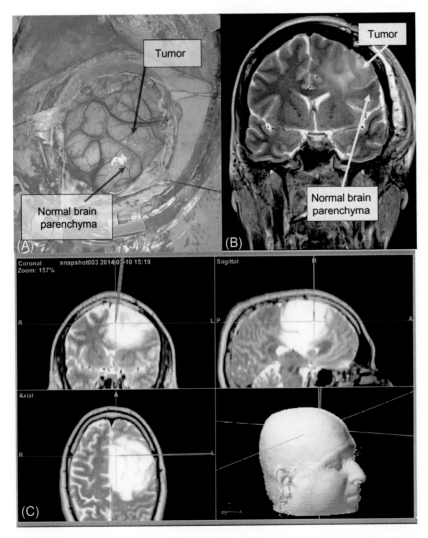

FIG. 2 An example of a left posterior frontal oligodendroglioma. (A) Intraoperative image highlighting gross visual similarity between tumor and normal adjacent cortex. (B) Preoperative coronal T2W image showing clearly abnormal hyperintense oligodendroglioma superiorly relative to normal isointense brain parenchyma inferiorly separated by a sulcus. (C) Intraoperative neuronavigation snapshot showing location of a navigated probe *(blue)* in the brain during surgery to guide the surgeon's resection based on the T2W-fused images obtained earlier that day. The navigation probe allows the surgeon to know whether the tissue adjacent to the probe is tumor or normal parenchyma based on the radiographic appearance of the tumor generally with millimeter accuracy.

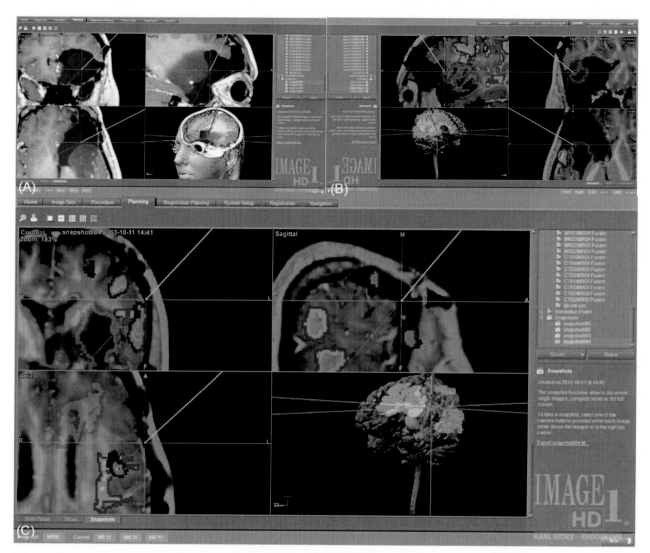

FIG. 3 Utilization of multiple modalities simultaneously with neuronavigation. In this patient with recurrent oligodendroglioma, tumor recurrence occurred posterior to the previous resection. (A) T1W-Gad-enhanced images show a hypodense mass involving the patient's dominant frontal and temporal lobes. As a result of a previous resection in that region, brain shift due to tumor, and neural pathway reorganization, the normal landmarks for identifying language and motor regions may not be reliable. To minimize this risk, preoperative magnetoencephalography, fMRI and DTI were performed to localize language and motor functions relative to the mass and previous resection cavity. In this example, T1W with GAD images are imported and used with the neuronavigation system to identify tumor relative to gliotic scar and normal brain parenchyma. (B) and (C) These intraoperative snapshots, demonstrate the fusion of MEG *(red)* for language generation, fMR *(blue)* for motor generation, DTI *(green)* for language tracts, and tumor *(purple)*. The fusion of these modalities into the neuronavigation system then allows the surgeon to approach the tumor, while avoiding these critical functional structures.

or registration via mask devices that can be done with or without head fixation. After a sufficient number of landmarks have been selected and registered to the system, the preoperative image should be accurately associated with the patient's anatomy such that placing the registration probe tip on anatomical landmarks, such as the lateral canthus or inside the external auditory canal, should result in the same location being shown on the imaging. If the images are not acceptably co-localized after the initial registration, additional methods such as surface refinement may be used to obtain better co-localization.

Once the neuronavigation has been registered satisfactorily, the imaging can be used to plan the incision. Planning the incision should take into account basic factors such as the size of the tumor and its medial, lateral, anterior, and posterior extent in order to have an incision or scalp flap large enough to allow for adequate exposure. The incision should also take into account prior incisions or scars, the potential need for conversion to a larger incision and craniotomy in the event of malignant cerebral edema or other severe complication, and even cosmetic factors such as the patient's hairline.

Once the incision has been planned, the patient's hair may be clipped, the surgical site prepped with sterile cleaning solution, and the operation begun.

TABLE 1 Advantages and limitations of imaging adjuncts

Imaging modality	Advantages	Limitations
Functional MRI	• Noninvasive • Localizes functional cortex • Can be used with neuronavigation	• Does not visualize subcortical paths • Variable correlation with intraoperative mapping
Magnetoencephalogram	• Noninvasive • Localizes functional cortex • Can be used with neuronavigation	• Does not visualize subcortical paths • Variable correlation with intraoperative mapping
High definition fiber tracking	• High resolution • Visualizes cortical and subcortical connections • Volumetric analysis possible • Can be used with neuronavigation	• May not fully detect fiber tracts, especially when invaded by tumor
Stereotactic neuronavigation	• Preoperative localization of tumor • Assistance with minimizing craniotomy size • Can be performed on all patients	• Intraoperative accuracy decreased due to brain shift
Intraoperative ultrasound	• Comparative pre- and postresection imaging • Minimal added time to utilize • Low cost	• Tissue manipulation can cause changes in echogenicity, making postresection interpretation difficult
Intraoperative MRI	• Increased extent of resection • Allows re-registration of neuronavigation to adjust for brain shift • Facilitates LITT	• Increased operating room time • High upfront cost • Requires a custom OR

Anesthesia and systemic considerations for brain surgery

When considering patients for surgical intervention, the goals of surgery must incorporate the patient's preoperative status, including their co-morbidities, performance status, level of preoperative neurologic dysfunction, and the lesion's involvement of eloquent structures. With respect to patients with significant medical co-morbidities, they may be unable to tolerate a prolonged surgical intervention from the standpoint of severe cardiovascular disease and risk of myocardial infarction. Alternatively, significant pulmonary disease may confer difficulty with extubation and weaning from the ventilator postoperatively. Both of these and other co-morbid considerations are reasons for the preoperative involvement of the anesthesia team to identify these risks and take steps preoperatively to minimize their impact on the surgery and the patient's outcome. For oligodendroglioma surgery, a patient's coagulation status needs to be essentially normal with an INR < 1.3 and a platelet count of > 100,000. We routinely ask patients to be off oral anticoagulation at least 5 days prior to surgery and check their coagulation status on the day of surgery. If a patient is at high risk for a thromboembolic event off anticoagulation, they are brought into the hospital once the oral anticoagulant is discontinued and they are bridged with IV heparin until the day of surgery. We discontinue the heparin 6 h prior to surgery. For patients on antiplatelet therapy, we ask that they discontinue their antiplatelet medications, including fish oils and herbal medications, 7–10 days prior to surgery. We then check their platelet function the day prior to, or day of surgery to confirm that no persistent platelet inhibition exists prior to bringing them to the operating room.

Prior to placing any patient under anesthesia, good communication with the anesthesiologist and intraoperative neurophysiologist is vital to ensure appropriate inhalational and intravenous agents are utilized throughout the case. For oligodendroglioma surgery, agents used should not increase intracranial pressure (ICP), lower seizure threshold, or interfere with the modalities necessary to appropriately monitor the brain from a neurophysiological perspective [EEG, somatosensory evoked potential (SSEP), motor evoked potential (MEP), cortical, and subcortical mapping]. Once the patient has been safely positioned for tumor resection, all pressure points are inspected and appropriately padded to prevent pressure sores or compressive neuropathies. Neurophysiological monitoring leads are then placed, and baseline studies are performed. Although such monitoring may not guide the resection of a tumor if located in a nonsensory or motor region, it can provide the surgeon other information that may be beneficial during a cranial procedure. For example, the use of intraoperative SSEP neuro-monitoring, including the ulnar, median, and tibial nerves, can alert the surgeon of a potential compressive neuropathy prior to the development of a permanent unintentional injury. Such intraoperative information allows for limb repositioning during the procedure to avoid a postoperative deficit. Intraoperative electroencephalography (EEG) can also provide information to the surgeon and anesthesiologist regarding depth of sedation. Such information can prevent a patient from becoming "too light" and thus avoiding patient movement, increases in ICP, or spikes in blood pressure during a procedure. For procedures over an hour, a urinary Foley catheter is routinely placed, arterial access is obtained for blood pressure monitoring, and lower extremity compression hose are placed to minimize the risk of deep venous thromboses. For cases where the venous sinuses will be exposed and the risk for potential air emboli exists, central venous access is obtained and precordial dopplers are used.

Surgical resection

The goal of surgery for oligodendrogliomas is a maximal safe resection, or a maximal tumor resection without incurring new neurologic deficits. The need for minimizing new postoperative deficits is based on the outcomes for glioblastoma resections where new language or motor deficits confer a high degree of morbidity and 1-year mortality.[6] While this has not been shown in oligodendrogliomas, the basic principle remains the same. In patients in whom gross total resection of the lesion is not possible, consideration of tumor debulking for symptomatic relief is an option. Alternatively, in patients who present with disease that is not resectable, due to the involvement of deep or eloquent structures, biopsy for diagnostic confirmation followed by up-front chemotherapy and radiation also is a reasonable treatment option.

Standard surgical resection through the use of a craniotomy relies on an adequate skin incision to create a bone flap that allows for adequate exposure of the cortex overlying the tumor. In both the creation of the skin incision and the craniotomy, the use of neuronavigation helps create the minimum exposure necessary to have adequate access to the area for surgical resection. At our institution, the craniotomy is created using a high-speed drill with either a perforator bit or a small matchstick-type drill bit to create burr holes and then connecting those holes together using a B1 drill bit with the footplate. Prior to the connecting the burr holes with the B1 drill bit, particular attention is paid to the dura to make sure that it is well separated from the inner table of the calvarium. Failure to do this can result in dural tears, cortical lacerations, venous injuries, and other complications.

Once the craniotomy has been completed and the bone flap removed, hemostasis is obtained and the dura may be palpated to determine if the brain is full and additional maneuvers are required to reduce ICP, so that the brain does not herniate through the dural opening that is about to be made. These maneuvers include elevation of the head of the bed to increase venous return to the heart, the administration of mannitol, an osmotic diuretic, and the use of transient hyperventilation to cause cerebrovascular vasoconstriction. Once the ICP has been lowered satisfactorily, the dura can be opened safely and surgery can proceed. It is worth noting that in brains that are particularly full, as the surgical resection proceeds and tumor is removed, this will also contribute to a lowering of the ICP. In cases where the brain is "full" in spite of these measures, initial tumor debulking should be expeditious to minimize the deleterious effects of brain herniation out of the dural and bony defect. Prolonged brain herniation out a bony defect can result in ischemic damage to normal tissues at the bony margin by direct compression adjacent to the herniation and venous engorgement and parenchymal bleeding due to vessels being compressed on the cortical surface.

During surgical resection of a tumor, normal tissue is protected often with cottonoid patties and retraction of normal tissues is minimized to avoid damage to such tissues as well as to minimize the effect of shift on neuronavigation accuracy. Visualization of tumor removal is done while wearing surgical magnifying loupes, using endoscopic magnification with a high-definition endoscope and monitor, or under an operating three-dimensional (3D) microscope. Ultrasonic aspirators and CO_2 lasers can be used in conjunction with microscopic suction to debulk and resect tumor tissues efficiently. Resection is guided by direct tissue visualization and observed differences between the tumor tissue and adjacent brain parenchyma, differences in tissue feel and consistency, neuronavigation, and neuromonitoring. Hemostasis is maintained utilizing bipolar electrocautery, hemostatic agents such as thrombin-soaked gel foam and powder, and by tamponading bleeding tissues with soaked peroxide cotton balls or cottonoid patties.

Of note, once the craniotomy has been completed, surgical adjuncts such as 5-aminolevulenic acid, intraoperative ultrasound, intraoperative MRI, and others may be used to help with the localization of the tumor and the assessment of the extent of resection. These surgical adjuncts will be discussed more fully in the chapter on Advanced Techniques in Surgery.

After completion of surgical resection of the tumor, the dura should be re-approximated and the bone flap should be plated using low-profile titanium plates and screws to hold it in place. Copious antibiotic irrigation is used at this point and the incision is then closed. The patient is removed from the cranial fixation device and extubated when safe to do so. Within

48 h of completing surgery, a postoperative MRI (or CT scan if the patient is unable to obtain MRIs) is obtained in order to establish a new baseline for comparison with long-term follow-up imaging and to detect tumor progression or recurrence.[7] We generally obtain this scan the morning following surgery and prior to the patient being sent to the neurosurgical floor from the intensive care or step down units.

Summary

The surgical management of oligodendrogliomas consists of maximal safe total resection for the purposes of obtaining a tissue diagnosis, relieving the symptoms of local mass effect, and improving survival. Surgical adjuncts such as neuronavigation enable the creation of appropriate operative corridors through the smallest incision and craniotomy. This in turn creates smaller wounds to heal and a lower risk of wound-related complications. After completion of the surgery, appropriate follow-up imaging should be obtained for both the determination of the postoperative baseline and the determination of tumor recurrence or progression in the future.

References

1. Mørk SJ, Lindegaard KF, Halvorsen TB, et al. Oligodendroglioma: incidence and biological behavior in a defined population. *J Neurosurg.* 1985;63(6):881–889. https://doi.org/10.3171/jns.1985.63.6.0881.

2. Smits M. Imaging of oligodendroglioma. *Br J Radiol.* 2016;89(1060). https://doi.org/10.1259/bjr.20150857.

3. Spampinato MV, Smith JK, Kwock L, et al. Cerebral blood volume measurements and proton MR spectroscopy in grading of oligodendroglial tumors. *AJR Am J Roentgenol.* 2007;188(1):204–212. https://doi.org/10.2214/AJR.05.1177.

4. Ribom D, Smits A. Baseline ^{11}C-methionine PET reflects the natural course of grade 2 oligodendrogliomas. *Neurol Res.* 2005;27(5):516–521. https://doi.org/10.1179/174313213X13789811969265.

5. Danchaivijitr N, Waldman AD, Tozer DJ, et al. Low-grade gliomas: do changes in rCBV measurements at longitudinal perfusion-weighted MR imaging predict malignant transformation? *Radiology.* 2008;247(1):170–178. https://doi.org/10.1148/radiol.2471062089.

6. Rahman M, Abbatematteo J, De Leo EK, et al. The effects of new or worsened postoperative neurological deficits on survival of patients with glioblastoma. *J Neurosurg.* 2017;127(1):123–131. https://doi.org/10.3171/2016.7.JNS16396.

7. Weller M, van den Bent M, Tonn JC, et al. European Association for Neuro-Oncology (EANO) guideline on the diagnosis and treatment of adult astrocytic and oligodendroglial gliomas. *Lancet Oncol.* 2017;18(6):e315–e329. https://doi.org/10.1016/S1470-2045(17)30194-8.

Chapter 17

Oligodendrogliomas: Advanced techniques in surgery

Melvin Field* and William C. Newman[†]

*Orlando Neurosurgery, University of Central Florida College of Medicine, Orlando, FL, United States; [†]Department of Neurological Surgery, University of Pittsburgh Medical Center, Pittsburgh, PA, United States

Introduction

Oligodendrogliomas are a well-differentiated class of slow-growing, infiltrative cortical and subcortical tumors classified as the World Health Organization (WHO) grade II, with a more aggressive anaplastic counterpart classified as the WHO grade III. In the United States, the incidence of oligodendrogliomas is 4500–5000 new diagnoses per year, accounting for approximately 5% of malignant primary brain and other central nervous system (CNS) malignancies.[1] Recent changes in the WHO classifications reflect the growing importance of tumor molecular biology for diagnosis and subsequent management. Current classification guidelines require identification of a 1p/19q chromosomal co-deletion and isocitrate dehydrogenase (IDH) mutation for diagnosis of oligodendroglioma.[2] In contrast to other WHO grade II neoplasms, such as diffuse astrocytic gliomas, oligodendrogliomas tend to have a more favorable response to chemotherapy resulting in improved progression-free survival, a finding that has been attributed to the 1p/19q co-deletion.[3,4] The current standard of care for oligodendrogliomas, consisting of maximal safe resection with adjuvant radiotherapy and chemotherapy, has resulted in median overall survival times of 10–15 years for oligodendrogliomas and approximately 5–9 years for anaplastic oligodendrogliomas.[1,5,6] Importantly, these assessments stem from diagnostic criteria that predates the more homogenizing WHO 2016 updates.

Surgical management

For oligodendrogliomas, the first principle of surgical management is the procurement of diagnostic tissue. This can be obtained through several different options: frame-based stereotactic biopsy, frameless stereotactic biopsy, open surgical biopsy, or in the context of a surgical resection.

Biopsy

While imaging techniques may narrow the differential, definitive diagnosis must be established with tissue sampling and pathologic analysis with molecular testing. The two main ways to obtain tissue for diagnosis are either via biopsy or during surgical resection. Surgical resection is preferred, except in situations where the patient is too unstable or critically ill to tolerate the procedure, or if the tumor predominantly involves an eloquent area that precludes safe resection. In these cases, stereotactic biopsy can be pursued. When obtaining a biopsy, intraoperative pathology consultation is helpful to confirm that diagnostic abnormal tissue has been obtained and that there is a sufficient quantity for molecular testing. Stereotactic biopsy can be divided into two classes: frame based and frameless, each with their own advantages and disadvantages. Frame-based biopsy using the Leksell frame (Elekta, Stockholm, Sweden) or the Cosman-Roberts-Wells (CRW) frame (Integra Life Sciences, Plainsboro, NJ, USA) has several advantages: a high degree of accuracy in a fixed coordinate system, a small stab incision that can be closed with a single stitch and does not impede starting subsequent radiation or chemotherapy, the ability to perform the procedure under sedation, and the absence of brain shift-associated distortion of stereotactic accuracy after dural opening. Disadvantages of frame-based biopsies include the need for imaging the day of surgery with the frame in place, the need to plan a trajectory on the day of surgery, and potentially longer operating room times.[7]

Frameless biopsies allow for traditional three-pin cranial fixation with registration of the patient's surface cranial anatomy to a recently obtained magnetic resonance imaging (MRI). In some cases, such biopsies can even be performed without any fixation by utilizing mask navigated reference systems. Advantages of frameless biopsies include shorter operating room times, flexibility in technique, and use of different image guidance software based on the surgeon preference. Disadvantages of frameless biopsies include decreased accuracy compared to frame-based biopsies, especially for deeper lesions (i.e., brainstem, thalamus), and possibly a larger incision to accommodate stereotactic bases that are fixed to the exposed calvarium. This larger incision is accompanied by wound healing concerns for patients who need to promptly start chemotherapy and radiation after diagnosis.

Biopsy technique

The goals and principles for both frame-based and frameless stereotactic brain biopsies are identical. The goals include obtaining an adequate diagnostic specimen reflecting the true biologic characteristics of the lesion in question and minimizing the risks of complication—the most important of which are neurological deficit and bleeding. To improve diagnostic yield, biopsy target selection should avoid areas of radiographic necrosis and aim for regions of enhancement, if present. To avoid complications of bleeding, biopsy trajectory and systemic risk factors need to be considered. Discontinuing antiplatelet and anticoagulant medications prior to biopsy, and demonstrating normal platelet and coagulation function prior to biopsy is integral to minimizing the risk of hemorrhagic complications. In addition, periprocedural blood pressure control can also decrease the risk of postoperative bleeding related to brain biopsy. Biopsy trajectory should avoid vascular structures to minimize bleeding. This includes avoiding sulci, cisterns, and regions where large vessels travel. Ideally, the biopsy trajectory should traverse only a single pial surface to reach the target. The trajectory should be along the long axis of the mass, if possible, so that additional specimens can be obtained in different regions of the mass along that trajectory without redirecting the biopsy needle to ensure a diagnostic specimen is obtained. Finally, minimizing the number of biopsies taken decreases the risk of bleeding and neurological complication, but can also increase the risk of a nondiagnostic specimen. To avoid too few or too many biopsies, the authors send each biopsy specimen for an intraoperative frozen section to evaluate tissue volume adequacy and whether diagnostic tissue is present. If the answer is yes to both no further biopsies are obtained and if not, additional biopsies are acquired until both criteria are met. Each biopsy is obtained using a Sedan-style side-aspirating needle that generates a 1 cm long by 1–2 mm diameter core biopsy. After each biopsy is obtained, the needle's outer stylet is kept in place at the biopsy target and the biopsy channel is kept open. By doing so, any bleeding that results from the sample acquisition is allowed to drain out the needle chamber and not collect at the biopsy site. If bleeding is seen, warm saline is used to flush the needle intermittently to prevent occlusion, until the bleeding stops. In addition to the above considerations, to decrease the risk of neurological deficit the trajectory chosen to reach the target should avoid as many areas of eloquent tissue as possible. For example, a lesion just in front of the motor strip would be more safely approached from an anterior trajectory extending posteriorly to the lesion than a posterior trajectory crossing the motor strip and extending anteriorly to biopsy the lesion.

The incorporation and fusion of fiber tract, functional cortical, and anatomical imaging when generating a neuronavigated biopsy plan allows the surgeon to create a maximally safe trajectory when performing a brain biopsy.

Open surgical resection
Preoperative planning

Imaging

With recent advances in neuroimaging, preoperative planning has come to involve not only imaging to localize the tumor but also studies to localize regions of eloquent cortex in areas surrounding the tumor. Diagnostic imaging tools such as functional MRI (fMRI) and magnetoencephalography (MEG) have enabled preoperative noninvasive cortical mapping of critical functional regions, including language, motor, somatosensory, or visual processing. By establishing the relationship between tumor and adjacent functional eloquent regions, surgeons are able to counsel patients and discuss the goals of surgery more specifically, define resectability, refine the choice of approach, and determine the need for awake craniotomy.

Given the heterogeneity of language localization,[5] functional imaging helps to determine the sideness (i.e., left or right) of language dominance as well as language centers that may be located in the abnormal regions. While fMRI and MEG are useful for localizing cortical regions of brain function, they are not as useful for subcortical analysis of fiber tract function. For those details, high-definition fiber tracking (HDFT)/diffusion tensor imaging (DTI) can be employed. By combining these different imaging modalities, a clearer picture of the tumor and its relationship with eloquent cortical and subcortical structures can be obtained, resulting in safer and more complete resection (Table 1).

TABLE 1 Advantages and limitations of imaging adjuncts

Imaging modality	Advantages	Limitations
Functional MRI	• Noninvasive • Localizes functional cortex • Can be used with neuronavigation	• Does not visualize subcortical paths • Variable correlation with intraoperative mapping
Magnetoencephalogram	• Noninvasive • Localizes functional cortex • Can be used with neuronavigation	• Does not visualize subcortical paths • Variable correlation with intraoperative mapping
High-Definition Fiber Tracking	• High resolution • Visualizes cortical and subcortical connections • Volumetric analysis possible • Can be used with neuronavigation	• May not fully detect fiber tracts, especially when invaded by tumor
Stereotactic Neuronavigation	• Preoperative localization of tumor • Assistance with minimizing craniotomy size • Can be performed on all patients	• Intraoperative accuracy decreased due to brain shift
Intraoperative Ultrasound	• Comparative pre- and post-resection imaging • Minimal added time to utilize • Low cost	• Tissue manipulation can cause changes in echogenicity, making post-resection interpretation difficult
Intraoperative MRI	• Increased extent of resection • Allows re-registration of neuronavigation to adjust for brain shift • Facilitates LITT	• Increased operating room time • High upfront cost • Requires a custom OR
Chemiluminescence	• Noninvasive • Eliminates need for expensive intraoperative imaging • Does not require neuronavigation	• Requires surgical microscope or endoscope with specialized filters to visualize illuminated tissue • No proven benefit for low grade tumors including oligodendrioglioma

Magnetic resonance imaging

As most patients with oligodendrogliomas or anaplastic oligodendrogliomas present with new onset seizures,[8] initial imaging work-up typically includes a non-contrast computed tomography (CT) scan of the brain followed by an MRI. The MRI should include a thin slice post-contrast image for two reasons. First, this allows for detailed visualization of the tumor and its adjacent vasculature. Second, thin-slice images can be uploaded to image guidance systems and used to register the patient's surface anatomy to their preoperative MRI. In the operating room, this is helpful for planning the incision, localizing important structures, and ensuring that the operative approach will not be impeded by head positioning or cranial fixation. At our institution, we often use 1.5-mm-thick slices for our image guidance.

Given that oligodendrogliomas exhibit variable degrees of contrast enhancement, it is worth noting that thin-sliced T2-weighted imaging can also be obtained and utilized with image guidance. T2-weighted imaging may better elucidate some lower grade oligodendrogliomas that demonstrate more T2 signal abnormality than they do contrast enhancement. Some image guidance software will also permit for the fusion of two sequences, yielding an image that visualizes both the non-contrast-enhancing parts of the tumor and the contrast-enhancing portions.

Functional MRI

fMRI is an imaging technique to visualize cortical activity by detecting subtle alterations in blood flow that occur in response to stimuli or actions. These alterations are measured by tracking changes in blood flow, blood volume, or intravascular magnetic susceptibility.[9] This technique is based on the concept that cerebral blood flow and neuronal activation are coupled, and that increased neuronal activation and increased metabolic demand correlate with a commensurate increase in cerebral blood flow to meet that demand. The resulting cortical areas of brain activation are then color coded and can be viewed simultaneously with the area of the tumor.

Given the degree of heterogeneity in language localization,[5] fMRI can elucidate cortical language centers that are either near or distant to the lesion. This language localization may identify patients for whom awake craniotomies may provide additional benefit or in whom they are not necessary for surgical resection. fMRI, however, is limited by its need for patient cooperation and participation and may not be suitable for all patients. Specifically, fMRI requires the active participation of the patient to generate accurate functional data, and in some patients (i.e., young children, neurologically impaired patients, etc.), this may not be possible.

Prior studies have shown that for language, fMRI correlates with intraoperative mapping between 59% and 100% of the time and for motor function around 71% of the time.[10] Thus, while preoperative fMRI is a useful adjunct, it may produce false negatives in patients for whom awake craniotomy would be the optimal mode of resection. Therefore, the decision to pursue or not to pursue awake craniotomy should not reside solely on the preoperative functional imaging, but should also take into account the proximity of the lesion to eloquent structures, the value of potential positive and negative mapping for safe surgical resection[11] and the patient's existing neurological deficits.

Magnetoencephalogram

MEG is another type of noninvasive functional neuroimaging that maps the brain by recording magnetic fields generated by cortically produced electrical currents. Synchronized firing by neuronal currents produces a weak magnetic field that can be recorded by using a superconducting quantum interference detector—a highly sensitive magnetic field detector. Due to the low signal strength of the brain magnetic field (10^{-15} T) relative to Earth's magnetic field (10^{-4} T) of the surrounding environment, these devices must be housed inside a magnetically shielded room.[12] MEG is combined with MRI to produce the magnetic source image. The patient is positioned within the machine and either asked to listen to sounds for somatosensory localization, to perform motor tasks for motor cortex localization, or perform verbal tasks such as reading for speech localization (Fig. 1). While early studies demonstrated reproducible identification of the central sulcus and somatosensory cortex with MEG,[13] like fMRI, this study does not allow for the determination of subcortical fiber paths and should still be combined with intraoperative mapping for definitive identification of eloquent brain areas.

High-definition Fiber tracking

HDFT is an imaging technique that uses processing, reconstruction, and tractography to identify white matter fibers that go from the cortex to cortical and subcortical regions while allowing for the delineation of crossing fibers.[14] The predecessor to this technology was DTI, which allowed for the noninvasive mapping of neuronal pathways, and was used to demonstrate the spatial relationship of brain lesions to white matter pathways. It has even been suggested that DTI may improve the preservation of eloquent brain by providing insight into connectivity when planning for the surgical resection of complex lesions.[15] DTI, however, is limited in that it is unable to resolve crossing fibers or determine the origin and destinations of fibers with significant accuracy, leading to potential false tracts and significant susceptibility to artifact.[6] HDFT is able to provide millimeter and submillimeter resolution with the ability to detect crossing fibers as well as both their origins and destinations.

HDFT has a role in both in preoperative planning as well as intraoperative navigation. With respect to preoperative planning, HDFT not only allows the surgeon to identify the cortical termination of a tract (i.e., the primary motor region as identified with MEG testing), but also allows for the spatial localization of the descending fiber tracts with respect to the lesion (i.e., the descending pathway of the corticospinal tracts) (Fig. 2). This localization allows the surgeon to plan the site of initial corticectomy as well as to consider the trajectory for deep surgical resections based on the location of these descending fibers. During the operation, the HDFT imaging can be uploaded to an image guidance system and used to visualize the affected tracts intraoperatively. When combined with intraoperative subcortical stimulation during a craniotomy, this may improve the safety of surgical resection. In addition, pre- and postoperative fiber tract volumes can be compared to assess the impact of surgical resection.[16]

One notable consideration when using HDFT, however, is that while tracts that can be visualized are displayed with excellent details, negative tractography does not rule out the possible persistence of a fiber tract, especially when invaded by tumors like gliomas.[17]

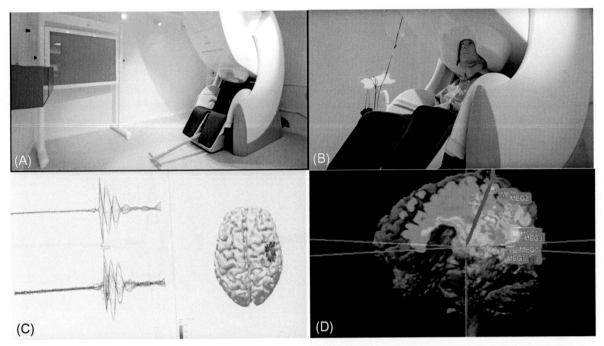

FIG. 1 Magnetoencephalography (MEG)—(A) shielded room housing an Elekta Neuromag TRIUX system. This specialized room has magnetic shielding composed of aluminum and mu-metal walls, floors, and ceilings to reduce high- and low-frequency noise including the Earth's own magnetic field. Such shielding is necessary to accurately measure the weak magnetic fields that synchronized neuronal currents create. These fields are measured in femtotesla (fT) and range from 10 fT for cortical activity to 1000 fT for alpha waves. Normal ambient fields in urban areas generally measure 100,000,000 fT which these rooms need to block in order to measure the ionic neuronal currents or the brain. (B) During a MEG study, the patient sits in the Neuromag unit with a set of head-positioning coils taped to the scalp. While performing various tasks to map language, motor, visual, auditory, or somatosensory cortical functional areas, the system uses hundreds of superconducting quantum interference devices (SQUIDs) to measure these activated magnetic signals that are less than one-billionth the strength of the Earth's own magnetic field. (C) Once the study is completed, MEG data are processed, and the recorded signals are used to determine where in the brain the activities originated. (D) This electrical functional data are then fused with MR image data allowing the physician to localize a patient's individual geographic anatomy with their associated function.

Additional testing

In addition to imaging, comprehensive preoperative planning includes a full neurological examination obtained by the neurosurgeon as well as, potentially, testing by a neuropsychologist. A complete neurological examination should detect gross neurologic abnormalities, such as speech dysfunction, motor weakness, or other signs of eloquent involvement. This can be helpful in preliminarily determining a patient's need for awake vs asleep craniotomy.

Neuropsychological testing provides a more detailed analysis of language, memory, and executive function. This type of testing can be used to detect subtler cognitive deficits not identified in a standard neurological examination. Neuropsychological testing is particularly useful in the evaluation of patients with tumors adjacent to or within the important regions of cognitive function including the frontal lobe and dominant temporal lobe. Studies have demonstrated that in many patients, extensive neurocognitive testing will reveal cognitive disturbances, particularly in working memory and executive function.[18] Identification of mild or subtle cognitive deficits may prompt the surgeon to adapt the surgical approach to account for these findings (i.e., awake craniotomy with functional mapping vs standard craniotomy under general anesthesia).

Preoperative neuropsychological testing is especially important in patients with lesions that appear to involve eloquent cortex as accurate pre- and postoperative evaluations allow for objective comparisons of clinical outcomes. Moreover, detailed testing may reveal evidence of neural plasticity with functional reorganization of brain networks, especially in slow-growing lesions like oligodendrogliomas. It may even reveal the potential reassignment of eloquent functions to regions adjacent to the lesion or even remote areas within the ipsilateral hemisphere or contralateral hemisphere, a form of neural plasticity.[19]

Careful preoperative assessment of a patient's neurologic baseline provides crucial comparison for intraoperative assessments during awake surgical procedures. Not only does it provide a context for specific tasks or questions used during the intraoperative examination, it may assist neuropsychologists in early detection of new neurologic deficits during resection.

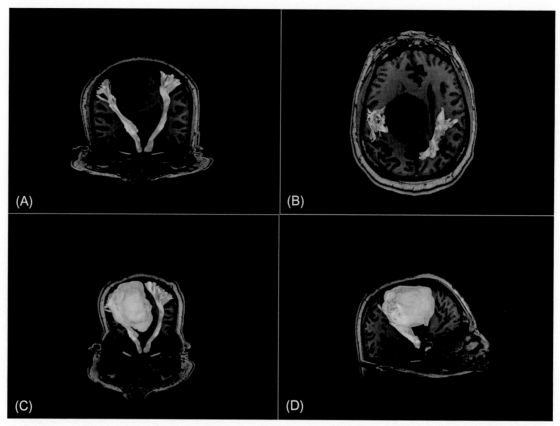

FIG. 2   An example of high-definition fiber tracking (HDFT) in a patient with a frontoparietal oligodendroglioma. (A) Preoperative T1W coronal imaging showing a large hypointense mass displacing the corticospinal tract laterally and inferiorly compared to its normal anatomic location as visualized on the contralateral side. (B) T1W axial imaging showing disorganized corticospinal fibers coalescing laterally relative to the mass and not being splayed diffusely by the tumor. (C) and (D) 3D computer remodeling and sequence fusion of HDFT data with T1W images and tumor volume reconstructions showing the relationship of the tumor relative to motor tracks. This information provides the surgeon with critical data allowing for a preoperative plan to resect the tumor while avoiding these tracks. In addition, this information demonstrates where subcortical or cortical stimulation will be most valuable to maximize the volume of disease resected while maintaining motor function. In this case, an anterior superomedial approach allows a safe corridor to the mass. As the surgeon, approaches the lateral and inferior margin of the tumor, subcortical stimulation will be essential to ensure preservation of the corticospinal track fibers as brain shift occurs.

Advanced operative techniques

Asleep craniotomy

A vast majority of craniotomies for oligodendroglioma are done with the patient asleep under general anesthesia. When tumors involve eloquent regions that cannot be monitored asleep, such as language, awake craniotomies are often performed. Neuronavigation is a standard tool used for asleep craniotomy and neurophysiological monitoring is often employed to monitor sensory and motor function during the procedure. Direct cortical and subcortical stimulation can be performed to guide a resection. Intraoperative imaging including ultrasound, CT, and MRI can be used to help the surgeon account for brain shift that occurs as tumor resection progresses. When resections are performed based on the navigation guidance and neurophysiological stimulation, as is often the case with oligodendroglioma, it is the authors' preference to define the margins of the tumor resection early in the surgery and before any significant brain shift has occurred. If a tumor comes close to an eloquent region, it is this region that the authors prefer to define first using cortical and subcortical stimulation. Once that boundary is defined, dissection using subcortical stimulating suction is performed to the depth of that tumor margin and before any significant tumor removal or brain retraction has occurred (Fig. 3). This maximizes the accuracy of the neuronavigation system at the location where it is most important and before significant brain shift has occurred. Once complete, the riskiest part of the surgery is now finished and further resection is guided away from the region of concern. Since the boundary to the eloquent region has already been established and preserved,

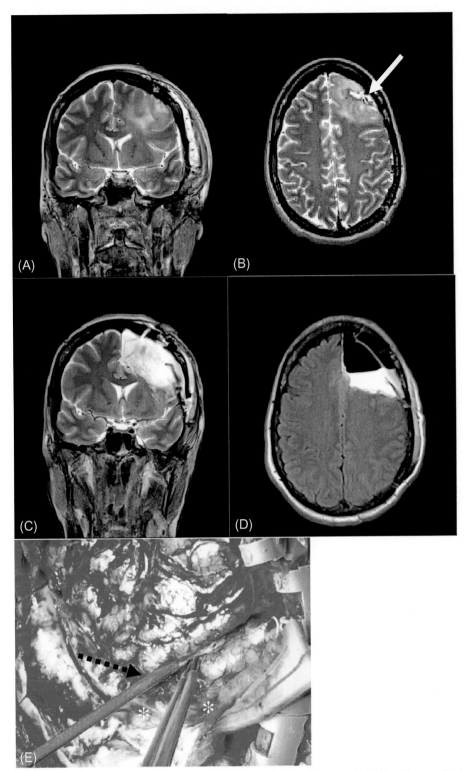

FIG. 3 A 35-year-old female with previous debulking resection via an asleep craniotomy. Pt presented initially with generalized seizure and postictal aphasia that resolved within 24 h of seizure onset. The patient's surgeon was concerned about causing a language deficit with an aggressive resection. Pathology came back as a grade II oligodendroglioma and tumor board recommended an awake craniotomy and maximally safe resection. fMR, MEG, and DTI evaluation suggested language localized within 1 cm of posterior and inferior margins of the flair margin of the tumor. Internal capsule motor fibers localized within 1 cm of the inferomedial margin. (A, B) Preoperative T2W images showing a left frontal mass extending to the corpus callosum medially. The previous resection cavity can be seen on the axial image (solid arrow). (C, D) Postoperative T2W images showing a complete radiographic resection of the mass. (E) Intraoperative photo showing tumor anteriorly (black) and cortex associated with expressive language posteriorly based on the preoperative functional data and intraoperative cortical stimulation resulting in speech arrest (white*). Once eloquent cortex is identified, initial resection is performed by dissecting along the margin of greatest concern before tumor debulking results in brain shift and distorts the accuracy of the neuronavigation system. In this photo, the surgeon is using a stimulating suction to dissect the posterior and inferior tumor margin (dashed arrow). Using subcortical stimulation, the surgeon guides the resection based on the clinical or neurophysiological changes in speech and motor function prior to coagulating, dissecting, or suctioning the underlying tissue. Once the margins adjacent to eloquent tissue are dissected, the surgeon is then free to debulk and resect more aggressively with confidence the remainder of the tumor.

there is less risk to the eloquent regions as the tumor bulk is resected away from that region and brain shift occurs. This approach also allows for a more aggressive, yet safe, resection as the surgeon is coming close to completing the resection. It also decreases the likelihood of finding unexpected residual disease once the intraoperative or postoperative MRI is completed.

Awake craniotomy

Traditional craniotomy for intra-axial supratentorial tumors is performed under general anesthesia with endotracheal intubation; a major limitation of this standard method is the inability to identify new neurologic deficits, particularly with language, until after the surgery is complete and the patient has recovered from anesthesia. In contrast, awake craniotomy is performed using only local anesthesia with small aliquots of pain medication, which allows for serial neurologic assessment in real time as the resection proceeds.

The goal of awake craniotomy is to facilitate maximal safe resection. This is especially important for glial neoplasms where the boundaries of the lesion may be difficult to fully discern or may abut critical structures. Real-time evaluation permits early detection of deficit and corresponding limits of safe surgical resection, thus minimizing risk of permanent postoperative neurologic deficit.[20] The importance of preventing new neurologic deficit was highlighted by studies reporting that in patients undergoing resection of high-grade gliomas, new language or motor deficits nullified the survival benefit gained from achieving a 95% extent of resection.[21] While awake craniotomy is a valuable tool, this technique is not appropriate for all patients. Factors including, but not limited to, lesion location, patient body habitus, poor airway, ability to tolerate laying in a single position for multiple hours, poorly controlled seizures, ability to consistently participate in neurologic testing, and overall level of anxiety are important to consider (Table 2). Despite its many benefits, awake surgical resection is only successful in select cases after careful consideration of appropriate tumor- and patient-related factors.

Anesthesia considerations

Patients for whom awake craniotomy is being considered should be screened for contraindications preoperatively by an experienced neuro-anesthesiologist. Considerations may include patient claustrophobia or anxiety, inability to lay still, cough, and co-morbidities such as obstructive sleep apnea or obese body habitus, that prevent safe maintenance of an unsecured airway.[22] In addition, it allows the anesthesiologist to discuss what the patient should expect with regard to emergence from anesthesia and re-sedation at the end of the case.

The airway evaluation is particularly important in planning for an awake craniotomy. During the procedure, the patient will not have a secured airway and contingency planning for airway compromise is critical. Plans for and conditions warranting elective or emergent conversion to endotracheal intubation should be determined preoperatively. Patients at particular risk of airway obstruction (i.e., morbidly obese, sleep apnea, etc.) should be identified and careful consideration given to the appropriateness of their candidacy. In the event, the patient has difficulty maintaining his or her oxygen saturation during an asleep portion of the case, several techniques may be employed: placement of a nasal trumpet,

---

TABLE 2 Relative and absolute contraindications to awake craniotomy

Absolute contraindications	Relative contraindications
Patient refusal	Anxiety disorders
Severe claustrophobia	Significant dysphasia
Inability to lay still for significant period of time	Chronic pain disorders
Inability to cooperate due to patient condition	Morbid obesity
Significant expressive or receptive speech deficit	Obstructive sleep apnea
	Expected difficult airway
	Uncontrolled coughing
	Dyspnea when lying flat
	Poorly controlled seizures

administration of supplemental oxygen, chin thrust maneuver, placement of a laryngeal mask airway, and lightening of sedation, if appropriate. The decision to convert to endotracheal intubation is one that should not be made lightly as this will terminate the possibility of a continuous patient assessment during awake surgical resection. Strong indications for conversion to general anesthesia with endotracheal intubation include uncontrolled seizures and high-risk patients who do not tolerate the awake portion of the procedure. In the event of seizures that do not respond to a combination of cold saline or lactated ringers applied to the brain, additional antiepileptics, or benzodiazepines, endotracheal intubation allows for the safe administration of additional sedatives that would otherwise risk airway compromise. In addition, in patients for whom airway management during the asleep portion is difficult and who are not tolerating the awake portion, a risk–benefit assessment may favor conversion to endotracheal intubation.

In the operating room, the anesthesia team should be involved in patient positioning after application of cranial fixation devices. From an airway perspective, it is important to ensure the patient is not placed in excessive flexion, as this may obstruct the airway. From a patient comfort perspective, it is important to ensure that the neck is not excessively extended or rotated, all pressure points are padded appropriately, and the patient is in a relatively comfortable, neutral position. In addition, patients should have a direct line of sight between themselves and either the anesthesiologist, neuropsychologist, or any other person performing the intraoperative neurocognitive testing. A heating blanket is also commonly used to prevent patient shivering and to further ensure comfort.

We routinely administer intravenous Fosphenytoin, Levetiracetam, or the patient's preexisting anticonvulsant upon entry into the operating room and prior to scalp block administration in an attempt to minimize the risk of intraoperative seizures. In addition, we keep sterile ice cold lactated ringers irrigation available for cortical lavage in the event an intraoperative seizure occurs during awake craniotomy. This has been shown to help extinguish seizure propagation from mechanical or electrical stimulation. Finally, we routinely use scopolamine patches and decadron intraoperatively to minimize nausea and emesis during surgery and thus preventing brain swelling and bleeding that can occur with such events.

Local anesthesia/scalp blocks

A circumferential scalp blockade using local anesthesia should be performed after sedation, but prior to placement of the cranial fixation device. At our institution, the anesthesia team administers total intravenous anesthesia with a combination of Propofol and remifentanil. We then assess for adequate sedation by brushing the eyelashes with a finger and observing any subsequent response. After confirming adequate sedation and planning the surgical incision, we begin the scalp blockade. A combination of short- and long-acting local anesthetic is used to address both the immediate pinning and future discomfort from surgical site dissection and retraction during the awake portion of the case. We generally use 0.5% lidocaine and 0.25% bupivacaine with 1:200,000 epinephrine. Adequate scalp blockade incorporates six nerves bilaterally—the supraorbital, supratrochlear, zygomaticotemporal, auriculotemporal, greater and lesser occipital, and great auricular[22]—as well as the planned incision (Fig. 4). Once circumferential blocks have been placed, the patient is placed in three-point cranial fixation with additional local anesthetic injected at the site of each of the three pins just prior to full engagement. The pins are then engaged; using 60–80 pounds of pressure. Neurophysiological monitoring leads are also placed at this time.

When administering local anesthetic, attention must be paid to the maximum total dose, based on the patient's lean body weight, to avoid systemic toxicities. These toxicities include, in order of increasing severity, the following: lingual numbness, lightheadedness, muscular twitching, hypotension, unconsciousness, convulsions, coma, respiratory arrest, and cardiac arrest.[23] In addition, it is important to aspirate prior to injection to ensure that you are not injecting intravascularly, as this can have significant cardiac effects.

Initial surgical steps (incision, craniotomy, and Dural opening)

Once the patient has been safely positioned, a urinary Foley catheter is placed, arterial access is obtained for blood pressure monitoring, and the patient's cranial surface anatomy is registered to an image guidance system. Adequate registration is then confirmed by applying the registration tool to known surface landmarks and visualizing that, when touched, the landmarks correlate with the image on the computer monitor. In our practice, we often use the lateral canthus and external auditory canal to confirm registration, then run the image guidance probe along the surface of the forehead and nose to assess for the accuracy of depth. If this accuracy is insufficient, reregistration or the use of additional registration points may be required.

When the patient is adequately registered, a surgical incision can be planned. This incision should take into account not just the size of the tumor but should be long enough to accommodate a craniotomy with appropriate exposure for speech or language mapping. In instances where positive mapping is the goal, a wider craniotomy may be required to exposure

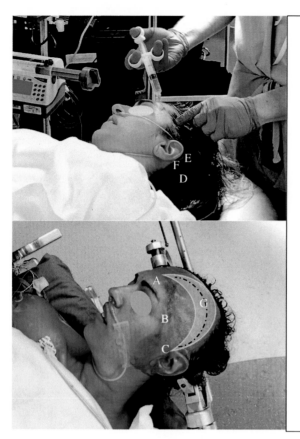

Scalp block locations for awake craniotomy

A. Supraorbital / Supratrochlear—Palpate the supraorbital notch and insert needle perpendicular to inject. Inject from the nasal root to the midpupillary line to cover both nerve distributions

B. Zygomatico-temporal—inject from the lateral edge of the supraorbital margin and proceed injecting to the distal zygomatic arch

C. Auriculo-temporal—inject over the zygomatic process and distal temporal artery. This is 1cm anterior to the tragus of the ear, just above the temporomandibular joint.

D. Greater Occipital—2.5cm lateral to median nuchal line and just medial to occipital artery. Inject just medial to the occipital artery to the midline along the nuchal line to the occipital protuberance.

E. Lesser Occipital—2.5cm lateral to greater occipital nerve injection site. Inject behind the ear from the top-down and then continue medially along the superior nuchal line to the greater occipital nerve injection site.

F. Greater Auricular—inject 2cm posterior to the ear at the level of the zygomatic root or tragus.

G. Peri-incisional—inject 1cm circumferentially around the planned incision

FIG. 4 Location for scalp block injections during awake craniotomy. 0.5% lidocaine and 0.25% bupivacaine with 1:200,000 epinephrine mixture is injected after induction of intravenous anesthesia with a combination of propofol and remifentanil to adequately sedate the patient. The block is done bilaterally and prior to placement of the headframe or neuromonitoring scalp electrodes.

adjacent or involved Rolandic cortex. Once marked, additional local anesthetic should be injected along the incision line. At this time, we generally ask the anesthesiologist to administer 0.5 g/kg intravenous mannitol and 10 mg intravenous Lasix to assist with brain relaxation. By administering these diuretics at this point in the surgery, we can assure maximal brain relaxation at the time of dural opening, usually 30–40 min later. In doing so, we minimize the risk of brain herniation out our opening, venous bleeding due to venous cortical compression, and ischemia due to compression at the bony margins.

The patient remains asleep during the opening incision, retraction of the scalp flap, and craniotomy. Once the dura is exposed, ultrasound may be used to further localize the tumor and serve as a form of real-time image guidance pre- and post-resection to determine if gross residual disease remains. Next, attention should be paid to how full the brain feels. If there is a concern for significant pressure, sedation should be decreased to allow for more rapid breathing, elimination of carbon dioxide, and subsequent brain relaxation. Additional reduction in intracranial pressure may be obtained by administering additional aliquots of mannitol. Lowering the intracranial pressure prior to dural incision reduces the risk of brain herniation through the dural opening. The dural incision is painful for the patient and not mitigated by any of the preoperative local anesthesia. To help minimize dural-related pain, we now apply 0.5% lidocaine with 0.25% bupivacaine soaked cottonoids onto the dural surface for 5 min prior to making our dural opening. The cottonoids are then removed and the dura is irrigated with saline to remove excess local anesthetic. The dura is opened, the patient's short-acting intravenous sedation is reduced, and the patient is allowed to awaken until conversant and oriented. When possible, the patient should only be aroused after completion of the durotomy to minimize conscious painful stimulation. The importance of having a team approach while performing an awake craniotomy cannot be understated. Inexperience with any member of the team can result in complications that negate any benefit of performing the procedure awake (Table 3).

Intraoperative mapping and stimulation/electrocorticography

Once the dura is open and the patient is awake, neuropsychological testing can begin. The patient is given tasks to make sure that they have returned to their baseline function (Fig. 5). While this is occurring, an electrocorticography (ECoG) strip

TABLE 3 Intraoperative complications of awake craniotomy

Anesthesia related	Surgical related
Hypoxia—desaturation	Bleeding
Airway obstruction	Brain swelling
Brain swelling—increased intracranial pressure	Brain shift
Hypotension—hypertension	Venous/arterial infarct—stroke
Bradycardia—tachycardia	Seizures
Pain	Expressive or receptive aphasia
Shivering	Motor/sensory/visual/memory deficits
Nausea—emesis	Pain
Confusion—agitation	Conversion to general anesthesia
Anesthetic toxicity	
Seizures	
Conversion to general anesthesia	

electrode or clover-leaf electrode (Ad-Tech, Frankfurt, Germany) is placed on the cortex in order to measure electrical activity and detect potential seizures from cortical stimulation before they become generalized convulsions. Direct electrical stimulation is then performed using a handheld bipolar stimulator to determine the patient's after discharge threshold. At our institution, we use the Ojemann Cortical Stimulator (Integra Neuroscience, New Jersey), starting with a current of 2 mA applied over 3–4 s. The current is progressively increased to elicit either speech arrest or motor arrest during the performance of a task ("positive mapping"). If an arrest occurs, the area is stimulated again to confirm it is a reproducible response and the area is marked. The absence of speech or motor arrest, so-called "negative mapping," is valuable as well, as it can signify an area of safe resection.[11]

During the course of stimulation, attention must be paid to the ECoG. If the neurophysiologist notices after-discharges—electrical discharges recorded after termination of the stimulus—stimulation should be halted and the stimulus current may need to be reduced. If an arrest is witnessed in conjunction with an after-discharge, it is important to reevaluate this area, as the arrest may be due to the after-discharge rather than the direct stimulation. If the ECoG shows signs of potential seizure, stimulation should be halted, ice cold saline or lactated ringers should be applied directly to the brain, and the surgery paused until the ECoG returns to normal. If the seizure persists, consideration of pharmacologic intervention in the form of additional antiepileptic doses, intravenous benzodiazepines, or other medications should be considered. Once the after-discharge threshold is determined, we then set our stimulation current for the tumor resection 2 mA below the after-discharge threshold but not to exceed a 12 mA maximum. As the resection proceeds, motor and language testing continues to monitor for new neurologic deficits. The surgeon can also elect to stimulate subcortically to investigate these regions for possible function. For subcortical mapping, we use the same level of stimulus as for cortical mapping, although the current may need to be increased slightly. As a general rule for white matter mapping, we find that for every 2–4 mA of stimulation with no neurophysiological response confers a region of safe resection around the stimulation site of 5–10 mm in diameter.

Positive and negative mapping

Traditionally, it was believed that large craniotomies incorporating definitive areas for positive mapping were needed to accurately delineate areas of safe surgical resection. However, more recent studies have shown that this is not universally the case. With respect to language function, negative mapping, or the absence of a stimulation-induced speech arrest, can identify areas in which surgical resection is safe and can be as useful as positive mapping.[11] The benefit of negative mapping is that smaller craniotomies can be used to expose relevant, adjacent brain tissue without the need to extend the craniotomy to find a site of speech arrest. The limitation of negative speech mapping is that it does not guarantee the absence of eloquent cortex and persistent postoperative deficits are still possible.[24] Overall, studies report rates upwards of 40% for transient postoperative neurological deficits with an approximately 5% rate of persistent neurologic deficits in patients with negative speech mapping.[25–27]

While negative mapping is commonly accepted for tumor resection when language function is of concern, for cases involving peri-Rolandic lesions in our practice we often increase the size of the craniotomy to find a site for positive motor

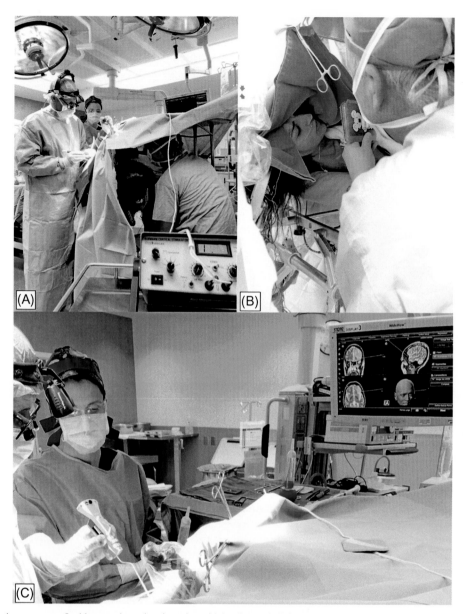

FIG. 5 Awake craniotomy setup. In this case, the patient is supine with head turned slightly to the right in a comfortable position and with a warming blanket to minimize shivering. (A) Neuropsychology is positioned adjacent to anesthesia to communicate with the patient and assess language while the surgeon uses the Ojemann cortical stimulation probe and subcortical suction probe to isolate eloquent brain tissue relative to the tumor to be resected. The neuromonitoring technician is positioned next to the Ojemann stimulator generator to adjust stimulation as necessary throughout the procedure and to provide neurophysiological feedback to the team as the surgery progresses. (B) Neuropsychology sits next to the patient and performs language paradigms observing facial expressions, movement, and responses to each task and question. (C) Neuronavigation is used to guide the tumor resection. In this photo, the surgeon is using the navigation probe to identify the inferior margin of the tumor based on the T2W images visualized on the neuronavigation monitor at the foot of the bed.

function mapping. Given the more homogeneous localization of motor function to the precentral gyrus, identifying an area of positive stimulation provides increased confidence in the safety of the planned resection.

For patients who are not candidates for awake craniotomy, but for whom there is concern for Rolandic or peri-Rolandic involvement, phase reversal can be used to localize primary motor and sensory cortex. Phase reversal utilizes intraoperative somatosensory-evoked potentials (SSEPs) to localize primary sensory and motor cortex. In this process, a strip electrode is placed on the cortex in an orientation perpendicular to the presumed orientation of the central sulcus. SSEP stimulation is performed while recording through the strip electrode. Phase reversal is identified when the negative N20 peak is replaced by a positive P20 peak between two electrodes on the contact array.[28]

Intraoperative adjuncts

Over the past 20 years, a multitude of highly specialized tools have been added to the surgeon's armamentarium to assist in maximizing the extent of tumor resection, minimize damage to adjacent normal structures, and to overcome the limitations that commonly occur with standard surgical techniques. These tools assist the surgeon with identifying anatomical structures and correlating them with radiographic findings. Neuronavigation and intraoperative imaging are now commonly used to assist with this. Chemiluminescence is another technique being increasingly utilized by neurosurgeons to assist in the visual discrimination of neoplastic tissue from normal brain parenchyma, allowing for a more complete tumor resection when visual discrimination under a microscope becomes difficult. And finally, newer intraoperative tools for tumor ablation are being investigated that may result in less invasive craniotomies while at the same time potentially minimizing the likelihood of tumor recurrence or progression. Laser interstitial thermal therapy (LITT) is one such tool currently being investigated for this purpose.

Neuronavigation

Image-guided stereotactic craniotomy is a term used routinely when discussing almost any form of surgery for a primary brain tumor in the developed world. Such surgery involves making an opening in the bone and biopsying, debulking, or removing an intracranial mass while utilizing imaging data that helps the surgeon to localize or "navigate" the region often with millimeter precision. Computerized and robotic systems that aid in this technique are known as neuronavigation systems. These systems use fiducial-based coordinates to fuse imaging data—CT, MR, angio, fMR, DTI to real-time fiducials that are fixed to the patient via a head holder, mask, or some other cranial fixation device that moves as the head moves. Cameras and sensors in the operating room that are integrated to the navigated systems then provide real-time feedback regarding patient positioning related to active and passive probes that tell the surgeon where in the head he is located using the fused imaging. Many people refer to this as a Global Positioning System (GPS) for the brain in that these systems show the surgeon a roadmap of where they are using preoperative and sometimes intraoperative imaging as a guide. For oligodendroglioma, such technology is integral to ensuring a maximally safe resection. Intraoperatively, such tumors may look indistinguishable from edematous surrounding brain parenchyma to the naked eye or under a microscope and yet look markedly different on MRI. This makes it very difficult for the surgeon to know where the tumor starts or ends. In such cases, the surgeon can use the navigation imaging tools to guide the resection of the mass while avoiding critical functional structures also seen on the navigation imaging. However, as the anatomy of the brain shifts during resection of a tumor or retraction of tissues relative to the fixed structures of the head occurs, such systems become unreliable as they are providing information based on the image sets created before any shift occurred. The larger the mass that is being removed or the more the brain needs to be retracted to perform the surgery, the more shift that occurs and the more inaccurate these systems become. The utilization of intraoperative imaging devices such as iMR, iCT, and iUS allows for new image sets to be obtained as brain shift occurs and recalibration/reregistration of the navigation systems to occur. This reregistration then fuses the new images to the system and now provides image guidance based on the new image sets that now account for the shift that has occurred with extreme accuracy. This allows the surgeon to overcome the limitations of brain shift and ensure an accurate resection prior to awakening the patient from anesthesia.

Chemiluminescence for resection of low-grade gliomas

In the past decade, there has been increasing interest in the use of intravenous chemiluminescent agents that are preferentially taken up by or bind to neoplastic cells relative to normal cells of the brain parenchyma. Such agents, in theory, when appropriately exposed to the proper wavelength of light preferentially "light up" tumor cells compared to normal brain cells allowing the surgeon to then remove the abnormally visualized tissue. Several agents are currently being evaluated, but 5-aminolevulinic acid (5-ALA) has been most widely studied, used, and accepted as an option when removing malignant primary brain tumors. 5-ALA is a naturally occurring substance in the human body that is a precursor in the heme biosynthetic pathway.[29] Exogenous administration of 5-ALA leads to the accumulation of protoporphyrin IX, a fluorescent compound, which selectively accumulates in malignant cells[30]; thus it has been used successfully in tumor surgery to induce tumor fluorescence for differentiation of pathologic and normal tissue. Its role in intraoperative fluorescence-guided resection of contrast enhancing malignant gliomas has been established, with reports demonstrating improved extents of resection and resultant increased progression-free survival.[31] While most of the evidence for the use of 5-ALA in gliomas pertains to high-grade lesions, there is evidence to suggest it may be a useful adjunct for lower-grade lesions with higher Ki-67 indices.[32,33] At this point, however, there is no significant data to support its consistent use in lower-grade lesions.

Intraoperative imaging adjuncts

Once the resection is felt to be complete, intraoperative imaging may provide additional confirmation. Image guidance tools may be used to assess the extent of resection, with the caveat that brain shift from tumor resection and cerebrospinal fluid egress will decrease its intraoperative accuracy. If ultrasound was used prior to the resection, it may again be employed to identify residual hyper- or hypoechoic regions relative to normal brain parenchyma that could denote residual tumor. This technique may be complicated by the fact that blood products and adjacent injured or contused brain can create hyperechoic regions that are sonographically difficult to distinguish from tumor.

Intraoperative MRI (iMRI) is another tool increasingly being embraced to assess extent of resection and, in some cases, to extend it when subtotal resection is identified. While most studies analyzing the impact of iMRI on patient outcomes have focused on glioblastoma—another infiltrative tumor frequently with poorly defined margins—these data are applicable to other infiltrating tumors such as grade II and grade III oligodendrogliomas. Current evidence suggests that use of iMRI improves extent of resection and prolongs survival after resection of glioblastomas.[34,35] For tumors that are well demarcated, iMRI has little benefit, as such technology is costly, increases operative time, requires MRI compatible OR equipment, and requires a specialized team experienced in the use of MRI in the operating room setting. However, for tumors with poorly defined margins, as is often the case with primary brain tumors including oligodendrogliomas, iMRI use can identify unintentional residual disease thought to have already been removed in up to one-third of cases (Fig. 6). The addition of iMRI allows the surgeon to identify the suboptimal resection prior to completing the craniotomy and then remove the residual disease at the time of the initial surgery. In addition, iMRI can be used to overcome the limitations of brain shift that occurs when retractors are used or as large masses are debulked. The use of iMRI allows for a reregistration of the neuronavigation system with updated imaging that is now accounting for such shift. In doing so, iMRI is allowing for a maximally safe resection based on the image guidance.

Laser interstitial thermal therapy

LITT is a treatment modality utilizing a small laser fiberoptic catheter to deliver heat precisely in conjunction with MR thermometry. This causes the destruction of tissue adjacent to the tip of the catheter. While the heat is being delivered, real-time MRI using MR thermometry is being performed with updates every few seconds to measure the temperature changes within the parenchyma. This allows for the determination of tissue that has been heated sufficiently to cause necrosis and cell death. Using predictive modeling, intraoperative MR thermometry imaging shows tissue volume death in the region of interest and allows the surgeon to determine when to stop lesion generation based on the thermal spread. In general, LITT remains a newer surgical tool with some early studies demonstrating its effectiveness in lesioning high-grade[36,37] and recurrent tumors,[38] as well as radiation necrosis (Fig. 7). This technique is minimally invasive and requires a small stab incision as well as an available iMRI or standalone MRI for the procedure. The main limitations of this technique with respect to glioma surgery are its absence of long-term outcomes compared to standard surgical resection as well as the inability, at this point, to be included in clinical trials after this procedure. As longer-term data regarding the efficacy of LITT become available, its role within the surgical management of glial neoplasms, including oligodendroglioma, will become more clearly defined.

Postoperative management

After surgical resection, there are several treatment algorithms. In patients with grade II oligodendrogliomas, up-front treatment with a combination of radiation and chemotherapy can be considered. Alternatively, due to the slow growing nature of these tumors, patients may be observed with serial imaging and neurological examinations with plan for radiation and chemotherapy at the time of disease progression. Decision making in this situation is based on the patient's functional status, tumor pathology, and amount of residual tumor. In those patients considered low-risk (i.e., high functional status, no visible residual, more benign tumor features), delaying cranial radiation and chemotherapy until recurrence is thought to spare patients from radiation-associated cognitive deficits for as long as possible.

In patients diagnosed with grade III oligodendrogliomas, standard postoperative treatment includes radiotherapy and chemotherapy. procarbazine, lomustine, vincristine (PCV), the main chemotherapeutic regimen for newly diagnosed oligodendrogliomas and anaplastic oligodendrogliomas, is a combination of procarbazine $60\,mg/m^2$, lomustine (also known as CCNU) $110\,mg/m^2$, and vincristine $1.4\,mg/m^2$ administered on a 29 day cycle repeated every 6 weeks.[39,40] Alternatively, oral temozolomide therapy in the doses of $75\,mg/m^2$ daily during radiotherapy followed by $150–200\,mg/m^2$ orally on days 1–5 every 4 weeks for six cycles of maintenance treatment can be given instead of PCV.[41] The benefits of temozolomide over PCV are its ease of administration and better patient tolerance. Long-term differences in outcome between the two agents are not yet available. In many ways, PCV is thus considered standard of care as it has been studied in two randomized

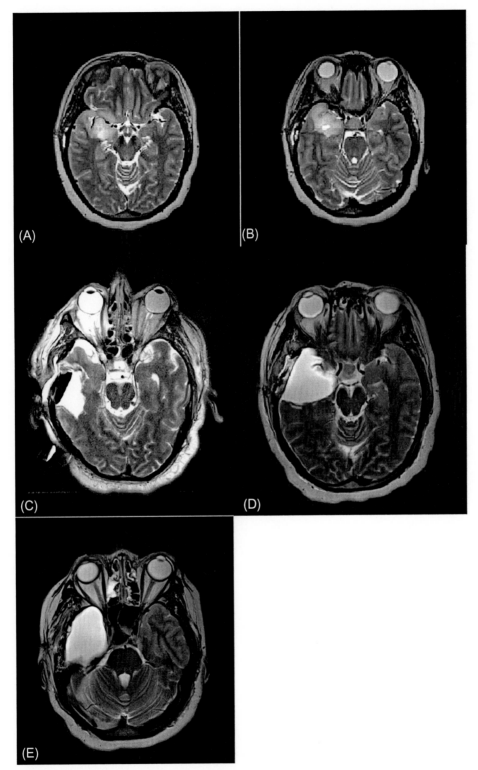

FIG. 6 Brain shift and iMRI. (A, B) Preoperative T2W axial images demonstrating a right temporal lobe tumor extending to the sylvian fissure super-omedially and tentorial edge inferomedially. During tumor resection, significant brain shift occurred related to tumor debulking and drainage of CSF. Based on the preoperative neuronavigation imaging, the resection resulted in a gross total resection of hyperintense tumor. (C). Intraoperative MRI (iMRI) showed residual disease missed anteriorly and medially. Neuronavigation was fused to the new intraoperative image data set and the surgeon proceeded to remove the residual disease. (D, E) Repeat imaging now shows a radiographic gross total resection.

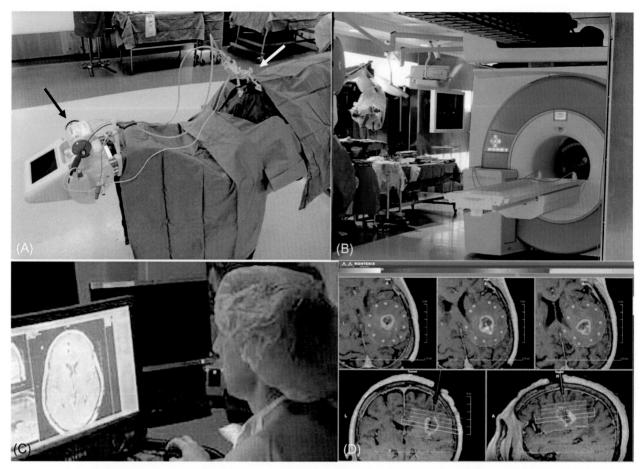

FIG. 7 Laser interstitial thermal therapy (LITT). (A) Using neuronavigation guidance, a burr hole is placed and a laser probed is advanced down to the tumor. Once at the predefined tumor target, the laser probe position is maintained using the cranial fixation device (white arrow). A robotic micromanipulator is used during the procedure to move the probe along the long axis of the tumor as needed while in the MRI to facilitate adequate thermal ablation throughout the tumor (black arrow). Using the micromanipulator, the laser also has a coolant system that allows for directional side firing of the laser. This helps with tumor contouring and directing thermal energy away from adjacent eloquent tissues. (B) Thermal ablation is performed while the patient is in MRI. (C) While in the MRI, the surgeon monitors and guides the ablation in a control room. In the control room, the surgeon can initiate and stop thermal heating of the tumor (ablation) with a foot pedal and direct the laser probe tip location in the tumor using the robotic micromanipulator. (D) MR thermometry predictive data are updated every 5 s, showing the surgeon real-time brain and tumor temperature information as well as ablation progress involving the regions of interest around the probe tip. As certain regions reach an ablative temperature, thus ensuring tumor cell death, the probe is redirected, advanced, or withdrawn to achieve a lethal thermal injury to the entire tumor. Probe tip repositioning is also used to minimize ablative temperatures to the surrounding normal brain parenchyma.

controlled trials (EORTC 26951 and RTOG 9402) that demonstrated increased survival with radiation therapy combined with PCV when compared to radiation therapy alone.[4,42] Of note, both of those studies involved anaplastic oligodendrogliomas. Radiotherapy is administered with external beam radiation as 59.4 Gy in 33 fractions (1.8 Gy each).[43] Hypofractionation with higher fractions and lower total doses may be appropriate for older patients and individuals with poor performance status.

Follow-up

General follow-up for patients with oligodendrogliomas and anaplastic oligodendrogliomas includes routine interval imaging with neurological examination. In the immediate postoperative period, patients should undergo an MRI of the brain (or CT if MRI is contraindicated) within 72 h to establish a new baseline for comparison with long-term follow-up imaging and diagnosis of recurrence.[44] Thereafter, follow-up can occur at 3-month intervals. If progression or recurrence is suspected or identified, subsequent treatment options should be based on a combination of Karnofsky Performance Scale score, prior treatment, and need for additional tissue sampling. Most studies analyzing the effectiveness of treatment in recurrent tumors have focused on anaplastic oligodendrogliomas, but even these studies are limited and report poor overall survival.[39,40]

Summary

Oligodendrogliomas and anaplastic oligodendrogliomas are an increasingly homogeneously defined tumor type based on the molecular mutations. Primary treatment of these lesions is with maximal safe resection, when possible, with the goal of obtaining diagnostic tissue and significant cytoreduction prior to chemotherapy and radiation therapy. Surgical adjuncts such as functional neuroimaging, stereotactic neuronavigation, awake craniotomy, and intraoperative chemiluminescence combine to increase the likelihood of maximal resection while minimizing permanent postoperative deficit.

References

1. Ostrom QT, Gittleman H, Xu J, et al. CBTRUS statistical report: primary brain and other central nervous system tumors diagnosed in the United States in 2009-2013. *Neuro-Oncol.* 2016;18(suppl_5):v1–v75. https://doi.org/10.1093/neuonc/now207.

2. Louis DN, Perry A, Reifenberger G, et al. The 2016 World Health Organization classification of tumors of the central nervous system: A summary. *Acta Neuropathol (Berl).* 2016;131(6):803–820. https://doi.org/10.1007/s00401-016-1545-1.

3. van den Bent MJ, Carpentier AF, Brandes AA, et al. Adjuvant procarbazine, lomustine, and vincristine improves progression-free survival but not overall survival in newly diagnosed anaplastic oligodendrogliomas and oligoastrocytomas: A randomized European Organisation for Research and Treatment of Cancer phase III trial. *J Clin Oncol Off J Am Soc Clin Oncol.* 2006;24(18):2715–2722. https://doi.org/10.1200/JCO.2005.04.6078.

4. Cairncross G, Wang M, Shaw E, et al. Phase III trial of chemoradiotherapy for anaplastic oligodendroglioma: Long-term results of RTOG 9402. *J Clin Oncol Off J Am Soc Clin Oncol.* 2013;31(3):337–343. https://doi.org/10.1200/JCO.2012.43.2674.

5. Ojemann G, Ojemann J, Lettich E, Berger M. Cortical language localization in left, dominant hemisphere. An electrical stimulation mapping investigation in 117 patients. *J Neurosurg.* 1989;71(3):316–326. https://doi.org/10.3171/jns.1989.71.3.0316.

6. Le Bihan D, Poupon C, Amadon A, Lethimonnier F. Artifacts and pitfalls in diffusion MRI. *J Magn Reson Imaging.* 2006;24(3):478–488. https://doi.org/10.1002/jmri.20683.

7. Dorward NL, Paleologos TS, Alberti O, Thomas DGT. The advantages of frameless stereotactic biopsy over frame-based biopsy. *Br J Neurosurg.* 2002;16(2):110–118.

8. Mørk SJ, Lindegaard KF, Halvorsen TB, et al. Oligodendroglioma: Incidence and biological behavior in a defined population. *J Neurosurg.* 1985;63(6):881–889. https://doi.org/10.3171/jns.1985.63.6.0881.

9. Logothetis NK, Pauls J, Augath M, Trinath T, Oeltermann A. Neurophysiological investigation of the basis of the fMRI signal. *Nature.* 2001;412(6843):150–157. https://doi.org/10.1038/35084005.

10. Giussani C, Roux F-E, Ojemann J, Sganzerla EP, Pirillo D, Papagno C. Is preoperative functional magnetic resonance imaging reliable for language areas mapping in brain tumor surgery? Review of language functional magnetic resonance imaging and direct cortical stimulation correlation studies. *Neurosurgery.* 2010;66(1):113–120. https://doi.org/10.1227/01.NEU.0000360392.15450.C9.

11. Sanai N, Mirzadeh Z, Berger MS. Functional outcome after language mapping for glioma resection. *N Engl J Med.* 2008;358(1):18–27. https://doi.org/10.1056/NEJMoa067819.

12. Singh SP. Magnetoencephalography: basic principles. *Ann Indian Acad Neurol.* 2014;17(Suppl 1):S107–S112. https://doi.org/10.4103/0972-2327.128676.

13. Sutherling WW, Crandall PH, Darcey TM, Becker DP, Levesque MF, Barth DS. The magnetic and electric fields agree with intracranial localizations of somatosensory cortex. *Neurology.* 1988;38(11):1705–1714.

14. Fernandez-Miranda JC, Pathak S, Engh J, et al. High-definition fiber tractography of the human brain: Neuroanatomical validation and neurosurgical applications. *Neurosurgery.* 2012;71(2):430–453. https://doi.org/10.1227/NEU.0b013e3182592faa.

15. Henry RG, Berman JI, Nagarajan SS, Mukherjee P, Berger MS. Subcortical pathways serving cortical language sites: initial experience with diffusion tensor imaging fiber tracking combined with intraoperative language mapping. *Neuroimage.* 2004;21(2):616–622. https://doi.org/10.1016/j.neuroimage.2003.09.047.

16. Fernandez-Miranda JC, Engh JA, Pathak SK, et al. High-definition fiber tracking guidance for intraparenchymal endoscopic port surgery. *J Neurosurg.* 2010;113(5):990–999. https://doi.org/10.3171/2009.10.JNS09933.

17. Leclercq D, Duffau H, Delmaire C, et al. Comparison of diffusion tensor imaging tractography of language tracts and intraoperative subcortical stimulations. *J Neurosurg.* 2010;112(3):503–511. https://doi.org/10.3171/2009.8.JNS09558.

18. Racine CA, Li J, Molinaro AM, Butowski N, Berger MS. Neurocognitive function in newly diagnosed low-grade glioma patients undergoing surgical resection with awake mapping techniques. *Neurosurgery.* 2015;77(3):371–379. discussion 379, https://doi.org/10.1227/NEU.0000000000000779.

19. Duffau H. Brain plasticity and tumors. *Adv Tech Stand Neurosurg.* 2008;33:3–33.

20. Eseonu CI, Rincon-Torroella J, ReFaey K, et al. Awake craniotomy vs craniotomy under general anesthesia for Perirolandic gliomas: Evaluating perioperative complications and extent of resection. *Neurosurgery.* 2017;81(3):481–489. https://doi.org/10.1093/neuros/nyx023.

21. Rahman M, Abbatematteo J, De Leo EK, et al. The effects of new or worsened postoperative neurological deficits on survival of patients with glioblastoma. *J Neurosurg.* 2017;127(1):123–131. https://doi.org/10.3171/2016.7.JNS16396.

22. Burnand C, Sebastian J. Anaesthesia for awake craniotomy. *Contin Educ Anaesth Crit Care Pain.* 2014;14(1):6–11. https://doi.org/10.1093/bjaceaccp/mkt024.

23. Rosenberg PH, Veering BT, Urmey WF. Maximum recommended doses of local anesthetics: a multifactorial concept. *Reg Anesth Pain Med.* 2004;29(6):564–575 [discussion 524].

24. Taylor MD, Bernstein M. Awake craniotomy with brain mapping as the routine surgical approach to treating patients with supratentorial intraaxial tumors: a prospective trial of 200 cases. *J Neurosurg.* 1999;90(1):35–41. https://doi.org/10.3171/jns.1999.90.1.0035.

25. Keles GE, Lundin DA, Lamborn KR, Chang EF, Ojemann G, Berger MS. Intraoperative subcortical stimulation mapping for hemispherical perirolandic gliomas located within or adjacent to the descending motor pathways: evaluation of morbidity and assessment of functional outcome in 294 patients. *J Neurosurg.* 2004;100(3):369–375. https://doi.org/10.3171/jns.2004.100.3.0369.

26. Carrabba G, Fava E, Giussani C, et al. Cortical and subcortical motor mapping in rolandic and perirolandic glioma surgery: impact on postoperative morbidity and extent of resection. *J Neurosurg Sci.* 2007;51(2):45–51.

27. Duffau H, Capelle L, Sichez J, et al. Intra-operative direct electrical stimulations of the central nervous system: the Salpêtrière experience with 60 patients. *Acta Neurochir.* 1999;141(11):1157–1167.

28. Cedzich C, Taniguchi M, Schäfer S, Schramm J. Somatosensory evoked potential phase reversal and direct motor cortex stimulation during surgery in and around the central region. *Neurosurgery.* 1996;38(5):962–970.

29. Bottomley SS, Muller-Eberhard U. Pathophysiology of heme synthesis. *Semin Hematol.* 1988;25(4):282–302.

30. Stummer W, Novotny A, Stepp H, Goetz C, Bise K, Reulen HJ. Fluorescence-guided resection of glioblastoma multiforme by using 5-aminolevulinic acid-induced porphyrins: a prospective study in 52 consecutive patients. *J Neurosurg.* 2000;93(6):1003–1013. https://doi.org/10.3171/jns.2000.93.6.1003.

31. Stummer W, Pichlmeier U, Meinel T, et al. Fluorescence-guided surgery with 5-aminolevulinic acid for resection of malignant glioma: a randomised controlled multicentre phase III trial. *Lancet Oncol.* 2006;7(5):392–401. https://doi.org/10.1016/S1470-2045(06)70665-9.

32. Valdés PA, Leblond F, Kim A, et al. Quantitative fluorescence in intracranial tumor: implications for ALA-induced PpIX as an intraoperative biomarker. *J Neurosurg.* 2011;115(1):11–17. https://doi.org/10.3171/2011.2.JNS101451.

33. Jaber M, Wölfer J, Ewelt C, et al. The value of 5-Aminolevulinic acid in low-grade gliomas and high-grade gliomas lacking glioblastoma imaging features: an analysis based on fluorescence, magnetic resonance imaging, 18F-Fluoroethyl tyrosine positron emission tomography, and tumor molecular factors. *Neurosurgery.* 2016;78(3):401–411. discussion 411, https://doi.org/10.1227/NEU.0000000000001020.

34. Kubben PL, ter Meulen KJ, Schijns OEMG, ter Laak-Poort MP, van Overbeeke JJ, van Santbrink H. Intraoperative MRI-guided resection of glioblastoma multiforme: a systematic review. *Lancet Oncol.* 2011;12(11):1062–1070. https://doi.org/10.1016/S1470-2045(11)70130-9.

35. Senft C, Bink A, Franz K, Vatter H, Gasser T, Seifert V. Intraoperative MRI guidance and extent of resection in glioma surgery: a randomised, controlled trial. *Lancet Oncol.* 2011;12(11):997–1003. https://doi.org/10.1016/S1470-2045(11)70196-6.

36. Barnett GH, Voigt JD, Alhuwalia MS. A systematic review and meta-analysis of studies examining the use of brain laser interstitial thermal therapy versus craniotomy for the treatment of high-grade tumors in or near areas of eloquence: an examination of the extent of resection and major complication rates associated with each type of surgery. *Stereotact Funct Neurosurg.* 2016;94(3):164–173. https://doi.org/10.1159/000446247.

37. Lee I, Kalkanis S, Hadjipanayis CG. Stereotactic laser interstitial thermal therapy for recurrent high-grade gliomas. *Neurosurgery.* 2016;79(Suppl 1): S24–S34. https://doi.org/10.1227/NEU.0000000000001443.

38. Tovar-Spinoza Z, Choi H. MRI-guided laser interstitial thermal therapy for the treatment of low-grade gliomas in children: a case-series review, description of the current technologies and perspectives. *Childs Nerv Syst ChNS Off J Int Soc Pediatr Neurosurg.* 2016;32(10):1947–1956. https://doi.org/10.1007/s00381-016-3193-0.

39. van den Bent MJ, Keime-Guibert F, Brandes AA, et al. Temozolomide chemotherapy in recurrent oligodendroglioma. *Neurology.* 2001;57 (2):340–342.

40. Chinot OL, Honore S, Dufour H, et al. Safety and efficacy of temozolomide in patients with recurrent anaplastic oligodendrogliomas after standard radiotherapy and chemotherapy. *J Clin Oncol Off J Am Soc Clin Oncol.* 2001;19(9):2449–2455. https://doi.org/10.1200/JCO.2001.19.9.2449.

41. Gorlia T, Delattre J-Y, Brandes AA, et al. New clinical, pathological and molecular prognostic models and calculators in patients with locally diagnosed anaplastic oligodendroglioma or oligoastrocytoma. A prognostic factor analysis of European organisation for research and treatment of cancer brain tumour group study 26951. *Eur J Cancer Oxf Engl.* 2013;49(16):3477–3485. https://doi.org/10.1016/j.ejca.2013.06.039.

42. Kouwenhoven MCM, Gorlia T, Kros JM, et al. Molecular analysis of anaplastic oligodendroglial tumors in a prospective randomized study: a report from EORTC study 26951. *Neuro-Oncol.* 2009;11(6):737–746. https://doi.org/10.1215/15228517-2009-011.

43. Triebels VHJM, Taphoorn MJB, Brandes AA, et al. Salvage PCV chemotherapy for temozolomide-resistant oligodendrogliomas. *Neurology.* 2004;63 (5):904–906.

44. Weller M, van den Bent M, Tonn JC, et al. European Association for Neuro-Oncology (EANO) guideline on the diagnosis and treatment of adult astrocytic and oligodendroglial gliomas. *Lancet Oncol.* 2017;18(6):e315–e329. https://doi.org/10.1016/S1470-2045(17)30194-8.

Chapter 18

Surgical management of 1p/19q co-deleted oligodendrogliomas WHO grade II and III

Niklas Thon, Friedrich-Wilhelm Kreth and Joerg-Christian Tonn
Department of Neurosurgery, Hospital of the University of Munich, Campus Grosshadern, Munich, Germany

Introduction

Over the last decade, advances in molecular and imaging-based biomarkers have induced a more versatile diagnostic classification and prognostic evaluation of glioma patients. This, in combination with new/improved treatment options enables increasingly individualized, risk-benefit-optimized treatment strategies. This approach to precision medicine in glioma patients requires that surgical procedures and related goals must be reevaluated in terms of indication, risk-benefit assessment, and prognostic impact in order to implement surgery within state-of-the-art management recommendations.

Isocitrate dehydrogenase mutated, 1p/19q co-deleted oligodendrogliomas WHO grade II and III

The 2016 revised World Health Organization (WHO) classification of central nervous system tumors has introduced a characteristic molecular pattern, namely the mutation in the gene encoding for isocitrate dehydrogenase (*IDH1/2*) and *1p/19q* co-deletion, as a prerequisite for the diagnosis of an oligodendroglioma grade II or III.[1,2] However, about 40%–65% of all grade II and III oligo-tumors show a heterogeneous histopathology including an astroglial tumor cell differentiation. This might have contributed to a high interobserver variability in histological diagnosis (up to 30% discordant findings), which is typical for former studies in grade II and III gliomas.[3–5] The molecular biomarker profile (i.e., *IDH* mutational and *LOH 1p/19q* status) has turned out to be homogeneously distributed throughout the WHO grade II and III gliomas.[6,7] Thus, the biomarker profile as detected from any viable part of the tumor can be regarded as representative for the entire tumor under consideration. In contrast, histopathological characteristics for anaplasia can sometimes be seen only focally, whereas major tumor parts exhibit low-grade characteristics.[8] Hence, accurate tumor grading relies on representative tissue sampling. Both the implementation of multimodal imaging methods for targeted tissue sampling and the introduction of molecular markers for integrated diagnoses has improved the diagnostic accuracy.

Tumor characteristics

1p/19q co-deleted oligodendrogliomas usually occur in middle-aged adults (35–50 years), but have also been observed in children. These tumors usually develop in the frontal, parietal, or temporal lobe of either hemisphere, but may also involve deep-seated parts of the brain.[9] The individual course of the disease is extremely variable and ranges from years of stable findings without specific symptoms to early uncontrolled progressive disease.[10] Overall, however, oligodendrogliomas tend to grow slowly and may, therefore, develop to a considerable size before specific symptoms such as seizures or focal neurological complaints lead to first diagnosis.[11] Focal calcifications may support the interpretation of slow tumor growth.

Preoperative diagnostic workup

In gliomas, the initial diagnosis is routinely based on the magnetic resonance imaging (MRI), which is increasingly being supplemented by anatomical, functional, and metabolic imaging data in terms of differential diagnosis, identification of intratumoral heterogeneity, and tumor extension in relation to function-relevant brain areas.

Oligodendroglioma. https://doi.org/10.1016/B978-0-12-813158-9.00018-9

Magnetic resonance imaging

Conventional MRI is the gold standard for differential diagnosis, treatment planning, and monitoring of treatment response in diffuse gliomas. Determination of true tumor extensions can be challenging as single tumor cells invade into the surrounding brain tissue far beyond MRI-defined tumor margins.[12] Lesions with sharp borders and well-matched tumor volumes in both T_1- and T_2-weighted sequences are clinically classified as circumscribed tumors, probably suitable for localized surgical treatment concepts. Lesions with significantly larger size in T2/fluid-attenuated inversion recovery (FLAIR) images as compared to the corresponding T1-weighted sequences are clinically classified as more diffuse lesions.[13] The latter implies a highly infiltrative growing pattern with tumor cells and normal brain tissue being intimately mixed with each other, which makes a clear separation difficult or even impossible. Diffuse tumors are often not completely resectable and alternative surgical strategies such as minimal invasive biopsy procedures have to be considered (instead of open tumor resection).[14] Conventional MRI suffers from low diagnostic sensitivity and specificity in oligodendrogliomas. Sharp delineations on MRI might be more common in oligodendroglial tumors, but do not exclude broad infiltration of adjacent brain areas.[15] Moreover, intratumoral calcifications and a patchy contrast enhancement may be seen; the latter does not necessarily exclude low-grade histology.[16] Conversely, non-enhancing tumors might turn out to be anaplastic tumors, which could be observed more frequently in the elderly subpopulation (~50% above 40 years of age).[8,17]

Advanced imaging techniques

Over the last years, advanced imaging modalities such as MR spectroscopy, MR perfusion analysis, and amino acid positron emission tomography (PET) have been shown to improve diagnostic accuracy and are increasingly used for noninvasive glioma evaluation.[8,15,18–36] These techniques can improve differential diagnosis and may detect infiltrative tumor tissue beyond conventional MRI-defined borders, indicating "true" biological tumor volumes. Proton MR spectroscopic and perfusion MR imaging have the potential to provide strong information about histological diagnosis, may help to identify local brain invasion outside of the volume readily identifiable by conventional MRI, and may indicate more aggressive behavior.[22,23,37] Increase in relative cerebral blood volume (rCBV), as being assessed by dynamic susceptibility contrast MRI (DSC-MRI), has been shown to predict high-grade transformation in astrocytic tumors before gadolinium enhancement occurs.[38] In oligodendroglial tumors, this distinction is less reliable as these tumors inherently exhibit higher rCBV values.[39] Diffusion tensor imaging (DTI) can be used to differentiate normal white matter and oedematous brain tissues from neoplastic tissues and may help to determine if fiber tracts are displaced, infiltrated, or disrupted by tumor formations.[40] Functional MRI (fMRI) can be used for noninvasive delineation of eloquent cortical areas and might be helpful to further estimate the tumors' resectability.[41,42] However, due to limits and loss of accuracy of spatial resolution in the process of data fusion of fMRI with the anatomical imaging data set, this information has to be used only very cautiously for intraoperative identification of functional areas.

In PET imaging oligodendroglial tumors are characterized by a higher uptake of radiolabeled amino acid tracers such as ^{11}C-methionine ([11]MET) and more importantly O-(2-[^{18}F] fluoroethyl)-L-tyrosine ([18]FET).[33,35,43–45] [18]FET-PET allows better delineation of true glioma extensions and has been shown to improve the differential diagnosis between pure astrocytic and oligo-tumors.[32,33,35] Moreover, the pattern of intratumoral [18]FET uptake kinetics can be exploited to identify focal anaplasia in MRI non-enhancing astrocytomas and oligo-tumors, and to define distinct biological subgroups with different clinical courses in suspected WHO grade II gliomas.[8]

The surgical perspective

Open tumor resection is considered one of the mainstays within the glioma treatment algorithm.[11,46,47] Neurosurgical resection is particularly important for symptomatic tumors with considerable mass effect and displacement of important brain structures. Besides clinical stabilization, there is good evidence that extensive resections improve survival in diffuse gliomas. Tumor characteristics support an initial surgical strategy, as radiographic complete resections may be achieved more frequently and with a lower risk in oligodendrogliomas.[48] Intraoperatively, the procedure of tumor resection can often be visually controlled and facilitated due to differences in color and consistency between tumor and brain tissues. This can be particularly helpful at the brain-tumor interface in the vicinity of functionally relevant cortical/subcortical structures. However, the inherent infiltrative growing pattern within complex neural circuits explains why a complete resection on the cellular level cannot be achieved. Accordingly, no patient can be cured by surgical means alone. Moreover, there is some evidence that the prognostic benefit from MRI-defined complete resection may be less pronounced than in astrocytic tumors.[49] The oncological benefit of incomplete resection—compared to a pure biopsy or even careful observation—has not yet been systematically evaluated. Since oligodendroglial gliomas exhibit increased sensitivity to radio/chemotherapy (as compared to astrocytomas), high-risk resections should be avoided. In patients with unresectable tumors, valid histological and molecular genetic analyses can be achieved using minimal invasive stereotactic biopsy techniques.[3,50–52]

Microsurgical tumor resection

On a population basis, about 70%–80% of diffuse glioma patients undergo open tumor resection as initial treatment. In oligodendroglial tumors, this number may be even higher, as these tumors frequently have lobar localization, may appear with sharp borders, particularly in T_2-weighted MRI sequences, may cause clinical symptoms due to large tumor volumes, and usually affect young patients with little or no co-morbidity. In asymptomatic patients with an incidental finding of a grade II oligodendroglioma, the timing of surgery remains controversial, as these tumors can be stable for years without any specific applied therapy.[53] However, MRI studies have shown that an annual growth rate of 3–4 mm in diameter must be considered even in clinically stable patients.[54] The rate of growth and extent of infiltration correlate with the degree of differentiation of the tumor according to the WHO classification, which may also be true for oligodendroglial tumors.[1] Ultimately, diffuse gliomas, including those with an oligodendroglial phenotype, may undergo more or less delayed malignant transformation with rapid clinical decline later on. It remains yet unclear, to which extent early resections can delay this fate.[55,56]

Prognostic impact of surgery

There is good evidence that extensive resection in diffuse gliomas is associated with improved overall survival. This is certainly correct for those cases where complete resection (as being defined by postoperative MRI criteria) can be achieved.[57,58] Some authors even promote a concept of "supratotal" resection for non-eloquent gliomas in order to account for the highly infiltrative growing behavior beyond MRI-defined volumes.[59] One cohort study from Norway, however, showed a significant prognostic advantage for any kind of resection compared to biopsy only.[60] Other data support that a volume reduction of at least 70%–80% should be attempted in order to improve outcome.[13,61] A similar significant correlation could be identified for the residual tumor volume.[62] In addition, any reduction of tumor burden appears to reduce the risk of a malignant transformation.[63,64] It should be emphasized, however, that all these data are based on the studies on diffuse WHO grade II gliomas and that no elaborate analysis for the subpopulation of *1p/19q* co-deleted/oligo-tumors has been published so far. Accordingly, the prognostic impact of partial resection in grade II and III oligodendrogliomas has not been systematically analyzed yet. Post-hoc analyses of prospective randomized trials (originally designed to analyze the prognostic relevance of chemotherapy and radiotherapy in the WHO grade III oligo-tumors) and numerous retrospective observational studies point toward a prognostic impact of extensive resections in oligodendrogliomas.[3,50,51] These data, however, should be regarded cautiously due to their inherent uncontrolled heterogeneity: patients receiving incomplete tumor resection or biopsy only usually were older, had a lower pretreatment clinical score, suffered more often from deep-seated and/or poorly delineated tumors, and cannot be compared with those receiving complete resection.[9,46,47,58,65] There is further indication that the impact of surgical resection may differ between anaplastic gliomas harboring an oligodendroglial phenotype vs astrocytic tumors. In one study, the impact of gross total resection was less pronounced in grade III oligoastrocytomas as compared to pure astrocytic tumors, although gross total resection was achieved more frequently.[49] This was also confirmed in grade II gliomas: increased extent of resection (EOR) resulted in better progression-free survival (PFS) in diffuse astrocytomas but not oligodendrogliomas.[66] The authors concluded that if surgery resulted in EOR <90%, patients with astrocytoma should undergo second-look surgery whereas patients with oligodendroglioma or oligoastrocytoma should rather be transferred to chemotherapy. Results from another retrospective analysis are in line with this interpretation.[53] In this report, the impact of surgery has been reevaluated according to the novel molecularly defined low-grade glioma subgroups. A strong prognostic impact of residual tumor volumes was particularly seen in *IDH*-mutated astrocytoma patients, also supporting a second look surgical strategy (if safely feasible), whereas no such conclusions were drawn for *1p/19q* co-deleted tumors. Another study on malignant gliomas also supported a relevant impact of molecular profiles on the prognostic relevance of EOR[67]: a positive association between gross total resection (of both enhancing and non-enhancing components) and overall survival was selectively observed for *IDH* mutant but not *IDH* wild-type tumors. Accordingly, this was analyzed for molecularly defined oligodendrogliomas—complete resection was important for astroglial tumors with *IDH1* mutation only; this could not be confirmed for patients with additional *1p/19q* co-deletion nor for patients with an *IDH* wild-type status.[68]

In case of surgical accessible and completely resectable tumors, open tumor resection continues to be a valuable and important treatment modality. Otherwise, for complex located tumors with lack of space occupying effects, a stereotactic biopsy should be considered as the initial step within an individualized management cascade.

The risk of surgery

In low-grade gliomas, a thorough risk/benefit estimation of the surgical procedure is mandatory, as treatment options potentially range from watchful waiting, local treatment concepts (i.e., surgical resection, interstitial brachytherapy, etc.) to aggressive strategies that advocate postoperative chemoradiation in high-risk patients in order to improve outcome.[11,69,70] Overall, approximately 5%–8% of periprocedural morbidity must be expected if surgical resection is planned in diffuse glioma patients. In eloquently located tumors, a markedly higher rate of complications may be seen.[71] By utilization of modern technical developments including pre- and intraoperative functional assessment, neuronavigation, and intraoperative imaging techniques, the surgery-associated morbidity may be reduced below 5% while increasing the proportion of complete resection to almost 90%.[13,67,72–77] A better patient selection upfront has certainly contributed to these favorable results. Taking into account the limited availability and high costs, it should be emphasized, however, that virtually none of the neuronavigation and intraoperative imaging techniques have undergone a systematic proof of clinical benefit.[19] The only randomized study on the use of intraoperative MRI, for example, comes from the surgical treatment of high-grade gliomas.[78] In here, the proportion of patients with complete removal of the solid tumor tissue was higher in the iMRI cohort. Similarly, the use of iMRI may improve EOR in low-grade gliomas.[79–81]

In general, the issue of perioperative complications is of fundamental relevance as it may worsen the patients' physical integrity, quality of life, and adjuvant treatment options. As a consequence negative effects on the patient's overall prognosis must be considered. This underscores the relevance of a risk-adapted concept which embeds any decision in favor of a high-risk open surgery into a thorough prognostic evaluation.[82] Notably, the procedural risk of minimal-invasive biopsy procedures, which may be considered upfront before active treatment commences, has been shown to be <1%, including very eloquent tumor sites such as the primary motor cortex or the brain stem.[52,83]

Stereotactic biopsy

In the majority of cases, histopathological diagnosis and molecular profiling is performed with the use of tissue samples from open tumor resection. A versatile neuropathological evaluation, however, can routinely be obtained from tissue samples which may be as small as the head of a match.[52] These tumor specimens can be harvested from any location within the brain (e.g., including the brain stem) with little morbidity by means of stereotactic biopsy procedures, which enables a comprehensive diagnostic investigation. Both microsurgical and stereotactic neurosurgeons have to ensure that tissue samples are representative of the tumors biological character, and have been harvested from the biologically most active and prognostically relevant parts of the tumor. This should particularly be considered in case of partial resections with uncharacterized residual tumor volumes left in situ. By utilization of advanced functional and metabolic imaging data for targeted sampling procedures, the risk of undergrading, misdiagnosis, and consecutive undertreatment can be reduced.[8,28] Since the most frequently requested genetic markers (*IDH1/2* mutation, *LOH 1p/19q*, *MGMT* promoter methylation, *p53* mutation) are homogeneously distributed in diffuse gliomas, the probability of obtaining a "false-negative" result or misclassification of the molecular profile is very low.[6] This also concerns heterogeneous tumor compositions with intratumoral hot spots of anaplasia.[8] Moreover, the clinically most important molecular genetic markers appear to remain stable in the majority of tumors during the course of the disease.[84] Given the powerful prognostic/predictive importance, for example, of *1p/19q* co-deletion, biopsy can be regarded as an important element for tailored treatment concepts particularly in complex located oligo-tumors.[85]

Interstitial iodine-125 brachytherapy

Interstitial implantation of a radioactive source such as iodine-125 is designed to deliver a high radiation dose to a well-defined tumor volume while minimizing the dose to the surrounding normal brain.[86,87] Low-dose rate interstitial iodine-125 brachytherapy can be successfully used for highly selected patients with small and circumscribed grade II diffuse gliomas in any location of the brain, either as initial treatment instead of high-risk open tumor resection, or for small-sized recurrences after previously performed surgery and/or multimodal treatment.[88] The impact of *1p/19q* co-deletion and oligodendroglial phenotype on the efficacy of interstitial iodine-125 brachytherapy has not been systematically evaluated yet. Tumor size is considered the most important risk factor of this treatment option: tumors with a diameter > 3.5 cm are usually not suitable for brachytherapy. In the case of larger complex located circumscribed tumors brachytherapy should, therefore, be combined with open tumor resection.[89] In these complex cases, usually a planned low-risk partial resection is performed as the initial step and iodine-125 implantation is then initiated 3 months later. The combined approach, which has been analyzed in low-grade gliomas, has been shown to be highly effective and not associated with permanent morbidity; it might be also an attractive treatment strategy for large and complex located grade II oligo-tumors. Generally, it can be assumed that brachytherapy is similarly effective as open tumor resection. A systematic comparative analysis of these two surgical treatment modalities, however, is still lacking. Given this uncertainty iodine-125 brachytherapy should be preserved for those tumors not accessible for safe open tumor resection.

Future perspectives

The overwhelming contribution of molecular markers for diagnostics, prognostic evaluation, and treatment considerations requires a reassessment of surgical strategies in the context of increasingly complex, risk, and benefit-optimized management considerations. Microsurgical resection should be performed if complete resection of entire tumor volumes can be safely achieved. In the case of unclear differential diagnosis and/or an unfavorable risk-benefit ratio for microsurgery, the molecular stereotactic biopsy can be a useful alternative as the first step within the management cascade. If the molecular profile indicates increased chemo- and/or radiation resistance, surgical resection may become more important even at increased risk. Conversely, neoadjuvant chemotherapy and primary irradiation, for example, for a *1p/19q* co-deleted tumor in a functionally relevant localization, can be considered in order to enable a second-stage resection in case of tumor regression (Fig. 1).[90] This individualized surgical strategy away from a classic sequential concept

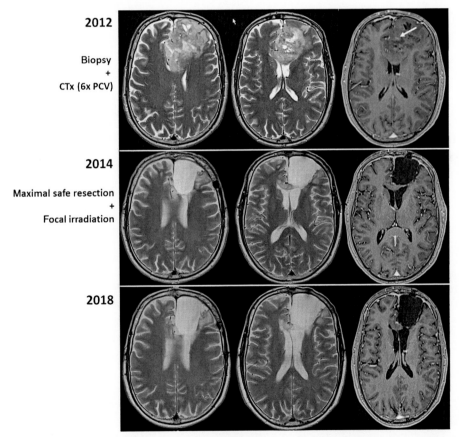

FIG. 1   Case of a 46-years-old male patient suffering from a left frontal (biopsy-proven) *IDH* mutant, *1p/19q* co-deleted oligodendroglioma WHO grade II (focal Ki67 up to 10%) with significant mass effect, critical involvement of the corpus callosum, and some patchy contrast enhancement (arrow) (Fig. 1, *first row*). At this point, neurocognitive testing showed no sustained abnormality, the patient was able to work as an engineer without any restrictions. The patient was initially reluctant to undergo primary resection and, in accordance with the recommendation of the interdisciplinary tumor board, primary chemotherapy (six cycles of procarbazine, CCNU and vincristin) was initiated. This led to a stable tumor formation but without significant volume reduction over 2 years (Fig. 1, *second row*). Ultimately, a maximum safe resection followed by percutaneus irradiation was performed. According to this individualized treatment concept, a stable tumor status without any new focal neurological deficit exists now for 4 years (Fig. 1, *last row*). *Legend: CT, chemotherapy; PCV, procarbazin plus CCNU and Vincristin.*

and toward a more dynamic approach certainly requires a future evaluation. Moreover, especially in symptomatic complex located low-grade glioma patients suffering from pharmacoresistant epilepsy, surgery may be indicated in order to reduce the burden from uncontrolled seizures. In here, a qualified diagnostic evaluation in highly specialized epilepsy centers may be indicated.[14,46,91] In individual cases, an invasive recording, for example, with stereotactically implanted deep electrodes, may be performed in order to determine one or multiple foci for targeted resections in the context of epilepsy surgery (Fig. 2). Mostly due to ethical considerations, it seems to be rather unlikely that class I evidence will be available on the impact of EOR on outcome measurements. However, an improved prognostic evaluation that also includes the

Progressive IDH mut, 1p/19q codeleted oligodendroglioma WHO grade II

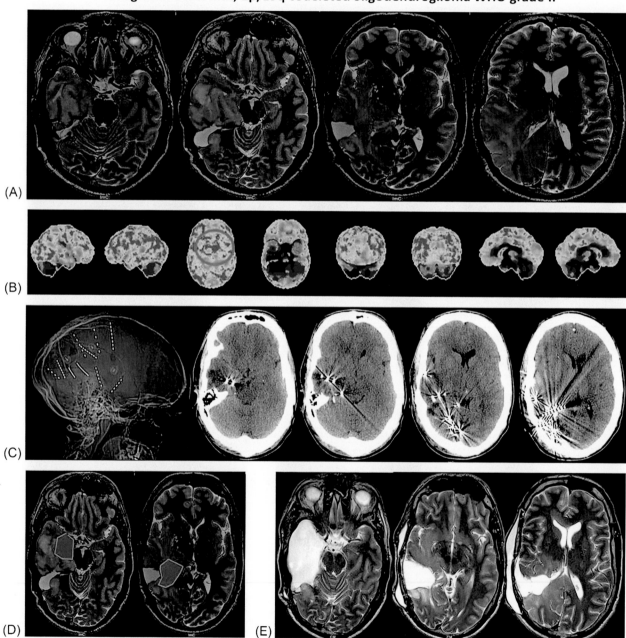

FIG. 2 (A) Case of a 35-years-old male patient with a progressive, open biopsy-proven *IDH* mutant, *1p/19q* co-deleted oligodendroglioma WHO grade II and pharmacoresistent seizures. (B) Ictal SPECT (Tc-99 m) with some increased uptake (green circle) but low spatial resolution. (C) Stereotactic implantation of multiple deep electrodes for invasive EEG monitoring. (D) Intralesional epileptogenic foci according to the invasive EEG evaluation (blue areas). (E) Epilepsy-tailored surgical strategy.

emerging field of molecular, metabolic, and imaging-based biomarkers will certainly help to identify the surgical procedure that fits most to the needs and limits of the individual patient within a multimodal risk—/benefit-optimized oncological management. Recent developments in targeted therapy will also push treatment concepts toward "molecular neuro-surgery," for example, by the use of conjugated immunotoxins that specifically bind to characteristic surface markers for glioma cells, which may be delivered by convection-enhanced delivery, stem cells, or interstitial radiosurgery. Emerging experimental therapies will certainly influence future management consideration and readjust the place of surgery in diffuse gliomas. Moreover, multimodal imaging systems with existing and new contrast agents, molecular tracers, technological advances, and advanced data analyses will serve as disease relevant biomarkers that will improve disease management and patient care.

Conclusion

Open tumor resection continues to be one of the mainstays of treatment in case of completely resectable oligodendrogliomas WHO grade II and III. For complex located circumscribed and small-sized tumors iodine-125 brachytherapy can be an attractive alternative treatment option. Molecular stereotactic biopsy is a valuable tool to obtain histology and the molecular signature of tumors which are not good candidates for a safe and complete resection. An elaborated diagnostic and prognostic evaluation is a fundamental prerequisite for individualized, multimodal treatment regimes. More data are necessary to support this concept of precision medicine for oligodendrogliomas.

Disclaimers

The authors are responsible for the topicality, correctness, completeness, and quality of the article.

Competing interests/disclosure

There is no competing interest and no funding to be disclosed.

References

1. Louis DN, Perry A, Reifenberger G, von Deimling A, Figarella-Branger D, Cavenee WK, et al. The 2016 World Health Organization classification of tumors of the central nervous system: a summary. *Acta Neuropathol.* 2016;131(6):803–820.
2. van den Bent MJ, Smits M, Kros JM, Chang SM. Diffuse infiltrating Oligodendroglioma and astrocytoma. *J Clin Oncol.* 2017;35(21):2394–2401.
3. Cairncross G, Wang M, Shaw E, Jenkins R, Brachman D, Buckner J, et al. Phase III trial of chemoradiotherapy for anaplastic oligodendroglioma: long-term results of RTOG 9402. *J Clin Oncol.* 2013;31(3):337–343.
4. van den Bent MJ. Interobserver variation of the histopathological diagnosis in clinical trials on glioma: a clinician's perspective. *Acta Neuropathol.* 2010;120(3):297–304.
5. Kros JM, Gorlia T, Kouwenhoven MC, Zheng PP, Collins VP, Figarella-Branger D, et al. Panel review of anaplastic oligodendroglioma from European organization for research and treatment of Cancer trial 26951: assessment of consensus in diagnosis, influence of 1p/19q loss, and correlations with outcome. *J Neuropathol Exp Neurol.* 2007;66(6):545–551.
6. Thon N, Eigenbrod S, Grasbon-Frodl EM, Ruiter M, Mehrkens JH, Kreth S, et al. Novel molecular stereotactic biopsy procedures reveal intratumoral homogeneity of loss of heterozygosity of 1p/19q and TP53 mutations in World Health Organization grade II gliomas. *J Neuropathol Exp Neurol.* 2009;68(11):1219–1228.
7. Thon N, Eigenbrod S, Kreth S, Lutz J, Tonn JC, Kretzschmar H, et al. IDH1 mutations in grade II astrocytomas are associated with unfavorable progression-free survival and prolonged postrecurrence survival. *Cancer.* 2012;118(2):452–460.
8. Kunz M, Thon N, Eigenbrod S, Hartmann C, Egensperger R, Herms J, et al. Hot spots in dynamic (18)FET-PET delineate malignant tumor parts within suspected WHO grade II gliomas. *Neuro-Oncology.* 2011;13(3):307–316.
9. Pignatti F, van den Bent M, Curran D, Debruyne C, Sylvester R, Therasse P, et al. Prognostic factors for survival in adult patients with cerebral low-grade glioma. *J Clin Oncol.* 2002;20(8):2076–2084.
10. Lang FF, Gilbert MR. Diffusely infiltrative low-grade gliomas in adults. *J Clin Oncol Off J Am Soc Clin Oncol.* 2006;24(8):1236–1245.
11. Weller M, van den Bent M, Tonn JC, Stupp R, Preusser M, Cohen-Jonathan-Moyal E, et al. European Association for Neuro-Oncology (EANO) guideline on the diagnosis and treatment of adult astrocytic and oligodendroglial gliomas. *Lancet Oncol.* 2017;18(6). e315–e29.
12. Sahm F, Capper D, Jeibmann A, Habel A, Paulus W, Troost D, et al. Addressing diffuse glioma as a systemic brain disease with single-cell analysis. *Arch Neurol.* 2012;69(4):523–526.
13. Ius T, Isola M, Budai R, Pauletto G, Tomasino B, Fadiga L, et al. Low-grade glioma surgery in eloquent areas: volumetric analysis of extent of resection and its impact on overall survival. A single-institution experience in 190 patients: clinical article. *J Neurosurg.* 2012;117(6):1039–1052.

14. Ius T, Pauletto G, Isola M, Gregoraci G, Budai R, Lettieri C, et al. Surgery for insular low-grade glioma: predictors of postoperative seizure outcome. *J Neurosurg.* 2014;120(1):12–23.

15. Pallud J, Varlet P, Devaux B, Geha S, Badoual M, Deroulers C, et al. Diffuse low-grade oligodendrogliomas extend beyond MRI-defined abnormalities. *Neurology.* 2010;74(21):1724–1731.

16. Zulfiqar M, Dumrongpisutikul N, Intrapiromkul J, Yousem DM. Detection of intratumoral calcification in oligodendrogliomas by susceptibility-weighted MR imaging. *AJNR Am J Neuroradiol.* 2012;33(5):858–864.

17. Barker 2nd FG, Chang SM, Huhn SL, Davis RL, Gutin PH, McDermott MW, et al. Age and the risk of anaplasia in magnetic resonance-nonenhancing supratentorial cerebral tumors. *Cancer.* 1997;80(5):936–941.

18. Weller M, van den Bent M, Tonn JC, Stupp R, Preusser M, Cohen-Jonathan-Moyal E, et al. Evidence-based management of adult patients with diffuse glioma—Authors' reply. *Lancet Oncol.* 2017;18(8). e430-e1.

19. Jenkinson MD, Barone DG, Bryant A, Vale L, Bulbeck H, Lawrie TA, et al. Intraoperative imaging technology to maximise extent of resection for glioma. *Cochrane Database Syst Rev.* 2018;1.

20. Choi C, Ganji SK, DeBerardinis RJ, Hatanpaa KJ, Rakheja D, Kovacs Z, et al. 2-Hydroxyglutarate detection by magnetic resonance spectroscopy in IDH-mutated patients with gliomas. *Nat Med.* 2012;18(4):624–629.

21. Elkhaled A, Jalbert LE, Phillips JJ, Yoshihara HAI, Parvataneni R, Srinivasan R, et al. Magnetic resonance of 2-hydroxyglutarate in IDH1-mutated low-grade gliomas. *Sci Transl Med.* 2012;4(116):116ra5.

22. Zonari P, Baraldi P, Crisi G. Multimodal MRI in the characterization of glial neoplasms: the combined role of single-voxel MR spectroscopy, diffusion imaging and echo-planar perfusion imaging. *Neuroradiology.* 2007;49(10):795–803.

23. Law M, Yang S, Wang H, Babb JS, Johnson G, Cha S, et al. Glioma grading: sensitivity, specificity, and predictive values of perfusion MR imaging and proton MR spectroscopic imaging compared with conventional MR imaging. *AJNR Am J Neuroradiol.* 2003;24(10):1989–1998.

24. Suchorska B, Giese A, Biczok A, Unterrainer M, Weller M, Drexler M, et al. Identification of time-to-peak on dynamic 18F-FET-PET as a prognostic marker specifically in IDH1/2 mutant diffuse astrocytoma. *Neuro-Oncology.* 2018;20(2):279–288.

25. Filss CP, Albert NL, Boning G, Kops ER, Suchorska B, Stoffels G, et al. O-(2-[18F]fluoroethyl)-L-tyrosine PET in gliomas: influence of data processing in different centres. *EJNMMI Res.* 2017;7(1):64.

26. Dunet V, Pomoni A, Hottinger A, Nicod-Lalonde M, Prior JO. Performance of 18F-FET versus 18F-FDG-PET for the diagnosis and grading of brain tumors: systematic review and meta-analysis. *Neuro-Oncology.* 2016;18(3):426–434.

27. Albert NL, Winkelmann I, Suchorska B, Wenter V, Schmid-Tannwald C, Mille E, et al. Early static (18)F-FET-PET scans have a higher accuracy for glioma grading than the standard 20–40 min scans. *Eur J Nucl Med Mol Imaging.* 2016;43(6):1105–1114.

28. Thon N, Kunz M, Lemke L, Jansen NL, Eigenbrod S, Kreth S, et al. Dynamic 18F-FET PET in suspected WHO grade II gliomas defines distinct biological subgroups with different clinical courses. *Int J Cancer.* 2015;136(9):2132–2145.

29. Rundle-Thiele D, Day B, Stringer B, Fay M, Martin J, Jeffree RL, et al. Using the apparent diffusion coefficient to identifying MGMT promoter methylation status early in glioblastoma: importance of analytical method. *J Med Radiat Sci.* 2015;62(2):92–98.

30. Jansen NL, Suchorska B, Wenter V, Eigenbrod S, Schmid-Tannwald C, Zwergal A, et al. Dynamic 18F-FET PET in newly diagnosed astrocytic low-grade glioma identifies high-risk patients. *J Nucl Med.* 2014;55(2):198–203.

31. Galldiks N, Stoffels G, Ruge MI, Rapp M, Sabel M, Reifenberger G, et al. Role of O-(2-18F-fluoroethyl)-L-tyrosine PET as a diagnostic tool for detection of malignant progression in patients with low-grade glioma. *J Nucl Med.* 2013;54(12):2046–2054.

32. Jansen NL, Graute V, Armbruster L, Suchorska B, Lutz J, Eigenbrod S, et al. MRI-suspected low-grade glioma: is there a need to perform dynamic FET PET? *Eur J Nucl Med Mol Imaging.* 2012;39(6):1021–1029.

33. Jansen NL, Schwartz C, Graute V, Eigenbrod S, Lutz J, Egensperger R, et al. Prediction of oligodendroglial histology and LOH 1p/19q using dynamic [(18)F]FET-PET imaging in intracranial WHO grade II and III gliomas. *Neuro-Oncology.* 2012;14(12):1473–1480.

34. Arbizu J, Tejada S, Marti-Climent JM, Diez-Valle R, Prieto E, Quincoces G, et al. Quantitative volumetric analysis of gliomas with sequential MRI and (1)(1)C-methionine PET assessment: patterns of integration in therapy planning. *Eur J Nucl Med Mol Imaging.* 2012;39(5):771–781.

35. la Fougere C, Suchorska B, Bartenstein P, Kreth FW, Tonn JC. Molecular imaging of gliomas with PET: opportunities and limitations. *Neuro-Oncology.* 2011;13(8):806–819.

36. Pallud J, Capelle L, Taillandier L, Fontaine D, Mandonnet E, Guillevin R, et al. Prognostic significance of imaging contrast enhancement for WHO grade II gliomas. *Neuro-Oncology.* 2009;11(2):176–182.

37. Saito T, Yamasaki F, Kajiwara Y, Abe N, Akiyama Y, Kakuda T, et al. Role of perfusion-weighted imaging at 3T in the histopathological differentiation between astrocytic and oligodendroglial tumors. *Eur J Radiol.* 2012;81(8):1863–1869.

38. Danchaivijitr N, Waldman AD, Tozer DJ, Benton CE, Brasil Caseiras G, Tofts PS, et al. Low-grade gliomas: do changes in rCBV measurements at longitudinal perfusion-weighted MR imaging predict malignant transformation? *Radiology.* 2008;247(1):170–178.

39. Cha S, Tihan T, Crawford F, Fischbein NJ, Chang S, Bollen A, et al. Differentiation of low-grade oligodendrogliomas from low-grade astrocytomas by using quantitative blood-volume measurements derived from dynamic susceptibility contrast-enhanced MR imaging. *AJNR Am J Neuroradiol.* 2005;26(2):266–273.

40. Witwer BP, Moftakhar R, Hasan KM, Deshmukh P, Haughton V, Field A, et al. Diffusion-tensor imaging of white matter tracts in patients with cerebral neoplasm. *J Neurosurg.* 2002;97(3):568–575.

41. Rutten GJ, Ramsey NF, van Rijen PC, Noordmans HJ, van Veelen CW. Development of a functional magnetic resonance imaging protocol for intraoperative localization of critical temporoparietal language areas. *Ann Neurol.* 2002;51(3):350–360.

42. Vlieger EJ, Majoie CB, Leenstra S, Den Heeten GJ. Functional magnetic resonance imaging for neurosurgical planning in neurooncology. *Eur Radiol.* 2004;14(7):1143–1153.

43. Saito T, Maruyama T, Muragaki Y, Tanaka M, Nitta M, Shinoda J, et al. 11C-methionine uptake correlates with combined 1p and 19q loss of heterozygosity in oligodendroglial tumors. *AJNR Am J Neuroradiol.* 2013;34(1):85–91.

44. Singhal T, Narayanan TK, Jacobs MP, Bal C, Mantil JC. 11C-methionine PET for grading and prognostication in gliomas: a comparison study with 18F-FDG PET and contrast enhancement on MRI. *J Nucl Med.* 2012;53(11):1709–1715.

45. Grosu AL, Astner ST, Riedel E, Nieder C, Wiedenmann N, Heinemann F, et al. An interindividual comparison of O-(2-[18F]fluoroethyl)-L-tyrosine (FET)- and L-[methyl-11C]methionine (MET)-PET in patients with brain gliomas and metastases. *Int J Radiat Oncol Biol Phys.* 2011;81 (4):1049–1058.

46. Soffietti R, Baumert BG, Bello L, von Deimling A, Duffau H, Frenay M, et al. Guidelines on management of low-grade gliomas: report of an EFNS-EANO task force. *Eur J Neurol.* 2010;17(9):1124–1133.

47. Stupp R, Tonn JC, Brada M, Pentheroudakis G, Group EGW. High-grade malignant glioma: ESMO clinical practice guidelines for diagnosis, treatment and follow-up. *Ann Oncol.* 2010;21(Suppl 5):v190–v193.

48. Mueller W, Hartmann C, Hoffmann A, Lanksch W, Kiwit J, Tonn J, et al. Genetic signature of oligoastrocytomas correlates with tumor location and denotes distinct molecular subsets. *Am J Pathol.* 2002;161(1):313–319.

49. Snyder LA, Wolf AB, Oppenlander ME, Bina R, Wilson JR, Ashby L, et al. The impact of extent of resection on malignant transformation of pure oligodendrogliomas. *J Neurosurg.* 2014;120(2):309–314.

50. Wick W, Hartmann C, Engel C, Stoffels M, Felsberg J, Stockhammer F, et al. NOA-04 randomized phase III trial of sequential radiochemotherapy of anaplastic glioma with procarbazine, lomustine, and vincristine or temozolomide. *J Clin Oncol.* 2009;27(35):5874–5880.

51. van den Bent MJ, Brandes AA, Taphoorn MJ, Kros JM, Kouwenhoven MC, Delattre JY, et al. Adjuvant procarbazine, lomustine, and vincristine chemotherapy in newly diagnosed anaplastic oligodendroglioma: long-term follow-up of EORTC brain tumor group study 26951. *J Clin Oncol Off J Am Soc Clin Oncol.* 2013;31(3):344–350.

52. Eigenbrod S, Trabold R, Brucker D, Eros C, Egensperger R, La Fougere C, et al. Molecular stereotactic biopsy technique improves diagnostic accuracy and enables personalized treatment strategies in glioma patients. *Acta Neurochir.* 2014;156(8):1427–1440.

53. Wijnenga MMJ, French PJ, Dubbink HJ, Dinjens WNM, Atmodimedjo PN, Kros JM, et al. The impact of surgery in molecularly defined low-grade glioma: an integrated clinical, radiological, and molecular analysis. *Neuro-Oncology.* 2018;20(1):103–112.

54. Pallud J, Blonski M, Mandonnet E, Audureau E, Fontaine D, Sanai N, et al. Velocity of tumor spontaneous expansion predicts long-term outcomes for diffuse low-grade gliomas. *Neuro-Oncology.* 2013;15(5):595–606.

55. Bourne TD, Schiff D. Update on molecular findings, management and outcome in low-grade gliomas. *Nat Rev Neurol.* 2010;6(12):695–701.

56. Potts MB, Smith JS, Molinaro AM, Berger MS. Natural history and surgical management of incidentally discovered low-grade gliomas. *J Neurosurg.* 2012;116(2):365–372.

57. Smith JS, Chang EF, Lamborn KR, Chang SM, Prados MD, Cha S, et al. Role of extent of resection in the long-term outcome of low-grade hemispheric gliomas. *J Clin Oncol.* 2008;26(8):1338–1345.

58. Keles GE, Lamborn KR, Berger MS. Low-grade hemispheric gliomas in adults: a critical review of extent of resection as a factor influencing outcome. *J Neurosurg.* 2001;95(5):735–745.

59. Yordanova YN, Moritz-Gasser S, Duffau H. Awake surgery for WHO grade II gliomas within "noneloquent" areas in the left dominant hemisphere: toward a "supratotal" resection. *J Neurosurg.* 2011;115(2):232–239.

60. Jakola AS, Skjulsvik AJ, Myrmel KS, Sjavik K, Unsgard G, Torp SH, et al. Surgical resection versus watchful waiting in low-grade gliomas. *Ann Oncol.* 2017;28(8):1942–1948.

61. Majchrzak K, Kaspera W, Bobek-Billewicz B, Hebda A, Stasik-Pres G, Majchrzak H, et al. The assessment of prognostic factors in surgical treatment of low-grade gliomas: a prospective study. *Clin Neurol Neurosurg.* 2012;114(8):1135–1144.

62. Shaw EG, Berkey B, Coons SW, Bullard D, Brachman D, Buckner JC, et al. Recurrence following neurosurgeon-determined gross-total resection of adult supratentorial low-grade glioma: results of a prospective clinical trial. *J Neurosurg.* 2008;109(5):835–841.

63. Ahmadi R, Dictus C, Hartmann C, Zurn O, Edler L, Hartmann M, et al. Long-term outcome and survival of surgically treated supratentorial low-grade glioma in adult patients. *Acta Neurochir.* 2009;151(11):1359–1365.

64. Chaichana KL, McGirt MJ, Laterra J, Olivi A, Quinones-Hinojosa A. Recurrence and malignant degeneration after resection of adult hemispheric low-grade gliomas. *J Neurosurg.* 2010;112(1):10–17.

65. Daniels TB, Brown PD, Felten SJ, Wu W, Buckner JC, Arusell RM, et al. Validation of EORTC prognostic factors for adults with low-grade glioma: a report using intergroup 86-72-51. *Int J Radiat Oncol Biol Phys.* 2011;81(1):218–224.

66. Nitta M, Muragaki Y, Maruyama T, Iseki H, Ikuta S, Konishi Y, et al. Updated therapeutic strategy for adult low-grade glioma stratified by resection and tumor subtype. *Neurol Med Chir.* 2013;53(7):447–454.

67. De Witt Hamer PC, Robles SG, Zwinderman AH, Duffau H, Berger MS. Impact of intraoperative stimulation brain mapping on glioma surgery outcome: a meta-analysis. *J Clin Oncol.* 2012;30(20):2559–2565.

68. Kawaguchi T, Sonoda Y, Shibahara I, Saito R, Kanamori M, Kumabe T, et al. Impact of gross total resection in patients with WHO grade III glioma harboring the IDH 1/2 mutation without the 1p/19q co-deletion. *J Neuro-Oncol.* 2016;129(3):505–514.

69. Xia L, Fang C, Chen G, Sun C. Relationship between the extent of resection and the survival of patients with low-grade gliomas: a systematic review and meta-analysis. *BMC Cancer.* 2018;18(1):48.

70. Gao Y, Weenink B, Van den Bent M, Erdem-Eraslan L, Kros JM, Sillevis Smitt PA, et al. Expression-based intrinsic glioma subtypes are prognostic in low-grade gliomas of the EORTC22033-26033 clinical trial. *Eur J Cancer.* 2018;94:10.

71. Satoer D, Visch-Brink E, Dirven C, Vincent A. Glioma surgery in eloquent areas: can we preserve cognition? *Acta Neurochir.* 2016;158 (1):35–50.

72. Chang EF, Clark A, Smith JS, Polley MY, Chang SM, Barbaro NM, et al. Functional mapping-guided resection of low-grade gliomas in eloquent areas of the brain: improvement of long-term survival. *J Neurosurg.* 2011;114(3):566–573.

73. Barone DG, Lawrie TA, Hart MG. Image guided surgery for the resection of brain tumours. *Cochrane Database Syst Rev.* 2014;1.

74. Mohammadi AM, Sullivan TB, Barnett GH, Recinos V, Angelov L, Kamian K, et al. Use of high-field intraoperative magnetic resonance imaging to enhance the extent of resection of enhancing and nonenhancing gliomas. *Neurosurgery.* 2014;74(4):339–350.

75. Sanai N, Mirzadeh Z, Berger MS. Functional outcome after language mapping for glioma resection. *N Engl J Med.* 2008;358(1):18–27.

76. Sollmann N, Wildschuetz N, Kelm A, Conway N, Moser T, Bulubas L, et al. Associations between clinical outcome and navigated transcranial magnetic stimulation characteristics in patients with motor-eloquent brain lesions: a combined navigated transcranial magnetic stimulation-diffusion tensor imaging fiber tracking approach. *J Neurosurg.* 2018;128(3):800–810.

77. Hervey-Jumper SL, Berger MS. Maximizing safe resection of low- and high-grade glioma. *J Neuro-Oncol.* 2016;130(2):269–282.

78. Senft C, Bink A, Franz K, Vatter H, Gasser T, Seifert V. Intraoperative MRI guidance and extent of resection in glioma surgery: a randomised, controlled trial. *Lancet Oncol.* 2011;12(11):997–1003.

79. Scherer M, Jungk C, Younsi A, Kickingereder P, Muller S, Unterberg A. Factors triggering an additional resection and determining residual tumor volume on intraoperative MRI: analysis from a prospective single-center registry of supratentorial gliomas. *Neurosurg Focus.* 2016;40(3).

80. Coburger J, Merkel A, Scherer M, Schwartz F, Gessler F, Roder C, et al. Low-grade Glioma surgery in intraoperative magnetic resonance imaging: results of a multicenter retrospective assessment of the German study Group for Intraoperative Magnetic Resonance Imaging. *Neurosurgery.* 2016;78(6):775–786.

81. Pala A, Brand C, Kapapa T, Hlavac M, Konig R, Schmitz B, et al. The value of intraoperative and early postoperative magnetic resonance imaging in low-grade Glioma surgery: a retrospective study. *World Neurosurg.* 2016;93:191–197.

82. Vogelbaum MA. Towards a genomic definition of completeness of resection? *Neuro-Oncology.* 2014;16(1):2–3.

83. Ragel BT, Ryken TC, Kalkanis SN, Ziu M, Cahill D, Olson JJ. The role of biopsy in the management of patients with presumed diffuse low grade glioma: a systematic review and evidence-based clinical practice guideline. *J Neuro-Oncol.* 2015;125(3):481–501.

84. Felsberg J, Thon N, Eigenbrod S, Hentschel B, Sabel MC, Westphal M, et al. Promoter methylation and expression of MGMT and the DNA mismatch repair genes MLH1, MSH2, MSH6 and PMS2 in paired primary and recurrent glioblastomas. *Int J Cancer.* 2011;129(3):659–670.

85. Tonn JC, Thon N, Schnell O, Kreth FW. Personalized surgical therapy. *Ann Oncol.* 2012;23(Suppl 10):x28–x32.

86. Schwarz SB, Thon N, Nikolajek K, Niyazi M, Tonn JC, Belka C, et al. Iodine-125 brachytherapy for brain tumours—a review. *Radiat Oncol.* 2012;7:30.

87. Kreth FW, Thon N, Siefert A, Tonn JC. The place of interstitial brachytherapy and radiosurgery for low-grade gliomas. *Adv Tech Stand Neurosurg.* 2010;35:183–212.

88. Kreth FW, Faist M, Grau S, Ostertag CB. Interstitial 125I radiosurgery of supratentorial de novo WHO grade 2 astrocytoma and oligoastrocytoma in adults: long-term results and prognostic factors. *Cancer.* 2006;106(6):1372–1381.

89. Schnell O, Scholler K, Ruge M, Siefert A, Tonn JC, Kreth FW. Surgical resection plus stereotactic 125I brachytherapy in adult patients with eloquently located supratentorial WHO grade II glioma—feasibility and outcome of a combined local treatment concept. *J Neurol.* 2008;255(10):1495–1502.

90. Taal W, van der Rijt CC, Dinjens WN, Sillevis Smitt PA, Wertenbroek AA, Bromberg JE, et al. Treatment of large low-grade oligodendroglial tumors with upfront procarbazine, lomustine, and vincristine chemotherapy with long follow-up: a retrospective cohort study with growth kinetics. *J Neuro-Oncol.* 2015;121(2):365–372.

91. Avila EK, Chamberlain M, Schiff D, Reijneveld JC, Armstrong TS, Ruda R, et al. Seizure control as a new metric in assessing efficacy of tumor treatment in low-grade glioma trials. *Neuro-Oncology.* 2017;19(1):12–21.

Chapter 19

Surgical results in anaplastic oligodendroglioma (AO) and anaplastic oligoastrocytoma (AOA)

Joshua L. Wang, Candice Carpenter, Ahmed Mohyeldin and J. Bradley Elder

Department of Neurological Surgery, The Ohio State University Wexner Medical Center, The James Cancer Hospital and Solove Research Institute, Columbus, OH, United States

Introduction

Despite improved radiation therapy protocols and chemotherapeutic regimens, neurosurgical intervention has remained the first step in effective management of oligodendrogliomas, of which 30% have anaplastic characteristics. Genome sequence profiling has ushered in a new era of cancer research that has led to the systematic clustering and cataloguing of oligodendrogliomas, not based on histopathology, but based on molecular phenotype. For patients with oligodendroglioma, treatment decisions and prognosis are predicated on these molecular findings, as well as other factors including location, size, grade, and extent of surgical resection. In addition to establishing a diagnosis and obtaining tissue for genome sequence profiling, surgery for accessible lesions provides immediate relief from increased intracranial pressure and neurologic symptoms secondary to tumor mass effect. Based on the most recent WHO guidelines, anaplastic oligoastrocytoma (AOA) is no longer a recommended diagnosis and can only be made on histopathology in the absence of adequate tissue for genomic characterization. With these recommendations noted, AOA are still discussed in this chapter in the context of data published prior to widespread routine genome analysis for these tumors.

Imaging

Magnetic resonance imaging (MRI) is considered the "gold standard" for diagnosing brain tumors, including anaplastic oligodendroglioma, and is the best noninvasive anatomic imaging modality available. Oligodendrogliomas are typically found in the cortical and/or subcortical white matter of the frontal or the temporal lobe. Radiographic appearance of WHO grade III anaplastic oligodendrogliomas (AO) is highly variable and appearance of the tumors using MRI can be similar to low-grade oligodendrogliomas, which do not reliably enhance with contrast, or to glioblastomas (GBM), which typically demonstrate irregular ring enhancement with nodularity with a central, nonenhancing region of necrosis. The AO tend to demonstrate more intense contrast enhancement, and may have more heterogeneity and infiltrative features, compared to low-grade gliomas. The heterogeneous radiographic appearance of AO may lead to different strategies in terms of surgical planning and goals of extent of resection (EOR), compared to their grade 2 or 4 counterparts. For example, for enhancing tumors that appear similar to grade 4 GBM, the objective of surgical resection is maximal removal of the enhancing lesion and central necrosis while minimizing potential for neurologic morbidity. Although there may be prognostic value in removing any T2/FLAIR high signal surrounding the enhancing component of the tumor, generally the surrounding FLAIR is presumed to be edema related to rapid tumor growth and is not targeted for resection. On the other hand, the radiographic appearance of AO may more closely mirror WHO grade II oligodendrogliomas, which are typically hypointense on T1-weighted MRI and hyperintense on T2-weighted MRI, often appearing fairly well demarcated (Fig. 1). For patients who present with nonenhancing tumors presumed to be grade 2 gliomas, surgical resection is targeted at the removal of all T2/FLAIR-hyperintense signal tissues. In this setting, complete resection of the T2/FLAIR signal combined with demographic factors and molecular findings may allow the patient to be classified as "low risk," and therefore avoid immediate postoperative adjuvant chemotherapy and radiation in lieu of serial imaging. The prognostic benefit related to resection of

Oligodendroglioma. https://doi.org/10.1016/B978-0-12-813158-9.00019-0

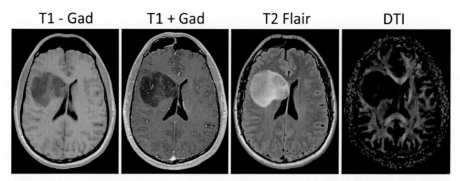

T1 - Gad   T1 + Gad   T2 Flair   DTI

FIG. 1 Representative T1-weighted (before and after gadolinium administration), FLAIR and DTI MRI sequences of a patient who presented with seizures due to a right frontal anaplastic oligodendroglioma. MRI features more closely resemble a grade 2 tumor rather than a grade 4 tumor. For example, the lesion appears to have distinct borders and white matter tracts are splayed around the tumor, rather than disrupted by the tumor.

anaplastic oligodendroglioma based on preoperative MRI findings remains unclear. As with GBM, resection of enhancing foci of tumor while minimizing neurologic morbidity is typically the primary goal of initial surgery. However, the prognostic benefit of resecting any surrounding FLAIR-hyperintense signal for enhancing tumors remains unclear, although extrapolation from other studies would suggest that the so-called "supramaximal" resection may offer benefit in the cases of AO.[1, 2] Ultimately, as with all histologic grades of oligodendroglioma, the invasive nature of AO makes total resection of all viable tumor cells difficult, if not impossible, despite complete resection of all radiographically apparent tumors.

Other MR imaging modalities may help with surgical planning, including diffusion tensor imaging (DTI) and functional MRI. The DTI typically shows white matter tracts splayed around oligodendrogliomas, rather than frank invasion. Peritumoral edema and mass effect are usually minimal or absent, regardless of the size of the tumor. The diagnosis of oligodendrogliomas is suggested preoperatively when intratumoral or peritumoral calcification is seen on neuroimaging (up to 90% cases). The use of advanced imaging modalities as surgical adjuncts is further discussed below.

Role of surgery

Ideally, the first step in effective management of AO/AOA is maximal safe surgical resection of the tumor. This provides extensive cytoreduction while preserving normal brain tissue and neurologic function. Surgical resection serves as both a diagnostic and therapeutic intervention by providing tissue for histopathological examination and genomic sequencing, which will guide adjuvant treatment decisions, and by rapidly relieving increased intracranial pressure and/or tumor mass effect. Other benefits of surgery may include improved seizure control and improvement in preoperative neurologic symptoms.

For some patients, factors such as medical co-morbidities and anatomic location of the tumor may preclude open surgery. All surgical intervention under general anesthesia is associated with perioperative risk, and patients with significant cardiovascular, pulmonary, or other medical co-morbidities may face an unacceptably high risk of medical morbidity or mortality due to the stress associated with undergoing anesthesia and a major surgery. Alternatively, patients who are healthy enough to tolerate surgery may not be candidates for open resection due to the anatomic location of their tumor. For example, multifocal and multicentric tumors may preclude maximal resection due to the morbidity inherent with multiple craniotomies and multiple surgical corridors within the brain. A more extreme, and uncommon, example of this is gliomatosis due to anaplastic oligodendroglioma.[3–5] A more common scenario that may preclude open surgery is tumor located within an eloquent area of the brain such as the primary motor cortex, Wernicke's area, or Broca's area. Although brain-mapping techniques can help maximize tumor resection in eloquent areas, in some cases tumors extensively involve an eloquent region of the brain and aggressive surgery may result in unacceptable permanent neurologic deficits such as aphasia or paralysis. Deep-seated tumors, such as those in the thalamus or brain stem, where access to the tumor requires traversing a significant distance through normal brain tissue, may not merit consideration for surgical resection due to associated morbidity of the surgical approach and eloquence of the target anatomy. In each of these anatomic scenarios, stereotactic biopsy is typically recommended with the goal of achieving diagnosis and providing tissue for molecular analysis, while minimizing the perioperative risk to the patient and preserving eloquent brain (Fig. 2). Additionally, patients may undergo biopsy initially to achieve diagnosis when the radiographic differential diagnosis is uncertain, with potential for open surgical resection if biopsy confirms tumor. Surgical risks of stereotactic biopsy are low, with reported mortality rates generally less than 1%. A downside of stereotactic biopsy is the potential for sampling error, due to the potential

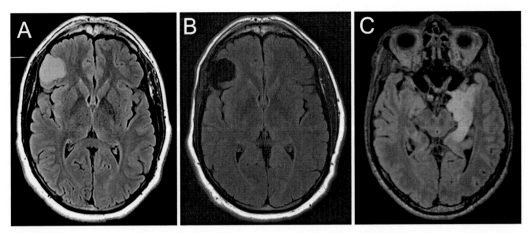

FIG. 2 FLAIR MRI sequences demonstrating how the anatomic location of the tumor helps determine the surgical strategy. The patient with the right frontal anaplastic oligodendroglioma (A) underwent gross total resection of the tumor (B), whereas the patient with the left temporal anaplastic oligoastrocytoma (C) underwent stereotactic biopsy only given the eloquent location and potential for significant neurologic morbidity with resection.

heterogeneous nature of these tumors, which may lead to "under-grading" or "over-grading" of the tumor. Because anaplastic oligodendroglioma may have clinical and radiographic features similar to either of the more commonly diagnosed grade 2 oligodendroglioma or glioblastoma, the risk of misdiagnosis due to sampling error may be higher. Sampling from multiple sites within the tumor may help mitigate this risk.

Surgical outcomes

Maximally safe resection is the goal of initial surgical intervention.[6] Extent of resection has been identified as a prognostic factor in several large clinical trials for low-grade and high-grade gliomas, though there are no randomized trials that focus on the role of extent of surgery specifically for AO. An analysis of the European Organisation for Research and Treatment of Cancer (EORTC) Trial 26951 data of AO patients ($n = 368$) showed that extensive and confirmed gross total resection (GTR) of AO ($n = 109$ of 368) without residual enhancing tumor on postoperative imaging had a significantly positive prognostic value for both progression-free survival [PFS, hazard ratio (HR) 0.437, 95% confidence interval (CI) 0.298–0.639, $P < .0001$) and overall survival (OS, HR 0.408, 95% CI 0.274–0.607, $P < .0001$), in comparison to both resection with residual enhancing tumor ($n = 208$) and biopsy alone ($n = 51$) (median PFS 43.5 vs 17.2 vs 6.9 months, respectively; median OS 70.9 vs 34.4 vs 16.2 months, respectively).[7, 8] Similarly, the Radiation Therapy Oncology Group (RTOG) 9402 trial demonstrated debulking surgery for AO (any nonbiopsy-only surgical resection) as initial intervention was independently associated with improved OS in a multivariate Cox regression compared to biopsy alone.[9] Wick and colleagues reported in the NOA-04 phase III trial of 318 patients with anaplastic gliomas [39 AO, 91 AOA, 144 anaplastic astrocytoma (AA)] that extent of resection (complete, incomplete, or biopsy alone) was a significant prognostic factor for PFS ($P = .0006$).[10] Complete versus incomplete resection trended toward improved PFS (HR 1.6, 95% CI 0.9–3.0), whereas complete and incomplete resection versus biopsy alone were both associated with significantly improved PFS (HR 2.1, 95% CI 1.1–4.0; and HR 3.5, 95% CI 1.8–7.0, respectively). By contrast, Puduvalli and colleagues analyzed data from 106 AO patients and did not find statistically significant differences in outcomes such as OS and PFS when stratifying based on extent of resection. The authors observed a trend toward longer survival with greater extent of resection, and felt that the lack of statistical significance was primarily due to the heterogeneity of the postoperative adjuvant treatments, which included chemotherapy or radiotherapy alone, chemotherapy followed by radiotherapy, radiotherapy followed by chemotherapy, and others.[11]

A retrospective analysis of extent of resection for 122 WHO grade III gliomas (81 AA or AOA, 41 AO) utilizing intraoperative MRI (iMRI) demonstrated a significant survival advantage with resection of >53% of the preoperative T2-weighted high-signal intensity volume in patients with AA or AOA (HR 3.03, 95% CI 1.20–7.21, $P = .021$) but was not able to establish a similar threshold for patients with AO.[12] The survival benefit associated with increased extent of resection was not demonstrated at lower extent of resection thresholds, suggesting that a minimum extent of resection is necessary to observe increasing survival benefit. When considering all WHO grade III gliomas together, resection of >60% of the preoperative T2-weighted high-signal intensity volume was associated with improved OS (HR 2.80, 95% CI 1.27–5.92, $P = .012$).

Much of the literature guiding surgical management of AO refers to or borrows from high-grade glioma studies (HGG), which may include AO, AOA, AA, and GBM, depending on the study, from which treatment guidelines are extrapolated about AO and AOA specifically. For the most part, higher quality studies suggest that maximal safe surgical resection is central to optimizing outcomes and maximizing survival for patients with HGG in general.[13, 14] Mounting retrospective data suggests that the extent of HGG resection is a significant prognostic factor for predicting OS, though there remains a lack of high-quality prospective trials evaluating extent of resection. In one of the larger retrospective series, Sanai and colleagues established an extent of resection threshold for 500 consecutive patients with newly diagnosed GBM.[15] They reported a significant survival advantage with ≥78% extent of resection ($P < .0001$, median OS 12.5 months) and stepwise improvement in survival with increasing extent of resection, even in the 90%–100% (median OS 13.8–16 months) range. Although this study does not specifically evaluate patients with AO or AOA, it is one of the larger and often referenced manuscripts and serves as a template for studying and understanding surgical outcomes in glioma patients in general. Other studies examining the role of extent of resection in prognosis of HGG are briefly described in Table 1, and further discussion in this chapter will focus on the few prospective reports.

In a prospective study of 361 consecutive patients with newly diagnosed GBM, Oszvald and colleagues reported that increasing age was a negative prognosticator in patients undergoing biopsy but not in patients receiving surgical resection, and concluded that resection should not be withheld from patients on the basis of age alone.[22] Although AO patients are typically younger than GBM patients, this study underscores the importance of surgical resection even in patients with characteristics typically associated with poorer prognoses.

Buckner and colleagues demonstrated in a phase III multicenter randomized controlled trial of 383 patients with newly diagnosed HGG (18 AOA, 49 AA, 293 GBM) that greater extent of surgery was a favorable prognostic variable in their Cox multivariate regression models.[28]

Three studies have examined the role of supratotal resection, which is typically defined as resection of all T1 contrast-enhancing tissue as well as resection of at least a portion of the surrounding, nonenhancing, T2/FLAIR-hyperintense tissue, in patients with HGG (primarily GBM). These three studies comprise a total of 58 patients who had supratotal resection and 361 nonsupratotal controls.[2, 34, 35] In Eyupoglu and colleague's series, 30 patients prospectively selected to receive supratotal resection had significantly increased median OS of 18.5 months (HR 0.449, 95% CI 0.289–0.696, $P = .0004$)

TABLE 1 Summary of studies discussing extent of resection for management of high-grade gliomas

Study	Design	n	Tumor type(s)	Key results and conclusions
Fuji et al. (2018)[12]	Retrospective, single center	122	Newly diagnosed AA or AOA (81), AO[16]	EOR of T2-hyperintense volume was one of the most important prognostic factors in AA and AOA, but not AO, patients; significant survival advantage with 53% T2-EOR
Im et al. (2018)[17]	Retrospective, single center	73	AO,[18] AA (48)	Maximal surgical resection associated with favorable prognosis
Padwal et al. (2016)[19]	SEER data	24717	AA (2755), GBM (21962)	Gross total resection associated with improved survival for both AA and GBM, but the survival advantage is greater in AA patients
Narita et al. (2015)[20]	Retrospective, multicenter	13431	Primary brain tumors: AO (126), AOA (106), AA (513), GBM (1489)	EOR ≥50% in AA, AO, AOA, and GBM patients significantly prolonged OS
Panageas et al. (2014)[21]	Retrospective, single center	587	AO (337), AOA (250)	Extent of resection (debulking surgery vs biopsy) was not a prognostic factor
Oszvald et al. (2012)[22]	Prospective, consecutive patients, single center	361	GBM	Increasing age is a negative prognostic factor in patients undergoing biopsy, but not in those receiving surgical resection
Sanai et al. (2011)[15]	Retrospective, consecutive patients, single center	500	Newly-diagnosed supratentorial GBM	EOR predictive of survival; significant survival advantage with 78% EOR, with stepwise improvement even in 95%–100% EOR range

TABLE 1 Summary of studies discussing extent of resection for management of high-grade gliomas—cont'd

Study	Design	n	Tumor type(s)	Key results and conclusions
McGirt et al. (2009)[23]	Retrospective, single center	1215	AOA (82), AA (167), GBM (700)	Increasing EOR associated with improved OS in primary and recurrent malignant astrocytomas, independent of age, performance status, WHO grade, or subsequent therapy
Wick et al. (2009)[10]	Phase III RCT, multicenter	318	Anaplastic gliomas	EOR was a significant prognostic factor for OS
Nomiya et al. (2007)[24]	Retrospective, single center	170	AA	Greater EOR (total or subtotal resection vs partial resection or biopsy) was the most important prognostic factor associated with increased OS and survival rate
Keles et al. (2006)[25]	Retrospective, single center	102	Hemispheric AA	Residual tumor volume associated with PFS and OS
RTOG et al. (2006)[9]	Phase III RCT, multicenter	289	AO, AOA	Debulking surgery was associated with increased OS
Stummer et al. (2006)[26]	Phase III RCT, multicenter	322	Malignant gliomas	5-ALA was associated with increased rates of total resection, which was associated with increased PFS
van de Bent et al. (2006)[8]	Phase III RCT, multicenter	368	AO, AOA	Confirmed resection of all enhancing tumor was a positive prognostic factor for PFS and OS
Pope et al. (2005)[27]	Retrospective, single center	153	Grade III[13] gliomas, GBM (110)	EOR "was not a statistically meaningful predictor of survival"
Ushio et al. (2005)[16]	Retrospective, consecutive patients, single center	105	Hemispheric GBM	Gross total resection was associated with longer PFS and OS compared to partial resection or biopsy; no difference between partial resection and biopsy
Puduvalli et al. (2003)[11]	Retrospective, multicenter	106	AO	Greater EOR showed a trend towards longer survival
Buckner et al. (2001)[28]	Phase III RCT, multicenter	383	Newly-diagnosed HGG	Greater EOR was a favorable prognostic variable in Cox multivariate regression model
Lacroix et al. (2001)[29]	Retrospective, consecutive patients, single center	416	GBM	Gross total resection was associated with longer OS
Keles et al. (1999)[30]	Retrospective, single center	92	Hemispheric GBM	EOR and residual tumor volume were both associated with PFS and OS
Kowalczuk et al. (1997)[31]	Retrospective, single center	75	AA,[32] GBM (52)	Aggressive surgical resection not associated with increased OS
Nitta and Sato (1995)[33]	Retrospective, single center	101	Supratentorial malignant gliomas	Total resection was associated with higher rate of survival compared to subtotal or partial resection; no survival difference between subtotal and partial resection

5-ALA, 5-aminolevulinic acid; *AA*, anaplastic astrocytoma; *AO*, anaplastic oligodendroglioma; *AOA*, anaplastic oligoastrocytoma; *EOR*, extent of resection; *GBM*, glioblastoma; *HGG*, high-grade gliomas; *OS*, overall survival; *RCT*, randomized controlled trial; *SEER*, surveillance, epidemiology, and end results database.

compared to retrospective controls receiving "only" GTR with median OS of 14 months. Pessina and colleagues reported 21 patients receiving supratotal resection in a retrospective series of 282 newly diagnosed GBM and found that extent of resection was associated with improved OS (supratotal resection 28.6 months vs GTR 16.2 months). Additionally, they identified a FLAIR removal threshold of 45% that was associated with improved OS at 2 years (54% with FLAIR resection volume >45% vs 12% with FLAIR resection volume <45%). Similarly, Glenn and colleagues reported that supratotal resection was associated with improved PFS [median 15 months vs 7 months for GTR and 6 months for subtotal resection (STR); HR 0.093; 95% CI, 0.01–0.89; $P = .039$) and OS (median 24 months vs 11 months GTR and 9 months STR; HR, 0.169; 95% CI, 0.05–0.57; $P < .004$) in their series of 32 consecutive temporal lobe GBM patients, of which 7 received supratotal resection. Of note, these studies excluded patients with tumors in eloquent or eloquent-adjacent areas, due to the risk of new neurologic deficit associated with supratotal resection in such locations. These three studies are potentially important for patients with AO or AOA, given the heterogeneous nature of MRI findings at presentation, and suggest a role for aggressive surgical resection of not only the contrast-enhancing material, but also the surrounding T2/FLAIR signal in patients with suspected AO/AOA.

A meta-analysis of biopsy versus surgical resection for malignant gliomas (31 WHO grade III, 1073 grade IV) showed that surgical resection provided superior OS (HR 0.61, 95% CI, 0.52–0.71, $P < .0001$), and further suggested improved quality of life in patients receiving surgical resection. However, in this study PFS was not impacted by extent of resection.[36]

Despite maximal intervention for newly diagnosed AO, roughly half of the AO patients have disease recurrence following initial therapy (Fig. 3).[11] Although re-resection of recurrent HGG certainly provides relief of progressive symptoms, evidence of impact on OS varies widely and there is no consensus on the role of reoperation for recurrent HGG. Moreover, much of the data is focused on GBM, and no prospective randomized studies have addressed the utility of reoperation for recurrent AO or AOA. A recent systematic review examining the benefit of repeat surgery for recurrent GBM analyzed 33 studies (2717 patients). From these studies, the authors reported an average median OS of 9.9 months following reoperation. This compared favorably to published data reporting a range of median OS of 5.8–8.1 months after disease recurrence, although only 20 out of 30 included studies noted a survival benefit of reoperation.[37] Additionally, 16 studies examined the role of extent of re-resection, of which 10 identified an associated survival benefit of greater extent of resection. Of note, Bloch and colleagues reported in their study of 107 patients, the survival disadvantage associated with incomplete resection at initial surgery could be salvaged at the time of disease recurrence. They classified patients based on extent of resection at first occurrence and second occurrence as either GTR or STR. Median OS for these four groups was: GTR/GTR 20.4 months, GTR/STR 18.4 months, STR/GTR 19.0 months, STR/STR 15.9 months; $P = .004$ between STR/GTR and STR/STR, no significant difference between GTR/STR and STR/GTR).[38] Oppenlander and colleagues identified a significant survival advantage associated with as little as 80% extent of resection (median OS 20.0 months for those

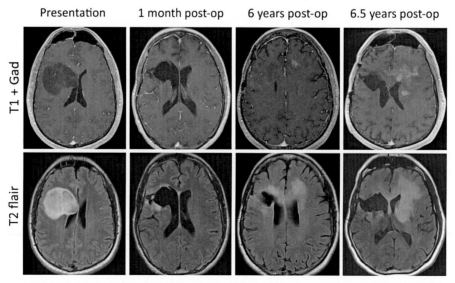

| Presentation | 1 month post-op | 6 years post-op | 6.5 years post-op |

FIG. 3 Representative T1-weighted, postgadolinium and FLAIR MRI sequences for the patient from Fig. 1. Difficulty in demonstrating postoperative edema from residual tumor is demonstrated in comparing the preoperative MRI to the one month postoperative MRI. Serial imaging was largely stable for nearly 6 years until the patient developed new seizures and an MRI revealed FLAIR signal emanating across the corpus callosum and a new area of gadolinium enhancement in the left frontal lobe. Stereotactic biopsy of this enhancing lesion confirmed recurrent anaplastic oligodendroglioma. Despite adjuvant temozolomide, subsequent imaging revealed further radiographic progression and the patient died 7 years after the initial surgery.

receiving at least 81% resection vs 19.0 months for the entire cohort of 170 patients) and stepwise improvement in OS through even in the 95%–100% resection range (97% resection equated to median OS of 30.0 months).[18] Thus, although prospective trials are lacking that could more definitively address the question regarding the value of aggressive surgical management at the time of recurrence, retrospective data increasingly supports the survival advantage associated with surgery. Like other surgical data for GBM, this is likely directly applicable to anaplastic gliomas such as AO and AOA.

Surgical adjuncts

Numerous surgical adjuncts, such as image-guided frameless navigation, iMRI, intraoperative ultrasound, and fluorescent tumor marking with 5-aminolevulinic acid (5-ALA), have been developed to assist in maximizing extent of resection while minimizing potential for neurologic morbidity associated with surgery. Current guidelines recommend maximal safe resection, with the use of surgical adjuncts such as 5-ALA intraoperatively to increase the rate of complete resection.[39]

Preoperative planning and risk assessment can be performed based on advanced MRI techniques that allow for the localization of eloquent speech and motor cortices (fMRI) and visualization of critical white matter tracts that may be splayed by the tumor (DTI). Specifically, a surgical exposure and trajectory toward the tumor can be planned such that eloquent cortical areas and white matter tracts are avoided. In 2007, Wu and colleagues performed a prospective randomized controlled trial to evaluate the impact of DTI versus standard neuronavigation in patients with gliomas.[40] The use of DTI in operative planning for high-grade gliomas (10 AO, 19 AA, 49 GBM) allowed for increased extent of resection (74.4% vs 33.3%, $P < .001$), higher Karnofsky Performance Scores at 6 months, and improved OS (median 21.2 vs 14.0 months; HR 0.570, 95% CI 0.325–1.003, $P = 0.05$).

Intraoperative mapping techniques for tumors in or near functionally eloquent cortex can help minimize neurologic morbidity and improve extent of resection. Awake surgery for mapping of language pathways may be required for tumors in the dominant temporal lobe, angular gyrus, or frontal operculum. Motor and sensory mapping can be performed awake or asleep. Although patients almost invariably prefer to be asleep if given a choice, mapping is more sensitive and accurate if the patient is awake and able to provide real-time feedback. Cortical and subcortical stimulation mapping can detect functional areas and pathways to help minimize motor and sensory morbidity. In a series of 611 consecutive patients undergoing brain tumor surgery, Hervey-Jumper and colleagues reported lower than previously published failure rates, defined as the inability to complete cortical stimulation mapping and tumor resection (0.5% vs 2.3%–6.4%) and complication rates (10% vs 14%–32%).[41] The main causes of failure in patients undergoing awake craniotomy include intraoperative seizures caused by localization stimulation, lack of effective intraoperative communication with the patient, and emotional intolerance.[32, 42] While rates of new neurologic deficit following awake craniotomy have been reported between 15% and 38%, Trinh and colleagues found that 67% of intraoperative deficits resolved within 3 months in their series of 214 patients (75 AO or AA, 87 GBM or gliosarcoma, 52 low-grade tumors), resulting in a 6% rate of persistent neurologic deficit at 3 months.[43]

5-ALA is a nonfluorescent orally administered prodrug that elicits synthesis and accumulation of fluorescent porphyrins in various epithelial and cancerous tissues, including in malignant gliomas. Porphyrin fluorescence can be visualized intraoperatively with a neurosurgical microscope with fluorescence capabilities and can aid in the detection of residual malignant glioma during resection. Stummer and colleagues reported in a phase III multicenter randomized controlled trial of 322 patients that the use of the surgical adjunct 5-ALA for intraoperative identification and resection of gliomas in comparison to conventional microsurgical techniques resulted in significantly improved rates of total resection of contrast-enhancing tumor, which translated into prolonged PFS (41.0% at 6 months vs 21.1%, median PFS 5.1 vs 3.6 months), but no improvement in OS.[26] Of note, most patients in this trial had pathologic diagnoses of GBM, and only three patients had AO or AOA. Della Pappa and colleagues also prospectively studied the use of 5-ALA in 31 patients with HGG (2 AO, 4 AA, 25 GBM) located in eloquent areas and found that complete resection of contrast-enhancing tumor was achieved in 74% of patients when 5-ALA and intraoperative monitoring were used. Currently, there are continuing clinical trials examining the role of 5-ALA in the resection of AO and other HGG.

Of the previously mentioned surgical adjuncts, frameless navigation has the most widespread use and is found in neurosurgical operating rooms worldwide. However, it is inherently limited due to intraoperative brain shifts during the normal course of surgery, which cannot be represented on the preoperative navigational scan. Intraoperative ultrasound and iMRI are less susceptible to intraoperative brain shifts, can distinguish tumor from surrounding edema or normal brain during surgery, and may improve extent of resection, especially for tumors that visually are difficult to distinguish from normal brain. iMRI was initially developed in 1997 to treat a wide range of intracranial lesions[44] and has since evolved in machine and workflow design and image quality. While the majority of the data surrounding the role of iMRI in glioma treatment has been obtained in patients with GBM and low-grade glioma, these data can likely be extrapolated to apply to AO patients as well. Senft and colleagues performed the only randomized controlled trial evaluating the benefit of iMRI compared to conventional neuronavigation in glioma surgery in 49 patients (1 AO, 1 AA, 44 GBM, 3 other).[45] They found that patients randomized to the iMRI group ($n = 24$) had a higher likelihood of GTR [23 iMRI patients received GTR (96%) vs 17 control patients (68%); OR 0.092, 95% CI 0.014-0.602, $P = .023$). Rates of new neurologic deficit were similar between the two groups and extent of resection was associated with improved OS; however, being in the iMRI group did not result in improved survival. Additionally, in a series of 170 consecutive GBM surgeries using iMRI, Coburger and colleagues reported improved rates of GTR and OS and decreased complication rates when compared to historical controls.[46]

Conclusions

For patients with anaplastic oligodendroglioma or anaplastic oligoastrocytoma, specific data regarding the value of aggressive surgical resection is sparse. Surgical resection remains the mainstay in initial treatment of AO, and existing data regarding AO/AOA combined with extrapolation from the literature regarding HGG and LGG suggests that maximizing extent of resection at diagnosis while minimizing neurologic morbidity leads to the best survival outcomes for these patients. Surgical adjuncts that can assist with this goal include intraoperative MRI, mapping of eloquent brain anatomy, and use of 5-ALA.

References

1. de Leeuw CN, Vogelbaum MA. Supratotal resection in glioma: a systematic review. *Neuro-Oncology*. 2018;.
2. Glenn CA, Baker CM, Conner AK, et al. An examination of the role of supramaximal resection of temporal lobe glioblastoma multiforme. *World Neurosurg*. 2018;114:e747–e755.
3. Glas M, Bahr O, Felsberg J, et al. NOA-05 phase 2 trial of procarbazine and lomustine therapy in gliomatosis cerebri. *Ann Neurol*. 2011;70:445–453.
4. Levin N, Gomori JM, Siegal T. Chemotherapy as initial treatment in gliomatosis cerebri: results with temozolomide. *Neurology*. 2004;63:354–356.
5. Sega S, Horvat A, Popovic M. Anaplastic oligodendroglioma and gliomatosis type 2 in interferon-beta treated multiple sclerosis patients. Report of two cases. *Clin Neurol Neurosurg*. 2006;108:259–265.
6. Sanai N, Berger MS. Glioma extent of resection and its impact on patient outcome. *Neurosurgery*. 2008;62:753–764. discussion 264–756.
7. Gorlia T, Delattre JY, Brandes AA, et al. New clinical, pathological and molecular prognostic models and calculators in patients with locally diagnosed anaplastic oligodendroglioma or oligoastrocytoma. A prognostic factor analysis of European Organisation for Research and Treatment of Cancer Brain Tumour Group Study 26951. *Eur J Cancer*. 2013;49:3477–3485.
8. van den Bent MJ, Carpentier AF, Brandes AA, et al. Adjuvant procarbazine, lomustine, and vincristine improves progression-free survival but not overall survival in newly diagnosed anaplastic oligodendrogliomas and oligoastrocytomas: a randomized European Organisation for Research and Treatment of Cancer phase III trial. *J Clin Oncol*. 2006;24:2715–2722.
9. Intergroup Radiation Therapy Oncology Group T, Cairncross G, Berkey B, et al. Phase III trial of chemotherapy plus radiotherapy compared with radiotherapy alone for pure and mixed anaplastic oligodendroglioma: Intergroup Radiation Therapy Oncology Group Trial 9402. *J Clin Oncol*. 2006;24:2707–2714.
10. Wick W, Hartmann C, Engel C, et al. NOA-04 randomized phase III trial of sequential radiochemotherapy of anaplastic glioma with procarbazine, lomustine, and vincristine or temozolomide. *J Clin Oncol*. 2009;27:5874–5880.
11. Puduvalli VK, Hashmi M, McAllister LD, et al. Anaplastic oligodendrogliomas: prognostic factors for tumor recurrence and survival. *Oncology*. 2003;65:259–266.
12. Fujii Y, Muragaki Y, Maruyama T, et al. Threshold of the extent of resection for WHO Grade III gliomas: retrospective volumetric analysis of 122 cases using intraoperative MRI. *J Neurosurg*. 2018;129:1–9.
13. Watts C. Surgical management of high-grade glioma: a standard of care. *CNS Oncol*. 2012;1:181–192.
14. Watts C, Price SJ, Santarius T. Current concepts in the surgical management of glioma patients. *Clin Oncol (R Coll Radiol)*. 2014;26:385–394.
15. Sanai N, Polley MY, McDermott MW, Parsa AT, Berger MS. An extent of resection threshold for newly diagnosed glioblastomas. *J Neurosurg*. 2011;115:3–8.
16. Ushio Y, Kochi M, Hamada J, Kai Y, Nakamura H. Effect of surgical removal on survival and quality of life in patients with supratentorial glioblastoma. *Neurol Med Chir (Tokyo)*. 2005;45:454–460. discussion 460–451.
17. Im JH, Hong JB, Kim SH, et al. Recurrence patterns after maximal surgical resection and postoperative radiotherapy in anaplastic gliomas according to the new 2016 WHO classification. *Sci Rep*. 2018;8(777).
18. Oppenlander ME, Wolf AB, Snyder LA, et al. An extent of resection threshold for recurrent glioblastoma and its risk for neurological morbidity. *J Neurosurg*. 2014;120:846–853.
19. Padwal JA, Dong X, Hirshman BR, Hoi-Sang U, Carter BS, Chen CC. Superior efficacy of gross total resection in anaplastic astrocytoma patients relative to glioblastoma patients. *World Neurosurg*. 2016;90:186–193.
20. Narita Y, Shibui S. Committee of brain tumor registry of Japan supported by the Japan Neurosurgical S: trends and outcomes in the treatment of gliomas based on data during 2001-2004 from the Brain Tumor Registry of Japan. *Neurol Med Chir (Tokyo)*. 2015;55:286–295.

21. Panageas KS, Reiner AS, Iwamoto FM, et al. Recursive partitioning analysis of prognostic variables in newly diagnosed anaplastic oligodendroglial tumors. *Neuro-Oncology*. 2014;16:1541–1546.

22. Oszvald A, Guresir E, Setzer M, et al. Glioblastoma therapy in the elderly and the importance of the extent of resection regardless of age. *J Neurosurg*. 2012;116:357–364.

23. McGirt MJ, Chaichana KL, Gathinji M, et al. Independent association of extent of resection with survival in patients with malignant brain astrocytoma. *J Neurosurg*. 2009;110:156–162.

24. Nomiya T, Nemoto K, Kumabe T, Takai Y, Yamada S. Prognostic significance of surgery and radiation therapy in cases of anaplastic astrocytoma: retrospective analysis of 170 cases. *J Neurosurg*. 2007;106:575–581.

25. Keles GE, Chang EF, Lamborn KR, et al. Volumetric extent of resection and residual contrast enhancement on initial surgery as predictors of outcome in adult patients with hemispheric anaplastic astrocytoma. *J Neurosurg*. 2006;105:34–40.

26. Stummer W, Pichlmeier U, Meinel T, et al. Fluorescence-guided surgery with 5-aminolevulinic acid for resection of malignant glioma: a randomised controlled multicentre phase III trial. *Lancet Oncol*. 2006;7:392–401.

27. Pope WB, Sayre J, Perlina A, Villablanca JP, Mischel PS, Cloughesy TF. MR imaging correlates of survival in patients with high-grade gliomas. *AJNR Am J Neuroradiol*. 2005;26:2466–2474.

28. Buckner JC, Schomberg PJ, McGinnis WL, et al. A phase III study of radiation therapy plus carmustine with or without recombinant interferon-alpha in the treatment of patients with newly diagnosed high-grade glioma. *Cancer*. 2001;92:420–433.

29. Lacroix M, Abi-Said D, Fourney DR, et al. A multivariate analysis of 416 patients with glioblastoma multiforme: prognosis, extent of resection, and survival. *J Neurosurg*. 2001;95:190–198.

30. Keles GE, Anderson B, Berger MS. The effect of extent of resection on time to tumor progression and survival in patients with glioblastoma multiforme of the cerebral hemisphere. *Surg Neurol*. 1999;52:371–379.

31. Kowalczuk A, Macdonald RL, Amidei C, et al. Quantitative imaging study of extent of surgical resection and prognosis of malignant astrocytomas. *Neurosurgery*. 1997;41:1028–1036 discussion 1036–1028.

32. Nossek E, Matot I, Shahar T, et al. Intraoperative seizures during awake craniotomy: incidence and consequences: analysis of 477 patients. *Neurosurgery*. 2013;73:135–140. discussion 140.

33. Nitta T, Sato K. Prognostic implications of the extent of surgical resection in patients with intracranial malignant gliomas. *Cancer*. 1995;75:2727–2731.

34. Eyupoglu IY, Hore N, Merkel A, Buslei R, Buchfelder M, Savaskan N. Supra-complete surgery via dual intraoperative visualization approach (DiVA) prolongs patient survival in glioblastoma. *Oncotarget*. 2016;7:25755–25768.

35. Pessina F, Navarria P, Cozzi L, et al. Maximize surgical resection beyond contrast-enhancing boundaries in newly diagnosed glioblastoma multiforme: is it useful and safe? A single institution retrospective experience. *J Neuro-Oncol*. 2017;135:129–139.

36. Tsitlakidis A, Foroglou N, Venetis CA, Patsalas I, Hatzisotiriou A, Selviaridis P. Biopsy versus resection in the management of malignant gliomas: a systematic review and meta-analysis. *J Neurosurg*. 2010;112:1020–1032.

37. Robin AM, Lee I, Kalkanis SN. Reoperation for recurrent glioblastoma multiforme. *Neurosurg Clin N Am*. 2017;28:407–428.

38. Bloch O, Han SJ, Cha S, et al. Impact of extent of resection for recurrent glioblastoma on overall survival: clinical article. *J Neurosurg*. 2012;117:1032–1038.

39. Stupp R, Brada M, van den Bent MJ, Tonn JC, Pentheroudakis G, Group EGW. High-grade glioma: ESMO clinical practice guidelines for diagnosis, treatment and follow-up. *Ann Oncol*. 2014;25(Suppl. 3):iii93–iii101.

40. Wu JS, Zhou LF, Tang WJ, et al. Clinical evaluation and follow-up outcome of diffusion tensor imaging-based functional neuronavigation: a prospective, controlled study in patients with gliomas involving pyramidal tracts. *Neurosurgery*. 2007;61:935–948 discussion 948–939.

41. Hervey-Jumper SL, Li J, Lau D, et al. Awake craniotomy to maximize glioma resection: methods and technical nuances over a 27-year period. *J Neurosurg*. 2015;123:325–339.

42. Nossek E, Matot I, Shahar T, et al. Failed awake craniotomy: a retrospective analysis in 424 patients undergoing craniotomy for brain tumor. *J Neurosurg*. 2013;118:243–249.

43. Trinh VT, Fahim DK, Shah K, et al. Subcortical injury is an independent predictor of worsening neurological deficits following awake craniotomy procedures. *Neurosurgery*. 2013;72:160–169.

44. Black PM, Moriarty T, Alexander 3rd E, et al. Development and implementation of intraoperative magnetic resonance imaging and its neurosurgical applications. *Neurosurgery*. 1997;41:831–842. discussion 842–835.

45. Senft C, Bink A, Franz K, Vatter H, Gasser T, Seifert V. Intraoperative MRI guidance and extent of resection in glioma surgery: a randomised, controlled trial. *Lancet Oncol*. 2011;12:997–1003.

46. Coburger J, Wirtz CR, Konig RW. Impact of extent of resection and recurrent surgery on clinical outcome and overall survival in a consecutive series of 170 patients for glioblastoma in intraoperative high field magnetic resonance imaging. *J Neurosurg Sci*. 2017;61:233–244.

Chapter 20

Surgical management of pediatric oligodendroglioma

Christine K. Lee* and Liliana C. Goumnerova[†]

*Department of Neurosurgery, Massachusetts General Hospital, Boston, MA, United States; [†]Department of Neurosurgery, Boston Children's Hospital, Boston, MA, United States

Introduction

Pediatric oligodendroglioma is a surgically treated disease, more so than its adult counterpart. Since the first report of oligodendrogliomas in children in 1978,[1] there have only been a small number of studies specific to these tumors, in part due to their rarity as a subtype of glioma in children. They may be classified as low-grade (WHO grade II) oligodendroglioma versus anaplastic (WHO grade III) oligodendroglioma, with significant differences in the natural history of disease, prognosis, and accordingly, surgical and medical management. There are certain distinguishing features of oligodendrogliomas in children, compared to those in adult patients, which have direct bearing on the surgical approach to these tumors as well as preoperative diagnostic evaluation and postoperative adjuvant treatments.

Overall survival outcomes are better in children with low-grade oligodendrogliomas compared to adults,[2, 3] with a reported 5-year survival rate as high as 90%. This survival advantage of children over adults does not appear to hold for anaplastic oligodendrogliomas.[2, 4] Pediatric oligodendrogliomas do not commonly harbor the 1p/19q co-deletion,[5, 6] which is an established molecular signature of adult oligodendrogliomas.[7] When present, this co-deletion, which has been associated with a favorable response to adjuvant therapy in adult patients with anaplastic oligodendrogliomas,[8, 9] does not have similar prognostic value for pediatric patients.[10] In general, pediatric oligodendrogliomas are less responsive to adjuvant treatments of chemotherapy and radiation than adult tumors, and aggressive surgical resection is the key to achieving a favorable long-term outcome.

This chapter reviews the key aspects of surgical management of pediatric oligodendrogliomas, including preoperative workup and preparation for surgery, surgical approaches and techniques, and postoperative adjuvant radiation, and medical oncologic treatments. We discuss technological adjuncts to surgery that are now the standard of care for the safe and reliable resection of these infiltrative tumors. Finally, we identify ongoing research efforts and clinical trials on the surgical management of pediatric oligodendrogliomas and other gliomas and discuss how the role of surgery is expected to change within the shifting landscape of new medical therapeutics such as immunotherapy and targeted therapies.

Epidemiology and presentation

Brain tumors are the most common solid tumor of childhood and a leading cause of pediatric cancer-related mortality.[11] Unlike adults, for whom high-grade gliomas account for the majority of primary brain tumors, children tend to have low-grade gliomas of which the most common type is cerebellar pilocytic astrocytoma. Of supratentorial hemispheric tumors in children, 60% are low-grade gliomas, with an incidence of approximately five cases per 1 million children per year.[12, 13] The majority of these supratentorial low-grade gliomas are low-grade astrocytomas, and the remainder are mixed gliomas, oligodendrogliomas, gangliogliomas, and other less common lesions such as pleomorphic xanthoastrocytomas, dysembryoplastic neuroepithelial tumors, and desmoplastic infantile gangliogliomas.[14]

Oligodendrogliomas have two incidence peaks: a smaller peak at 6–12 years and a larger peak at 35–44 years.[15–17] Only about 6%–7.5% of all oligodendrogliomas are diagnosed in children, and overall oligodendrogliomas represent a small proportion (~1%) of childhood brain tumors.[3, 16, 18, 19] In the United States, the majority of pediatric oligodendrogliomas were diagnosed in Caucasians (approximately 70% of all cases), similar to the adult patient population.[4] Oligodendrogliomas are more common among males than females in pediatric and adult patients, with a ratio of 1.2:1.[4]

Oligodendroglioma. https://doi.org/10.1016/B978-0-12-813158-9.00020-7

The presenting symptoms of pediatric oligodendrogliomas are generally similar to those of adult oligodendrogliomas, including seizures, headaches, and nausea and vomiting.[11, 20, 21] These symptoms are shared by other pediatric brain tumors in general, with the most common symptoms and signs at diagnosis being headache, nausea and vomiting, abnormalities of gait and coordination, and papilledema.[21] Young children under the age of four with intracranial tumors have different presenting symptoms, with the most common being macrocephaly, nausea and vomiting, irritability, and lethargy.[21] Thus, the differential diagnosis of a pediatric oligodendroglioma includes other intracranial glial tumors such as low-grade astrocytoma, mixed tumors such as oligoastrocytoma, mixed neuronal-glial tumors such as ganglioglioma or dysembryoplastic neuroepithelial tumors, and medulloblastoma or ependymoma for posterior fossa tumors. Nonneoplastic etiologies, such as an abscess or inflammatory lesion should always be considered, though are typically readily ruled out by imaging studies and clinical history.

Pediatric oligodendrogliomas are usually hemispheric but can arise elsewhere including in the cerebellum and posterior fossa,[22] brainstem, thalamus/basal ganglia, and spinal cord.[23] Adult oligodendrogliomas occur primarily in the frontal lobes (53%–55%),[2, 24] and second most commonly in the temporal lobes. In contrast, pediatric oligodendrogliomas arise less frequently in the frontal lobe than in adult oligodendrogliomas (22% vs 53%), but are more common in the temporal lobe than in adults (32% vs 18%) and extracortical regions (19% vs 5%).[2]

Seizures are present in approximately 10%–13% of the pediatric brain tumor patient population at the time of diagnosis,[11, 21, 25] with certain factors predisposing for seizures, such as supratentorial cortical involvement rather than subcortical, deep midline or posterior fossa involvement.[11] Interestingly, oligodendrogliomas that manifest with seizures in pediatric or adult patients often have an indolent disease course, while lesions that present with intracranial hypertension or severe neurologic deficits may have a more aggressive course.[3, 26, 27] While adult patients with low-grade oligodendrogliomas are significantly more likely to be associated with seizures compared to other tumors such as low-grade astrocytomas,[28] this propensity for oligodendrogliomas to present with seizures has not been borne out in the pediatric literature, and therefore, the presenting symptom of seizures in a pediatric patient should be considered nonspecific with regards to inferring the tumor subtype.

Molecular and histologic classification and relevance to surgery

Knowing the molecular and histopathologic classification of pediatric oligodendroglioma subtypes and the relevant data on the association between these subtypes and postsurgical outcomes is essential for the neurosurgeon to make intraoperative decisions regarding the extent of resection. Rapid advancements in our understanding of the molecular underpinnings of central nervous system tumors have led to a major restructuring of the classification of these tumors, including gliomas, as demonstrated by the updated 2016 World Health Organization (WHO) Classification of Tumors of the Central Nervous System.[29] Previously, histological phenotype exclusively determined the diagnosis of glioma subtypes, as in the 2007 WHO Classification which defined two types of "oligodendroglial tumors," "oligodendroglioma," and "anaplastic oligodendroglioma."[30] With the updated classification, the diagnostic categories are now "oligodendroglioma, IDH-mutant and 1p/19q co-deleted" and "anaplastic oligodendroglioma, IDH-mutant and 1p/19q co-deleted." However, the differentiation between these two categories (i.e., the grading of oligodendroglial tumors) still relies on histology alone.[30] As such, intraoperative, preliminary diagnoses of low-grade oligodendroglioma versus anaplastic oligodendroglioma are still done based on histologic analysis of frozen tissue samples. The histopathological features of pediatric and adult oligodendrogliomas are similar, even though their underlying genetics and clinical behavior are distinct. Classic histologic features of oligodendroglioma include round monomorphous cells, often with a "chicken-wire" background due to a reticular network of vessels, with round nuclei and perinuclear halos which create a "clear-cell" appearance, microcysts, and microcalcifications.[5] Anaplastic oligodendroglioma show prominent mitotic activity, microvascular changes, focal endothelial cell proliferation, and necrosis. Given the distinctive histologic features of oligodendroglioma, frozen surgical pathology specimens frequently yield a reliable intraoperative diagnosis.

Pediatric oligodendrogliomas appear to have a distinct molecular makeup compared to adult oligodendrogliomas,[5, 6] without a unifying molecular signature that is known at the present time. While 40%–80% of low-grade oligodendrogliomas in adults show co-deletion of 1p/19q,[31, 32] this co-deletion is rare in pediatric patients, and nearly nonexistent in the first decade of life.[5, 33] There does appear to be an "adult-type oligodendroglioma" which can present in childhood, particularly in older children or adolescents, and is characterized by the genetic hallmarks of adult oligodendroglioma including co-deletion of 1p/19q and IDH-mutation.[5, 6, 34] Currently, there is no reliable intraoperative genetic-based test that can be used on frozen specimens to diagnose and grade pediatric oligodendroglioma, though this continues to be an active area of research.

Mixed tumors such as oligoastrocytoma are rare in children, and it appears that mixed histology is a poor prognostic marker compared to pure oligodendroglioma.[19, 35] In adults, mixed anaplastic oligoastrocytoma may have a similar genotypic and phenotypic profile to glioblastoma and may be better characterized as "glioblastoma with oligodendroglial phenotype,"[36] though this has yet to be clearly described in children.

Radiographic features

Obtaining preoperative knowledge of the grade and subtype of a newly diagnosed glioma is valuable for surgeons and other clinicians, as knowing whether the tumor will be responsive to postoperative adjuvants such as chemotherapy or radiation will greatly impact discussions on the risks and benefits of surgery. As discussed above, the clinical presentation of pediatric oligodendroglioma is nonspecific and similar to that of other brain tumors and thus, is not useful in preoperative differentiation between oligodendroglioma and other tumors. Are there features of oligodendrogliomas which may help in early identification of these tumors and inform surgical approach and treatment plans?

General radiographic features of pediatric oligodendrogliomas have some similarities to those of adult oligodendrogliomas. The majority of pediatric oligodendrogliomas are hypodense on computed tomography (CT), hypointense relative to white matter on T1-weighted magnetic resonance (MR) images, and hyperintense on T2-weighted MR images.[3] These features are similar to adult oligodendrogliomas, which additionally also have marked heterogeneity on T2-weighted MR images and often appear well demarcated.[17, 37] Cystic degeneration and hemorrhage are not common findings. These tumors generally do not enhance, but patchy multifocal enhancement with a lacy pattern has been reported in a subset of cases and may correlate with the reticular network of vessels seen on histopathologic examination.[37] Some studies on adult patients have suggested that the presence of peri-tumoral edema, enhancement, cystic degeneration and hemorrhage may indicate a more aggressive tumor such as an anaplastic oligodendroglioma,[17, 37–39] while other studies debate this.[27, 40]

However, oligodendrogliomas in children and adolescents do differ from those in adults, in that calcifications, enhancement, and edema are seen less frequently.[3] Some reports indicate 38% of pediatric oligodendrogliomas with calcifications,[3] which stands in contrast to reports of up to 90% of adult oligodendrogliomas with coarse calcifications.[17] While calcifications are nonspecific as they can occur in other tumors such as astrocytomas or gangliogliomas or other tumors with dystrophic calcification from old hemorrhage,[38] interestingly, there are reports that calcification on CT is highly predictive of 1p/19q loss of heterozygosity in adult oligodendroglioma.[41]

Although there are no reliable radiologic features of pediatric oligodendrogliomas using conventional CT or MR imaging which unambiguously distinguish them from other gliomas,[42] some features of newer advanced imaging techniques such as MR spectroscopy, MR perfusion, or specific positron emission tomography (PET) imaging are useful for the diagnosis of tumor subtypes or differentiation of tumor grades. High-grade components of oligodendrogliomas often demonstrate elevated peaks of choline, N-acetyl-aspartate, and lactate, as well as an increased choline-to-creatine ratio in MR spectroscopy.[37] In addition, the degree of MR perfusion may be able to aid the differentiation of low-grade glioma from high-grade gliomas,[43] while 11C-methionine uptake on PET imaging may correlate with 1p/19q loss of heterozygosity[44] and thus help predict tumor grade.

Preoperative care and preparation

Preoperative steps include stabilization of the patient, acquisition of imaging and laboratory studies, and surgical planning. The timing of surgery depends on the patient's clinical condition and can range from emergent for life-threatening tumors to elective for smaller tumors, which often present with pharmacologically controllable seizures.[14] Patients with a large or centrally located tumor can present in extremis, obtunded, with respiratory depression due to brainstem compression and may require emergent intubation and cardiovascular support followed by urgent surgical resection. If obstructive hydrocephalus is contributing to clinical deterioration, an external ventricular drain may need to be urgently placed for cerebrospinal fluid (CSF) diversion. Such patients, as well as the more clinically stable patients harboring large tumors with associated edema and significant mass effect on the brain, should be given corticosteroids expeditiously to decrease peri-tumoral edema. If corticosteroids are started perioperatively, these are typically tapered during a period of several days after surgical resection has taken place. If patients present with seizures, anticonvulsants are begun preoperatively and continued through the perioperative period. Patients who have not had seizures may also be given anticonvulsants prophylactically to decrease seizure risk, particularly if intraoperative neuromonitoring involving cortical or subcortical stimulation is planned.

Preoperative radiographic studies should include a MRI brain with and without gadolinium contrast, which may need to be obtained with sedation for children unable to remain still voluntarily. Basic sequences include T1-weighted pre- and postcontrast studies, T2-weighted/FLAIR to evaluate edema, susceptibility to determine presence of hemorrhage or calcifications, and diffusion restriction (Fig 1A–C). If planning to use stereotactic navigation, which can be useful for resection of pediatric oligodendrogliomas, which are frequently infiltrative and difficult to distinguish macroscopically from surrounding brain tissue, MRI sequences should be optimized for use with the navigation device. Gadolinium contrast studies such as a MR arteriogram and/or venogram may be obtained to visualize vessels in greater detail for operative planning, particularly if the tumor encases or abuts major vessels. Alternatively, a CT arteriogram or venogram with iodinated contrast can be obtained and may show greater detail and have shorter acquisition times, but subjects the child to ionizing radiation. Additionally, for patients with epilepsy, preoperative mapping of the brain with functional MRI (fMRI) may be used to identify eloquent brain regions (Fig 1D–E). In children who cannot perform task-based fMRI or have difficulty communicating, newer technologies such as resting-state fMRI[45, 46] may be useful for preoperative preparation as well as to assess changes in connectivity and epileptic networks postoperatively. Additional studies such as MR spectroscopy may be used to gain further preoperative information on tumor subtype (Fig 1F).

CSF dissemination of intracranial tumors to the spinal cord, known as drop metastases, have infrequently been reported in pediatric and adult patients with oligodendrogliomas.[47, 48] Pediatric brain tumors with a higher propensity for drop metastases, such as medulloblastoma or other neuroepithelial tumors can be difficult to differentiate from oligodendroglioma on the basis of preoperative imaging alone. As such, it is often useful to obtain a screening spine MRI in a child with a newly discovered intracranial lesion to evaluate for drop metastases. Metastasis of oligodendroglioma outside of the

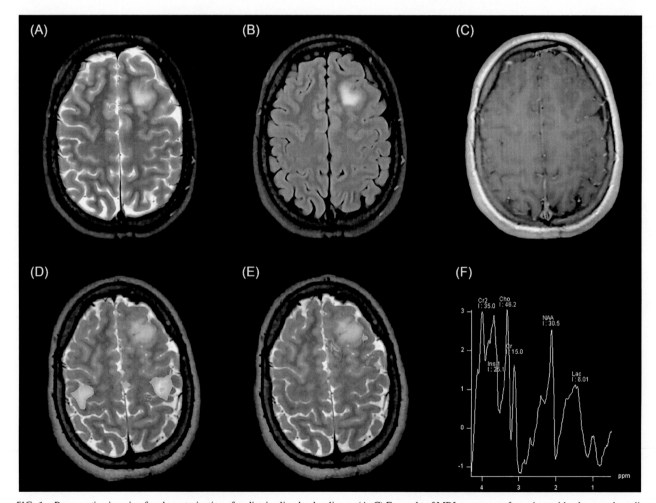

FIG. 1 Preoperative imaging for characterization of pediatric oligodendroglioma. (A–C) Example of MRI sequences of a patient with a low-grade pediatric oligodendroglioma, showing hyperintense T2 lesion (A) which is more prominently displayed on T2 FLAIR sequence (B) and notably is nonenhancing as seen on postcontrast T1 sequence (C). (D–E) Example of functional MRI (fMRI) studies for the same patient, used to delineate eloquent cortex, with sequences such as those demonstrating active regions during finger tapping (D) or verb generation (E). (F) Example of MR spectroscopy of another pediatric patient with oligodendroglioma, demonstrating elevated peaks of *N*-acetylaspartate and choline.

central nervous system, while very rare,[49, 50] may occur more frequently than other glioma types and has a propensity for bone and bone marrow involvement.[51] Therefore, unusual focal symptoms warrant further radiographic investigation to rule out extraneural disease.

Goals of surgery

The dual goals of surgery are arriving at an accurate diagnosis and achieving a durable therapeutic benefit. For supratentorial hemispheric tumors in noneloquent areas, both of these goals may be pursued with relative ease. However, in considering lesions which are difficult to resect completely, such as those in eloquent areas or in subcortical structures, the safe acquisition of an adequate amount of tissue should still be attempted for accurate histopathological diagnosis and grading, due to significant prognostic and therapeutic implications. In adult gliomas, with the addition of genetic criteria to the WHO grading classification system,[29] molecular testing (e.g., for IDH mutations or 1p/19q co-deletion) can compensate for diagnostic uncertainty and decreased accuracy of grading of small tumor specimens[52, 53] such as those obtained by stereotactic biopsy.[54, 55] However, as there are no reliable diagnostic molecular markers of pediatric oligodendrogliomas at present, accurate histopathological diagnosis remains critical, which in turn depends on the acquisition of an adequate amount of tumor specimen. Furthermore, although the genetic makeup of pediatric oligodendrogliomas is currently unclear, future research is bound to elucidate the genomic landscape of these tumors, and retrospective genetic testing of previously obtained surgical specimens may prove valuable in determining the long-term management paradigm of postsurgical patients who present years later with recurrent or progressive disease.[56]

The therapeutic benefit of surgical resection of oligodendrogliomas include both short-term relief of present or imminent symptoms (such as seizures, headaches or nausea due to elevated intracranial pressures, cranial nerve defects from direct compression, and lethargy or life-threatening dysautonomia due to brainstem compression), as well as long-term and durable alterations in the natural course of the disease and improvement in survival and quality of life. Regarding seizures, for example, the extent of surgical resection of temporal lobe tumors in children correlates with the probability of seizure freedom,[57–59] and pediatric patients are more likely than adult patients to achieve seizure freedom after resection of low-grade temporal lobe gliomas.[59] Studies on pediatric brain tumors in other locations have also shown that subtotal tumor resection is a predictor of seizure recurrence,[11] supporting the general surgical goal of gross total resection in order to achieve seizure freedom.

The correlation between complete resection of tumor and survival outcome is unclear in pediatric oligodendroglioma patients (see Section "Outcomes"), unlike in adults for whom there are data supporting the survival benefit of complete resection for low-grade oligodendroglioma.[17] Of the studies which have analyzed prognostic factors in pediatric oligodendroglioma,[1, 2, 5, 19, 20, 23, 35, 60] several report that extent of resection is a significant predictor of outcome,[2, 35, 61] while others assert that there is no correlation.[19, 20, 42, 60] One reason for this lack of consensus may be the confounding question of whether a complete resection is directly responsible for improved outcomes, or whether tumors that lend themselves to complete resection (by virtue of their location or separability from surrounding brain tissue) represent an inherently more favorable subtype of oligodendroglioma. Regardless, given that pediatric oligodendrogliomas are not sensitive to adjuvant chemotherapy or radiation, and as the cumulative effects of toxicity from these adjuvant treatments can be higher in children for whom overall prognosis and duration of survival are favorable, complete resection of the tumor should remain the goal if the risks of surgery are tolerable.

Surgical approaches and techniques

The surgical plan for resection of a suspected oligodendroglioma in a child is akin to that of any brain tumor and begins with a comprehensive operative plan agreed on by all teams: neurosurgery, anesthesia, scrub technicians, circulating nurses, and neuro-monitoring. Preoperative planning should include a discussion about the type of neuro-monitoring planned (which can affect the anesthetic agents to be given), patient positioning, antibiotic selection, brain relaxation agents (such as mannitol or furosemide), corticosteroids and anticonvulsants, arterial line placement, venous access through peripheral intravenous versus central lines, and urinary catheter placement.

Surgeries are performed under general anesthesia, although in select cases for older children or adolescents, awake craniotomies may be performed for lesions near eloquent areas. The positioning of the patient and surgical approach are determined by the location of the tumor and the operative trajectory planned. For supratentorial tumors, the patient is placed either supine (with gel or foam padding under the patient if the patient needs to be tilted to allow the head to be rotated further) or in lateral decubitus. For infratentorial tumors, the patient is typically placed in the prone position. Pinned head fixation may be used for children older than 2–3 years, while younger children may be held in position with a horseshoe head holder. If stereotactic navigation is to be employed, the system may be set up at this time. A strip of hair is shaved along the planned incision line, and a standard skin preparation is completed with chlorhexidine, or for infants, a combination of betadine and alcohol.

The skin incision is determined by the location and size of the tumor, with linear or C-shaped incisions for posterior frontal, parietal, or occipital lesions; a bicoronal incision for an anterior frontal lesion; and a question mark, linear or C-shaped incision for temporal lesions. After elevating the scalp flap, the craniotomy and dural opening are performed in a standard fashion. The tumor is then approached, using stereotactic navigation or ultrasound guidance if necessary. Once the tumor is identified, a specimen is typically sent to pathology for a frozen section diagnosis early in the process of resection, as preliminary knowledge of the tumor subtype and grade may influence the extent to which a total resection is pursued. For example, if the tumor is anaplastic rather than low-grade, this may inspire the surgeon to pursue a more aggressive resection, knowing that pediatric anaplastic oligodendrogliomas are not very responsive to chemoradiation. The tumor is then resected, typically by attempting to define a tumor-brain interface circumferentially around the tumor, or by an inside-out technique with debulking of the tumor from within. Meticulous hemostasis is achieved and a watertight dural closure is pursued if possible, followed by replacement of the bone flap and a multilayer closure of the superficial tissues and skin.

To aid the goal of complete tumor resection, modern neurosurgery includes techniques such as frameless stereotactic navigation, intraoperative neurophysiology, electrocorticography (ECoG), intraoperative MRI (iMRI), intraoperative ultrasound, and fluorescence-guided resection. These technical adjuncts have been employed for over twenty years[62] and are now part of the standard of care for safe and effective tumor resection, particularly for tumors in eloquent brain regions. There are specific considerations for pediatric patients regarding these technologies, owing to differences in efficacy as well as anatomical considerations. For infants who cannot tolerate pinned head fixation, stereotactic navigation using pinless electromagnetic tracking technology can be useful. Intraoperative ECoG can be helpful for surgical resection of brain tumors in pediatric patients with seizures,[63] and its use has been associated with long-term seizure control, particularly for low-grade glioma patients with seizures refractory to medical therapy.[64] Intraoperatively, grid electrodes can be placed on the cortex to help localize epileptogenic foci prior to resecting the tumor and then can be set aside (Fig 2A). After resection, the grid can again be placed over the region to determine whether epileptogenic foci have been satisfactorily removed.

For frequently infiltrative tumors such as pediatric oligodendrogliomas, iMRI can be very helpful to determine whether any residual tumor remains and needs to be further resected (Fig 2B). In both pediatric[65] and adult patients,[66, 67] iMRI has been shown to contribute to greater extent of resection during glioma surgery, not only for high-grade gliomas which often have enhancing components[66, 67] but also for nonenhancing tumors including low-grade gliomas (Fig 2C–E).[68, 69] In addition, intraoperative ultrasonography, both conventional and contrast-enhanced, may be useful for pediatric low-grade glioma resection.[70] Ultrasonography has the advantage of being a real-time imaging modality, in contrast to stereotactic navigation, which typically relies on preoperative imaging and becomes increasingly less accurate as the dissection proceeds and the tissue is distorted. Interestingly, it has been reported that low-grade gliomas which did not show contrast enhancement on preoperative MRI demonstrated contrast enhancement on ultrasound,[70, 71] supporting the use of ultrasound in improving surgical resection.

Fluorescence-guided surgery, such as the use of the endogenous fluorescent biomarker protoporphyrin IX (PpIX) induced by exogenously administered 5-aminolevulinic acid (5-ALA) or indocyanine green, is being used increasingly commonly in adult patients. While there have been reports of improved progression-free survival with 5-ALA guided resection,[72] the application of 5-ALA in children remains an off-label use and the utility of fluorescence-guided resection of pediatric tumors is debated. A recent study discussed 120 cases of 5-ALA application in pediatric brain tumor resection,[73] of which a handful of cases involved pediatric oligodendroglioma. Positive fluorescence was seen for anaplastic oligodendroglioma, but only faint heterogeneous fluorescence for low-grade oligodendrogliomas.[73, 74] In contrast, 86% of pediatric glioblastoma demonstrated avid fluorescence,[73] consistent with other reports that 5-ALA fluorescence guided resection is most useful for high-grade gliomas.[75] While quantitative methods of fluorescence have reported increased accuracy of 5-ALA-guided resection of adult low-grade gliomas compared to visual fluorescence,[76] these technologies are in development and have not yet been widely tested in pediatric patients.

Postoperative care

Standard postoperative care includes recovery in an intensive care unit, step-down unit, or regular unit, wherever the patient's neurologic exam can be carefully monitored. Unexpected exam findings of somnolence or altered mental status, pain or nausea out of proportion to the postanesthetic state, or focal neurologic deficits should prompt an immediate evaluation with a noncontrast head CT or a rapid MRI which includes targeted axial T2, diffusion-weighted and susceptibility sequences to evaluate for postoperative hemorrhage, infarct, or other acute abnormalities. Other explanations for such

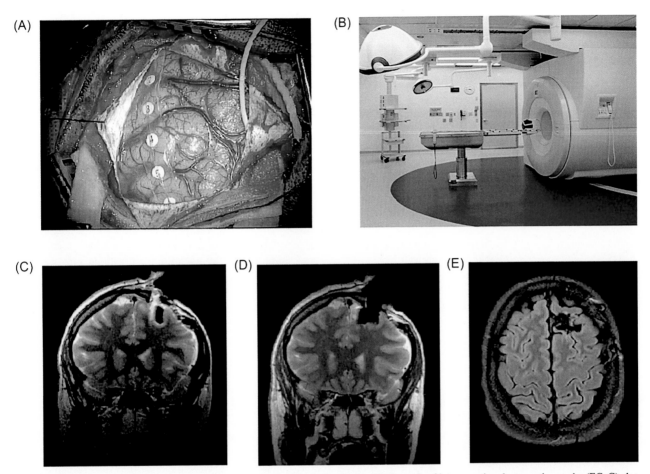

FIG. 2   Technologies to optimize surgical resection of pediatric oligodendroglioma. (A) Example of intraoperative electrocorticography (ECoG) electrode grids overlying a frontal lobe tumor. (B–E) Use of intraoperative MRI for optimal surgical resection. An MRI scan obtained intraoperatively (B) after resection demonstrated a small rim of residual lesion (C), with further resection yielding a gross total resection demonstrated on a second intraoperative MRI (D), and follow-up scan after one year showing no residual or recurrent tumor (E).

symptoms include seizures, over-narcotization, medication reactions, toxic-metabolic etiologies, and each possibility should be ruled out systematically.

Patients who have seizures preoperatively will be susceptible to seizures postoperatively, so anticonvulsants may need to be up-titrated in the perioperative setting. Postoperative steroid taper over several days can help decrease postoperative edema. Antibiotics may be given postoperatively. Surgical dressings may be kept in place for up to one week, especially for young children who may be unable to refrain from manipulating the surgical wound. Early mobilization is recommended to promote recovery and physical and occupation therapy evaluations are central to determining whether the child may be safely discharged home or may need transfer to a rehabilitation center. Child life support and social work are often very helpful in normalizing the operative experience for children and supporting the parents and other family through the stressors of surgery.

All pediatric brain tumor patients should be managed by a multidisciplinary team consisting of neurologists, neurosurgeons, neuro-oncologists, and radiation therapists. Ideally, children are evaluated preoperatively by each component of this team in order to help the patient and the family begin to prepare for a multistep treatment plan, of which surgical resection is the first major step. Certainly there should be a postoperative evaluation by the entire team, and management decisions are to be made with critical input from all members of the team once surgical pathology is available and a final diagnosis is reached.

Adjuvant therapy

Postoperative adjuvant treatments in use today are chemotherapy and radiation. Factors to consider when deciding which regimen to use in children may include extent of tumor resection, tumor location, anticipated toxicity, genetic alterations, and patient/family preferences. For pediatric oligodendroglioma patients, who generally have an excellent prognosis with an indolent disease course and delayed recurrence, the risks of adjuvant treatments must be heavily weighed against the potential benefits. At our institutions, as at many others, children who have undergone gross total resection of a low-grade oligodendroglioma are managed postoperatively with observation and surveillance scans. This is partly due to the previous studies reporting that in children with oligodendrogliomas, neither postoperative chemotherapy nor radiation therapy are associated with improved outcomes.[35] As mentioned previously, the presence of the 1p/19q co-deletion, which has been associated with a favorable response to adjuvant therapy in adult patients with anaplastic oligodendrogliomas[8, 9] does not have similar prognostic value for pediatric patients.[10] For MGMT-methylated pediatric anaplastic oligodendrogliomas, durable postoperative control by solitary treatment with temozolamide has been reported.[77]

More data exists in the adult patient realm. For adult anaplastic oligodendroglioma with 1p/19q co-deletion and/or IDH mutations, postoperative treatment with adjuvant procarbazine, lomustine, vincristine (PCV) chemotherapy is now the standard of care, based on multiple studies[8, 9, 78] including phase 3 trials RTOG 9402[79, 80] and EORTC 26951,[36, 81] demonstrating that patients who received chemotherapy plus radiation had greater progression-free and overall survival than those who received radiation alone. Adult anaplastic oligodendroglioma patients without either the 1p/19q co-deletion or IDH mutations do not seem to benefit from the addition of chemotherapy to radiation.[82] Recent reports have also demonstrated that for young adult patients 18–40 years old with low-grade oligodendroglioma, administering PCV chemotherapy with radiation therapy provides a survival benefit over radiation therapy alone, especially for patients with IDH1 R132H mutations.[83] In adults, the difference in therapeutic benefit between PCV and temozolamide are unclear.[8, 82] There are few data on immunotherapy, tumor vaccines, or other targeted therapies for oligodendroglioma, or other low-grade glioma in children, or adults at the present time.[84]

Radiation therapy has been utilized less frequently over time for pediatric low-grade oligodendrogliomas, likely due to conflicting results regarding the survival benefit of radiation and a risk of cognitive decline conferred by radiation as seen in a proportion of long-term adult survivors.[85, 86] For pediatric patients with subtotally resected low-grade oligodendroglioma, some have reported a significant increase in progression-free survival in those who underwent additional adjuvant therapy,[87] while others reported comparable progression-free survival with subtotal resection alone.[60] Other studies reported that postoperative radiation therapy failed to improve overall survival for pediatric oligodendroglioma patients, though these studies did not stratify based on the extent of resection of the tumor.[19, 35] Overall, given insufficient data and the risks of treating children with radiation, it would be reasonable to defer radiation for patients with gross totally resected oligodendrogliomas. Radiation therapy and chemotherapy may be considered for patients with subtotally resected oligodendrogliomas, disseminated disease to leptomeninges or spine, or recurrent disease. Larger studies aimed at identifying additional prognostic markers and predicting response to chemotherapy, radiation therapy, and newer therapeutics such as immunotherapy, targeted agents and other biologics will be needed to improve outcomes.

Outcomes

The mean overall survival in pediatric oligodendroglioma exceeds that of the adult population, with 5-year overall survival rates ranging between 65% and 94%.[1, 2, 5, 19, 35, 60] This survival advantage is apparent for low-grade oligodendroglioma, but may not hold for anaplastic oligodendroglioma. For example, a recent, large-scale retrospective analysis of pediatric and adult oligodendroglioma patients demonstrated that survival duration was significantly longer for pediatric patients than adult patients in all categories of cases matched for gender, race, tumor size, tumor location, use of radiotherapy, and extent of surgery.[2] However, this was not true for cases matched by tumor grade. While pediatric patients demonstrated better outcomes than adults for low-grade oligodendrogliomas, there was no survival advantage in pediatric patients with high-grade oligodendrogliomas over adult patients.

Several studies have examined predictors of clinical outcome in pediatric oligodendroglioma. A recent example was an individual-patient-data meta-analysis on predictors of clinical outcome in pediatric oligodendroglioma,[61] which demonstrated that subtotal resection, mixed pathology tumor, initial presentation of headache, and a parietal lobe tumor location were predictors of both tumor progression/recurrence and death. Age at diagnosis of pediatric oligodendroglioma patients has been variably reported as being prognostic for survival by some groups, who report age less than 3 years[35] or less than 12 years at diagnosis[19] as correlating with overall survival, while others suggest that age is not a prognostic factor.[20] While the co-deletion of 1p/19q and mutations in IDH-1 or IDH-2 are strong independent prognostic factors for overall and progression free survival in adult oligodendroglioma patients,[8, 36, 79, 81, 82] this contrasts sharply with data suggesting that 1p/19q-co-deleted oligodendrogliomas in pediatric patients may have an equally or even more aggressive clinical course than non-co-deleted tumors.[5]

Whether the extent of resection is correlated with outcome for pediatric oligodendroglioma patients is contested. Of the handful of studies which have analyzed prognostic factors in pediatric oligodendroglioma,[1, 2, 5, 19, 20, 23, 35, 60] several report that extent of resection is a significant predictor of outcome.[2, 35, 61] In one study on 37 pediatric patients with oligodendroglioma, those who underwent gross total resection experienced a significantly higher 5-year progression-free survival rate of 100% compared to 28.8% in patients treated with subtotal resection or biopsy alone.[35] In contrast, some studies report that the completeness of resection was not a significant predictor of outcome.[19, 20, 42, 60] It has also been reported that pediatric oligodendroglioma patients with incomplete tumor resection have a low incidence of disease progression,[3, 87] raising the question of how important it actually is to achieve a radiologically complete tumor resection at the expense of increased surgical risk.

There are a few reasons for this disagreement over the effects of completeness of resection. One reason is the confounding question of whether a complete resection is directly causative for improved outcomes, or whether tumors that lend themselves to complete resection (by virtue of their location or separability from surrounding brain tissue) represent an inherently more favorable subtype of oligodendroglioma.[14] Perhaps deep-seated, more central gliomas have worse outcomes than more superficial hemispheric lesions, not because these tumors are difficult to access surgically and resect completely, but because they constitute a biologically more aggressive group with poor disease outcomes. The second reason for the lack of consensus regarding whether extent of resection is a predictor of improved outcomes is that much of the published data has not been stratified by tumor grade. For example, in the adult literature, there is a survival benefit of complete resection in adult patients with oligodendrogliomas,[17] and a nonsignificant trend toward longer survival with a greater extent of resection in adult patients with anaplastic oligodendrogliomas.[88] In general, extensive surgical resection of low-grade supratentorial gliomas in adult patients has been associated with prolonged recurrence-free survival,[89, 90] and thus, complete resection is the standard of care for all low-grade gliomas in adults. Recent research, however, asserts that the extent of resection was not correlated with overall survival for adult anaplastic oligodendroglioma,[91] possibly due to the fact that the majority of these anaplastic oligodendrogliomas are 1p/19q co-deleted tumors which are highly responsive to chemotherapy. Further studies with careful matching of variables such as tumor location and tumor size may help elucidate this question in pediatric patients.

Studies examining the characteristics of children with brain tumors and seizures have shown that subtotal resection of tumor is a predictor of seizure recurrence,[11] and the extent of surgical resection of temporal lobe tumors in children correlates with seizure freedom.[57–59] The adult literature reports that a long preoperative history of uncontrolled seizures and simple partial seizure type predispose to poor postoperative control of seizures in patients with low-grade gliomas.[28] The factors associated with freedom from seizures were gross-total resection, preoperative seizure history of less than 1 year, and nonsimple partial seizure type. In all, 98% of patients with preoperative seizures were seizure free at 6 months when all of these three factors were present, compared with only 44% when none were present.[28] Pediatric patients are more likely than adult patients to achieve seizure freedom after resection of low-grade temporal lobe gliomas,[59] supporting the general surgical goal of total resection in order to achieve seizure freedom and improve quality of life, even if uncertainty may persist about the impact of total resection on survival duration.

Surveillance and recurrence

Surveillance imaging for tumor recurrence can be challenging, particularly in low-grade gliomas. Low-grade oligodendrogliomas and other gliomas are frequently nonenhancing and infiltrate brain parenchyma, and the radiographic progression of such tumors can be subtle, such as the distortion of the brain tissue with relative preservation of the boundaries of gray and white matter.[56] Immediate postoperative changes along the margins of the resection cavity can appear similar to residual tumor, and the evolution of these postoperative changes—including disruption of adjacent neural tissue and subsequent gliosis—can also be difficult to distinguish from tumor progression. Radiation therapy can result in radiographic changes occurring on timescales between acute (days) to delayed (months to years), with edema or contrast-enhancing necrosis seen on CT and MRI. In addition, surgery and adjuvant treatment can lead to "pseudo-progression," which is the phenomenon of new and progressive contrast-enhancing lesions on MRI either within the prior resection cavity or elsewhere in the brain that mimics true tumor progression but turns out to be necrotic tissue and inflammation with vascular changes.[92] This phenomenon has been well characterized in patients who receive combination radiation and chemotherapy in high-grade gliomas[93] and can also occur in patients with low-grade gliomas who receive adjuvant treatment.[94, 95]

The decision to reoperate on a recurrent lesion can be complex, particularly for pediatric low-grade oligodendroglioma patients who have a good prognosis in the setting of their indolent disease and for whom the morbidity of a second surgery could be high. On the other hand, rapid progression of a new or recurrent lesion raises the specter of histopathologic transformation of a low-grade oligodendroglioma into an anaplastic oligodendroglioma and early initiation of additional treatment may be necessary. If the lesion is readily accessible by a repeat craniotomy, a reoperation may be favored since resection of the lesion would provide diagnostic information that could inform subsequent treatment and prognosis. However, if the risks of surgery are high, nonoperative treatment with radiation and/or chemotherapy would be reasonable to pursue.

The limitations of conventional MR imaging may be overcome by advanced imaging techniques to aid in tumor surveillance. MR perfusion, which signals abnormal hemodynamic changes related to increased angiogenesis and vascular permeability[43] may be used to differentiate pseudo-progression from true tumor progression.[95, 96] MR spectroscopy is used to measure regional variations in neurochemistry and the concentration of various brain metabolites and indices such as choline/creatine or choline/N-acetyl-aspartate may be used to distinguish between radiation necrosis and recurrent tumor, though there is not yet consensus on the specific reference values for oligodendroglioma.[37] MR spectroscopy may even be used intraoperatively to differentiate residual tumor from nontumoral changes around the resection cavity.[97] Dynamic susceptibility contrast (DSC) imaging, which can report relative cerebral blood volume which may be elevated in malignant transformation far before other radiographic or clinical signs become apparent,[98] can also be used to track treatment effects as well as to distinguish pseudo-progression from true progression in oligodendrogliomas and other tumors.[37, 99]

Surgical clinical trials and research frontiers

Greater research efforts and clinical trials are needed for the surgical management of pediatric oligodendrogliomas, particularly as the role of surgery is expected to evolve within the shifting landscape of new medical therapeutics including immunotherapy or biological therapies. Research efforts relevant to the neurosurgical approach to these tumors include: (1) preoperative diagnostic tools, (2) intraoperative technological adjuncts to improve tumor resection, and (3) novel therapeutics deployed during surgery.

Preoperative knowledge of glioma grade and subtype is valuable for the clinical team, as knowing whether the tumor will be responsive to postoperative adjuvant such as chemotherapy or radiation will greatly impact the discussion on risk/benefit of surgery. For pediatric oligodendrogliomas, which are not sensitive to chemoradiation, being able to identify these tumors and differentiate them from other glioma subtypes could very well change the operative strategy. Advanced imaging techniques are increasingly being developed to aid the identification of the molecular subtype of the tumor as well as tumor grade, as well as to differentiate between tumor recurrence and radiation necrosis, but no unequivocal method or combination of methods are available at the present time. Functional imaging techniques also aim to visualize and quantify various aspects of the tumor, including microstructure (by diffusion-weighted imaging), perfusion (by MR perfusion imaging), metabolites (by MR spectroscopy), and metabolism (by PET).[37] Currently, the most widely used parameter for glioma grading is relative cerebral blood volume (rCBV), as angiogenic activity of high-grade tumors causes increased microvascular density and slow-flowing collateral vessels, resulting in an increased rCBV. While it has been shown that high-grade gliomas can be differentiated from low-grade glioma with 95% sensitivity using a specific rCBV threshold value,[100] this has not been applicable to oligodendroglioma which may have high rCBV even for low-grade tumors and in which a reliable distinction between high and low grade oligodendrogliomas cannot be made.[37, 101, 102]

Given the importance of optimal surgical resection for improved outcomes, there is a focus on technologies to improve surgical precision and accuracy. Some have been discussed above (see "Surgical approaches and techniques"). Recent areas of research include the impact of MR spectroscopy guided resection on prognosis, the use of 5-ALA or fluorescein to improve resection of tumors, comparison of neuronavigation techniques such as intraoperative MR, intraoperative ultrasound, and ^{18}F-FDOPA PET. Studies must be pursued to determine the safety of these intraoperative techniques as well as to optimize them for practical use in the OR. Noninvasive molecular imaging of tumors may be used increasingly in the future for preoperative accurate diagnosis, assessment of prognosis, and selection of optimal treatment, possibly obviating the need for surgery altogether in certain cases.[53]

As strides are made in elucidating the genetic underpinnings of pediatric oligodendrogliomas and other low-grade gliomas,[6] novel therapeutics which need to be delivered during surgery or via a procedure are also being developed. In adults, for example, chimeric antigen receptor T-cells and other targeted therapies often need to be delivered intraventricularly due to the blood-brain barrier precluding the use of intravenous therapies. Intraoperative delivery of these therapeutics as well as implantation of devices (e.g., reservoirs, pumps) for serial treatments will increasingly require neurosurgeons to be involved in performing these procedures safely and efficiently.

Conclusion

Pediatric oligodendroglioma remains an understudied tumor type, with its clear and distinctive histopathologic features contrasting with its nebulous genetic underpinnings. While survival trends of adult oligodendroglioma patients have evolved,[103] with a significant improvement in overall survival of anaplastic oligodendroglioma patients partly due to developments in adjuvant therapies,[79, 81] the care of pediatric oligodendroglioma patients has largely remained static over the years with a management strategy that is primarily surgical. Further studies will be needed to identify new surgical techniques and medical treatments that hold the potential to improve survival and quality of life in this challenging patient population.

References

1. Dohrmann GJ, Farwell JR, Flannery JT. Oligodendrogliomas in children. *Surg Neurol.* 1978;10:21–25.
2. Goel NJ, Abdullah KG, Lang SS. Outcomes and prognostic factors in pediatric oligodendroglioma: a population-based study. *Pediatr Neurosurg.* 2018;53:24–35.
3. Tice H, Barnes PD, Goumnerova L, et al. Pediatric and adolescent oligodendrogliomas. *AJNR Am J Neuroradiol.* 1993;14:1293–1300.
4. Achey RL, Khanna V, Ostrom QT, et al. Incidence and survival trends in oligodendrogliomas and anaplastic oligodendrogliomas in the United States from 2000 to 2013: a CBTRUS report. *J Neuro-Oncol.* 2017;133:17–25.
5. Rodriguez FJ, Tihan T, Lin D, et al. Clinicopathologic features of pediatric oligodendrogliomas: a series of 50 patients. *Am J Surg Pathol.* 2014;38(8):1058–1070.
6. Zhang J, Wu G, Miller CP, et al. Whole-genome sequencing identifies genetic alterations in pediatric low-grade gliomas. *Nat Genet.* 2013;45:602–612.
7. Reifenberger J, Reifenberger G, Liu L, et al. Molecular genetic analysis of oligodendroglial tumors shows preferential allelic deletions on 19q and 1p. *Am J Pathol.* 1994;145:1175–1190.
8. Lassman AB, Iwamoto FM, Cloughesy TF, et al. International retrospective study of over 1000 adults with anaplastic oligodendroglial tumors. *Neuro-Oncology.* 2011;13:649–659.
9. Ino Y, Betensky RA, Zlatescu MC, et al. Molecular subtypes of anaplastic oligodendroglioma: implications for patient management at diagnosis. *Clin Cancer Res.* 2001;7:839–845.
10. Pollack IF, Finkelstein SD, Burnham J, et al. Association between chromosome 1p and 19q loss and outcome in pediatric malignant gliomas: results from the CCG-945 cohort. *Pediatr Neurosurg.* 2003;39:114–121.
11. Ullrich NJ, Pomeroy SL, Kapur K, et al. Incidence, risk factors, and longitudinal outcome of seizures in long-term survivors of pediatric brain tumors. *Epilepsia.* 2015;56:1599–1604.
12. Pollack IF. Brain tumors in children. *N Engl J Med.* 1994;331:1500–1507.
13. Berger MS, Keles GE, Geyer JR. Cerebral hemispheric tumors of childhood. *Neurosurg Clin N Am.* 1992;3:839–852.
14. Albright AL, Pollack IF, Adelson PD. *Principles and Practice of Pediatric Neurosurgery.* 3rd ed. New York: Thieme; 2014.
15. Chin HW, Hazel JJ, Kim TH, et al. Oligodendrogliomas. I A clinical study of cerebral oligodendrogliomas. *Cancer.* 1980;45:1458–1466.
16. Mørk SJ, Lindegaard KF, Halvorsen TB, et al. Oligodendroglioma: incidence and biological behavior in a defined population. *J Neurosurg.* 1985;63:881–889.
17. Engelhard HH, Stelea A, Mundt A. Oligodendroglioma and anaplastic oligodendroglioma: clinical features, treatment, and prognosis. *Surg Neurol.* 2003;60:443–456.
18. Lau CS, Mahendraraj K, Chamberlain RS. Oligodendrogliomas in pediatric and adult patients: an outcome-based study from the surveillance, epidemiology, and end result database. *Cancer Manag Res.* 2017;9:159–166.
19. Razack N, Baumgartner J, Bruner J. Pediatric oligodendrogliomas. *Pediatr Neurosurg.* 1998;28:121–129.
20. Bowers DC, Mulne AF, Weprin B, et al. Prognostic factors in children and adolescents with low-grade oligodendrogliomas. *Pediatr Neurosurg.* 2002;37:57–63.
21. Wilne S, Collier J, Kennedy C, et al. Presentation of childhood CNS tumours: a systematic review and meta-analysis. *Lancet Oncol.* 2007;8:685–695.
22. Furtado SV, Venkatesh PK, Ghosal N, et al. Clinical and radiological features of pediatric cerebellar anaplastic oligodendrogliomas. *Indian J Pediatr.* 2011;78:880–883.
23. Wang KC, Chi JG, Cho BK. Oligodendroglioma in childhood. *J Korean Med Sci.* 1993;8:110–116.
24. Ludwig CL, Smith MT, Godfrey AD, et al. A clinicopathological study of 323 patients with oligodendrogliomas. *Ann Neurol.* 1986;19:15–21.
25. Sjörs K, Blennow G, Lantz G. Seizures as the presenting symptom of brain tumors in children. *Acta Paediatr.* 1993;82:66–70.
26. Rizk T, Mottolèse C, Bouffet E, et al. Cerebral oligodendrogliomas in children: an analysis of 15 cases. *Childs Nerv Syst.* 1996;12:527–529.
27. Mirsattari SM, Chong JJ, Hammond RR, et al. Do epileptic seizures predict outcome in patients with oligodendroglioma. *Epilepsy Res.* 2011;94:39–44.
28. Chang EF, Potts MB, Keles GE, et al. Seizure characteristics and control following resection in 332 patients with low-grade gliomas. *J Neurosurg.* 2008;108:227–235.
29. Louis DN, Perry A, Reifenberger G, et al. The 2016 World Health Organization Classification of Tumors of the Central Nervous System: a summary. *Acta Neuropathol.* 2016;131:803–820.
30. Louis DN, Ohgaki H, Wiestler OD, et al. The 2007 WHO classification of tumours of the central nervous system. *Acta Neuropathol.* 2007;114:97–109.
31. Ohgaki H, Kleihues P. Genetic profile of astrocytic and oligodendroglial gliomas. *Brain Tumor Pathol.* 2011;28:177–183.
32. Riemenschneider MJ, Jeuken JW, Wesseling P, et al. Molecular diagnostics of gliomas: state of the art. *Acta Neuropathol.* 2010;120:567–584.
33. Suri V, Jha P, Agarwal S, et al. Molecular profile of oligodendrogliomas in young patients. *Neuro-Oncology.* 2011;13:1099–1106.

34. Schwartzentruber J, Korshunov A, Liu XY, et al. Driver mutations in histone H3.3 and chromatin remodelling genes in paediatric glioblastoma. *Nature.* 2012;482:226–231.

35. Creach KM, Rubin JB, Leonard JR, et al. Oligodendrogliomas in children. *J Neuro-Oncol.* 2012;106:377–382.

36. Kouwenhoven MC, Gorlia T, Kros JM, et al. Molecular analysis of anaplastic oligodendroglial tumors in a prospective randomized study: a report from EORTC study 26951. *Neuro-Oncology.* 2009;11:737–746.

37. Smits M. Imaging of oligodendroglioma. *Br J Radiol.* 2016;89:20150857.

38. Lee YY, Van Tassel P. Intracranial oligodendrogliomas: imaging findings in 35 untreated cases. *AJR Am J Roentgenol.* 1989;152:361–369.

39. Daumas-Duport C, Varlet P, Tucker ML, et al. Oligodendrogliomas. Part I: patterns of growth, histological diagnosis, clinical and imaging correlations: a study of 153 cases. *J Neuro-Oncol.* 1997;34:37–59.

40. White ML, Zhang Y, Kirby P, et al. Can tumor contrast enhancement be used as a criterion for differentiating tumor grades of oligodendrogliomas. *AJNR Am J Neuroradiol.* 2005;26:784–790.

41. Saito T, Muragaki Y, Maruyama T, et al. Calcification on CT is a simple and valuable preoperative indicator of 1p/19q loss of heterozygosity in supratentorial brain tumors that are suspected grade II and III gliomas. *Brain Tumor Pathol.* 2016;33:175–182.

42. Puget S, Boddaert N, Veillard AS, et al. Neuropathological and neuroradiological spectrum of pediatric malignant gliomas: correlation with outcome. *Neurosurgery.* 2011;69:215–224.

43. Abrigo JM, Fountain DM, Provenzale JM, et al. Magnetic resonance perfusion for differentiating low-grade from high-grade gliomas at first presentation. *Cochrane Database Syst Rev.* 2018;1.

44. Saito T, Maruyama T, Muragaki Y, et al. 11C-methionine uptake correlates with combined 1p and 19q loss of heterozygosity in oligodendroglial tumors. *AJNR Am J Neuroradiol.* 2013;34:85–91.

45. Hart MG, Price SJ, Suckling J. Functional connectivity networks for preoperative brain mapping in neurosurgery. *J Neurosurg.* 2017;126:1941–1950.

46. Boerwinkle VL, Vedantam A, Lam S, et al. Connectivity changes after laser ablation: resting-state fMRI. *Epilepsy Res.* 2017;.

47. Ng HK, Sun DT, Poon WS. Anaplastic oligodendroglioma with drop metastasis to the spinal cord. *Clin Neurol Neurosurg.* 2002;104:383–386.

48. Fayeye O, Sankaran V, Sherlala K, et al. Oligodendroglioma presenting with intradural spinal metastases: an unusual cause of cauda equina syndrome. *J Clin Neurosci.* 2010;17:265–267.

49. Bruggers C, White K, Zhou H, et al. Extracranial relapse of an anaplastic oligodendroglioma in an adolescent: case report and review of the literature. *J Pediatr Hematol Oncol.* 2007;29:319–322.

50. Macdonald DR, O'Brien RA, Gilbert JJ, Cairncross JG. Metastatic anaplastic oligodendroglioma. *Neurology.* 1989;39(12):1593–1596.

51. Zustovich F, Della Puppa A, Scienza R, et al. Metastatic oligodendrogliomas: a review of the literature and case report. *Acta Neurochir.* 2008;150:699–702.

52. Kim BY, Jiang W, Beiko J, et al. Diagnostic discrepancies in malignant astrocytoma due to limited small pathological tumor sample can be overcome by IDH1 testing. *J Neuro-Oncol.* 2014;118:405–412.

53. Cahill DP, Sloan AE, Nahed BV, et al. The role of neuropathology in the management of patients with diffuse low grade glioma: a systematic review and evidence-based clinical practice guideline. *J Neuro-Oncol.* 2015;125:531–549.

54. Jackson RJ, Fuller GN, Abi-Said D, et al. Limitations of stereotactic biopsy in the initial management of gliomas. *Neuro-Oncology.* 2001;3:193–200.

55. Glantz MJ, Burger PC, Herndon JE, et al. Influence of the type of surgery on the histologic diagnosis in patients with anaplastic gliomas. *Neurology.* 1991;41:1741–1744.

56. Chi AS, Cahill DP, Larvie M, Louis DN. Case 38-2016. A 52-year-old woman with recurrent oligodendroglioma. *N Engl J Med.* 2016;375:2381–2389.

57. Cataltepe O, Turanli G, Yalnizoglu D, et al. Surgical management of temporal lobe tumor-related epilepsy in children. *J Neurosurg.* 2005;102:280–287.

58. Khajavi K, Comair YG, Wyllie E, et al. Surgical management of pediatric tumor-associated epilepsy. *J Child Neurol.* 1999;14:15–25.

59. Englot DJ, Han SJ, Berger MS, et al. Extent of surgical resection predicts seizure freedom in low-grade temporal lobe brain tumors. *Neurosurgery.* 2012;70:921–928.

60. Peters O, Gnekow AK, Rating D, Wolff JE. Impact of location on outcome in children with low-grade oligodendroglioma. *Pediatr Blood Cancer.* 2004;43:250–256.

61. Wang KY, Vankov ER, Lin DDM. Predictors of clinical outcome in pediatric oligodendroglioma: meta-analysis of individual patient data and multiple imputation. *J Neurosurg Pediatr.* 2018;21:153–163.

62. Berger MS. The impact of technical adjuncts in the surgical management of cerebral hemispheric low-grade gliomas of childhood. *J Neuro-Oncol.* 1996;28:129–155.

63. Fallah A, Weil AG, Sur S, et al. Epilepsy surgery related to pediatric brain tumors: Miami Children's Hospital experience. *J Neurosurg Pediatr.* 2015;16:675–680.

64. Berger MS, Ghatan S, Haglund MM, et al. Low-grade gliomas associated with intractable epilepsy: seizure outcome utilizing electrocorticography during tumor resection. *J Neurosurg.* 1993;79:62–69.

65. Giordano M, Samii A, Lawson McLean AC, et al. Intraoperative magnetic resonance imaging in pediatric neurosurgery: safety and utility. *J Neurosurg Pediatr.* 2017;19:77–84.

66. Senft C, Bink A, Franz K, et al. Intraoperative MRI guidance and extent of resection in glioma surgery: a randomised, controlled trial. *Lancet Oncol.* 2011;12:997–1003.

67. Kubben PL, ter Meulen KJ, Schijns OE, et al. Intraoperative MRI-guided resection of glioblastoma multiforme: a systematic review. *Lancet Oncol.* 2011;12:1062–1070.

68. Mohammadi AM, Sullivan TB, Barnett GH, et al. Use of high-field intraoperative magnetic resonance imaging to enhance the extent of resection of enhancing and nonenhancing gliomas. *Neurosurgery.* 2014;74:339–348.

69. Martin C, Alexander E, Wong T, et al. Surgical treatment of low-grade gliomas in the intraoperative magnetic resonance imager. *Neurosurg Focus.* 1998;4.

70. Mattei L, Prada F, Legnani FG, et al. Neurosurgical tools to extend tumor resection in hemispheric low-grade gliomas: conventional and contrast enhanced ultrasonography. *Childs Nerv Syst.* 2016;32:1907–1914.

71. Prada F, Mattei L, Del Bene M, et al. Intraoperative cerebral glioma characterization with contrast enhanced ultrasound. *Biomed Res Int.* 2014;2014:484261.

72. Stummer W, Pichlmeier U, Meinel T, et al. Fluorescence-guided surgery with 5-aminolevulinic acid for resection of malignant glioma: a randomised controlled multicentre phase III trial. *Lancet Oncol.* 2006;7:392–401.

73. Roth J, Constantini S. 5ALA in pediatric brain tumors is not routinely beneficial. *Childs Nerv Syst.* 2017;33:787–792.

74. Stummer W, Rodrigues F, Schucht P, et al. Predicting the "usefulness" of 5-ALA-derived tumor fluorescence for fluorescence-guided resections in pediatric brain tumors: a European survey. *Acta Neurochir (Wien).* 2014;156:2315–2324.

75. Preuß M, Renner C, Krupp W, et al. The use of 5-aminolevulinic acid fluorescence guidance in resection of pediatric brain tumors. *Childs Nerv Syst.* 2013;29:1263–1267.

76. Valdés PA, Jacobs V, Harris BT, et al. Quantitative fluorescence using 5-aminolevulinic acid-induced protoporphyrin IX biomarker as a surgical adjunct in low-grade glioma surgery. *J Neurosurg.* 2015;123:771–780.

77. Sorge C, Li R, Singh S, et al. Complete durable response of a pediatric anaplastic oligodendroglioma to temozolomide alone: case report and review of literature. *Pediatr Blood Cancer.* 2017;64.

78. Cairncross G, Macdonald D, Ludwin S, et al. Chemotherapy for anaplastic oligodendroglioma. National Cancer Institute of Canada Clinical Trials Group. *J Clin Oncol.* 1994;12:2013–2021.

79. Cairncross G, Wang M, Shaw E, et al. Phase III trial of chemoradiotherapy for anaplastic oligodendroglioma: long-term results of RTOG 9402. *J Clin Oncol.* 2013;31:337–343.

80. Cairncross JG, Wang M, Jenkins RB, et al. Benefit from procarbazine, lomustine, and vincristine in oligodendroglial tumors is associated with mutation of IDH. *J Clin Oncol.* 2014;32:783–790.

81. van den Bent MJ, Brandes AA, Taphoorn MJ, et al. Adjuvant procarbazine, lomustine, and vincristine chemotherapy in newly diagnosed anaplastic oligodendroglioma: long-term follow-up of EORTC brain tumor group study 26951. *J Clin Oncol.* 2013;31:344–350.

82. Lecavalier-Barsoum M, Quon H, Abdulkarim B. Adjuvant treatment of anaplastic oligodendrogliomas and oligoastrocytomas. *Cochrane Database Syst Rev.* 2014;2014.

83. Buckner JC, Shaw EG, Pugh SL, et al. Radiation plus procarbazine, CCNU, and vincristine in low-grade glioma. *N Engl J Med.* 2016;374:1344–1355.

84. Sloan AE, Okada H, Ryken TC, et al. The role of emerging therapy in the management of patients with diffuse low grade glioma. *J Neuro-Oncol.* 2015;125:631–635.

85. Douw L, Klein M, Fagel SS, et al. Cognitive and radiological effects of radiotherapy in patients with low-grade glioma: long-term follow-up. *Lancet Neurol.* 2009;8:810–818.

86. Habets EJ, Taphoorn MJ, Nederend S, et al. Health-related quality of life and cognitive functioning in long-term anaplastic oligodendroglioma and oligoastrocytoma survivors. *J Neuro-Oncol.* 2014;116:161–168.

87. Pollack IF, Claassen D, al-Shboul Q, et al. Low-grade gliomas of the cerebral hemispheres in children: an analysis of 71 cases. *J Neurosurg.* 1995;82:536–547.

88. Puduvalli VK, Hashmi M, McAllister LD, et al. Anaplastic oligodendrogliomas: prognostic factors for tumor recurrence and survival. *Oncology.* 2003;65:259–266.

89. Shaw EG, Berkey B, Coons SW, et al. Recurrence following neurosurgeon-determined gross-total resection of adult supratentorial low-grade glioma: results of a prospective clinical trial. *J Neurosurg.* 2008;109:835–841.

90. Smith JS, Chang EF, Lamborn KR, et al. Role of extent of resection in the long-term outcome of low-grade hemispheric gliomas. *J Clin Oncol.* 2008;26:1338–1345.

91. Fujii Y, Muragaki Y, Maruyama T, et al. Threshold of the extent of resection for WHO Grade III gliomas: retrospective volumetric analysis of 122 cases using intraoperative MRI. *J Neurosurg.* 2017;e20171–e20179.

92. Taal W, Brandsma D, de Bruin HG, et al. Incidence of early pseudo-progression in a cohort of malignant glioma patients treated with chemoirradiation with temozolomide. *Cancer.* 2008;113:405–410.

93. Brandsma D, Stalpers L, Taal W, et al. Clinical features, mechanisms, and management of pseudoprogression in malignant gliomas. *Lancet Oncol.* 2008;9:453–461.

94. Lin AL, White M, Miller-Thomas MM, et al. Molecular and histologic characteristics of pseudoprogression in diffuse gliomas. *J Neuro-Oncol.* 2016;130:529–533.

95. Meyzer C, Dhermain F, Ducreux D, et al. A case report of pseudoprogression followed by complete remission after proton-beam irradiation for a low-grade glioma in a teenager: the value of dynamic contrast-enhanced MRI. *Radiat Oncol.* 2010;5:9.

96. Oborski MJ, Laymon CM, Lieberman FS, et al. Distinguishing pseudoprogression from progression in high-grade gliomas: a brief review of current clinical practice and demonstration of the potential value of ^{18}F-FDG PET. *Clin Nucl Med.* 2013;38:381–384.

97. Pamir MN, Özduman K, Yıldız E, et al. Intraoperative magnetic resonance spectroscopy for identification of residual tumor during low-grade glioma surgery: clinical article. *J Neurosurg.* 2013;118:1191–1198.

98. Danchaivijitr N, Waldman AD, Tozer DJ, et al. Low-grade gliomas: do changes in rCBV measurements at longitudinal perfusion-weighted MR imaging predict malignant transformation. *Radiology*. 2008;247:170–178.

99. Lacerda S, Law M. Magnetic resonance perfusion and permeability imaging in brain tumors. *Neuroimaging Clin N Am*. 2009;19:527–557.

100. Law M, Yang S, Wang H, et al. Glioma grading: sensitivity, specificity, and predictive values of perfusion MR imaging and proton MR spectroscopic imaging compared with conventional MR imaging. *AJNR Am J Neuroradiol*. 2003;24:1989–1998.

101. Chawla S, Oleaga L, Wang S, et al. Role of proton magnetic resonance spectroscopy in differentiating oligodendrogliomas from astrocytomas. *J Neuroimaging*. 2010;20:3–8.

102. Lev MH, Ozsunar Y, Henson JW, et al. Glial tumor grading and outcome prediction using dynamic spin-echo MR susceptibility mapping compared with conventional contrast-enhanced MR: confounding effect of elevated rCBV of oligodendrogliomas. *AJNR Am J Neuroradiol*. 2004;25:214–221.

103. Brandel MG, Alattar AA, Hirshman BR, et al. Survival trends of oligodendroglial tumor patients and associated clinical practice patterns: a SEER-based analysis. *J Neuro-Oncol*. 2017;133:173–181.

Section E

Radiation therapy

Chapter 21

Basic principles of brain tumor radiotherapy

Katarina Petras*, Samir Sejpal†, Sean Sachdev‡ and Minesh P. Mehta§,¶

*Department of Radiation Oncology, Robert H. Lurie Comprehensive Cancer Center, Northwestern University, Feinberg School of Medicine, Chicago, IL, United States; †University of Central Florida, Florida Hospital Cancer Institute, Orlando, FL, United States; ‡Department of Radiation Oncology, Lou and Jean Malnati Brain Tumor Institute, Robert H. Lurie Comprehensive Cancer Center, Northwestern University, Feinberg School of Medicine, Chicago, IL, United States; §Department of Radiation Oncology, Florida International University, Miami, FL, United States; ¶Miami Cancer Institute, Miami, FL, United States

Introduction

Brain neoplasms, a diverse group of tumors with heterogeneous behavior patterns and markedly different clinical outcomes, remain challenging to treat despite numerous recent advances in therapeutic options. This chapter will focus on the principles of radiotherapy for the treatment of brain neoplasms, with an emphasis on external beam radiotherapy (EBRT) and adult gliomas.

Primary CNS neoplasms

Primary glial neoplasms can arise from cells in any part of the central nervous system (CNS) from an intracranial origin or from the spinal cord. The presence of the blood-brain-barrier (BBB) makes the CNS unique in its patterns of progression, and spread of a primary CNS tumor via blood or lymphatics to another part of the body is relatively uncommon. Inside the CNS itself, tumors can spread to adjacent or semidistant parts of the brain via white-matter pathways such as the corpus callosum, along the subependymal lining, or along the leptomeninges.

The Central Brain Tumor Registry of the United States (CBTRUS) is an organization that compiles data from 12 tumor registries to report on the incidence of malignant and benign CNS lesions in the United States. From 2008 to 2012, the US incidence of primary brain and nervous system tumors in adults aged 20 and older was estimated to be 28.6 cases per 100,000 persons.[1] About one-third of these tumors were malignant, while the remaining two-thirds were borderline malignant or benign. The incidence rate for children aged 0–19 was only 5.6 per 100,000, but a larger portion (65%) were malignant.[2]

Low-grade gliomas (LGGs) include grade I tumors (e.g., pilocytic astrocytoma) and grade II tumors (e.g., astrocytoma, oligodendroglioma, and oligoastrocytoma). They account for about 10% of all primary brain tumors. Ten-year recurrence free survival for grade I gliomas following resection is approximately 95%.[3] Patients with grade II gliomas can have excellent long-term clinical outcomes[4] (Fig. 1). The most recent WHO classification now uses molecular features to categorize tumors, and the lower grade gliomas more commonly possess mutations in *IDH1* or *2* genes, and a subset of these tumors also possess the pericentromeric unequal translocation, 1p/19q (co-deleted), which imparts a favorable prognosis.

High-grade glioma (HGG), which includes anaplastic astrocytoma (grade III, G3) and glioblastoma (grade IV, G4), are the most common primary malignant brain tumors in adults (Fig. 2). Using the recent WHO molecular classification, these are more frequently *IDH* wild-type (and the *IDH* wild-type tumors are almost always non-co-deleted at 1p/19q). Respectively, they account for 6% and 54% of all gliomas.[1] Glioblastoma carries a dismal prognosis. Only about half of patients are alive 1-year after diagnosis, and less than 5% live beyond 5 years. However, those with promoter methylation of the *MGMT* gene, treated with combination chemoradiotherapy, experience significantly superior 5-year survival. Only about a quarter (27%) of G3 anaplastic astrocytoma patients are expected to be alive at 5-years.[1] The most important prognostic

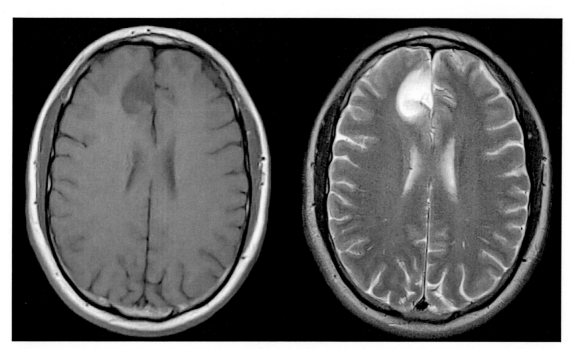

FIG. 1   Axial images of WHO grade II glioma. The tumor does not enhance on a T1 postcontrast sequence (left) but demonstrates clear abnormal intensity on T2 imaging (right).

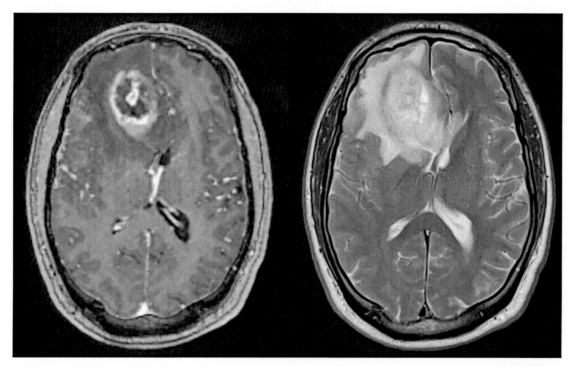

FIG. 2   Axial images of WHO grade IV glioblastoma. A centrally necrotic mass with peripheral enhancement can be seen on the T1 postcontrast image (left). The tumor demonstrates marked enhancement on the T2 sequence (right).

factors associated with survival include age >50, histology, performance status (PS), mental status changes, and symptom duration ≥3 months.[5] However, the G3 anaplastic oligodendroglioma, characterized by *IDH* mutations and also frequently expressing co-deletion of 1p/19q, have significantly superior survival, especially when treated with combination chemoradiotherapy.

There are many types of benign primary CNS neoplasms including meningiomas, vestibular schwannomas, and pituitary adenomas. Meningiomas account for about 30% of primary CNS tumors.[6] Benign tumors are generally managed with primary surgical resection, or primary radiation therapy and/or a combination of these modalities.

Although some hereditary conditions and exposure to ionizing radiation[7] are known to be associated with the development of brain tumors, the majority of cases are thought to arise sporadically. More recently, cellular phone use has been investigated as a potential risk factor for brain tumor development, but so far the results are largely inconclusive and unconvincing.[8, 9]

Radiation therapy

Radiation therapy is used for the curative treatment of CNS malignancies, either in an adjuvant or definitive setting, but it can also be used with palliative intent. The difference between these two approaches lies mainly in the doses used and techniques of treatment delivery.

The most commonly used form of radiation is EBRT. This is commonly delivered using a linear accelerator or "LINAC" (Fig. 3) a device that accelerates electrons onto a heavy metal (tungsten) target. Bombardment of the target causes production of high energy x-rays (photons), which are channeled through the treatment head toward the patient. Photons are produced through a process known as *brehmsstrahlung* or "breaking radiation" which occurs when charged particles are decelerated by an atomic nucleus. The LINAC can also produce a beam of primary electrons for treatment. Electrons have different physical properties than photons, and thus different advantages and limitations depending on the clinical situation; given their preferential superficial dose deposition, their role in the treatment of CNS malignancies is limited.

EBRT is the most common form of RT used in the treatment of gliomas. It is generally delivered as a daily treatment over 4–6 weeks with or without chemotherapy. It is utilized either in the adjuvant setting following surgery, or in the definitive setting following biopsy, limited surgery, or in cases of patient ineligibility (i.e., the patient is not a surgical candidate due to medical comorbidities or tumor location).

Proton therapy is an alternative form of EBRT utilizing heavy charged particles. Protons are produced and accelerated by the delivery system's cyclotron. Protons behave differently than traditional photon therapy in that they travel a certain distance within the patient, depositing a low radiation dose until they reach their termination destination, where along a narrow distance known as the Bragg Peak they deposit the bulk of their energy. Unlike photon therapy, protons have minimal exit dose, thus limiting the exposure of distal normal tissues to radiation.

Protons are believed to have slightly different biological effect, with slightly higher potency per equivalent dose compared to photons. This "relative biological effectiveness" (RBE) is generalized to a value of 1.1 compared to photons, but there is growing understanding that there is variability in this, with RBE values possibly as high as 1.3 immediately beyond the Bragg peak.[10] Increased RBE can pose the potential for increased normal tissue toxicity, particularly around sensitive structures such as the optic nerves/chiasm and brainstem. Indeed, such toxicities have been reported, particularly in the pediatric setting,[11] and are of concern as we learn more about utilizing protons as a clinical treatment modality. Proton therapy has been used extensively in pediatric brain tumors due to its advantage of limiting dose (and volume of lower

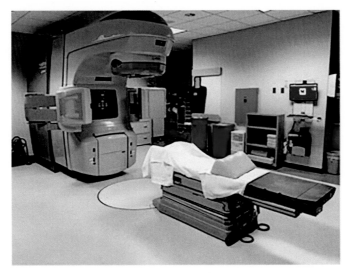

FIG. 3   A modern linear accelerator used for patient treatment.

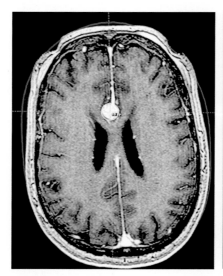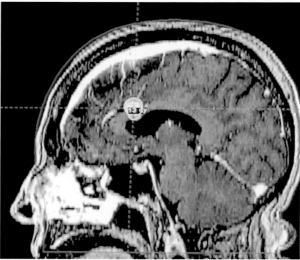

FIG. 4 Stereotactic radiosurgery (SRS) in the treatment of a benign meningioma. This lesion was treated to 13 Gy in one fraction. The 50% isodose line is demonstrated in *yellow*.

dose exposure) of adjacent critical structures. This property of protons remains especially relevant in the pediatric setting, as these patients, if they are long term survivors, can be at higher risk of numerous secondary effects including radiotherapy-induced functional deficits (such as neurocognitive or endocrine dysfunction) and secondary malignancies.[12]

Stereotactic radiosurgery, or SRS, is a highly conformal type of EBRT. In this technique, the target volume receives a highly concentrated dose of radiation with sharp falloff outside of the target volume. SRS can be delivered using a traditional linear accelerator with a specialized/advanced immobilization set-up designed for this purpose. When SRS is utilized for intracranial disease, specialized systems are used including a LINAC adaption such as the CyberKnife. Another system, the Gamma Knife, utilizes Cobalt-60 radioisotope-generated gamma rays for treatment. Both devices deliver radiation therapy with extreme precision (submillimetric). Examples of use for SRS in the treatment of benign neoplasms are shown in Fig. 4, with a highly conformal treatment delivered in a single fraction.

Radiation can also be delivered using a technique called brachytherapy, wherein small radioactive sources (or seeds) are placed temporarily in the body, usually via indwelling catheters. It can also be done through direct instillation of radioactive solutions where balloon-like devices are surgically implanted within a tumor cavity and subsequently filled percutaneously with the solution. Brachytherapy forms a "dose cloud" inside the patient, with sharp fall off of dose outside the cloud. While highly conformal, the process is labor intensive and requires optimal geographic distribution of the dose-depositing source. Risks include infection or brain necrosis.[13] Because of these issues, brachytherapy is used uncommonly in CNS disease.[14]

Foundations of radiotherapy: The 5 "R's"

There are five biological principles of tumor cell behavior that lie at the heart of radiotherapy fractionation schemata. These are: repair, reassortment, repopulation, reoxygenation, and radiosensitivity; or the "R's" of radiobiology.[15, 16]

1. Repair—DNA repair mechanisms are activated in response to radiotherapy-induced damage. In general, tumors are less efficient at repairing this damage than most normal tissues, with certain notable exceptions.
2. Reassortment—cells exhibit varying degrees of radiosensitivity depending on the cell cycle phase they are in. In general, mitosis is the most sensitive cell cycle phase and late S-phase is the most resistant. Fractionation ensures that over a protracted course of treatment, almost all cycling cells will be exposed to radiation while in the sensitive phases.
3. Repopulation—clonogenic tumor cells that survive radiation exposure may repopulate the remaining tumor environment quickly. Fractionation allows repeated opportunities to target these repopulating cells.
4. Reoxygenation—radiation sensitivity increases with increasing oxygen in the tumor microenvironment. In general, tumors under 1 mm are considered to be fully oxygenated, but larger tumors will develop regions of hypoxia as they outgrow their blood supply. Fractionation tends to eliminate the radiosensitive well-oxygenated cells first, thereby permitting the less-oxygenated cells to access the available oxygen in greater concentration, rendering these cells now more radiosensitive to subsequent fractions of radiation.

5. Radiosensitivity—different tumor and normal tissue histologies have varying degrees of inherent sensitivity to radiation treatments. The most radiosensitive, or easily damaged cells, are those that divide rapidly, are undifferentiated, and are mitotically active. This, however, is a rather simplistic summary of multiple inherent molecular processes that balance pro-survival versus pro-death signal cascades following initial DNA damage.

These various biological properties are taken into account when designing a treatment regimen for any tumor. The efficacy and safety depends on the size and the type of tumor being treated, the tolerance of the adjacent tissue, and on fraction size. Standard fractionation (SF) is considered to be a dose of 1.8–2 Gy per fraction. Any single dose greater than 2 Gy is generally associated with lowering the total number of fractions used, hence decreasing overall treatment time, and is therefore considered a form of hypofractionation (HF).

In adult brain tumors, radiotherapy is usually given as a daily treatment over 2 to 6 weeks. In the case of gliomas, RT is delivered at SF to total doses of 50.4–54 Gy (LGG) or 59.4–60 Gy (HGG) over 6 weeks. Of late, hypofractionated schedules for HGG are being increasingly utilized in certain patient subsets. RT may be given concurrently with chemotherapy to potentiate radiation sensitivity. This is done most commonly for HGGs, but sometimes also for low-grade disease. In adult glioma patients older than 65–70 or with poor PS, EBRT can be delivered using a HF schedule of 2.66 Gy daily over 3 weeks to a total dose of 40 Gy (or one of several other possible hypofractionated schedules).[17]

Important, too, is the consideration of whether the treatment target and adjacent critical tissue is a parallel versus serial organ.[18] Parallel organs, such as the liver, have tissue redundancy. Thus when part of the organ is damaged, the remaining healthy subunits can still yield normal function of the organ. On the other hand, the spinal cord is a classic serial organ. Loss of function occurs even if a small part of the structure is permanently damaged. Location of critical tissues near target volume may affect the total dose and fractionation scheme chosen for treatment of the target volume.

Initial diagnosis and surgical management of brain tumors

Although multiple MRI sequences will likely be utilized to establish a preliminary diagnosis, the most helpful studies often include the postcontrast TI and T2 as well as fluid attenuation inversion recovery (FLAIR) images. The postcontrast T1 study will clarify the dimensions and characteristics of a contrast-enhancing primary tumor (HGGs enhance after the administration of contrast). The T2 and FLAIR sequences will better characterize nonenhancing tumor or peritumoral edema, which has been found to house microscopic tumor cells. For primary brain neoplasms, maximal safe surgical resection, when feasible, remains the cornerstone of treatment.[19] Surgical removal is important for both diagnostic and therapeutic purposes, particularly where symptomatic mass effect exists. Needle biopsies are generally reserved for inoperable brain lesions, such as those in vital brain regions. Biopsies can be unrepresentatively misleading from a pathological standpoint due to sampling bias and heterogeneity of tumor tissue. When surgery is performed adjacent to a critical portion of the brain, such as the sensory/motor cortex or speech region, functional mapping can be performed intraoperatively to help determine the optimal limits of the surgical resection.[20, 21] Following surgery, important variables such as the extent or resection and pathologic findings, will help determine the appropriate combination of adjuvant radiation therapy with or without chemotherapy.

Steps in modern radiotherapy (Table 1)
Patient simulation

Modern radiotherapy requires many steps after an initial consultation to ensure that treatments are delivered accurately and safely. The ultimate goal is to be able to perform accurate imaging-based planning to tailor treatments to an individual's unique anatomy. In order to do this, patients undergo a process known as CT simulation.

The goal of the CT simulation is to create a reliable and comfortable position for the patient to assume during their daily treatments. In the case of intracranial disease, this involves the creation of a thermoplastic mask that fits over the patient's head and neck. The patient is first placed in the supine position with their chin at a neutral angle. A flat piece of malleable plastic with fishnet texture is immersed in a warm water bath and allowed to become flexible. This is then quickly stretched over the patient's face and neck and secured via pegs to the treatment table underneath. As the thermoplastic mask dries and cools, it hardens and takes on a permanent shape of the patient's facial contour. At the end of this process, the mask will allow for reliable daily rigid immobilization for reproducibility, prior to treatment (Fig. 5).

TABLE 1 Steps in modern radiotherapy planning

Planning step	Substeps
(1) Patient simulation	• Create thermoplastic mask • Perform CT scan of head and neck • Place isocenter in middle of target volume • Mark 3D coordinates on mask surface
(2) Treatment planning	• Fuse relevant pre-op/post-op MRIs • Contour critical OARs • Contour GTV and CTV • Expand to create PTV • Decide on beam arrangement/treatment technique
(3) Plan evaluation	• Does the target have adequate coverage? • Are dose constraints for critical OARs met?
(4) Quality assurance	• Dose delivery verification by medical physicist
(5) Treatment initiation	• Align patient on treatment couch • Obtain port films (kV or CBCT) • Perform shifts in necessary directions • If alignment correct, deliver treatment
(6) On-treatment	• Treating physician approves daily set-up films • Physician and patient meet weekly to address treatment progress and toxicities
(7) Follow-up	• Patients initially seen at 4–8 weeks post-RT • Then seen quarterly on a yearly basis • Monitor clinical and radiographic response • Address any long-term toxicities

OARs, organs at risk; *GTV*, gross tumor volume; *CTV*, clinical target volume; *PTV*, planning target volume; *CBCT*, cone beam CT.

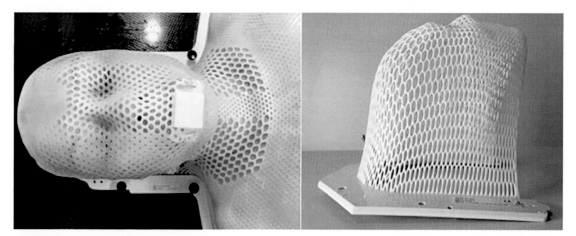

FIG. 5   A thermoplastic mask used for immobilization of the patient's head during external beam radiotherapy.

Once the mask has hardened, CT images of the head and upper neck are obtained. A virtual point is placed near the middle of the intracranial target volume, which will correspond to the geometric central axis of the treatment machine known as the "isocenter" (Fig. 6). Leveling lasers inside the CT scanner room are turned on, and 3D coordinates of this isocenter point are obtained. These are then marked on the patient's mask and will be used, in addition to the mask itself, to line the patient up for daily treatments.

This entire simulation process takes approximately one hour. Radiation therapy will generally commence 1–2 weeks after this is complete. The next steps involve treatment planning where relevant preoperative and postoperative MRI scans are fused to the planning CT scan to better delineate the extent of intracranial disease and peritumoral edema.

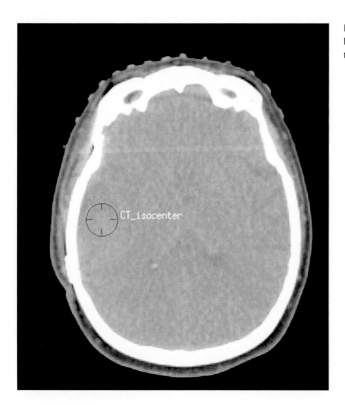

FIG. 6 Axial planning CT slice with the virtual isocenter point that will be used for dose calculations during treatment planning. This is placed in the center of the target volume during the patient simulation process.

Treatment planning

Once the preoperative and postoperative MRI scans are fused to the planning scan, one begins to three-dimensionally define target volumes as well as adjacent critical normal structures. Starting on axial planning CT slices, critical CNS structures near the intracranial target volume are contoured (outlined) so that the dose to their 3D volume during treatment can be calculated (Fig. 7). This ensures that the known tolerance for each individual structure is not exceeded. With CNS disease, organs-at-risk (OARs) include the right/left orbit, lenses, optic nerves, optic chiasm, hippocampi, cochleae, temporal lobes, eyes, lacrimal gland, pituitary gland, brainstem, and upper cervical spinal cord. Critical structures and their respective dose tolerances are outlined in Table 2.

The determination of target volumes in radiotherapy is based on the concept of three different volume subtypes (Fig. 8). These include gross tumor volume (GTV), clinical target volume (CTV), which accounts for subclinical disease, and planning target volume (PTV), accounting for daily set-up errors and unmeasurable errors. For gliomas, GTV includes any residual disease, T1 enhancement, T2 or FLAIR signal, and the postoperative tumor bed. The CTV, which accounts for microscopic tumor infiltration, is created by expanding the prior volume uniformly by 1–2.5 cm depending on histological subtype. The CTV is appropriately manipulated in shape off anatomical barriers to tumor spread such as the dura, ventricles, falx, and tentorium cerebelli. A PTV margin of 3–5 mm is then added to the CTV.

Once target structures and OARs have been delineated, the planning process begins. In general, there are two types of radiotherapy planning, forward and inverse planning. During forward planning, the planner selects beam arrangements in the treatment planning system that will provide adequate coverage of the target volumes while reducing dose to OARs. Based on the selected beam energy and geometric arrangement as well as shape, the treatment planning system calculates the three dimensional shape of the dose distribution allowing evaluation and further manual input in modification of beams. Forward planning is rarely utilized in the CNS due to the need to minimize high dose to adjacent areas and critical structures.

Instead, inverse planning is more commonly utilized. In this situation, the planner assigns OARs various importance weight factors. The treatment planning system then selects beam arrangements that give priority to avoiding the OARs while covering the target volumes as best as possible. Intensity-modulated radiotherapy (IMRT) is a commonly utilized technique based on inverse planning. A variation of IMRT known as volumetric-modulated arc therapy (VMAT) delivers treatment with continuous motion of the LINAC head, allowing for nonceasing beam modulation. VMAT is becoming

FIG. 7 Different critical organs-at-risk (OARs) are outlined on a planning CT slice so that they can be avoided during treatment of the target volume.

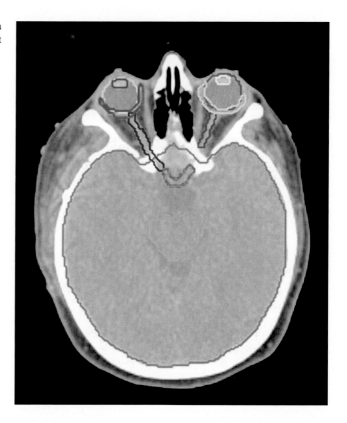

TABLE 2 Normal tissue dose tolerances

EBRT using 1.8–2.0 Gy/fractions	SRS Max point dose
Whole brain 50 Gy	Brainstem 12 Gy
Partial brain 60 Gy	Optic nerve and chiasm 8–10 Gy
Brainstem 54 Gy	Cranial nerves 12 Gy
Spinal cord 45 Gy	
Chiasm 50–54 Gy	
Retina 45 Gy	
Lens 10 Gy	
Cochlea 30–35 Gy	

EBRT, external beam radiotherapy; *SRS*, stereotactic radiosurgery.

more commonly utilized in radiation oncology as it has potential dose-distribution benefits and can minimize treatment duration. However, it should be remembered that multiple solutions can fit the constraints imposed by the planner, and selection between these can be a challenge. These "optimal" solutions are generally described as residing along a mathematical plane, the Paredo plane, and because of the possibility of more than one "correct" solution, artificial-intelligence based programs still require human guidance in seeking an optimal plan for a given patient.

Dose reporting

Once a suitable plan is generated, the radiation oncologist will carefully evaluate the plan to ensure that the target volume is receiving adequate dose and that constraints for adjacent critical structures are met. If dose objectives are not satisfactory, the plan will be iteratively modified until a suitable plan is achieved. Modern treatment planning systems (TPS) allow

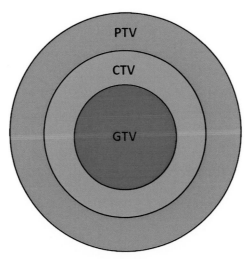

FIG. 8   Schematic representation of gross tumor volume (GTV), clinical target volume (CTV), and planning target volume (PTV).

multicriteria optimization, wherein the user can alter the dose to a specific structure while visualizing the consequential impact of this on a change in dose to all other structures. The final objectives and dose constraints are evaluated and reported.

Different energies of ionizing radiation have different characteristic patterns of energy deposition in tissue, and this allows for the biologic dose distribution to be estimated. Isodose lines and planes are two- and three-dimensional representations on a treatment plan that represent planar and volumetric variations in absorbed dose. When high-energy photons enter a patient, they cause the production of high-speed electrons at the surface and in subsequent tissue layers. These electrons deposit their energy at a slight distance from where they are produced. The further the photon travels in tissue, the more energy it loses, and therefore the less likely it is to produce new electrons. As a result, electron dose builds with depth until a maximum dose depth is reached. After this, dose decreases with further depth. Isodose curves represent a volume receiving (at minimum) an equivalent dose; this is often evaluated in absolute dose values or percentage dose values (Fig. 9).

Radiation oncologists often evaluate a dose-volume-histogram (DVH), which graphically depicts volume and dose deposited. This allows one to evaluate proper target coverage and safe dose to OARs. The DVH is a visual representation of maximum, minimum, and average doses to varying subvolumes of a structure. The physician can evaluate the DVH of an individual structure, or group of structures, when assessing treatment plan quality (Fig. 10).

Quality assurance

Once the physician has approved a patient's plan, a trained medical radiation physicist must verify that the calculated dose is what will actually be delivered to the patient at the time of treatment. This is done at least a day prior to patient arrival as part of quality assurance measures. The physicist will use a tissue phantom and dose verification mechanisms to compare the dose delivered during a test treatment to what was calculated by the planning system. If within sufficient range, the treatment plan is deemed safe to be delivered to the actual patient.

Initiation of treatment

On the first day of treatment, the patient is aligned in the thermoplastic mask on the treatment couch. Imaging is obtained just prior to treatment (either from lower-energy beams emanating from the treatment head or specialized attachments, or using megavoltage beam energies). This allows bony (and also soft-tissue) anatomy to be compared to reconstructed images from the patient's CT-simulation scan (Fig. 11). If the patient is not aligned properly, shifts can be made in the x, y, and z (horizontal, vertical, and lateral) directions to correct the misalignment. Modern day couch-tops also permit angular correction. A second set of films is then taken to confirm that the shifts resulted in correct alignment, which the physician approves prior to treatment delivery.

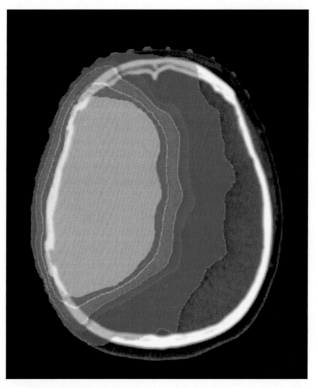

FIG. 9 Isodose curve lines representing volumes receiving (at minimum), a certain percentage of the prescribed dose. The *yellow* volume represents a high dose region, while the outer *dark blue* volume represents a low dose region.

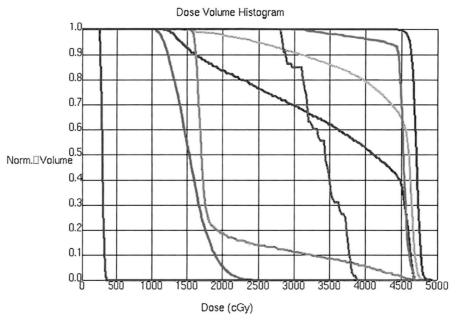

FIG. 10 A dose-volume-histogram (DVH) which plots dose to volume of specified OARs and target structures.

Many linear accelerators have the ability to yield cone-beam CT (CBCT) scans utilizing kilovoltage beams and a special attachment. These CBCTs can be compared to the planning CT images (Fig. 12). Other devices can produce megavoltage CT scans (MvCTs), which have the added benefit of minimizing artifacts induced by metallic implants/devices.

After the positioning (referred to as set-up) is approved, trained radiation therapists will deliver the radiation treatments, allowing the treatment system to replicate beams that were formerly planned and evaluated. Modern treatment machines operate using record-and-verify systems, which ensure that all subsystems and steps are in appropriate sequence,

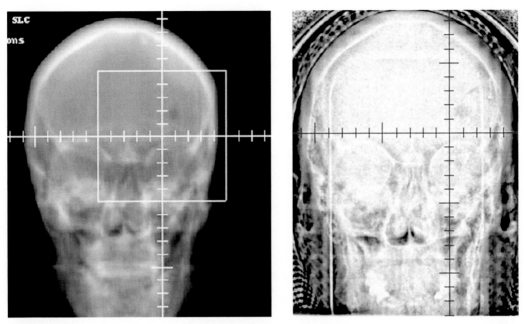

FIG. 11   Kilovoltage images, also known as port films, are taken at the start of treatment and daily to verify a patient's position on the treatment table. The real-time image (right) is compared to a 3D reconstruction from the patient's CT simulation (left). Bony landmarks are used to verify set-up position, and images are taken from at least two angles.

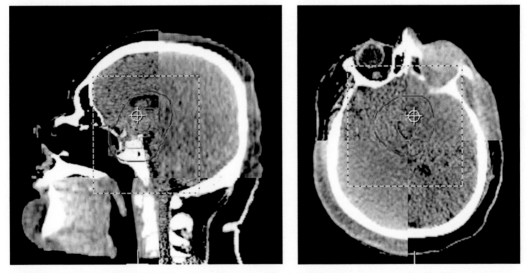

FIG. 12   Cone-beam CTs (CBCTs) allow for real-time comparison of a patient's soft tissue images to the planning CT scan.

minimizing the likelihood of error. Regular interval imaging is obtained to ensure appropriate target coverage and proper alignment. This is often done daily when treatment intent is curative, the treatment is complex, or if the likelihood of damage to an OAR is higher than desired.

Sometimes during the course of treatment a patient may no longer be optimally set-up as planned. For example, a patient may gain or lose weight, or develop facial swelling due to corticosteroids, and the thermoplastic mask may not fit. In these cases, adjustments to the plan or immobilization equipment may be needed. Often patients are replanned in a new mask, and a new treatment plan is created.

On-treatment visits and follow-up

While physicians see patients as often as needed during the treatment course, they are scheduled for at least a once weekly clinical appointment called an on-treatment visit, or OTV, so that their clinical course can be closely evaluated. The goal of weekly visits is to monitor patient progress, such as clinical response during treatment, assess acute treatment related toxicity, and address questions or concerns that patient and/or family members may have. These visits also serve the important purpose of reinforcing treatment goals and objectives, which many patients often require given the stressful situation of dealing with a tumor. Following completion of treatment, patients are seen in follow-up initially at 4–8 weeks and then at predefined intervals based on diagnosis to assess radiographic and clinical response and attend to any long-term treatment related toxicities.

Specific treatment techniques

Described below are target volume expansions, target doses, and dose constraints for specific brain neoplasms.

Low-grade infiltrative astrocytomas and oligodendrogliomas

WHO grade II tumors, which include infiltrating low-grade astrocytomas, oligodendrogliomas, and mixed oligoastrocytomas, are a rare and diverse group of tumors. Features associated with a higher risk of rapid postoperative relapse include age ≥40, astrocytoma histology, tumor size ≥5 cm, tumor that crosses midline, subtotal resection, and neurological deficit prior to resection; molecular features associated with poorer prognosis are *IDH* wild type and 1p-19q non-co-deletion status.[22, 23] Surgical resection remains the mainstay of treatment, and the role of postoperative radiotherapy continues to be more optimally defined. About 70%–80% of LGGs will transform to high-grade disease.[24]

Timing of adjuvant RT was examined in EORTC 22845.[24] Adult patients were randomized to either early versus delayed RT of 54 Gy. The 5-year analysis showed that disease-free survival was better with early adjuvant RT (44% vs 37%, $P = .02$) but without a statistically significant difference in overall survival (OS). Delayed RT also resulted in poorer seizure control in patients. Thoughtful deferment of adjuvant RT for future treatment, in appropriate patients, was deemed an acceptable option following this trial, a recommendation that now requires reconsideration in the light of modern results.

When adjuvant RT is delivered, margins are often defined as the following: gross target volume (GTV) = tumor bed and residual disease seen on T1 but more importantly T2-weighted and/or FLAIR MRI images. These tumors rarely demonstrate enhancement on T1 sequences, and almost always demonstrate abnormal T2 or FLAIR intensity.

CTV is defined as GTV plus a 1–2 cm margin, which allows for coverage of the infiltrative tumor border that is not adequately visualized on MR imaging. In recent EORTC trials, there is a trend toward decreasing this CTV margin for low-grade tumors based on the contention that it would likely reduce the volume of normal brain being irradiated. Given that these patients are now generally treated with combination chemo-radiotherapy, the rationale is that chemotherapy will "mop-up" any microscopic disease missed by decreasing CTV margins (note that there is no firm evidence for this contention, as of yet).

A final PTV margin of approximately 5 mm is added to the CTV (Fig. 13) to account for day-to-day variability in reproducibly aligning the patient for treatment. With modern-day daily volumetric imaging, margins as low as 3 mm can be used but require very diligent daily review by the physician. The standard dose range is 45–54 Gy at 1.8–2.0 Gy per fraction. There has been no proven benefit to dose escalation in these patients.[25, 26]

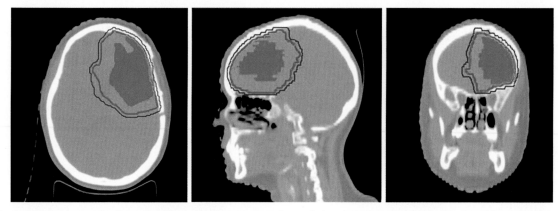

FIG. 13   An adjuvant low-grade glioma plan. The *red* volume represents the surgical cavity and residual T2 FLAIR signal. The patient was treated to 54 Gy as demonstrated by the *green* CTV and *blue* PTV lines.

RTOG 9802 examined the role of adjuvant combination chemo-radiotherapy in patients with LGG.[27] A total of 251 patients with LGG were stratified into low-risk (age < 40, gross total resection) and high-risk (age > 40, subtotal resection/biopsy) groups and the high-risk group was randomized to adjuvant RT alone versus RT followed by 6 cycles of combination chemotherapy (procarbazine, CCNU, and vincristine or PCV). At 12 years follow-up, the RT + chemotherapy arm had improved median OS (13.3 vs 7.8 years) and median progression-free survival (10.4 vs 4.0 years).

These results require very serious introspection regarding the prior recommendation of "watch-and-wait" for some high-risk LGG patients. Because the early versus delayed radiotherapy trials did not alter OS, it was felt that there was nothing lost by waiting. However, these results categorically demonstrate that the combination chemoradiotherapy arm, which yielded a median survival in excess of 13 years, now represents the modern recommendation for *all* high-risk low-grade glioma patients. Whether chemotherapy alone followed by salvage radiotherapy would yield similar results remains unproven, and in single institution series where this practice is adopted, the median survival is much shorter than the 13.3 years observed on RTOG 9802.[28, 29] Furthermore, a randomized EORTC trial failed to show superiority of adjuvant temozolamide to adjuvant radiotherapy.[30]

Low-risk LGG patients are still considered appropriate for resection followed by observation, but an important caveat here is to be wary of the "slippery-slope" effect. Some practitioners have started designating all *IDH* mutated, or all *IDH*-mutated and 1p-19q co-deleted G2 gliomas (which are mostly G2 oligodendroglioma) as "low-risk." Post hoc subset analysis of RTOG 9802 demonstrate that these patients categorically benefit from combined chemo-radiotherapy, and therefore caution is advised at attempts to "convert" high-risk LGG to so-called low-risk on the basis of molecular features, for the purposes of treatment deferral.

High-grade gliomas

For grade III anaplastic astrocytomas and grade IV glioblastomas, radiation is usually delivered following biopsy or resection in the adjuvant setting. The radiation volumes and margins are often defined using one of at-least two different philosophical approaches: the so-called EORTC approach or the NRG/RTOG approach.

In the EORTC approach, GTV = tumor bed and residual disease seen on T1 images, using the larger volume seen on either the pre-op versus post-op MRI. CTV = GTV = a 2–3 cm margin to include the FLAIR abnormality, but trimmed off anatomic barriers to tumor spread. PTV = CTV = 3–5 mm. The typical total dose delivered is 59.4–60 Gy in 1.8–2.0 Gy per fraction.

An alternative method is the two-step NRG/RTOG approach. This utilizes a larger lower-dose volume treated to 46 Gy (GTV1 = FLAIR or T2; CTV1 = GTV1 = 2 cm with appropriate "anatomic trimming"; PTV1 = CTV1 = 3–5 mm). This is followed by a sequential boost bringing the total dose up to 59.4–60 Gy (GTV2 = enhancing tumor and tumor bed; CTV2 = GTV2 = 2.5 cm, with appropriate "anatomic trimming"; PTV2 = CTV2 = 3–5 mm) (Fig. 14).

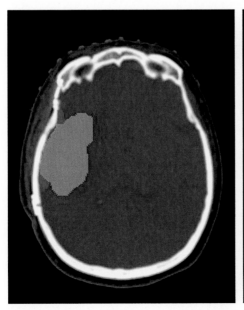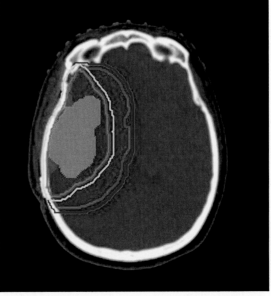

FIG. 14 An adjuvant high-grade glioma plan. The postsurgical cavity and residual T2 enhancement are first contoured using a fusion of MR images to the planning CT scan (*red* and *orange* volumes on left). The volume receiving 46 Gy is represented by the *green* CTV and *blue* PTV lines, while the boost volume receiving 60 Gy is represented by the *pink* CTV and *yellow* PTV lines (right).

Glioblastoma

Radiotherapy is given concurrently with the alkylating agent temozolomide (TMZ) in the case of glioblastoma. HGGs that harbor promoter-methylation of the *MGMT* gene have increased sensitivity to the effects of TMZ. In 2004, the EORTC and NCIC performed a randomized trial among adult glioblastoma patients comparing standard radiotherapy, 60 Gy in 30 fractions, to the same radiation schedule with concurrent TMZ followed by 6 cycles of adjuvant TMZ. At 5-years, there was improved OS with TMZ compared to RT alone (9.8% vs 1.9%, $P < .0001$). *MGMT* promoter methylation was the strongest predictor of benefit and improved outcome from TMZ chemotherapy.[31] Five-year OS in methylated patients treated with combination therapy was 13.8% versus 8.3% in unmethylated patients; with RT alone this dropped to 5.2% and 0%, respectively.

In patients older than 65–70 or those with a poor PS, shortened courses of radiotherapy are acceptable. Potential dosing schemes include 34 Gy in 10 fractions,[32] 40 Gy in 15 fractions,[17] 50 Gy in 20 fractions,[33] or 25 Gy in 5 fractions.[34] A very recent NCIC-C trial has established that at least in the *MGMT*-methylated elderly glioblastoma patients, hypofractionated RT plus TMZ produces superior survival to hypofractionated RT alone; although for *MGMT*-unmethylated patients, the survival difference was statistically nonsignificant.[35]

A very recent development has been the improved survival observed for glioblastoma patients treated with chemoradiotherapy followed by the alternating electrical field device, Optune (Novo-TTF).[36] The device is applied in the maintenance phase and does not impact delivery of radiotherapy. However, data for concomitant device use with radiotherapy are essentially nonexistent and given the higher likelihood of skin toxicities, concurrent use with radiotherapy is not routine.

Anaplastic astrocytoma

Historically, G3 astrocytomas were lumped with G4 tumors in clinical trials. Because of this, dose and margin recommendations are similar to those used for glioblastoma. The recently reported EORTC CATNON trial[37] for G3 astrocytoma has demonstrated categorical survival advantage with the addition of TMZ following radiotherapy, making this the new standard of care. The RTOG 9813 trial demonstrated that survival was comparable with radiotherapy combined with either TMZ or a nitrosourea,[38] although the nitrosourea regimen yielded more toxicity.

Two randomized trials, EORTC 26951[39] and RTOG 9402[40] demonstrated the optimal therapy for maximizing OS in G3 oligodendroglioma, at the very least, for the molecularly favorable 1p-19q co-deleted tumors. This is combination radiotherapy and PCV chemotherapy (four cycles, given either pre or post-RT).

In EORTC 26951, 368 patients underwent surgery, radiotherapy (58.4 Gy/33 fractions), and were then randomized to observation versus six cycles of PCV. In the published 11-year updated results, OS was prolonged in the RT=PCV arm (42.3 vs 30.6 months). 1p-19q co-deleted patients had better PFS and OS than patients with non-co-deleted tumors (PFS, 76 vs 11 months; OS, 123 vs 23 months). PFS and OS were also better in patients with methylated *MGMT* promoter status and *IDH*-mutated tumors. These genetically favorable patients, or those with confirmed G3 histology, seemed to derive more benefit from the addition of PCV.

In RTOG 9402, 289 patients underwent maximal surgical resection and were then randomized to four cycles of PCV prior to RT versus RT alone (59.4 Gy/33 fractions). Co-deleted patients had improved MS of 14.7 years with PCV=RT compared to 7.3 years with RT alone ($P = .03$). The addition of PCV did not benefit non co-deleted tumors (MS 2.6 and 2.7 years, $P = .39$). In multivariate analysis that accounted for co-deletion status, the adjusted OS for all patients was prolonged by PCV=RT (HR=0.67, $P = .01$).

Given the legacy nature of these trials, field designs for G3 tumors are similar to those used for G4 tumors, and dose/fractionation schedules also reflect glioblastoma standards. Whether treatment de-intensification for favorable subsets of patients or substitution of TMZ for PCV can be undertaken remains to be seen. Currently, an ongoing randomized trial, CODEL,[41] is attempting to answer the TMZ versus PCV question.

Radiation side effects

EBRT to the brain, like other oncologic treatments, is associated with its own set of short-term and long-term side effects. Common acute side effects include alopecia, headaches, nausea or vomiting, skin erythema, and fatigue. The extent of alopecia depends generally on the volume of scalp within the radiation field, and is usually partial in the treatment of gliomas. Efforts can be made to spare a strip of scalp superior to the target lesion to spare these patients the added toxicity of alopecia. Hair growth usually resumes in the months after radiotherapy but can be delayed or limited depending on multiple factors, including other treatments the patient may be receiving. Symptoms of headaches are generally managed with steroids or oral analgesics. Symptoms of fatigue and lethargy generally improve 2–6 weeks after radiation therapy is completed, but can be prolonged, particularly in elderly patients. Patients are frequently concerned about the cognitive effects of brain radiotherapy. The quantification of true neurologic deficit after brain radiotherapy can be difficult to determine due to competing clinical factors such as postsurgical tissue injury or disease-related neurologic dysfunction.

Surveillance imaging

Regular clinical and radiographic follow-up is a crucial part of patient management. Sequential comparison of new imaging studies to the initial post-RT scan is used to determine whether disease has remained controlled or is progressing. Two important imaging findings, pseudo-progression and radionecrosis, can pose challenges for the treating physicians.

Pseudo-progression

Both surgery and radiotherapy contribute to breakdown of the BBB. As a result, BBB dysfunction can mimic tumor progression by increasing contrast enhancement, T2-weighted abnormalities, and mass effect during posttreatment surveillance imaging.[42] This entity, known as pseudo-progression, is seen in 15%–30% of patients after chemoradiation for HGGs, and typically occurs within 3 months of completing chemoradiation. It is more common in those with unmethylated *MGMT* promoter status.[43]

Distinguishing pseudo-progression from early progression can be challenging. Patients with pseudo-progression are more often asymptomatic compared to those with true progression. While specialized imaging such as MR spectroscopy or FDG-PET may help distinguish pseudo-progression from true progression, surgery may be required to establish a definitive diagnosis. Pseudo-progression is frequently identified retrospectively, after pooling the results of serial imaging and a patient's clinical response to interventions.

Radionecrosis

Cerebral radionecrosis can occur months to years following intracranial radiotherapy. It results from direct vascular injury, oligodendrocyte and white matter damage, and immune-mediated damage.[44] Necrotic brain tissue may exert mass effect on surrounding normal brain parenchyma causing neurological dysfunction.

MR spectroscopy, MR perfusion, and FDG-PET[45] may be utilized to help in the diagnosis. In general, the process can be managed conservatively in asymptomatic patients. Patients with neurological sequelae are managed with oral steroids to reduce cerebral edema. Antiangiogenic medications such as bevacizumab or even resection may be used in patients who do not respond to conservative measures.

Recurrent glioma

Unfortunately, the majority of HGG patients will relapse after completing radiation and chemotherapy, and some LGG patients will recur with transformation to higher-grade disease. If recurrent disease is confirmed, further treatment decisions should be tailored to patient preferences, as any further therapy is palliative and not curative. Retreatment poses the risk of increased neurotoxicity and subsequent negative impact on quality of life. Additionally, there are no randomized studies that directly compare intervention with best supportive care. Depending on the clinical situation, progressive disease can be managed with surgery, radiation therapy, or systemic therapy.

The Response Assessment in Neuro-Oncology (RANO) working group proposed a set of guidelines to help define response after definitive treatment. Progressive disease is defined by changes in the volume of T1 gadolinium enhancing disease and T2/FLAIR signal on MR imaging, presence of a new lesion, and decline in clinical status.[46] The RANO recommendation is that an increase in corticosteroid dose alone should not solidify a diagnosis of progression when no persistent clinical decline is present.

The potential management strategies for recurrent HGG are numerous, but largely hinge on patient PS. In patients with good PS, intervention depends on the pattern of recurrence/progression. Most recurrences are local.[47, 48] If the area of local recurrence is amenable to total/gross-total resection, reoperation[49] followed by systemic therapy can be considered. If surgery is not feasible, systemic therapy, reirradiation, or alternating electric fields remain treatment options. Ideally retreatment would be conducted in the setting of a clinical trial.

Reirradiation fields are typically smaller and may utilize hypofractionated dosing regimens.[50–53] RT may be combined with bevacizumab,[54, 55] but we do not yet know if this is superior to bevacizumab alone.[56–58] The most commonly employed single agent systemic treatments are bevacizumab, lomustine, or TMZ[59] but unfortunately, no individual agent has been shown to improve OS. Given the similar efficacies of these agents, the choice of one over another is frequently influenced by prior patient exposure, side effect profile, extent of disease/edema, and steroid tolerance. The role of immunotherapy for recurrent glioma is currently under investigation.

In patients with good PS but a diffuse or multifocal recurrence/progression pattern, systemic therapy or alternating electric fields are the most viable interventions. For patients with poor PS, the underlying cause of their clinical condition must be examined. If the poor PS is due to peritumoral edema and/or corticosteroid side effects, they may benefit from a trial of bevacizumab. If the poor PS cannot be explained by these causes, best supportive care should be pursued. The European Association for Neuro-Oncology (EANO) recently published a comprehensive set of guidelines to aid practitioners in selecting the optimal palliative care approach for terminal adult glioma patients.[60] This includes management of patient symptoms (such as pain, headache, seizures, fatigue, and mood/behavior disorders), as well as addressing complex psychosocial caregiver needs.

Future directions and conclusion

The field of Neuro-Oncology continues to push for improvement in and advancement of treatment of adult brain neoplasms. There are ongoing studies in glioblastoma examining the benefit of additional chemotherapy, dose-escalation with photons or protons, and tumor-targeted vaccines. A phase 2 national trial (NRG-BN001) is currently accruing patients with GBM to hypofractionated dose escalated photon therapy or proton therapy to a dose of 75 Gy in 30 fractions versus conventionally dosed photon therapy of 60 Gy in 30 fractions with concurrent and adjuvant TMZ. Together, radiation oncologists, neurosurgeons, and neuro-oncologists are working to overcome the unique hurdles posed by adult gliomas to improve outcomes for patients battling a challenging diagnosis.

References

1. Ostrom QT, Gittleman H, Fulop J, et al. CBTRUS statistical report: primary brain and central nervous system tumors diagnosed in the United States in 2008–2012. *Neuro-Oncology.* 2015;17:iv1–iv62.
2. Kohler BA, Ward E, McCarthy BJ, et al. Annual report to the Nation on the Status of Cancer, 1975-2007, featuring tumors of the brain and other nervous system. *J Natl Cancer Inst.* 2011;103(9):714–736.
3. Watson G, Kadota R, Wisoff J. Multidisciplinary management of pediatric low-grade gliomas. *Semin Radiat Oncol.* 2001;11(2):152–162.
4. Claus E, Black P. Survival rates and patterns of care for patients diagnosed with supratentorial low-grade gliomas: data from the SEER program, 1973–2001. *Cancer.* 2006;106(6):1358.
5. Curran W, Scott C, Horton J, Al E. Recursive partitioning analysis of prognostic factors in three Radiation Therapy Oncology Group malignant glioma trials. *J Natl Cancer Inst.* 1993;85(9):704–710.
6. Adamson D, Rasheed B, McLendon R, Bigner D. Central nervous system. *Cancer Biomark.* 2010;9(16):193–210.
7. Braganza MZ, Kitahara CM, de González AB, Inskip PD, Johnson KJ, Rajaraman P. Ionizing radiation and the risk of brain and central nervous system tumors: a systematic review. *NeuroOncology.* 2012;14(11):1316–1324.
8. Little MP, Rajaraman P, Curtis RE, et al. Mobile phone use and glioma risk: comparison of epidemiological study results with incidence trends in the United States. *BMJ.* 2012;344(1147):1–16.
9. Deltour I, Auvinen A, Feychting M, et al. Mobile phone use and incidence of glioma in the nordic countries 1979–2008. *Epidemiology.* 2012;23(2):301–307.
10. Jones B. Proton radiobiology and its clinical implications. *Ecancermedicalscience.* 2017;11(777).
11. Indelicato D, Flampouri S, Rotondo R, Al E. Incidence and dosimetric parameters of pediatric brainstem toxicity following proton therapy. *Acta Oncol (Madr).* 2014;53:1298–1304.
12. Kumar S. Second malignant neoplasms following radiotherapy. *Int J Environ Res Public Health.* 2012;9(12):4744–4759.
13. Sneed PK, Lamborn KR, Larson DA, et al. Demonstration of brachytherapy boost dose-response relationships in glioblastoma multiforme. *Int J Radiat Oncol Biol Phys.* 1996;35(1):37–44.
14. Kickingereder P, Hamisch C, Suchorska B, et al. Low-dose rate stereotactic iodine-125 brachytherapy for the treatment of inoperable primary and recurrent glioblastoma: single-center experience with 201 cases. *J Neuro-Oncol.* 2014;120(3):615–623.
15. Pajonk F, Vlashi E, McBride W. Radiation resistance of cancer stem cells: the 4 R's of radiobiology revisited. *Stem Cells.* 2010;28(4):639–648.
16. Steel G, McMillan T, Peacock J. The 5Rs of radiobiology. *Int J Radiat Biol.* 1989;56(6):1045–1048.
17. Roa W, Brasher PMA, Bauman G, et al. Abbreviated course of radiation therapy in older patients with glioblastoma multiforme: a prospective randomized clinical trial. *J Clin Oncol.* 2004;22(9):1583–1588.
18. Brashears J. *Encyclopedia of Radiation Oncology.* Berlin Heidelberg: Springer-Verlag; 2013.
19. Pichlmeier U, Bink A, Schackert G, Stummer W. Resection and survival in glioblastoma multiforme: an RTOG recursive partitioning analysis of ALA study patients. *Neuro-Oncology.* 2008;10(6):1025–1034.
20. Hervey-Jumper SL, Li J, Lau D, et al. Awake craniotomy to maximize glioma resection: methods and technical nuances over a 27-year period. *J Neurosurg.* 2015;123(2):325–339.
21. Simpson J, Horton J, Scott C, Al E. Influence of location and extent of surgical resection on survival of patients with glioblastoma multiforme: results of three consecutive radiation therapy oncology group (RTOG) clinical trials. *Int J Radiat Oncol Biol Phys.* 1993;26(2):239–244.

22. Gorlia T, Wu W, Wang M, et al. New validated prognostic models and prognostic calculators in patients with low-grade gliomas diagnosed by central pathology review: a pooled analysis of EORTC/RTOG/NCCTG phase III clinical trials. *Neuro-Oncology.* 2013;15(11):1568–1579.

23. Pignatti BF, Van Den BM, Curran D, et al. Prognostic factors for survival in adult patients with cerebral low-grade glioma. *J Clin Oncol.* 2015;20 (8):2076–2084.

24. van den Bent M, Afra D, de Witte O, et al. Long-term efficacy of early versus delayed radiotherapy for low-grade astrocytoma and oligodendroglioma in adults: the EORTC 22845 randomised trial. *Lancet.* 2005;366(9490):985–990.

25. Karim A, Maat B, Hatlevoll R, Al E. A randomized trial on dose-response in radiation therapy of low-grade cerebral glioma: European Organization for Research and Treatment of Cancer (EORTC) Study 22844. *Int J Radiat Oncol Biol Phys.* 1996;36(3):549–556.

26. Shaw E, Arusell R, Scheithauer B, Al E. Prospective randomized trial of low- versus high-dose radiation therapy in adults with supratentorial low-grade glioma: initial report of a North Central Cancer Treatment Group/Radiation Therapy Oncology Group/Eastern Cooperative Oncology Group study. *J Clin Oncol.* 2002;20(9):2267–2276.

27. Buckner JC, Shaw EG, Pugh SL, et al. Radiation plus procarbazine, CCNU, and vincristine in low-grade glioma. *N Engl J Med.* 2016;374 (14):1344–1355.

28. Bauman G, Fisher B, Watling C. Adult supratentorial low-grade glioma: long-term experience at a single institution. *Int J Radiat Oncol Biol Phys.* 2009;75(5):1401–1407.

29. Youland R, Brown P, Giannini C. Adult low-grade glioma 19-year experience at a single institution. *Am J Clin Oncol.* 2013;36(6):612–619.

30. Baumert B, van den Bent M, von Deimling A. Temozolomide chemotherapy versus radiotherapy in high-risk low-grade glioma (EORTC 22033-26033): a randomised, open-label, phase 3 intergroup study. *Lancet Oncol.* 2016;17:1521–1532.

31. Stupp R, Hegi M, Mason W, Al E. Effects of radiotherapy with concommitant and adjuvant temozolomide versus radiotherapy alone on survival in glioblastoma in a randomised phase III study: 5-year analysis of the EORTC-NCIC trial. *Lancet Oncol.* 2009;10(5):459–466.

32. Malmström A, Grønberg BH, Marosi C, et al. Temozolomide versus standard 6-week radiotherapy versus hypofractionated radiotherapy in patients older than 60 years with glioblastoma: the Nordic randomised, phase 3 trial. *Lancet Oncol.* 2012;13(9):916–926.

33. Slotman B, Kralendonk J, van Alphen H, Al E. Hypofractionated radiation therapy in patients with glioblastoma multiforme: results of treatment and impact of prognostic factors. *Int J Radiat Oncol Biol Phys.* 1996;34(4):895–898.

34. Guedes de Castro D, Matiello J, Roa W, Al E. Survival outcomes with short-course radiation therapy in elderly patients with glioblastoma: data from a randomized phase 3 trial. *Int J Radiat Oncol Biol Phys.* 2017;98(4):931–938.

35. Perry J, Laperriere N, O'Callaghan C. Short-course radiation plus temozolomide in elderly patients with glioblastoma. *NEJM.* 2017;376:1027–1037.

36. Stupp R, Taillibert S, Kanner A. Maintenance therapy with tumor-treating fields plus temozolomide vs temozolomide alone for glioblastoma: a ran-domized clinical trial. *JAMA.* 2015;314(23):2535–2543.

37. van den Bent M, Baumert B, Erridge S. Interim results from the CATNON trial (EORTC study 26053-22054) of treatment with concurrent and adjuvant temozolomide for 1p/19q non-co-deleted anaplastic glioma: a phase 3, randomised, open-label intergroup study. *Lancet.* 2017;390 (10103):1645–1653.

38. Chang S, Zhang P, Cairncross J. Phase III randomized study of radiation and temozolomide versus radiation and nitrosourea therapy for anaplastic astrocytoma: results of NRG Oncology RTOG 9813. *Neuro-Oncology.* 2017;19(2):252–258.

39. van den Bent M, Brandes A, Taphoorn M. Adjuvant procarbazine, lomustine, and vincristine chemotherapy in newly diagnosed anaplastic oligoden-droglioma: long-term follow-up of EORTC brain tumor group study 26951. *J Clin Oncol.* 2013;31(3):344–350.

40. Cairncross G, Wang M, Shaw E. Phase III trial of chemoradiotherapy for anaplastic oligodendroglioma: long-term results of RTOG 9402. *J Clin Oncol.* 2013;31(3):337–343.

41. Radiation Therapy With Concomitant and Adjuvant Temozolomide Versus Radiation Therapy With Adjuvant PCV Chemotherapy in Patients With Anaplastic Glioma or Low Grade Glioma. https://clinicaltrials.gov/ct2/show/NCT00887146.

42. O'Brien B, Colen R. Post-treatment imaging changes in primary brain tumors. *Curr Oncol Rep.* 2014;16:397.

43. Brandes A, Franceschi E, Tosoni A. MGMT promoter methylation status can predict the incidence and outcome of pseudoprogression after con-comitant radiochemotherapy in newly diagnosed glioblastoma patients. *J Clin Oncol.* 2008;26(13):2192–2197.

44. Miyatake S, Nonoguchi N, Furuse M, Al E. Pathophysiology, diagnosis, and treatment of radiation necrosis in the brain. *Neurol Med Chir.* 2015;55 (1):50–59.

45. Shah R, Vattoth S, Jacob R, Al E. Radiation necrosis in the brain: imaging features and differentiation from tumor recurrence. *RadioGraphics.* 2012;32:1343–1359.

46. Wen P, Macdonald D, Reardon D. Updated response assessment criteria for high-grade gliomas: response assessment in neuro-oncology working group. *J Clin Oncol.* 2010;28(11):1963–1972.

47. Halperin E, Bentel G, Heinz E, Burger P. Radiation therapy treatment planning in supratentorial glioblastoma multiforme: an analysis based on post mortem topographic anatomy with CT correlations. *Int J Radiat Oncol Biol Phys.* 1989;17(6):1347–1350.

48. Wallner K, Galicich J, Krol G. Radiation therapy treatment planning in supratentorial glioblastoma multiforme: an analysis based on post mortem topographic anatomy with CT correlations. *Int J Radiat Oncol Biol Phys.* 1989;16(6):1405–1409.

49. Hervey-Jumper S, Berger M. Reoperation for recurrent high-grade glioma: a current perspective of the literature. *Neurosurgery.* 2014;75(5):491–499.

50. Bauman G, Sneed P, Wara W. Reirradiation of primary CNS tumors. *Int J Radiat Oncol Biol Phys.* 1996;36:433–441.

51. Combs S, Widmer V, Thilmann C. Stereotactic radiosurgery (SRS): treatment option for recurrent glioblastoma multiforme (GBM). *Cancer.* 2005;104:2168–2173.

52. Shrieve D, Alexander EI, Wen P. Comparison of stereotactic radiosurgery and brachytherapy in the treatment of recurrent glioblastoma multiforme. *Neurosurgery.* 1995;36:275–282.

53. Combs S, Tilmann C, Edler L. Efficacy of fractionated stereotactic reirradiation in recurrent gliomas: long-term results in 172 patients treated in a single institution. *J Clin Oncol*. 2005;23:8863–8869.
54. Cabrera A, Cuneo K, Desjardins A. Concurrent stereotactic radiosurgery and bevacizumab in recurrent malignant gliomas: a prospective trial. *Int J Radiat Oncol Biol Phys*. 2013;86(5):873.
55. Gutin P, Iwamoto F, Beal K. Safety and efficacy of bevacizumab with hypofractionated stereotactic irradiation for recurrent malignant gliomas. *Int J Radiat Oncol Biol Phys*. 2009;75(1):156–163.
56. Wick W, Gorlia T, Bendszus M. Lomustine and bevacizumab in progressive glioblastoma. *NEJM*. 2017;377(20):1954.
57. Friedman H, Prados M, Wen P. Bevacizumab alone and in combination with irinotecan in recurrent glioblastoma. *J Clin Oncol*. 2009;27(28):4733.
58. Brandes A, Finocchiaro G, Zagonel V. AVAREG: a phase II, randomized, noncomparative study of fotemustine or bevacizumab for patients with recurrent glioblastoma. *Neuro-Oncology*. 2016;18(9):1304.
59. Perry J, Belanger K, Mason W. Phase II trial of continuous dose-intense temozolomide in recurrent malignant glioma: RESCUE study. *J Clin Oncol*. 2010;28(12):2051.
60. Pace A, Dirven L, Johan A. European association for neuro-oncology (EANO) guidelines for palliative care in adults with glioma. *Lancet Oncol*. 2017;18:e330–e340.

Chapter 22

Standard external beam radiation therapy for oligodendroglioma

Gustavo Nader Marta*,†, Fabio Y. Moraes*,‡, Erin S. Murphy§ and John H. Suh§

*Department of Radiation Oncology, Hospital Sírio-Libanês, Sao Paulo, Brazil; †Department of Radiology and Oncology, Division of Radiation Oncology, Instituto do Câncer do Estado de São Paulo (ICESP), Faculdade de Medicina da Universidade de São Paulo, Sao Paulo, Brazil; ‡Radiation Medicine Program, Princess Margaret Cancer Centre, University Health Network, Toronto, ON, Canada; §Department of Radiation Oncology, Cleveland Clinic, Taussig Cancer Institute, Cleveland, OH, United States

Introduction

The management of patients diagnosed with gliomas has advanced due to improvements in surgery, neuroimaging, and the understanding of genetic and molecular biomarkers, which can be predictive and/or prognostic. Radiation therapy (RT) has an important role in the management of glioma patients. Since most randomized clinical trials particularly in grade 2 gliomas often pooled together astrocytic and oligodendroglial histological types, the results do not specifically focus on oligodendroglial tumors. When available, subgroup outcomes analyses based on specific tumor histologies were performed post hoc. The available data from clinical trials indicate that oligodendroglioma with 1p/19q co-deletion is more responsive to treatment than astrocytoma and normally correlates with better clinical outcomes.[1, 2]

Practice of RT

In general, upfront postoperative RT, following surgical resection or biopsy, is considered as a treatment option patients with grade 2 or 3 glioma. The timing of RT is controversial given the concerns regarding quality of life and cognitive function for patients receiving radiation treatments. Tumor and patient characteristics along with physician preference are the main factors that guide whether to proceed with RT.

Tumor features include tumor size, extent of resection (gross total resection vs partial resection vs biopsy), tumor grade (grade 2 vs grade 3), presence of mass effect, and genetic and molecular features of the tumor. Patient characteristics include performance status, age, clinical presentation at the time of diagnosis, and symptoms following resection.[3]

Low-grade gliomas: observation versus RT

The European Organisation for Research and Treatment of Cancer (EORTC) Radiotherapy and Brain Tumor Groups conducted a randomized phase III trial (EORTC 22845) of 314 low-grade glioma patients (14% oligodendroglioma; 51% astrocytoma; 13% mixed oligoastrocytoma) to compare immediate postoperative RT with observation (RT was performed at the time of tumor progression). Most patients underwent gross or partial resection (42% and 35%, respectively). A total dose of 54 Gy in 30 fractions (1.8 Gy per fraction) was used with computed tomography (CT)-based radiation field targeting preoperative tumor volume. The median progression-free survival was statistically significantly higher in the immediate postoperative RT group (5.3 years vs 3.4 years; $P < .0001$). However, no difference in overall survival rates was observed between groups (median survival 7.4 years vs 7.2 years; $P = .872$). In the observation arm, 65% of patients underwent salvage RT at progression. Although quality of life was not formally studied, better tumor control was related to improved seizure control at 1 year (41% for the RT group versus 25% for the observation group; $P = .0329$).[4]

This trial proposed that immediate postoperative RT can provide benefit for patients with neurological symptoms (i.e., seizures). However, the lack of improvement in overall survival indicated that delaying RT until tumor progression was a reasonable management choice.

Oligodendroglioma. https://doi.org/10.1016/B978-0-12-813158-9.00022-0

Low-grade gliomas: temozolomide alone

A multiinstitutional phase II trial evaluated adjuvant temozolomide alone for 120 low-grade glioma patients who underwent less than gross total resection.[5] Patients received up to 12 cycles of temozolomide the until time of progression. Primary end point was radiographic response: the overall response rate was only 6%, however, 87% had stable or improved disease during temozolomide therapy. After a median follow-up of 7.5 years, median progression-free survival was 3.8 years and overall survival was 9.7 years. The rate of grade 2–4 hematologic toxicity was 10%. A total of 97 patients had molecular status analyzed. Results showed that histology, molecular subtype, and pretreatment tumor volume (68cc) were significant for both progression—free survival and overall survival. The rate of progression during temozolomide was 0% for 1p/19q co-deleted patients, 8% for IDH mutated, and 56% for IDH wild-type patients.

Low-grade gliomas: RT versus chemotherapy

The effectiveness of chemotherapy in the postoperative scenario (adjuvant and salvage setting) for grade 2 and 3 gliomas has been shown. The most commonly used chemotherapy agents are PCV (procarbazine, CCNU, and vincristine) and temozolomide.[6, 7] Since phase II trials have demonstrated that postoperative chemotherapy alone can result in satisfactory local control and survival rates, the issue of whether low-grade glioma patients should be treated with postoperative chemotherapy alone to avoid the side effects of RT has been raised.[5, 8–10]

In the EORTC 22033-26033 trial, 477 patients with low-grade (WHO grade 2) glioma (41% oligodendroglioma, 35% astrocytoma, and 24% oligoastrocytoma) with at least one high-risk feature (tumor size >5 cm, tumor crossing the midline, age >40 years, neurological symptoms or progressive disease) were randomized to conformal RT (up to 50.4 Gy in 28 fractions of 1.8 Gy/fraction delivered over 6 weeks) or to dose-dense oral temozolomide [75 mg/m^2 daily for 21 days, repeated every 28 days (one cycle)—maximum of 12 cycles]. For this EORTC trial, the RT target volume was focused on the postoperative surgical bed. The trial reported no statistically significant difference in progression-free survival between the groups (median progression-free survival 39 months for the temozolomide group versus 46 months for the RT group; $P = .22$). In a post hoc analysis of tumor biomarkers [1p/19q co-deletion, O^6-methylguanine DNA-methyltransferase (MGMT) promoter methylation status, and isocitrate dehydrogenases (IDH1/IDH2) mutations], patients with IDH wild-type tumors had the worst prognosis independent of treatment, whereas patients with IDH mutated or 1p/19q co-deleted tumors had the best prognosis regardless of treatment modality.[11] In addition, the effect of postoperative RT or temozolomide on global cognitive functioning and health-related quality of life were not different ($P = .98$).[12] Of note, on exploratory analysis, patients with IDHmt/1p/19q non-co-deleted tumors had longer PFS if treated with RT than with temozolomide.

The Neuro-oncology Working Group of the German Cancer Society randomized 318 anaplastic glioma patients on the NOA-04 trial, into three groups: (1) conventional RT (focal irradiation to gross tumor volume plus a 2-cm margin over 6-weeks using 1.8- to 2-Gy fractions to a total 60-Gy dose based on preoperative magnetic resonance imaging), (2) PCV [four 8-week cycles of lomustine (110 mg/m^2 on day 1), vincristine (2 mg on days 8 and 29), and procarbazine (60 mg/m^2 on days 8 through 21)], and (3) temozolomide [eight 4-week cycles of temozolomide (200 mg/m^2 on days 1 through 5)]. With follow-up of 9.5 years, the median time to treatment failure (primary end point), progression-free survival, and overall survival were equivalent for all groups. Methylation of the MGMT promoter, mutations of IDH1, and oligodendroglial histology were associated with lower risk of progression.[13, 14]

Based on these trials, there is no standard postoperative approach. Either PCV or temozolomide have been shown to be effective. Since temozolomide is better tolerated with lower grade 3 and 4 toxicity rates compared to PCV, many clinical practices prefer the use of temozolomide. This strategy can be considered to avoid immediate postoperative RT, particularly in patients diagnosed with favorable molecular features tumors, particularly 1p/19q co-deletion.[3, 15]

Low- and high-grade gliomas: RT and chemotherapy

The use of postoperative RT in combination with chemotherapy for the treatment of patients with grade 2 and 3 gliomas has been evaluated for many years.[16] We will highlight and summarize the studies and results that corroborate this treatment paradigm.

Grade 2 gliomas and PCV chemotherapy

The RTOG 9802 study assessed supratentorial WHO grade 2 glioma (including astrocytoma, oligodendroglioma, or mixed oligoastrocytoma) patients with high-risk characteristics (subtotal resection/biopsy, or age ≥40 years regardless of extent of resection) and randomized them into two groups: postoperative RT alone (54 Gy in 30 fractions) and postoperative

RT (54 Gy in 30 fractions) followed by six cycles of PCV (combined modality). Between 1998 and 2002, 251 patients were accrued. The first publication of RTOG 9802 demonstrated that progression-free survival but not overall survival was improved with combined postoperative management.[1] The long-term results showed that patients who underwent combined modality (postoperative RT plus PCV) had improved progression-free survival (10-year progression-free survival: 51% vs 21%; $P < .05$) as well as median overall survival (13.3 vs 7.8 years; $P = .003$) and 10-year overall survival rates (60% vs 40%; $P < .05$).[17]

The long-term results of this study demonstrated that postoperative RT followed by PCV is associated with a better overall survival rate compared to postoperative RT alone. This improvement was not just statistically significant, but also clinically relevant indicating that if the decision is to perform postoperative treatment, the sequential approach with RT followed by PCV should be considered, particularly for high-risk disease.

Grade 3 gliomas and PCV chemotherapy

The EORTC 26951 randomized trial demonstrated that RT (59.4 Gy/33 fractions) followed by six cycles of PCV increased progression-free survival and overall survival in anaplastic oligodendroglioma tumors compared to postoperative RT alone. Those patients with 1p/19q co-deleted tumors benefit from adjuvant PCV compared with patients with 1p/19q nondeleted tumors.[2] The major side effects of PCV on health-related quality of life are loss of appetite, vomiting/náusea, and lethargy during and shortly after treatment.[18]

The RTOG 9402 trial randomized 289 anaplastic oligodendroglioma and anaplastic oligoastrocytoma patients to postoperative RT alone (59.4 Gy in 33 fractions) or combination therapy (PCV followed by RT). The median survival time was comparable between groups (4.7 years RT alone vs 4.9 years PCV followed by RT; $P = .26$). Progression-free survival was better in the combined therapy group (1.7 years RT alone vs 2.6 years PCV followed by RT; $P = .04$). However, the addition of PCV was associated with significant toxicity (65% of patients had grades 3 and 4 toxicities; one death was observed). The median survival of those with co-deleted tumors treated with PCV plus RT was 14.7 versus 7.3 years for those treated with RT alone.[19, 20] However, for those with non-co-deleted tumors, there was no benefit from the addition of chemotherapy.

Temozolomide with RT

Temozolomide is the standard chemotherapy option for patients with glioblastoma due to the radiosensitizing properties and survival benefit in combination with radiotherapy. This is a well-tolerated drug and its benefits were formally proven through a randomized clinical trial.[21] Nonetheless, the benefit of adding temozolomide to radiation for low-grade gliomas is not as well established.

RTOG 0424 was a phase II trial that evaluated 129 low-grade gliomas patients with three or more risk features for relapse (astrocytoma histology, age $\geq$40 years, bihemispherical tumor, preoperative neurological function status of >1 or preoperative tumor $\geq$6 cm). The treatment consisted of postoperative RT (54 Gy in 30 fractions) concurrent with low-dose daily temozolomide followed by adjuvant temozolomide for up to 12 cycles. The 3-year overall survival was 73.1%, which was significantly superior, compared to historical control of 54% from EORTC trials ($P < .001$).[22]

The recently published CATNON trial (EORTC study 26053-22054) assessed the addition of temozolomide during and/or after RT compared to RT alone in the treatment of anaplastic glioma patients without 1p/19q co-deletion.[23] CATNON trial is a 2×2 factorial design where patients were randomized to receive RT alone (59.4 Gy in 33 fractions) or with adjuvant temozolomide; or to receive RT with concurrent temozolomide with or without adjuvant temozolomide. The interim results from the CATNON trial (EORTC 26053-22054) demonstrated that the use of adjuvant temozolomide was correlated to a significant survival benefit compared to patients who did not receive adjuvant temozolomide (5-year overall survival: 55.9% vs 44.1%; $P < .05$). Additional evaluation of the role of concurrent temozolomide management and molecular tumor issues is expected from this study.[24]

Ongoing trials are assessing the role of temozolomide associated with RT for the treatment of nonglioblastoma gliomas.[23, 25–27] The Alliance for Clinical Trials in Oncology Groups is evaluating the use of RT with concomitant and adjuvant temozolomide compared to RT with adjuvant PCV in patients with high-risk low-grade and anaplastic gliomas.[25] Eastern Cooperative Oncology Group trial E3F05, which was a trial studying postoperative RT with or without temozolomide for low-grade glioma patients, was suspended in 2014 based on the findings from RTOG 98-02. NRG-BN005 is a phase II study that is assessing cognitive outcomes in patients treated with intensity modulated radiation therapy (IMRT) or proton therapy. Adjuvant temozolomide is given 4 weeks after completion of radiation.[27]

RT fractionation and dose

Based on the finding that the relapses are mainly local and within the radiation target, dose escalation was a strategy which was investigated to improve clinical outcomes in patients with gliomas. However, no published trial proved any benefit in local control or survival rates after the use of higher doses of RT for gliomas (Table 1).

The EORTC 22844 study randomized 379 patients with low-grade gliomas (22% oligodendroglioma; 69% astrocytoma; 9% mixed oligoastrocytoma) to receive postoperative RT (45 Gy/25 fractions or 59.4 Gy/33 fractions). The study reported no statistically significant differences in 5-year progression-free survival (47% for 45 Gy arm vs 50% for 59.4 Gy arm) or overall survival (58% for 45 Gy arm vs 59% for 59.4 Gy arm) between the groups.[28]

Similarly, the North Central Cancer Treatment Group/RT Oncology Group/Eastern Cooperative Oncology Group randomized study compared the clinical outcomes of low-grade gliomas patients treated with 50.4 Gy/28 fractions or 64.8 Gy/36 fractions. There were no significant differences in 2-year and 5-year overall survival rates in patients treated with 50.4 Gy or 64.8 Gy. A higher rate of radiation necrosis (grade 3–5) was observed in the 64.8 Gy group compared to the 50.4-Gy group (5% vs 2.5%).[29]

RT: simulation, target volume, dose, treatment planning, and techniques

To allow for reproducible setups, patients are immobilized in a thermoplastic mask during CT simulation and treatment.

The target volume comprises the surgical bed and any observable remaining tumor. Postoperative magnetic resonance imaging (MRI), in which tumor appears hyperintense on the FLAIR sequence, is commonly used to define gross tumor volume (GTV). Clinical target volume (CTV) is created by expanding GTV. The GTV to CTV expansion ranges from 1 to 2 cm and respects natural anatomic barriers such as the falx and bone. An extra geometric margin of 3–5 mm is generated to create the planning target volume (PTV) as shown in Fig. 1A. Furthermore, organs at risk (optic nerves, chiasm, retina, lenses, lacrimal glands, pituitary, cochlea, brainstem, and spinal cord) are delineated and dose limits (constraints) should be respected throughout the RT planning course (Table 2). For low-grade gliomas, the prescribed dose ranges from 45 to 54 Gy in 1.8 Gy per fraction. Current studies use a total dose of 54 Gy in 30 fractions.

Regarding RT techniques, most institutions are using three-dimensional conformal RT (3D-CRT), IMRT, or volumetric arc therapy (VMAT) as shown in Fig. 1B. As mentioned earlier, an ongoing randomized phase II trial, NRG-BN005, is randomizing IDH-mutant low-grade gliomas to protons or photons. Commonly, ≥95% of the PTV should be covered by 100% of the total prescribed dose. The maximum dose to any point must be reduced to 105% of the prescription dose and preferably situated in the GTV. Table 2 shows suggestions for normal tissue constraints.[30–33]

Toxicities of RT

Common acute reactions include fatigue, radiation dermatitis, alopecia, headaches, nausea, vomiting, and worsening of preexisting neurologic symptoms.[34–37] Symptoms of fatigue usually occur during the last few weeks of RT and gradually resolve over several weeks after completion of radiation. Most patients experience at least grade 1 fatigue and almost 50% of them can develop mild to moderate symptoms.[34] Alopecia (partial or total) appears where the radiation beams enter the scalp. The extent and duration of alopecia is associated with scalp dose. Similarly, regions of skin that received high doses of irradiation might show dry or moist desquamation.[35] Mild headaches during the RT treatment can occur and are frequently minor or transitory enough that patients do not need any specific intervention. However, headaches might also

TABLE 1 Target volumes and dose for low-grade oligodendroglioma patients

Target volumes	
GTV	FLAIR or T2 weighted abnormalities on MRI. After surgical resection, the resection bed and residual FLAIR or T2 abnormalities are included
CTV	GTV plus 1–2 cm except at natural barriers such as inner calvarium, falx, and ventricles
PTV	CTV plus 3–5 mm
Dose[a]	

[a]NRG-BN 005 uses dose of 54 Gy.

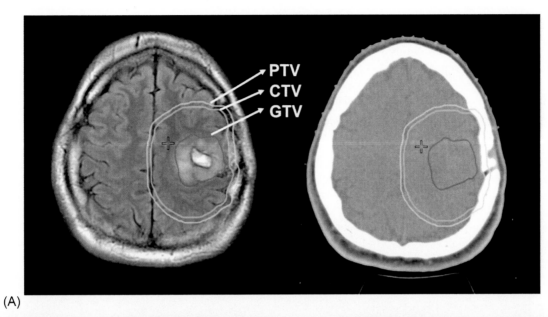

(A)

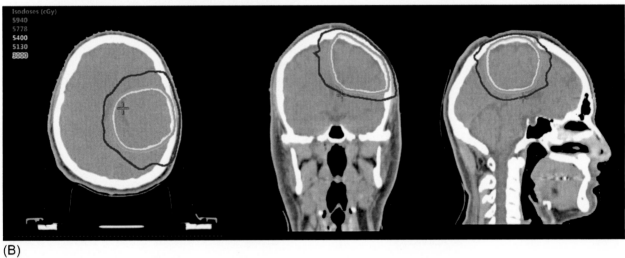

(B)

FIG. 1 (A) Target volume: gross tumor volume (GTV), clinical target volume (CTV), and planning target volume (PTV). (B) Radiation therapy planning: volumetric arc therapy (VMAT) technique.

TABLE 2 Normal tissue dose constraints for low-grade and high-grade oligodendroglioma patients

Critical structures	Constraints
Brainstem	Max < 55 Gy for low-grade/max < 60 Gy for high-grade
Cochleae	Max < 45 Gy (mean 35 Gy)
Lacrimal glands	Max < 35 Gy
Lenses	Max < 7 Gy
Optic chiasm	Max < 55 Gy
Optic nerves	Max < 55 Gy
Pituitary gland	Max < 45 Gy
Retina	Max < 45 Gy

be a consequence of mass effect from tumor: simultaneous evolution of neurologic deficits with worsening headaches, somnolence, and new or worsening nausea and vomiting can be associated with tumor progression or peritumoral edema. Acute encephalopathy and transient focal neurologic symptoms are uncommon acute reactions.[36, 37]

Late complications of cranial irradiation are represented by new or persistent symptoms after 6 months from the completion of RT. These include neurocognitive sequelae, cerebrovascular effects (stroke and occlusive vascular disease, intracranial hemorrhage), neuroendocrine dysfunction, secondary tumor formation, and radiation necrosis.[38–45] Radiation necrosis often occurs 12–36 months after cranial irradiation, though this event has also described more than 10 years after treatment. The incidence of radiation necrosis in patients with low-grade glioma is almost 3% and normally develops within high-dose regions of the PTV.[38, 39] Partial cranial irradiation can result in long-term neurocognitive deficits, but the risk is meaningfully less than with whole brain irradiation, and the observed sequelae might be due to multiple aspects such as surgery, tumor-related damage, and/or antiseizure drug therapy.[40, 41] Of note, 20 patients who were treated on RTOG 9110 underwent an extensive battery of neurocognitive testing prior to radiotherapy and every 18 months up to 5 years.[46] These patients experienced stable cognitive function at 5 years postradiotherapy.

Malignant transformation is an event that can occur for low-grade gliomas, however, the causes and risk factors are not well known. Data from EORTC 22845 in which patients were randomized to immediate or delayed radiotherapy at progression demonstrate a 66%–72% risk of malignant transformation regardless of radiotherapy timing.[4] Although these were biopsy-proven events, they were not confirmed with central pathology review and these patients were diagnosed and treated with CTs. There is more recent data from a large modern retrospective series of 599 patients that suggests a crude malignant transformation incidence of 21% (of which 61.3% were biopsy proven).[47]

References

1. Shaw EG, Wang M, Coons SW, Brachman DG, Buckner JC, Stelzer KJ, et al. Randomized trial of radiation therapy plus procarbazine, lomustine, and vincristine chemotherapy for supratentorial adult low-grade glioma: initial results of RTOG 9802. *J Clin Oncol.* 2012;30:3065–3070.

2. van den Bent MJ, Brandes AA, Taphoorn MJ, Kros JM, Kouwenhoven MC, Delattre JY, et al. Adjuvant procarbazine, lomustine, and vincristine chemotherapy in newly diagnosed anaplastic oligodendroglioma: long-term follow-up of EORTC brain tumor group study 26951. *J Clin Oncol.* 2013;31:344–350.

3. Chung C, Laperriere N. Radiation therapy and grade II/III oligodendroglial tumors. *CNS Oncol.* 2015;4:325–332.

4. van den Bent MJ, Afra D, de Witte O, Ben Hassel M, Schraub S, Hoang-Xuan K, et al. Long-term efficacy of early versus delayed radiotherapy for low-grade astrocytoma and oligodendroglioma in adults: the EORTC 22845 randomised trial. *Lancet.* 2005;366:985–990.

5. Wahl M, Phillips JJ, Molinaro AM, Lin Y, Perry A, Haas-Kogan DA, et al. Chemotherapy for adult low-grade gliomas: clinical outcomes by molecular subtype in a phase II study of adjuvant temozolomide. *Neuro Oncol.* 2017;19:242–251.

6. Brada M, Stenning S, Gabe R, Thompson LC, Levy D, Rampling R, et al. Temozolomide versus procarbazine, lomustine, and vincristine in recurrent high-grade glioma. *J Clin Oncol.* 2010;28:4601–4608.

7. Buckner JC, Gesme Jr. D, O'Fallon JR, Hammack JE, Stafford S, Brown PD, et al. Phase II trial of procarbazine, lomustine, and vincristine as initial therapy for patients with low-grade oligodendroglioma or oligoastrocytoma: efficacy and associations with chromosomal abnormalities. *J Clin Oncol.* 2003;21:251–255.

8. Kesari S, Schiff D, Drappatz J, LaFrankie D, Doherty L, Macklin EA, et al. Phase II study of protracted daily temozolomide for low-grade gliomas in adults. *Clin Cancer Res.* 2009;15:330–337.

9. van den Bent MJ, Taphoorn MJ, Brandes AA, Menten J, Stupp R, Frenay M, et al. Phase II study of first-line chemotherapy with temozolomide in recurrent oligodendroglial tumors: the European Organization for Research and Treatment of Cancer Brain Tumor Group Study 26971. *J Clin Oncol.* 2003;21:2525–2528.

10. Quinn JA, Reardon DA, Friedman AH, Rich JN, Sampson JH, Provenzale JM, et al. Phase II trial of temozolomide in patients with progressive low-grade glioma. *J Clin Oncol.* 2003;21:646–651.

11. Baumert BG, Hegi ME, van den Bent MJ, von Deimling A, Gorlia T, Hoang-Xuan K, et al. Temozolomide chemotherapy versus radiotherapy in high-risk low-grade glioma (EORTC 22033-26033): a randomised, open-label, phase 3 intergroup study. *Lancet Oncol.* 2016;17:1521–1532.

12. Reijneveld JC, Taphoorn MJ, Coens C, Bromberg JE, Mason WP, Hoang-Xuan K, et al. Health-related quality of life in patients with high-risk low-grade glioma (EORTC 22033-26033): a randomised, open-label, phase 3 intergroup study. *Lancet Oncol.* 2016;17:1533–1542.

13. Wick W, Hartmann C, Engel C, Stoffels M, Felsberg J, Stockhammer F, et al. NOA-04 randomized phase III trial of sequential radiochemotherapy of anaplastic glioma with procarbazine, lomustine, and vincristine or temozolomide. *J Clin Oncol.* 2009;27:5874–5880.

14. Wick W, Roth P, Hartmann C, Hau P, Nakamura M, Stockhammer F, et al. Long-term analysis of the NOA-04 randomized phase III trial of sequential radiochemotherapy of anaplastic glioma with PCV or temozolomide. *Neuro Oncol.* 2016;18:1529–1537.

15. Weller M, van den Bent M, Tonn JC, Stupp R, Preusser M, Cohen-Jonathan-Moyal E, et al. European Association for Neuro-Oncology (EANO) guideline on the diagnosis and treatment of adult astrocytic and oligodendroglial gliomas. *Lancet Oncol.* 2017;18:e315–e329.

16. Fine HA, Dear KB, Loeffler JS, Black PM, Canellos GP. Meta-analysis of radiation therapy with and without adjuvant chemotherapy for malignant gliomas in adults. *Cancer.* 1993;71:2585–2597.

17. Buckner JC, Shaw EG, Pugh SL, Chakravarti A, Gilbert MR, Barger GR, et al. Radiation plus procarbazine, CCNU, and vincristine in low-grade glioma. *N Engl J Med.* 2016;374:1344–1355.

18. Taphoorn MJ, van den Bent MJ, Mauer ME, Coens C, Delattre JY, Brandes AA, et al. Health-related quality of life in patients treated for anaplastic oligodendroglioma with adjuvant chemotherapy: results of a European Organisation for Research and Treatment of Cancer randomized clinical trial. *J Clin Oncol.* 2007;25:5723–5730.

19. Intergroup Radiation Therapy Oncology Group T, Cairncross G, Berkey B, Shaw E, Jenkins R, Scheithauer B, et al. Phase III trial of chemotherapy plus radiotherapy compared with radiotherapy alone for pure and mixed anaplastic oligodendroglioma: Intergroup Radiation Therapy Oncology Group Trial 9402. *J Clin Oncol.* 2006;24:2707–2714.

20. Cairncross G, Wang M, Shaw E, Jenkins R, Brachman D, Buckner J, et al. Phase III trial of chemoradiotherapy for anaplastic oligodendroglioma: long-term results of RTOG 9402. *J Clin Oncol.* 2013;31:337–343.

21. Stupp R, Mason WP, van den Bent MJ, Weller M, Fisher B, Taphoorn MJ, et al. Radiotherapy plus concomitant and adjuvant temozolomide for glioblastoma. *N Engl J Med.* 2005;352:987–996.

22. Fisher BJ, Hu C, Macdonald DR, Lesser GJ, Coons SW, Brachman DG, et al. Phase 2 study of temozolomide-based chemoradiation therapy for high-risk low-grade gliomas: preliminary results of Radiation Therapy Oncology Group 0424. *Int J Radiat Oncol Biol Phys.* 2015;91:497–504.

23. U.S. National Library of Medicine. https://clinicaltrials.gov/ct2/show/NCT00626990?cond=NCT00626990&rank=1.

24. van den Bent MJ, Baumert B, Erridge SC, Vogelbaum MA, Nowak AK, Sanson M, et al. Interim results from the CATNON trial (EORTC study 26053-22054) of treatment with concurrent and adjuvant temozolomide for 1p/19q non-co-deleted anaplastic glioma: a phase 3, randomised, open-label intergroup study. *Lancet.* 2017;390:1645–1653.

25. U.S. National Library of Medicine. https://clinicaltrials.gov/ct2/show/NCT00887146?term=NCT00887146&rank=1.

26. U.S. National Library of Medicine. https://clinicaltrials.gov/ct2/show/NCT00978458?term=NCT00978458&rank=1.

27. U.S. National Library of Medicine. https://clinicaltrials.gov/ct2/show/NCT03180502.

28. Karim AB, Maat B, Hatlevoll R, Menten J, Rutten EH, Thomas DG, et al. A randomized trial on dose-response in radiation therapy of low-grade cerebral glioma: European Organization for Research and Treatment of Cancer (EORTC) Study 22844. *Int J Radiat Oncol Biol Phys.* 1996;36:549–556.

29. Shaw E, Arusell R, Scheithauer B, O'Fallon J, O'Neill B, Dinapoli R, et al. Prospective randomized trial of low- versus high-dose radiation therapy in adults with supratentorial low-grade glioma: initial report of a North Central Cancer Treatment Group/Radiation Therapy Oncology Group/Eastern Cooperative Oncology Group study. *J Clin Oncol.* 2002;20:2267–2276.

30. Marks LB, Yorke ED, Jackson A, Ten Haken RK, Constine LS, Eisbruch A, et al. Use of normal tissue complication probability models in the clinic. *Int J Radiat Oncol Biol Phys.* 2010;76:S10–S19.

31. Gondi V, Pugh SL, Tome WA, Caine C, Corn B, Kanner A, et al. Preservation of memory with conformal avoidance of the hippocampal neural stem-cell compartment during whole-brain radiotherapy for brain metastases (RTOG 0933): a phase II multi-institutional trial. *J Clin Oncol.* 2014;32:3810–3816.

32. Gilbert MR, Dignam JJ, Armstrong TS, Wefel JS, Blumenthal DT, Vogelbaum MA, et al. A randomized trial of bevacizumab for newly diagnosed glioblastoma. *N Engl J Med.* 2014;370:699–708.

33. Emami B, Lyman J, Brown A, Coia L, Goitein M, Munzenrider JE, et al. Tolerance of normal tissue to therapeutic irradiation. *Int J Radiat Oncol Biol Phys.* 1991;21:109–122.

34. Powell C, Guerrero D, Sardell S, Cumins S, Wharram B, Traish D, et al. Somnolence syndrome in patients receiving radical radiotherapy for primary brain tumours: a prospective study. *Radiother Oncol.* 2011;100:131–136.

35. Lawenda BD, Gagne HM, Gierga DP, Niemierko A, Wong WM, Tarbell NJ, et al. Permanent alopecia after cranial irradiation: dose-response relationship. *Int J Radiat Oncol Biol Phys.* 2004;60:879–887.

36. Armstrong C, Ruffer J, Corn B, DeVries K, Mollman J. Biphasic patterns of memory deficits following moderate-dose partial-brain irradiation: neuropsychologic outcome and proposed mechanisms. *J Clin Oncol.* 1995;13:2263–2271.

37. Armstrong CL, Corn BW, Ruffer JE, Pruitt AA, Mollman JE, Phillips PC. Radiotherapeutic effects on brain function: double dissociation of memory systems. *Neuropsychiatry Neuropsychol Behav Neurol.* 2000;13:101–111.

38. Strenger V, Lackner H, Mayer R, Sminia P, Sovinz P, Mokry M, et al. Incidence and clinical course of radionecrosis in children with brain tumors. A 20-year longitudinal observational study. *Strahlenther Onkol.* 2013;189:759–764.

39. Janny P, Cure H, Mohr M, Heldt N, Kwiatkowski F, Lemaire JJ, et al. Low grade supratentorial astrocytomas. Management and prognostic factors. *Cancer.* 1994;73:1937–1945.

40. Correa DD, DeAngelis LM, Shi W, Thaler HT, Lin M, Abrey LE. Cognitive functions in low-grade gliomas: disease and treatment effects. *J Neurooncol.* 2007;81:175–184.

41. Kiehna EN, Mulhern RK, Li C, Xiong X, Merchant TE. Changes in attentional performance of children and young adults with localized primary brain tumors after conformal radiation therapy. *J Clin Oncol.* 2006;24:5283–5290.

42. Marta GN, Murphy E, Chao S, Yu JS, Suh JH. The incidence of second brain tumors related to cranial irradiation. *Expert Rev Anticancer Ther.* 2015;15:295–304.

43. Murphy ES, Xie H, Merchant TE, Yu JS, Chao ST, Suh JH. Review of cranial radiotherapy-induced vasculopathy. *J Neurooncol.* 2015;122:421–429.

44. Taphoorn MJ, Heimans JJ, van der Veen EA, Karim AB. Endocrine functions in long-term survivors of low-grade supratentorial glioma treated with radiation therapy. *J Neurooncol.* 1995;25:97–102.

45. Collet-Solberg PF, Sernyak H, Satin-Smith M, Katz LL, Sutton L, Molloy P, et al. Endocrine outcome in long-term survivors of low-grade hypothalamic/chiasmatic glioma. *Clin Endocrinol (Oxf)*. 1997;47:79–85.

46. Laack NN, Brown PD, Ivnik RJ, Furth AF, Ballman KV, Hammack JE, et al. Cognitive function after radiotherapy for supratentorial low-grade glioma: a North Central Cancer Treatment Group prospective study. *Int J Radiat Oncol Biol Phys*. 2005;63:1175–1183.

47. Murphy ES, Leyrer CM, Parsons M, Suh JH, Chao ST, Yu J, et al. Risk factors for malignant transformation of low grade glioma. *Int J Radiat Oncol Biol Phys*. 2018; [Ref ROB_24660/ROB_ROB-D-17-01036, accepted 12.11.17].

Chapter 23

Stereotactic radiosurgery in the management of oligodendroglioma

Fabio Y. Moraes*, Gustavo Nader Marta[†,‡], Erin S. Murphy[§] and John H. Suh[§]

*Division of Radiation Oncology, Department of Oncology, Kingston Health Sciences Centre, Queen's University, Kingston, ON, Canada; [†]Department of Radiation Oncology, Hospital Sírio-Libanês, Sao Paulo, Brazil; [‡]Department of Radiology and Oncology, Division of Radiation Oncology, Instituto do Câncer do Estado de São Paulo (ICESP), Faculdade de Medicina da Universidade de São Paulo, Sao Paulo, Brazil; [§]Department of Radiation Oncology, Cleveland Clinic, Taussig Cancer Institute, Cleveland, OH, United States

Introduction

For the past 10 years, we have experienced a revolution in the assessment and classification of gliomas. By definition, gliomas are primary central nervous system tumors, consisting primarily of aberrantly growing astrocytes and oligodendrocytes (glial cells). Gliomas are graded, as per World Health Organization (WHO), into grades I–IV, mainly based on histological characteristics (as astrocytoma, oligodendroglioma, or oligoastrocytoma).[1] Grades I and II are low-grade gliomas (LGGs) and grades III and IV are high-grade gliomas (HGGs). The LGGs are generally less aggressive than HGG. Typically, LGG patients have median survival of 5–10 years with a range of up to 15 years for certain subtype.[2,3] Given the wide variation in survival, better classification systems were needed.

Advances in genomic characterization has provided a robust method to categorize and prognosticate glial tumors (Table 1).[1,4] For instance, isocitrate dehydrogenase (IDH) mutations were found to be present in the majority of the LGGs, regardless of histologic subtype.[5] Tumors of oligodendroglial lineage (oligodendrogliomas) exhibit allelic loss of chromosome 1p and 19q (1p/19q co-deletion) in addition to the IDH mutation, and this allelic loss is associated with a favorable prognosis and improved overall survival (OS).[6,7] Also, mutations in the telomerase reverse transcriptase (TERT) promoter were shown to be frequent in LGG.[8] Table 2 illustrates the median progression-free survival and OS of some seminal trials including radiotherapy (RT) for LGG with molecular classification.[9,10]

Oligodendrogliomas represent the third most common type of glioma, comprising about 10% of all gliomas and can be classified into grade II (LGG) and grade III (HGG—anaplastic characteristics), according to the available literature.[11] Roughly 30% of oligodendroglial tumors have anaplastic features, and the formal diagnosis of oligodendroglioma and anaplastic oligodendroglioma requires the demonstration of both an IDH gene family mutation and combined whole-arm losses of 1p and 19q—1p/19q co-deletion, a strong prognostic factor and a powerful predictor of prolonged survival.[1]

It is well established that surgical resection is the first treatment of choice for resectable oligodendrogliomas. Complete resection directly correlates with a significant increase in OS compared to biopsy alone or subtotal resection.[12,13] However, as infiltrative glial cells tend to extend beyond the margins of the tumor, complete resection is not always possible.[14] For many years, RT has been considered standard of care following best tumor resection for selected cases. The indication, timing, and dose of RT remain controversial, especially for grade II oligodendroglioma, because of the potential cognitive effects, radiation necrosis risk, and minimal impact on OS.[15–19] For grade III oligodendroglioma, two well-performed randomized clinical trials (EORTC 26951 and RTOG 9402) provide evidence that patients managed with procarbazine, lomustine, and vincristine chemotherapy (before or after RT), have longer OS compared with patients treated with RT alone, especially in a subset of patients with 1p/19q co-deletion.[20,21]

In the cases of unresectable or subtotally resected tumors, RT and chemotherapy are important treatment modalities to improve patient outcomes. In addition, several reports in recent years have studied the role of SRS in the treatment of LGG and HGG (which include oligodendrogliomas) and early data show favorable safety and efficacy profiles.

Oligodendroglioma. https://doi.org/10.1016/B978-0-12-813158-9.00023-2

TABLE 1 Relative survival according to low-grade glioma molecular classification

Molecular classification	Histologic subtype	Relative survival	
TERT mutation (+) IDH mutation (+) 1p19q codeletion (+)	Oligodendroglial	Highest	
TERT mutation (+) IDH mutation (+) 1p19q codeletion (−)	Astrocytic (ATRX mutation and TP53 mutation)	Highest	
TERT mutation (−) IDH mutation (+) 1p19q codeletion (−)	Astrocytic (ATRX mutation and TP53 mutation)	Intermediate	
TERT mutation (+) IDH mutation (−) 1p19q codeletion (−)	Glioblastoma-like (transformation)	Low	

TERT, telomerase reverse transcriptase; *IDH*, isocitrate dehydrogenase; *ATRX*, α-thalassemia/mental-retardation-syndrome-X-linked gene.

TABLE 2 Median progressive free survival (years) and overall survival (years) of selected trials including RT for low-grade gliomas

Study	Population	PFS (years)	*P* value	OS (years)	*P* value
Buckner (2016) RTOG 9802 RT+PCV vs RT[a]	Overall	10.4 (RT+PCV) vs 4 (RT)	0.002	13.3 (RT+PCV) vs 7.8 (RT)	0.03
	IDH wild type vs mutant	1.5 vs 7.6		5.1 vs 13.1	0.02
Baumert (2016) EORTC 22033-26033 (2016) RT vs TMZ[a]	Overall	3.9 (RT) vs 3.3 (TMZ)	0.23	Not reached (RT+PCV) vs 6.2	0.55
	1p intact	3.4 (RT) vs 2.5 (TMZ)	0.06	NR	NR
	1p deleted	4.8 (RT) vs 4.6 (TMZ)	0.95	NR	NR
	IDH Mutant/1p19q co-deleted	5.1 (RT) vs 4.6 (TMZ)	0.85	NR	NR
	IDH Mutant/1p19q non-co-deleted	4.6 (RT) vs 3.0 (TMZ)	0.004	NR	NR
	IDH wild type	1.6 (RT) vs 2.0 (TMZ)	0.67	NR	NR

[a]*Grade 2 astrocytoma, oligodendroglioma, and oligoastrocytoma; RT, radiotherapy; TMZ, temozolomide; PFS, progression free survival; OS, overall survival; IDH, isocitrate dehydrogenase; NR, not reported.*

Basics of radiosurgery

Conventional RT when indicated in high-risk and residual and/or progressive LGG achieves acceptable long-term local control, but is tempered with the risk of developing long-term neurocognitive and negative quality of life (QoL) impact. As the molecular assessment of tumors and treatments evolve, RT continues to progress based on imaging, accurate positioning, treatment techniques, and precision of delivery.

Evidence in applying SRS to brain tumors is vast but largely based on the treatment of single or multiple brain metastases.[22] Regarding brain metastases, the use of SRS has been correlated with better local control compared to whole brain RT alone, and associated with improved neurocognitive and QoL outcomes.[23,24] As technologies have expanded and become more available, the applications of SRS to neuro-oncology have increased to include multiple brain metastases, vestibular schwannomas, intracranial meningiomas, cerebral arteriovenous malformation, trigeminal neuralgia, and gliomas.[25]

The term SRS was first described by Leksell over five decades ago.[26] Currently, SRS is defined as the delivery of highly conformal radiation to a specific target using precise stereotactic localization systems with sharp dose fall-off to surrounding structures in single session or multiple sessions (commonly up to five fractions).[24,27] The SRS involves the use of multiple beamlets of radiation aimed precisely at a static (or minimally movable) target to deliver high ablative dose of radiation. A high dose is deposited at the juncture of these numerous beamlets, with a steep dose fall-off outside the target. With this approach, SRS can maximize dose to the target while lessening the dose to the surrounding brain. In this context, the use of SRS for the treatment of LGG could be an option with the intent to reduce potential sides effects associated with conventional RT modalities, which includes fatigue, radiation necrosis, neurocognitive decline, and impact on QoL.

An amalgamation of factors including advances in imaging techniques and immobilization systems combined with high-powered computer systems have facilitated the use and development of SRS. Currently, there are three main categories of SRS treatment devices: (1) cobalt-60-based system; Gamma Knife radiosurgery (GKRS) (Elekta), (2) robotic multi-leaf collimator (MLC) based system; CyberKnife system (Accuray), and (3) high-definition MLC-based linear accelerator system (Elekta, Varian, BrainLAB).

Clinical use of radiosurgery for LGG

SRS: upfront approach

The data for the potential positive neurocognitive and QoL impact with the use of SRS for oligodendroglioma is limited to small and retrospective series (Table 3).[28–32] The lack of widespread use of SRS for these tumors may be related to the infiltrative nature of these tumors, which suggests microscopic disease beyond the gross tumor requiring a treatment volume larger than the target itself (i.e., clinical target volume, CTV), which is not commonly used for SRS target delineation. In addition, concern arises regarding the changes to standard treatment approaches as even modest improvement in outcomes would take several years to establish.

Although surgical resection remains the standard treatment for most oligodendrogliomas, these tumors may grow in eloquent areas, making gross total resection impossible. In addition, the biological properties of some gliomas, specifically relating to their oxygenation, may make them good targets for SRS. Typically, mammalian tumor cell survival curves have a characteristic shape or shoulder after a single dose of irradiation. At a specific point, there is an evolution in cell viability as aerated cells are nonviable and hypoxic cells dominate the response to radiation. It is well described that higher dose of radiation per fraction (as occurs with SRS) increases the lethality of hypoxic cells due to greater interference with intrinsic cellular repair mechanisms during irradiation.[33] Thus, oligodendrogliomas in adverse location or with tumor recurrence or progression despite surgery may benefit from the use of additional treatment modalities such as SRS.

In a study of 39 patients with LGG (minority of them oligodendroglioma) who received GKRS as primary treatment (30.8%) or salvage following previous treatment (69.2%), the authors concluded that GKRS was effective and safe. The mean tumor volume was $2.7\,cm^3$. The median marginal dose was $15\,Gy$.[32] At a mean follow-up of 60.5 months, the actuarial progression-free survival and OS at 5 years was 39.1% and 91.8%, respectively. The authors also reported that at last follow-up, volume reduction was documented in 57.7% of cases and that GKRS improved patients' functional performance and QoL. Regarding QoL assessment, the authors reported significant improvement between pretreatment and post-treatment Karnofsky performance status, Eastern Cooperative Oncology Group scale, and modified Rankin Scale score with positive impact on functional status.

In another series, the authors reported on 49 LGG patients (96% astrocytoma) treated with GKRS for tumors in eloquent brain, residual tumor postsurgery, or for late progression after surgery. Treatment median volume was $2.4\,cm^3$, and median

TABLE 3 Selected series on stereotactic radiosurgery for low-grade gliomas

Study	# Patients	Histology	Median size (mm or cc)	Dose (Gy)	Follow up months	Survival	Late toxicity
Kida (2000)	31	Astrocytoma	25.4 mm	15.7	27.6	NR	50%
Souhami (2003)	21	Mixed	14 mm	NR max 37.5–45.5	86.4	72% at five years	14%
Hadjipayanis (2002)	37	Astrocytoma	0.42–45.1 cc.	15	28	85% at 32 months after SRS	NR
Heppner (2006)	49	Astrocytoma	2.4 cc	15	63	86%	8%
Park (2011)	25	Astrocytoma	3.7 cc	14	65	54.1% PFS at 5 years	4%
Gagliardi (2017)	39	Astrocytoma	2.7 cc	15	60.5	52.8% PFS at 5 years 91.8% OS at 5 years	4%

#, number; NR, not reported; PFS, progression free survival; OS, overall survival; SRS, stereotactic radiosurgery.

peripheral dose was 15 Gy. At a median follow-up of 63 months, the median clinical progression-free survival was 44 months and only 4 patients had treatment-related toxicity. Complete radiological remission was seen in 14 patients (29%), and mortality due to tumor progression occurred in 7 patients (14%).[31]

The previously presented studies in addition to the ones summarized in Table 3 provide limited evidence on safety and efficacy of SRS for oligodendrogliomas. Thus, in special situations, including tumors in eloquent brain areas where maximal safe resection is not possible or with meaningful residual tumor, the use of SRS could provide satisfactory disease control in patients with acceptable side effects.

SRS for recurrent LGG (reirradiation)

Oligodendrogliomas are slow-growing tumors, but the majority of these tumors ultimately show progression and/or malignant degeneration independent of first-line treatment modality.[34] The failure to control oligodendrogliomas correlates with progressive neurological deficits and decrease of QoL, causing eventual death in most patients. Second or third time total surgical resection is often impossible without morbidity when the infiltrative gliomas progress.[35,36] Systemic chemotherapy, especially temozolomide at the time of progression, can offer some benefit for a selected group of patients.[37–39] However, this benefit is often not durable. Due to its high conformality and minimally invasive nature, the use of reirradiation with SRS for recurrences might be considered.

Again, most of the evidence is composed of retrospective series, which highlights the use of SRS reirradiation in selected patients with lesions of limited size. Table 4 lists selected studies on the use of SRS in the recurrent/reirradiation setting.[40–50] Typically, only the area of contrast uptake, which could represent an area of malignant transformation, is defined as the target volume with median survival times between 7 and 12 months after SRS reirradiation.[51–53] Fig. 1 demonstrates a case of a patient with recurrent LGG treated with radiosurgery.

In a retrospective review that aimed to assess safety and efficacy of SRS in the management of patients with oligodendrogliomas or mixed oligoastrocytomas, 18 patients (21 tumors) who had failed to respond to conventional therapies were reported. Tumor grades at the most recent operation were grade 1 (1), grade 2 (1), grade 3 (12) and grade 4 (7 patients). A total of 17 patients had undergone previous RT. The authors described a median tumor volume and margin dose of 8.2 cm³ (range 1.9–47.7 cm³) and 15 Gy (range 12–20 Gy), respectively. In this mixed cohort, survival after SRS was 44% at 48 months, and toxicity was acceptable.[54]

In a series of 63 patients with LGG (53 astrocytoma; 6 oligodendroglioma, and 4 others) effectiveness of reirradiation in recurrent LGG was assessed.[55] All included patients were treated with surgical resection of the primary tumor followed by RT. The median RT total dose was 60 Gy (range 30–80 Gy). Using stereotactic technique for reirradiation, a median total

TABLE 4 Selected series (>25 patients) of single fraction radiosurgery (SRS) or fractionated stereotactic RT (FSRT) with or without of systemic therapy for recurrent low and high-grade gliomas

	Therapeutic approach (N)	Median target volume (cc)	Prescription dose (N of fractions)	Median OS after re-RT (months)	Toxicity[a]
Elliott et al. (2011)	SRS (26)	1.22	15 Gy (1 fx)	13.5	RN 7.7%
Skeie et al. (2012)	SRS (51)	NA	12.2 (1 fx); 24–25 Gy (3–5 fx)	12	NA
McKenzie et al. (2013)	FSRT (35)	8.5	30 Gy (5 fx)	8.6	RTOG (G3–G4) 9%
Martínez-Carrillo et al. (2014)	SRS (87)	6	18 Gy (1 fx)	10	None
Pinzi et al. (2015)	SRS (42) FSRT (67)	SRS 2 FSRT 11	23–24 Gy (3 fx)	11.5	RN 6%
Minniti et al. (2011)	FSRT + TMZ (36)	–	37.5 Gy (15 fx)	9.7	NA
Cuneo et al. (2012)	SRS/FSRT ± BEV (63)	–	18 Gy (1 fx) 25 Gy (5 fx)	11	RTOG G3 = 11%; RN 10%
Niyazi et al. (2012)	FSRT ± BEV (30)	–	36 Gy (18 fx)	Not reached	RTOG G3 = 1; G4 = 1; RN 2
Greenspoon et al. (2014)	SRS + TMZ (31)	–	25–35 Gy (5 fx)	9	RTOG G3 = 3, G4 = 1
Combs et al. (2005)	FSRT (172)	44	36 Gy (variable)	23	No grade 4 events
Sutera et al. (2017)	SRS (65)	65		68.4% at 1 years for LGG	No grade 3 or 4 events

[a]*RN: radionecrosis or RTOG grade 3 or 4.*
ReRT, reirradiation; SRS, radiosurgery; FSRT, fractionated stereotactic radiotherapy; HGG, high grade glioma; RTOG, radiation therapy oncology group; G, toxicity grade; NA, not available; RN, radionecrosis; fx, fraction(s).

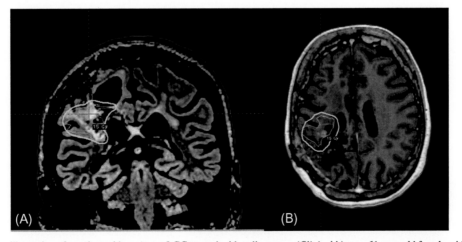

FIG. 1   Clinical case illustration of a patient with recurrent LGG treated with radiosurgery (*Clinical history*: 61-year-old female with right parietal low-grade oligodendroglioma and multiple recurrences): (A) coronal T2/flair MR imaging and (B) axial T1 MR imaging. First resections (gross total) was performed in August 2000. Patient had a recurrence and repeat resection performed in September 2011 (pathology consistent with oligodendroglioma (grade II, 1p/19q co-deletion, EGFR not amplified, Ki-67 19%–20%). Patient progressed again in August 2013 and was treated on the ECOG E3F05 [randomized to RT alone (5040 cGy/28 fractions)]. She had a new progression and received temozolomide from January 2016 to February 2017. MRI brain in November 2017 revealed progressive disease of the right fronto-parietal region (A and B). The new progressive disease was treated with radiosurgery 1500 cGy in a single fraction prescribed to 57% isodose line *(yellow line)*, which covered 100% of the target. The plan utilized 51 shots using 8 mm, 4 mm, composite sectors. Target volume = 11.3 cc; maximum dose = 2630.0 cGy; maximum diameter = 3.8 cm; MD/PD = 1.753; PIV/TV = 1.735; gradient index = 2.8; number of fractions = 1. Treatment volume was based on T1 and FLAIR MRI scans *(green line)*.

dose of 36 Gy (range 15–62 Gy) was delivered to a planning target volume (PTV). The median PTV (contrast-enhancing lesion in T1-weighted and the hyperintense areas in T2-weighted MRI scans, plus a 1-cm safety margin) size was 44 cm^3. Reirradiation was well tolerated, and no grade 4 or 5 side effects occurred. From the time of reirradiation, median survival and PFS was 23 and 12 months, respectively.

A single-institution prospective cohort reported on 114 recurrent malignant glioma (HGG) patients who received salvage SRS.[56] The median tumor volume was 10.6 cm^3 (roughly a 2.7-cm diameter tumor) and the median prescription dose was 16 Gy. The SRS was well tolerated with grade 1 or 2 radiation necrosis reported in 24.4% of patients [graded as grade 1 asymptomatic; clinical or diagnostic observations only; intervention not indicated. Grade 2 moderate symptoms; corticosteroids indicated. Grade 3 severe symptoms; medical intervention indicated. Grade 4 life-threatening consequences; urgent intervention indicated. Grade 5 death as per the *Common Terminology Criteria for Adverse Events (CTCAE), v4.0*]. In a large institutional validation analysis of 199 patients treated for recurrent glioma, different schemes of reirradiation were assessed.[49,57] The RT was performed with stereotactic technique. The clinical treatment volume (CTV) included the macroscopic tumor and a safety margin of about 5 mm; PTV included an additional margin of 1–2 mm. Median PTV volume was 62.0 cm^3 (range 0.4–480.6 cm^3) and >50% of the patients received SRS or dose fractions >5 Gy. With median follow-up after re-RT of 2.5 months, the median OS was 13.6 months for HGG. Univariate analyses confirmed HGG ($P < .001$), time between primary RT and reirradiation ($P < .001$), age > 50 years old ($P = .002$) as factors correlated with worse survival.

In another report, salvage SRS for recurrent glioma was assessed in 65 patients with 76 lesions (55 high-grade, 21 low-grade). At a median follow-up of 8.5 months [interquartile range (IQR): 3.9–15.8)], the 1-year OS was 68.4% and 38.7% for LGG and HGG, respectively. No clinically meaningful toxicity was reported.[58] Finally, a systematic review on the management of patients with recurrence of LGG reported with level 3 evidence (based on three case series, one including SRS)[49,55,59] that reirradiation at recurrence should be considered due to possible benefit in disease control.[60] It is important to highlight that case selection is the key and recurrent volume should be limited to 10 cm^3 and 2.5 cm in maximum diameter.

Radiosurgery treatment planning, delivery, and techniques

Studies of SRS for LGG have used variable target volumes and dose prescription with variable margins intended to treat residual and/or microscopic disease. Definition of the target volume has varied among the studies, but the postoperative residual tumor with minimal margin is mostly utilized. Dose prescription also is quite variable among studies with different scenarios (no previous RT or previous RT) and technique (SRS or fractionated stereotactic RT). Future studies will help refine the role of SRS in the upfront and recurrent/reirradiation setting.

References

1. Louis DN, Perry A, Reifenberger G, et al. The 2016 World Health Organization Classification of Tumors of the Central Nervous System: a summary. *Acta Neuropathol (Berl)*. 2016;131(6):803–820. https://doi.org/10.1007/s00401-016-1545-1.
2. Delgado-López PD, Corrales-García EM, Martino J, Lastra-Aras E, Dueñas-Polo MT. Diffuse low-grade glioma: a review on the new molecular classification, natural history and current management strategies. *Clin Transl Oncol Off Publ Fed Span Oncol Soc Natl Cancer Inst Mex*. 2017;19(8):931–944. https://doi.org/10.1007/s12094-017-1631-4.
3. Wessels PH, Weber WEJ, Raven G, Ramaekers FCS, Hopman AHN, Twijnstra A. Supratentorial grade II astrocytoma: biological features and clinical course. *Lancet Neurol*. 2003;2(7):395–403.
4. Eckel-Passow JE, Lachance DH, Molinaro AM, et al. Glioma groups based on 1p/19q, IDH, and TERT promoter mutations in tumors. *N Engl J Med*. 2015;372(26):2499–2508. https://doi.org/10.1056/NEJMoa1407279.
5. Turcan S, Rohle D, Goenka A, et al. IDH1 mutation is sufficient to establish the glioma hypermethylator phenotype. *Nature*. 2012;483(7390):479–483. https://doi.org/10.1038/nature10866.
6. Kraus JA, Koopmann J, Kaskel P, et al. Shared allelic losses on chromosomes 1p and 19q suggest a common origin of oligodendroglioma and oligoastrocytoma. *J Neuropathol Exp Neurol*. 1995;54(1):91–95.
7. von Deimling A, Louis DN, von Ammon K, Petersen I, Wiestler OD, Seizinger BR. Evidence for a tumor suppressor gene on chromosome 19q associated with human astrocytomas, oligodendrogliomas, and mixed gliomas. *Cancer Res*. 1992;52(15):4277–4279.
8. Killela PJ, Reitman ZJ, Jiao Y, et al. TERT promoter mutations occur frequently in gliomas and a subset of tumors derived from cells with low rates of self-renewal. *Proc Natl Acad Sci U S A*. 2013;110(15):6021–6026. https://doi.org/10.1073/pnas.1303607110.
9. Buckner JC, Shaw EG, Pugh SL, et al. Radiation plus procarbazine, CCNU, and vincristine in low-grade glioma. *N Engl J Med*. 2016;374(14):1344–1355. https://doi.org/10.1056/NEJMoa1500925.

10. Baumert BG, Hegi ME, van den Bent MJ, et al. Temozolomide chemotherapy versus radiotherapy in high-risk low-grade glioma (EORTC 22033-26033): a randomised, open-label, phase 3 intergroup study. *Lancet Oncol.* 2016;17(11):1521–1532. https://doi.org/10.1016/S1470-2045(16)30313-8.

11. Koeller KK, Rushing EJ. From the archives of the AFIP: oligodendroglioma and its variants: radiologic-pathologic correlation. *Radiogr Rev Publ Radiol Soc N Am Inc.* 2005;25(6):1669–1688. https://doi.org/10.1148/rg.256055137.

12. Smith JS, Chang EF, Lamborn KR, et al. Role of extent of resection in the long-term outcome of low-grade hemispheric gliomas. *J Clin Oncol Off J Am Soc Clin Oncol.* 2008;26(8):1338–1345. https://doi.org/10.1200/JCO.2007.13.9337.

13. Ius T, Isola M, Budai R, et al. Low-grade glioma surgery in eloquent areas: volumetric analysis of extent of resection and its impact on overall survival. A single-institution experience in 190 patients. *J Neurosurg.* 2012;117(6):1039–1052 [clinical article], https://doi.org/10.3171/2012.8.JNS12393.

14. Yordanova YN, Moritz-Gasser S, Duffau H. Awake surgery for WHO Grade II gliomas within "noneloquent" areas in the left dominant hemisphere: toward a "supratotal" resection. *J Neurosurg.* 2011;115(2):232–239 [clinical article], https://doi.org/10.3171/2011.3.JNS101333.

15. Douw L, Klein M, Fagel SS, et al. Cognitive and radiological effects of radiotherapy in patients with low-grade glioma: long-term follow-up. *Lancet Neurol.* 2009;8(9):810–818. https://doi.org/10.1016/S1474-4422(09)70204-2.

16. Sarmiento JM, Venteicher AS, Patil CG. Early versus delayed postoperative radiotherapy for treatment of low-grade gliomas. *Cochrane Database Syst Rev.* 2015;6:CD009229. https://doi.org/10.1002/14651858.CD009229.pub2.

17. Karim AB, Maat B, Hatlevoll R, et al. A randomized trial on dose-response in radiation therapy of low-grade cerebral glioma: European Organization for Research and Treatment of Cancer (EORTC) Study 22844. *Int J Radiat Oncol Biol Phys.* 1996;36(3):549–556.

18. Shaw E, Arusell R, Scheithauer B, et al. Prospective randomized trial of low- versus high-dose radiation therapy in adults with supratentorial low-grade glioma: initial report of a North Central Cancer Treatment Group/Radiation Therapy Oncology Group/Eastern Cooperative Oncology Group study. *J Clin Oncol Off J Am Soc Clin Oncol.* 2002;20(9):2267–2276. https://doi.org/10.1200/JCO.2002.09.126.

19. van den Bent MJ, Afra D, de Witte O, et al. Long-term efficacy of early versus delayed radiotherapy for low-grade astrocytoma and oligodendroglioma in adults: the EORTC 22845 randomised trial. *Lancet Lond Engl.* 2005;366(9490):985–990. https://doi.org/10.1016/S0140-6736(05)67070-5.

20. Cairncross G, Wang M, Shaw E, et al. Phase III trial of chemoradiotherapy for anaplastic oligodendroglioma: long-term results of RTOG 9402. *J Clin Oncol Off J Am Soc Clin Oncol.* 2013;31(3):337–343. https://doi.org/10.1200/JCO.2012.43.2674.

21. van den Bent MJ, Brandes AA, Taphoorn MJB, et al. Adjuvant procarbazine, lomustine, and vincristine chemotherapy in newly diagnosed anaplastic oligodendroglioma: long-term follow-up of EORTC brain tumor group study 26951. *J Clin Oncol Off J Am Soc Clin Oncol.* 2013;31(3):344–350. https://doi.org/10.1200/JCO.2012.43.2229.

22. Moraes FY, Taunk NK, Marta GN, Suh JH, Yamada Y. The rationale for targeted therapies and stereotactic radiosurgery in the treatment of brain metastases. *Oncologist.* 2016;21(2):244–251. https://doi.org/10.1634/theoncologist.2015-0293.

23. Brown PD, Jaeckle K, Ballman KV, et al. Effect of radiosurgery alone vs radiosurgery with whole brain radiation therapy on cognitive function in patients with 1 to 3 brain metastases: a randomized clinical trial. *JAMA.* 2016;316(4):401–409. https://doi.org/10.1001/jama.2016.9839.

24. Sahgal A, Ruschin M, Ma L, Verbakel W, Larson D, Brown PD. Stereotactic radiosurgery alone for multiple brain metastases? A review of clinical and technical issues. *Neuro-Oncol.* 2017;19(suppl 2):ii2–ii15. https://doi.org/10.1093/neuonc/nox001.

25. Pannullo SC, Fraser JF, Moliterno J, Cobb W, Stieg PE. Stereotactic radiosurgery: a meta-analysis of current therapeutic applications in neuro-oncologic disease. *J Neuro-Oncol.* 2011;103(1):1–17. https://doi.org/10.1007/s11060-010-0360-0.

26. Leksell L. The stereotaxic method and radiosurgery of the brain. *Acta Chir Scand.* 1951;102(4):316–319.

27. Pollock BE, Lunsford LD. A call to define stereotactic radiosurgery. *Neurosurgery.* 2004;55(6):1371–1373.

28. Kida Y, Kobayashi T, Mori Y. Gamma knife radiosurgery for low-grade astrocytomas: results of long-term follow up. *J Neurosurg.* 2000;93(Suppl 3):42–46. https://doi.org/10.3171/jns.2000.93.supplement.

29. Roberge D, Souhami L. Stereotactic radiosurgery in the management of intracranial gliomas. *Technol Cancer Res Treat.* 2003;2(2):117–125. https://doi.org/10.1177/153303460300200207.

30. Hadjipanayis CG, Kondziolka D, Flickinger JC, Lunsford LD. The role of stereotactic radiosurgery for low-grade astrocytomas. *Collections.* 2007;15(4):1–7. https://doi.org/10.3171/foc.2003.14.5.16@col.4.

31. Heppner PA, Sheehan JP, Steiner LE. Gamma knife surgery for low-grade gliomas. *Neurosurgery.* 2008;62(suppl 2):755–762. https://doi.org/10.1227/01.neu.0000316279.22371.b8.

32. Gagliardi F, Bailo M, Spina A, et al. Gamma knife radiosurgery for low-grade gliomas: clinical results at long-term follow-up of tumor control and patients' quality of life. *World Neurosurg.* 2017;101:540–553. https://doi.org/10.1016/j.wneu.2017.02.041.

33. Balagamwala EH, Chao ST, Suh JH. Principles of radiobiology of stereotactic radiosurgery and clinical applications in the central nervous system. *Technol Cancer Res Treat.* 2012;11(1):3–13. https://doi.org/10.7785/tcrt.2012.500229.

34. Wallner KE, Galicich JH, Krol G, Arbit E, Malkin MG. Patterns of failure following treatment for glioblastoma multiforme and anaplastic astrocytoma. *Int J Radiat Oncol Biol Phys.* 1989;16(6):1405–1409.

35. Dirks P, Bernstein M, Muller PJ, Tucker WS. The value of reoperation for recurrent glioblastoma. *Can J Surg J Can Chir.* 1993;36(3):271–275.

36. Harsh GR, Levin VA, Gutin PH, Seager M, Silver P, Wilson CB. Reoperation for recurrent glioblastoma and anaplastic astrocytoma. *Neurosurgery.* 1987;21(5):615–621.

37. van den Bent MJ, Taphoorn MJB, Brandes AA, et al. Phase II study of first-line chemotherapy with temozolomide in recurrent oligodendroglial tumors: the European Organization for Research and Treatment of Cancer Brain Tumor Group Study 26971. *J Clin Oncol Off J Am Soc Clin Oncol.* 2003;21(13):2525–2528. https://doi.org/10.1200/JCO.2003.12.015.

38. van den Bent MJ, Chinot O, Boogerd W, et al. Second-line chemotherapy with temozolomide in recurrent oligodendroglioma after PCV (procarbazine, lomustine and vincristine) chemotherapy: EORTC Brain Tumor Group phase II study 26972. *Ann Oncol Off J Eur Soc Med Oncol*. 2003; 14(4):599–602.

39. Jaeckle KA, Hess KR, Yung WKA, et al. Phase II evaluation of temozolomide and 13-cis-retinoic acid for the treatment of recurrent and progressive malignant glioma: a North American Brain Tumor Consortium study. *J Clin Oncol Off J Am Soc Clin Oncol*. 2003;21(12):2305–2311. https://doi.org/10.1200/JCO.2003.12.097.

40. Elliott RE, Parker EC, Rush SC, et al. Efficacy of gamma knife radiosurgery for small-volume recurrent malignant gliomas after initial radical resection. *World Neurosurg*. 2011;76(1-2):128–140 [discussion 61–62]. https://doi.org/10.1016/j.wneu.2010.12.053.

41. Skeie BS, Enger PØ, Brøgger J, et al. γ knife surgery versus reoperation for recurrent glioblastoma multiforme. *World Neurosurg*. 2012; 78(6):658–669. https://doi.org/10.1016/j.wneu.2012.03.024.

42. McKenzie JT, Guarnaschelli JN, Vagal AS, Warnick RE, Breneman JC. Hypofractionated stereotactic radiotherapy for unifocal and multifocal recurrence of malignant gliomas. *J Neuro-Oncol*. 2013;113(3):403–409. https://doi.org/10.1007/s11060-013-1126-2.

43. Martínez-Carrillo M, Tovar-Martín I, Zurita-Herrera M, et al. Salvage radiosurgery for selected patients with recurrent malignant gliomas. *Biomed Res Int*. 2014; https://doi.org/10.1155/2014/657953.

44. Pinzi V, Orsi C, Marchetti M, et al. Radiosurgery reirradiation for high-grade glioma recurrence: a retrospective analysis. *Neurol Sci Off J Ital Neurol Soc Ital Soc Clin Neurophysiol*. 2015;36(8):1431–1440. https://doi.org/10.1007/s10072-015-2172-7.

45. Minniti G, Armosini V, Salvati M, et al. Fractionated stereotactic reirradiation and concurrent temozolomide in patients with recurrent glioblastoma. *J Neuro-Oncol*. 2011;103(3):683–691. https://doi.org/10.1007/s11060-010-0446-8.

46. Cuneo KC, Vredenburgh JJ, Sampson JH, et al. Safety and efficacy of stereotactic radiosurgery and adjuvant bevacizumab in patients with recurrent malignant gliomas. *Int J Radiat Oncol Biol Phys*. 2012;82(5):2018–2024. https://doi.org/10.1016/j.ijrobp.2010.12.074.

47. Niyazi M, Ganswindt U, Schwarz SB, et al. Irradiation and bevacizumab in high-grade glioma retreatment settings. *Int J Radiat Oncol Biol Phys*. 2012;82(1):67–76. https://doi.org/10.1016/j.ijrobp.2010.09.002.

48. Greenspoon JN, Sharieff W, Hirte H, et al. Fractionated stereotactic radiosurgery with concurrent temozolomide chemotherapy for locally recurrent glioblastoma multiforme: a prospective cohort study. *OncoTargets Ther*. 2014;7:485–490. https://doi.org/10.2147/OTT.S60358.

49. Combs SE, Thilmann C, Edler L, Debus J, Schulz-Ertner D. Efficacy of fractionated stereotactic reirradiation in recurrent gliomas: long-term results in 172 patients treated in a single institution. *J Clin Oncol Off J Am Soc Clin Oncol*. 2005;23(34):8863–8869. https://doi.org/10.1200/JCO.2005.03.4157.

50. Sutera PA, Bernard ME, Gill BS, et al. Salvage stereotactic radiosurgery for recurrent gliomas with prior radiation therapy. *Future Oncol Lond Engl*. 2017;13(29):2681–2690. https://doi.org/10.2217/fon-2017-0226.

51. Cho KH, Hall WA, Gerbi BJ, Higgins PD, McGuire WA, Clark HB. Single dose versus fractionated stereotactic radiotherapy for recurrent high-grade gliomas. *Int J Radiat Oncol Biol Phys*. 1999;45(5):1133–1141.

52. Gaspar LE, Zamorano LJ, Shamsa F, Fontanesi J, Ezzell GE, Yakar DA. Permanent 125iodine implants for recurrent malignant gliomas. *Int J Radiat Oncol Biol Phys*. 1999;43(5):977–982.

53. Taunk NK, Moraes FY, Escorcia FE, Mendez LC, Beal K, Marta GN. External beam re-irradiation, combination chemoradiotherapy, and particle therapy for the treatment of recurrent glioblastoma. *Expert Rev Anticancer Ther*. 2016;16(3):347–358. https://doi.org/10.1586/14737140.2016.1143364.

54. Sarkar A, Pollock BE, Brown PD, Gorman DA. Evaluation of gamma knife radiosurgery in the treatment of oligodendrogliomas and mixed oligo-dendroastrocytomas. *J Neurosurg*. 2002;97(5 Suppl):653–656. https://doi.org/10.3171/jns.2002.97.supplement.

55. Combs SE, Ahmadi R, Schulz-Ertner D, Thilmann C, Debus J. Recurrent low-grade gliomas: the role of fractionated stereotactic re-irradiation. *J Neuro-Oncol*. 2005;71(3):319–323. https://doi.org/10.1007/s11060-004-2029-z.

56. Kong D-S, Lee J-I, Park K, Kim JH, Lim D-H, Nam D-H. Efficacy of stereotactic radiosurgery as a salvage treatment for recurrent malignant gliomas. *Cancer*. 2008;112(9):2046–2051. https://doi.org/10.1002/cncr.23402.

57. Kessel KA, Hesse J, Straube C, et al. Validation of an established prognostic score after re-irradiation of recurrent glioma. *Acta Oncol Stockh Swed*. 2017;56(3):422–426. https://doi.org/10.1080/0284186X.2016.1276621.

58. Sutera PA, Bernard ME, Gill BS, et al. Salvage stereotactic radiosurgery for recurrent gliomas with prior radiation therapy. *Future Oncol*. 2017; https://doi.org/10.2217/fon-2017-0226.

59. Shepherd SF, Laing RW, Cosgrove VP, et al. Hypofractionated stereotactic radiotherapy in the management of recurrent glioma. *Int J Radiat Oncol Biol Phys*. 1997;37(2):393–398.

60. Nahed BV, Redjal N, Brat DJ, et al. Management of patients with recurrence of diffuse low grade glioma. *J Neuro-Oncol*. 2015;125(3):609–630. https://doi.org/10.1007/s11060-015-1910-2.

Chapter 24

Proton beam therapy for oligodendroglioma

Vonetta M. Williams*, Simon S. Lo[†] and Lia M. Halasz[†]

*Department of Radiation Oncology, University of Washington School of Medicine, Seattle, WA, United States; [†] Department of Radiation Oncology and Neurological Surgery, University of Washington School of Medicine, Seattle, WA, United States

Introduction

With maximal safe surgical resection, radiation therapy, and chemotherapy, many patients with oligodendroglioma live for several years after diagnosis. The recent RTOG 9802 trial showed a median survival of 13.3 years for high-risk low-grade glioma (subtotal resection and/or age more than 40) patients treated with adjuvant radiation therapy to 54 Gy and procarbazine, CCNU, and vincristine (PCV) chemotherapy.[1] This trial included oligodendroglioma, astrocytoma, and oligoastrocytoma tumors; however, the greatest benefit from PCV was seen among patients with oligodendrogliomas. The RTOG 9402 trial similarly showed excellent 14.7 year overall survival for anaplastic oligodendrogliomas with 1p/19q co-deletion.[2] Thus, long-term late effects of treatment are important in maintaining a good quality of life for these patients. Several trials have sought to omit radiation therapy given long-term effects, including neurocognitive decline, hypopituitarism, and secondary malignancy. Additionally, technological advances in limiting the radiation dose to normal brain and adjacent organs have been important. One of these strategies has been the use of proton therapy, which has the ability to decrease the integral dose to the brain and adjacent organs.

Background on protons

Protons are charged particles that were discovered in 1919 by Ernest Rutherford. They were first suggested as a treatment for cancer by Robert Wilson, and the first patient was treated in 1954 in Berkley, California. The first clinical treatment center was built in Loma Linda in 1990. Since then, proton therapy has expanded. According to the National Association for Proton Therapy there are 28 active proton therapy centers and another 23 centers under construction or in development.

Protons, in contrast to the photons that are used for conventional external beam radiation therapy, are particles with mass and charge. They have a superior dose distribution with no exit dose that allows for the deposition of dose within the intended target, while sparing surrounding normal tissue. The characteristic proton Bragg peak is generated from the loss of proton energy within the last few millimeters of tissue penetration. Distal to this, no additional dose is deposited (see Fig. 1). Placement of the beam edge at precise locations is achieved by modulating the proton energy.[3] Studies indicate that in comparison to even modern techniques of delivery of photon therapy such as with IMRT, the dose to normal tissue can be decreased by as much as 50% using protons without sacrificing tumor coverage.[4]

The Role of Radiation Therapy for Oligodendroglioma

The cornerstones of treatment for oligodendroglioma for both adults and children consist of maximal safe resection, radiation therapy, and chemotherapy. Radiation therapy is often recommended in cases of unresectable disease, subtotal resection, or recurrent disease. Radiation therapy should also be considered for patients with anaplastic oligodendroglioma or low-grade oligodendroglioma with high-risk features. Pignatti defined these high-risk features as age ≥40, astrocytoma histology, tumors ≥6 cm, tumor crossing midline, and preoperative neurologic deficits based on EORTC data. The RTOG 9802 trial simplified these high-risk features as age >40 years and subtotal resection. Given the recent changes in WHO classification to include molecular markers of 1p1/9q deletion status and IDH mutation status, the categories of patients who require radiation therapy may change over the coming years.

Oligodendroglioma. https://doi.org/10.1016/B978-0-12-813158-9.00024-4

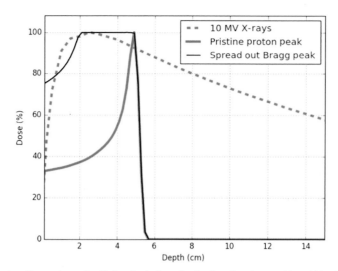

FIG. 1 Proton Bragg peak illustration. Comparison of radiation depth-dose distributions for photons (*dotted blue line*), a single Bragg peak (*red solid line*), and several Bragg peaks shifted in depth to create a spread out Bragg peak (*black solid line*). In comparison to photons, protons have no exit dose beyond the deposited dose.

The most common radiation modality for treatment of oligodendroglioma is photon therapy. Photon therapy involves the use of high-energy photon beams aimed at the tumor and surrounding brain at risk for tumor infiltration. Low-grade gliomas are generally treated with 45–54 Gy in either 1.8–2 Gy fractions. Treatment volumes consist of the tumor extent and resection bed based on MRI T1 and T2 postoperative imaging. The use of modern techniques of conformal radiotherapy and intensity-modulated therapy (IMRT) has improved the precision of delivery allowing for a lower dose of radiation to normal tissue. However, the delivery of dose from several angles still allows for exposure of normal tissue to low-dose radiation.

Historically, radiation therapy was shown to improve progression free survival for low-grade gliomas and overall survival for higher-grade gliomas.[5, 6] However, the cognitive side effects of radiation therapy have caused concern, especially for patients surviving more than 1 year. These include memory loss, difficulty with executive function, and attention deficit problems.[7, 8] In addition, brain tumor patients report more fatigue, depression, and anxiety in addition to cognitive limitations.[9] Klein and colleagues showed that long-term survivors (>6 years) with low-grade gliomas who received radiation upfront had worse cognitive function than those who had not received radiation therapy.[10, 11] These concerns have led many providers to delay or avoid radiation therapy. Multiple trials, including the NOA-04 trial[12] that randomized patients with anaplastic glioma to standard radiation therapy, PCV, or TMZ in a 2:1:1 ratio and the NCCTG/Mayo clinic trial that treated low-grade oligodendroglioma patients with PCV and only RT on progression have looked at avoiding radiation therapy as upfront treatment.[13]

For pediatric patients, long-term side effects may be an even greater concern. In 2005 there were over 50,000 estimated survivors of pediatric central nervous system (CNS) malignancies.[14] This highlights the importance of reducing treatment toxicity in such a large cohort with significant treatment related morbidity. Among childhood survivors of cancer 62% report at least one long-term side effect of treatment, many of which are sequelae from radiotherapy.[15] Given the excellent rate of overall survival in the treatment of low-grade glioma, long-term sequelae of treatment is of particular concern. Long-term radiation toxicity in the treatment of low-grade gliomas consist of risk of cataracts, vision loss, sensorineural hearing loss, cognitive changes, pseudo-progression, radiation necrosis, pituitary dysfunction, secondary malignancy, and in children, decrement in IQ, infertility, vasculopathy, and musculoskeletal anomalies. According to data from the Childhood Survivor study, which compared the health of children who received cranial radiation to the health of their siblings, these children have a seven-fold higher risk of severe long-term morbidity compared to their siblings.[15]

Initial experience with proton therapy

In 2001, Fitzek and colleagues from the Massachusetts General Hospital/Harvard Cyclotron Laboratory first published on the use of combined protons and photons for lower-grade gliomas (at the time this was according to the Daumas-Duport classification[16]).[17] They utilized protons with the goal of dose escalation to doses of 68.2 cobalt Gray equivalent for grade 2 lesions and 79.7 CGE for grade 3 lesions. Among the 20 patients in the study, 10 had mixed gliomas and one had oligodendroglioma. Overall, dose escalation failed to improve outcomes. Subsequent reports from the Heidelberg Ion Therapy Center, Massachusetts General Hospital, and University of Pennsylvania on proton therapy for treatment of low-grade gliomas focused on using more standard doses. Hauswald and colleagues published on 19 patients treated for low-grade glioma between 2010 and 2011 utilizing proton therapy to a median total dose of 54 Gy E. One patient had tumor progression after a median follow-up of 5 months. Alopecia was the predominant toxicity.[18] Maquilan and colleagues published on 23 patients treated for meningioma and low-grade glioma reporting that patients experienced mild fatigue, headaches, and insomnia similar to photon patients.[19] A more recent abstract from a database maintained by six proton centers in the United States confirmed that all side effects were grade 1 or 2 and similar to photon therapy for 58 patients treated with proton therapy for low-grade glioma.[18a]

Subsequently published data presented dosimetric comparisons between proton and photon therapy for low-grade gliomas. Fig. 2 illustrates a plan comparison between protons and photons for a low grade glioma patient. Dennis and colleagues from the Massachusetts General Hospital published a comparison of IMRT and proton therapy for 11 patients treated with fractionated proton therapy. The prescription dose to the CTV was 54 Gy (RBE). Based on the dosimetric data, they estimated risk of toxicity utilizing equivalent uniform dose (EUD) and normal tissue complication probability (NTCP) modeling. They estimated that the mean excess risk of secondary malignancy was 106 per 10,000 cases for IMRT versus 47 per 10,000 cases for proton plans. Of note there was no difference in the NTCP for both modalities.[20] Harrabi and colleagues from Heidelberg compared 3D conformal proton and photon plans for 74 patients with low-grade glioma. Overall, target volume coverage was comparable for proton and photon plans. However, protons allowed for decreased dose to organs at risk, especially in the contralateral brain. For instance, the contralateral hippocampus received a mean dose of 39.1 Gy E (standard deviation ±21.6 Gy E) with photon therapy versus 13.9 Gy E (standard deviation ±21.6 Gy E) with proton therapy.[21]

Another study looking at the potential benefit of proton therapy versus photons in the pediatric population modeled dose characteristics in four different CNS tumors. Data for 40 patients with optic pathway gliomas, craniopharyngiomas, ependymomas, and medulloblastoma were analyzed using dose-cognitive effects data. The outcomes were estimated over a 5-year period. They found better sparing of smaller OARs such as cochlea and hypothalamus particularly when they were further away from the PTV. Larger structures such as temporal lobes received less low- and intermediate-dose radiation. These reductions based on modeling would result in clinically significant IQ differences.[22] Dosimetric comparisons of three-dimensional conformal proton radiotherapy (3D-PRT) and intensity-modulated proton therapy (IMPT) compared with intensity-modulated radiotherapy (IMRT) in pediatric craniopharyngioma demonstrated that all modalities were able to provide adequate PTV coverage while meeting dose constraints, but proton therapy with both three-dimensional conformal proton radiotherapy (3D-PRT) and IMPT was able to decrease the hippocampal dose.[23]

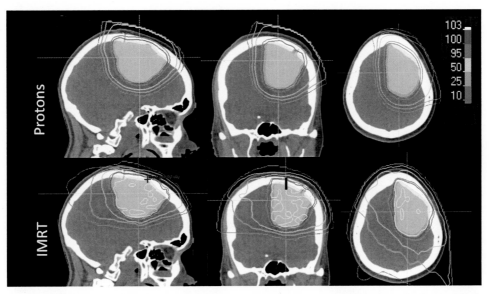

FIG. 2 Low-grade glioma comparison plan. Proton therapy plan is represented on the top row and IMRT photon plan is represented on the bottom row. CTV is solid green and PTV (3 mm expansion) is solid yellow. Dose lines display percentage of prescribed dose for each plan. The proton plan allows for less integral dose to the brain and more homogeneity within the target to achieve the same target dose coverage.

Neurocognitive outcomes

Shih and colleagues from the Massachusetts General Hospital performed the first prospective trial with detailed neurocognitive, neuroendocrine, and quality-of-life outcomes for patients with WHO grade 2 glioma (20% oligodendrogliomas) treated with proton therapy. They treated 20 patients with a median age of 37.5 years to 54 Gy (RBE). After a median follow up of 5.1 years for patients without progression and a median follow-up for neurocognitive measures of 3.2 years, they found that intellectual functioning was within the normal range for the study group and remained stable over time. This included tests of visuospatial ability, attention/working memory, and executive function. For eight patients there were baseline neurocognitive impairments observed in language, memory, and processing with no overall decline in cognitive function over time. Of note, the PFS for this group was 85% at 3 years and 40% at 5 years, and patients were removed from the study at progression. Given historical data on neurocognitive decline after radiation therapy, though it had a small number of patients, this prospective trial suggested proton therapy may be helpful to maintain neurocognitive function for these patients.[24, 25]

For pediatric patients, studies have demonstrated that proton therapy can reduce the volume of supratentorial brain that is irradiated. This reduction can in turn decrease decline in IQ observed in patients treated with photons. In a study of 32 patients with a median age of 7.4 years treated for low-grade glioma with proton therapy the cognitive effects 7 years after treatment were minimal.[26] A 2016 paper published in Pediatric Oncology by Kahalley et al. compared IQ over time in pediatric patients treated with proton beam versus external beam radiotherapy with photons for brain tumors. Their data demonstrated no decline in IQ in the proton beam group compared to an approximately 1 point per year decline in the XRT group, yet the clinical significance of this was unclear.[27] This is in contrast to the findings of a meta-analysis of 39 studies of survivors of pediatric CNS tumors, which demonstrated verbal and nonverbal cognitive deficits that were one full standard deviation below the mean.[28] Another study demonstrated a mean 12–14 point decline in IQ among survivors of pediatric tumors, a number which increased in prevalence and severity with time.[29] The rate of decline of IQ has been correlated to the volume of supratentorial brain that is radiated.[30] Of note, this is thought to be due to the lack of development of new knowledge and not a loss of existing knowledge. A 2004 Swedish population-based study of the cognitive effects of low-dose radiation such as that received for cutaneous hemangioma suggest that it was sufficient to have an impact on IQ in a study of over 3000 children who received radiation treatment under 18 months of age.[31]

Currently the NRG Oncology group is conducting a randomized phase II clinical trial (BN005) to study how well proton beam or intensity-modulated radiation therapy works in preserving brain function in patients with IDH mutant grade II or III glioma. The primary outcome is neurocognitive outcomes over time as measured by the clinical trial battery composite calculated from the Hopkins Verbal Learning Test Revised Total Recall, Delayed Recall, and Delayed Recognition; Controlled Oral Word Association Test; and the Trail Making Test. Additional outcomes measured will include quality of life, adverse effects, symptom inventory, and patterns of failure. Hopefully, this trial will provide us with the needed randomized trial data to truly test the efficacy and neurocognitive side effects of proton versus photon therapy. The estimated completion date of this study is 2022.

Neuroendocrine outcomes

Another common long-term sequelae of radiation for brain tumors is endocrine dysfunction. It is caused by radiation injury to the hypothalamus and pituitary. This results in decreases in the production of GH, ACTH, LH, FSH, and TSH. The deficiencies may be corrected with hormone replacement therapy, however this requires long-term follow up with endocrinology and adds to the overall cost of cancer treatment. Among childhood brain tumor survivors treated with RT the incidence of hormone deficiency over 40 years ranges from 72.4% for GH to 5.2% for ACTH.[32] The incidence of GH deficiency has been correlated with radiation dose, with a greater loss in adult height seen in patients who received over 24 Gy without hormone supplementation[33, 34]

A recent paper published in JCO by Vatner et al. on the results of a prospective study of approximately 200 pediatric and young adult patients treated with proton radiation for brain tumors demonstrated that the incidence of hormone deficiency was highest in those that received a mean hypothalamus and pituitary dose ≥40 Gy (CGE) and lowest in those receiving ≤20 CGE.[34a] In the prospective trial from MGH on 20 low-grade glioma adult patients, sparing of the pituitary depended on tumor location and the dose to the pituitary was associated with neuroendocrine outcomes overall.[25]

Secondary malignancy

Secondary malignancy is a concern after CNS irradiation for primary brain tumors, particularly in the pediatric population. The risk is increased in long-term survivors. A retrospective study of 558 patients treated with Proton therapy at Harvard Cyclotron was compared to a matched cohort from the national SEER database. It found no evidence of an increased incidence of secondary malignancies in the proton cohort.[35] In adults, the burden of secondary malignancy is 1.5 times the risk in the general population. It is postulated that proton beam therapy may half the risk of secondary malignancy compared to photon therapy due to the decreased dose of radiation that surrounding normal brain tissue is exposed to with proton radiation.[20] Overall, further clinical follow-up is needed to determine whether this decreased dose indeed leads to fewer secondary malignancies.

Pseudo-progression and radiation treatment effect

Pseudo-progression or radiation treatment effect is the development of contrast enhancing changes surrounding the treatment field which may be mistaken for tumor progression. Pseudo-progression is transient and occurs within 6 months after treatment with radiation. Radiation treatment effect, or radionecrosis, may be transient or permanent. The mechanism is thought to be injury to oligodentrocytes and vascular injury that leads to leaking of the blood-brain barrier allowing for contrast enhancement.[36] There is some concern that pseudo-progression and radiation treatment effect rates may be higher in brain tumor patients treated with proton therapy, however this has not been definitely shown, especially in adult cohorts.

The concern stems from small studies in pediatric cohorts. Gunther and colleagues from the MD Anderson Cancer Center found more common postradiation MRI changes in patients less than 3 years of age at diagnosis when treated with proton therapy versus IMRT for ependymoma.[37] A 2014 paper by Indelicato et al. seemed to indicate that the incidence of brainstem injury for pediatric patients with posterior fossa tumors may be higher among patients treated with proton versus photon radiation therapy using the same dose contraints.[38] However, after employing certain dose constraints, the investigators found that this incidence actually decreased. A review published in the IJROBP by the National Cancer Institute Workshop on Proton Therapy for Children examined the issue of brainstem injury with proton therapy treatment in over 600 patients treated at the three highest volume centers from 2006 to 2016. They established guidelines to allow for the safe delivery of proton radiotherapy utilizing certain planning constraints.[39]

For adults, Bronk and colleagues from MD Anderson compared 65 patients treated with photons with 34 patients treated with protons for low-grade and anaplastic glioma. Overall, there were no differences in the pseudo-progression rate between protons and photons. However, they demonstrated pseudo-progression rates were higher in patients with oligodendroglioma compared to astrocytoma patients.[40] Acharya and colleagues from Washington University performed a similar analysis with 123 patients treated with photons and 37 patients treated with protons. They reported that the 2-year cumulative incidence for significant radiation necrosis was 18.7% for protons and 9.7% for photons. However, given the small numbers this was not significantly different. They confirmed the MD Anderson data that co-deleted oligodendroglioma was a significant risk factor for radiation necrosis. They also showed that the volume receiving 60 Gy RBE was a significant dosimetric predictor of radionecrosis in oligodendrogliomas. They concluded there was insufficient evidence to conclude a difference in the incidence of radionecrosis between proton and photon therapy.

Technological improvements

Just as photon therapy has had multiple advances in radiation delivery, dose calculation, and image guidance throughout the years, the ability for accurate proton therapy delivery is constantly evolving. Though the initial series of proton therapy for gliomas utilized passive scattering techniques, centers are increasingly turning to pencil beam-scanning techniques. Passively scattered proton plans are similar to 3D photon plans and are forward planned. A pencil beam is created by the accelerator and delivered through a beam nozzle where it is scattered by a physical filter to generate a broad proton field; this field is then collimated by a custom physical aperture. A range modulator is used to form a spread out Bragg peak to achieve the desired depths with a compensator to adjust the depth at the distal edge of the target.

Pencil beam scanning (PBS) is inversely planned utilizing an optimization algorithm with specified goals for target coverage and organs at risk. To deliver PBS, a pencil beam is created by the accelerator and transported to the beam nozzle where two sets of magnets are used to deflect the beam in two directions perpendicular to the beam's direction. Before it hits the nozzle, the beam energy is adjusted to achieve the desired range in tissue. Utilizing the energy and magnets, the beam's Bragg Peak can be directed to any location. This is known as the beam spot. Each beam spot has an individual monitor unit setting. This allows for more precise conformality on both the proximal and distal edges of the targets, thereby allowing for more conformal plans, better control of where the hot spots are placed, and dose painting. Given the precision of this technique, it is important to evaluate the robustness of the treatment delivery by analyzing how shifts in position in all directions or changes in the density of tissues penetrated may change the target coverage and dose to the objects at risk.

For oligodendroglioma patients, PBS may allow for better conformality that will further decrease excess radiation dose to normal brain. In addition, it allows for better target coverage while respecting the dose constraints of the optic nerves and brainstem, even for complex shapes. Finally, it allows for better sparing of the scalp as the proximal aspect of the treatment can be better controlled. However, PBS (and proton therapy in general) can be sensitive to changes in tissue density and thickness. Thus, in cases where tissue thickness may change (such as a changing pseudo-meningocele) it is important to ensure that the dose distribution has not changed.

Cost

Proton therapy is more costly than photon therapy, so much of the debate on protons versus photons lies around the cost-benefit analysis of this treatment. Important to this debate will be the accrual on randomized trials to quantify the advantages and disadvantages of proton therapy for gliomas. It may be that only in the cases where a large clinical benefit of protons is shown, that it will be cost effective. According to the National Cancer Institute Workshop on Proton Therapy for Children in a consensus statement from May 2018, they recommend that protons be used in pediatric patients with curable brain tumors in order to avoid injury and long-term damage to critical central nervous system structures.[39] Studies have also indicated that there may be a cost benefit to the use of proton therapy for CNS disease in children.[41] The cost of management of the long- term seqeulae of CNS radiation such as Endocrinology for management of GH and other pituitary hormone deficits can be significantly decreased if the incidence of these is decreased with more sparing of normal tissue.

Conclusions

With treatment, patients with oligodendroglioma can achieve prolonged life expectancy. Long-term sequelae of treatment in both pediatric and adult survivors is of particular concern. Endocrine dysfunction, neurocognitive deficits and risk of secondary malignancy significantly impact quality of life. Several dosimetric analyses have demonstrated that proton plans can spare normal tissue better than photon therapy plans even with the use of IMRT technology. Randomized data comparing protons versus photons in the treatment of CNS tumors is lacking even in the pediatric population. Data for proton therapy in the treatment of low-grade gliomas is sparse given the relative incidence of these tumors and the sparsity of proton centers. However, trials looking at endpoints such as neurocognition are currently underway comparing protons versus photon therapy. Overall, given the improvement in overall survival for patients with oligodendroglioma established by recent trials, proton therapy represents one promising way of reducing the long-term toxicities of radiation therapy.

References

1. Buckner JC, Shaw EG, Pugh SL, et al. Radiation plus procarbazine, CCNU, and vincristine in low-grade glioma. *N Engl J Med.* 2016;374:1344–1355.
2. Cairncross G, Wang M, Shaw E, et al. Phase III trial of chemoradiotherapy for anaplastic oligodendroglioma: long-term results of RTOG 9402. *J Clin Oncol.* 2013;31:337–343.
3. Hsi WC, Moyers MF, Nichiporov D, et al. Energy spectrum control for modulated proton beams. *Med Phys.* 2009;36:2297–2308.
4. Ladra MM, Edgington SK, Mahajan A, et al. A dosimetric comparison of proton and intensity modulated radiation therapy in pediatric rhabdomyosarcoma patients enrolled on a prospective phase II proton study. *Radiother Oncol.* 2014;113:77–83.
5. van den Bent MJ, Afra D, de Witte O, et al. Long-term efficacy of early versus delayed radiotherapy for low-grade astrocytoma and oligodendroglioma in adults: the EORTC 22845 randomised trial. *Lancet.* 2005;366:985–990.
6.. Walker MD, Alexander E, Jr., Hunt WE, et al. Evaluation of BCNU and/or radiotherapy in the treatment of anaplastic gliomas. A cooperative clinical trial. J Neurosurg 1978;49:333-43.
7. Gregor A, Cull A, Traynor E, Stewart M, Lander F, Love S. Neuropsychometric evaluation of long-term survivors of adult brain tumours: relationship with tumour and treatment parameters. *Radiother Oncol.* 1996;41:55–59.
8. Surma-aho O, Niemela M, Vilkki J, et al. Adverse long-term effects of brain radiotherapy in adult low-grade glioma patients. *Neurology.* 2001;56:1285–1290.
9. Calvio L, Feuerstein M, Hansen J, Luff GM. Cognitive limitations in occupationally active malignant brain tumour survivors. *Occup Med (Lond).* 2009;59:406–412.

10. Klein M, Heimans JJ, Aaronson NK, et al. Effect of radiotherapy and other treatment-related factors on mid-term to long-term cognitive sequelae in low-grade gliomas: a comparative study. *Lancet*. 2002;360:1361–1368.

11. Douw L, Klein M, Fagel SS, et al. Cognitive and radiological effects of radiotherapy in patients with low-grade glioma: long-term follow-up. *Lancet Neurol*. 2009;8:810–818.

12. Wick W, Roth P, Hartmann C, et al. Long-term analysis of the NOA-04 randomized phase III trial of sequential radiochemotherapy of anaplastic glioma with PCV or temozolomide. *Neuro Oncol*. 2016;18:1529–1537.

13. Buckner JC, Gesme D, O'Fallon JR, et al. Phase II trial of procarbazine, lomustine, and vincristine as initial therapy for patients with low-grade oligodendroglioma or oligoastrocytoma: efficacy and associations with chromosomal abnormalities. *J Clin Oncol*. 2003;21:251–255.

14. Mariotto AB, Rowland JH, Yabroff KR, et al. Long-term survivors of childhood cancers in the United States. *Cancer Epidemiol Biomarkers Prev*. 2009;18:1033–1040.

15. Oeffinger KC, Mertens AC, Sklar CA, et al. Chronic health conditions in adult survivors of childhood cancer. *N Engl J Med*. 2006;355:1572–1582.

16. Daumas-Duport C, Scheithauer B, O'Fallon J, Kelly P. Grading of astrocytomas. A simple and reproducible method. *Cancer*. 1988;62:2152–2165.

17. Fitzek MM, Thornton AF, Gt H, et al. Dose-escalation with proton/photon irradiation for Daumas-Duport lower-grade glioma: results of an institutional phase I/II trial. *Int J Radiat Oncol Biol Phys*. 2001;51:131–137.

18. Hauswald H, Rieken S, Ecker S, et al. First experiences in treatment of low-grade glioma grade I and II with proton therapy. *Radiat Oncol*. 2012;7:189.

18a. Wilkinson B, Morgan H, Gondi V, Larson GL, Hartsell WF, Laramore GE, Halasz LM, Vargas C, Keole SR, Grosshans DR, Shih HA, Mehta MP. Low levels of acute toxicity associated with proton therapy for low-grade glioma: a proton collaborative group study. *Int J Radiat Oncol Biol Phys*. 2016;96 (2S):E135. https://doi.org/10.1016/j.ijrobp.2016.06.930.

19. Maquilan G, Grover S, Alonso-Basanta M, Lustig RA. Acute toxicity profile of patients with low-grade gliomas and meningiomas receiving proton therapy. *Am J Clin Oncol*. 2014;37:438–443.

20. Dennis ER, Bussiere MR, Niemierko A, et al. A comparison of critical structure dose and toxicity risks in patients with low grade gliomas treated with IMRT versus proton radiation therapy. *Technol Cancer Res Treat*. 2013;12:1–9.

21. Harrabi SB, Bougatf N, Mohr A, et al. Dosimetric advantages of proton therapy over conventional radiotherapy with photons in young patients and adults with low-grade glioma. *Strahlenther Onkol*. 2016;192:759–769.

22. Merchant TE, Hua CH, Shukla H, Ying XF, Nill S, Oelfke U. Proton versus photon radiotherapy for common pediatric brain tumors: comparison of models of dose characteristics and their relationship to cognitive function. *Pediatric Blood Cancer*. 2008;51:110–117.

23. Boehling NS, Grosshans DR, Bluett JB, et al. Dosimetric comparison of three-dimensional conformal proton radiotherapy, intensity-modulated proton therapy, and intensity-modulated radiotherapy for treatment of pediatric craniopharyngiomas. *Int J Radiat Oncol Biol Phys*. 2012;82:643–652.

24. Sherman JC, Colvin MK, Mancuso SM, et al. Neurocognitive effects of proton radiation therapy in adults with low-grade glioma. *J Neurooncol*. 2016;126:157–164.

25. Shih HA, Sherman JC, Nachtigall LB, et al. Proton therapy for low-grade gliomas: results from a prospective trial. *Cancer*. 2015;121:1712–1719.

26. Greenberger BA, Pulsifer MB, Ebb DH, et al. Clinical outcomes and late endocrine, neurocognitive, and visual profiles of proton radiation for pediatric low-grade gliomas. *Int J Radiat Oncol Biol Phys*. 2014;89:1060–1068.

27. Kahalley LS, Ris MD, Grosshans DR, et al. Comparing intelligence quotient change after treatment with proton versus photon radiation therapy for pediatric brain tumors. *J Clin Oncol*. 2016;34:1043.

28. Robinson KE, Kuttesch JF, Champion JE, et al. A quantitative meta-analysis of neurocognitive sequelae in survivors of pediatric brain tumors. *Pediatr Blood Cancer*. 2010;55:525–531.

29. Mulhern RK, Hancock J, Fairclough D, Kun L. Neuropsychological status of children treated for brain tumors: a critical review and integrative analysis. *Med Pediatr Oncol*. 1992;20:181–191.

30. Reddick WE, Mulhern RK, Elkin TD, Glass JO, Merchant TE, Langston JW. A hybrid neural network analysis of subtle brain volume differences in children surviving brain tumors. *Magn Reson Imaging*. 1998;16:413–421.

31. Hall P, Adami HO, Trichopoulos D, et al. Effect of low doses of ionising radiation in infancy on cognitive function in adulthood: Swedish population based cohort study. *Brit Med J*. 2004;328:19–21.

32. Chemaitilly W, Li Z, Huang S, et al. Anterior hypopituitarism in adult survivors of childhood cancers treated with cranial radiotherapy: a report from the St Jude Lifetime Cohort study. *J Clin Oncol*. 2015;33:492–500.

33. Merchant TE, Goloubeva O, Pritchard DL, et al. Radiation dose-volume effects on growth hormone secretion. *Int J Radiat Oncol Biol Phys*. 2002;52:1264–1270.

34. Melin AE, Adan L, Leverger G, Souberbielle JC, Schaison G, Brauner R. Growth hormone secretion, puberty and adult height after cranial irradiation with 18 Gy for leukaemia. *Eur J Pediatr*. 1998;157:703–707.

34a. Vatner RE, Niemierko A, Misra M, Weyman EA, Goebel CP, Ebb DH, Jones RM, Huang MS, Mahajan A, Grosshans DR, Paulino AC, Stanley T, MacDonald SM, Tarbell NJ, Yock TI. Endocrine deficiency as a function of radiation dose to the hypothalamus and pituitary in pediatric and young adult patients with brain tumors. *J Clin Oncol*. 2018;36(28):2854–2862. https://doi.org/10.1200/JCO.2018.78.1492.

35. Chung CS, Yock TI, Nelson K, Xu Y, Keating NL, Tarbell NJ. Incidence of second malignancies among patients treated with proton versus photon radiation. *Int J Radiat Oncol*. 2013;87:46–52.

36. Rider WD. Radiation damage to the brain—a new syndrome. *J Can Assoc Radiol*. 1963;14:67–69.

37. Gunther JR, Sato M, Chintagumpala M, et al. Imaging changes in pediatric intracranial ependymoma patients treated with proton beam radiation therapy compared to intensity modulated radiation therapy. *Int J Radiat Oncol Biol Phys*. 2015;93:54–63.

38. Indelicato DJ, Flampouri S, Rotondo RL, et al. Incidence and dosimetric parameters of pediatric brainstem toxicity following proton therapy. *Acta Oncol.* 2014;53:1298–1304.

39. Haas-Kogan D, Indelicato D, Paganetti H, et al. National cancer institute workshop on proton therapy for children: considerations regarding brainstem injury. *Int J Radiat Oncol Biol Phys.* 2018;101:152–168.

40. Bronk JK, Guha-Thakurta N, Allen PK, Mahajan A, Grosshans DR, McGovern SL. Analysis of pseudoprogression after proton or photon therapy of 99 patients with low grade and anaplastic glioma. *Clin Transl Radiat Oncol.* 2018;9:30–34.

41. Mailhot Vega R, Kim J, Hollander A, et al. Cost effectiveness of proton versus photon radiation therapy with respect to the risk of growth hormone deficiency in children. *Cancer.* 2015;121:1694–1702.

Chapter 25

Interstitial brachytherapy treatment for oligodendrogliomas

Ibrahim Abu-Gheida, Sarah Sittenfeld, Samuel T. Chao and John H. Suh

Department of Radiation Oncology, Cleveland Clinic, Taussig Cancer Institute, Cleveland, OH, United States

INTRODUCTION

Standard treatment of malignant brain gliomas consists of surgical resection followed by external beam radiation therapy (EBRT) and chemotherapy. Unfortunately, recurrence rates remain very high, with most recurrences occurring within 2 cm of the original contrast enhancing tumor.[1] Adjuvant EBRT prolongs progression-free survival (PFS) and overall survival (OS) in patients with gliomas. Early adjuvant radiation therapy post maximal safe resection has been shown to improve PFS but not OS in patients with low-grade gliomas (LGG) of which 13.5% were oligodendrogliomas (OD).[2] The addition of chemotherapy to radiation has also been shown to improve outcomes in both high- and low-grade gliomas including OD.[3, 4] Since these patients have a better prognosis with a median OS of 16.7 years, any benefit of radiotherapy must be balanced against risk of long-term side effects; mainly radiation-induced necrosis and neurocognitive function decline.[5] Brachytherapy has been proposed as an alternative method of delivering radiation for these tumors by the virtue of its low energy and steep dose falloff, which limits the radiation exposure to surrounding healthy brain tissue. Finally, by decreasing the number of visits for radiation treatments, this also improves the patient's quality of life.[6] This chapter will focus mainly on the indications, risks and benefits, and early and late side effects of brain glioma brachytherapy, with a focus on outcomes of oligodendroglioma patients.

HISTORY OF BRAIN BRACHYTHERAPY
Procedure

Brain tumor brachytherapy procedures were first reported back in the early 20th century. In 1914, Dr. Charles H. Frazier, from Philadelphia, reported on the implantation of radium radiation sources into cerebral parenchymal tumors using 50–100 mg of radium implanted for up to 20 h.[7] His procedure, which was commonly accompanied with EBRT, resulted in encouraging outcomes.[7] Meanwhile, radium loaded probes were also being used for treatment of pituitary adenomas and craniopharyngiomas.[8] Breakthrough in brain brachytherapy came in the late 1940s when Dr. Leksell updated the stereotactic frame that ultimately revolutionized stereotactic approaches for the treatment of brain malignancies.[9] His apparatus and modified versions of his frame are still the most popular and widely used of all stereotactic head frames (Fig. 1). It relies on a set of three coordinates (*x*, *y*, and *z*) in an orthogonal frame for reference, or alternatively, an angle, depth, and anteroposterior (or axial) location coordinates for a cylindrical system. The mechanical frame has head-holding clamps and bars which put the head in a fixed position in reference to the coordinate system; zero or origin device (Fig. 1). Guide bars in the *x*, *y*, and *z* directions, or alternately in the polar coordinate holder are fitted with high precision Vernier scales (Fig. 2), allowing the neurosurgeon to position the point of a probe containing radioactive seeds inside the brain in the desired location through a small hole in the skull. By utilizing this technique, radioactive source implantation became more accurate and possible in deep inoperable tumors[10] (Fig. 2).

Isotopes

Different isotopes have been studied to deliver brain brachytherapy. By 1961, Mundinger's group from Germany pioneered implanting different radioactive isotopes in 261 patients including radioactive Gold (Au-198), Cobalt (Co-60), Phosphorus

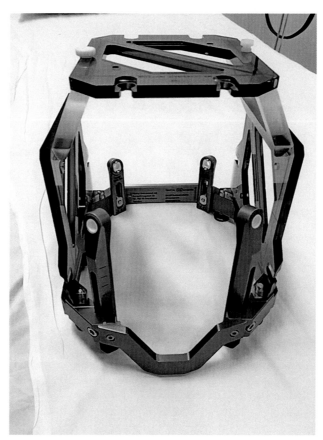

FIG. 1   Today common use head stereotactic frame, developed from the original Leksell model. *(Copyrights from Cleveland Clinic.)*

(P-32), Tantalum (Ta-182), and Iridium (Ir-192).[11] They also developed the "gamma-med," which is a brain intraoperative radiotherapy machine that utilizes a hot Ir-192 source in an afterloading technique for treatment of intracavitary tumor-beds.[11, 12] This machine is still commonly used in various brachytherapy centers for different body sites for Ir-192 after-loading procedures. Therefore, many radionuclides have been used for clinical brachytherapy against malignancies at various body sites (Table 1). An ideal source would be identified based on its radiobiophysical properties, which is discussed further in the "Rationale of Brachytherapy" section. Radioactive Iodine (I-125), which was used in prostate and head and neck brachytherapy, was introduced for brain brachytherapy in the 1970s.[13, 14] Having a low average energy and therefore, a low half-value range (distance by which dose is reduced by 50%), I-125 became the most commonly used source in the brain brachytherapy because of its rapid fall-off in tissues resulting in minimal exposure to outlying normal brain. Furthermore, I-125 brachytherapy was shown to reduce histological features that are prognostic for tumor progression, that is, cellularity, pleomorphism, vessel hyperplasia, and degree of mitosis, and lowers the proliferating cell nuclear antigen, a marker for late G1- and S-phases of the cell cycle.[15] Permanent low activity I-125 seeds, temporary high activity I-125 seeds, and temporary high activity I-125 Iotrex solution have been used. Table 2 summarizes the different radiobiological characteristic of I-125 used as sources for brain brachytherapy.[16]

Imaging advancement

It was not until the introduction of computed tomography (CT) imaging in the early 1970s that tumors were localized and doses planned on the basis of films obtained by angiography, nuclear scanning, and pneumoencephalography.[11, 17] The CT scans allowed for better target delineation and more accurate dosimetric calculation. Furthermore, CT scans accompanied with stereotactic head frames have made the accurate placement of radioactive sources into tumors more reliable.[17]

FIG. 2 (A–D) Polar coordinate guide bars fitted with high precision Vernier scales. Probes containing radioactive seeds are placed directly through the scales to allow for precise-point positioning. (E) Antero-posterior X-ray post brain brachytherapy implant. Seeds are seen and numbered on the left temporal lobe. *((A–D) Copyrights from Cleveland Clinic. (E) Copyrights from Cleveland Clinic.)*

TABLE 1 Radioisotopes used for interstitial brachytherapy in different body sites

Isotope	Half-life	Alpha decay	Beta decay	Gamma (energy in keV)
Radium-226	1064 years	+	+	180–2200
Radon-222	3.8 days	+	+	220–2200
Gold-198	2.7 days	−	+	410
Yttrium-90	2.7 days	−	+	−
Phosphorus-32	14.4 days	−	+	−
Cesium-137	30.0 years	−	+	660
Cobalt-60	5.3 years	−		1170–1330
Iridium-192	74.2 days	−	+	300–610
Iodine-125	60.3 days	−	−	28–35

TABLE 2 Comparison of temporary high activity and permanent low I-125 seeds with temporary high activity I-125 GliaSite Implants[16]

	Temporary seeds	Permanent seeds	Temporary GliaSite
Activity	10–20 mCi per seed source	0.6–1.0 mCi per seed source	19–30 mCi/mL of diluted Iotrex
Dose rate	40–60 cGy/h	7–20 cGy/h initially	50 cGy/h
Median number of sources	Eight seeds	54 seeds	Median balloon diameter, 3 cm (range 2–6 cm)
Hospital cost per procedure—1998 costs	NA	$3200	$11,000
Median of total dose	53 Gy	210 Gy	60 Gy
Time of total dose	4–5 days	Infinity	3–7 days (the calculated dwell time)

Key: *mCi*, milliCurie; *cGy/h*, centiGray per hour; *Gy*, Gray; *Yr*, year.

RATIONALE OF BRACHYTHERAPY

Currently, there are few centers in the world that are still actively engaged in brain tumor brachytherapy, either using Ir-192 High Dose Rate (HDR), Au-198, or more commonly I-125 seeds, via low dose permanent seeds or high dose temporary seeds or solution. The main advantage of Iodine seeds, commonly used in extracranial cancers such as prostate and head and neck, is that its generated gamma rays have low energies of 28–34 keV with a half value layer of 2 cm. This results in a sharp dose fall off and minimal exposure of normal surrounding brain tissue to radiation. Furthermore, with a half-life of 60.2 days, I-125 results in a less complicated dosimetric calculation when compared to Au-198, which has a half-life of only 2.7 days.

Since seeds are directly placed into the tumor, this site usually receives a very high dose of radiation therapy, with a very sharp dose fall off that limits toxicity to surrounding normal brain tissue. This results in a very focal and conformal radiation therapy plan. The dose gradient that is achievable at the target surface using I-125 seeds is even greater than that which can be achieved with Gamma Knife radiosurgery or linear accelerator radiosurgery.[18] Finally, as repopulation and redistribution during treatment are of minor importance in the therapy of LGGs, low dose rate seed implantation appears to be a rational therapeutic strategy. With low dose rate treatment, sublethal damage can be repaired and long-term side effects of late-responding tissues can be avoided, which is especially important in the periphery of the target volume. In the center of the treated volume, highly focused necrotizing intratumoral doses with a steep dose decrease from the center to the periphery can be achieved. Therefore, seeds combine the advantage of fractionated radiotherapy (repair of surrounding normal tissue) and radiosurgery (tumor cell death irrespective of radiosenstivity) in one modality.[19] Moreover, permanent seed placement is a one-day procedure, allowing it to be a more convenient option for health care providers and patients.

BRACHYTHERAPY PROCEDURE

The procedure has significantly developed over many decades from the initial immediate postoperative placement of radio-active seeds to the development of stereotactic radiosurgery frames by Leksell.[9] Assisted imaging modality improvements, more accurate target visualization, dosimetric calculation, and more accurate planning algorithms are all utilized for improvement of treatment delivery. Finally, improved surgical techniques for better isotope delivery allow for more reliable outcomes from brain tumor brachytherapy.

At present, patients receiving seed brachytherapy (either permanent or temporary implants) undergo implantation under general anesthesia, using a stereotactic frame. Stereotactic CT examination is done and fused with contrast enhancing axial T1 and T2 weighted MR images or PET data performed 1–3 days before treatment. Stereotactic 3D planning is performed. Entry and target points of catheters are determined taking into account both the optimal dose distribution and the safest trajectory. Therapeutic isodose curves are superimposed on the surface of the tumor.

The stereotactic device is set up and a burr hole(s) is drilled. Outer nylon catheters are placed stereotactically and loaded with an inner catheter in which the seeds had been placed at position(s) previously determined during planning. For catheters with a single seed, the coordinates of the seed center can be used to determine the realized target point. For catheters with several seeds, the catheter trajectory can be realized by subtracting the planned coordinates (x, y, and z) from the realized coordinates (x, y, and z), squaring the results and adding them together to the power of ½. This determines the spatial, lateral, and depth displacement of the realized catheter trajectory (Fig. 3). After verification by intraoperative orthogonal stereotactic X-ray, catheters are fixed in the burr hole and the skin is sutured.

After the procedure is done, the following parameters should be documented for each brachytherapy case[20]:

- Prescribed total dose
- Dose rate
- Minimum tumor dose
- Percentage of tumor receiving less than the prescribed dose
- Maximum dose in "normal" brain at 1 cm from tumor border
- Number of seeds used
- Activity of the seeds
- Volumes of the tumor and surrounding tissue irradiated to various total doses and dose rates

Finally, as a safety measure, postimplant applied dose distribution is calculated by using the coordinates of the realized target points and seed position in the catheters. A comparison of the realized dose distribution is then done against the

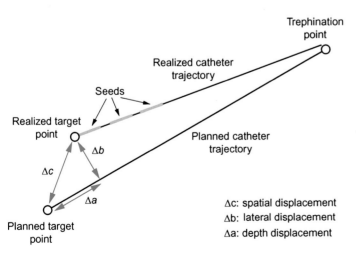

FIG. 3   Determining the spatial, lateral, and depth displacement of the realized multi-seed catheter trajectory. Subtract the planned (x, y, z) coordinates from the realized (x, y, z) coordinates. Square the results Δa (depth), Δb (lateral), and Δc (spatial) then add them together to the power of ½. *From Treuer H, Klein D, Maarouf M, et al. Accuracy and conformity of stereotactically guided interstitial brain tumour therapy using 1–25 seeds.* Radiother. Oncol. *2005;77:202–209. Copyrights 2005 Elsevier Ltd (approval granted from publishers by Kristi Anderson).*

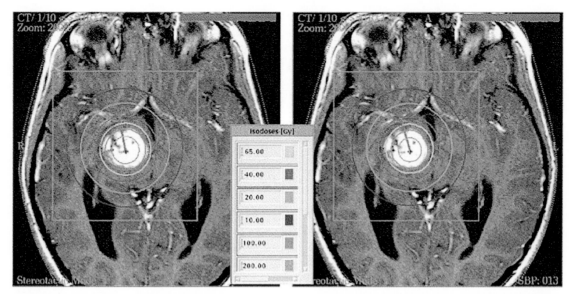

FIG. 4 Brain brachytherapy isodose line distribution of the realized applied dose distribution from the realized target points. Comparison to the planned dose should lead to an ideal ratio of 1. *From Treuer H, Klein D, Maarouf M, et al. Accuracy and conformity of stereotactically guided interstitial brain tumour therapy using 1–25 seeds.* Radiother. Oncol. *2005;77:202–209. Copyrights 2005 Elsevier Ltd (approval granted from publishers by Kristi Anderson).*

planned dose (Fig. 4). The ideal ratio should be 1; meaning that the realized dose is very similar to the planned dose required to adequately cover the tumor, while keeping the safest dose to normal brain tissue.[21]

BRACHYTHERAPY IN GLIOMAS

Data on brain glioma brachytherapy mostly originates from institutional series.[22–31] Utilized either in the primary, adjuvant, or recurrent setting, glioma brachytherapy has been reported for low-grade as well as for high-grade gliomas. Mean OS has ranged from 8 to almost 16 months in some series. Toxicity ranges also varied from zero up to 35% severe toxicity in some series.[26] Patients' performance status, age, preimplant treatment volume, and use of adjuvant chemotherapy have been shown to be important predictive factors for high-grade gliomas treated with seed brachytherapy. Table 3 summarizes the most recent studies and outcomes of brain brachytherapy for high-grade gliomas. Brachytherapy has also been used either as a sole method of radiation delivery or as a boost post EBRT. However, two randomized prospective trials, from Canada and the Brain Tumor Cooperative Group (BTCG) 8701, showed no survival advantage of brachytherapy boost.[32, 33] The rate of necrosis was about 24% with significant steroid dependence in the brachytherapy treatment arm.[32]

BRACHYTHERAPY IN OLIGODENDROGLIOMAS

Since this procedure is not commonly used worldwide, and given the low numbers of oligodendroglioma patients, data regarding the use of brachytherapy in OD arises mostly from retrospective single institution series. Data from LGG series as well as from grade III studies that included patient with oligo-containing pathology are summarized in Table 4.

As shown by Table 4, the reported median and 5 year PFS and OS varies significantly between different series, whether brachytherapy was offered in the primary or recurrent setting. This is likely attributed to the wide variability in patient grouping and outcomes reporting. However, despite inconsistent results between various studies, baseline performance status (KPS), tumor volume, and tumor pathology tend to predict overall oncologic outcomes in these patients.[6, 31–40] Furthermore, brachytherapy following microsurgery for tumor located in eloquent locations seemed to provide adequate local control comparable to those with tumors located in noneloquent areas where a more complete craniotomy resection can be done.[36] Schatz et al. reported their outcomes of 67 patients treated with LDR brachytherapy for unresectable, LGGs limited to the insula and paralimbic regions. Of those, 4 were OD and 11 were oligoastrocytomas.[36] Survival outcomes were similar for those after surgery and radiation therapy for resectable tumors with low perioperative mortality and morbidity. The only significant predictive factor of outcome was baseline KPS >90.[36] Furthermore, another pilot study by Schnell et al.[39] assessed the feasibility, risk, and effectiveness of microsurgery with I-125 brachytherapy for tumors >4 cm or

TABLE 3 Outcomes of brain brachytherapy for high grade gliomas

Author	WHO grade (n)	Setting	BT type	BT dose (Gy)	Chemotherapy	PFS (months)	OS (months)
Simon et al.[22]	IV (42)	Recurrent	HDR	50	NR	NR	12.5
Larson et al.[23]	IV (38)	Recurrent	LDR	300	NR	4	13
Tselis et al.[24]	IV (84)	Recurrent	HDR	40	NR	NR	9.25
Julow et al.[25]	IV (6)	Recurrent	LDR	60	NR	NR	7.6
Darakchiev et al.[26]	IV (34)	Recurrent	LDR	120	Yes	11	14.75
Fabrini et al.[27]	III (3) IV (18)	Recurrent	HDR	18	NR	NR	8
Archavlis et al.[28]	IV (50)	Recurrent	HDR	40	NR	8	9.25
Kickingereder et al.[29]	IV (98)	Primary	LDR	60	NR	5.9	10.4
Archavlis et al.[30]	IV (17)	Recurrent	HDR	40	NR	7	8
Schwartz et al.[31]	III (28) IV (40)	Recurrent	LDR	50	NR	8.3	NR

Key: *BT*, brachytherapy; *Gy*, Gray; *N*, number of patients; *LDR*, low dose rate; *HDR*, high dose rate; *NR*, not reported; *PFS*, progression free survival; *OS*, overall survival.

I-125 brachytherapy alone for tumors less than 4 cm. In that study, complete, partial, and stable diseases were seen in 8, 9, and 14 out of a total of 31 patients enrolled, respectively. Later, the German group reported the outcomes of 58 LGG patients treated with brachytherapy alone or combined with microsurgery.[43] The 5-year PFS and OS was 87% and 95%, respectively. However, it is important to note that the differential pathological subtype (oligodendroglioma vs. oligoastrocytomas) was not specified in that study. Finally, the Italian group reported the outcomes of 36 patients with unresectable low-grade brain gliomas treated with brachytherapy (11 patients had oligodendroglioma pathology).[35] The median survival time was 110 months for the entire cohort, and no significant differences were found between different pathological subtypes of tumors, similar to what the German group reported.[36] Yet patients' performance status and volume of the target were significant predictive factors for outcomes.[35, 36]

On the other hand, pathology did influence outcomes from a larger series by the German group.[41] For low-grade tumors, the 5-year PFS of 75% for OD was significantly higher than the 51% for oligoastrocytomas.[41] Interestingly, the same group later reported the outcomes of treating 20 patients with high-grade oligoastrocytoma and OD treated with I-125 brachytherapy[42] and showed these patients had significantly better outcomes than patients with high-grade anaplastic astrocytomas.

Finally, a recently published series by El Majdoub et al. described 63 patients, with pure oligodendroglioma, treated with EBRT and I-125 brachytherapy seeds.[6] This series probably represents the only report of seed brachytherapy for patient with pure OD in the modern era. It included patients treated in the primary, adjuvant, and salvage setting, with a surface dose between 50 and 65 Gy with 25 Gy EBRT and showed encouraging 2, 5, and 10 year outcomes.[6]

BRACHYTHERAPY FOR PEDIATRIC PATIENTS

Since OD can also present in the pediatric age group, some studies specifically evaluated the safety, efficacy, and outcomes of LGGs in pediatric patients treated with brachytherapy. Peraud et al.[44] published outcomes of a pilot study done in Germany on 11 pediatric patients with WHO I/II astrocytoma treated with Iodine radiosurgery ± microsurgery for larger tumors and showed the safety and efficacy of this procedure.

In 2011, Ruge and colleagues[45] published the outcomes of 147 patients, less than 20 years of age, prospectively enrolled with primary or recurrent LGGs treated with permanent I-125 seeds with a prescribed dose of 65 Gy, or 50 Gy if previously

TABLE 4 Brachytherapy for non-WHO grade IV gliomas

Author	Pathology (n)	Total N	Setting	EBRT (at rec.)	EBRT dose	BT dose to margin	5-Year PFS	5-Year OS
Ostertag et al.[34]	OD (29) OA (44)	539	Prim. (485) Rec. (54)	Neoadj. (23) Adj. at rec. (24)	NR	60 Gy	NR	OD (58%) OA (80%)
Scerrati et al.[35]	Oligo (11)	36	Prim.	Adj. at rec. (19)	30–54 Gy	25–70 Gy (temporary) 60–130 Gy (permanent)	NR	Med. OS: 9.3 year
Schatz et al.[36]	OD (4) OA (11)	67	Prim.	Neoadj. (6)	NR	60 Gy	NR	OD (54%) OA (38%)
Kreth et al.[37]	OD (27) OA (60)	455	Prim.	NR	NR	60 Gy (temporary) 100 Gy (permanent)	NR	OD (50%) OA (49%)
Kreth et al.[38]	OA (52)	239	Prim.	None	NA	60 Gy (temporary) 100 Gy (permanent)	45%	56%
Schnell et al.[39]	Oligo (4)	31	Prim. (18) Rec. (13)	After (1)	NR	54 Gy	Rec. (62%) Prim. (72%)	92%
Korinthenberg et al.[40]	OD (5) OA (4)	94	Prim. (93) Rec. (1)	Neoadj. (3) Adj. at rec. (4)	NA	50–60 Gy	NR	97%
Suchorska et al.[41]	OD(12) OA (14)	95	Prim. (43) Rec. (52)	Neoadj. (3) Adj. at rec. (19)	56 Gy	50–65 Gy	43.4% OA (51.1%) OD (75%)	Med. OS: 19.8 years
Majdoub et al.[6]	OD (63)	63	Prim. (35) Rec. (28)	Both[a]	Neoadj (NR) Adj (25.2 Gy)	50–65 Gy	OD WHO II (93.8 %) OD WHO III (58.4%)	OD WHO II (97%) OD WHO III (77%)
Suchorka et al.[42]	OD (48)	172	Prim. (99) Rec. (73)	Concomitant	25 Gy	50–60 Gy	Med. rec. (32.6 months) Med. prim. (21.4 months)	Med. rec. (49.4 months) Med. prim (28.9 months)

[a]Both: Treated with adjuvant EBRT despite no progression.
OD, oligodendroglioma; OA, oligoastrocytoma; Oligo, oligo-component; nonspecified; N, number; Prim, primary; Rec., recurrence; EBRT, external beam radiation therapy; Gy, Gray; NR, not reported; NA, not available; Neoadj., neoadjuvant; Adj. at rec., adjuvant at recurrence; WHO, World Health Organization grade; PFS, progression free survival; OS, overall survival; Med., median.

treated with EBRT or if adjacent to an optic nerve, the optic chiasm, or internal capsule. Of those, only 5 had oligoden-droglioma pathology (2 had WHO grade II and 3 had WHO grade III). The results of this study were grouped and reported with 5, 10, and 15 year PFS of 92%, 74%, and 67%, respectively. Complete response was seen in 24.6% of the patients, 31% had partial responses, and 30% had stable disease. Within the same year, another group from Germany reported the long-term outcomes of pediatric patients with LGGs treated with Iodine brachytherapy.[40] The 5- and 10-year survival was 97% and 92%, respectively; interestingly, treatment response did not correlate with histology (only five had OD histology). Finally, Kunz et al. exclusively included LGG pediatric patients who were treated with LDR brachytherapy with or without combined microsurgery.[43] Outcomes were again comparable to complete resection. Unfortunately, pathology subtype was not specified in this patient population.[43]

GLIASITE BRACHYTHERAPY

Another brachytherapy treatment approach for brain metastasis and gliomas was introduced in 2001 with the GliaSite Radi-ation Therapy System. This technique, offered by IsoRay Inc., was more commonly used in the United States as compared to radioactive seeds, which were more commonly used in Europe. It consisted of an expandable balloon positioned in the resection cavity in its uninflated status. Multilumen silicone catheter shafts connected the balloon portion of the applicator to the infusion port, which was placed subcutaneously under the scalp. After the patient had recovered from surgery, the infusion port was filled with an aqueous solution containing I-125 and saline (Iotrex) (Fig. 5). The diluted Iotrex solution remained within the balloon for the duration of brachytherapy (3–7 days) and was then withdrawn from the infusion port and the GliaSite device was removed.

GLIASITE EFFICACY AND SAFETY (RECURRENT SETTING)

Early efficacy and safety results on the use of GliaSite brachytherapy originated from a multicenter North American series on 21 patients with recurrent brain gliomas.[46] Device implantation, delivery of radiation, and the catheter removal were well tolerated. Doses of 40–60 Gy were delivered to all tissues within the target volume. There were no serious adverse device related events during brachytherapy and no symptomatic radiation necrosis identified over 21.8 patient years of follow-up.

In a subsequent study from Johns Hopkins Hospital, a median survival of 9.3 months was noted in good performance status (KPS ≥ 70) recurrent GBM patients compared to RTOG RPA database historical controls. This study also confirmed the safety and tolerability of this treatment modality.[47]

A larger multi-institutional study of 95 patients with recurrent grade III or grade IV gliomas, median KPS of 80, con-firmed that good KPS was the only significant predictor for survival, and that those patients with the initial diagnosis of GBM had a worse median survival than those of non-GBM pathology (35.9 vs 43.6 weeks, respectively).[48] Out of the 95 treated patients, 3 developed pathological radiation necrosis, but otherwise this procedure was well tolerated with the delivery of 60 Gy to 1 cm depth and median dose rate of 52.3 Gy/h.

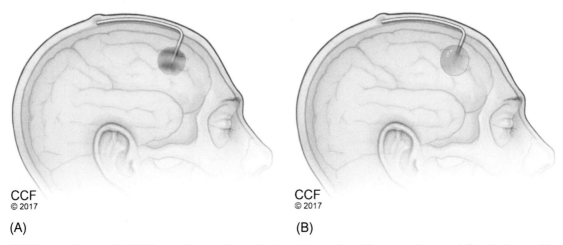

CCF
© 2017

(A)

CCF
© 2017

(B)

FIG. 5 GliaSite brachytherapy. (A) Multilumen silicone catheter placed under the scalp and is connected to an un-inflated balloon positioned in the resection cavity. (B) Radioactive I-125 saline infused into the balloon providing radiation for 3–7 days. *(Copyrights from Cleveland Clinic.)*

In an attempt to determine the optimal Gliasite RT dose, the Adult Brain Tumor Consortium (ABTC) initiated NABTT 2106 in patients with recurrent high-grade glioma.[49] This trial was closed as too few patients were assessable to allow dose escalation. Therefore no conclusion regarding maximal tolerated dose (MTD) could be found. It is interesting to note that the median survival for this group was 12.8 months.

GLIASITE IN THE UPFRONT SETTING

Wernicke and colleagues reported the feasibility and safety of treating 10 consecutive patients with recurrent, primary, and metastatic brain malignancies treated with upfront maximal safe resection and GliaSite brachytherapy combined with EBRT.[50] With no acute or late grade 3 or 4 toxicities in their prospective cohort, the group was able to report their outcomes and show GliaSite brachytherapy may be considered as an option for dose escalation in patients with localized CNS malignancy in the upfront setting.

The ABTC launched a trial in an attempt to determine the MTD for Gliasite brachytherapy followed by conventional EBRT in newly diagnosed GBM (NABTT 2105). Only 10 patients were enrolled on this trial, and it was closed early due to poor outcomes in this patient population. This trial failed to show the MTD of this treatment approach because of the high incidence of early progression that prevented assessment of treatment related toxicity. Median survival of that study cohort was 15.3 months (20 months for patients with no residual vs 9 months for patients with residual disease).[49]

GLIASITE CONCLUSION

Given the results of the NABTT 2105 and 2106 trials, the ABTC abandoned investigation of the use of the Gliasite device as a single therapeutic modality, as well as a combination method with other systemic or locally infused agents. Following this, GliaSite brachytherapy production was stopped and this treatment is currently not clinically used. Therefore, IsoRay Inc. who previously marketed GliaSite has discontinued that line of production and is currently focusing on Cesium-131 brachytherapy for brain metastasis.

BRAIN BRACHYTHERAPY TOXICITY AND LIMITATIONS

Like any procedure, brain brachytherapy has its own risks and potential side effects. Despite the low mortality rate of <1%, morbidity related to brain brachytherapy ranges from 4% to 11% even in large volume high experience centers.[6, 37] Catheter placement errors, despite advances in stereotactic technique approaches, remains highly dependent on center volume and experience. Moreover, extended duration of surgery and prolonged percutaneous access to the brain at each insertion site might be associated with increased risk of surgical site infection.[51] Finally, percutaneous catheter-derived brachytherapy may be associated with an increased incidence of extraneural metastatic gliomas of the scalp, skull, and cervical nodes reaching up to 4.3%.[52] This is especially important in the setting of multiple procedures, which would not be the case when a single seed implant procedure or GliaSite is used.[53]

The most commonly reported toxicity across the studies was the risk of brain necrosis. From the high-grade glioma series, the rate of severe toxicity ranged from 2 up to 35%.[22–31] The highest rate of severe toxicity was seen when permanent I-125 seeds were used in combination with BCNU wafers, with most related to symptomatic radiation necrosis requiring intervention.[26]

In studies excluding WHO IV tumors, symptomatic radiation necrosis was reported from 0% to 25%, with higher rates of necrosis reported after a course of EBRT.[32, 35] Tumor volume and mode of implantation were other risk factors predictive of symptomatic radiation necrosis; critical volume cutoff value of 32 cc and dose of 100 and 50 Gy for permanent and temporary implants, respectively, were also reported as strong predictors for radiation necrosis.[35, 37] Evaluation of brain necrosis after permanent seed implantation showed necrotic areas within the 100 Gy isodose line and damage to the blood-brain barrier within the 50 Gy isodose line. Therefore, Kreth et al. proposed the following guidelines to reduce the risk for side effects: use of temporary implants, dose rates around 10 cGy/h and activities <20 mCi, high dose zones >150 Gy should not be located within normal tissue or next to vessels, and volumes >4 cm should not be implanted.[19] Finally, on long term follow-up, the malignant transformation rate was 33%, 54%, and 67% at 5, 10, and 15 years, respectively, for patients with WHO II oligoastrocytoma or pure astrocytoma pathology, which are similar to rates expected when patients undergo open tumor resection alone.[38]

CONCLUSION

Brain brachytherapy has been shown to be a safe and effective method in treating both adults and pediatric patients with gliomas, including patients with OD. Long-term follow-up data is available and encouraging. Its current use, however, remains mostly limited to a few centers. It appears to be an appealing alternative for patients with gliomas in eloquent brain locations where surgery cannot be safely done, as results have shown comparable outcomes to those able to undergo surgical resection. It may also have a role as salvage therapy for a selected group of patients with recurrent lesions. Data regarding OD specific patients is limited as this pathology presents its own unique outcomes. Further prospective studies using clinical and survival end points, including quality of life measurements, are warranted to study brain brachytherapy. In addition to technical improvement, more attention should be paid to understanding the biologic events involved in late radiation complications and the molecular factors responsible for radioresistance of CNS malignancies.

REFERENCES

1. Harsh GR. Management of recurrent gliomas. In: Berger MS, Wilson CB, eds. *The Gliomas*. Philadelphia: WB Saunders; 1996:649–659.
2. Van Den Bent MJ, Afra D, De Witte O, et al. Long-term efficacy of early versus delayed radiotherapy for low-grade astrocytoma and oligodendroglioma in adults: the EORTC 22845 randomised trial. *Lancet*. 2005;366(9490):985–990.
3. Buckner JC, Shaw EG, Pugh SL, et al. Radiation plus procarbazine, CCNU, and vincristine in low-grade glioma. *N Engl J Med*. 2016; 374(14):1344–1355.
4. Cairncross JG, Wang M, Jenkins RB, et al. Benefit from procarbazine, lomustine, and vincristine in oligodendroglial tumors is associated with mutation of IDH. *J Clin Oncol: Off J Am Soc Clin Oncol*. 2014;32(8):783–790.
5. Olson JD, Riedel E, Deangelis LM. Long-term outcome of low-grade oligodendroglioma and mixed glioma. *Neurology*. 2000;54:1442–1448 (0028-3878 (Print)).
6. El Majdoub F, Neudorfer C, Blau T, et al. Stereotactic interstitial brachytherapy for the treatment of oligodendroglial brain tumors. *Strahlenther Onkol*. 2015;191(12):936–944.
7. Frazier CH. The effects of radium emanations upon brain tumors. *Surg Gynecol Obstet*. 1920;31:236–239.
8. Hirsch O. Symptoms and treatment of pituitary tumors. *Arch Otolaryngol Head Neck Surg*. 1952;55(3):268–306.
9. Leksell L. The stereotaxic method and radiosurgery of the brain. *Acta Chir Scand*. 1951;102:316–319.
10. D'Andrea F, de Divitiis E, Signorelli CD, Cerillo A. Stereotaxic implantation of radio-isotopes in brain tumours. *Neurochirurgia*. 1972;15:86–92.
11. Bernstein M, Guti PH. Interstitial irradiation of brain tumors: a review. *Neurosurgery*. 1981;9(6):741–750.
12. Mundinger F, Riechert T. Stereotaxic irradiation-procedure of brain tumors and pituitary adenomas by means of radio-isotopes and its results. *Confin Neurol*. 1962;22:190–203.
13. Kim JH, Hilaris B. Iodine 125 source in interstitial tumor therapy: clinical and biologic considerations. *AJR*. 1975;123:163–169.
14. Krishnaswamy V. Dose distribution around an 125I seed source in tissue. *Radiology*. 1978;126:489–491.
15. Siddiqi SN, Provias J, et al. Effects of Iodine-125 brachytherapy on proliferative capacity and histopathological features of glioblastoma recurring after initial therapy. *Neurosurgery*. 1997;40(5):910–917.
16. McDermott MW, Sneed PK, Gutin PH. Interstitial brachytherapy for malignant brain tumors. *Semin Surg Oncol*. 1998;14:79–87.
17. MacKay A, Gutin PG, Hosobuchi Y, Norman D. CT stereotaxis and interstitial radiation for brain tumors. In: Moss AA, Goldberg HI, eds. *Interventional Radiological Techniques: Computed Tomography and Ultrasonography*. San Francisco: University of California Printing Department; 1981:93–99.
18. Flickinger JC, Levine G, Maitz A, Lundsford LD. Radioisotopes and radiophysics of brachytherapy. In: Gildenberg PL, Tasker RR, eds. *Textbook of Stereotactic and Functional Neurosurgery*. New York: McGraw-Hill; 1998:561–568.
19. Kreth FW, Thon N, et al. The place of interstitial brachytherapy and radiosurgery for low grade gliomas. *Adv Tech Stand Neurosurg*. 2010;35:183–212.
20. Eddy M, et al. Quality assurance for I-125 brain implants program description and preliminary results. *Med Phys (Woodbury)*. 1991;18(3):605.
21. Treuer H, Klein D, Maarouf M, et al. Accuracy and conformity of stereotactically guided interstitial brain tumour therapy using 1-25 seeds. *Radiother Oncol*. 2005;77:202–209.
22. Simon JM, Cornu P, Boisserie G, et al. Brachytherapy of glioblastoma recurring in previously irradiated territory: predictive value of tumor volume. *Int J Radiat Oncol Biol Phys*. 2002;53:67–74.
23. Larson DA, Suplica JM, Chang SM, et al. Permanent iodine 125 brachytherapy in patients with progressive or recurrent glioblastoma multiforme. *Neuro-Oncology*. 2004;6:119–126.
24. Tselis N, Kolotas C, Birn G, et al. CT-guided interstitial HDR brachytherapy for recurrent glioblastoma multiforme. Long-term results. *Strahlenther Onkol*. 2007;183:564–570.
25. Julow J, Viola A, Balint K, et al. Image fusion-guided stereotactic iodine-125 interstitial irradiation of inoperable and recurrent gliomas. *Prog Neurol Surg*. 2007;20:303–311.
26. Darakchiev BJ, Albright RE, Breneman JC, Warnick RE. Safety and efficacy of permanent iodine-125 seed implants and carmustine wafers in patients with recurrent glioblastoma multiforme. *J Neurosurg*. 2008;108:236–242.
27. Fabrini MG, Perrone F, De Franco L, et al. Perioperative high-dose-rate brachytherapy in the treatment of recurrent malignant gliomas. *Strahlenther Onkol*. 2009;185:524–529.

28. Archavlis E, Tselis N, Birn G, et al. Survival analysis of HDR brachytherapy versus reoperation of recurrent glioblastoma multiforme. *BMJ*. 2013; e002262. Open 3.

29. Kickingereder P, Hamisch C, Suchorka B, et al. Low-dose rate stereotactic iodine-125 brachytherapy for the treatment of inoperative primary and recurrent glioblastoma: single-center experience with 201 cases. *J Neuro-Oncol*. 2014;120:615–623.

30. Archavlis E, Tselis N, Birn G, et al. Salvage therapy for recurrent glioblastoma multiforme: a multimodal approach combining fluorescence-guided resugery, interstitial irradiation and chemotherapy. *Neurol Res*. 2014;26:1047–1055.

31. Shwartz C, Romagna A, Thon N, et al. Outcome and toxicity profile of salvage low-dose-rate iodine-125 stereotactic brachytherapy in recurrent high-grade gliomas. *Acta Neurochir*. 2015;157:1757–1764.

32. Laperriere NJ, Leung PM, McKenzie S, et al. Randomized study of brachytherapy in the initial management of patients with malignant astrocytoma. *Int J Radiat Oncol*. 1998;41(5):1005–1011.

33. Selker RG, Shapiro WR, Burger P, et al. The brain tumor cooperative group NIH trial 87-01: a randomized comparison of surgery, external radiotherapy and carmustine versus surgery, interstitial radiotherapy boost, external radiation therapy and carmustine. *Neurosurgery*. 2002;51:343–357.

34. Ostertag CB, Kreth FW. Iodine-125 interstitial irradiation for cerebral gliomas. *Acta Neurochir*. 1992;119(1-4):53–61.

35. Scerrati M, Montemaggi P, Iacoangeli M, Roselli R, Rossi GF. Interstitial brachytherapy for low-grade cerebral gliomas: Analysis of results in a series of 36 cases. *Acta Neurochir*. 1994;131(1-2):97–105.

36. Schätz CR, Kreth FW, Faist M, Warnke PC, Volk B, Ostertag CB. Interstitial 125-iodine radiosurgery of low-grade gliomas of the insula of Reil. *Acta Neurochir*. 1994;130(1-4):80–89.

37. Kreth FW, Faist M, Warnke PC, et al. Interstitial radiosurgery of low-grade gliomas. *J Neurosurg*. 1995;82:418–429.

38. Kreth FW, Faist M, Grau S, Ostertag CB. Interstitial 125I radiosurgery of supratentorial de novo WHO Grade 2 astrocytoma and oligoastrocytoma in adults: Long-term results and prognostic factors. *Cancer*. 2006;106(6):1372–1381.

39. Schnell O, Schöller K, Ruge M, Siefert A, Tonn JC, Kreth FW. Surgical resection plus stereotactic 125I brachytherapy in adult patients with eloquently located supratentorial WHO grade II glioma—feasibility and outcome of a combined local treatment concept. *J Neurol*. 2008;255(10):1495–1502.

40. Korinthenberg R, Neuburger D, Trippel M, Ostertag C, Nikkhah G. Long-term results of brachytherapy with temporary iodine-125 seeds in children with low-grade gliomas. *Int J Radiat Oncol Biol Phys*. 2011;79(4):1131–1138.

41. Suchorska B, Ruge M, Treuer H, Sturm V, Voges J. Stereotactic brachytherapy of low-grade cerebral glioma after tumor resection. *Neuro-Oncology*. 2011;13(10):1133–1142.

42. Suchorska B, Hamisch C, Treuer H, et al. Stereotactic brachytherapy using iodine 125 seeds for the treatment of primary and recurrent anaplastic glioma WHO° III. *J Neuro-Oncol*. 2016;130(1):123–131.

43. Kunz M, Nachbichler SB, Ertl L, et al. Early treatment of complex located pediatric low-grade gliomas using iodine-125 brachytherapy alone or in combination with microsurgery. *Cancer Med*. 2016;5(3):442–453.

44. Peraud A, Goetz C, Siefert A, Tonn JC, Kreth FW. Interstitial iodine-125 radiosurgery alone or in combination with microsurgery for pediatric patients with eloquently located low-grade glioma: a pilot study. *Childs Nerv Syst*. 2007;23(1):39–46.

45. Ruge MI, Simon T, Suchorska B, et al. Stereotactic brachytherapy with iodine-125 seeds for the treatment of inoperable low-grade gliomas in children: Long-term outcome. *J Clin Oncol*. 2011;29(31):4151–4159.

46. Tatter SB, Shaw EG, Rosenblum ML, et al. An inflatable balloon catheter and liquid 125I radiation source (GliaSite Radiation Therapy System) for treatment of recurrent malignant glioma: multicenter safety and feasibility trial. *J Neurosurg*. 2003;99:297–303.

47. Chan TA, Weingart JD, Parisi M, et al. Treatment of recurrent glioblastoma multiforme with gliasite brachytherapy. *Int J Radiat Oncol Biol Phys*. 2005;62(4):1133–1139.

48. Gabayan AJ, Green SB, Sanan A, et al. Gliasite brachytherapy for treatment of recurrent malignant gliomas: a retrospective multi-institutional analysis. *J Neurosurg*. 2006;58:701–709.

49. Kleinberg LR, Stieber V, Mikkelsen T, et al. Outocme of adult brain tumor Consortium (ABTC) prospective dose-finding trials of I-125 balloon brachytherapy in high-grade gliomas: challenges in clinical trial design and technology development when MRI treatment effect and recurrence appear similar. *J Radiat Oncol*. 2015;4(3):235–241.

50. Wernicke AG, Sherr DL, Schwartz TH, et al. Feasibility and safety of GliaSite brachytherapy in treatment of CNS tumors following neurosurgical resection. *J Can Res Ther*. 2010;6:65–74.

51. Leong G, Wilson J, Charlett A. Duration of operation as a risk factor for surgical site infection: comparison of English and US data. *J Hosp Infect*. 2006;63(3):255–262.

52. Houston SC, Crocker IR, et al. Extraneural metastatic glioblastoma after interstitial brachytherapy. *Int J Radiat Oncol Biol Phys*. 2000;48(3):831–836.

53. Monroe JI, Dempsy JF, Dorton JA, et al. Experimental validation of dose calculation algorithms for the GliaSite RTS, a Novel 125-I liquid filled balloon brachytherapy applicator. *Med Phys*. 2001;28(1):73–85.

Chapter 26

The role of radiation in pediatric oligodendrogliomas

Joelle P. Straehla* and Karen J. Marcus[†,‡]

*Department of Pediatric Hematology/Oncology, Dana-Farber/Boston Children's Cancer and Blood Disorders Center, Boston, MA, United States
[†]Department of Radiation Oncology, Dana-Farber/Brigham and Women's Hospital, Boston, MA, United States; [‡]Dana Farber/Boston Children's Cancer and Blood Disorders Center, Boston, MA, United States

Introduction

Oligodendrogliomas are glial tumors, typically arising within the white matter of the hemispheres. They have been reported to comprise 5%–20% of glial tumors in adults, but in the pediatric population, oligodendrogliomas are much more rare, accounting for about 1% of pediatric brain tumors according to the surveillance, epidemiology and end result (SEER) database (with pediatric patients defined as 19 years of age and younger).[1] These tumors occur slightly more often in male patients, with a male-to-female ratio of 1.2:1 in pediatric patients. In all, 66% of tumors are located in the temporal and frontal locations and clinically, pediatric patients tend to present with seizures or hydrocephalus.[1,2] On magnetic resonance imaging (MRI), tumors are generally hypointense on T1 and hyperintense on T2, and contrast enhancement is intermittent.[3]

Historically, oligodendrogliomas have been defined by World Health Organization (WHO) histologic grade and can be categorized as either grade II or grade III tumors, with grade III tumors also known as anaplastic oligodendrogliomas. The WHO grading was updated in 2016 to incorporate molecular characterization for these tumors, such that a tumor with co-deletion of 1p/19q in combination with IDH1 or IDH2 mutations is defined an oligodendroglioma regardless of histologic appearance.[4] This has resulted in reclassification of some tumors that were previously called oligoastrocytomas in the 2007 classification, such that the terms oligoastrocytoma and anaplastic oligoastrocytoma are now strongly discouraged. For adult patients with oligodendrogliomas, multiple retrospective reviews have demonstrated prognostic significance of 1p/19q co-deletion and IDH mutations,[5–7] but in the pediatric population, the molecular characteristics differ. In the largest series of pediatric oligodendrogliomas (50 cases), Rodriguez et al. reported that only 25% had 1p/19q co-deletion and 18% had the most common IDH1 mutation (R132H); nearly all of those with 1p/19q co-deletion were over 16 years of age.[2] These findings have been confirmed in other case series as well, establishing that the genetic landscape of pediatric oligodendrogliomas is distinct from their adult counterparts.[8–10] Thus, it is difficult to attribute prognostic significance to molecular characteristics in pediatric patients, and these data are not used to determine therapy at this time.

Outcomes in pediatric oligodendrogliomas are generally favorable, with SEER data showing overall survival (OS) of 94% at 1 year and 85% at 5 year, though it should be noted that 90% (410 of 455) of pediatric oligodendrogliomas in the SEER database were grade II.[10] In the 50-patient case series by Rodriguez et al., 76% of tumors were grade II and 24% were grade III.[2] The rarity of anaplastic oligodendrogliomas in children make estimation of survival in this subpopulation more difficult, but multiple case series show a trend of decreased OS for patients with grade III disease.[2,11] Pediatric patients as a whole fare better than adult patients with oligodendrogliomas, with 5-year OS for adults at 66%.[1] Recurrence is common in pediatric patients with oligodendrogliomas, with approximately a third of patients experiencing progressive disease at a median of 2.5 years after diagnosis in one study.[11] This is fitting with other pediatric low-grade gliomas (defined as gliomas that are WHO grade I or II), in that recurrent or progressive disease is common, but tumor-related mortality is low.

Treatment of pediatric oligodendrogliomas

The development of a treatment plan for pediatric patients with oligodendrogliomas is complex and must involve a multidisciplinary team of pediatric oncologists, neuroradiologists, neurosurgeons, radiation oncologists, and neuropathologists.

Oligodendroglioma. https://doi.org/10.1016/B978-0-12-813158-9.00026-8

The treating team is initially only privy to radiologic findings and the clinical examination and must take into account the location of the tumor and age of the patient in order to make a preliminary treatment plan. In order to obtain histologic and molecular pathology data, a biopsy or resection must first be undertaken. The role of surgery in pediatric oligodendrogliomas is paramount, and is more fully addressed in Chapter 20. In short, whenever possible from the standpoint of surgical risk, an up-front resection is performed, and the extent of surgical resection achieved greatly impacts survival. In a case series of 37 patients, Creach et al. reported a 5-year PFS of 100% for pediatric patients achieving a gross total resection (GTR, defined as >90% resection based on immediate postoperative imaging) compared to 29% for those achieving less than GTR.[11] Thus, maximal surgical resection is the primary means of treatment whenever possible.

The role of chemotherapy in pediatric oligodendrogliomas is primarily for those patients with residual or recurrent disease after primary resection. A secondary role is in an effort to halt tumor growth in order to delay radiation for young children. This is in contrast to adult patients with oligodendrogliomas, in whom adjuvant chemotherapy with temozolomide or the PCV [procarbazine, lomustine (CCNU), vincristine] regimen has been shown to be beneficial in patients who require radiation and is now a standard of care.[12] Responsiveness to these chemotherapeutic regimens can be predicted by 1p/19q co-deletion in adult patients,[6,7,13] but this has not held true in pediatric patients. The presence of 1p and/or 19q loss is rare in pediatric patients, but its presence actually appeared to predict a decreased response to chemotherapy in one cohort.[9] Chemotherapy practices for pediatric patients are further discussed in ***Chapter 33, and in general the decision to use chemotherapy is made on a case-by-case basis.

Clinical trial data—Grade II oligodendrogliomas

There have been no randomized trials conducted to study radiation use in pediatric patients with grade II oligodendrogliomas. These patients have been eligible for therapeutic trials for low-grade gliomas, however, and there have been two prospective randomized controlled trials (RCTs) completed to date, both focusing on the use of chemotherapy in patients with residual or progressive disease after surgical resection. Patients with grade II oligodendrogliomas were eligible for the Children's Oncology Group (COG) trial A9952, but only two patients with the diagnosis were enrolled out of 274 patients. This trial compared two chemotherapy regimens (carboplatin and vincristine versus thioguanine, procarbazine, dibromodulcitol, lomustine, and vincristine) in patients with residual or progressive low-grade (WHO grade I or II) gliomas, with the goal of using chemotherapy as an alternative to radiation.[14] The event-free survival was not statistically different between the two regimens, and there were not enough patients with oligodendrogliomas to make any conclusions in this subgroup. The second RCT including patients with grade II oligodendrogliomas was run in Europe by the Society of Pediatric Oncology—Low-Grade Glioma (SIOP LGG) Committee and also compared two chemotherapy regimens in 497 patients with residual or recurrent disease, though patients with residual disease also had to have evidence of "severe tumor-related symptoms" such that stable residual disease was not an indication for randomization. The two treatment regimens studied were vincristine and carboplatin (VC) versus VC plus etoposide (VCE); similar to COG A9952, there was no statistical difference in PFS or OS.[15] The number of enrolled patients with oligodendrogliomas was not reported, though is assumed to be >5% based on the predominance of other named tumor types. In these two RCTs, some patients who progressed during the follow-up period did receive radiation, but these data were not collected in such a way as to be able to comment on efficacy of RT.

There are numerous clinical trials that include adult patients with grade II oligodendrogliomas, which are more fully discussed in other sections of this book. In short, these have established that surgery should be the primary treatment for low-grade gliomas in adults, but contrary to what is seen in children; ancillary treatments such as radiotherapy (RT) and chemotherapy are nearly always required due to eventual progression of disease. Patients' risk factors are taken into account when deciding between immediate postoperative RT and/or chemotherapy versus a "watch-and-wait" approach. For patients with low-grade gliomas, some of the risk factors for early progression include age >40 years, tumor size >5 cm, tumor crossing the midline, neurologic symptoms, and lack of 1p/19q co-deletion.[16,17] Of the trials focusing on the optimal timing and dose of radiation,[18–21] it was determined that early versus delayed postoperative RT improved PFS but did not affect OS, favoring a "watch-and-wait" approach for the asymptomatic patient. When a lower dose of radiation was compared with a higher dose (45 Gy versus 59.4 Gy in one study and 50.4 Gy versus 60.8 Gy in another) there was no statistical difference in OS between the arms, but patients receiving the higher dose in each study experienced more treatment-related toxicity.[18,20] More recently, there has been one large RCT comparing RT alone (up to 50.4 Gy) versus temozolomide monotherapy in 477 patients with low-grade gliomas and one or more high-risk feature.[17] While the data has not yet matured to be able to comment on differences in OS, the PFS is not statistically different between the treatment arms.

In total, the majority of what is known about the use of radiation in pediatric and adult grade II oligodendrogliomas is extrapolated from RCTs studying the collective group of low-grade gliomas, of which oligodendrogliomas comprise only

5%–20%. It is clear that oligodendrogliomas are radiosensitive, but the most appropriate setting for RT in pediatric population remains unclear. Certainly RT must be considered for the treatment of residual or progressive disease, though chemotherapy may be comparable in efficacy if data from low-grade gliomas trials can be extrapolated to this population. For patients with surgically resected disease or stable residual disease who remain symptom free, there is currently no role for radiotherapy.

Clinical trial data—Grade III anaplastic oligodendrogliomas

Pediatric patients with anaplastic oligodendrogliomas were historically eligible for prospective RCTs focused on high-grade gliomas, but very few patients were enrolled. For example, the Children's Cancer Group (CCG) 945 trial was designed to compare two chemotherapy regimens given in addition to RT for patients with high-grade gliomas. Four patients with anaplastic oligodendrogliomas were enrolled, however, a subsequent central review confirmed only one of the four to actually have an anaplastic oligodendroglioma.[22] This very low agreement rate not only precluded any subgroup analysis for anaplastic oligodendrogliomas, but also highlights the difficulty in diagnosing these tumors on histology alone (this study was prior to the advent of standard molecular profiling). More recently, as oligodendrogliomas are considered to have a 'more favorable' prognosis, they have been excluded from high-grade gliomas trials,[23] thus there is a true paucity of data for pediatric patients with grade III oligodendrogliomas.

Adult patients with anaplastic oligodendrogliomas have been included in numerous studies, which are more fully discussed in Sections 5 and 6. However, it is worth remarking on three notable trials that reported on the use of RT in a randomized, controlled fashion. First, the European Organization for Research and Treatment of Cancer (EORTC) 26951 was an RCT including 368 patients over 18 years of age with anaplastic oligodendrogliomas or anaplastic oligoastrocytomas, randomized to receive RT alone versus RT plus the PCV regimen. All patients received RT dosed at 45 Gy with a boost of 14.4 Gy (cumulative dose of 59.4 Gy) to the tumor bed, given in 33 fractions and started within 6 weeks of surgery. One of the primary findings reported from the EORTC 26951 trial by van den Bent et al. was improved PFS with the use of adjuvant PCV, though it should be noted that OS was not statistically different between the groups. Of the 183 patients enrolled, 131 (72%) in the RT alone arm experienced tumor progression, and the majority of these patients then received chemotherapy. Post hoc analyses for molecular markers (the significance of which had not been known at the onset of the study) showed that patients with 1p/19q co-deletion experienced a greater improvement in PFS than those without.[24,25] Second, the Radiation Therapy Oncology Group (RTOG) 9402 trial included 289 patients with anaplastic oligodendrogliomas or anaplastic oligoastrocytomas who were similarly randomized to receive RT alone (cumulative dose of 59.4 Gy) versus RT plus PCV regimen. Their results, reported by Cairncross et al. mirrored those of EORTC 26951, with improved PFS in the group receiving adjuvant chemotherapy, but similar OS between the groups.[12] Again, the subgroup analysis showed those patients with 1p/19q co-deletion had significantly improved PFS compared to those without. Third, the Neuro-Oncology Working Group (NOA) of the German Cancer Society conducted an RCT (NOA-04) that varied from the others in studying the effects of changing the sequence of therapy. In NOA-04, 274 patients with anaplastic astrocytoma, anaplastic oligodendroglioma, or anaplastic oligoastrocytoma randomized in a 2:1:1 fashion to postsurgical treatment with RT alone (60 Gy in 33 fractions) versus PCV regimen alone versus temozolomide alone. As expected, the majority of patients had progressive disease during the study period and at that time patients initially treated with RT were randomized 1:1 to the chemotherapy regimens; patients initially treated with chemotherapy were treated with RT. The primary end point reported by Wick et al. was time to treatment failure (TTF), with secondary end points of PFS and OS. The median TTF was not statistically different between the groups: 42.7 months in patients receiving RT followed by chemotherapy, and 43.8 months in patients receiving chemotherapy followed by RT. On subgroup analysis, the use of PCV versus temozolomide as the chemotherapy regimen did not alter TTF regardless of sequence. Similarly, PFS and OS were not affected by sequence of treatment or chemotherapy regimen.[26]

These three landmark trials in adult patients with anaplastic oligodendrogliomas have set the precedent for standard treatment in adults: maximally tolerated resection followed by a combination of radiation therapy and chemotherapy, although the optimal order and timing continue to be debated. It should also be noted that practice varies in regard to the optimal chemotherapy regimen for patients with 1p/19q co-deletion, and there is an ongoing clinical trial in this subset of patients (ClinicalTrials.gov identifier NCT00887146) comparing PCV versus single-agent temozolomide side by side.

Radiotherapy considerations

From the clinical data available, it is clear that oligodendrogliomas are sensitive to radiation, but given its many short- and long-term side effects, it is considered carefully in pediatrics and best practice is currently to limit its use to the most aggressive cases. Given that oligodendrogliomas are uncommon in the pediatric population—and anaplastic oligodendrogliomas even more rare—there are relatively few children who receive radiation as part of their treatment plan. To put into perspective, over 40 years there were 455 pediatric patients with oligodendrogliomas in the SEER database; the majority of these (319 or 70.6%) received surgery as their only treatment modality. Only 103 patients (22.8%) received radiation either in addition to surgery (19.7%) or radiation alone (3.1%).[1] As noted above, there are very limited clinical trial data regarding the use of radiation in pediatric oligodendrogliomas, thus the guiding principles of radiation use in pediatric oligodendrogliomas are extrapolated from the treatment of other pediatric gliomas and from clinical trials of adult patients.

Radiation therapy techniques

When radiation therapy is used in the management of pediatric oligodendrogliomas, the radiation modality and dosing utilized depends on the size and location of the tumor. Treatment plans must also take into account patient age and functional status, as well as any prior therapy received and future therapy anticipated.

The preferred radiation modality in recent decades has been intensity modulated radiation therapy (IMRT) as it is available at most cancer centers and allows for very precise delivery of photon radiation to a target volume while limiting the dose to nearby critical structures. Compared with three-dimensional (3D) conformal radiotherapy (CRT), which was used commonly in the 1990s, IMRT allows for increased control over the radiation field and sparing of normal structures to a much greater degree. This is accomplished by modifying the radiation beam, using collimation within each beam path and combining multiple beams. This allows for targeted areas to receive a much higher dose of radiation than the surrounding tissues. This does not eliminate radiation dosing to normal structures, and in fact there is a larger area receiving lower irradiation doses in order to deliver the treatment dose in a more targeted manner (Fig. 1).[27–29]

Stereotactic radiation therapy (SRT) and stereotactic radiosurgery (SRS) are used increasingly in pediatric patients, as these techniques allow for an even more precise delivery of photon radiation. The use of moving arcs as opposed to static beams allows for exquisite accuracy to within 1 mm. However, this accuracy requires intensive immobilization, which is a limiting factor in the pediatric population, as sedation is often required even for older children. Stereotactic radiation can be delivered in fractionated doses (in the case of SRT) or as one single dose (linear accelerator-based SRS or with Gamma Knife), but in both cases tumors must be relatively small, less than 4 cm for SRT and less than 3 cm for SRS.[28] The role for SRT or SRS in pediatric oligodendrogliomas could be considered for recurrence after prior radiotherapy or small unresectable recurrence of high-grade disease.

More recently, the advent of charged particle radiation has changed the field of radiation oncology, with the ability to deliver radiation via a particle with mass and charge as opposed to via photons. With particle radiation, there is a minimal

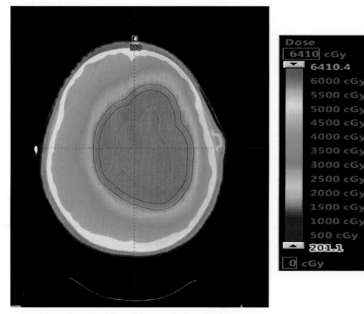

FIG. 1 Example of isodose distribution for patient receiving IMRT for a CNS glioma.

exit dose of radiation, thus allowing for near complete tissue sparing beyond the target volume.[27] Proton therapy is the most widely used form of charged particle radiation in pediatric patients, and is especially appealing for tumors located in close proximity to the brainstem, skull base, or spinal cord. Dosimetric studies have shown reduced integral radiation dose of approximately 50% as well as a significant reduction in the doses to critical brain structures with proton therapy over 3D photon therapy (Fig. 2).[30]

While there are clear advantages with proton therapy in terms of dosimetry and theoretical advantages in regard to expected long-term toxicity, there is a paucity of clinical trial evidence for increased efficacy or decreased late effects over photon radiation. There is also some suggestion of increased short-term effects such as radiation necrosis.[31] This lack of trial data is in part due to the relative newness of this treatment modality, as cohorts currently being followed have not matured. As late effects data is gathered for pediatric patients receiving proton therapy and the technology becomes more readily available and cost effective, its use will likely become more widespread.

Regarding radiation dose, there are no set guidelines for the treatment of pediatric oligodendrogliomas or even for low- or high-grade gliomas in general. There is, however, a de facto standard in the field of pediatric radiation oncology of delivering doses of 50.4–54.0 Gy, conventionally fractionated in 1.8 Gy per fraction, when treating pediatric central nervous system (CNS) gliomas with curative intent. This allows for a boost to the tumor bed of 5.4–9.0 Gy such that the usual maximum dose is 59.4 Gy. In the rare situation of disseminated spread of disease, craniospinal irradiation could be considered. The dose recommended is 36 Gy with additional treatment directed at the primary tumor of 45–54 Gy depending on the location.[27,29]

As oligodendroglioma is rare in children and radiation therapy is used infrequently for pediatric oligodendrogliomas, there are no standard guidelines. The chief indication is high-risk histology. When it is indicated, the complex decisions should be made on a case-by-case basis, including the choice of radiation modality and dose. Therefore, it is imperative that a multidisciplinary team at a tertiary cancer center be involved in both the decision to treat and in the development and execution of a treatment plan for pediatric patients undergoing radiation therapy for oligodendrogliomas.

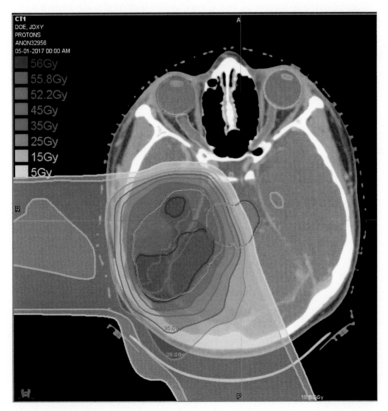

FIG. 2   Example of isodose distribution for patient receiving proton therapy for a CNS gliomas.

Late effects of radiation

The long-term side effects of radiation to the CNS are well described in review articles[32–35] and textbooks,[27,29,36] but are worth briefly reviewing here with a focus on pediatric patients as they significantly impact its use for oligodendrogliomas. Radiation injury to normal tissues is the foundation for long-term effects, impairing growth, development, and function of brain tissues. Late effects of radiation can be broadly categorized into neurocognitive, neuroendocrine, and neurovascular effects. In addition, radiation injury can lead to secondary cancers occurring many years or even decades after treatment.

Neurocognitive impairment is one of the chief concerns when using radiation in young children, as childhood is a period of remarkable brain development. In particular for those under the age of 5 years, there is a pronounced drop in IQ after receiving radiation to the CNS, with higher doses of radiation correlating with increased neurocognitive dysfunction.[34,35] Hearing loss can also contribute to decreased scholastic achievement in these patients. The cochlea is exquisitely sensitive to radiation,[37] and hearing impairment appears to have an additive effect on neurocognitive functioning through an effect on classroom learning.[38] Consequently, the use of RT in children school-aged or younger is only undertaken after careful consideration of any alternative treatment options.

The effects of radiation on neuroendocrine function are also significant in children, as damage to the hypothalamus and pituitary can lead to several endocrinopathies that affect growth and development.[34,39] It is estimated that 40%–80% of childhood brain tumor survivors suffer from endocrine dysfunction, with higher doses of radiation correlated with increased severity (Table 1). For example, growth hormone deficiency is the most common endocrinopathy and can be seen with any amount of radiation exposure; however, doses of more than 50 Gy invariably lead to growth hormone deficiency and are also associated with high incidences of panhypopituitarism.[32] It is thus imperative that children who receive CNS radiation have close follow-up with an endocrinologist (ideally within a survivorship clinic setting) to address these issues.

The effect of radiation on the cerebral vasculature has been increasingly recognized as a source of morbidity in patients undergoing cranial radiation, and pediatric survivors are affected disproportionately given they are more likely than their adult counterparts to outlive the relatively long latency period. Cerebral vascular disease can be in the form of small- or large-vessel disease including atherosclerotic changes, cavernous malformations, microhemorrhages, and moyamoya disease. The common end point for all these vasculopathies is stroke (either ischemic or hemorrhagic), the incidence of which increases over time and can lead to significant morbidity or mortality. The median latency for the development of moyamoya after cranial radiation is 40 months, and the median time to the detection of cavernous malformations is 12 years,[32] again highlighting the need for long-term specialty follow-up. The risks of cerebral vasculopathy increase with higher doses of radiation, especially in patients receiving over 50 Gy. Location also effects vasculopathy risk, with radiation to the region of the Circle of Willis conveying the highest risk of moyamoya. For pediatric patients already at risk of neurocognitive impairment, the added morbidity of cerebral vascular disease can be devastating.

Secondary cancers are a risk to both pediatric and adult patients receiving radiation, but given the longer life expectancy of a pediatric patient cured of disease versus an adult, second cancers are a much larger burden for patients treated at a young age. In one cohort of patients treated for CNS tumors, the 15-year incidence of second neoplasms was 4%, with the majority being high-grade gliomas or meningiomas.[40] A systematic review by Bowers et al. reported that pediatric patients treated with cranial radiation have substantial excess risk of a subsequent CNS tumor, with standardized incidence ratios of 9.5–52.3.[41] Second brain tumors appear to be especially deadly, with SEER data showing a 10-year OS of only 13.6% for pediatric patients with a CNS tumor after receiving treatment for another solid tumor.

TABLE 1 Threshold doses of radiation to the hypothalamic axis leading to specific hormone abnormalities

Hormone abnormality	Threshold dose to hypothalamic axis
Growth hormone deficit	18–25 Gy
ACTH deficit	40 Gy
TRH/TSH deficit	40 Gy
Precocious puberty	20 Gy
LH/FSH deficit	40 Gy
Hyperprolactinemia	40 Gy

ACTH, corticotropin; *LH/FSH*, leutinizing hormone/follicle-stimulating hormone; *TRH/TSH*, thyrotropin-releasing hormone/thyrotropin.
Reprint permission requested from Leibel and Phillips Textbook of Radiation Oncology (Elsevier) via RightsLink on 11/20/17.

It should be noted that the above-referenced late effects data were primarily from patients who received 3D conformal radiation, which had been standard radiation delivery technique used in the last several decades. There are limited data on late effects for patients receiving newer, more directed modalities (expanded upon above), though early studies suggest increased ability to spare normal tissues.[30] In sum, the long-term effects of cranial irradiation are particularly important to consider in pediatric patients. Especially given the overall low mortality of pediatric oligodendrogliomas and the relatively high morbidity of treatment, radiotherapy is only undertaken in cases that cannot be managed by surgical means or those with high-grade histology.

Current best practices for radiotherapy in pediatric oligodendrogliomas

Based on the favorable prognosis for pediatric patients with grade II oligodendrogliomas, treatment for these patients mirrors that for other low-grade gliomas, with surgical resection as the primary treatment modality. For patients with residual disease after surgery or for the significant proportion with subsequent tumor progression, consideration is given to chemotherapy or radiation therapy. In general, due to the numerous late effects of radiation therapy in children, chemotherapy is preferred as a disease stabilizer in this population. Based on data from the two randomized clinical trials referenced above for low-grade gliomas, treatment with carboplatin and vincristine has been the first line for patients with residual or progressive disease. However, the field is rapidly changing as molecular sequencing is more routinely performed and targeted therapies become available; this is further discussed in Chapter 33. Radiation is rarely used for grade II oligodendrogliomas, primarily for patients who progress despite chemotherapy or older patients with particularly recurrent unresectable tumors.

In contrast, radiation therapy is indicated in pediatric patients with grade III or anaplastic oligodendrogliomas. The clinical trial data referenced above for anaplastic oligodendrogliomas in adults have been extrapolated for pediatric patients with anaplastic oligodendrogliomas such that maximal surgery followed by radiation therapy is the standard of care. Given the lack of trial evidence regarding adjuvant chemotherapy in the pediatric populations and the distinct molecular pattern of these tumors, the type of adjuvant chemotherapy is considered on a case-by-case basis. When radiation therapy is used for anaplastic oligodendrogliomas, IMRT or proton beam RT are recommended, though the specifics of a particular case (tumor location, surgical outcome, patient age, and other medical comorbidities) as well as access to treatment will determine the choice of modality and dosing.

In summary, the current best practice for radiation in the management of pediatric oligodendrogliomas is to reserve its use for patients with anaplastic oligodendrogliomas or those with grade II oligodendrogliomas with residual or progressive disease after surgery and chemotherapy.

References

1. Lau CS, Mahendraraj K, Chamberlain RS. Oligodendrogliomas in pediatric and adult patients: an outcome-based study from the surveillance, epidemiology, and end result database. *Cancer Manag Res.* 2017;9:159–166.
2. Rodriguez FJ, Tihan T, Lin D, et al. Clinicopathologic features of pediatric oligodendrogliomas: a series of 50 patients. *Am J Surg Pathol.* 2014;38(8):1058–1070.
3. Koeller KK, Rushing EJ. From the archives of the AFIP: oligodendroglioma and its variants: radiologic-pathologic correlation. *Radiographics.* 2005;25(6):1669–1688.
4. Louis DN, Perry A, Reifenberger G, et al. The 2016 World Health Organization classification of tumors of the central nervous system: a summary. *Acta Neuropathol.* 2016;131(6):803–820.
5. Mariani L, Deiana G, Vassella E, et al. Loss of heterozygosity 1p36 and 19q13 is a prognostic factor for overall survival in patients with diffuse WHO grade 2 gliomas treated without chemotherapy. *J Clin Oncol.* 2006;24(29):4758–4763.
6. Jenkins RB, Blair H, Ballman KV, et al. A t(1;19)(q10;p10) mediates the combined deletions of 1p and 19q and predicts a better prognosis of patients with oligodendroglioma. *Cancer Res.* 2006;66(20):9852–9861.
7. Kouwenhoven MCM, Kros JM, French PJ, et al. 1p/19q loss within oligodendroglioma is predictive for response to first line temozolomide but not to salvage treatment. *Eur J Cancer.* 2006;42(15):2499–2503.
8. Kreiger PA, Okada Y, Simon S, Rorke LB, Louis DN, Golden JA. Losses of chromosomes 1p and 19q are rare in pediatric oligodendrogliomas. *Acta Neuropathol.* 2005;109(4):387–392.
9. Raghavan R, Balani J, Perry A, et al. Pediatric oligodendrogliomas: a study of molecular alterations on 1p and 19q using fluorescence in situ hybridization. *J Neuropathol Exp Neurol.* 2003;62(5):530–537.
10. Nauen D, Haley L, Lin MT, et al. Molecular analysis of pediatric oligodendrogliomas highlights genetic differences with adult counterparts and other pediatric gliomas. *Brain Pathol.* 2016;26(2):206–214.
11. Creach KM, Rubin JB, Leonard JR, et al. Oligodendrogliomas in children. *J Neurooncol.* 2012;106(2):377–382.

12. Cairncross G, Berkey B, Shaw E, et al. Phase III trial of chemotherapy plus radiotherapy compared with radiotherapy alone for pure and mixed anaplastic oligodendroglioma: intergroup radiation therapy oncology group trial 9402. *J Clin Oncol.* 2006;24(18):2707–2714.

13. Jaeckle KA, Ballman KV, Rao RD, Jenkins RB, Buckner JC. Current strategies in treatment of oligodendroglioma: evolution of molecular signatures of response. *J Clin Oncol.* 2006;24(8):1246–1252.

14. Ater JL, Zhou T, Holmes E, et al. Randomized study of two chemotherapy regimens for treatment of low-grade glioma in young children: a report from the Children's Oncology Group. *J Clin Oncol.* 2012;30(21):2641–2647.

15. Gnekow AK, Walker DA, Kandels D, et al. A European randomised controlled trial of the addition of etoposide to standard vincristine and carboplatin induction as part of an 18-month treatment programme for childhood (≤16 years) low grade glioma—a final report. *Eur J Cancer.* 2017;81:206–225.

16. Sepúlveda-Sánchez JM, Muñoz Langa J, Arráez M, et al. SEOM clinical guideline of diagnosis and management of low-grade glioma. *Clin Transl Oncol.* 2018;20:3. https://doi.org/10.1007/s12094-017-1790-3.

17. Baumert BG, Hegi ME, van den Bent MJ, et al. Temozolomide chemotherapy versus radiotherapy in high-risk low-grade glioma (EORTC 22033-26033): a randomised, open-label, phase 3 intergroup study. *Lancet Oncol.* 2016;17(11):1521–1532.

18. Karim AB, Maat B, Hatlevoll R, et al. A randomized trial on dose-response in radiation therapy of low-grade cerebral glioma: European Organization for Research and Treatment of Cancer (EORTC) Study 22844. *Int J Radiat Oncol Biol Phys.* 1996;36(3):549–556.

19. Karim AB, Afra D, Cornu P, et al. Randomized trial on the efficacy of radiotherapy for cerebral low-grade glioma in the adult: European Organization for Research and Treatment of Cancer Study 22845 with the Medical Research Council study BRO4: an interim analysis. *Int J Radiat Oncol Biol Phys.* 2002;52(2):316–324.

20. Shaw E, Arusell R, Scheithauer B, et al. Prospective randomized trial of low- versus high-dose radiation therapy in adults with supratentorial low-grade glioma: initial report of a North Central Cancer Treatment Group/Radiation Therapy Oncology Group/Eastern Cooperative Oncology Group study. *J Clin Oncol.* 2002;20(9):2267–2276.

21. van den Bent MJ, Afra D, de Witte O, et al. Long-term efficacy of early versus delayed radiotherapy for low-grade astrocytoma and oligodendroglioma in adults: the EORTC 22845 randomised trial. *Lancet.* 2005;366(9490):985–990.

22. Hyder DJ, Sung L, Pollack IF, et al. Anaplastic mixed gliomas and anaplastic oligodendroglioma in children: results from the CCG 945 experience. *J Neurooncol.* 2007;83(1):1–8.

23. Jakacki RI, Cohen KJ, Buxton A, et al. Phase 2 study of concurrent radiotherapy and temozolomide followed by temozolomide and lomustine in the treatment of children with high-grade glioma: a report of the Children's Oncology Group ACNS0423 study. *Neuro Oncol.* 2016;18(10):1442–1450.

24. van den Bent MJ, Carpentier AF, Brandes AA, et al. Adjuvant procarbazine, lomustine, and vincristine improves progression-free survival but not overall survival in newly diagnosed anaplastic oligodendrogliomas and oligoastrocytomas: a randomized European Organisation for Research and Treatment of Cancer phase III trial. *J Clin Oncol.* 2006;24(18):2715–2722.

25. van den Bent MJ, Brandes AA, Taphoorn MJ, et al. Adjuvant procarbazine, lomustine, and vincristine chemotherapy in newly diagnosed anaplastic oligodendroglioma: long-term follow-up of EORTC brain tumor group study 26951. *J Clin Oncol.* 2013;31(3):344–350.

26. Wick W, Hartmann C, Engel C, et al. NOA-04 randomized phase III trial of sequential radiochemotherapy of anaplastic glioma with procarbazine, lomustine, and vincristine or temozolomide. *J Clin Oncol.* 2009;27(35):5874–5880.

27. Hoppe RT, Phillips TL, Roach MI. *Leibel and Phillips Textbook of Radiation Oncology.* 3rd ed. Philadelphia, PA: Elsevier Saunders; 2010.

28. Hoffman KE, Yock TI. Radiation therapy for pediatric central nervous system tumors. *J Child Neurol.* 2009;24(11):1387–1396.

29. Gunderson L, Tepper J. *Clinical Radiation Oncology.* 4th ed. Philadelphia, PA: Elsevier; 2016.

30. Eaton BR, Yock T. The use of proton therapy in the treatment of benign or low-grade pediatric brain tumors. *Cancer J.* 2014;20(6):403–408.

31. Leroy R, Benahmed N, Hulstaert F, Van Damme N, De Ruysscher D. Proton therapy incChildren: a systematic review of clinical effectiveness in 15 pediatric cancers. *Int J Radiat Oncol Biol Phys.* 2016;95(1):267–278.

32. Roddy E, Mueller S. Late effects of treatment of pediatric central nervous system tumors. *J Child Neurol.* 2016;31(2):237–254.

33. Thorp N. Basic principles of paediatric radiotherapy. *Clin Oncol (R Coll Radiol).* 2013;25(1):3–10.

34. Merchant TE, Conklin HM, Wu S, Lustig RH, Xiong X. Late effects of conformal radiation therapy for pediatric patients with low-grade glioma: prospective evaluation of cognitive, endocrine, and hearing deficits. *J Clin Oncol.* 2009;27(22):3691–3697.

35. Armstrong GT, Conklin HM, Huang S, et al. Survival and long-term health and cognitive outcomes after low-grade glioma. *Neuro Oncol.* 2011;13(2):223–234.

36. Cox J, Ang KK. *Radiation Oncology: Rationale, Technique, Results.* 9th ed. Philadelphia, PA: Elsevier; 2010.

37. Bass JK, Hua CH, Huang J, et al. Hearing loss in patients who received cranial radiation therapy for childhood cancer. *J Clin Oncol.* 2016;34(11):1248–1255.

38. Orgel E, O'Neil SH, Kayser K, et al. Effect of sensorineural hearing loss on neurocognitive functioning in pediatric brain tumor survivors. *Pediatr Blood Cancer.* 2016;63(3):527–534.

39. Chemaitilly W, Armstrong GT, Gajjar A, Hudson MM. Hypothalamic-pituitary axis dysfunction in survivors of childhood CNS tumors: importance of systematic follow-up and early endocrine consultation. *J Clin Oncol.* 2016;34(36):4315–4319.

40. Broniscer A, Ke W, Fuller CE, Wu J, Gajjar A, Kun LE. Second neoplasms in pediatric patients with primary central nervous system tumors: the St. Jude Children's Research Hospital experience. *Cancer.* 2004;100(10):2246–2252.

41. Bowers DC, Nathan PC, Constine L, et al. Subsequent neoplasms of the CNS among survivors of childhood cancer: a systematic review. *Lancet Oncol.* 2013;14(8):e321–e328.

Section F

Chemotherapy and immunotherapy

Chapter 27

Basic principles of brain tumor chemotherapy

Kester A. Phillips* and David Schiff†

*Inova Medical Group Hematology Oncology, Inova Fairfax Hospital, Falls Church, VA, United States; †Neuro-Oncology Center, University of Virginia Health System, Charlottesville, VA, United States

Introduction

In an era of modern medicine, knowledge of the molecular underpinnings of tumorigenesis has led to substantial improvements in the outcome of cancer patients. Though patient survivorship is on the rise, the development of effective antineoplastic agents against some of the most menacing forms of brain tumors remains a formidable task. Nowadays, clinical trial research methodology utilizing scientific model-based statistical tests has outweighed the historical empiricism of drug development. However, when compared to the strides made in chemotherapy for systemic tumors, the brain tumor armamentarium is disappointingly inadequate and barriers to drug advancement remain a dire plight. Glioblastoma is the epitome of chemotherapy-resistant brain tumors and despite a flurry of investigational agents against this tumor entity, the pace of growth over the last 15 years has been stagnant. Currently, there are only four drugs and one treatment device approved by the FDA for the management of newly diagnosed glioblastoma (Table 1), while there is no consensus for second-line treatment for disease progression or recurrence.

Lately, brain tumor research has been heavily steeped in the investigation of molecularly targeted anticancer agents against a wide spectrum of adult and pediatric brain tumors (Table 2). In 2009 the pharmaceutical company Genentech received accelerated approval from the FDA for the use of bevacizumab against recurrent glioblastoma, but since then studies aimed at expanding the treatment landscape for relapsed high-grade glioma (HGG) has produced a paucity of results. Regrettably, the efficacy of targeted therapy against both adult and childhood primary brain tumors have been limited and remains experimental. On the upside, this shift in the treatment paradigm has shown promise for brain metastases with actionable mutations such as BRAF V600E mutant melanoma, and *EGFR*-mutant and *ALK*-rearranged non-small cell lung cancer (NSCLC). Most recently, cancer immunotherapy has made headway in the neuro-oncology arena, and preliminary data suggest a benefit of immune checkpoint inhibitors against some secondary brain tumors, particularly

TABLE 1 FDA approved drugs/device for the treatment of several brain tumors

Year	Agent/device	Class	Indication
1970s	Lomustine	Nitrosoureas	Newly diagnosed glioblastoma and recurrent glioma
1997	Gliadel	Nitrosoureas	Newly diagnosed glioblastoma and recurrent glioma
1999	Temozolomide	Methylating	Recurrent anaplastic astrocytoma
2005	Temozolomide	Methylating	Newly diagnoses glioblastoma
2009	Bevacizumab	VEGF inhibitor	Recurrent glioblastoma
2010	Everolimus	mTOR inhibitor	Subependymal giant cell astrocytoma
2011	Optune device	N/A	Recurrent glioblastoma
2015	Optune device	N/A	Newly diagnosed glioblastoma

TABLE 2 Molecularly targeting agents for brain tumors

Agent	Pathway	Target	Clinical application
Gefitinib, Erlotinib, Lapatinib	RTK/RAS/PI3K	EGFR	NSCLC, HER2+ BC BM
Bevacizumab	VEGF pathway	VEGF-A	GBM, meningioma, schwannoma
Sunitinib	VEGF pathway	VEGFR	Meningioma, RCC BM
Crizotinib	ALK pathway	c-Met/(HGFR)TK	EML4-*ALK* fusion NSCLC
Everolimus	PI3K/AKT/mTOR	mTOR	SEGA, meningioma, GBM
Dabrafenib	RAS/RAF/MEK/ERK	BRAF	Melanoma BM
Trametinib	RAS/RAF/MEK/ERK	MEK ½	Melanoma BM
Vemurafenib	RAS/RAF/MEK/ERK	BRAF	Melanoma BM
Pazopanib	VEGF pathway	VEGFR	RCC BM

brain metastasis from melanoma. For patients with meningioma, surgical resection remains the gold standard of treatment and a role for antineoplastic agents remains to be defined. Additionally, there are no FDA approved agents for adult low-grade glioma (LGG) and the off-label use of nitrosoureas and temozolomide for this brain tumor entity is widely accepted. Furthermore, chemotherapy has not been standardized and remains investigational for the treatment of pediatric brain tumors such as primitive neuroectodermal tumor (e.g., medulloblastoma), astrocytoma, and ependymoma.

The road map to the discovery of new brain tumor therapies over the past decade has been fraught with many bumps and dead ends that have counteracted the translation of novel therapies. Most notably, it is well recognized that brain tumor entities are vast and represent a biologically distinct group of diseases with complex tumor heterogeneity and genomic landscapes that govern growth rates and response to treatment, so the "one-size-fits-all" approach to therapy is impractical. Moreover, certain brain tumor subtypes carry inherent or acquired genotypic signatures such as mutations and/or epigenetic changes that confer drug resistance. Additionally, drug delivery presents a unique challenge specific to brain tumors because of the impact of the blood-brain barrier (BBB), and methods to circumvent this hurdle are desperately needed. Furthermore, issues with trial design, patient selection, low accrual rates, and endpoint analysis further thwart the advancement of novel treatment options. For instance, overall survival (OS) is typically the primary endpoint in phase III trials in patients with brain metastasis. However, most patients with brain metastases succumb to their disease because of systemic progression independently of intracranial disease control.[1,2] Likewise, the introduction of molecular profiling of brain tumors coupled with advancements in pharmacogenomic techniques has opened the door to personalized medicine. Consequently, new trials testing targeted therapies require the recruitment of patients with tumor-specific mutations, which makes recruitment for large-scale randomized trials problematic since patients with rare brain tumors must share similar cytogenetics. Lastly, the pitfalls in neuroimaging to reliably discriminate viable tumor from treatment-related change represent a major conundrum in the assessment of response to treatment. In this chapter, we provide a comprehensive, broad overview of chemotherapy in neuro-oncology and highlight the major challenges of drug development and ongoing research aimed at addressing these difficulties. A thorough discussion of the conventional and molecularly targeted agents is beyond the scope of this chapter, but a summary of these agents and emerging treatment strategies is provided.

Tumor burden and cell kill

Tumor burden refers to the number of tumor cells and size of tumor present in any organ or organ system at any given point in time. The original experiments on tumor growth and therapeutic regression involved studies of murine leukemia models by Skipper and colleagues, who injected different doses of tumor into animals to determine the fatal tumor volume. The results of these studies popularized the well-known log-kill model, based on the observation that the tumors grew exponentially at a constant doubling time until the tumor load becomes lethal.[3] Implicit in this study was an inverse relationship between tumor burden and curability. The group theorized that cytotoxic agents kill tumor cells by first-order kinetics, where a constant fraction of cells are killed rather than an absolute number of cells.[3] Most chemotherapy agents target actively dividing cells and based on first-order kinetics each chemotherapeutic cycle kills a fixed percentage (e.g., 90%) of all proliferating tumor cells resulting in a reduction of the cell number by a factor of 10. Therefore, a 2-log-kill

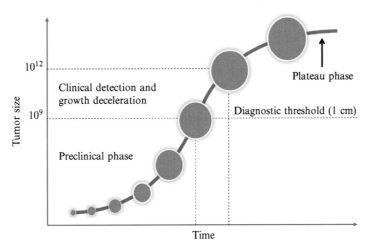

FIG. 1 Gompertzian curve and tumor growth.

is equivalent to 99% and 3-log-kill is equivalent to 99.9%, and so on. This model predicts high cure rates at chemotherapy cycles of 4–6; however, while the model helped explain the kinetics of leukemogenesis, there was divergence between the theory and actual observation, particularly in studies with solid tumors.[4,5] A major criticism of the group's work is that they considered a homogeneously sensitive population of cancer cells, which, in fact, is not realistic—especially for brain tumors. We now know from the success of cancer cytopathology that the malignant phenotype of gliomas, for example, is comprised of a heterogeneous tumor microenvironment with both infiltrative and proliferative subpopulations of tumor cells with inherent and acquired mechanisms for drug resistance. In fact, the invasive tumor cells are less likely to undergo proliferation.[6]

Gompertzian growth

Solid tumors do not grow exponentially because they do not have a fixed doubling time throughout their life cycle. The Gompertzian model (Fig. 1) was adapted from the theory behind the Skipper-Schabel-Wilcox model to explain the growth dynamics of solid tumors. In this model, the early phase of tumor growth is exponential-like, but as the tumor increases in size, there is growth deceleration until a plateau phase is reached. The kinetics is contrasted from the exponential growth model because the doubling times are not fixed but increases progressively as the tumors grow larger. Historically, this growth trajectory has been widely accepted in tumor biology, and it has been hypothesized that the growth rate of solid tumors is a function of the availability of nutrients to fuel cellular metabolism and proliferation. In the early phase, small tumor cells will have the largest growth fraction but as the tumor grows larger, cells become more quiescent at the other end of the graph. Both ends of the graph are separated by the clinical phase (slope of the line), where it is purported that tumors become clinically detectable and where growth rates decelerate. When the concept of cell kill and tumor burden is applied in this setting, we can expect higher chemosensitivity of tumor cells during the early phase—the subclinical phase, which has implications for timely diagnosis and the initiation of therapy. Conversely, during the clinical phase, the non-dividing and slowly growing cells vary in morphology and are less sensitive to drug treatment.[7] The idea of nutrient competition has been challenged on the grounds that large tumors can induce neovascularization, allowing them to grow to large sizes.[8] Yet, others posit that there is competition for space between the tumor and the host, that is, proliferative cells are restricted to the border of the tumor and exert pressure on the inner cells and restrict their proliferation because of the lack of space. It is suggested that chemotherapy would kill all the cells at the periphery while the inner cells would escape the therapeutic effect. They would, therefore, survive to become the new peripheral, proliferative layer wherein chemotherapy administration simply repeats the process.[9]

Drug resistance

The resistance of tumor cells to chemotherapy is one of the most vexing issues associated with therapeutic failure and poor patient outcome. Mechanisms underlying drug resistance can be extrinsic, where resistance is induced by environmental signals, or intrinsic, such as by aberrant gene expression and epigenetic alterations.[10] These changes can be acquired

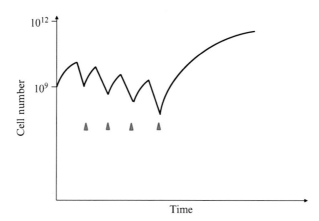

FIG. 2 Concept of drug resistance from spontaneous mutations. Blue arrows indicate chemotherapy administration.

secondary to drug administration, or they may be innate. Mathematical models have also been used to provide a framework for understanding drug resistance. In 1979, Goldie and Coldman proposed the first model (Fig. 2) to relate the chemosensitivity of tumors to their own spontaneous mutations that bestow resistance to one or more chemotherapeutic agents. They concluded that the mechanisms underlying drug resistance do not require exposure to chemotherapy, but resistant clones develop spontaneously because of genetic instability of the tumor cells resulting in point mutations, gene amplifications, or other epigenetic changes. It was hypothesized that chemotherapy exerts selective pressure in favor of these resistant clones, which continue to proliferate so that with time the proportion of cells will increase unless exposed to a drug to which they are sensitive (Fig. 3). It has been postulated that the administration of multiple non-cross-resistant drugs (concurrently or in an alternating fashion) as early as possible may offer the best chance of eradicating malignant tumors. This practice is widely utilized in clinical oncology to stave off the emergence of resistant clones, thus increasing the odds of cure. However, in neuro-oncology, survival end points are used as a barometer for patient outcome rather than cure rates. Though chemotherapy resistance represents one of the emblematic features of HHG, other tumors such as PCNSL, oligodendroglioma, IDH1-mutant LGG, intracranial germ cell tumors, and brain metastasis from small cell lung cancer are exquisitely chemosensitive.

Cell-cycle specificity of chemotherapy agents

The hallmark of cancer is an accelerated proliferative rate. Most traditional antineoplastic drugs exert their cytotoxic effects by exploiting components of the cell cycle (Fig. 4) that induce growth arrest and trigger apoptotic signal pathways. Tumor cells are vulnerable to these agents because they divide at rapid growth rates when compared to normal cells. Most standard anticancer drugs developed over the past 50 years either damage DNA, interfere with cell metabolism, or alter the stability and/or function of microtubules.[11] For example, alkylating agents create DNA base damage that induces single strand breaks and arrests the replication fork machinery leading to cell death. The cell cycle is a complex, well-choreographed, series of events culminating in DNA duplication and the birth of two genetically identical daughter cells. One cycle is defined as the interval between the midpoint of mitosis in the parent cell to the midpoint of mitosis in one of the daughter cells.[12] Each cycle consists of five phases: G0, G1, S, G2, and M phase. DNA replication occurs during synthesis phase

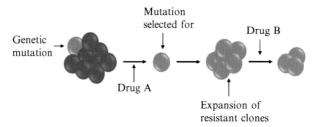

FIG. 3 Schematic representation of tumor resistance to chemotherapy due to gene mutation. Green indicated mutated cell that was selected for after chemotherapy. Red indicates normal cells.

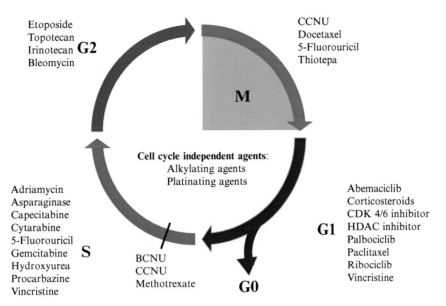

FIG. 4 Cell cycle phases and chemotherapy.

(S phase), which is followed by mitosis phase (M phase) where chromosome segregation and cell division occur. Most cells require extra time to grow and to monitor environmental conditions before embarking on DNA synthesis and mitosis. Henceforth, gap phases (G phase) are integrated into most cell cycles—G1 is the pause between M phase and S phase, while G2 is the pause between S phase and M phase. If environmental conditions are unsuitable, cells may enter a quiescent phase known as G0 (G zero), where they may remain indefinitely before proliferation recommences. A mammalian cell requires less than an hour to complete mitosis.[13]

Chemotherapeutic agents designed to intercept and halt mechanisms of cell division during the cell cycle can unleash cytotoxic effects against normally dividing host cells. To avoid this blanketed effect, newer drugs targeting molecular defects in cell cycle checkpoints are being developed to achieve selective killing of cancer cells—a concept termed "cyclotherapy."[14] For example, an evolving approach to selective killing of cancer cells by cycle-dependent chemotherapy involves the use of novel agents that target DNA-damage repair mechanisms and the cell cycle regulatory kinases, cyclin-dependent kinase 4/6 (CDK4/6). Mammalian cells are kept in senescence by retinoblastoma protein (pRb), a tumor suppressor protein. When cells are ready to replicate CDK4/6 phosphorylates pRb, rendering the protein inactive and drives cell cycle progression. CDK4/6 inhibitors, ribociclib and palbociclib are used in the treatment of metastatic breast cancer and are currently being tested in brain tumors. Furthermore, recent studies to elucidate the epigenetic regulatory mechanisms of tumor cells have clarified the role of histone deacetylases (HDACs) in the interplay between the remodeling of chromatin architecture and the stimulation of gene transcription. As it turns out, the acetylation and deacetylation of histones represent an important epigenetic regulatory mechanism of gene expression.[15] Histone acetylation is accomplished by histone acetylases (HATs) which promote the relaxation of chromatin structure through cancellation of electrostatic affinity between histones and DNA.[16] Histone acetylation enhances transcription while histone deacetylation promotes condensation of the nucleosomal fiber and represses transcription. Homeostatic imbalances in favor of HDACs foster the inactivation of tumor suppressing and cell-cycle regulatory genes and incite tumorigenesis.[17] Additionally, HDACs have also been implicated in the downregulation of c-myc expression leading to stimulation of cell proliferation and inhibition of apoptosis.[18] Thus, by reversing gene silencing, HDAC inhibition serves as another potential tactic for improving the therapeutic outcome of patients with brain tumors—by mechanisms that may also limit the collateral damage incurred by bystander host cells. HDAC inhibitors remain investigational in patients with brain tumors. Preclinical data have shown activity of the HDAC antagonist vorinostat against pediatric HGGs and myc-driven medulloblastoma.[19,20] Vorinostat was also established as a potent radiosensitizer in human glioblastoma cell lines.[21] However, the addition of vorinostat to gold standard therapy did not improve the survival of patients with newly diagnosed glioblastoma.[22] Valproic acid, a modest HDAC inhibitor, was evaluated for a possible survival benefit in patients with glioblastoma treated with this anticonvulsant at the start of standard radiotherapy and temozolomide in a trial. However, a pooled analysis of several key studies upended the belief that treatment with valproic acid yielded a survival advantage in patients with newly diagnosed glioblastoma.

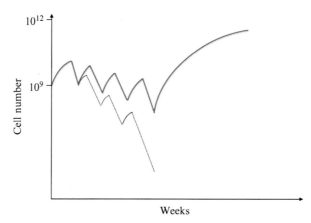

FIG. 5   Concept of dose-dense chemotherapy (black line). Red line equals regular chemotherapy schedule.

Dose intensity

The concept of dose-dense chemotherapy (Fig. 5) rests on the theory of the Gompertzian growth curve. High-dose chemotherapy assaults rapidly dividing cells during the early phase of the growth curve; however, between cycles, tumor cells are not exposed to chemotherapy and can regrow. Dose-dense therapeutic agents scheduled with shorter intervals between cycles provide a knockout punch to populations of cells that may be on the brink of replication. Standard treatment for patients with newly diagnosed glioblastoma is comprised of the administration of daily temozolomide at 75 mg/m^2 during involved-field cranial irradiation for 6 weeks. Following chemoradiotherapy, six cycles of 150–200 mg/m^2 of adjuvant temozolomide are administered for 5 days of a 28-day cycle. The addition of temozolomide to radiotherapy was shown to improve median OS by 2.5 months.[23] Furthermore, those patients with O^6-methylguanine-DNA methyltransferase (MGMT) promoter hypermethylated tumors derived a benefit from temozolomide, whereas those with wild-type promoter status did not.[24] Thus, MGMT promoter status is a predictive biomarker for the response of glioblastoma to alkylating agents. Based on these results several approaches to exhaust MGMT activity have been studied. As it turns out, a regimen composed of alternating schedules of daily temozolomide for 7 days at 75–175 mg/m^2 followed by a 7-day rest period was shown to decrease MGMT activity in peripheral blood mononuclear cells. Likewise, a 21- day continuous schedule of temozolomide at 85–125 mg/m^2 also depleted MGMT activity.[25] Given these positive results, it was hypothesized that dose-dense temozolomide would deplete MGMT activity thus sensitizing tumors to the effects of alkylating agents. Disappointingly, when compared to standard adjuvant temozolomide, patients with newly diagnosed glioblastoma did not derive a survival benefit from cumulative doses of adjuvant temozolomide for 21 days of a 28-day cycle.[26] In another study, dose-dense temozolomide scheduled for 7 days on/7 days off was not superior to conventional doses for glioma patients.[27]

Protracted, daily low-dose chemotherapy or metronomic chemotherapy (generally 50 mg/m^2/day) has also been extensively studied. A few studies have reported antitumor effects of metronomic chemotherapy via mechanisms that fortify response to anti-VEGF therapies.[28,29] Some studies reported a modest benefit of metronomic temozolomide for patients with progressive HGG;[30,31] however, in another study, daily metronomic schedules of oral etoposide or temozolomide administered in combination with bevacizumab did not show a therapeutic benefit in heavily pretreated glioblastoma patients.[32] A study of a multi-kinase inhibitor plus dose-dense or metronomic temozolomide for patients with recurrent HGG is currently ongoing (NCT02942264). The administration of dose-dense chemotherapy has improved outcome for patients with atypical teratoid/rhabdoid tumor,[33] ependymoma,[34] medulloblastoma,[35] and adenohypophysial tumors.[36] Dose-dense temozolomide has failed to improve survival for pediatric patients diagnosed with DIPG.[37]

Adjuvant and neoadjuvant chemotherapy

Adjuvant chemotherapy targets tumor cells after the completion of primary treatment, such as surgery or radiotherapy. As mentioned earlier, the rationale for this approach stems from early mathematical models of tumor growth kinetics wherein the preclinical phase of tumor growth represents the largest growth fraction with the highest sensitivity to chemotherapy. Additionally, in the preclinical phase, microscopic tumor may be less likely to harbor cytotoxic resistant cells. Therefore, adjuvant chemotherapy should be initiated early after primary therapy to prevent the growth of residual tumor cells and to help retard the appearance of resistant clones. In neuro-oncology, however, adjuvant chemotherapy is initiated after radiotherapy despite the residual tumor burden. In those patients treated with craniospinal axis radiation, the initiation of adjuvant chemotherapy is delayed by 4–6 weeks to allow recovery of bone marrow function in the vertebral bodies.

Some tumors are exquisitely chemosensitive. Therefore, the upfront treatment (neoadjuvant) with chemotherapy may shrink these tumors rendering them more amenable to surgery. Neoadjuvant therapy against chemosensitive tumors may also spare patients of the neurotoxic side effects of radiotherapy. In some tumor types that respond well to chemotherapy, such as CNS germinoma, neoadjuvant chemotherapy permits reduced-dose involved field cranial irradiation therapy to patients with localized disease. Neoadjuvant chemotherapy has also been shown to reduce tumor vascularity in some pediatric brain tumors and allow gross total resection without significant intraoperative hemorrhaging.[38]

Combination therapy

Antineoplastic polytherapy is a hallmark of cancer treatment and was first conceived by Frei et al. for the treatment of pediatric patients with acute lymphocytic leukemia.[39] In principle, this treatment paradigm employs multiple drugs with different mechanisms of action and non-overlapping toxicities that have shown a therapeutic efficacy in phase II clinical trials. Multimodal therapy improves the therapeutic efficacy because multiple key pathways of tumorigenesis are targeted in a synergistic or additive manner. Thus, treatment responses can be achieved with fewer cycles. This approach can potentially reduce treatment resistance while working in concert to provide either cytotoxic or cytostatic antineoplastic effects. It should be noted, however, that antineoplastic polytherapy can equally produce undesirable side effects. Therefore, a careful evaluation of drug interactions prior to creating a combination regimen is prudent.

In patients with glioma the benefit of this approach is unproven. A few trials have investigated the efficacy of combined therapy to monotherapy. Levin et al. found that PCV (procarbazine, CCNU [1,3-*cis*(2-chloroethyl)-1-nitrosourea] and vincristine) was superior to BCNU monotherapy for the treatment of patients with anaplastic glioma.[40] Conversely, a trial showed no significant differences in treatment effect when HGG patients were treated with PCV or radiotherapy alone.[41] Moreover, data from a recent randomized phase III trial showed no difference in survival in HGG patients treated with PCV vs temozolomide monotherapy.[42] In clinical practice, both CCNU and procarbazine can produce clinically significant delayed myelosuppression, with the nadir in blood counts occurring 5–6 weeks after administration. Additionally, these agents are emetogenic and patients may require powerful antiemetic medication. Quite often, the side-effect profile of PCV may interrupt the 6-week treatment course and at times may lead to discontinuation of planned treatment. These challenges make temozolomide monotherapy more appealing to treating physicians.

For patients with PCNSL, the optimal treatment protocol has not been standardized but nowadays the use of an immunochemotherapy combination regimen (Table 3), with methotrexate as the backbone, has been widely adopted. Historically, treatment with WBRT and methotrexate monotherapy has yielded a median OS of 12.2 months[43] and 25 months,[44] respectively. However, Morris and colleagues demonstrated a median OS of 6.6 years in patients with newly diagnosed PCNSL treated with rituximab, methotrexate, procarbazine, and vincristine followed by consolidation reduced-dose WBRT and cytarabine.[45] In this cohort of patients with symptomatic vasogenic edema, the incorporation of corticosteroids in the treatment protocol also carries lymphocytotoxic effects.

TABLE 3 Brain tumor specific chemotherapeutic agents

High-grade glioma	Temozolomide; carboplatin; PCV; BEV/irinotecan; BCNU; BEV/carboplatin
Low-grade glioma	Temozolomide; carboplatin, PCV
Oligodendroglioma	Temozolomide; PCV
Primary CNS Lymphoma	R-MPV; MTX; rituximab/temozolomide; MTX/cytarabine; MPV; ifosfamide
Germinoma	Carboplatin or cisplatin/etoposide
Medulloblastoma/PNET	Vincristine/cisplatin/CCNU or cyclophosphamide Carboplatin or cisplatin/etoposide/cyclophosphamide

Radiotherapy with concomitant chemotherapy is often employed for local tumor control. Combination chemotherapy with and without radiotherapy has improved patient outcome in other tumor types such as medulloblastoma, ependymoma, LGG, and atypical teratoid rhabdoid tumor. More recently, the combination of dabrafenib and trametinib has shown improved survival of patients with extracranial metastatic melanoma with BRAF V600E mutation.[46–48] Two clinical trials will clarify the efficacy of this approach in patients with brain metastasis (NCT02039947, NCT02537600).

Clinical trials

The goal of clinical trials (Table 4) is to answer scientific questions through experiments or observations by recruiting and treating patients in a uniform manner to measure a primary outcome or end point. The data generated in clinical research are rigorously analyzed using scientific model-based statistical tests to measure response to therapy. Responses can be an evaluation of drug safety, toxicity, or direct/indirect measures of antineoplastic efficacy. Common end points used in clinical trials include OS, progression free survival (PFS), time to progression (TTP), and objective response rates (ORR). The clinical trial algorithm primarily includes four phases of drug development.

TABLE 4 Several investigational agents against a variety of brain tumors

Agent	Target	Tumor histology	Phase	Clinical trial
Veliparib	PARP	Glioma, meningioma, PCNSL ependymoma	Phase II	NCT03220646
		Glioma	Phase II/III	NCT02152982
Letrozole	Estrogen receptor	Glioma	Phase 0/1	NCT03122197
Erlotinib	EGFR	NSCLC	Phase II	NCT00871923
Cabozantinib	c-MET, VGFR	BC BM Glioma	Phase II Pilot Study	NCT02260531, NCT02885324
Abemaciclib	CDK4/6	Glioma	Phase I	NCT02644460
IDH1 Peptide Vaccine	IDH1 mutation	Glioma	Phase I	NCT02193347
CUDC-907	HDAC and PI3K	Pediatric brain tumors	Phase I	NCT02909777
CAR transduced PBL	EGFRvIII	Glioma	Phase I/II	NCT01454596
Trastuzumab-emtansine	HER2	BC BM	Phase I/II	NCT03190967
TG02	Multiple kinases	Glioma	Phase II	NCT02942264
ABT-414	EGFR	Glioma	Phase II	NCT02343406
Sorafenib Tosylate	VEGFR, PDGFR, RAF/MEK/ERK	Glioma	Phase II	NCT01817751
Vemurafenib	BRAF	Glioma	Phase 0	NCT01748149
Marizomib	Proteasome	Glioma	Phase 1b	NCT02903069
Dabrafenib	BRAF	Glioma	Phase I/IIa	NCT01677741
Selinexor	XPO1	Glioma	Phase I	NCT02323880
Tucatinib	HER2	BC BM	Phase I	NCT01921335
Icotinib	EGFR	NSCLC	Phase II	NCT02726568
Vismodegib	Smoothened	Meningioma	Phase II	NCT01239316
Palbociclib	CDK4/6	Solid tumor BM	Phase II	NCT02896335
Osimertinib	EGFR T790M mutation	NSCLC	Phase II	NCT02856893
Vemurafenib	BRAF	Craniopharyngioma	Phase II	NCT03224767

Phase I trials

Phase I clinical trials are conducted to evaluate the toxicity and pharmacokinetic profiles of new chemotherapeutic agents, as well as to establish the MTD using a specified schedule or route of administration. The MTD is defined as the peak dose of a drug that provides a therapeutic efficacy but with reduced rates of untoward side effects. The optimal dose is established by dose incrementation; however, for some molecular targeting agents with a calculable biological effect, the desired effect can be achieved at doses below the MTD. Therefore, in these cases a dose escalation until deleterious effects arises is unnecessary. By convention, the sample size (N) in each dose level of phase I studies is typically small to limit the number of patients subjected to both extremes of the drug dose (i.e., low and high).

The translation of targeted therapy into clinical application for brain tumor treatment has been faced with many challenges. Notably, there have been insufficient early studies to evaluate the biological activity of many of these molecular targets. Furthermore, most comparative studies were performed retrospectively and there has been a paucity of prospective studies with baseline genomic profiling of tissue to identify those subsets of patients who may derive a clinical benefit. There has also been vacillation among research investigators regarding which surrogate end points are most suitable for studies involving molecularly targeted agents. To help bridge the gap between scientific knowledge and clinical research for glioma, an early phase study design (phase 0) has been integrated into the standard clinical research paradigm. This trial design evaluates both the pharmacokinetic and pharmacodynamic profiles of investigational agents in a small group of patients to capture relevant translational data applicable to clinical end points. The approach also includes tissue acquisition at time of diagnosis or at disease relapse to study the biological effects of target modulation using laboratory models such as xenograft models or cultured stem cells. The hope is that in early phase studies the optimal biological dose of targeted therapies can be achieved before patients proceed to conventional study designs. For example, in an ongoing phase 0/II study (NCT02933736), patients with Rb positive recurrent glioblastoma and meningioma scheduled for surgery will be administered ribociclib, pre-surgically to identify relevant pharmacokinetic and pharmacodynamics end points for phase II studies.

Phase II trials

Phase II protocols evaluate the efficacy and safety of the drug dose established in the phase I trial to determine whether further evaluation in a large-scale confirmatory phase III study would be worthwhile. Typically, efficacy is measured by radiographic and clinical response metrics and/or laboratory endpoints. A large homogeneous cohort is preferred for randomization into one of the following three classifications: (1) randomization to parallel non-comparative single-arm experimental regimens each with an independent decision rule; (2) randomized selection designs for selecting the most promising experimental regimen among several similar experimental regimens; and (3) randomized screening design for comparing an experimental regimen to standard of care.[49] In the former, volunteers are randomized to ≥2 experimental treatment arms; each designed as an independent phase II study to determine therapeutic potential when compared against historical controls. In the selection designs phase II trial, patients are randomized to ≥2 "rivaling" experimental regimen and the final results are then organized by superiority. The investigational therapy with the most superior outcome is then selected for further investigation against standard of care in a larger phase III trial. Lastly, randomized screening designs compare one or more investigational regimes against the standard of care treatment. To illustrate, when untreated patients with metastatic NSCLC were randomly assigned to either low- or high-dose bevacizumab combined with carboplatin and paclitaxel (two experimental arms) or to carboplatin and paclitaxel alone (control group), higher response rates were observed in the high-dose bevacizumab group (15 mg/kg). This observation led to the pivotal phase III trial that confirmed the efficacy of bevacizumab plus carboplatin and paclitaxel in this cohort of patients. Unfortunately, brain tumor phase II trials are often plagued by low patient accrual yielding insufficient data. This scenario leads to approximations of the true effect of the investigational agent—with a wide confidence interval around the observed end point—that tend to generate nebulous conclusions. Another important limitation of small studies is that they can result in false-positive results or over-estimations of the "observed" response rate which can, faultily, qualify a study drug for a more thorough evaluation in a large-scale definitive phase III study.

Phase III trials

These trials are considered the definitive phase of drug development in the clinical trial algorithm. If an agent/regimen shows therapeutic efficacy and tolerability in a phase II trial the experimental agent(s) are tested in a large-scale phase III confirmatory study. Classically, these are multicenter studies where several hundred to thousands of patients are randomly allocated to receive either the investigational agent or standard therapy. In those cases where a standard treatment is nonexistent, a placebo-controlled, double-blinded trial is favored over an open-label study. These trials are the longest of the clinical trials and can remain ongoing for over a decade. For instance, the key European study that demonstrated improved survival in patients with anaplastic oligodendroglioma treated with a drug regimen after radiotherapy took 17 years to complete.[50] These prolonged phase III studies frequently suffer from issues with noncompliance, protocol deviations, and high patient dropout rates, which might create important prognostic differences among treatment groups and may also lead to an inadequate statistical power.[51] To counteract this effect the concept of intent to treat analysis was introduced. This method gives an unprejudiced estimate of the treatment effect in the final analysis because it disregards events that occur after randomization—results are based on the initially assigned treatment rather than the treatment received.

Phase IV trials

Phase IV after-marketing studies are performed to capture any long-term sequelae (risks and benefits) of the new chemotherapeutic agent. Phase IV evaluations usually have several thousand participants.

Assessment of response to chemotherapy

The main objective of phase II and phase III clinical trials in the assessment of chemotherapy is to measure tumor response to treatment. Typically, survival end points (OS being "the gold standard" for HGG trials) are analyzed in phase III trials, whereas imaging-based end points, such as an overall radiographic response (either by 2-dimensional or 1-dimensional measurements) and PFS is assessed in phase II trials. Of note, several new chemotherapeutic agents are cytostatic and are less likely to produce tumor regression. Additionally, most traditional cytotoxic agents do not produce substantial overall radiographic response rates but may show utility in terms of PFS and OS when given to newly diagnosed HGG patients as upfront therapy (e.g., temozolomide). For these agents, the ability to achieve stable disease is the end goal, thus another endpoint, PFS at 6-months, is appropriate. Moreover, given increased patient survival coupled with the recent advances in neuroimaging methods as well as in the assessment of the functional status and well-being of patients, there has been a budding interest in the integration of ancillary end points into clinical trials. End points such as cognitive function, quality of life, and seizure control in brain tumor-related epilepsy have been suggested as metrics in future trials, but await validation.

An objective assessment of tumor response to treatment can be problematic because the spatial extent of tumors at baseline as well as during therapy can be obscure on MRI. For example, post-radiation treatment effects in HGG patients are quite often indistinguishable from viable tumor and may even mimic tumor progression. Furthermore, the T1 post-gadolinium enhancement on MRI correlates with regions of tumor microvascular disruption and is not a true reflection of tumor dimension or border. Likewise, the abnormal T2 signal on MRI may reflect infiltrating tumor or vasogenic edema, which should be considered during the assessment of response to treatment. Equally important, antiangiogenic agents such as bevacizumab, reduces vascular permeability; therefore, the disappearance of contrast enhancement on MRI in tumors under the influence of these agents may be interpreted as a pseudo-response. However, despite these shortcomings, MRI remains the standard imaging modality in the assessment of radiological response to treatment. Because of the impact of corticosteroids on neuroimaging, both the Response Assessment for Neuro-Oncology and Macdonald criteria require patients with HGG to be free of steroid use or on stable doses to permit accurate radiographic assessments of the efficacy of novel therapies. To resolve the issue of haphazard tumor measurements and estimates in neuro-oncology, it has been agreed upon that a response status be defined as a $\geq 50\%$ reduction in the tumor size (bi-dimensionally) following treatment. However, such a definition implies that for any changes in tumor size of $<50\%$ the chemotherapeutic agent under investigation may be ineffective when they may, in fact, possess antitumor properties. For this reason, time-dependent patient variables such as the period of clinically stable disease, is a useful gauge for response to therapy. Equally effective, a tumor shrinkage of 30% uni-dimensionally also defines objective tumor response,[52] but this approach has not been widely adopted.

The importance of pharmacokinetic studies such as drug clearance and steady state characteristics of investigational drugs during early phase testing was a learned lesson in the North American Brain Tumor Consortium (NABTC) phase II studies during the period of 1994–98. At that time, the NABTC used dosing schedules adopted from phase I studies of cytotoxic agents against different systemic solid tumors. The coadministration of agents that alter the pharmacokinetics of the investigational drugs represented an Achilles heel in these trials. Hepatic enzyme-inducing drugs such as phenytoin, carbamazepine, phenobarbital, and corticosteroids can induce the metabolism of therapeutic agents through the cytochrome

P450 system, wherein the standard dose used for other systemic tumors becomes subtherapeutic in brain tumor patients. To illustrate this point, results of a phase II study of paclitaxel showed that the toxicity profile seen in a cohort of brain tumor patients was remarkably lower than what was expected based on the reported toxicity in studies for the patients without brain tumors.[53] Also, in a phase I study, enzyme inducing antiepileptic drugs increased the clearance of irinotecan and its metabolites and required subsequent dose escalations until the MTD was achieved.[54] Drug doses were later stratified according to EIAD use in the follow-up phase II study.[55]

More advanced neuroimaging techniques have been investigated for their use in treatment planning as well as for the assessment of response to treatment. Methods such as volumetric imaging, MR diffusion-weighted, perfusion-weighted, and metabolic imaging modalities can provide more reliable tumor measurements. Additionally, these advanced techniques offer valuable data regarding the physiological and metabolic properties of the tumor microenvironment (e.g., tumor cellularity, hypoxia, disruption of normal tissue architecture, changes in vascular density, and vessel permeability) that can be used as proxies for the biological effects in clinical trials. Furthermore, an evolving approach to cancer diagnosis and monitoring of tumor response to treatment involves techniques for analyzing peripheral blood and CSF to identify circulating tumor cells and tumor biomarkers, termed a "liquid biopsy."

Challenges of drug delivery
Blood-brain barrier

The presence of the BBB is at the core of the major impediments of drug delivery to the CNS. This topic has garnered widespread research attention since the early 20th century, but while most issues have been resolved there are existing unanswered questions and unsettled controversies. However, it is understood that the BBB is an extraordinarily specialized and well-regimented interface composed of brain endothelial cells, a highly specialized basal membrane, astrocytic end feet, pericytes, and an extracellular matrix that physiologically separates the systemic circulation and CNS. The initial study that paved the way to the concept of the BBB was carried out by Ehrlich[56] who, in search of potential chemotherapeutic agents, observed that several intravenously injected dyes selectively stained tissues outside the brain. Later on, experiments by Berlin physician Lewandowski concluded that "brain capillaries must hold back certain molecules from entering the brain."[57] Despite these discoveries, researchers were befuddled about the discriminative properties of the BBB, but it was not until the advent of electron microscopy that the integrity and components of the BBB were identified.[58] We now know that the physical and metabolic integrity of the BBB is maintained by adherent junctions that are composed of vascular endothelium, cadherin, actinin, and catenin.[59] Moreover, the integral membrane proteins—occludin, claudin, junction adhesion molecules, and cytoplasmic accessory proteins—collectively form tight junctions which adjoin the cytoskeletons of adjacent brain microvascular endothelial cells and are thus primarily responsible for the permeability of the BBB.[60] The brain is considered a sanctuary site so there is restriction of the passage of certain compounds into the CNS. To this end, the transendothelial passage of endogenous compounds is achieved by passive diffusion, carrier-mediated transport (CMT), endocytosis via receptor-mediated transport (RMT), and active transport. These complex intercellular tight junctions prevent paracellular movement of molecules and restrict the passive diffusion of molecules into the brain. Furthermore, the BBB selectively inhibits agents with high molecular weight ($> 600\,Da$), strongly negatively charged compounds, and compounds with low lipophilicity.

The integrity of the BBB varies across different histologic types of tumor, and it is well known that the inter-endothelial scaffolding in HGG is functionally altered because of the presence of leaky tight junctions. Though the exact mechanism is not clear, it is thought that the aberrant expression of occludins and claudins in microvessels may contribute to the permeability of the BBB in brain tumors.[61] Also, the secretion of VEGF by high-grade astrocytoma has been implicated in increasing endothelial cell permeability.[62] Interestingly, the BBB is heterogeneously disrupted in glioblastoma with a more intact interface at the leading edge of the infiltrative tumor area where it is thought that the invasive tumor cells populate.[63,64] In this situation, passive accumulation of chemotherapeutic drugs within the region of the infiltrating tumor cells may be limited,[65,66] resulting in a diminished therapeutic effect. This may also be true for LGG which is less infiltrative and does not wreak havoc on the normal astrocytes, thus leaving a much more intact BBB. The BBB is also breached in metastatic brain tumors and even plays a significant role in their formation, but the disruption is negligible in smaller aggregates of metastatic tumor cells.[61] Consequently, microscopic tumors grow unperturbed despite the administration of efficacious systemic chemotherapy, underscoring the urgent need for rapid clinical translation of next-generation therapies with the ability to penetrate the BBB.

Efflux pumps

The BBB also safeguards the CNS against neurotoxins. This is accomplished by a variety of endothelial ATP-binding cassette transporters that actively pump toxic compounds and therapeutic agents out of the brain. P-glycoprotein (P-gp), breast cancer resistance protein (BCRP), and members of the multidrug resistance (MDR) related proteins are three ATP-driven drug efflux pumps that are highly expressed in the BBB and blood-cerebrospinal fluid barrier (BCSFB).[67] These proteins are localized at the luminal surface of brain capillaries and actively prevent a large variety of lipophilic molecules, xenobiotics, potentially toxic metabolites, and drugs from reaching the brain via the transcellular route.[61,68] As a result, intracellular drug concentrations are reduced to subtherapeutic levels.

One of the most widely studied efflux transport proteins, P-glycoprotein, is a 170-kDa transmembrane ATP-driven drug efflux transporter encoded by MDR genes[69] that actively excretes its substrates, unidirectionally, out of the cells.[70] This discovery was revolutionary because it helped unravel the mystery of the mechanisms underlying treatment failure due to drug resistance. P-gp has been identified in primary brain tumors and has been accepted as a crucial transporter that confers resistance to many anticancer drugs, such as taxanes, vinca-alkaloids, etoposide and analogs, anthracyclines, lonafarnib, imatinib, topotecan, and palbociclib.[71–73] These findings carried huge clinical implications, and the resistant mechanisms represented an exciting avenue for exploitation to augment the delivery of CNS therapeutic agents. Thus, considerable research effort over the past two decades has focused on screening of agents for their potential role in modulating the expression or activity of P-gp transport as a strategy to improve brain delivery of anticancer drugs. Many brain tumor cells express P-gp[74]; however, in HGG P-gp expression is disproportionately localized at the endothelial cells of the tumor blood vessels. In relation to the BBB, P-gp activity is disrupted at the necrotic core of glioblastoma, but it is preserved at the tumor border.[75] This finding is clinically relevant with respect to the postoperative administration of chemotherapy in patients with HGG. The residual border cells with an intact BBB along with potential P-gp overexpression can significantly impede drug delivery leading to relapse into larger and more aggressive tumors.[76] To illustrate this point, in a study to evaluate the regional tumor distribution of erlotinib in a xenograft model of glioblastoma, Agarwal and colleagues found higher erlotinib concentrations at the necrotic core (where P-gp disruption occurs) and significantly low erlotinib concentrations around the tumor rim. However, after inhibition of P-gp and BCRP the concentration of erlotinib in the tumor core increased fourfold, while that in the brain around the tumor and normal brain increased >12-fold compared with the control group.[64] These results suggest that the pharmacological inhibition of P-gp and BCRP markedly enhanced erlotinib delivery, especially to the tumor periphery and normal brain tissue, where the BBB is intact. To date, preclinical findings of modulators of these CNS gatekeepers have not been translated into clinical application, possibly because of unfavorable pharmacokinetic and toxicity profiles as well as adverse drug interactions.

DNA repair enzymes

Eukaryotic cells have evolved several mechanisms to correct errors made during DNA replication, but a greater propensity of genetic instability exists in tumor cells making them more prone to the accumulation of DNA damage. Malignant tumors inflicted with DNA damage in the context of radiation and chemotherapy can escape cell death by activating de novo or acquired DNA damage response signaling events. For example, the cytotoxic effect of alkylating agents is induced by the formation of cross-links with DNA at the O^6 or N^7 position of guanine. However, preservation of the genomic integrity can be accomplished by MGMT, a repair protein found in normal cells, but with higher activity in a variety of brain tumors.[77] This protein removes the alkyl groups from the O^6-position of guanine, and reverses the cytotoxic effect of alkylating agents. High levels of MGMT in many tumor cells render them resistant to drugs such as nitrosourea, procarbazine, and temozolomide. Conversely, tumor cells lacking this protein are extremely sensitive to alkylating agents in vitro.[77] MGMT activity has been detected in primary brain tumors, including meningioma, HGG, astrocytoma, oligodendroglioma, medulloblastoma, and ependymoma.[78] Thus, the inhibition of this DNA damage response maintains the genotoxic forms of DNA and enhances response rates to cytotoxic chemotherapy.

In theory, this approach to cancer therapy seems straightforward but several studies utilizing an agent to modify resistance to nitrosoureas and temozolomide ultimately proved to be negative—illustrating the difficulty of clinical trials in this arena. Inhibition of the DNA damage response by O^6-benzylguanine, a methylating agent, was evaluated in several clinical trials. Preclinical data indicated that O^6-benzylguanine was effective at counteracting the effects of MGMT.[79,80] A phase I trial of carmustine (BCNU) plus O^6-benzylguanine defined the toxicity and maximum-tolerated dose (MTD) of BCNU in conjunction with the pre-administration of O^6-benzylguanine in patients with recurrent or progressive HGG. However, a phase II trial did not produce tumor regression in patients with nitrosourea-resistant HGG, while 67% of the patients experienced significant hematopoietic toxicity.[81] In another study, temozolomide sensitivity was not restored in patients with temozolomide-resistant glioblastoma treated with O^6-benzylguanine.[82]

The MGMT gene can be silenced by promoter methylation resulting in reduced MGMT protein expression. Thus, patients with a hypermethylated MGMT promoter status are more sensitive to temozolomide. Since the landmark study by Hegi,[24] there has been a cornucopia of data confirming the prognostic and predictive value of MGMT methylation status for patients with glioblastoma. Interestingly, recent data have shown that the bi-functional alkylating agent, VAL-083, can unleash potent cytotoxic effects against temozolomide resistant glioblastoma stem cells by creating DNA cross-links at N^7-guanine. Thus, the activity of VAL-083 is not influenced by MGMT. A pivotal phase III trial is ongoing (NCT03149575).

Poly ADP-ribose polymerase (PARP) enzyme represents a third DNA-damage repair mechanism that plays an essential role in single strand break DNA repair. Currently, there are several FDA approved PARP inhibitors marketed for systemic tumors, but these agents are still under investigation for their role as chemo- and radiosensitizers in brain tumor patients. Following DNA insult, homologous recombinational repair proteins BRCA1 and BRCA2, are summoned to repair the DNA damage. Tumor cells that harbor mutations in the homologous recombinational repair proteins become dependent on PARP1, which is recruited to single strand breaks and orchestrate base-excision repair. Thus, the inhibition of PARP1 results in multiple double strand breaks leading to cell death, since the deficiencies in the DNA-damage repair mechanisms preclude efficient DNA repair. In normal cells, the homologous recombinational repair proteins are competent; therefore, DNA damage repair can be efficiently achieved. In addition, PARP inhibition potentially enhances the sensitivity of tumor cells to DNA damaging agents, including radiotherapy.[83] More recently, data has suggested that PARP inhibition may be beneficial in the treatment of IDH1-mutant LGG because of deficiencies in DNA repair mechanisms specific to this molecularly favorable glioma subtype. IDH1-mutant LGG has known vulnerability to chemotherapy when compared to tumors with wild-type IDH1, possibly because of defects in oxidative metabolism and the production of the oncometabolite 2-hydroxyglutarate.[84] In the presence of PARP inhibitors, IDH1-mutant glioma cells become exquisitely sensitive to temozolomide.[85] Furthermore, the inhibition of PARP in combination with radiotherapy demonstrated preclinical efficacy in the treatment of MGMT unmethylated glioblastoma.[86] A modest benefit was derived from the combination of irinotecan plus the PARP inhibitor iniparib in the treatment of progressive brain metastasis from triple negative breast cancer.[87] A clinical trial to determine the efficacy of veliparib in combination with radiation therapy and temozolomide in pediatric patients with diffuse pontine gliomas (DIPG) is ongoing (NCT01514201). At the time of this manuscript submission, a phase I trial of olaparib and temozolomide for treating patients with relapsed glioblastoma was also in progress (NCT01390571). The role of DNA mismatch repair proteins in relation to cancer immunotherapy will be discussed later.

Strategies for improved CNS drug delivery
Osmotic and chemical disrupt of the BBB

Various chemical agents have been utilized for the sole purpose of opening the BBB to increase delivery of chemotherapeutic agents to the tumor niche. Mannitol, bradykinin, and the bradykinin analog RMP-7 are the most extensively studied agents. Mannitol is the most commonly used agent in preclinical and clinical trials. The mechanism of hyperosmolar BBB disruption entails the intra-arterial infusion of a hyperosmotic agent into the main feeding artery of the tumor, creating an osmotic gradient allowing the seepage of water out of the endothelial cells. Subsequently, there is shrinkage of the cerebrovascular endothelial cells, which, reversibly, alters the configuration of the inter-endothelial tight junctions and consequently, increases BBB permeability.[88] Typically, a standard dose of 10 mL of 1.4 M mannitol is infused over 2 min, prior to the chemotherapeutic agent of choice.[89] The intra-arterial mannitol infusion combined with methotrexate for the treatment of PCNSL has demonstrated increased BBB penetration by up to 100-fold with improved response rates.[90–93] Likewise, the delivery of carboplatin and the liposomal formulations of cisplatin and oxaliplatin into glioma cells improved with the administration of intra-arterial mannitol.[94] Osmotic agents have been estimated to yield an increase in drug delivery by 10–100 times.[95] The barrier remains open for up to 2–3 h.[96] This approach is not used in standard practice partly because it requires repeated hospitalizations, frequently necessitating general anesthesia. It has also been shown to increase the risk of seizures and strokes. A phase II trial will evaluate the efficacy of intra-arterial cetuximab after BBB disruption by mannitol in patients with HGG (NCT02800486).

MRI-guided focused ultrasound

Pioneering work by Hynynen et al. at the turn of the 21st century introduced the concept of BBB disruption via focused ultrasound (FUS).[97] Successive studies have established FUS as an innovative approach to establishing a focal and transient breach of the BBB.[98,99] In principle, FUS utilizes an acoustic lens to concentrate multiple intersecting beams of ultrasound on a region of interest. Convergence of these beams at the targeted site creates high intensity focused ultrasound energy resulting in tissue ablation.[100] Studies have suggested that FUS may promote BBB disruption by producing shear stress in cells or by activation of signaling pathways involved in the regulation of permeability.[101,102] Disruption of tight junction-related proteins by FUS have also been suggested.[103] This approach is advantageous when compared to other strategies because of the ease of repeatability, noninvasiveness, negligible side effects, and precise targeting in conjunction with MRI. Worth mentioning, a limitation of FUS is signal attenuation and distortion from the skull.[104] Nonetheless, ultrasound-induced BBB disruption has potentiated the delivery of several drugs into the CNS.[105] More studies are needed to assess the safety and efficacy of this technique in the treatment of patients with brain tumors.

Radiation-induced BBB disruption

The permeability of the BBB increases during radiotherapy,[106] and the degree of disruption is directly proportional to the radiation dose.[107] The pathophysiology of radiation-induced vascular permeability is unclear but studies have postulated that it may be a result of inflammatory damage to astrocytes.[108] It has been shown that the permeability is increased with daily fractions of 3 Gy up to a total dose of 30 Gy in brain regions bordering the irradiation field.[109] However, fractionation with a single daily dose of 2 Gy has been suggested to reduce long-term neurotoxicity.[110] Qin et al. demonstrated that the fractionation dose of 2 Gy used to alter the BBB permeability can potentiate the effect of chemotherapy.[107] Consequently, the concomitant administration of chemotherapy with cranial irradiation as a therapeutic strategy has been a topic of interest for neuro-oncological therapy. Several agents with poor CNS penetration were evaluated in the presence of targeted cranial radiotherapy for the treatment of glioma.[111–114] Although the results of these studies are promising, the transient opening of the BBB may present a limited window of opportunity for the accumulation of therapeutic doses of chemotherapy in the tumor cells. For patients with brain metastasis, the role of chemotherapy is still evolving. Recent studies and observations have reported objective responses with systemic chemotherapy in several tumor types, particularly when coadministered with WBRT.[115–117] Jiang et al. reported that the optimal time window for interventions with chemotherapy is during and within 2 weeks following 20 and 40 Gy WBRT.[118]

Laser interstitial thermotherapy-induced BBB disruption

Laser interstitial thermotherapy (LITT) is a novel technique whereby tumors are ablated via the insertion of a fiber-optic laser applicator. This modality is minimally invasive, and like FUS, the tissue is obliterated through the administration of thermal energy. The application is used in many tissue types, including glioblastoma, although at present there is a paucity of data to justify its role as a treatment modality. Nonetheless, LITT may have clinical utility in the disruption of the BBB to increase delivery of chemotherapy locally. Preclinical data suggest that laser ablation may open the BBB for up to several days following treatment.[119] Future studies are needed to clarify its use in this context.

BBB circumvention
Drug modification

At present, the most efficacious drugs for brain tumor therapies are temozolomide, procarbazine, carmustine, and lomustine given the favorable molecular weights of these agents for BBB penetration (Table 5). Other chemotherapeutics such as doxorubicin, paclitaxel, cisplatin, and methotrexate have limited BBB penetration.[65] Given this problem, new cancer therapeutics with the ability to circumvent the BBB are desperately needed to improve patient survival. As mentioned earlier, a drug's capacity to traverse the BBB is influenced by several factors such as molecular size (<600 Da), low hydrogen bonding capacity, and high lipophilicity. Thus, in drug modification, the biochemical structure of an existing drug can be chemically altered to establish the favorable pharmacological properties that grant entry into the brain. Methods such as lipidization, structural modifications to enhance stability, and transformation to a prodrug have been investigated.[120] For instance, linoleic acid conjugated to paclitaxel showed higher cellular uptake and increased antitumor effect in a preclinical glioma model.[121] In another study, higher objective response rates were observed in breast cancer patients with brain metastases treated with a liposomal doxorubicin-cyclophosphamide combination.[122] The biochemical structures of drugs may also be tailored to mimic endogenous carrier-mediated substrates (e.g., sugars, amino acids, nucleosides) and are thus transported, veiled, by CMT. The hope in this approach is to avoid possible side effects of the drug construct and to preserve the biological function and the pharmacological activity of CMT. This approach has been applied to small molecules but not

TABLE 5 Common chemotherapeutic agents used in the treatment of brain tumors

Drug	Mol. Wt. (Da)	Route	Mechanism of action	Common side effects
Bevacizumab	149,000	IV	Anti-VEGF-A	Hypertension, proteinuria, GI perforation, poor wound healing, hemorrhage
Carboplatin	373	IV	Platinating agent	Myelosuppression (Plt), N&V, neuropathy, alopecia, hypersensitivity reaction
Carmustine	214	IV	Alkylating agent	Myelosuppression, pulmonary fibrosis, N&V, renal insufficiency
Cisplatin	300	IV	Platinating agent	Myelosuppression, alopecia, N&V, ototoxicity, renal insufficiency
Cyclophosphamide	261	IV	Alkylating agent	Myelosuppression, cystitis, water retention, cardiomyopathy
Cytarabine	243	SC, IV, IT	Antimetabolite	Myelosuppression, hepatic injury, N&V, mucositis
Etoposide	588	PO, IV	Topoisomerase II inhibitor	Myelosuppression, alopecia, N&V, infertility, headache
Everolimus	958	PO	mTOR inhibitor	Rash, acne, mucositis, diarrhea, hypertriglyceridemia, hypercholesterolemia
Ifosfamide	261	IV	Alkylating agent	Myelosuppression, encephalopathy, N&V, alopecia, cystitis, cardiomyopathy
Irinotecan	586, 623,[a] 677[b]	IV	Topoisomerase I inhibitor	Diarrhea, N&V, myelosuppression
Lomustine	233	PO	Alkylating agent	Myelosuppression, N&V, pulmonary fibrosis
Methotrexate	454	PO, IM, SC	Dihydrofolate reductase inhibitor	Mucositis, rash, diarrhea, hepatotoxicity, leukopenia
Procarbazine	221	PO	Methylating agent	Myelosuppression, N&V, decreased appetite
Temozolomide	194	PO, IV	Methylating agent	Myelosuppression, constipation, headache, fatigue
Vincristine	824	IV	Antimicrotubule agent	Hair loss, neuropathy, constipation, taste changes

N&V = nausea and vomiting.
[a]*HCl.*
[b]*HCl trihydrate.*

high molecular weight biological molecules.[120] Furthermore, local delivery methods such as the packaging of chemotherapeutic agents into nanocarriers can be combined strategically with biochemically modified drug formulations to elude the BBB. For example, liposomes targeting glutathione transporters (2B3–101) have been shown to enhance doxorubicin delivery to the brain extracellular space by 4.8-fold compared with nontargeted liposomal doxorubicin.[123] 2B3–101 has demonstrated clinical activity in a phase II clinical trial involving patients with brain metastasis from breast cancer and glioblastoma. The drug (now called 2×-111) will be developed and commercialized in the Netherlands for further clinical trial evaluations.

Targeting receptor-mediated transporters

The novel approach to RMT-targeting involves conjugating a targeting ligand (e.g., antibody or antibody fragment, growth factors, therapeutic antibiotics, peptide, and natural ligand) that has an affinity for an endocytic cell-surface receptor expressed in brain endothelial cells to the drug or to a drug-loaded nanocarrier.[124] In essence, the binding and clustering of the targeted receptor on the cell surface trigger intracellular signaling cascades that mediate the internalization of membrane-bound intracellular vesicles via endocytosis. These vesicles then undergo intracellular trafficking that culminates into their accumulation into lysosomes (where they are digested) or into endosomes where a fraction of the content of the vesicles is expelled into the side of the cell by exocytosis. It is presumed that nanomedicines may proceed to therapeutic sites within the brain parenchyma following exocytosis.[125] Penetrating-targeting drug delivery has been investigated in other receptors, including the transferrin receptor, lactoferrin receptor, low-density lipoprotein receptor, folate receptor, insulin receptor, and leptin receptor.[126–131]

Cancer-directed gene therapy

Gene therapy represents an emerging and exciting area of research in the field of neuro-oncology, particularly with respect to therapy for glioblastoma given the unmet need for agents able to extend OS. This treatment paradigm rests on the capability of delivering genes directly to the tumor niche to yield antineoplastic effects. Several different gene therapy approaches are currently being examined for their utility in treating glioblastoma. One of the most studied strategies, suicide gene therapy, exploits enzyme-encoding genes that can convert inert prodrugs into cytotoxic compounds. The herpes simplex virus (HSV) type 1 thymidine kinase (tk)/ganciclovir (GCV) system and the cytosine deaminase (CD)/5-fluorocytosine (5-FC) system are two well-known suicide gene therapies that have been investigated.[132] In both constructs, delivery of a gene (tk or CD) to the tumor cells generates enzymes that catalyze the conversion of a systemically administered prodrug (GCV or 5-FC, respectively) to an activated chemotherapeutic agent (ganciclovir triphosphate or 5-fluorouracil, respectively). The latter approach, currently in clinical trial, has shown preliminary signs of clinical efficacy. Results from a recent clinical trial involving a retroviral replicating vector, Toca 511 (*vocimagene amiretrorepvec*), and a prodrug, Toca FC (5-fluorocytosine), have demonstrated a survival benefit in patients with relapsed HGG.[133] In this methodology, the retroviral replicating vector is engineered with yeast cytosine deaminase, a prodrug-activator gene, which generates the cytotoxic compound 5-fluorouracil within the infected tumor cells from its precursor 5-fluorocytosine. Researchers have also discovered that the drug kills immune suppressive cells within the tumor microenvironment, triggering an immune response against the residual tumor. An accelerated phase III trial, Toca 5, (NCT02414165) is currently underway. The delivery of immunomodulatory genes, tumor-suppressor genes, and oncolytic virotherapy are other types of gene therapies currently being investigated. Other viral vehicles for gene therapy include adenovirus, adeno-associated virus-2, reovirus, H1 parvovirus, measles virus, and poliovirus.[134]

Convection-enhanced delivery

Locally administered macromolecules and bioactive agents in the brain have been shown to have restricted penetration secondary to inadequate diffusion.[135] This hurdle can be overcome by the convection-enhanced delivery (CED) platform, which has been shown to enhance the distribution, and consequently, the effectiveness of locally administered drugs in the brain.[136] In this apparatus, the drug of choice is infused through a cannula that is inserted into the region of interest. In principle, a constant pressure difference between the cannula tip and the surrounding tissue during interstitial infusion allows the administered drug to be transported into the tissue via convective flow. Deeper tissue penetration can be achieved via CED, resulting in a large volume of distribution relative to the infused volume at drug concentrations orders of magnitude greater than systemic levels.[137] The co-infusion of imaging contrast material, such as radiotracers, iodinated compounds, or paramagnetic materials into the infusion volume permit the use of diagnostic imaging platforms for real-time surveillance of CED. Additionally, with an image-guided design, the ratio of the volume of distribution to the infusion volume can be estimated. CED has been utilized to improve the delivery of several chemotherapeutic agents,[138] and may limit the side effects generally observed in systemic chemotherapy.

Implantable devices

Implantable drug depots represent another method for local delivery of antineoplastics. In this technique, a polymer-controlled drug delivery system is placed intratumorally or near the resection margin during neurosurgical intervention, thereby permitting the release of drugs into neighboring tumor cells. Gliadel, a biodegradable polymeric wafer loaded with BCNU (1,3-bis (2-chloroethyl)-1-nitrosourea), was the inaugural implantable depot tested clinically against primary brain tumors,[139] and remains the only FDA approved drug-polymer device to date. Single-agent Gliadel or in combination with radiotherapy was shown to extend the median survival of patients with primary or relapsing glioblastoma by 2–3 months.[140,141] However, given the modest survival advantage with BCNU other implantable device models enriched with other agents (e.g., temozolomide and doxorubicin) are being tested.[142–144] The approach is also being developed for group 3 medulloblastoma.[145] The most efficacious implantable device should be designed with an agent with controlled and predictable release kinetics and, like in CED, has the capacity to diffuse away from the supply to achieve a clinically meaningful volume of distribution. Although implantable depots, in principle, have several advantages over conventional drug delivery, their therapeutic yield may be narrow due to an inability to reload the drug source, change its location or modify the drug release kinetics without an invasive surgical procedure.[124]

Intracerebrospinal fluid chemotherapy

Intra-CSF chemotherapy requires the intrathecal infusion of agents either into the lumbar subarachnoid space or lateral ventricle, typically via a surgically implanted subcutaneous apparatus, such as an Ommaya reservoir.[146] Treatment is palliative and is used for non-nodular and non-bulky leptomeningeal carcinomatosis from systemic solid cancers, as well as hematologic neoplasms. Intrathecal administration of conventional agents such as methotrexate, cytarabine, thiotepa, topotecan, and etoposide has been employed for the treatment of neoplastic meningitis. Lesion thickness influences treatment decision-making. It has been hypothesized that thick leptomeningeal lesions are well vascularized and may respond favorably to systemic chemotherapy, whereas intra-CSF chemotherapy penetrates only 2–3 mm into carcinomatous meningitis.[147] The intrathecal approach is appealing because it yields high drug concentrations within the CSF. For example, intrathecally administered methotrexate reaches drug concentration levels 10-fold higher than systemically administered routes.[148] Furthermore, when administered via an Ommaya reservoir the therapeutic effect is retained for at least 24 h.[148] This strategy is also advantageous because of negligible drug degradation by enzymes as well as the enhanced availability of unbound drugs in the CSF. The drug distribution is highly dependent on CSF flow dynamics; therefore, CSF flow studies should precede intrathecally administered chemotherapy to avoid neurotoxicity secondary to drug accumulation.

Therapeutic activity of intrathecal interleukin-2[149] against melanoma and lorlatinib[150] against *ALK* positive lung cancer has been reported but awaits validation. Intrathecal trastuzumab, a monoclonal antibody against human epidermal growth receptor 2 (HER2), has shown efficacy for the management of carcinomatous meningitis in HER2-amplified breast cancer patients,[151] and has reached clinical trial testing (NCT01373710, NCT01325207, NCT02598427). Intrathecal administration of the CD20 antibody rituximab, was shown to be both feasible and efficacious for the treatment of leptomeningeal lymphomatosis[152,153]; but more studies are needed to verify a clinical benefit. Leptomeningeal seeding confers poor prognosis in patients with HGG. A few small studies have reported a palliative efficacy of intrathecal chemotherapy in these patients,[154–156] but generally a multimodal approach to treatment is performed in clinical practice.

References

1. Suh JH, Stea B, Nabid A, et al. Phase III study of efaproxiral as an adjunct to whole-brain radiation therapy for brain metastases. *J Clin Oncol.* 2006;24:106–114.
2. Huang CF, Kondziolka D, Flickinger JC, Lunsford LD. Stereotactic radiosurgery for brainstem metastases. *J Neurosurg.* 1999;91(4):563–568.
3. Skipper HE, Schabel Jr FM, Wilcox WS. Experimental evaluation of potential anticancer agents. XIII. On the criteria and kinetics associated with "curability" of experimental leukemia. *Cancer Chemother Rep.* 1964;35:1–111.
4. Skipper HE, Schabel Jr FM, Mellett LB, et al. Implications of biochemical, cytokinetic, pharmacologic, and toxicologic relationships in the design of optimal therapeutic schedules. *Cancer Chemother Rep.* 1970;54(6):431–450.
5. Shackney SE, McCormack GW, Cuchural Jr GJ. Growth rate patterns of solid tumors and their relation to responsiveness to therapy: an analytical review. *Ann Intern Med.* 1978;89(1):107–121.
6. Eidel O, Burth S, Neumann JO, et al. Tumor infiltration in enhancing and non-enhancing parts of glioblastoma: a correlation with histopathology. *PLoS One.* 2017;12(1).
7. Ebadi M. Taxol (paclitaxel) and cancer chemotherapy. In: *Pharmacodynamic Basis of Herbal Medicine.* 1st ed. CRC Press; 2010:643.
8. Folkman J, Shing Y. Angiogenesis. *J Biol Chem.* 1992;267(16):10931–10934.
9. Bru A, Albertos S, Luis Subiza J, Garcia-Asenjo JL, Bru I. The universal dynamics of tumor growth. *Biophys J.* 2003;85(5):2948–2961.
10. Lavi O, Gottesman MM, Levy D. The dynamics of drug resistance: a mathematical perspective. *Drug Resist Updat.* 2012;15(1–2):90–97.
11. Zhuang Z, Lu J, Lonser R, Kovach JS. Enhancement of cancer chemotherapy by simultaneously altering cell cycle progression and DNA-damage defenses through global modification of the serine/threonine phospho-proteome. *Cell Cycle.* 2009;8:3303–3306.
12. Hill BT, Baserga R. The cell cycle and its significance for cancer treatment. *Cancer Treat Rev.* 1975;2(3):159–175.
13. Alberts B, Johnson A, Lewis J, Raff M, Roberts K, Walter P. *Molecular Biology of the Cell.* 4th ed. New York: Garland Science; 2002.
14. Blagosklonny MV, Pardee AB. Exploiting cancer cell cycling for selective protection of normal cells. *Cancer Res.* 2001;61(11):4301–4305.
15. Peart MJ, Smyth GK, van Laar RK, et al. Identification and functional significance of genes regulated by structurally different histone deacetylase inhibitors. *Proc Natl Acad Sci U S A.* 2005;102(10):3697–3702.
16. Eyupoglu IY, Savaskan NE. Epigenetics in brain tumors: HDACs take center stage. *Curr Neuropharmacol.* 2016;14(1):48–54.
17. Marks PA, Richon VM, Miller T, Kelly WK. Histone deacetylase inhibitors. *Adv Cancer Res.* 2004;91:137–168.

18. Hideshima T, Cottini F, Ohguchi H, et al. Rational combination treatment with histone deacetylase inhibitors and immunomodulatory drugs in multiple myeloma. *Blood Cancer J*. 2015;5.

19. Williams MJ, Singleton WG, Lowis SP, Malik K, Kurian KM. Therapeutic targeting of histone modifications in adult and pediatric high-grade glioma. *Front Oncol*. 2017;7:45.

20. Ecker J, Oehme I, Mazitschek R, et al. Targeting class I histone deacetylase 2 in MYC amplified group 3 medulloblastoma. *Acta Neuropathol Commun*. 2015;3:22.

21. Shi W, Lawrence YR, Choy H, et al. Vorinostat as a radiosensitizer for brain metastasis: a phase I clinical trial. *J Neurooncol*. 2014;118(2):313–319.

22. Galanis E, Anderson SK, Miller CR, et al. Phase I/II trial of vorinostat combined with temozolomide and radiation therapy for newly diagnosed glioblastoma: final results of alliance N0874/ABTC 02. In: *Neuro Oncol*; England 2017.

23. Stupp R, Mason WP, van den Bent MJ, et al. Radiotherapy plus concomitant and adjuvant temozolomide for glioblastoma. *N Engl J Med*. 2005;352:987–996.

24. Hegi ME, Diserens AC, Gorlia T, et al. MGMT gene silencing and benefit from temozolomide in glioblastoma. *N Engl J Med*. 2005;352:997–1003.

25. Tolcher AW, Gerson SL, Denis L, et al. Marked inactivation of O6-alkylguanine-DNA alkyltransferase activity with protracted temozolomide schedules. *Br J Cancer*. 2003;88(7):1004–1011.

26. Gilbert MR, Wang M, Aldape KD, et al. Dose-dense temozolomide for newly diagnosed glioblastoma: a randomized phase III clinical trial. *J Clin Oncol*. 2013;31(32):4085–4091.

27. Taal W, Segers-van Rijn JM, Kros JM, et al. Dose dense 1 week on/1 week off temozolomide in recurrent glioma: a retrospective study. *J Neurooncol*. 2012;108(1):195–200.

28. Kerbel RS, Kamen BA. The anti-angiogenic basis of metronomic chemotherapy. *Nat Rev Cancer*. 2004;4:423–436.

29. Laquente B, Lacasa C, Ginesta MM, et al. Antiangiogenic effect of gemcitabine following metronomic administration in a pancreas cancer model. *Mol Cancer Ther*. 2008;7:638–647.

30. Omuro A, Chan TA, Abrey LE, et al. Phase II trial of continuous low-dose temozolomide for patients with recurrent malignant glioma. *Neuro Oncol*. 2013;15(2):242–250.

31. Kesari S, Schiff D, Doherty L, et al. Phase II study of metronomic chemotherapy for recurrent malignant gliomas in adults. *Neuro Oncol*. 2007;9(3): 354–363.

32. Da DAVJJ R, et al. Metronomic chemotherapy with daily, oral etoposide plus bevacizumab for recurrent malignant glioma: a phase II study. *Br J Cancer*. 2009;101(12):1986 [-1987-1994].

33. Slavc I, Chocholous M, Leiss U, et al. Atypical teratoid rhabdoid tumor: Improved long-term survival with an intensive multimodal therapy and delayed radiotherapy. The Medical University of Vienna experience 1992–2012. *Cancer Med*. 2014;3(1):91–100.

34. Ruda R, Bosa C, Magistrello M, et al. Temozolomide as salvage treatment for recurrent intracranial ependymomas of the adult: a retrospective study. *Neuro Oncol*. 2016;18(2):261–268.

35. Gajjar A, Chintagumpala M, Ashley D, et al. Risk-adapted craniospinal radiotherapy followed by high-dose chemotherapy and stem-cell rescue in children with newly diagnosed medulloblastoma (St Jude Medulloblastoma-96): long-term results from a prospective, multicentre trial. *Lancet Oncol*. 2006;7:813–820.

36. Ortiz LD, Syro LV, Scheithauer BW, et al. Temozolomide in aggressive pituitary adenomas and carcinomas. *Clinics (Sao Paulo)*. 2012;67(suppl 1): 119–123.

37. Bailey S, Howman A, Wheatley K, et al. Diffuse intrinsic pontine glioma treated with prolonged temozolomide and radiotherapy—results of a United Kingdom phase II trial (CNS 2007 04). *Eur J Cancer*. 2013;49(18):3856–3862.

38. Iwama J, Ogiwara H, Kiyotani C, et al. Neoadjuvant chemotherapy for brain tumors in infants and young children. *J Neurosurg Pediatr*. 2015;15(5): 488–492.

39. Frei 3rd E, Karon M, Levin RH, et al. The effectiveness of combinations of antileukemic agents in inducing and maintaining remission in children with acute leukemia. *Blood*. 1965;26(5):642–656.

40. Levin VA, Silver P, Hannigan J, et al. Superiority of post-radiotherapy adjuvant chemotherapy with CCNU, procarbazine, and vincristine (PCV) over BCNU for anaplastic gliomas: NCOG 6G61 final report. *Int J Radiat Oncol Biol Phys*. 1990;18:321–324.

41. Medical Research Council Brain Tumor Working Party. Randomized trial of procarbazine, lomustine, and vincristine in the adjuvant treatment of high-grade astrocytoma: a Medical Research Council trial. *J Clin Oncol*. 2001;19(2):509–518.

42. Wick W, et al. EORTC 26101 phase III trial exploring the combination of bevacizumab and lomustine in patients with first progression of a glioblastoma. *J Clin Oncol*. 2016;34. (Suppl):Absract 2001.

43. Nelson DF, Martz KL, Bonner H, et al. Non-Hodgkin's lymphoma of the brain: can high dose, large volume radiation therapy improve survival? Report on a prospective trial by the Radiation Therapy Oncology Group (RTOG): RTOG 8315. *Int J Radiat Oncol Biol Phys*. 1992;23:9–17.

44. Herrlinger U, Kuker W, Uhl M, et al. NOA-03 trial of high-dose methotrexate in primary central nervous system lymphoma: final report. *Ann Neurol*. 2005;57(6):843–847.

45. Morris PG, Correa DD, Yahalom J, et al. Rituximab, methotrexate, procarbazine, and vincristine followed by consolidation reduced-dose whole-brain radiotherapy and cytarabine in newly diagnosed primary CNS lymphoma: final results and long-term outcome. *J Clin Oncol*. 2013;31 (31):3971–3979.

46. Long GV, Stroyakovskiy D, Gogas H, et al. Combined BRAF and MEK inhibition versus BRAF inhibition alone in melanoma. *N Engl J Med*. 2014;371(20):1877–1888.

47. Long GV, Stroyakovskiy D, Gogas H, et al. Dabrafenib and trametinib versus dabrafenib and placebo for Val600 BRAF-mutant melanoma: a multi-centre, double-blind, phase 3 randomised controlled trial. *Lancet*. 2015;386:444–451.

48. Robert C, Karaszewska B, Schachter J, et al. Improved overall survival in melanoma with combined dabrafenib and trametinib. *N Engl J Med.* 2015;372(1):30–39.

49. Mandrekar SJ, Sargent DJ. Randomized phase II trials: time for a new era in clinical trial design. *J Thorac Oncol.* 2010;5(7):932–934.

50. van den Bent MJ, Brandes AA, Taphoorn MJ, et al. Adjuvant procarbazine, lomustine, and vincristine chemotherapy in newly diagnosed anaplastic oligodendroglioma: long-term follow-up of EORTC brain tumor group study 26951. *J Clin Oncol.* 2013;31:344–350.

51. Gupta SK. Intention-to-treat concept: a review. *Perspect Clin Res.* 2011;2(3):109–112.

52. Eisenhauer EA, Therasse P, Bogaerts J, et al. New response evaluation criteria in solid tumours: revised RECIST guideline (version 1.1). *Eur J Cancer.* 2009;45:228–247.

53. Prados MD, Schold SC, Spence AM, et al. Phase II study of paclitaxel in patients with recurrent malignant glioma. *J Clin Oncol.* 1996;14(8): 2316–2321.

54. Prados MD, Yung WK, Jaeckle KA, et al. Phase 1 trial of irinotecan (CPT-11) in patients with recurrent malignant glioma: a north American brain tumor consortium study. *Neuro Oncol.* 2004;6(1):44–54.

55. Prados MD, Lamborn K, Yung WK, et al. A phase 2 trial of irinotecan (CPT-11) in patients with recurrent malignant glioma: a north American brain tumor consortium study. *Neuro Oncol.* 2006;8(2):189–193.

56. Ehrlich P. Uber die Beziehungen von chemische Constitution, Vertheilung, und Pharmakologischer Wirkung. In: *Collected Studies in Immunity.* New York: Repr and Transl Wiley; 1906:567–595.

57. Lewandowsky M. Zur Lehre der Cerebrospinalflussigkeit. *Z Klin Med.* 1900;40:480–494.

58. Reese TS, Karnovsky MJ. Fine structural localization of a blood-brain barrier to exogenous peroxidase. *J Cell Biol.* 1967;34(1):207–217.

59. Vorbrodt AW, Dobrogowska DH. Molecular anatomy of intercellular junctions in brain endothelial and epithelial barriers: electron microscopist's view. *Brain Res Brain Res Rev.* 2003;42:221–242.

60. Zlokovic BV. The blood-brain barrier in health and chronic neurodegenerative disorders. *Neuron.* 2008;57:178–201.

61. Bhowmik A, Khan R, Ghosh MK. Blood brain barrier: a challenge for effectual therapy of brain tumors. *Biomed Res Int.* 2015;2015:320941.

62. Davies DC. Blood-brain barrier breakdown in septic encephalopathy and brain tumours. *J Anat.* 2002;200(6):639–646.

63. Wen PY, Kesari S. Malignant gliomas in adults. *N Engl J Med.* 2008;359:492–507.

64. Agarwal S, Manchanda P, Vogelbaum MA, Ohlfest JR, Elmquist WF. Function of the blood-brain barrier and restriction of drug delivery to invasive glioma cells: findings in an orthotopic rat xenograft model of glioma. *Drug Metab Dispos.* 2013;41(1):33–39.

65. Juillerat-Jeanneret L. The targeted delivery of cancer drugs across the blood-brain barrier: chemical modifications of drugs or drug-nanoparticles? *Drug Discov Today.* 2008;13:1099–1106.

66. Noell S, Mayer D, Strauss WS, Tatagiba MS, Ritz R. Selective enrichment of hypericin in malignant glioma: pioneering in vivo results. *Int J Oncol.* 2011;38(5):1343–1348.

67. Begley DJ. ABC transporters and the blood-brain barrier. *Curr Pharm Des.* 2004;10(12):1295–1312.

68. Abbott NJ. Blood-brain barrier structure and function and the challenges for CNS drug delivery. *J Inherit Metab Dis.* 2013;36(3):437–449.

69. Gottesman MM, Hrycyna CA, Schoenlein PV, Germann UA, Pastan I. Genetic analysis of the multidrug transporter. *Annu Rev Genet.* 1995;29:607–649.

70. Sharom FJ. The P-glycoprotein efflux pump: How does it transport drugs? *J Membr Biol.* 1997;160(3):161–175.

71. Schinkel AH, Smit JJ, van Tellingen O, et al. Disruption of the mouse mdr1a P-glycoprotein gene leads to a deficiency in the blood-brain barrier and to increased sensitivity to drugs. *Cell.* 1994;77:491–502.

72. Tsuji A. P-glycoprotein-mediated efflux transport of anticancer drugs at the blood-brain barrier. *Ther Drug Monit.* 1998;20(5):588–590.

73. Parrish KE, Pokorny J, Mittapalli RK, Bakken K, Sarkaria JN, Elmquist WF. Efflux transporters at the blood-brain barrier limit delivery and efficacy of cyclin-dependent kinase 4/6 inhibitor palbociclib (PD-0332991) in an orthotopic brain tumor model. *J Pharmacol Exp Ther.* 2015;355 (2):264–271.

74. Regina A, Demeule M, Laplante A, et al. Multidrug resistance in brain tumors: roles of the blood-brain barrier. *Cancer Metastasis Rev.* 2001;20 (1–2):13–25.

75. Deeken JF, Loscher W. The blood-brain barrier and cancer: transporters, treatment, and Trojan horses. *Clin Cancer Res.* 2007;13:1663–1674.

76. Cheshier SH, Kalani MY, Lim M, Ailles L, Huhn SL, Weissman IL. A neurosurgeon's guide to stem cells, cancer stem cells, and brain tumor stem cells. *Neurosurgery.* 2009;65:237–249. discussion 249–250; quiz N236.

77. Wiestler O, Kleihues P, Pegg AE. O6-alkylguanine-DNA alkyltransferase activity in human brain and brain tumors. *Carcinogenesis.* 1984;5(1): 121–124.

78. Mineura K, Izumi I, Watanabe K, Kowada M. O6-alkylguanine-DNA alkyltransferase activity in human brain tumors. *Tohoku J Exp Med.* 1991; 165(3):223–228.

79. Dolan ME, Stine L, Mitchell RB, Moschel RC, Pegg AE. Modulation of mammalian O6-alkylguanine-DNA alkyltransferase in vivo by O6-benzylguanine and its effect on the sensitivity of a human glioma tumor to 1-(2-chloroethyl)-3-(4-methylcyclohexyl)-1-nitrosourea. *Cancer Commun.* 1990;2(11):371–377.

80. Felker GM, Friedman HS, Dolan ME, Moschel RC, Schold C. Treatment of subcutaneous and intracranial brain tumor xenografts with O6-benzylguanine and 1,3-bis(2-chloroethyl)-1-nitrosourea. *Cancer Chemother Pharmacol.* 1993;32(6):471–476.

81. Quinn JA, Pluda J, Dolan ME, et al. Phase II trial of carmustine plus O(6)-benzylguanine for patients with nitrosourea-resistant recurrent or progressive malignant glioma. *J Clin Oncol.* 2002;20(9):2277–2283.

82. Quinn JA, Jiang SX, Reardon DA, et al. Phase II trial of temozolomide plus o6-benzylguanine in adults with recurrent, temozolomide-resistant malignant glioma. *J Clin Oncol.* 2009;27(8):1262–1267.

83. van Vuurden DG, Hulleman E, Meijer OL, et al. PARP inhibition sensitizes childhood high grade glioma, medulloblastoma and ependymoma to radiation. *Oncotarget*. 2011;2(12):984–996.

84. Lu Y, Kwintkiewicz J, Liu Y, et al. Chemosensitivity of IDH1-mutated gliomas due to an impairment in PARP1-mediated DNA repair. *Cancer Res*. 2017;77(7):1709–1718.

85. Sulkowski PL, Corso CD, Robinson ND, et al. 2-Hydroxyglutarate produced by neomorphic IDH mutations suppresses homologous recombination and induces PARP inhibitor sensitivity. *Sci Transl Med*. 2017;9(375).

86. Jue TR, Nozue K, Lester AJ, et al. Veliparib in combination with radiotherapy for the treatment of MGMT unmethylated glioblastoma. *J Transl Med*. 2017;15(1):61.

87. Anders C, Deal AM, Abramson V, et al. TBCRC 018: Phase II study of iniparib in combination with irinotecan to treat progressive triple negative breast cancer brain metastases. *Breast Cancer Res Treat*. 2014;146(3):557–566.

88. Rapoport SI, Robinson PJ. Tight-junctional modification as the basis of osmotic opening of the blood-brain barrier. *Ann N Y Acad Sci*. 1986;481:250–267.

89. Burkhardt JK, Riina H, Shin BJ, et al. Intra-arterial delivery of bevacizumab after blood-brain barrier disruption for the treatment of recurrent glioblastoma: progression-free survival and overall survival. *World Neurosurg*. 2012;77(1):130–134.

90. Angelov L, Doolittle ND, Kraemer DF, et al. Blood-brain barrier disruption and intra-arterial methotrexate-based therapy for newly diagnosed primary CNS lymphoma: a multi-institutional experience. *J Clin Oncol*. 2009;27(21):3503–3509.

91. Neuwelt EA, Frenkel EP, Diehl JT, et al. Osmotic blood-brain barrier disruption: a new means of increasing chemotherapeutic agent delivery. *Trans Am Neurol Assoc*. 1979;104:256–260.

92. Neuwelt EA, Bauer B, Fahlke C, et al. Engaging neuroscience to advance translational research in brain barrier biology. *Nat Rev Neurosci*. 2011;12(3):169–182.

93. Doolittle ND, Abrey LE, Shenkier TN, et al. Brain parenchyma involvement as isolated central nervous system relapse of systemic non-Hodgkin lymphoma: an International Primary CNS Lymphoma Collaborative Group report. *Blood*. 2008;111:1085–1093.

94. Charest G, Sanche L, Fortin D, Mathieu D, Paquette B. Optimization of the route of platinum drugs administration to optimize the concomitant treatment with radiotherapy for glioblastoma implanted in the Fischer rat brain. *J Neurooncol*. 2013;115(3):365–373.

95. Miller G. Drug targeting. Breaking down barriers. *Science*. 2002;297:1116–1118.

96. Chi OZ, Wei HM, Lu X, Weiss HR. Increased blood-brain permeability with hyperosmolar mannitol increases cerebral O2 consumption and O2 supply/consumption heterogeneity. *J Cereb Blood Flow Metab*. 1996;16(2):327–333.

97. Hynynen K, McDannold N, Vykhodtseva N, Jolesz FA. Noninvasive MR imaging-guided focal opening of the blood-brain barrier in rabbits. *Radiology*. 2001;220(3):640–646.

98. Hynynen K, McDannold N, Sheikov NA, Jolesz FA, Vykhodtseva N. Local and reversible blood-brain barrier disruption by noninvasive focused ultrasound at frequencies suitable for trans-skull sonications. *Neuroimage*. 2005;24:12–20.

99. Hynynen K, McDannold N, Vykhodtseva N, Jolesz FA. Non-invasive opening of BBB by focused ultrasound. *Acta Neurochir Suppl*. 2003;86:555–558.

100. Coluccia D, Fandino J, Schwyzer L, et al. First noninvasive thermal ablation of a brain tumor with MR-guided focused ultrasound. *J Ther Ultrasound*. 2014;2:17.

101. VanBavel E. Effects of shear stress on endothelial cells: possible relevance for ultrasound applications. *Prog Biophys Mol Biol*. 2007;93:374–383.

102. Jalali S, Huang Y, Dumont DJ, Hynynen K. Focused ultrasound-mediated bbb disruption is associated with an increase in activation of AKT: experimental study in rats. *BMC Neurol*. 2010;10:114.

103. Shang X, Wang P, Liu Y, Zhang Z, Xue Y. Mechanism of low-frequency ultrasound in opening blood-tumor barrier by tight junction. *J Mol Neurosci*. 2011;43(3):364–369.

104. McDannold N, Clement GT, Black P, Jolesz F, Hynynen K. Transcranial magnetic resonance imaging-guided focused ultrasound surgery of brain tumors: Initial findings in 3 patients. *Neurosurgery*. 2010;66(2):323–332 [discussion 332].

105. Lamsam L, Johnson E, Connolly ID, Wintermark M, Hayden GM. A review of potential applications of MR-guided focused ultrasound for targeting brain tumor therapy. *Neurosurg Focus*. 2018;44(2).

106. d'Avella D, Cicciarello R, Angileri FF, Lucerna S, La Torre D, Tomasello F. Radiation-induced blood-brain barrier changes: pathophysiological mechanisms and clinical implications. *Acta Neurochir Suppl*. 1998;71:282–284.

107. Qin D, Zheng R, Ma J, Xiao J, Tang Z. Influence of radiation on the blood-brain barrier and optimum time of chemotherapy. *Zhongguo Yi Xue Ke Xue Yuan Xue Bao*. 1999;21(4):307–310.

108. Zawaski JA, Gaber MW, Sabek OM, Wilson CM, Duntsch CD, Merchant TE. Effects of irradiation on brain vasculature using an in situ tumor model. *Int J Radiat Oncol Biol Phys*. 2012;82(3):1075–1082.

109. Diserbo M, Agin A, Lamproglou I, et al. Blood-brain barrier permeability after gamma whole-body irradiation: an in vivo microdialysis study. *Can J Physiol Pharmacol*. 2002;80(7):670–678.

110. Kraemer DF, Fortin D, Neuwelt EA. Chemotherapeutic dose intensification for treatment of malignant brain tumors: recent developments and future directions. *Curr Neurol Neurosci Rep*. 2002;2(3):216–224.

111. Baumann BC, Kao GD, Mahmud A, et al. Enhancing the efficacy of drug-loaded nanocarriers against brain tumors by targeted radiation therapy. *Oncotarget*. 2013;4(1):64–79.

112. Bow H, Hwang LS, Schildhaus N, et al. Local delivery of angiogenesis-inhibitor minocycline combined with radiotherapy and oral temozolomide chemotherapy in 9L glioma. *J Neurosurg*. 2014;120(3):662–669.

113. Barth RF, Yang W, Huo T, et al. Comparison of intracerebral delivery of carboplatin and photon irradiation with an optimized regimen for boron neutron capture therapy of the F98 rat glioma. *Appl Radiat Isot.* 2011;69:1813–1816.

114. Khatri A, Gaber MW, Brundage RC, et al. Effect of radiation on the penetration of irinotecan in rat cerebrospinal fluid. *Cancer Chemother Pharmacol.* 2011;68(3):721–731.

115. Abboud M, Saghir NS, Salame J, Geara FB. Complete response of brain metastases from breast cancer overexpressing Her-2/neu to radiation and concurrent Lapatinib and Capecitabine. *Breast J.* 2010;16(6):644–646.

116. Walbert T, Gilbert MR. The role of chemotherapy in the treatment of patients with brain metastases from solid tumors. *Int J Clin Oncol.* 2009; 14(4):299–306.

117. Zeng YD, Liao H, Qin T, et al. Blood-brain barrier permeability of gefitinib in patients with brain metastases from non-small-cell lung cancer before and during whole brain radiation therapy. *Oncotarget.* 2015;6(10):8366–8376.

118. Jiang J, Wei WH, Feng YL, et al. Application of 99mTc-DTPA in evaluation of blood-brain barrier permeability in patients receiving whole brain irradiation. *Nan Fang Yi Ke Da Xue Xue Bao.* 2010;30(2):329–330.

119. Nakagawa M, Matsumoto K, Higashi H, Furuta T, Ohmoto T. Acute effects of interstitial hyperthermia on normal monkey brain-magnetic resonance imaging appearance and effects on blood-brain barrier. *Neurol Med Chir (Tokyo).* 1994;34:668–675.

120. Witt KA, Gillespie TJ, Huber JD, Egleton RD, Davis TP. Peptide drug modifications to enhance bioavailability and blood-brain barrier permeability. *Peptides.* 2001;22:2329–2343.

121. Ke XY, Zhao BJ, Zhao X, et al. The therapeutic efficacy of conjugated linoleic acid - paclitaxel on glioma in the rat. *Biomaterials.* 2010;31:5855–5864.

122. Linot B, Campone M, Augereau P, et al. Use of liposomal doxorubicin-cyclophosphamide combination in breast cancer patients with brain metastases: A monocentric retrospective study. *J Neurooncol.* 2014;117(2):253–259.

123. Birngruber T, Raml R, Gladdines W, et al. Enhanced doxorubicin delivery to the brain administered through glutathione PEGylated liposomal doxorubicin (2B3–101) as compared with generic Caelyx,((R))/Doxil((R))—a cerebral open flow microperfusion pilot study. *J Pharm Sci.* 2014;103:1945–1948.

124. Papademetriou IT, Porter T. Promising approaches to circumvent the blood-brain barrier: progress, pitfalls and clinical prospects in brain cancer. *Ther Deliv.* 2015;6(8):989–1016.

125. Wang S, Meng Y, Li C, Qian M, Huang R. Receptor-mediated drug delivery systems targeting to glioma. *Nanomaterials (Basel).* 2015;6(1).

126. Cui Y, Xu Q, Chow PK, Wang D, Wang CH. Transferrin-conjugated magnetic silica PLGA nanoparticles loaded with doxorubicin and paclitaxel for brain glioma treatment. *Biomaterials.* 2013;34:8511–8520.

127. Su Z, Xing L, Chen Y, et al. Lactoferrin-modified poly(ethylene glycol)-grafted BSA nanoparticles as a dual-targeting carrier for treating brain gliomas. *Mol Pharm.* 2014;11(6):1823–1834.

128. Hayavi S, Halbert GW. Synthetic low-density lipoprotein, a novel biomimetic lipid supplement for serum-free tissue culture. *Biotechnol Prog.* 2005;21(4):1262–1268.

129. Chen YC, Chiang CF, Chen LF, Liang PC, Hsieh WY, Lin WL. Polymersomes conjugated with des-octanoyl ghrelin and folate as a BBB-penetrating cancer cell-targeting delivery system. *Biomaterials.* 2014;35:4066–4081.

130. Pardridge WM, Kang YS, Buciak JL, Yang J. Human insulin receptor monoclonal antibody undergoes high affinity binding to human brain capillaries in vitro and rapid transcytosis through the blood-brain barrier in vivo in the primate. *Pharm Res.* 1995;12(6):807–816.

131. Liu Y, Li J, Shao K, et al. A leptin derived 30-amino-acid peptide modified pegylated poly-L-lysine dendrigraft for brain targeted gene delivery. *Biomaterials.* 2010;31:5246–5257.

132. Fischer U, Steffens S, Frank S, Rainov NG, Schulze-Osthoff K, Kramm CM. Mechanisms of thymidine kinase/ganciclovir and cytosine deaminase/5-fluorocytosine suicide gene therapy-induced cell death in glioma cells. *Oncogene.* 2005;24:1231–1243.

133. Cloughesy TF, Landolfi J, Hogan DJ, et al. Phase 1 trial of vocimagene amiretrorepvec and 5-fluorocytosine for recurrent high-grade glioma. *Sci Transl Med.* 2016;8. 341ra375.

134. Wollmann G, Ozduman K, van den Pol AN. Oncolytic virus therapy for glioblastoma multiforme: Concepts and candidates. *Cancer J.* 2012;18(1): 69–81.

135. Strasser JF, Fung LK, Eller S, Grossman SA, Saltzman WM. Distribution of 1,3-bis(2-chloroethyl)-1-nitrosourea and tracers in the rabbit brain after interstitial delivery by biodegradable polymer implants. *J Pharmacol Exp Ther.* 1995;275(3):1647–1655.

136. Lonser RR, Sarntinoranont M, Morrison PF, Oldfield EH. Convection-enhanced delivery to the central nervous system. *J Neurosurg.* 2015;122(3): 697–706.

137. Bobo RH, Laske DW, Akbasak A, Morrison PF, Dedrick RL, Oldfield EH. Convection-enhanced delivery of macromolecules in the brain. *Proc Natl Acad Sci U S A.* 1994;91(6):2076–2080.

138. Zhan W, Wang CH. Convection enhanced delivery of chemotherapeutic drugs into brain tumour. *J Control Release.* 2018;271:74–87.

139. Brem H, Gabikian P. Biodegradable polymer implants to treat brain tumors. *J Control Release.* 2001;74:63–67.

140. Brem H, Piantadosi S, Burger PC, et al. Placebo-controlled trial of safety and efficacy of intraoperative controlled delivery by biodegradable polymers of chemotherapy for recurrent gliomas. The Polymer-brain Tumor Treatment Group. *Lancet.* 1995;345:1008–1012.

141. Westphal M, Hilt DC, Bortey E, et al. A phase 3 trial of local chemotherapy with biodegradable carmustine (BCNU) wafers (Gliadel wafers) in patients with primary malignant glioma. *Neuro Oncol.* 2003;5(2):79–88.

142. Scott AW, Tyler BM, Masi BC, et al. Intracranial microcapsule drug delivery device for the treatment of an experimental gliosarcoma model. *Biomaterials.* 2011;32:2532–2539.

143. Upadhyay UM, Tyler B, Patta Y, et al. Intracranial microcapsule chemotherapy delivery for the localized treatment of rodent metastatic breast adenocarcinoma in the brain. *Proc Natl Acad Sci U S A*. 2014;111(45):16071–16076.

144. Masi BC, Tyler BM, Bow H, et al. Intracranial MEMS based temozolomide delivery in a 9L rat gliosarcoma model. *Biomaterials*. 2012;33(23): 5768–5775.

145. Jonas O, Calligaris D, Methuku KR, et al. First in vivo testing of compounds targeting group 3 Medulloblastomas using an implantable microdevice as a new paradigm for drug development. *J Biomed Nanotechnol*. 2016;12(6):1297–1302.

146. Ommaya AK. Subcutaneous reservoir and pump for sterile access to ventricular cerebrospinal fluid. *Lancet*. 1963;2:983–984.

147. Niwinska A, Rudnicka H, Murawska M. Breast cancer leptomeningeal metastasis: propensity of breast cancer subtypes for leptomeninges and the analysis of factors influencing survival. *Med Oncol*. 2013;30(1):408.

148. Shapiro WR, Young DF, Mehta BM. Methotrexate: distribution in cerebrospinal fluid after intravenous, ventricular and lumbar injections. *N Engl J Med*. 1975;293(4):161–166.

149. Glitza IC, Rohlfs M, Guha-Thakurta N, et al. Retrospective review of metastatic melanoma patients with leptomeningeal disease treated with intrathecal interleukin-2. *ESMO Open*. 2018;3(1).

150. Hochmair MJ, Schwab S, Prosch H. Complete remission of intrathecal metastases with lorlatinib therapy in a heavily pretreated ALK-positive lung cancer patient. *Anticancer Drugs*. 2017;28(8):928–930.

151. Zagouri F, Sergentanis TN, Bartsch R, et al. Intrathecal administration of trastuzumab for the treatment of meningeal carcinomatosis in HER2-positive metastatic breast cancer: A systematic review and pooled analysis. *Breast Cancer Res Treat*. 2013;139(1):13–22.

152. Rubenstein JL, Fridlyand J, Abrey L, et al. Phase I study of intraventricular administration of rituximab in patients with recurrent CNS and intraocular lymphoma. *J Clin Oncol*. 2007;25:1350–1356.

153. Villela L, Garcia M, Caballero R, Borbolla-Escoboza JR, Bolanos-Meade J. Rapid complete response using intrathecal rituximab in a patient with leptomeningeal lymphomatosis due to mantle cell lymphoma. *Anticancer Drugs*. 2008;19(9):917–920.

154. Witham TF, Fukui MB, Meltzer CC, Burns R, Kondziolka D, Bozik ME. Survival of patients with high grade glioma treated with intrathecal thio-triethylenephosphoramide for ependymal or leptomeningeal gliomatosis. *Cancer*. 1999;86:1347–1353.

155. Chamberlain MC. Combined-modality treatment of leptomeningeal gliomatosis. *Neurosurgery*. 2003;52(2):324–329 [discussion 330].

156. Dardis C, Milton K, Ashby L, Shapiro W. Leptomeningeal metastases in high-grade adult glioma: Development, diagnosis, management, and outcomes in a series of 34 patients. *Front Neurol*. 2014;5:220.

Chapter 28

The use of PCV chemotherapy in oligodendrogliomas

Michael W. Ruff* and Jan C. Buckner[†]

*Department of Neurology, Department of Medical Oncology, Rochester, MN, United States; [†]Department of Medical Oncology, Rochester, MN, United States

Background

In 2016, the World Health Organization classification of brain tumors, including diffuse gliomas, was redefined to include molecular criteria, often supplanting tumor morphology.[1] Prognosis has been shown to be more closely associated with molecular diagnosis than with morphology; however, grade remains prognostically important.[2] Oligodendrogliomas are defined by the presence of whole-arm co-deletion in chromosomal arms 1p and 19q. The presence of a mutation in IDH is typical of 1p/19q co-deleted tumors, and may precede the 1p/19q co-deletion teleologically.[3]

The treatment paradigm for oligodendroglioma has shifted, owing to new diagnostic criteria and phase III clinical trial evidence. Patients benefit more from chemotherapy (CT) combined with radiation therapy compared to radiation therapy alone. Specifically, treatment with radiation therapy plus CT with procarbazine, lomustine, and vincristine (PCV) results in the prolongation of both progression-free survival (PFS) and overall survival (OS).

Evidence that led to definitive trials

Several trials led to the Radiation Therapy Oncology Group (RTOG) 9802 and 9402, as well as The European Organization for Research and Treatment of Cancer (EORTC) 26,951 phase III clinical trials.[4–6] At the time of initiation of the RTOG 9802 trial, there had been previous descriptions of benefit in recurrent low-grade glioma, and specifically oligodendroglioma, with the use of PCV CT.[7,8] Other combinatorial regimens including carmustine plus interferon, as well as mechlorethamine, vincristine, and procarbazine, showed potential benefit in patients with recurrent low-grade glioma, or those with anaplastic oligodendroglial components.[9,10]

In an early prospective phase II trial by Cairncross et al., patients with newly diagnosed or recurrent anaplastic oligodendroglioma received up to six cycles of PCV, 75% of patients experienced radiologic responses.[8] Based on these responses, it was declared that, "anaplastic oligodendrogliomas are chemosensitive brain cancers which respond predictably, durably, and often completely to PCV..."

As with recurrent and anaplastic oligodendroglioma, the utility of PCV in treating low-grade glioma was demonstrated in a small case series involving nine symptomatic patients with low-grade glioma treated with a standard and intensified formulation of PCV.[11] Eight of these patients were treated with CT at presentation and one at recurrence after radiotherapy (RT) had failed, with none of the patients having received prior CT. All patients improved by clinical or imaging criteria with PCV CT and no patient deteriorated during therapy. Responses were sustained for a median of 35 months (range 22–45 months). Five patients received a standard PCV regimen: CCNU $110 \, mg/m^2$ on day 1; vincristine $1.4 \, mg/m^2$ on days 8 and 29 (maximum daily dose 2 mg); and procarbazine $60 \, mg/m^2$ per day from day 8 to 29 every 8 weeks. Four patients received an intensified PCV regimen: CCNU $130 \, mg/m^2$ on day 1; vincristine $1.4 \, mg/m^2$ on days 8 and 29 (no maximum); and procarbazine $75 \, mg/m^2$ per day from day 8 to 21 every 6 weeks. All patients developed myelosuppression, but only those receiving the intensified regimen required dose reduction or premature discontinuation of treatment. Treatment with standard regimen PCV was well tolerated; however, intensive PCV CT was shown to be significantly more toxic than the standard regimen and was not completed by any patient in this series.

A subsequent phase II trial further demonstrated the promise of PCV CT to treat oligodendroglial tumors.[12] After biopsy or subtotal resection, patients received up to six cycles of PCV, with subsequent radiation therapy (59.4 or 54.0 Gy) within

10 weeks of completing CT or immediately if there was evidence of tumor progression on PCV. In all 8 of 28 (29%) and 13 of 25 (52%) eligible patients demonstrated tumor regression as assessed by the treating physician and a blinded central neuroradiology reviewer, respectively. The authors concluded that PCV produced tumor regressions in a meaningful percentage of patients with low-grade oligodendrogliomas.

Retrospective insights

In a retrospective study examining 16 newly diagnosed patients with oligodendroglioma and oligoastrocytoma, as well as 5 patients with recurrent disease ($n = 7$ with 1p/19q co-deletion) treated with PCV, 3 of 5 patients with recurrent tumors responded, and 13 of 16 newly diagnosed patients responded to treatment. The median time to disease progression was >24 months. Only one patient experienced disease progression while receiving CT. Several patients showed a significant clinical improvement despite only a modest improvement of the tumor on the MRI (magnetic resonance imaging) scans.[13]

In a large retrospective study, published after RTOG 9402 and RTOG 9802 had commenced but prior to their publication, Lassman et al. reported outcomes in 1013 patients (which spanned 1981–2007) with histologically diagnosed anaplastic oligodendroglioma.[14] Patients had received RT alone, CT alone, or a combination as an upfront strategy. The OS for the entire cohort was 6.3 years, and the overall time to progression was 3.1 years. Median follow-up was 5.2 years among 502 surviving patients.

There was substantial crossover of modalities in this retrospective review, as it is in the real-world setting, and the majority of patients who progressed received both CT and RT during their disease course. Efficacy was not recorded for salvage treatments.

Median OS of all patients treated with upfront CT alone was 7.0 years; for patients treated with CT and RT (sequential or concurrent) median OS was 7.3 years. The difference between these two groups was not significant ($p = 0.84$); however, median OS of patients treated with RT alone upfront was 4.4 years, and this was a statistically significant difference in median OS between patients treated with CT or CT and RT ($p = 0.0001$). In patients receiving both CT and RT ($n = 528$), there was no statistically significant difference between patients who received PCV vs temozolomide (TMZ) in either time to progression or OS ($P = 0.26$ for TTP, $P = 0.62$ for OS).

Patients with 1p/19q co-deletion treated with upfront chemo- RT ($n = 133$) had a prolonged time to progression (7.2 years), compared to CT alone ($n = 93$, 3.9 years) and RT alone ($n = 54$, 2.5 years). However, differences in OS were not statistically significant (mOS 8.4 year following CT + RT, 10.5 year following CT alone, 8.7 year following RT alone) (Fig. 1). Of note, 49% of patients with 1p/19q co-deletion who received RT/PCV were alive up to 10 years (95% CI 34–68), compared to 15% (95% CI 1–48) of patients who were treated with RT/TMZ.[14]

In patients treated with upfront CT alone, PCV provided longer disease control than TMZ, with a longer time to progression following treatment with PCV alone ($n = 21$, 7.6 years) vs TMZ alone ($n = 58$, 3.3 years, $P = 0.019$). Notably, the discrepancy between TMZ and PCV in regard to time to progression in patients with co-deletion remained significant when accounting for potential confounders on multivariate analysis, including extent of resection, KPS (Karnofsky Performance Scale), age, and histology. In patients treated with upfront CT alone, median OS trended toward favoring PCV (10.5 vs 7.2 year, $P = 0.16$). Median follow-up for survivors was 7 years among patients receiving PCV and 3.6 years for TMZ. It is possible that there have been too few events to reliably assess a difference in OS because follow-up time was not long enough to demonstrate a clear superiority to one regimen or another.

Of note, in patients with histologically diagnosed anaplastic "oligodendrogliomas" with IDH mutations but without 1p/19q co-deletion (IDH mutant astrocytomas, per 2016 WHO classification) there was a benefit with chemo-RT manifesting as prolonged time to progression, as well as an increased OS when compared to treating with CT or RT. Median time to progression was longer following CT and RT (3.1 year) than CT (0.9 year, $P = 0.0124$) or RT (1.1 year, $P < 0.0001$) alone among these patients.[15]

Definitive trials

In the EORTC 22033 trial, patients with low-grade glioma (WHO grade II, pre-2016 revision) were treated with single-agent RT vs primary TMZ CT in patients with low-grade glioma who were felt to be at high risk (≥ 40 years of age and had radiological tumor progression, worsening neurological symptoms, or refractory seizures).[16] A total of 477 patients were randomized to receive either RT ($n = 240$) or TMZ ($n = 237$). At a median follow-up of 48 months, median PFS was 39 months in the TMZ group and 46 months in the RT group (HR 1.16, 95% CI 0.9–1.5, $p = 0.22$). Median OS has not been reached. There was no significant difference in PFS in patients with IDH mutated co-deleted patients. In IDH mutant astrocytomas, patients treated with RT had a longer PFS than those treated with TMZ. There was no difference in PFS with IDH

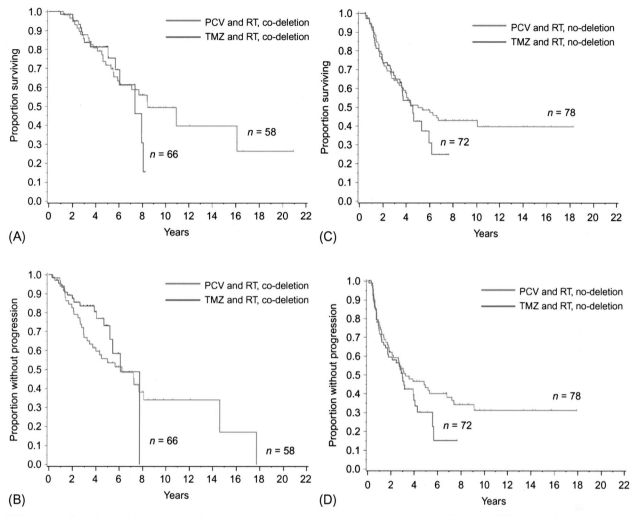

FIG. 1 Overall survival and time to progression by treatment and deletion status. *(Lassman, AB, Neuro-Oncology. 2011 supplemental data).*

wild-type astrocytomas. The practice of many treatment providers who advocate TMZ as an initial therapy following surgery in order to defer radiation treatment and potentially defer radiation-induced cognitive decline does not seem to be warranted based on this data.

The RTOG 9802 trial was a pivotal demonstration of the efficacy of adjuvant CT in the management of low-grade glioma, with the most substantial improvement in PFS demonstrated in 1p/19q co-deleted tumors.[1] This trial was preceded by RTOG 9402 and EORTC 26951, which also demonstrated substantial improvement in OS in grade III 1p/19q co-deleted tumors. The PCV CT in addition to radiation has a clear benefit in patients with oligodendroglioma and IDH mutated astrocytoma.

In the RTOG 9402 study 291 patients with histologically defined anaplastic oligodendroglioma, or histologically defined anaplastic oligoastrocytoma were randomized to PCV plus RT ($n = 148$) vs RT alone ($n = 143$), with the primary end point of OS.[5] For the entire cohort there was no difference in median OS by treatment (4.6 years PCV/RT vs 4.7 RT). However, in those patients with 1p/19q co-deleted tumors (molecular oligodendrogliomas) treatment with RT plus PCV doubled median OS (14.7 vs 7.3 years, HR 0.59, with 95% CI 0.37–0.95, $P = 0.03$). The IDH mutation was balanced between the two arms, but not explicitly reported in the 2006 publication of the RTOG 9402 study. For those patients with non-co-deleted tumors, there was no difference in median survival by treatment arm (2.6 vs 2.7 years, HR 0.85, 95% CI 0.58–1.23; $P = 0.39$). The authors concluded that for the subset of patients with 1p/19q co-deleted anaplastic oligodendroglioma, and anaplastic oligoastrocytoma, PCV plus RT may be an especially effective treatment, though the observation was derived from an unplanned analysis for which it was underpowered. Additionally the authors noted that the histologic criteria were not reliable in determining which patients would harbor the 1p/19q co-deletion, as 29% of anaplastic oligodendrogliomas had preservation of 1p and 19q, whereas 24% of anaplastic oligoastrocytomas did harbor 1p/19q co-deletion. Additionally, the doubling of survival in the subset of patients with 1p/19q co-deletion was not detectable in 2006, when the median survival was 5 years. The implication that combinatorial chemo-RT was superior to RT emerged only with mature follow-up.

In 2014, further analysis of the RTOG 9402 patient data demonstrated that in 210 (of the 291 on study) patients who were able to be evaluated for IDH mutation status, 154 (74%) had IDH mutations.[15] In patients with IDH wild-type tumors, combinatorial therapy did not prolong median survival (1.3 vs 1.8 years, HR 1.14, 95% CI 0.63–2.04, $p = 0.67$) or 10-year survival (6% in the combinatorial group, 4% in the RT only group), however, patients without co-deletion but with IDH mutation did live longer after combinatorial therapy than with RT alone (5.5 vs 3.3 years, HR 0.56, 95% CI 0.32–0.99, $p < 0.5$). The authors concluded that 1p/19q co-deletion and IDH mutational status identified patients with histologically diagnosed oligodendroglial tumors who benefited from combinatorial chemo-RT, as demonstrated in Fig. 2.

EORTC 26951, a phase III clinical trial in patients with newly diagnosed anaplastic oligodendroglioma, randomized 368 patients to receive either radiation therapy alone (59.4 Gy) or the same RT followed by six cycles of adjuvant PCV.[6] Results demonstrated a significantly longer OS in the RT/PCV arm vs RT-alone arm (42.3 vs 30.6 months, HR 0.75; 95% CI, 0.60–0.95). In 80 patients with a 1p/19q co-deletion, OS was increased, with a trend toward more benefit from adjuvant PCV (OS not reached in the RT/PCV group vs 112 months in the RT group; HR, 0.56; 95% CI, 0.31–1.03). The authors concluded that the addition of six cycles of PCV after 59.4 Gy of RT increases both OS in anaplastic oligodendroglial tumors and that 1p/19q co-deleted tumors derived more benefit from adjuvant PCV compared with non-1p/19q co-deleted tumors. IDH mutation was also prognostically significant.

As noted, oligodendrogliomas had been demonstrated to respond exquisitely to PCV, with the response rate to PCV as an initial therapy ranging from 52% to 100%.[11–13] RTOG 9802 was a phase III trial in adults with "high-risk disease" defined as those patients >40 years old, or those with less than a GTR (gross total resection) (as estimated by the treating neurosurgeon), regardless of age. Patients were randomized and received either radiation therapy alone or radiation plus PCV CT. The radiation dose was 54 Gy in 30 fractions. Patients assigned to receive CT were treated with six cycles of procarbazine (60 mg/m^2 orally per day on days 8 through 21 of each cycle), lomustine (110 mg/m^2 orally on day 1 of each cycle), and vincristine [1.4 mg/m^2 (maximum 2 mg)] intravenously on days 8 and 29 of each cycle. The cycle length was 8 weeks. Median PFS was 10.4 vs 4.0 years for chemoradiation vs for radiation therapy alone, with a 10-year PFS of 21% and 51%, respectively. The OS was also superior in the chemoradiation group. Median survival was 7.8 years vs 13.3 years for radiation therapy alone versus chemoradiation, and 10-year survival was 40% vs 60%, respectively. The hazard ratios in favor of chemoradiation were 0.43 for patients with oligodendroglioma and 0.42 for IDH mutant tumors, respectively.

In the RTOG 9802 study, 97 patients had sufficient DNA for profiling. Of these, 33 (34%) were *IDH* mutant 1p/19q co-deleted patients. This subgroup had superior PFS with PCV + RT than those treated with RT alone [oligodendroglioma (molecular) HR = 0.16, 95% CI 0.05–0.058, $p = 0.002$].[13a]

The OS was not significantly different among this group, owing to the lack of patient deaths (events). The OS benefit may take even more time to be born out due to such long survival within the two groups. The IDH mutant astrocytomas (IDH mutant without 1p/19q co-deletion) were significantly correlated with better PFS and OS with the addition of PCV. There was no significant difference observed with the addition of PCV in the IDH non-mutant subgroup for either OS or PFS.

The RTOG 9802 trial was a pivotal demonstration of the efficacy of adjuvant CT in the management of low-grade glioma, with the most substantial improvement in PFS demonstrated in 1p/19q co-deleted tumors.[1] This trial was preceded by RTOG 9402 and EORTC 26951 which also demonstrated substantial improvement in OS in grade III 1p/19q co-deleted tumors, in essence doubling OS.

The PCV CT in addition to radiation has a clear benefit in patients with oligodendroglioma and IDH mutated astrocytoma. The PCV CT as an adjuvant therapy has clear benefit in grade II and III oligodendroglioma. There is also benefit in patients with IDH mutated astrocytoma. The benefit of adjuvant PCV CT is less clear in patients with non-IDH mutated astrocytoma. Whether TMZ has a similar benefit in oligodendroglioma is less convincing from the available data, however, an ongoing trial, CODEL, will further explore this.

Toxicity management per the RTOG 9802 trial[4]:

Hematologic toxicity: The dose of CCNU and procarbazine were reduced based on blood count nadir of the previous cycle: If the absolute granulocyte nadir is <0.5 × x 10^9 (< 500) or the platelet count nadir is <50 × 10^9 (<50,000), then the dose should be reduced by 25% (of the previous cycle's dose).

Neurotoxicity: Vincristine was stopped for grade 3 or grade 4 neurosensory or neuromotor toxicity. CCNU and procarbazine continue as per protocol. For severe abdominal or jaw pain, reduce vincristine dose by 50% on all subsequent doses.

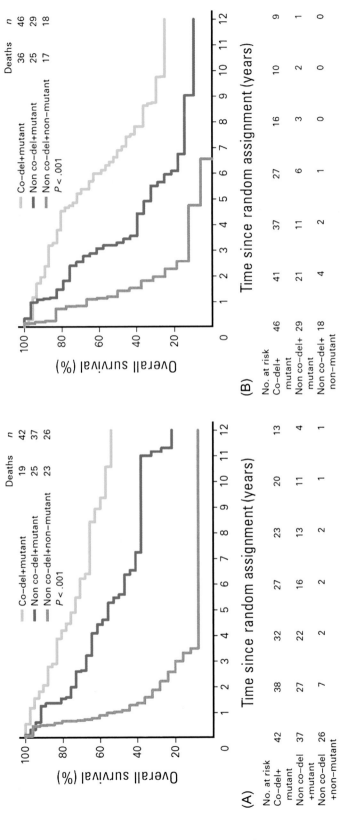

FIG. 2 Survival based on mutational status and combinatorial therapy vs RT alone in the RTOG 9402 trial.

Nausea/vomiting: If grade 3 or grade 4 nausea or vomiting persist despite antiemetics, the dose of CCNU and procarbazine may be reduced (as per other toxicity, below).

Skin Toxicity: Procarbazine will be discontinued should any urticarial rash develop from PCV. CCNU and vincristine will continue per protocol.

Pulmonary toxicity: CCNU will be stopped if cough, shortness of breath, or other pulmonary symptoms develop and if the DLCO is <60% of predicted. Procarbazine and vincristine will continue per protocol.

Hepatic toxicity: Hold all drugs for AST (aspartate aminotransferase also referred to as aspartate transaminase) or ALT (alanine aminotransferase also referred to as alanine transaminase) >3 × normal and resume with 25% dose reduction when AST or ALT is ≤2 × normal.

Other toxicity: Doses will be reduced by 25% for grade 3 toxicity, and 50% for grade 4 toxicity. The PCV may be discontinued for any grade 4 toxicity.

Further discussion: Treatment of recurrent oligodendroglioma: Temozolomide or PCV? Either?

In a prospective phase II study, 43 patients with LGG (29 with histologic astrocytoma, 4 with oligodendroglioma, and 10 with oligoastrocytoma) were treated with TMZ at the time of progression.[17] Of these, 30 patients had previously received RT, and 16 patients had received prior CT. A total of 4 patients had a complete response (9%), while 16 had a partial response (37%). In all 17 patients had stabilization of disease (39.5%), and 6 patients continued to have disease progression in spite of treatment (13%). Median response duration was 10 months, with a 76% rate of PFS at 6 months, and a 39% rate of PFS at 12 months.

In a separate phase II trial of TMZ in 46 patients with progressive low-grade glioma, an objective response rate occurred in 61% (24% CR and 37% PR), with 35% of patients demonstrating stable disease.[18] Median PFS was 22 months, with a 6-month PFS of 98%, and a 12-month PFS of 76%.

In the NOA-04 trial, patients with newly diagnosed anaplastic gliomas were randomized to initial RT followed by CT (PCV or TMZ) at progression or intolerance of unacceptable toxicity, or the inverse sequence with initial CT followed by radiation therapy at progression.[19] The study involved 274 patients in the intention to treat analysis. There was no statistically significant difference in activity of primary CT (PCV or TMZ) vs RT in any subgroup of anaplastic glioma, with time to treatment failure as defined by the study; however, there was a statistically significant difference in PFS with patients treated with TMZ having a shorter PFS (4.46 year) than PCV (9.4 year) and RT (8.67 year) ($p = 0.02$).

The RTOG 9802, RTOG 9402, and the EORTC 26951 studies demonstrated superiority of chemo-RT over RT alone, whereas the result of NOA-04 did not show a difference between initial monotherapy with either CT (TMZ or PCV) or RT, and did not support the use of primary CT as monotherapy in patients with oligodendroglioma. Patients with molecular oligodendrogliomas (defined in NOA-O4 as CIMP, 1p/19q co-deleted patients) treated with TMZ had a worse outcome, reaching median OS (8.09 years), while the PCV as initial monotherapy group, and similarly the RT as initial monotherapy group, did not reach median OS with a median overall follow-up of 9.5 years. NOA-04 demonstrated that primary monotherapy with CT is not superior to primary RT, however, the authors concluded that this trial does support the use of primary chemo-RT with PCV.[19]

Prior to the publication of the RTOG 9802, RTOG 9402, and EORTC 26951 studies, TMZ had largely replaced the use of PCV in the treatment of gliomas; however, PCV was demonstrated retrospectively to likely be superior in this tumor group (molecular oligodendroglioma).[14] Additionally, in a case series of patients with low-grade oligodendroglioma, Johnson et al. performed genome sequence analysis of initial and recurrent human gliomas, with and without exposure to TMZ (notably there was no PCV arm).[20] Tumors treated with TMZ were hyper-mutated at recurrence in a greater proportion of patients who were treated with TMZ alone than of those who received no interval treatment. The resulting genetic alterations resemble those found in glioblastoma (disruption of the retinoblastoma-associated protein tumor suppressor pathway and activation of the Akt–mTOR pathway), raising the possibility that TMZ monotherapy could accelerate tumor transformation to a more aggressive phenotype.

Ongoing studies

The "co-deleted Tumors" (CODEL) trial randomizes patients with oligodendroglioma (1p/19q co-deleted tumors only, both grade II and grade III) to RT followed by PCV or to RT plus TMZ followed by adjuvant TMZ, to determine if PCV has superior disease control when compared to TMZ.[21,22] Additionally, this trial will compare quality of life and neurologic function following initial treatment regimens. Importantly, tissue collection will permit detailed genomic analyses to determine variability in outcomes based on differences in genetic alterations. Of note, in the original trial design, there was a TMZ-alone arm; however; randomization was halted by the Data Safety Monitoring Committee owing to more frequent tumor progression in the TMZ-alone arm ($n = 6/12$; 50%) in comparison to the radiation therapy arms ($n = 2/24$; 8%; $P = 0.002$), after a median follow-up of 3.4 years. Additionally, 3 of 12 patients who progressed on TMZ died, whereas 1 of 2 patients in the RT-alone arm died of disease progression. These data are consistent with the results of the NOA-04 and EORTC 22033 trials, demonstrating the inferiority of CT as initial therapy.

Deferred treatment

Owing to the indolent growth rate of low-grade gliomas, and the lack of symptoms in many (seizure free) patients balanced against the morbidity of surgery, radiation, and CT, and the evidence that early radiation therapy does not alter life expectancy, as it was shown in EORTC 22845, a trial to gauge the benefit of the "wait and see" approach, delayed RT at progression vs upfront early RT.[23] Postoperatively patients with low-grade astrocytoma, oligodendroglioma, mixed oligoastrocytoma, and incompletely resected pilocytic astrocytoma were randomized to early RT (54 Gy in fractions of 1.8 Gy) or deferred RT until the time of progression. A total of 157 patients were assigned to early RT, and 157 to RT at progression. Median PFS was 5.3 years in the early RT group and 3.4 years in the control group (HR 0.59, 95% CI 0.45–0.77; $p < 0.0001$). The OS was similar between groups: median survival in the RT group was 7.4 years compared with 7.2 years in the control group (HR 0.97, 95% CI 0.71–1.34; $p = 0.872$). It was concluded that RT could be deferred for patients with low-grade glioma who are in a good condition, provided they are carefully monitored. Accordingly, patients are often counseled by their treating oncologists to defer any additional therapy or surgery until symptoms become bothersome or MRI demonstrates that the tumor is growing and will cause impending neurologic injury in the future. The EORTC 22845 data and the deleterious late neurologic effects of RT upon cognition have led many physicians to recommend delaying treatment until there is radiographic or clinical worsening.[24–28] In EORTC 22033, patients did not begin treatment until they had become symptomatic.[16] Median time from initial diagnosis to treatment, with either radiation therapy or CT, was only about 6 months, and the maximum time to treatment was only about 3 years. Perhaps the data reflect a pattern of practice in which lesions thought to represent oligodendroglioma or indolent astrocytoma are not biopsied at the time of initial detection by MRI, but only when the clinician is intending to begin treatment. It could also suggest that the majority of patients do develop symptoms warranting treatment rather quickly after histologic confirmation of disease, and that a minority of patients can be watched for a meaningful period of time without treatment intervention.

Nevertheless, a proportion of patients may be able avoid the adverse effects of therapy for many months to years without compromising the expectations for their survival. Some patients have such indolent disease that immediate treatment with chemoradiation is not necessary; however, prospectively identifying these patients has previously remained challenging. Given the recent reclassification of oligodendroglioma in the WHO 2016 reclassification, and the subsequent analysis of the RTOG 9802 data, one could infer that (molecular) oligodendrogliomas may fall into this group more frequently than other tumor types.

The RTOG 9802 and EORTC 22033 trials used tumor size, tumor histology, and patient age to assess the risk of tumor recurrence and death. These risk factors are now supplanted by molecular diagnoses. Specifically, patients with IDH mutations, with or without the 1p/19q co-deletion, have a median life expectancy that exceeds a decade. Counterintuitively, in RTOG 9802, patients with the best prognosis had the most notable prolongation of life with combinatorial treatment (oligodendroglioma: HR = 0.15; IDH1-Arg132His mutations: HR = 0.42). Median survival of patients with oligodendroglioma was 10.8 years with radiation treatment alone, and was not reached (>13.1 years) during the follow-up of patients who received chemoradiation; 10-year survival rates were 57% and 79%, respectively. Median survival of patients with IDH1-Arg132His mutations was 10.1 years with radiation treatment alone and was not reached (>11.3 years) during the follow-up of patients who received chemoradiation; 10-year survival rates were 53% and 75%, respectively.

Deferral of radiation therapy could logically be posited to preserve cognitive function and maintain the known prolongation of life with chemoradiation, though no prospective clinical trial data currently confirm this hypothesis.

No prospective data are currently available regarding observation vs immediate chemoradiation for patients under 40 years of age who have undergone gross total tumor resection. Consequently, treatment recommendations for these patients must be individualized on the basis of the known risks of chemoradiation vs the potential risks of delayed therapy, such as the greater toxicity associated with larger volumes of brain being included in future radiation fields, or development of interim mutational events that might render the tumor more resistant to chemoradiation in the future (e.g., TMZ).[28]

Future directions

Questions remain as to why oligodendrogliomas, now defined by 1p/19q co-deletion typically with IDH mutation, are chemosensitive. An interesting theory as to the chemosensitivity incurred by the 1p/19q co-deletion was recently published.[29] The answer may be the reduced formation of tumor microtubules (TM), as 1p/19q co-deletion has been shown to be inversely correlated with TM formation. Astrocytoma cells have been demonstrated to extend ultra-long membrane protrusions termed TM, and use them as routes for brain invasion, proliferation, and to interconnect over long distances. These tumors via TM may act as a functional syncytium to resist CT and radiation, ship materials to adjacent cells, and potentially communicate. The TM-connected astrocytoma cells in a mouse model have demonstrated resistance to RT and CT mediated cell death by buffering calcium, and resupplying damaged cells with organelles such as mitochondria and nuclei, whereas 1p/19q co-deleted tumor did not demonstrate these abilities, and when TM were present in 1p/19q co-deleted tumors they were less robust and shorter.

The IDH was recently characterized as leading to loss of gene insulator function and resulting in gliomagenesis. The IDH mutations disrupt how the genome folds, bringing together disparate genes, and disrupt structural regulatory controls to spur cancer growth.[30] How this results in increased tumor sensitivity to adjuvant treatment remains unclear.

As we sort out the appropriate timing of initiating chemoradiation, it is important to consider strategies to reduce radiation-related toxicity, such as reducing dose of radiation therapy. Prior studies have demonstrated that 65 Gy is not superior to 50 Gy and 60 Gy is not superior to 45 Gy.[31,32] Treatment plans suggest that there is less normal white matter exposure to radiation with protons compared with IMRT (intensity-modulated radiation therapy). In NRG-005 (Proton Beam or IMRT in Preserving Brain Function in Patients with IDH Mutant Grade II or III Glioma), patients with grade II or III IDH mutated tumors were randomized to either proton beam or photon beam radiation.[33] The primary end point of the trial is time to cognitive decline as measured by specific neuropsychometric tests.

Conclusion

Clinical trials, including RTOG 9402, EORTC 26951, and RTOG 9802, in contrast to NOA-04 and EORTC 22033, have conclusively demonstrated that chemoradiation is superior to either radiation therapy alone or CT alone for patients with oligodendroglioma. Regardless of grade, PCV CT, in addition to radiation, has a clear benefit in patients with oligodendroglioma and IDH mutated astrocytoma. TMZ may be less effective in the treatment of oligodendroglioma, but clearly has a benefit in astrocytoma. The superiority of PCV or TMZ in the treatment of co-deleted glioma will be addressed in the CODEL trial. Although survival time can exceed 15 years in many patients, the majority of patients eventually develop tumor recurrence, and die prematurely. We clearly need to explore better initial therapies that may ultimately lead to cure, improve outcomes through novel therapies at recurrence, and reduce toxicity of both radiation therapy and CT so that patients not only live longer, but better as well.

References

1. Louis D, Perry A, Reifenberger G, et al. The 2016 World Health Organization classification of tumors of the central nervous system: a summary. *Acta Neuropathol.* 2016;131(6):803–820.
2. Eckel-Passow JE, Lachance DH, Molinaro AM, et al. Glioma groups based on 1p/19q, IDH, and TERT promoter mutations in tumors. *N Engl J Med.* 2015;372(26):2499–2508.
3. Cahill DP, Louis DN, Cairncross JG. Molecular background of oligodendroglioma: 1p/19q, IDH, TERT, CIC and FUBP1. CNS Oncol 2015; 4(5):287–94.
4. Buckner JC, Shaw EG, Pugh SL, et al. Radiation plus Procarbazine, CCNU, and vincristine in low-grade glioma. *N Engl J Med.* 2016;374(14):1344–1355.
5. Cairncross G, Berkey B, Shaw E, et al. Phase III trial of chemotherapy plus radiotherapy compared with radiotherapy alone for pure and mixed anaplastic oligodendroglioma: Intergroup radiation therapy oncology group trial 9402. *J Clin Oncol.* 2006;24:2707–2714.
6. van den Bent MJ, Brandes AA, Taphoorn MJ, et al. Adjuvant procarbazine, lomustine, and vincristine chemotherapy in newly diagnosed anaplastic oligodendroglioma: Long-term follow-up of EORTC brain tumor group study 26951. *J Clin Oncol.* 2013;31(3):344–350.
7. Levin VA, Edwards MS, Wright DC, et al. Modified procarbazine, CCNU, and vincristine (PCV 3) combination chemotherapy in the treatment of malignant brain tumors. *Cancer Treat Rep.* 1980;64:237–244.
8. Cairncross G, Macdonald D, Ludwin S, et al. Chemotherapy for anaplastic oligodendroglioma. *J Clin Oncol.* 1994;12:2013–2021.
9. Buckner JC, Brown LD, Kugler JW, et al. Phase II evaluation of recombinant interferon alpha and BCNU in recurrent glioma. *J Neurosurg.* 1995;82:430–435.
10. Galanis E, Buckner JC, Burch PA, et al. Phase II trial of nitrogen mustard, vincristine, and procarbazine in patients with recurrent glioma: North central Cancer treatment group results. *J Clin Oncol.* 1998;16:2953–2958.

11. Mason WP, Krol GS, DeAngelis LM. Low-grade oligodendroglioma responds to chemotherapy. *Neurology*. 1996;46(1):203–207.

12. Buckner JC, Gesme Jr D, O'Fallon JR, et al. Phase II trial of procarbazine, lomustine, and vincristine as initial therapy for patients with low-grade oligodendroglioma or oligoastrocytoma: efficacy and associations with chromosomal abnormalities. *J Clin Oncol*. 2003;21:251–255.

13. Stege EM, Kros JM, de Bruin HG, et al. Successful treatment of low-grade oligodendroglial tumors with a chemotherapy regimen of procarbazine, lomustine, and vincristine. Cancer2005;103:802–809.

13a. Bell EH, Zhang P, Shaw EG, Buckner JC, Barger GR, Coons SW, Bullard DE, Mehta MP, Gilbert MR, Brown PD, Stelzer KJ, Fleming J, McElroy JP, Timmers CD, Becker AP, Salavaggione AL, Liu Z, Aldape K, Brachman DG, Gertler SZ, Murtha AD, Schultz CJ, Johnson D, Shu H-K, Chakravarti A. ACTR-37. Predictive significance of *IDH1/2* mutation and 1p/19q co-deletion status in a post-hoc analysis of NRG oncology/RTOG 9802: a phase III trial of RT vs RT + PCV in high risk low-grade gliomas. *Neuro-Oncol*. 2017;19(suppl_6):vi8. https://doi.org/10.1093/neuonc/nox168.028.

14. Lassman AB, Iwamoto FM, Cloughesy TF, et al. International retrospective study of over 1000 adults with anaplastic oligodendroglial tumors. *Neuro Oncol*. 2011;13(6):649–659.

15. Cairncross JG, Wang M, Jenkins RB, et al. Benefit from procarbazine, lomustine, and vincristine in oligodendroglial tumors is associated with mutation of IDH. *J Clin Oncol*. 2014;32(8):783–790.

16. Baumert BG, Hegi ME, van den Bent MJ, et al. Temozolomide chemotherapy versus radiotherapy in high-risk low-grade glioma (EORTC 22033-26033): a randomised, open-label, phase 3 intergroup study. *Lancet Oncol*. 2016;17(11):1521–1532.

17. Pace A, Vidiri A, Galiè E, et al. Temozolomide chemotherapy for progressive low-grade glioma: Clinical benefits and radiological response. *Ann Oncol*. 2003;14(12):1722–1726.

18. Quinn JA, Reardon DA, Friedman AH, et al. Phase II trial of temozolomide in patients with progressive low-grade glioma. *J Clin Oncol*. 2003;21(4):646–651.

19. Wick W, Roth P, Hartmann C, et al. Long-term analysis of the NOA-04 randomized phase III trial of sequential radiochemotherapy of anaplastic glioma with PCV or temozolomide. *Neuro Oncol*. 2016;18(11):1529–1537.

20. Johnson BE, Mazor T, Hong C, et al. Mutational analysis reveals the origin and therapy-driven evolution of recurrent glioma. *Science*. 2014;343(6167):189–193.

21. Jaeckle K, et al. ATCT-16CODEL (ALLIANCE-N0577; EORTC-26081/2208; NRG-1071; NCIC-CEC-2): phase III randomized study of RT versus RT + TMZ versus TMZ for newly diagnosed 1p/19q-codeleted anaplastic glioma. Analysis of patients treated on the original protocol design. *Neuro Oncol*. 2015;17(suppl 5):v4–v5.

22. Clinicaltrials.Gov, Radiation Therapy With Concomitant And Adjuvant Temozolomide Versus Radiation Therapy With Adjuvant PCV Chemotherapy In Patients With Anaplastic Glioma Or Low Grade Glioma, Full Text View, Clinicaltrials.Gov." Clinicaltrials.gov. N. p., 2017. Web. 24 Oct. 2017.

23. van den Bent MJ, Afra D, de Witte O, et al. Long-term efficacy of early versus delayed radiotherapy for low-grade astrocytoma and oligodendroglioma in adults: the EORTC 22845 randomised trial. *Lancet*. 2005;366(9490):985–990.

24. Peterson H, DeAngelis LM. Weighing the benefits and risk of radiation therapy for low-grade glioma. *Neurology*. 2001;56:1255–1256.

25. Postma TJ, Klein M, Verstappen CCP, et al. Radiotherapy induced cerebral abnormalities in patients with low-grade glioma. *Neurology*. 2002;59:121–123.

26. Surma-aho O, Niemela M, Vikki J, et al. Adverse long-term effects of brain radiotherapy in adult low-grade glioma patients. *Neurology*. 2001;56:1285–1290.

27. Packer RJ, Metha M. Neurocognitive sequelae of cancer treatment. *Neurology*. 2002;59:8–10.

28. Buckner J, Giannini C, Eckel-Passow J, et al. Management of diffuse low-grade gliomas in adults - use of molecular diagnostics. *Nat Rev Neurol*. 2017;13(6):340–351.

29. Osswald M, Jung E, Sahm F, et al. Brain tumor cells interconnect to a functional and resistant network. *Nature*. 2015;528(7580):93–98.

30. Flavahan WA, Drier Y, Liau BB, et al. Insulator dysfunction and oncogene activation in IDH mutant gliomas. *Nature*. 2016;529(7584):110–114.

31. Shaw E, Arusell R, Scheithauer B, et al. Prospective randomized trial of low- versus high-dose radiation therapy in adults with supratentorial low-grade glioma: Initial report of a north central Cancer treatment group/radiation therapy oncology group/eastern cooperative oncology group study. *J Clin Oncol*. 2002;20(9):2267–2276.

32. Karim AB, Maat B, Hatlevoll R, et al. A randomized trial on dose-response in radiation therapy of low-grade cerebral glioma: European Organization for Research and Treatment of Cancer (EORTC) study 22844. *Int J Radiat Oncol Biol Phys*. 1996;36(3):549–556.

33. Clinicaltrials.Gov, Proton Beam or Intensity-Modulated Radiation Therapy in Preserving Brain Function in Patients with IDH Mutant Grade II or III Glioma—Full Text View—Clinicaltrials.Gov. [Clinicaltrials.gov. N. p., 2018. Web. 23 Feb. 2018].

Chapter 29

Temozolomide chemotherapy for oligodendroglial tumors

Herbert B. Newton[*,†,‡] and Nina A. Paleologos[§,¶]

[*]Neuro-Oncology Center, AdventHealth Cancer Institute, AdventHealth Medical Group, Orlando, FL, United States; [†]CNS Oncology Program, AdventHealth Cancer Institute, AdventHealth Medical Group, Orlando, FL, United States; [‡]Division of Neuro-Oncology, Esther Dardinger Endowed Chair in Neuro-Oncology James Cancer Hospital & Solove Research Institute, Wexner Medical Center at the Ohio State University, Columbus, OH, United States; [§]Advocate Medical Group, Advocate Healthcare, Chicago, IL, United States; [¶]Rush University Medical School, Chicago, IL, United States

Introduction

Oligodendrogliomas are an uncommon form of diffuse glioma that can be of pure or mixed histology, and are classified as World Health Organization (WHO) grade II or III.[1–4] Molecular phenotyping is also necessary nowadays to finalize the diagnosis of an oligodendroglioma–with the requirement for 1p/19q co-deletion to be present.[4] They typically occur in young to middle-aged adults (peak age 35–45 years) with a history of seizures, within the white matter of the frontal and temporal lobes.

In the late 1980s, it was demonstrated by Cairncross and Macdonald that oligodendrogliomas were chemosensitive tumors, and that they were especially responsive to the combination chemotherapy regimen of procarbazine, 1-(2-chloroethyl)-3-cyclohexyl-1-nitrosourea (CCNU), and vincristine (PCV).[5] For the next 12–15 years, PCV became the mainstay of treatment for recurrent and progressive low-grade and anaplastic oligodendrogliomas (see Chapter 28).[2, 3, 6] However, during and after the FDA approval process of temozolomide (TMZ) for recurrent and newly diagnosed glioblastoma multiforme (GBM; finalized in 2005), clinicians began to apply TMZ to recurrent and progressive oligodendroglial tumors as well (Table 1).

TMZ is an imidazotetrazine derivative of the alkylating agent dacarbazine with activity against systemic and CNS malignancies.[7–9] The drug undergoes chemical conversion at physiological pH to the active species 5-(3-methyl 1-triazeno) imidazole-4-carboxamide (MTIC) (see Fig. 1). TMZ exhibits schedule-dependent antineoplastic activity by interfering with DNA replication through the process of methylation. The methylation of DNA is dependent on the formation of a reactive methyldiazonium cation, which interacts with DNA at the following sites: N^7-guanine (70%), N^3-adenine (9.2%), and O^6-guanine (5%).

The cytotoxicity of TMZ can be modulated by the degree of activity of three DNA-repair enzyme systems: DNA mismatch repair, O^6-alkylguanine-DNA alkyltransferase (AGT), and poly (ADP-ribose) polymerase.[7, 9–12] AGT, also known as O^6-methylguanine DNA methyltransferase (MGMT), is a protein encoded by the O^6-methylguanine DNA methyltransferase (*MGMT*) gene.[13] MGMT repairs the naturally occurring mutagenic DNA lesion O^6-methylguanine back to guanine and in doing so prevents mismatch and errors during DNA replication and transcription.

DNA-mismatch repair pathways must be at a normal functional capacity to confer TMZ cytotoxicity. The mechanism may involve initiation of apoptosis in those cells that cannot repair the methylated sites.[14] Tumor cells with mutations in, or phenotypically low expression of, DNA-mismatch repair genes are more resistant to TMZ and other methylating agents. Conversely, high expression of MGMT confers resistance to TMZ by removing methyl groups from DNA before cell injury and death can occur.[15]

TMZ for low-grade oligodendroglial tumors

TMZ was first used for treatment of newly diagnosed and progressive low-grade oligodendrogliomas in the late 1990s and early 2000s. The earliest studies used TMZ for low-grade gliomas, including grade II oligodendrogliomas, as primary treatment after surgical biopsy or resection.[16, 17] At the time of the initial presentation of this data in 2000 at a

TABLE 1 Summary of TMZ regimens and results used for low-grade and anaplastic oligodendrogliomas

Author/Ref	N	TMZ Dose	CR/PR/MR	PFS	OS
Low-grade Oligodendroglioma					
Brada/17	13	200 mg/m^2/day	2/0/3	PFS-2, 76%	OS-2, 87%
Quinn/18	25	200	6/8/0	mPFS, 22 mo	N/A
Pace/19	14	150-200	2/4/0	N/A	N/A
van den Bent/21	30	150-200	3/6/0	PFS-6, 44%	N/A
Costanza/22	25	150-200	1/6/5	mPFS, 6 mo	N/A
van den Bent/23	39	150-200	10/10/0	PFS-6, 71%	N/A
van den Bent/24	32	150	7/0/0	PFS-6, 29%	mOS 12.3 mo
Hoang-Xuan/25	25	200	10/0/8	PFS-12, 73.4%	N/A
Levin/26	28	200	0/10/7	PFS-12, 89%	N/A
Kesari/27	38	75	0/9/0	N/A	mOS > 72 mo
		7 wks on/4 wks off			
Wahl/28	77	200	0/7/0	N/A	mOS 10.8 yrs
Baumert/29	310	75	N/A	mPFS 39 mo	N/A
		3 wks on/1 wk off			
Anaplastic Oligodendroglioma					
Chinot/31	48	150-200	8/13/0	mPFS 6.7 mo	mOS 10 mo
Gwak/32	72	150-200	8/9/0	mPFS 8 mo	OS-2, 63%
Taliansky-Aronov/33	20	200	0/15/0	mPFS 24 mo	N/A
Gan/35	40	200	15/6/0	mPFS 21 mo	mOS 43 mo
Ahluwalia/36	62	150	2/0/0	mPFS 27.2 mo	mOS 105.8 mo
		7 days on/7 days off			
Vogelbaum/37	39	150	2/8/0	N/A	N/A
		7 days on/7 days off; chemo-RT			
Thomas/38	41	200	N/A	mPFS 62.8 mo	OS-5, 93.4%
		TMZ induction		1p/19q co-deleted	
		High-dose chemo with			
		Thiotepa/Busulfan, ASCT			
Ducray/39	44	150-200	0/13/0	mPFS 6.9 mo	mOS 12.1 mo

Abbreviations: *PFS-2*, PFS rate at 2 years; *PFS-5*, PFS rate at 5 years; *mPFS*, median PFS; *OS-2*, OS rate at 2 years; *mo*, months; *PFS-6*, PFS rate at 6 months; *N/A*, not available; *mOS*, median OS; *yrs*, years; *wks*, weeks; *ASCT*, autologous stem cell transplant.

Neuro-Oncology meeting, there were a total of 25 patients in the cohort.[16] When the final manuscript was published by Brada and coworkers, a total of 30 patients had been entered into the phase II trial.[17] All patients had surgically verified WHO grade II gliomas, including 17 astrocytomas, 11 oligodendrogliomas, and 2 oligoastrocytomas; 29 patients were eligible for response. None of the patients had received prior treatment with radiotherapy (RT) or chemotherapy. Treatment consisted of TMZ 200 mg/m^2/day orally for 5 days every 28 days; a maximum of 12 cycles was planned per the protocol. The primary endpoint of the trial was the response rate during follow-up imaging with magnetic resonance imaging (MRI). The median age for the cohort was 40 years, with a range of 25–68 years; 17 patients were male and 13 were female. A total

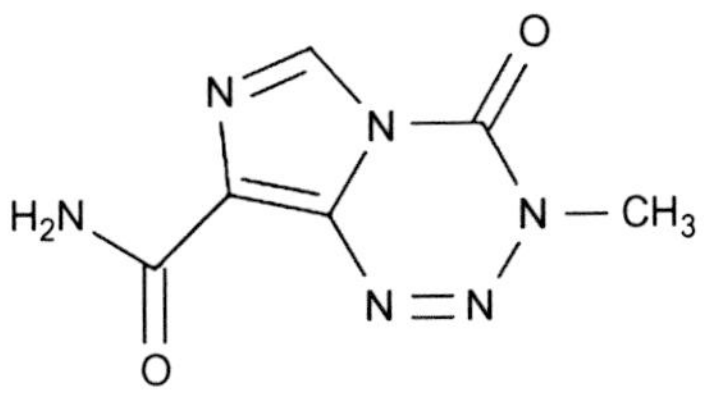

Temozolomide

FIG. 1   Chemical structure of Temozolomide.

of 324 cycles of TMZ were administered; 24 of the 29 evaluable patients completed 12 months of treatment. The overall imaging responses included 3 partial responses (PR), 14 minor responses (MR), 11 with stable disease (SD), and 1 with progressive disease (PD). For the oligodendroglioma subgroup, there were 2 PR (20%), 3 MR (30%), and 5 with SD (50%). Nine patients had PD while on study; 3 while on chemotherapy and 6 after the completion of chemotherapy. The 2-year progression-free survival (PFS) was 76%, with a 3-year PFS of 66%. The 2-year and 3-year actuarial survival rates were 87% and 82%, respectively. In the subgroup of patients with active seizures, there was a 53% response rate in terms of reduction in seizure frequency.

A similar phase II study was reported by the Duke group, who evaluated the response to TMZ in a cohort of 46 patients with progressive low-grade gliomas, including 20 patients with oligodendrogliomas and 5 patients with mixed oligoastrocytomas.[18] The median age of the patients was 41 years, with a range of 7–61 years and a male predominance (59%). The majority of patients had undergone a biopsy or subtotal resection; most of the patients had not had any RT or chemotherapy. At the time of entry onto study, most of the patients had documented MRI progression, while a few had only neurological progression. All of the patients were placed on TMZ 200 mg/m^2/day for 5 consecutive days per month, for a maximum of 12 cycles of treatment. For the entire cohort, there was an overall objective response rate of 61% (28 of 46 patients). In the oligodendroglioma subgroup, the MRI follow-up demonstrated 5 complete responses (CR; 25%), 7 PR (35%), and 8 with SD (40%). Patients with mixed oligoastrocytomas had 1 CR (20%), 1 PR (20%), and 2 with SD (40%). The median PFS for the oligodendroglioma subgroup was 22 months, with a 6-month PFS rate and 12-month PFS rate of 100% and 79%, respectively.

Pace and colleagues also reported on the use of TMZ for progressive low-grade gliomas.[19] In their study, 43 patients with progressive low-grade gliomas, including 14 patients with oligodendroglial tumors (4 pure oligodendroglioma, 10 mixed oligoastrocytoma), were treated with TMZ 200 mg/m^2/day × 5 days per month (unless had failed prior PCV chemotherapy, then 150 mg/m^2/day). The median age of the cohort was 39.5 years (range 21–69 years), with a median Karnofsky Performance Status (KPS) of 90. On prestudy MRI, 26 patients had some enhancement, while 17 had no evidence for enhancement. A total of 30 patients had had prior RT, while 16 patients had tried and failed prior PCV chemotherapy. For the oligodendroglioma and oligoastrocytoma subgroups, the median number of TMZ cycles was 8.5, with a range of 4–17. MRI follow-up demonstrated 2 CR, 4 PR, 6 SD, and 3 PD. Overall, the imaging response rate was lower in the patients with non-enhancing tumors (29%; $N = 17$) vs those with enhancing tumors (58%; $N = 26$). There was clinical evidence of uncontrolled seizure activity in 31 of the patients prior to onset of TMZ. After the start of TMZ, improved seizure control was noted in 15 patients (6 complete seizure control and 9 partial seizure control), without any change in the anticonvulsant regimen. The improvement in seizure control was more robust in patients with non-enhancing tumors in comparison to those with enhancing tumors (81% vs 30%; $P < .05$).

Several early publications have also reported the use of TMZ for patients with recurrent oligodendrogliomas.[20, 21] In a European study using TMZ (150–200 mg/m^2/day for 5 consecutive days, every 28 days) for salvage therapy in patients with recurrent and progressive oligodendrogliomas, van den Bent and colleagues reported their results in a cohort of 30 patients, which had a mix of anaplastic, low-grade, and oligoastrocytoma histologies. The median age of the patients was 45 years, with a range of 23–67 years. More than half of the cohort had had more than one surgical resection, and all of them had undergone a course of RT. Most of the patients had had prior chemotherapy, mainly with PCV ($N = 27$), as well as a few other drugs (e.g., BCNU, cisplatin, and carboplatin). On MRI follow-up, there were 3 CR (10%), 6 PR (20%), 8 SD (27%), and 12 PD (40%). In the subgroup of patients with previous PCV chemotherapy, there were 7 of 27 with an objective response (26%). In addition, 2 of the 3 chemotherapy naïve patients had excellent responses (both CR). The median time to progression in responding patients was 13 months, with a range of 5–22 months. The 6-month PFS rate and 12-month PFS rate were 44% and 27%, respectively. Median survival for the responding group was 7 months. There was no predictive relationship noted between the initial response to PCV and subsequent responses to TMZ. In the subgroup of patients that had a first relapse within 2 years of completing the course of RT, there was a less sensitive response to TMZ (Fisher's exact test; $P = .04$).

A similar study was reported by Costanza and an Italian group, in a cohort of 32 patients with recurrent oligodendroglial tumors, after they had all failed initial PCV chemotherapy.[22] The patients had a mix of low-grade and anaplastic oligodendrogliomas ($N=25$), as well as a subgroup of oligoastrocytomas ($N=7$). The median age of the group was 48 years (range 23–68), with 20 women and 12 men, and a median KPS of 90. All of the patients had undergone a biopsy or surgical resection, and most had had a course of RT. TMZ was administered on the standard schedule at 150–200 mg/m^2/day. A median of 5 cycles of TMZ was administered during the study, with a range of 1–16 cycles. During follow-up imaging with MRI, there was 1 CR (3%), 6 PR (19%), 15 SD (47%; including 5 with an MR), and 10 with PD (31%). The median time to progression was 6 months for the entire cohort, and 8 months for the responding subgroup. All of the responding patients had pure oligodendrogliomas.

Soon after, the European Organization for Research and Treatment of Cancer (EORTC) published their data from a phase II study of first-line use of TMZ (150–200 mg/m^2/day) in patients with recurrent oligodendrogliomas after failing surgical resection and RT.[23] A total of 39 patients were enrolled into the study from eight institutions, with a median age of 48.6 years (range 21–64 years); 32 of the patients were WHO performance status 0 or 1. The pathology was pure oligodendroglioma in 24 patients and mixed oligoastrocytoma in 15 patients. A total of 34 of the patients had undergone a course of RT prior to starting TMZ. The median number of cycles of TMZ was 10, with a range of 1–14 cycles; the full treatment plan of 12 cycles was achieved in 17 patients. There was an overall objective response rate of 53%, with 10 of the 39 patients responding–10 CR and 10 PR, along with 12 SD and 6 PD. The median time to progression was 10.4 months overall, and 13.2 months in the responding subgroup. The 6-month and 12-month PFS rates were 71% and 40%, respectively. Median survival for the entire cohort had not been reached. The authors concluded that the use of TMZ should be considered as the first-line treatment option for patients with recurrent or progressive oligodendrogliomas.

A separate parallel phase II study by the EORTC evaluated the use of second-line TMZ in the setting of patients with recurrent or progressive oligodendrogliomas that had failed prior to PVC chemotherapy.[24] A total of 32 patients were enrolled into the study, with a median age of 43.6 years (range 27–56 years); 20 of the patients were WHO performance status 0 or 1. The pathology was pure oligodendroglioma in 17 patients and mixed oligoastrocytoma in 11 patients. There were 12 objective responses to prior PCV chemotherapy in the cohort: 5 CR and 7 PR. The median interval since prior PCV chemotherapy for the entire cohort was 6.5 months (range 1–55 months). All patients received TMZ at 150 mg/m^2/day for 5 consecutive days every 28 days. Due to prior PCV chemotherapy, there was no dose escalation to 200 mg/m^2/day after cycle #1. The median number of cycles of TMZ was 5, with a range of 1 to 16 cycles; in 84% of the patients, 90% or more of the intended dose intensity was delivered. There was an overall objective response rate of 22% (7 of 32 patients). However, in the pure oligodendroglial subgroup the rate was 25% (7 of 28 patients). The median time to progression for the responding cohort was 8.0 months. The PFS-6 and PFS-12 rates for the entire cohort were 29% and 11%, respectively; in these nonprogressive surviving patients, the median overall survival was 12.3 months. The authors concluded that TMZ should be considered the drug of choice for patients with progressive oligodendrogliomas during or after PCV failure.

In a study by Hoang-Xuan and a French group, patients with low-grade oligodendrogliomas and oligoastrocytomas that were progressive after surgery were treated with TMZ as initial therapy, and also had correlation with 1p deletion status.[25] None of the patients had undergone RT or any form of chemotherapy. A total of 60 consecutive patients were enrolled into the study; the median age was 43 years (range 24–72 years), with a median KPS of 90 at the onset of treatment. There were 49 patients with pure grade II oligodendrogliomas (82%) and 11 with grade II mixed oligoastrocytomas. The median time from surgical resection to starting TMZ chemotherapy was 3 months (range 0.1–108 months). All patients were placed on TMZ at 200 mg/m^2/day for 5 days per month for at least 12 cycles if stable and doing well, and up to 24 cycles in selected cases. The median number of cycles of TMZ was 11, with a range of 1–20 cycles. There was an objective response rate by MRI criteria of 31% (18 of 49 patients), including 10 PR (17%) and 8 MR (14%), along with 36 patients with SD (61%) and 5 with PD (8%). The median time to the onset of radiographic response was 4 months, with some responses being delayed up to 6–10 cycles. In addition, there were some patients who had purely clinical responses, mainly a reduction in seizure frequency noted in 30 patients (51%). The PFS-12 rate was 73.4%. Paired tumor and blood DNA samples were available for molecular genetic analysis in 26 patients. Loss of heterozygosity (LOH) of 1p (alone or in combination with LOH of 19q) was detected in 12 patients (46%). LOH of 1p was present in 6 of the patients with objective responses (3 PR, 3 MR) and in 6 patients with SD, but not in any of the patients with PD. There was a significant association between LOH of 1p and a response to chemotherapy with TMZ ($P < .004$).

A report from Levin and co-workers used TMZ for treatment of adults with progressive low-grade oligodendrogliomas (WHO grade II), who were still RT naive.[26] The results were correlated with 1p/19q deletion status and expression of MGMT protein in the tumor tissue. A total of 28 patients were enrolled into the study (17 males and 11 females), with a median age of 38 years (range 17–77 years). The median time from diagnosis to tumor progression was 33.5 months, with a range of 1–133 months. At the time of enrollment into the study, 16 patients demonstrated tumor progression on neuro-imaging studies, while 18 patients had clinical and neurological progression. The forms of clinical and neurological progression included severe seizure exacerbation (12 patients), progressive focal neurological deficits (4 patients), and significant cognitive decline (2 patients). Each patient was placed on TMZ 200 mg/m^2 per day for 5 consecutive days every month; treatment cycles were repeated every 28 days. The median number of TMZ cycles was 12.5, with a range of 2–24 cycles; 71% of the cohort received at least 10 or more treatment cycles. Marked clinical improvement was noted in 15 patients (53%), including 9 patients who were refractory to antiepileptic therapy and noted a significant reduction in seizure frequency. Clinical stability was noted in 10 patients (36%), while 3 patients (11%) had progressive clinical deterioration. On follow-up with MRI, there were 17 objective responses (61%; 10 PR, 7 MR), while 10 patients were able to maintain SD and 1 had rapid PD. The median TTP was 31 months, with a PFS rate at 12 months and 24 months of 89% and 70%, respectively. Six of the patients died (21%) secondary to PD within a median follow-up time of 29 months. In 15 patients, tumor and blood DNA pairs were available for 1p/19q co-deletion analysis. Co-deletion of 1p and 19q was noted in 10 patients (66%). Of these 10 patients, 9 were noted to have a radiographic response (PR 5, MR 4). The association between 1p deletion and radiographic response was statistically significant ($P < .003$). High MGMT protein expression levels (>50%) were observed in patients with non-deleted 1p. In this same group of patients, the best radiographic response was SD. In contrast, for patients with deletion of 1p, the expression of MGMT protein was low to intermediate. In addition, the patients with deletion of 1p had objective responses on MRI. Overall, there was a significant correlation between deletion of 1p, MGMT protein expression, and radiographic responses to TMZ ($P < .047$).

In an effort to reduce the concentration of tumor cell MGMT and increase overall Dose Intensity, Kesari and coworkers used a protracted daily regimen of TMZ in a phase II study of patients with newly diagnosed and recurrent low-grade gliomas (mainly oligodendrogliomas).[27] There were 15 newly diagnosed oligodendrogliomas and 3 newly diagnosed oligoastrocytomas; the newly diagnosed patients had to have a MIB-1 labeling index of >5% to be included in the study. There were also 11 patients with recurrent oligodendroglioma, 9 with recurrent oligoastrocytoma, and 6 with recurrent astrocytoma. Patients received TMZ 75 mg/m^2/day in 11-week cycles of 7 weeks on and 4 weeks off, for a total of 6 cycles or until progression or unacceptable toxicity. The clinical results were correlated with MGMT promoter methylation and 1p/19q co-deletion status. There were a total of 44 patients entered into the study; the median age was 43 years (range 20–68 years); 30 of the patients were male. The median KPS was 90, with a range of 70–100. The median number of treatment cycles of TMZ was 4.7 (range of 0.2–6.0; equivalent to 12.5 months). During MRI follow-up there were 0 CR, 9 PR (20%), 33 SD (75%), and 2 with PD (5%)–giving an overall positive response rate of 20%. The median overall survival was >72 months, with OS rates at 1-, 3-, and 5-years of 98%, 81%, and 73%, respectively. Patients with oligodendroglioma lived significantly longer than those with astrocytomas (HR = 0.14; $P = .018$). The overall median PFS was 38 months, with 1-, 3-, and 5-year PFS rates of 91%, 57%, and 34%, respectively. PFS did not differ statistically by tumor type. Patients with methylated MGMT had a longer OS ($P = .008$). Deletion of either 1p or 19q was also predictive of longer OS (HR = 0.17; log-rank $P = .02$). There was a significant co-segregation of MGMT methylation and deletion of 1p and/or 19q, with an Odds Ratio of 16 ($P = .09$). The authors concluded that the protracted course of TMZ was well tolerated and seemed to produce effective tumor control in newly diagnosed and recurrent low-grade gliomas.

In a phase II study, Wahl and the UCSF group used TMZ as the primary adjuvant therapy in a cohort of newly diagnosed patients with WHO grade II gliomas after surgical biopsy or partial resection.[28] A total of 120 patients were enrolled into the study; 57 had pure oligodendrogliomas, 20 had oligoastrocytomas, and 43 had astrocytomas. The median age of the group was 39 years, with a range of 19–71 years. The majority of patients had a KPS of 90–100; 56% of the patients were male. A subtotal resection was performed in 77% of the patients, while the other 23% only had a biopsy. TMZ was administered at a dose of 200 mg/m^2/day for 5 consecutive days, for up to 12 cycles. The mean number of cycles of TMZ was 10, with a range of 1–12 cycles; 72% of the patients completed the entire course of 12 cycles. On MRI follow-up, an overall objective response rate of only 6% was noted (0 CR, 7 PR). The objective response rate (ORR) was 7% for the oligodendroglioma subgroup, 5% for the oligo-astrocytoma subgroup, and 4% for the astrocytoma subgroup. There was SD noted in 97 patients (81%)–yielding an 87% rate of stable or improved disease during treatment. When analyzed by molecular subgroups, there were 5 responders that were 1p/19q co-deleted (11% response rate), 1 responder that was IDH-1 mutant (3% response rate), and 0 responders that were IDH-1 wild type. Progression rates were significantly different by molecular subtype (i.e., IDH-1 mutant vs IDH-1 wild type and 1p/19q co-deleted vs non-co-deleted; $P < .001$). The overall median PFS was 3.8 years, including 4.6 years in the oligodendroglioma subgroup and 2.7 years in the oligo-astrocytoma subgroup; the PFS was not significantly different depending on the histology ($P < .08$). The overall median overall survival was 9.7 years, including 10.8 years in the oligodendroglioma subgroup and 5.7 years in the oligoastrocytoma subgroup; the OS was significantly different depending on the histology ($P < 0.007$). The authors concluded that although the study had a low objective response rate, the high rate of stable or improved disease (86%) on TMZ therapy was still significant, and was equivalent to similar outcomes reported for adjuvant RT (e.g., RTOG 9802 and RTOG 0424). The use of TMZ in this group of patients was able to meaningfully delay the need for RT.

The EORTC recently performed a randomized phase III intergroup collaborative study, addressing the question of whether initial TMZ chemotherapy conferred an advantage in outcome in comparison to standard RT, for patients with high-risk low-grade glioma.[29] Patients had to have histologically confirmed supratentorial diffusely infiltrating WHO grade II gliomas, and required further active treatment other than surgical resection. Patients were entered into the study for radiological progression (63%), clinical progression other than seizures (22%), and refractory seizure activity (13%). After randomization, patients would receive 3D-conformal RT up to 5040 cGy in 28 fractions over 5–6 weeks vs TMZ 75 mg/m^2/day on a 21 day on and 7 day off schedule, for up to 12 cycles or PD or unacceptable toxicity. Molecular phenotyping was performed on all available tumor samples, including IDH-1/2 mutation status, 1p/19q deletion status, and MGMT promoter methylation status. A total of 477 patients were randomized to receive either RT ($N = 240$) or TMZ ($N = 237$), including pure oligodendrogliomas (40%), oligoastrocytomas (25%), and astrocytomas (35%). The median age of the entire cohort was 45 years (range 18–75 years); the patients were male in 58% of the cohort; and 96% were WHO Performance Status 0 or 1. Prior surgical status of the group consisted of complete resection in 17%, partial resection in 43%, and biopsy in 40%. The overall median time between the last surgical procedure and start of treatment on the protocol was 4.8 months (range of 0.0–151.5 months). The RT plan was completed in 91% of the patients, while the TMZ plan of 12 cycles was completed in 75% of the patients. The median PFS was 46 months in the RT subgroup and 39 months in the TMZ subgroup; the unadjusted HR for TMZ vs RT was 1.16 (log-rank $P = .22$). After adjusted HR were computed, the results were similar and also non-significant. No meaningful analyses of OS were possible at the time of publication, since only 25% of patients had died. On molecular analysis, IDH-1 or IDH-2 mutations were detected in 83% of the tumor specimens. Co-deletions of 1p and 19q were identified in 33% of the tumors, while MGMT promoter methylation was present in 90% of the tumors. MGMT promoter methylation was present in all of the tumors that were IDH-1/2 mutant and 1p/19q co-deleted (45/45; 100%). In patients with IDH-1/2 mutant and 1p/19q co-deleted tumors, the median PFS was very favorable at 62 months. The median PFS was also quite good for patients with IDH-1/2 mutant and 1p/19q non-co-deleted patients—at 48 months. The median PFS was only 20 months for patients with tumors that were IDH-1/2 wild type and 1p/19q non-co-deleted. The authors concluded that there was no significant difference in outcome between TMZ and RT for progressive low-grade gliomas. In addition, the option of initial chemotherapy with TMZ alone should be considered in good-prognosis patients with IDH mutant and 1p/19q co-deleted tumors.

A recent report by Izquierdo and colleagues analyzed the impact of up front TMZ treatment on the growth kinetics of diffuse low-grade 1p/19q co-deleted gliomas.[30] Using serial MRI, they followed the mean tumor diameter (MTD) in 36 progressive tumors that received TMZ treatment. After the onset of TMZ, 94.4% of patients had an initial decrease in MTD, that lasted for a median duration of 23 months (range 3–114 months). MTD re-growth occurred in 10 patients (27%) during TMZ treatment and in 22 patients (66%) after TMZ was stopped. In this subgroup, the median time to MTD re-growth after TMZ discontinuation was 12 months (range 1–88 months). The rate of MTD re-growth after TMZ onset at 3 and 5 years was 77% and 94%, respectively. The TTP based on volumetric analysis was shorter than the TTP based on Response Assessment in Neuro-Oncology (RANO) bidimensional criteria (23 months vs 35 months; $P = .05$). The authors concluded that volumetric analysis of tumors was a more precise method to assess chemotherapy efficacy in 1p/19q co-deleted low-grade gliomas.

TMZ for anaplastic oligodendroglial tumors

At the same time that clinicians were evaluating the use of TMZ in low-grade pure and mixed oligodendrogliomas, it was also being applied to anaplastic oligodendroglial tumors. A seminal paper by Chinot and a French group looked at the use of TMZ in the setting of patients with anaplastic oligodendrogliomas that had failed surgery, RT, and standard first-line PCV chemotherapy.[31] In this phase II study, TMZ was administered at 150-200 mg/m^2/day for 5 continuous days every 28 days, up to a maximum of 24 cycles, unless there was tumor progression or unacceptable toxicity. A total of 48 patients were entered into the study, with a median age of 41 years (range 21–65 years); 71% of the patients had a KPS of 70–80. The tumor types were pure anaplastic oligodendroglioma in 39 patients (81%) and anaplastic oligoastrocytoma in 9 patients (19%). In 35% of the patients ($N = 17$), the tumor had evolved from a lower grade tumor. All of the patients had received prior RT, while 47 patients had received chemotherapy with PCV ($N = 44$) or carmustine ($N = 3$). The overall median time from diagnosis to recurrence was 31.4 months. A median of 6 cycles of TMZ was administered to the cohort, with a median time to response of 8.13 months (range 1–12 months). There were 21 patients who developed an objective response on MRI follow-up, including 8 with a CR (16.7%) and 13 with a PR (27.1%), for an overall ORR of 43.8%. In addition, there were also 19 patients (39.6%) with SD and 8 with PD (16.7%). The overall median PFS for the entire cohort was 6.7 months. The

PFS-6 and PFS-12 rates were 50.5% and 25.4%, respectively. The overall median OS for the entire cohort was 10 months (range 2–36+ months), with OS-6 and OS-12 rates of 77.1% and 45.8%, respectively. In the subgroup with the objective responses, the PFS was 11.8+ months for the CR cohort and 8.9 months for the PR cohort, while the OS was 26+ months and 11 months, respectively. There was a significant correlation noted between patients having an objective response and experiencing longer PFS and OS ($r = 0.8937$; $P < .0001$). It was also noted that in patients that had had an objective response to prior PCV, there was a similar objective response achieved in 44%–46% of this group with TMZ. Patients with pure anaplastic oligodendroglioma responded better than patients with anaplastic oligoastrocytoma to TMZ treatment, with a higher ORR (48.7% vs 22.2%), median PFS (7.3 months vs 5.6 months), and median OS (9.9 months vs 8.7 months).

A similar more recent study from Gwak and colleagues used TMZ for treatment in a group of 72 patients with recurrent anaplastic oligodendroglioma and oligoastrocytoma.[32] There were 51 patients (71%) with pure anaplastic oligodendroglioma and 21 patients (29%) with anaplastic oligoastrocytoma; the tumors were derived from a lower grade neoplasm in 21% of the cases. The mean age of the group was 47.8 years, with a range of 11–76 years; slightly over half of the patients were male. The ECOG performance status of the cohort was 0 or 1 in 73% of the patients. Over half the cohort had undergone prior surgical resection, RT, and chemotherapy (56%), while surgery and RT was noted for 36%, surgery and radiosurgery for 4%, and surgery and chemotherapy for 4%. Prior chemotherapy was mainly PCV (70%), but also included a few patients that had received TMZ, bevacizumab and irinotecan, ACNU-cisplatin, and BCNU-cisplatin. TMZ was administered using standard dosing of 150–200 mg/m^2/day for 5 consecutive days per month until progression or unacceptable toxicity. The cohort received a median of 5.3 cycles of TMZ (range 1–41 cycles). There were a total of 17 objective responses on neuroimaging follow-up (24% ORR), including 8 CR (11%) and 9 PR (13%). In addition, there were also 23 with SD (32%) and 32 with PD (44%). The overall tumor stabilization rate for the cohort was 56% while on TMZ. The patients with pure anaplastic oligodendroglioma demonstrated a trend for improved responses in comparison to the patients with anaplastic oligoastrocytoma, but did not reach statistical significance (CR 14% vs 4.8%; PR 14% vs 9.5%). The overall median PFS for the group was 8 months, with a range of 6–10 months. The PFS-6 and PFS-12 rates were 52% and 27%, respectively. The median OS was not reached for the entire cohort, after median follow-up of 11.5 months (range of 2–66 months). The estimated 1-year and 2-year survival rates were 81% and 63%, respectively.

Several authors have reported studies of using TMZ as the primary treatment modality in patients with newly diagnosed anaplastic oligodendroglioma, holding off on any form of RT until after progression or recurrence.[33–36] An initial report by Taliansky-Aronov and co-workers reviewed their experience from a group of 20 patients with anaplastic oligodendroglioma; 10 patients had secondary tumors that evolved from a lower grade oligodendroglioma.[33] The median age of the cohort was 47 years (range 26–65 years), with a median KPS of 70% (range 30%–90%). Surgical procedures prior to TMZ treatment included debulking surgery in 9 patients and a biopsy in 11 patients. Three of the patients had received prior therapy at the time of diagnosis of a low-grade tumor, including RT in two patients and PCV chemotherapy in one patient. The median time from diagnosis of anaplastic oligodendroglioma to the first cycle of TMZ was 43 days (range 9–130 days). Patients were placed on TMZ 200 mg/m^2/day for 5 consecutive days every 28 days and received 24 cycles unless there was tumor progression or unacceptable toxicity. The median number of cycles of TMZ was 14, with a range of 3–24. There was clinical improvement documented in 12 patients (60%), as noted by KPS and neurologic function scales, while five remained stable and nine progressed. The objective responses as documented by MRI follow-up were 0 CR and 15 PR, for an ORR of 75%. The median time to reach a maximal objective response was 6 months (range 3–12 months). The overall PFS was 24 months (range 3–34 months). The presence of loss of 1p was correlated with longer PFS; the PFS-24 rate was 100% for patients with intact 1p and only 20% for those with 1p deletion ($P = .057$). The authors concluded that newly diagnosed anaplastic oligodendrogliomas demonstrated a high rate of response to initial therapy with TMZ, which was very similar to the responses reported for PCV combination therapy.

In a report by Mikkelsen et al., they treated two cohorts of patients with newly diagnosed anaplastic oligodendroglioma using different regimens, stratified by co-deletion or non-co-deletion of 1p/19q (i.e., loss of heterozygosity; LOH).[34] The co-deleted subgroup received "up front" TMZ (150–200 mg/m^2/day for 5 consecutive days, every 28 days), without any RT. The 1p/19q non-co-deleted group received from 2–4 cycles of similar "up front" TMZ, followed by a course of standard chemo-radiotherapy (6000 cGy) with concomitant daily TMZ (75 mg/m^2/day) over 6 weeks, followed by additional cycles of adjuvant TMZ. A total of 48 patients were entered into the study (pure anaplastic oligodendroglioma–39 [81.3%]; anaplastic oligoastrocytoma–8 [16.7%], anaplastic mixed–1 [2.1%]); 36 with 1p/19q co-deletion and 12 with 1p/19q non-co-deletion. The median age of the cohort was 45.5 years (range 18–81 years); 60% of the patients were male, with a median KPS of 90. The median number of cycles of TMZ was significantly higher in the 1p/19q co-deleted subgroup in comparison to the non-co-deleted cohort (12 cycles [range 2–24] vs 8 cycles [range 1–12]; $P = .013$). Overall, there were 45 objective responses, including 4 CR (8.9%) and 22 PR (48.9%), along with 14 SD (31.1%) and 5 (11.1%) with PD. The median time to radiographic response was 5.3 months (range 2–18). The percentage of patients demonstrating objective responses was similar between the 1p/19q co-deleted and non-co-deleted groups: 21 (3 CR, 18 PR; 60%) vs 6 (0 CR, 6 PR; 60%), respectively. The overall median PFS for the entire cohort was 28.3 months, with PFS-6 and PFS-12 rates of 89.4% and 72.0%, respectively. The median OS was 77.3 months, with OS-12 and OS-24 rates of 95.2% and 88.8%, respectively. The median PFS was longer in the 1p/19q co-deleted subgroup in comparison to the non-co-deleted subgroup (28.7 months vs 13.5 months), but not reach statistical significance. However, the OS-12 and OS-24 rates were statistically significantly longer in the patients with 1p/19q co-deletion in comparison to those with non-co-deletion (97.2% vs 83.3%, 90.1% vs 83.3%; $P = .016$). Patients with pure anaplastic oligodendroglioma had longer median OS than those with mixed anaplastic tumors (77.3 months vs 44.6 months; $P = .017$). The authors concluded that patients with newly diagnosed 1p/19q co-deleted anaplastic oligodendrogliomas can be safely treated with primary TMZ, and that they did not experience earlier or more frequent tumor progression.

In the phase II trial by Gan and colleagues, TMZ was used as primary therapy for a cohort of 40 newly diagnosed patients with anaplastic oligodendroglial tumors.[35] Eleven of the patients (28%) were pure anaplastic oligodendrogliomas, while 29 (72%) were mixed anaplastic oligodendrogliomas. Prior surgical intervention included complete resection in 9 patients (22%), subtotal resection in 27 patients (68%), and biopsy only in 4 patients (10%). The median age of the group was 43 years (range 18–71 years), with the majority of patients in the 0–1 ECOG performance status range. 1p/19q co-deletion was noted in 18 patients (45%), 1p deletion was present in 3 patients (7%), and 19q deletion was demonstrated in 5 patients (13%). MGMT promoter status included 10 patients with methylated tumors (25%) and 11 patients with unmethylated tumors (28%). TMZ was administered at 200 mg/m^2/day for 5 consecutive days every 4 weeks for 6 total cycles. On MRI follow-up, the objective responses included 15 CR (38%) and 6 PR (15%)–for an overall ORR of 53%, as well as 9 patients (23%) with disease stabilization. The median overall PFS was 21 months (range 3–39 months); the PFS-6 rate was 77%. The median OS was 43 months, with a range of 20–66 months. The authors concluded that TMZ was active in newly diagnosed anaplastic oligodendrogliomas, and should be considered as an alternative to RT.

Using a different approach for treatment of newly diagnosed anaplastic oligodendroglial tumors, Ahluwalia and coworkers administered a dose-intense TMZ regimen to 62 patients in a multi-center phase II trial.[36] TMZ was administered at 150 mg/m^2/day on days 1–7 and 15–21, every 28 days for 8 cycles. There were 41 patients with pure anaplastic oligodendrogliomas (66%) and 21 with mixed anaplastic oligoastrocytomas (34%). The median age of the cohort was 43 years (range 18–83 years), with 81% having a KPS of 90–100, and a fairly even male to female ratio. Prior surgery included a gross total resection in 18 patients (30%), partial resection in 19 patients (31%), and a biopsy only in 11 patients (18%). In evaluable patients, 1p loss was detected in 36 patients (58%), while 19q loss was noted in 23 patients (55%); co-deletion of 1p/19q was present in 20 patients (48%). In patients with measurable disease ($N = 25$), there were only a few objective responses, including 2 CR (8%) and no PR, while 56% had SD and 36% had tumor progression. The estimated overall median PFS was 27.2 months, with a range of 11.9–36.3 months. The estimated overall median OS was 105.8 months, with a range of 51.5 to N/A months. Loss of 1p and co-deletion of 1p/19q were significant prognostic factors for PFS ($P < .001$) and OS ($P < .001$). Co-deletion of 1p/19q was also suggestive of a better response to chemotherapy ($P = .007$). The authors concluded that the results supported the further study of first-line TMZ monotherapy as an alternative to RT in patients with newly diagnosed anaplastic oligodendrogliomas.

Vogelbaum and colleagues decided on a more aggressive strategy when they were planning the RTOG BR0131 phase II trial for newly diagnosed anaplastic oligodendrogliomas, using preradiotherapy dose intense TMZ, followed by chemo-radiotherapy with daily concurrent TMZ.[37] The primary end point of the study was the response rate during the 6-month, preradiotherapy chemotherapy. The histology was pure anaplastic oligodendroglioma in 13 patients (33%) and various types of mixed anaplastic oligoastrocytoma in 26 patients (67%). Each patient received preradiotherapy TMZ dosed at 150 mg/m^2/day given on a 7-day on/7-day off schedule (days 1–7 and 15–21) for up to 6 cycles, followed by standard chemo-radiotherapy(6000 cGy; in patients with residual measurable disease after initial TMZ) over 6 weeks in combination with daily oral TMZ (75 mg/m^2/day). A total of 39 evaluable patients were enrolled into the study; the median age of the cohort was 45 years (range 18–71 years), with a slight male preponderance (54%). Prior surgical therapy included biopsies in 4 patients (11%), partial resection in 18 patients (46%), and a complete resection in 17 patients (44%). The majority of patients had minor to moderate neurological symptoms (61%) at the time of entry into the study, with normal mental status in 82%. In 17 of 28 (60.7%) evaluable cases, there was co-deletion of 1p and 19q. In 16 of 20 (80%) evaluable cases, MGMT promoter methylation was present. Of the 31 patients that were evaluable for neuro-imaging review, there were 10 objective responses (32%), including 2 CR (6%) and 8 PR (26%), as well as 16 SD (52%) and 3 PD (10%). All of the patients with tumors that had 1p/19q co-deletion and MGMT promoter methylation were free from progression during the neoadjuvant TMZ phase of the study (i.e., PFS-6 rate of 100%). The overall rate of progression during the preradiotherapy neoadjuvant phase of TMZ was 10%, which compared favorably with the prior experience with preradiotherapy chemotherapy using PCV (RTOG 9402)–that had a rate of progression of 20%.

As a potential strategy to delay or defer RT in newly diagnosed anaplastic oligodendroglioma patients, Thomas et al. report their experience using induction TMZ, followed by myeloablative high-dose chemotherapy with autologous stem-cell transplantation.[38] A total of 41 patients were entered into the study; 36 (88%) had pure anaplastic oligodendroglioma, while 5 (12%) were mixed anaplastic oligoastrocytomas. In 80% of the cohort (33 of 41), the tumors were co-deleted for 1p and 19q. Residual measurable disease (i.e., >10 mm) was present in 11 patients (27%), while the majority of patients did not have any significant residual disease (26 patients; 63%). The median age of the cohort was 44 years (range 30–66 years), with a median KPS of 90 and a male to female ratio of almost 2/1. The induction regimen consisted of TMZ 200 mg/m^2/day for 5 consecutive days in 28-day cycles, for a total of 6 cycles if the patient achieved an objective response or SD. All patients achieving an objective response or SD while on TMZ were eligible for the high-dose chemotherapy regimen consisting of intravenous Thiotepa 250 mg/m^2/day for 3 consecutive days, on days −8, −7, and −6 prior to transplant, followed by intravenous Busulfan 3.2 mg/kg/day on days −5, −4, and −3. Peripheral blood stem cells were re-infused 2 days (48 hours) after the completion of the Busulfan infusion, on day 0. After the completion of induction chemotherapy, 9 patients (25.7%) had PD, leaving 26 patients eligible for transplant. The median PFS and median OS had not been reached in the transplanted cohort after a median follow-up of 65.7 months. The 2-year and 5-year PFS rates were 85.7% and 60.4%, respectively. At the time of publication, none of the transplanted patients had died, with a 5-year OS rate of 100%. 1p and 19q co-deletion was a predictor of PFS ($P = .03$) and OS ($P < .001$). In the subgroup of patients with 1p and 19q co-deletion ($N = 33$), the median PFS was 62.8 months, with a 2-year PFS rate of 75.9% and a 5-year PFS rate of 50.3%. The median OS for this cohort has not been reached; the 5-year OS rate was 93.4%. The authors concluded that induction chemotherapy with TMZ followed by high-dose chemotherapy and autologous stem cell transplant was active and safe in patients with newly diagnosed anaplastic oligodendrogliomas, and had promising PFS and OS data.

In the report from Ducray and colleagues, they evaluated the use of TMZ in elderly patients with anaplastic oligodendrogliomas.[39] A cohort of 44 consecutive patients above 70 years of age were enrolled in the study of "up front" use of TMZ, dosed at 150–200 mg/m^2/day for 5 consecutive days on a 28-day cycle. The median age of the group was 74.4 years (range 70–90 years), with a median KPS of 70; roughly 60% of the cohort was male. The tumors were pure anaplastic oligodendroglioma in 30 patients (68%) and anaplastic oligoastrocytoma in 14 patients (32%). MGMT methylation status was assessable in 38 patients and was present in 19 patients (50%). The median number of TMZ cycles administered during the protocol was 5 (range 1–23 cycles). During MRI follow-up, the responses included 0 CR, 13 PR (32%), 17 SD (41%), and 11 with PD (27%). Clinical improvement was noted in 13 patients (30%). The overall median PFS was 6.9 months (range 1–24.9 months); the median PFS in patients with a PR or SD was 12.6 months and 7.1 months, respectively. The overall median OS was 12.1 months (range 2–38 months), with a median OS of 18 months for patients with a PR and 14.2 months for those with SD. Patients with tumors that demonstrated MGMT promoter methylation had longer PFS (8.7 months vs 5.7 months; $P = .01$) and OS (16.2 months vs 12.4 months; $P = .05$). The objective response rate did not differ in terms of MGMT methylation status (38% vs 31%; $P = 1.0$); however, there was a significant difference in the duration of response in those with methylated tumors (16.1 months vs 9.6 months; $P = .0004$). The authors concluded that up-front TMZ was active in elderly patients with newly diagnosed anaplastic oligodendrogliomas, should be considered as a reasonable alternative to RT, and warranted further study.

In a case report by Michotte et al., a rare patient with primary leptomeningeal anaplastic oligodendrioma has been described, who responded to primary treatment with TMZ.[40] The patient was a 60-year old male who had a seizure 3 years earlier, and then was referred for further evaluation when he became depressed and had rapid cognitive decline. A brain MRI scan revealed a diffuse right parieto-occipital subarachnoid enhancing lesion without any intra-axial extension. A biopsy of the mass proved the presence of an anaplastic oligodendroglioma, that had co-deletion of 1p and 19q. There was rapid progression of the lesion, with spinal leptomeningeal metastases. He was placed on standard TMZ 150 mg/m^2/day for 5 consecutive days every 4 weeks. After 6 cycles of TMZ he had a significant clinical recovery, and also had near complete regression of the enhancing lesion on MRI. After cycle #8, he developed recurrent disease and had to undergo craniospinal RT.

TMZ vs PCV for oligodendrogliomas

There is an ongoing controversy regarding the appropriate therapy for newly diagnosed and recurrent low-grade and anaplastic oligodendrogliomas–which is the better regimen: PCV or TMZ?[3, 6, 41, 42] Based on the data reviewed above, it is obvious that TMZ has shown evidence for significant activity against low-grade and anaplastic oligodendrogliomas in the newly diagnosed and recurrent settings. However, it is also clear from the data reviewed in Chapter 28 that PCV has significant efficacy in these tumor types as well, and may be capable of inducing more durable responses than TMZ. Some investigators have suggested that since TMZ and PCV have similar efficacy against this group of tumors, and that TMZ is much less toxic and is also easier to use, that TMZ should remain the primary treatment option.[43] In contrast, other investigators feel that the superior outcomes obtained with PCV, in terms of PFS and OS, outweigh the issues related to toxicity and ease of administration, so that PCV should become the primary treatment option.[44] The ongoing intergroup "CODEL" trial (National Clinical Trials Identifier—NCT00887146; Alliance, EORTC, NRG, NCIC) will hopefully answer this question definitively.[6, 43] In this modified trial, newly diagnosed patients with 1p/19q co-deleted anaplastic oligodendroglial tumors will be treated with RT followed by PCV vs RT with concurrent and adjuvant TMZ vs TMZ alone.

References

1. Wesseling P, van den Bent M, Perry A. Oligodendroglioma: pathology, molecular mechanisms and markers. *Acta Neuropathol*. 2015;129:809–827.
2. van den Bent MJ. Oligodendrogliomas: a short history of clinical developments. *CNS Oncol*. 2015;4:281–285.
3. van den Bent JM, Bromberg JE, Buckner J. Low-grade and anaplastic oligodendroglioma. *Handb Clin Neurol*. 2016;134:361–380.
4. Louis DN, Perry A, Reifenberger G, et al. The 2016 World Health Organization classification of tumors of the central nervous system: a summary. *Acta Neuropathol*. 2016;131:803–820.
5. Cairncross JG, Mcdonald DR. Successful chemotherapy for recurrent malignant oligodendroglioma. *Ann Neurol*. 1988;23:360–364.
6. Drappatz J, Lieberman F. Chemotherapy of oligodendrogliomas. *Prog Neurol Surg*. 2018;31:152–161.
7. Stevens MFG, Newlands ES. From triazines and triazenes to temozolomide. *Eur J Cancer*. 1993;29:1045–1047.
8. Newlands ES, Stevens MFG, Wedge SR, et al. Temozolomide: a review of its discovery, chemical properties, pre-clinical development and clinical trials. *Cancer Treat Rev*. 1997;23:35–61.
9. Hvizdos KM, Goa KL. Temozolomide. *CNS Drugs*. 1999;12:237–243.
10. Liu L, Markowitz S, Gerson SL. Mismatch repair mutations override alkyltransferase in conferring resistance to temozolomide but not to 1,3-bis(2-chloroethyl)nitrosourea. *Cancer Res*. 1996;56:5375–5379.
11. Friedman HS, Johnson SP, Dong Q, et al. Methylator resistance mediated by mismatch repair deficiency in a glioblastoma multiforme xenograft. *Cancer Res*. 1997;57:2933–2936.
12. Friedman HS, McLendon RE, Kerby T, et al. DNA mismatch repair and O^6-alkylguanine-DNA alkyltransferase analysis and response to Temodal in newly diagnosed malignant glioma. *J Clin Oncol*. 1998;16:3851–3857.
13. Tano K, Shiota S, Collier J, et al. Isolation and structural characterization of a cDNA clone encoding the human DNA repair protein for O^6-alkylguanine. *Proc Natl Acad Sci USA*. 1990;87:686–690.
14. Ochs K, Kaina B. Apoptosis induced by DNA damage O^6-methylguanine is Bcl-2 and caspase-9/3 regulated and Fas/caspase-8 independent. *Cancer Res*. 2000;60:5815–5824.
15. Hegi ME, Liu L, Herman JG, et al. Correlation of O^6-methylguanine methyltransferase (MGMT) promoter methylation with clinical outcomes in glioblastoma and clinical strategies to modulate MGMT activity. *J Clin Oncol*. 2008;26:4189–4199.
16. Viviers L, Brada M, Hines F, et al. A phase II trial of primary chemotherapy with temozolomide in patients with low-grade cerebral gliomas (abstract). *Neuro-Oncol*. 2000;(Suppl 1):S43. July.
17. Brada M, Viviers L, Abson C, et al. Phase II study of primary temozolomide chemotherapy in patients with WHO grade II gliomas. *Ann Oncol*. 2003;14:1715–1721.
18. Quinn JA, Reardon DA, Friedman AH, et al. Phase II trial of temozolomide in patients with progressive low-grade glioma. *J Clin Oncol*. 2003;21:646–651.
19. Pace A, Vidiri A, Galie E, et al. Temozolomide chemotherapy for progressive low-grade glioma: clinical benefits and radiological response. *Ann Oncol*. 2003;14:1722–1726.
20. van den Bent MJ, Keime-Guibert F, Brandes AA, et al. Temozolomide chemotherapy in recurrent oligodendroglioma (abstract). *Neurology*. 2000;2 (Suppl 1):S26.
21. van den Bent MJ, Keime-Guibert F, Brandes AA, et al. Temozolomide chemotherapy in recurrent oligodendroglioma. *Neurology*. 2001;57:340–342.
22. Costanza A, Borgognone M, Nobile M, et al. Temozolomide in recurrent oligodendroglial tumors: A phase II study (abstract). *Neuro-Oncology*. 2001;3(Suppl 1):S66.
23. van den Bent MJ, Taphoorn MJB, Brandes AA, et al. Phase II study of first-line chemotherapy with temozolomide in recurrent oligodendroglial tumors: The European Organization for Research and Treatment of Cancer Brain Tumor Group Study 26971. *J Clin Oncol*. 2003;21:2525–2528.
24. van den Bent MJ, Chinot O, Boogerd W, et al. Second-line chemotherapy with temozolomide in recurrent oligodendroglioma after PCV (procarbazine, lomustine and vincristine) chemotherapy: EORTC Brain Tumor Group phase II study 26972. *Ann Oncol*. 2003;14:599–602.
25. Hoang-Xuan K, Capelle L, Kujas M, et al. Temozolomide as initial treatment for adults with low-grade oligodendrogliomas or oligoastrocytomas and correlation with chromosome 1p deletions. *J Clin Oncol*. 2004;22:3133–3138.
26. Levin N, Lavon I, Zelikovitsh B, et al. Progressive low-grade oligodendrogliomas: response to temozolomide and correlation between genetic profile and O6-methylguanine DNA methyltransferase protein expression. *Cancer*. 2006;106:1759–1765.
27. Kesari S, Schiff D, Drappatz J, et al. Phase II study of protracted daily temozolomide for low-grade gliomas in adults. *Clin Cancer Res*. 2009;15:330–337.
28. Wahl M, Phillips JJ, Molinaro AM, et al. Chemotherapy for adult low-grade gliomas: clinical outcomes by molecular subtype in a phase II study adjuvant temozolomide. *Neuro-Oncology*. 2017;19:242–251.

29. Baumert BG, Hegi ME, van den Bent MJ, et al. Temozolomide chemotherapy versus radiotherapy in high-risk low-grade glioma (EORTC 23033-26033): a randomized, open-label, phase 3 intergroup study. *Lancet Oncol.* 2016;17:1521–1532.

30. Izquierdo C, Alentorn A, Idbaih A, et al. Long-term impact of temozolomide on 1p/19q-codeleted low-grade glioma growth kinetics. *J Neuro-Oncol.* 2018;136:533–539.

31. Chinot OL, Honore S, Dufour H, et al. Safety and efficacy of temozolomide in patients with recurrent anaplastic oligodendrogliomas after standard radiotherapy and chemotherapy. *J Clin Oncol.* 2001;19:2449–2455.

32. Gwak HS, Yee GT, Park CK, et al. Temozolomide salvage chemotherapy for recurrent anaplastic oligodendroglioma and oligo-astrocytoma. *J Korean Neurosurg Soc.* 2013;54:489–495.

33. Taliansky-Aronov A, Bokstein F, Lavon I, Siegal T. Temozolomide treatment for newly diagnosed anaplastic oligodendrogliomas: a clinical efficacy trial. *J Neuro-Oncol.* 2006;79:153–157.

34. Mikkelsen T, Doyle T, Anderson J, et al. Temozolomide single-agent chemotherapy for newly diagnosed anaplastic oligodendroglioma. *J Neuro-Oncol.* 2009;92:57–63.

35. Gan HK, Rosenthal MA, Dowling A, et al. A phase II trial of primary temozolomide in patients with grade III oligodendroglial brain tumors. *Neuro-Oncology.* 2010;12:500–507.

36. Ahluwalia MS, Xie H, Dahiya S, et al. Efficacy and patient-reported outcomes with dose-intense temozolomide in patients with newly diagnosed pure and mixed anaplastic oligodendroglioma: a phase II multicenter study. *J Neuro-Oncol.* 2015;122:111–119.

37. Vogelbaum MA, Berkey B, Peereboom D, et al. Phase II trial of preirradiation and concurrent temozolomide in patients with newly diagnosed anaplastic oligodendrogliomas and mixed anaplastic oligoastrocytomas: RTOG BR0131. *Neuro-Oncology.* 2009;11:167–175.

38. Thomas AA, Abrey LE, Terziev R, et al. Multicenter phase II study of temozolomide and myeloablative chemotherapy with autologous stem cell transplant for newly diagnosed anaplastic oligodendroglioma. *Neuro-Oncology.* 2017;19:1380–1390.

39. Ducray F, del Rio MS, Carpentier C, et al. Up-front temozolomide in elderly patients with anaplastic oligodendroglioma and oligoastrocytoma. *J Neuro-Oncol.* 2011;101:457–462.

40. Michotte A, Chaskis C, Sadones J, et al. Primary leptomeningeal anaplastic oligodendroglioma with a 1p36-19q13 deletion: report of a unique case successfully treated with temozolomide. *J Neurol Sci.* 2009;287:267–270.

41. Villano JL, Wen PY, Lee EQ, et al. PCV for anaplastic oligodendrogliomas: back to the future or a step backwards? A point/counterpoint discussion. *J Neuro-Oncol.* 2013;113:143–147.

42. Lassman AB. Procarbazine, lomustine and vincristine or temozolomide: which is the better regimen? *CNS Oncol.* 2015;4:341–346.

43. Rinne ML, Wen PY. Treating anaplastic oligodendrogliomas and WHO grade 2 gliomas: PCV or temozolomide? The case for temozolomide. *Oncology (Willston Park).* 2015;29:275.

44. Levin VA. Treating anaplastic oligodendrogliomas and WHO grade 2 gliomas: PCV or temozolomide? The case for PCV. *Oncology (Willston Park).* 2015;29:264.

Chapter 30

Miscellaneous chemotherapy approaches to oligodendroglial tumors

Nina A. Paleologos*,† and Herbert B. Newton‡,§,¶

*Advocate Medical Group, Advocate Healthcare, Chicago, IL, United States; †Rush University Medical School, Chicago, IL, United States; ‡Neuro-Oncology Center, AdventHealth Cancer Institute, AdventHealth Medical Group, Orlando, FL, United States; §CNS Oncology Program, AdventHealth Cancer Institute, AdventHealth Medical Group, Orlando, FL, United States; ¶Division of Neuro-Oncology, Esther Dardinger Endowed Chair in Neuro-Oncology James Cancer Hospital & Solove Research Institute, Wexner Medical Center at the Ohio State University, Columbus, OH, United States

Oligodendrogliomas are much less commonly diagnosed than astrocytomas, and are thought to comprise approximately 5%–6% of primary brain tumors.[1] Oligodendrogliomas were first appreciated as a distinct clinical and histological entity in the 1920s.[2–4] It was initially suggested by Bailey and Hiller in 1924 that there might be a variety of glioma that was composed primarily of oligodendrocytes.[2] Bailey and Cushing then described oligodendrogliomas in a detailed manner in 1926, as part of their initial treatise on the classification of gliomas.[3] Subsequently, in 1929, Bailey and Bucy published the classic paper that characterized the clinical and pathological features of oligodendrogliomas, noting the predilection for involving the cerebral hemispheres, onset in adulthood, frequent calcification, and tendency for slow clinical evolution.[4]

In the decades that followed, it was observed that oligodendrogliomas had a more indolent and less aggressive clinical course than glioblastoma (GBM) or anaplastic astrocytoma (AA).[5–7] Several of the authors of these early reports noted extended preoperative periods, as well as long postoperative periods in many of the patients. For example, in the report from the Mayo Clinic, Earnest and colleagues recorded preoperative symptoms of up to 256 months, as well as several patients with very long-term survival after the onset of the first symptoms (i.e., > 20 years).[6] In addition, numerous case reports began to appear in the literature, describing patients with much extended survival times with oligodendroglial tumors.[8–10] This included a report from Freeman and Feigin of a patient with a 35-year survival, as well as a 15-year-old teenage boy who had lived with the tumor for 40 years after diagnosis, and was still living at the time of the publication.[8,10]

The early approaches to brain tumor chemotherapy in the 1960s and 1970s focused on newly diagnosed and progressive GBM and AA, and were mainly using single-agent nitrosourea drugs and nitrosourea-based combination regimens.[11,12] The most commonly used drugs at that time were 1,3-bis(2-chloro-ethyl)-1-nitrosourea (BCNU), 1-(2-chloroethyl)-3-cyclo-hexyl-1-nitrosourea (CCNU), procarbazine, and vincristine. The CCNU was sometimes used as a single agent, or more commonly used in combination with procarbazine and vincristine–the PCV (procarbazine, lomustine, and vincristine) regimen.[11,13] Oligodendrogliomas were not included too often in the reported series of high-grade glioma patients undergoing chemotherapy. When they were included, they were usually anaplastic oligodendroglial tumors that had progressed through surgery and radiotherapy. However, it was noted in many of these early papers that the cohorts of anaplastic oligodendroglioma patients tended to respond better to the various chemotherapy regimens than the GBM and AA patients, with more objective responses on neuroimaging, and longer progression-free survival (PFS) and overall survival (OS).

In the 1980s there were more reports of chemotherapy being used for the treatment of oligodendroglioma tumors; many of them containing positive responses to therapy. A report from a French group early on described their experience with 31 patients with oligodendrogliomas.[14] For 21 patients in the cohort, the only therapeutic approach was surgical resection, with no additional adjuvant treatment with radiotherapy or chemotherapy. The other 10 patients had surgical resection followed by a course of chemotherapy (CCNU, 60 mg/m²; teniposide, 60 mg/m²). There was a definite advantage in terms of tumor control for the patients who had chemotherapy added to the surgical resection. Using a different strategy, Lange and colleagues reported their results in 102 patients with inoperable high-grade gliomas (seen over a period of 65 months in a single clinic), after a stereotactic biopsy or subtotal resection.[15] The cohort consisted of 35% GBM, 30% AA, and 21% anaplastic oligodendroglioma. In the subgroup that had not received any form of radiotherapy (N = 75), the treatment plan consisted of three courses of concomitant radiotherapy and chemotherapy, with a 5-week interval between courses. During each course, the patient received daily external beam radiotherapy up to a maximum of 1500 cGy. At the same time, they received intravenous (IV) ifosfamide on days 1–5 at a dosage of $1.2 g/m^2$/day, as well as IV BCNU at a dosage of $30 mg/m^2$/day on days 1, 3, and 5. After the three courses of chemoradiotherapy (i.e., at a total RT dose of 4500 cGy), the tumor was then locally boosted to a total of 5000–6000 cGy. Using UICC (Union of International Cancer Control) criteria, the overall imaging responses included 12 complete responses (CR; 16.9%) and 35 partial responses (PR; 49.3%). For the anaplastic oligodendroglioma cohort, there were 2 CR (13.3%), 11 PR (73.%), and 2 with No Change (13.3%), with a mean duration of response of 16.7 months. In 21 cases (29.6%) that were actively growing, the progression was halted by the course of treatment. In a retrospective analysis of 39 cases of supratentorial oligodendroglioma from 1965 to 1984, a Japanese group noted 5- and 10-year survival rates of 73.1% and 32.2%, respectively.[16] Patients from the earlier years of their series were only treated with surgical resection, and had short survival times. More recent patients had undergone surgery followed by a combination of radiotherapy and chemotherapy, and had much longer PFS and OS. The radiotherapy was 5000–6000 cGy over 5–6 weeks, while the chemotherapy was a nitrosourea-based combination regimen using ACNU [1-(4-amino-2-methyl-5-primidinyl)-methyl-3-(2-chloroethyl)-3-nitrosourea], FT-207 [N1-(2 tetrahydrofuryl)-5-fluorouracil], and PSK (polysaccharide-K). A report from Levin and coworkers used a combination of eflornithine (irreversible inhibitor of ornithine decarboxylase) and mitoguazone (a competitive inhibitor of S-adenosylmethionine decarboxylase) in a phase I/II study of 33 patients with recurrent or progressive malignant primary brain tumors.[17] In the cohort there were 19 cases of anaplastic gliomas, 6 with GBM, and several other tumor types, including one with an oligodendroglioma. There was disease stabilization noted in over half of the group with anaplastic gliomas. However, the one patient with an oligodendroglioma did not respond and had progressive disease after one cycle. In 1988 Cairncross and Mcdonald published the seminal paper that established malignant oligodendrogliomas—and oligodendrogliomas in general—as chemosensitive tumors.[18] They treated a series of eight consecutive patients with recurrent malignant oligodendrogliomas with systemic chemotherapy, and had clinical and radiographic responses (cranial CT scans) in the entire cohort. The chemotherapy approaches consisted of PCV in six patients, BCNU in one patient, and diaziquone in one patient. There was one CR that lasted for 78 weeks and seven PRs that lasted from 30+ to 68+ weeks. In two of the partial responders who also had extraneural metastases, there was restabilization of the systemic disease after the initiation of chemotherapy.

In the 1990s, after the publication of the Cairncross and Macdonald paper, the focus of chemotherapy for progressive and recurrent oligodendroglial tumors was the use of PCV (see Chapter 28). However, there were a few other chemotherapy drugs and regimens that were attempted as well. In a study of melphalan (L-phenylalanine mustard) for recurrent oligodendrogliomas and astrocytomas, Brown and colleagues treated 67 patients in a phase II study.[19] The regimen consisted of IV melphalan dosed at $40 mg/m^2$, administered every 4 weeks. For the oligodendroglioma subgroup ($N = 11$ patients), there were six patients with a CR or PR (55%), three with stable disease (SD; 27%), and two with progressive disease (PD; 18%). The mean duration of response in the oligodendroglioma cohort was 26 weeks, with a range of 6–47+ weeks. The responses in the non-oligodendroglioma subgroup (19 GBM, 14 AA, 4 astrocytoma, and 5 ependymoma) were not as impressive, with 6 CR or PR (11%), 7 with SD (13%), and 22 with PD (42%). Dosage reductions due to bone marrow toxicity were necessary in 25 patients.

Another approach that was started in the 1990s was the concept of "myeloablative" chemotherapy, followed by hematopoietic reconstitution.[20] In this case report, Saarinen and coworkers described a 7-year-old male child who developed diplopia and headache, and was found on workup to have a left frontal enhancing mass with extension across the midline. A partial resection of the mass revealed a WHO grade III oligodendroglioma. The remaining tumor was treated with RT to 5000 cGy, followed by a course of chemotherapy with bleomycin and CCNU for 3 years. He soon had an aggressive recurrence of tumor, which required another course of RT to 3000 cGy, as well as additional chemotherapy with cisplatin and teniposide for 6 months. Within 4 months of stopping the new chemotherapy, he had another recurrence of disease that was mainly intraventricular, prompting treatment with the "8-in-1" chemotherapy regimen, which induced a near CR of the intraventricular enhancing tumor. The next year the intraventricular tumor recurred, with thick sub-ependymal enhancement. At this time he was started on intrathecal thiotepa, which helped to thin out the sub-ependymal tumor. Due to the response from intrathecal thiotepa, it was decided to proceed to IV high-dose thiotepa ($375 mg/m^2$/day) on three consecutive days, followed by an autologous bone marrow transplant. He tolerated the first course of high-dose thiotepa and transplant well with a CR of the enhancing sub-ependymal tumor on follow-up cranial CT scan. After a 2-month interval, he was treated with a second cycle of high-dose thiotepa and autologous bone marrow transplant for consolidative purposes. Due to complications of visceral candidiasis and renal dysfunction, he eventually passed away after an acute pulmonary hemorrhage. At autopsy, the neuropathological inspection revealed areas of necrotic tumor with only occasional tumor cell clusters of questionable viability. Another approach to myeloablative chemotherapy treatment of recurrent oligodendrogliomas was reported by Cairncross and colleagues.[21] In their study, patients with previously irradiated tumors that had measurable, enhancing disease and were behaving aggressively were treated with induction chemotherapy and then, depending on their response, would be considered for myeloablative chemotherapy and hematopoietic reconstitution.

The cohort consisted of 38 patients with recurrent tumors; all were treated with induction chemotherapy consisting of dose intense PCV or cisplatin plus etoposide. The patients with a CR or major PR (i.e., 75% or more reduction in tumor size) were then considered for high-dose thiotepa (300 mg/m^2/day × 3 days), followed by autologous reconstitution with bone marrow or peripheral blood stem cells. A total of 20 patients [10 men, 10 women; median age 46 years; median KPS (Karnofsky performance status) = 80) received the high-dose thiotepa regimen. For the entire high-dose thiotepa cohort, the median event-free survival time from recurrence was 17 months. The overall PFS and OS for the cohort were 20 and 49 months, respectively. Tumor control in excess of 2 years was noted in six patients (30%). Four patients were alive and free of tumor 27–77 months (median, 42 months) from the start of the induction therapy. However, there was significant toxicity from the treatment regimen, including four patients who had fatal treatment-related toxicities (20%; progressive encephalopathy, wasting syndrome, intratumoral hemorrhage). The authors concluded that high-dose thiotepa was not a good strategy for recurrent aggressive oligodendrogliomas, in spite of several durable responses, mainly because of the potential for severe—and sometimes fatal—toxicity and complications.

Several reports have come from Chamberlain and colleagues regarding the use of irinotecan (CPT-11) for salvage chemotherapy of recurrent oligodendrogliomas [22,23]. In a phase I study, 15 patients (age range 24–55 years; 10 males, 5 females) with recurrent tumors were treated with CPT-11.[22] All of the patients had failed prior PCV chemotherapy and were on P-450 enzyme-inducing anticonvulsants (e.g., phenytoin). The dosing of CPT-11 was escalated in four cohorts—three patients at 400 mg/m^2 every 3 weeks, three patients at 500 mg/m^2, six patients at 600 mg/m^2, and three patients at 700 mg/m^2. The median number of cycles of CPT-11 was 4, with a range of 2–12. There were two patients with PR, five patients with SD, and eight patients with PD. The duration of responses ranged from 1.5 to 9.0 months, with a median duration of 3.0 months. In the subgroup of patients with a PR or SD, the duration ranged from 4.5 to 9.0 months, with a median duration of response of 6.0 months. The PFS at 6 months (PFS-6) was 33%, while at 12 months the PFS was 0%. The median OS after the initiation of CPT-11 was 3 months, with a range from 2 to 12 months. However, in responders the median OS was 7 months, with a range from 6 to 12 months. The maximum tolerated dose (MTD) for CPT-11 was 600 mg/m^2. Significant toxicity included neutropenia, abdominal pain with or without diarrhea, nausea and emesis, and thrombocytopenia. In a more recent related study, Chamberlain and Glantz reported the results of a phase II study of CPT-11 in adults with recurrent TMZ refractory, 1p/19q co-deleted anaplastic oligodendrogliomas.[23] A total of 22 patients were enrolled in the study (11 men; 11 women), with a median age of 40 years (range 26–65 years). All patients had previously undergone surgery and involved-field RT, as well as an initial form of adjuvant chemotherapy (TMZ in 15; BCNU in 6). At the time of first recurrence, 15 patients had also tried an alternative form of chemotherapy, while 13 patients had undergone repeat surgical resection. CPT-11 was administered to the patients for every 3 weeks. A total of 141 cycles of were administered, with a median of 3 cycles (range 3–18 cycles). There were five patients with a PR (23%), eight patients with SD (36%), and nine patients with PD (41%) after completing three cycles of CPT-11. The median time to progression was 4.5 months, with a range of 2–13.5 months. The PFS-6 and PFS-12 were 33% and 4.5%, respectively, with a median OS of 5.5 months (range 2–21 months). Toxicity was similar to the phase I study and included diarrhea, neutropenia, fatigue, and delayed nausea and emesis.

Belanger and coworkers used topotecan, a camptothecin analog, in a multicenter phase II trial for salvage treatment of patients with recurrent or progressive anaplastic oligodendrogliomas and anaplastic mixed oligoastrocytomas.[24] A total of 16 patients (with an original plan for an $N = 30$) were enrolled in the study; all had failed prior chemotherapy with PCV. Each patient received topotecan 1.5 mg/m^2 for over 30 min daily for 5 consecutive days every 3 weeks. There were no objective responses documented for any of the patients, while 11 had SD for a median of 3.8 months, and 3 had PD. Significant toxicity was noted, with grade 3 or 4 neutropenia noted in 15 of 16 patients, leading to dosage reductions in almost half of the cycles. Due to the lack of objective responses and overall efficacy, the study was terminated early.

Platinum drug-containing regimens have been used for solid tumors for many years, including brain tumors.[11] Several reports of salvage treatment of progressive or recurrent oligodendrogliomas have included combination regimens with carboplatin.[25,26] In the first study, an Italian group used the combination of carboplatin and teniposide in a phase II trial for salvage therapy (i.e., third line) of patients with progressive oligodendroglioma or oligoastrocytoma after failure of surgery, RT, and chemotherapy with PCV and TZM.[25] The regimen consisted of carboplatin 350 mg/m^2 on day 1, and teniposide 50 mg/m^2 on days 1–3, every 4 weeks. A total of 23 patients were enrolled on study and received treatment with carboplatin and teniposide; a total of 103 cycles were administered (mean 4.4 cycles; range 1–9 cycles). Neuroimaging evaluation showed that 2 patients had PR (8.6%) and 12 patients had SD (52.17%). The PFS-6 was 34.8%, with a median time to progression of 19 weeks (range 11.4–35.0 weeks). At 12 months 51% of the cohort was still alive, with a median OS of 60.7 weeks. The toxicity profile was mild and mainly hematological, with a low risk for high-grade neutropenia and thrombocytopenia. Overall, the authors felt that the regimen of carboplatin and teniposide had moderate activity in heavily pretreated patients with oligodendroglial tumors, along with acceptable toxicity. In the second study, another Italian group used the combination of carboplatin and etoposide in a phase II trial for the treatment of patients with recurrent or progressive oligodendroglial tumors.[26] The treatment regimen consisted of carboplatin AUC-5 on day 1 and etoposide 120 mg/m^2 on days 1–3, every 28 days. A total of 32 patients were enrolled on the study and had the following diagnoses: 9 oligodendroglioma, 3 oligoastrocytoma, 11 anaplastic oligodendroglioma, and 9 anaplastic oligoastrocytoma. Neuroimaging responses demonstrated an overall objective response rate of 46.9%, including 5 patients with CR (15.6%), 10 patients with PR (31.3%), and 11 patients with SD (34.4%). The median time to progression for the entire cohort was 8 months. The PFS-6 and PFS-12 were 80% and 46.9%, respectively. Toxicity was mainly hematological, with grade 3–4 neutropenia in five (15.6%) patients. The authors concluded that carboplatin and etoposide had moderate activity against recurrent and progressive oligodendroglial tumors, and had an acceptable toxicity profile.

In a phase II study, Reardon and colleagues used imatinib and hydroxyurea for patients with progressive or recurrent low-grade gliomas.[27] The cohort consisted of 64 patients, including 32 each with astrocytoma and oligodendroglioma (28 oligodendroglioma, and 4 mixed oligoastrocytoma). The oligodendroglioma cohort had a median age of 46.2 years (range of 22.6–67.2 years), and was 60% male. Prior chemotherapy consisted of mainly TZM, as well as carboplatin and PCV. Each patient was placed on hydroxyurea 500 mg twice per day, as well as imatinib 400 mg per day (nonenzyme inducing anticonvulsant) or 500 mg twice per day (enzyme inducing anticonvulsant). There were no objective radiological responses in either histological group. For the oligodendroglial cohort, the median PFS was 43.3 weeks, with a PFS-12 and PFS-24 of 34.4% and 21.9%, respectively. The median OS was not estimable, but the survival rate at 12 and 24 months was 100% and 90.6%, respectively. The authors concluded that the regimen was well tolerated, but had negligible activity in recurrent or progressive low-grade gliomas.

References

1. Ostrum QT, Gittleman H, Kromer C, et al. CBTRUS statistical report: primary brain and other central nervous system tumors diagnosed in the United States in 2009–2013. *Neuro-Oncol.* 2016;18(Suppl_5). v1–v75.
2. Bailey P, Hiller G. The interstitial tissues of central nervous system: a review. *J Nerv Ment Dis.* 1924;59:337–361.
3. Bailey P, Cushing H. *A classification of the tumors of the glioma group on a histogenetic basis with a correlated study of prognosis.* Philadelphia: Lippincott Co; 1926. pp. 87–89.
4. Bailey P, Bucy PC. Oligodendrogliomas of the brain. *J Pathol.* 1929;32:735–751.
5. Shenkin HA, Grant FC, Drew JH. Post-operative period survival of 25 patients with oligodendroglioma of the brain. Report of 25 cases. *Arch Neurol Psych.* 1948;59:434–436.
6. Earnest F, Kernohan JW, Craig WM. Oligodendroglioma: a review of 200 cases. *Arch Neurol Psych.* 1950;63:964–976.
7. Reymond A, Oligodendroglioma RW. Anatomico-clinical study of 74 cases. *Schweitzer Arch Neurol Psych.* 1950;65:221–254.
8. Freeman L, Feigin I. Oligodendroglioma with 35 years survival. *J Neurosurg.* 1963;20:363–365.
9. Solitaire GB, Robinson F, Lamarche JB. Oligodendroglioma: recurrence following an exceptionally long postoperative symptom-free interval. *Canad Med Assoc J.* 1967;Vol. 97:862–865.
10. Roberts M, German WJ. Oligodendroglioma: a 40-year survival. Case report. *J Neurosurg.* 1969;31:355–357.
11. Newton HB. Chemotherapy of high-grade astrocytomas. In: Newton HB, ed. *Handbook of Brain Tumor Chemotherapy.* Amsterdam: Elsevier/Academic Press; 2006:347–363. Vol. 24.
12. Levin VA, Crafts DC, Wilson CB, et al. BCNU (NSC-409962) and procarbazine (NSC-77213) treatment for malignant brain tumors. *Cancer Treat Rep.* 1976;60:243–249.
13. Shapiro WR, Young DF. Chemotherapy of malignant glioma with CCNU alone and CCNU combined with vincristine sulfate and procarbazine hydrochloride. *Trans Am Neurol Assoc.* 1976;101:217–220.
14. Barbizet J, Caron JP, Comoy J, et al. Oligodendrogliomas: clinical and therapeutic study (French). *Sem Hop.* 1981;57:221–224.
15. Lange OF, Haase KD, Scheef W. Simultaneous radio- and chemotherapy of inoperable brain tumours. *Radiother Oncol.* 1987;8:309–314.
16. Kitihara M, Katakura R, Mashiyama S, et al. Results in oligodendroglioma: postoperative radiotherapy combined with chemotherapy (Japanese). *No Shinkei Geka.* 1987;15:397–403.
17. Levin VA, Chamberlain MC, Prados MD, et al. Phase I–II study of eflornithine and mitoguazone combined in the treatment of recurrent primary brain tumors. *Cancer Treat Rep.* 1987;71:459–464.
18. Cairncross JG, Mcdonald DR. Successful chemotherapy for recurrent malignant oligodendroglioma. *Ann Neurol.* 1988;23:360–364.
19. Brown M, Cairncross JG, Vick NA, et al. Differential response to recurrent oligodendrogliomas and astrocytomas to intravenous (IV) melphalan (abstract). *Neurol.* 1990;40(Suppl 1):397–398.
20. Saarinen UM, Pihko H, Makipernaa A. High-dose thiotepa with autologous bone marrow rescue in recurrent malignant oligodendroglioma: a case report. *J Neurooncol.* 1990;9:57–61.
21. Cairncross G, Swinnen L, Bayer R, et al. Myeloablative chemotherapy for recurrent aggressive oligodendroglioma. *Neuro-Oncol.* 2000;2:114–119.
22. Chamberlain MC. Salvage chemotherapy with CPT-11 for recurrent oligodendrogliomas. *J Neurooncol.* 2002;59:157–163.
23. Chamberlain MC, Glantz MJ. CPT-11 for recurrent temozolomide-refractory 1p19q co-deleted anaplastic oligodendroglioma. *J Neurooncol.* 2008;89:231–238.

24. Belanger K, MacDonald D, Cairncross G, et al. A phase II study of topotecan in patients with anaplastic oligodendroglioma or anaplastic mixed oligoastrocytoma. *Invest New Drugs*. 2003;21:473–480.

25. Brandes AA, Basso U, Vastola F, et al. Carboplatin and teniposide as third-line chemotherapy in patients with recurrent oligodendroglioma or oligoastrocytoma: a phase II study. *Ann Oncol*. 2003;14:1727–1731.

26. Scopece L, Franceschi E, Cavallo G, et al. Carboplatin and etoposide (CE) chemotherapy in patients with recurrent or progressive oligodendroglial tumors. *J Neurooncol*. 2006;79:299–305.

27. Reardon DA, Desjardins A, Vredenburgh JJ, et al. Phase II study of Gleevec plus hydroxyurea in adults with progressive or recurrent low-grade gliomas. *Cancer*. 2012;118:4759–4767.

Chapter 31

Molecular Therapy for Oligodendrogliomas

Julie J. Miller* and Patrick Y. Wen[†]

*Department of Neurology, Pappas Center for Neuro-Oncology, Massachusetts General Hospital, Boston, MA, United States; [†]Center for Neuro-Oncology, Dana-Farber Cancer Institute, Boston, MA, United States

Introduction

Oligodendroglial tumors are primary brain tumors that are believed to originate from glial precursor cells and account for slightly less than 10% of all adult gliomas.[1] These tumors appear to resemble oligodendrocytes histologically, with a characteristic round nucleus and swollen, empty-appearing cytoplasm lending to the "fried egg" description used by pathologists.[2] Oligodendrogliomas are classified as the World Health Organization (WHO) grade II if the tumor appears well differentiated while anaplastic oligodendrogliomas, WHO grade III, display more anaplasia, mitotic activity, and occasional necrosis.[2]

On a molecular level, oligodendrogliomas and anaplastic oligodendrogliomas are defined by mutations in the *isocitrate dehydrogenase 1 or 2 (IDH1* or *IDH2)* gene, as well as heterozygous whole arm loss of the long arm of chromosome 1 and the short arm of chromosome 19, known as 1p/19q co-deletion.[2-4] In addition, the majority of oligodendroglial tumors acquire a mutation in the promoter of the telomerase reverse transcriptase (*TERT*) gene, which encodes for the enzyme telomerase which maintains the ends of chromosomes.[5,6]

It has long been appreciated that oligodendroglial tumors typically exhibit a slower pattern of growth and less aggressive behavior compared to other types of gliomas, including astrocytomas.[7] Even in the absence of conventional treatment, overall survival of patients with these tumors can be greater than a decade.[8] Current conventional treatments include a combination or safe maximal resection, radiation, and chemotherapy including procarbazine, lomustine, and vincristine (PCV) and temozolomide (TMZ). The long natural history of these tumors has led to varying opinions concerning the optimal timing of treatment. Because patients with oligodendrogliomas tend to be young, with median age at diagnosis of 35,[9] and enjoy prolonged overall survival, there is much concern surrounding the long-term side effects of aggressive treatments, including cytotoxic chemotherapy, and impact on quality of life. Nevertheless, there is increasing evidence supporting the notion that radiation therapy in combination with chemotherapy leads to increased progression free and overall survival for both oligodendrogliomas and anaplastic oligodendrogliomas. While the only currently approved drugs for oligodendrogliomas and anaplastic oligodendrogliomas are chemotherapies, newer classes of agents, including targeted molecular therapies and immune modulating therapies, are under investigation in clinical trials. There is much hope that these agents, which are designed to specifically target pathways believed to be dysregulated in oligodendrogliomas, will be less toxic than traditional chemotherapies. Novel therapeutics under study in clinical trials and on the horizon will be the focus of this chapter.

The current status of systemic therapy for treatment of oligodendroglial tumors

Currently, the only approved systemic therapies for the treatment of oligodendroglioma and anaplastic oligodendroglioma are chemotherapies, with the most common regimens involving TMZ or the three-drug combination PCV. Although treatment with chemotherapy has been shown to lead to a significant prolongation of overall survival,[10-12] treatment with chemotherapy is not without risks. In addition to gastrointestinal symptoms and bone marrow suppression that occurs relatively commonly during a typical cycle of PCV or TMZ, these drugs can have long-term side effects, including increasing the risk of secondary malignancy, particularly hematologic disorders like leukemia and myelodysplastic syndromes.[13,14]

Oligodendroglioma. https://doi.org/10.1016/B978-0-12-813158-9.00031-1

As with other areas of oncology, there is much interest in the development of targeted therapies for oligodendrogliomas, as these therapies offer the promise of better efficacy coupled with lower side effects.

Several promising investigational agents are under study in clinical trials. While not specifically designed for the treatment of patients with oligodendroglial tumors, these drugs have been developed based on the underlying biology of *IDH1* and *IDH2* gliomas, thereby encompassing nearly all oligodendrogliomas.

IDH inhibitors

As mentioned previously, in addition to loss of chromosome arms 1p and 19q, the majority of oligodendroglial tumors have a mutation in either the *IDH1* or *IDH2* gene.[15] Wild-type IDH1 and IDH2 are metabolic enzymes that promote the conversion of isocitrate to alpha-ketoglutarate (α-KG), with IDH1 functioning in the cytoplasm and IDH2 active in the mitochondria. The recurrent mutation that has been observed in glioma occurs predominantly at the active site (arginine 132 in IDH1 and arginine 172 in IDH2)[16] and leads to the acquisition of additional enzymatic activity, resulting in the subsequent conversion of α-KG to the metabolite R-2-hydroxyglutarate (2-HG).[17] 2-HG accumulates in large amounts, and excess levels are thought to trigger tumorigenesis[17–19] by leading to a change in the epigenetic state of the cells and preventing differentiation (Fig. 1). The IDH mutation and excess 2-HG production has been correlated with elevated repressive histone methylation.[19] This appears to be related to the ability of 2-HG to inhibit the α-KG-dependent family of dioxygenases that have an important role in controlling the histone methylation state.[18,19] Further, the presence of the IDH mutation leads to a global DNA hypermethylation phenotype known as CpG island methylator phenotype (CIMP),[20] perhaps in part related to alterations in histone methylation patterns. Overall these changes result in an impaired differentiation state.

Acquisition of an *IDH* mutation appears to be an early driving event in glioma development[21,22] and *IDH* mutations persist during tumor recurrence and transformation of the tumor to a higher grade.[23] These characteristics render IDH mutant-specific inhibitors an attractive therapeutic strategy. In 2013, a compound designed to inhibit 2-HG production by mutant IDH1 was demonstrated to efficiently inhibit the growth of IDH1 R132H expressing cells in vitro as well as IDH mutant glioma xenografts in mouse models.[24] In contrast, the compound was ineffective in an IDH wild-type model.[24] In line with data suggesting that excess 2-HG leads to hypermethylation through inhibition of ten-eleven translocation (TET) hydrolases and Jumonji-C domain histone demethylases[19] (Fig. 2), profiling of cells treated with the IDH inhibitor showed decreased histone methylation and a consequent increase in glial differentiation markers.[24]

Since the publication of those data, a number of oral, small molecule IDH inhibitors that potently lower 2-HG levels have been investigated in early phase clinical trials for recurrent gliomas. Preliminary results from the phase I dose-escalation and expansion trial of the IDH1 mutant-specific inhibitor AG-120 (Ivosidenib, Agios Pharmaceuticals, NCT02073994) were presented at the Annual Meeting of the Society of Neuro-Oncology in both 2016 and 2017.[25,26] The agent was not effective for high-grade tumors but the investigators leading the study reported stable disease in 29 of 35 patients with non-enhancing, recurrent *IDH* mutant gliomas, with a median duration of treatment of 16 months. Using volumetric quantification to calculate tumor growth rates, they observed a slowing of tumor growth while on the treatment

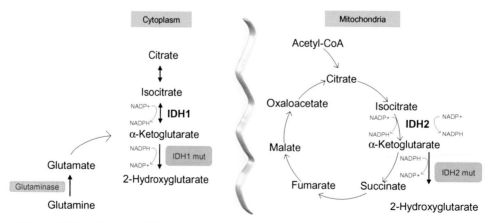

FIG. 1 Wild-type IDH1 and IDH2 catalyze the NADP+-dependent reduction of isocitrate to α-ketoglutarate in the cytoplasm and mitochondria, respectively. The mutant form of the enzymes (IDH1 mut, IDH2 mut) gain the additional ability to convert α-ketoglutarate to *D*-2-hydroxyglutarate (2-HG), resulting in the oxidation of NADPH. Glutamate, derived from glutamine, is a critical precursor to α-ketoglutarate production. *IDH: isocitrate dehydrogenase; NADP +/NADPH: nicotinamide adenine dinucleotide phosphate.*

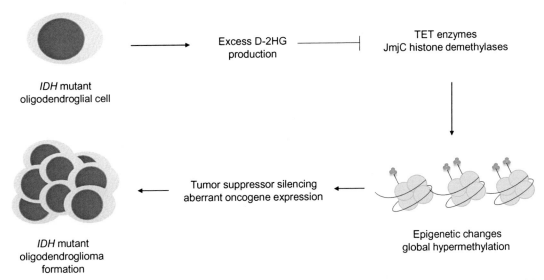

FIG. 2 The proposed role of *D*-2-hydroxyglutarate (2-HG) in gliomagenesis. Production of 2-HG by mutant IDH enzyme inhibits multiple α-ketoglutarate-dependent dioxygenases, including the ten-eleven translocation (TET) family of 5-methyl cytosine hydrolases and jumonji domain containing (JmjC) family of histone demethylases. This is believed to change the overall epigenetic state of the cell, leading to a pattern of global hypermethylation and subsequent aberrant expression of oncogenes and tumor suppressors. The altered transcriptional program results in the formation of an oligodendroglioma. IDH: isocitrate dehydrogenase; D-2HG: D-2-hydroxyglutarate; TET: ten-eleven translocation; JmjC: jumonji domain containing.

with AG-120 compared to the rate observed in the year leading up to trial enrollment. The most common side effects experienced by patients on trial included grade 1 or 2 headache (34%), diarrhea (26%), and nausea/vomiting (20%). The only significant adverse event was hypophosphatemia, experienced by two patients.[26]

In addition to AG-120 (ivosidenib), a combined IDH1 and IDH2 inhibitor, AG-881 is being tested in an ongoing phase I trial (NCT02481154) in a variety of *IDH* mutant tumors, including gliomas. IDH305 is another IDH1 mutant-specific inhibitor that may be tested in a phase II trial in progressive or recurrent grade II and III *IDH1* mutant gliomas (NCT02977689); however, the phase I trial was halted secondary to concerns around hepatotoxicity. In addition, the oral IDH1 inhibitor BAY1436032[27] is in a phase I clinical trial (NCT02746081) for advanced solid tumors with IDH1 mutations, including anaplastic gliomas and glioblastoma.

A major challenge to the development of IDH inhibitors is that if they merely stabilize disease, randomized trials using progression-free survival as an endpoint will be required for registration. For oligodendrogliomas, in particular, these types of trials will take a number of years to complete and pose a significant hurdle to obtaining results in a timely manner. Ongoing studies are underway to determine if these agents change the growth trajectory of the tumors and whether this can serve as a surrogate endpoint.

It is worth noting that the lack of complete and partial responses in this trial is in stark contrast to the striking results noted for a similar IDH mutant inhibitor that was tested in *IDH2* mutant acute myeloid luekemia[28] and now approved for clinical use (Enasidenib, Agios Pharmaceuticals). The lack of obvious clinical efficacy in glioma patients may be related to incompletely understood differences in IDH function between glial and hematologic cells. Although elevated 2-HG production from mutant IDH is generally well accepted as a fundamental event in the path to glioma formation, there is continued debate in the field as to whether decreasing 2-HG levels is useful once the tumor has been established. The epigenetic changes that are promoted by 2-HG presumably happen as an early oncogenic event and lead to a stable change in differentiation state[19,20] (Fig. 2). Once this change in state is established, it may not be easily reversed. Indeed, in some IDH mutant patient-derived cell lines grown in culture or used in xenograft models, IDH inhibitors have minimal effect on cell growth and viability.[29]

One major challenge to understanding why a differential effect exists is the lack of in vitro models that faithfully recapitulate the behavior of low-grade IDH mutant astrocytomas and oligodendrogliomas. Cell lines derived from tumor resected from patients with IDH mutant gliomas serve as the primary model being used to study this tumor type, but the majority of these lines have been derived from IDH mutant glioblastoma and, less commonly, anaplastic astrocytoma.[30–32] It remains possible that IDH inhibitors will only be effective at the earliest stages of glioma formation and are rendered ineffective after a certain point in the tumor development. If this hypothesis is true, then it is not surprising that IDH inhibitors have not shown wide efficacy in IDH mutant glioma models. This concept is further supported by work from the Pieper lab using a cell culture model in which normal human astrocytes lacking tumor suppressor 53 (p53) and retinoblastoma protein (Rb) were transformed by the expression of IDH1 R132H and TERT. They observed that the treatment with an IDH1 inhibitor prior to or concomitant with induction of IDH1 expression blocks IDH1-induced transformation, as detected by growth in soft agar. Delaying IDH1 inhibitor treatment by as little as 4 days after transformation, however, resulted in failure to halt glioma growth. Later administration of IDH inhibitor also failed to reverse the characteristic histone methylation observed after IDH1 mutant-induced transformation.[33] There may be a short window of time during which inhibition of 2-HG production by IDH inhibitors are effective, after which it may be too late for the epigenetic changes conferred by excess 2-HG to be reversed.

Overall, additional in-depth preclinical data as well as more clinical data with traditional endpoints are needed before true efficacy of this drug class can be fully assessed in the brain tumor population.

Everolimus

Many low-grade gliomas, including oligodendrogliomas, have been found to have activation of the phosphatidylinositol 3-kinase (PI3K) pathway.[34] One strategy to target aberrant activation of PI3K involves inhibition of the downstream target, mammalian target of rapamycin (mTOR). In a phase II, single-arm trial involving 58 patients with recurrent astrocytomas, oligodendrogliomas, or oligoastrocytomas treated with everolimus, the progression-free survival at 6 months was determined to be 84% (95% CI 71%–94%) for the subset of patients (47 of 58) with the WHO grade II gliomas at enrollment. The vast majority of these patients had *IDH* mutant tumors, 22 of which were oligodendroglioma with 1p/19q co-deletion. Many of these patients remained with stable disease and 48% stayed on treatment for at least 1 year.[35] The preliminary data are encouraging but additional, randomized trials with a larger, more uniform patient population will be required to determine the true efficacy of everolimus.

PARP inhibitors

Another line of investigation involves the use of poly (adenosine 5′-diphosphate ribose) polymerase (PARP) inhibitors for the treatment of grade II and grade III gliomas with *IDH* mutations. There is evidence that elevated levels of 2-HG produced by mutant IDH leads to deficits in homologous repair and DNA damage, at least in vitro. This deficiency can be exploited by concurrently shutting down a parallel DNA repair pathway that utilizes the PARP1 enzyme via small molecule, oral PARP inhibitors.[36] Several PARP inhibitors are currently in clinical use for other malignancies, including breast cancer and ovarian cancer. Based on the enthusiasm surrounding the preclinical data, an open-label, phase II trial using Olaparib has been initiated by the National Cancer Institute for treating patients with advanced solid tumors with *IDH1* and *IDH2* mutations, including gliomas (NCT03212274). The primary endpoint of this trial will be radiologic response rate, using Response Assessment in Neuro-Oncology (RANO) criteria[37] for patients with glioma. In addition, the American Brain Tumor Consortium (ABTC) is currently in the process of planning a trial using the PARP1/2 inhibitor BGB-290 (BeiGene USA, Inc.) administered concurrently with TMZ for recurrent or progressive IDH mutant tumors.

Glutaminase inhibition

α-KG is the required substrate for IDH mutant-mediated production of 2-HG (Fig. 1). Much of the α-KG in the cell is derived from glutamate, produced by the hydrolysis of glutamine through the activity of the enzyme glutaminase.[38] Therefore, IDH mutant tumors may be sensitive to a decrease in the pool of available glutamate. Indeed, inhibition of glutaminase by small interfering RNA (siRNA) or by small molecule inhibitor BPTES (bis-2-(5-phenylacetamido-1,2,4-thiadiazol-2-yl)ethyl sulfide) led to a selective slowing of growth in IDH mutant glioblastoma cell lines compared to IDH wild-type glioblastoma cell lines.[38] Similar IDH-dependent sensitivity to BPTES has been reported for IDH mutant acute myelogenous leukemia cells in vitro.[39] In addition, glutamate serves as a critical building block for production of nucleic acids, fatty acids, and proteins, all of which are crucial for cell proliferation, particularly in rapidly growing cancer cells. Blockade of glutamate supply through glutaminase inhibition potentiates DNA damage from ionizing radiation by decreasing the supply of nucleotides, leading to fewer precursors available to repair DNA.[40] There is also evidence that glutaminase inhibitors hypersensitize to alkylating agents, such as the chemotherapeutics used in the treatment of glioma.[41]

In light of these data and the alterations in metabolic flux caused by IDH mutations, there is much interest in the community in using glutaminase inhibitors for the treatment of IDH tumors. Calithera Biosciences, Inc. has developed a selective, oral glutaminase inhibitor CB-839 that was well tolerated in initial phase I studies.[42] It is now under further investigation in an extension of the phase I trial in a variety of solid tumor types, including tumors harboring IDH1 or IDH2 mutations (NCT02071862). In addition, Calithera is collaborating with the National Cancer Institute Cancer Therapy Evaluation Program (CTEP) to study CB-839 more in-depth in IDH mutant glioma patients, potentially with concomitant administration during radiation treatment.

Immunotherapy approaches

Given the successful, long-term remissions observed in patients with melanoma and lung cancer following the treatment with immune checkpoint inhibitors, a variety of agents designed to promote immune system activation for tumor targeting are being explored for the *IDH* mutant population. The most popular approach in grade II and III gliomas has thus far been the design of IDH mutant vaccines. Indeed, three different IDH1 R132H vaccine trials are underway, including two peptide vaccines (NCT02454634, NCT02193347) and one dendritic cell vaccine (NCT02771301). An IDH mutant peptide vaccine efficiently decreased tumor growth in mouse gliomas[43] and generated antitumor immunity,[44] offering promise for translation to the clinical setting. This strategy is quite attractive because these vaccines theoretically enhance the ability of the patient's immune system to recognize and attack tumor cells with the IDH1 R132H mutation; at the same time, this tumor-specific marker should not result in an autoimmune response elsewhere in the body.

The immune checkpoint inhibitor Avelumab, which targets the programmed death-1 receptor (PD-1) and renders the tumor more vulnerable to the host immune system, is being tested with hypofractionated radiation therapy in patients who have malignant *IDH* mutant tumors that have progressed to glioblastoma (NCT02968940). Enrolled patients must have been previously treated with chemotherapy regimens that included an alkylating agent, such as TMZ or lomustine, which has been correlated with the emergence of a hypermutated phenotype at recurrence.[23] Increased number of somatic mutations is believed to lead to a higher neoantigen load, which should theoretically lead to an enhanced response to immunotherapy. Although oligodendrogliomas and anaplastic oligodendrogliomas are not thought to progress to glioblastomas, the use of a PD-1 inhibitor in this setting suggests that other checkpoint inhibitors may be explored for *IDH* mutant tumors and possibly oligodendrogliomas in the future.

Notably, recent results suggest that 2-HG produced by mutant IDH enzyme creates an immunosuppressive environment. Introduction of mutant IDH1 leads to decreased STAT1 expression, thereby decreasing chemokine production and ultimately leading to a decrease in CD8+ T-cell infiltration into tumor sites. This impairment in CD8+ T-cell recruitment to tumor sites could be reversed by using an IDH1 inhibitor that decreases 2-HG levels. The authors further demonstrated that the efficacy of a peptide vaccine was enhanced with concurrent administration of an IDH inhibitor in a mouse syngeneic glioma model.[45] In agreement with these data, a separate group noted that increased tumor infiltrating lymphocytes and higher PD-L1 expression is observed in IDH wild-type grade II and grade III gliomas compared to IDH mutant tumors of the same grade.[46] These findings have prompted the initiation of trials, still in the planning stages, to combine immunotherapy strategies with IDH inhibition to overcome the immunosuppressive environment.

Other promising targets
DNA hypermethylation

As a group, *IDH* mutant glioma exhibits an altered epigenome and a hypermethylated DNA pattern known as the CIMP. The CIMP pattern is believed to be a result of elevated 2-HG levels, which inhibit α-KG-dependent dioxygenases, including the TET family of demethylases, which are believed to be critical for controlling levels of DNA methylation throughout the genome.[20] The 2-HG-induced altered methylation pattern is thought to result in aberrant expression of oncogenes and tumor suppressors.[20,47] Increased levels of methylation at insulator sites, which normally create boundaries and allow the creation of topographical domains within the genome leads to decreased binding of insulator regulatory proteins and allows changes in chromatin structure. Recent work has demonstrated that one consequence of decreased insulator binding is aberrant activation of PDGFRA expression in *IDH* mutant gliomas.[47] Platelet-derived growth factor receptor (PDGFR) is a receptor tyrosine kinase that promotes signal transduction following activation by PDGF, a growth factor that stimulates growth and survival; excess activation can lead to glioma development.[48] Disruption of topographical domains may lead to abnormal activation of other oncogenes as well. Therefore, an alternative treatment strategy for *IDH* mutant gliomas is to attempt to reverse the hypermethylated phenotype using agents that cause hypomethylation, such as decitabine or 5-azacytadine. This class of drugs has been shown to promote differentiation and decrease glioma proliferation in vitro and in mouse *IDH* mutant tumor models.[49,50] Demethylating agents also decrease PDGFRA expression in vitro, in line with the role of insulator function disruption in gliomagenesis.[47] While no glioma trials involving demethylating agents have yet been initiated, the concept has been explored for patients with *IDH* mutant acute myelogenous leukemia[51] and is likely to be tested in glioma in the near future. Alternatively, the aberrant activation of PDGFR could be targeted by using receptor tyrosine kinase inhibitors that are being explored in other tumor types, including lung cancer, glioblastoma, and sarcoma.[48]

TERT activation

The majority of oligodendroglial tumors with 1p/19q co-deletion also exhibit mutations in the *TERT* promoter, leading to telomere maintenance. This raises the possibility of targeting excess telomerase activity to halt tumor growth. Telomerase, however, is a difficult enzyme to target safely because the enzymatic activity is required by normal tissue, particularly stem cells.[52] Interestingly, the TERT promoter mutations commonly found in oligodendroglioma appear to recruit special transcription factors that are not normally involved in the regulation of TERT.[53] A deeper understanding of this abnormal recruitment and activation offers the potential for tumor-specific therapies in the future.

Nicotinamide phosphoribosyltransferase inhibition

IDH mutant gliomas have been shown to exhibit low basal levels of nicotinamide adenine dinucleotide (NAD).[29] This metabolic deficiency can be exploited by the use of drugs that inhibit the NAD biosynthesis enzyme nicotinamide phosphoribosyltransferase (NAMPT), which efficiently causes *IDH*-mutant-specific cytotoxicity both in vitro and in vivo in patient-derived cell lines.[29] Unfortunately, the clinical development of NAMPT inhibitors have been hampered by retinal and cardiac toxicity observed in rodents.[54,55] This treatment strategy may be viable if the exquisite sensitivity to NAMPT inhibition is coupled with mechanisms to deplete NAD pools by other means. For instance, preclinical studies have demonstrated that TMZ induces PARP-mediated consumption of NAD. The combination of TMZ with NAMPT inhibitor at low doses, which would likely avoid toxicity concerns, results in improved efficacy in an *IDH* mutant tumor model compared to single-drug treatment alone.[56]

Conclusions

In recent years, the treatment of oligodendrogliomas and anaplastic oligodendrogliomas has evolved to now routinely incorporate systemic chemotherapy into regimens containing maximal surgical resection and radiation therapy. Given the many side effects and toxicities associated with traditional chemotherapy agents, much effort is being put into the discovery of novel molecular therapies to treat these tumors, particularly as we gain more knowledge about the molecular characteristics defining oligodendrogliomas and the underlying mechanisms driving tumor development. The promising strategies on the horizon take advantage of this knowledge and include targeting excess 2-hydroxyglutarate production with direct IDH inhibitors, exploiting DNA damage deficiency with PARP inhibitors, and manipulating the mTOR pathway and capitalizing on glutamine dependence. These strategies are currently in early phase clinical trials but offer hope for enhanced antitumor efficacy combined with less toxicities. One challenge to developing novel agents for slow growing IDH-mutated tumors is that clinical trials using standard endpoints such as progression-free survival can take a very long time. There are efforts underway to evaluate novel endpoints such as change in tumor volume growth trajectory.

References

1. Ostrom QT, Gittleman H, Liao P, et al. CBTRUS Statistical Report: Primary brain and other central nervous system tumors diagnosed in the United States in 2010–2014. *Neuro Oncol.* 2017;19(suppl 5). v1-v88.
2. Louis DN, Ohgaki H, Wiestler O, et al. *WHO Classification of Tumours of the Central Nervous System.* 4th ed. International Agency for Research on Cancer: Lyon; 2016.
3. von Deimling A, Louis DN, von Ammon K, Petersen I, Wiestler OD, Seizinger BR. Evidence for a tumor suppressor gene on chromosome 19q associated with human astrocytomas, oligodendrogliomas, and mixed gliomas. *Cancer Res.* 1992;52(15):4277–4279.
4. Kraus JA, Koopmann J, Kaskel P, et al. Shared allelic losses on chromosomes 1p and 19q suggest a common origin of oligodendroglioma and oligoastrocytoma. *J Neuropathol Exp Neurol.* 1995;54(1):91–95.
5. Suzuki H, Aoki K, Chiba K, et al. Mutational landscape and clonal architecture in grade II and III gliomas. *Nat Genet.* 2015;47(5):458–468.
6. Eckel-Passow JE, Lachance DH, Molinaro AM, et al. Glioma groups based on 1p/19q, IDH, and TERT promoter mutations in tumors. *N Engl J Med.* 2015;372(26):2499–2508.
7. Ricard D, Kaloshi G, Amiel-Benouaich A, et al. Dynamic history of low-grade gliomas before and after temozolomide treatment. *Ann Neurol.* 2007;61 (5):484–490.
8. Lebrun C, Fontaine D, Ramaioli A, et al. Long-term outcome of oligodendrogliomas. *Neurology.* 2004;62(10):1783–1787.

9. Batchelor TT, Nishikawa R, Tarbell NJ, Weller M. *Oxford Textbook of Neuro-Oncology.* 1st ed. New York, New York, USA: Oxford University Press; 2017.

10. Cairncross G, Wang M, Shaw E, et al. Phase III trial of chemoradiotherapy for anaplastic oligodendroglioma: long-term results of RTOG 9402. *J Clin Oncol.* 2013;31(3):337–343.

11. van den Bent MJ, Brandes AA, Taphoorn MJ, et al. Adjuvant procarbazine, lomustine, and vincristine chemotherapy in newly diagnosed anaplastic oligodendroglioma: long-term follow-up of EORTC brain tumor group study 26951. *J Clin Oncol.* 2013;31(3):344–350.

12. Buckner JC, Shaw EG, Pugh SL, et al. Radiation plus Procarbazine, CCNU, and vincristine in low-grade glioma. *N Engl J Med.* 2016;374 (14):1344–1355.

13. Villano JL, Letarte N, Yu JM, Abdur S, Bressler LR. Hematologic adverse events associated with temozolomide. *Cancer Chemother Pharmacol.* 2012;69(1):107–113.

14. Momota H, Narita Y, Miyakita Y, Shibui S. Secondary hematological malignancies associated with temozolomide in patients with glioma. *Neuro Oncol.* 2013;15(10):1445–1450.

15. Yan H, Parsons DW, Jin G, et al. IDH1 and IDH2 mutations in gliomas. *N Engl J Med.* 2009;360(8):765–773.

16. Hartmann C, Meyer J, Balss J, et al. Type and frequency of IDH1 and IDH2 mutations are related to astrocytic and oligodendroglial differentiation and age: a study of 1,010 diffuse gliomas. *Acta Neuropathol.* 2009;118(4):469–474.

17. Dang L, White DW, Gross S, et al. Cancer-associated IDH1 mutations produce 2-hydroxyglutarate. *Nature.* 2009;462(7274):739–744.

18. Xu W, Yang H, Liu Y, et al. Oncometabolite 2-hydroxyglutarate is a competitive inhibitor of α-ketoglutarate-dependent dioxygenases. *Cancer Cell.* 2011;19(1):17–30.

19. Lu C, Ward PS, Kapoor GS, et al. IDH mutation impairs histone demethylation and results in a block to cell differentiation. *Nature.* 2012;483 (7390):474–478.

20. Turcan S, Rohle D, Goenka A, et al. IDH1 mutation is sufficient to establish the glioma hypermethylator phenotype. *Nature.* 2012;483(7390):479–483.

21. Watanabe T, Nobusawa S, Kleihues P, Ohgaki H. IDH1 mutations are early events in the development of astrocytomas and oligodendrogliomas. *Am J Pathol.* 2009;174(4):1149–1153.

22. Lai A, Kharbanda S, Pope WB, et al. Evidence for sequenced molecular evolution of IDH1 mutant glioblastoma from a distinct cell of origin. *J Clin Oncol.* 2011;29(34):4482–4490.

23. Johnson BE, Mazor T, Hong C, et al. Mutational analysis reveals the origin and therapy-driven evolution of recurrent glioma. *Science.* 2014;343 (6167):189–193.

24. Rohle D, Popovici-Muller J, Palaskas N, et al. An inhibitor of mutant IDH1 delays growth and promotes differentiation of glioma cells. *Science.* 2013;340(6132):626–630.

25. Mellinghoff IK, Touat M, Maher E, et al. ACTR-46. AG120, a first-in-class mutant IDH1 inhibitor in patients with recurrent or progressive IDH1 mutant glioma: results from the phase 1 glioma expansion cohorts. *Neuro-Oncology.* 2016;18(suppl_6).vi12-vi12.

26. Mellinghoff IK, Touat M, Maher E, et al. ACTR-46. AG-120, A first-in-class mutant IDH1 inhibitor in patients with recurrent or progressive IDH1 mutant glioma: updated results from the phase 1 non-enhancing glioma population. *Neuro-Oncology.* 2017;19(suppl_6).vi10-vi11.

27. Pusch S, Krausert S, Fischer V, et al. Pan-mutant IDH1 inhibitor BAY 1436032 for effective treatment of IDH1 mutant astrocytoma in vivo. *Acta Neuropathol.* 2017;133(4):629–644.

28. Stein EM, DiNardo CD, Pollyea DA, et al. Enasidenib in mutant-IDH2 relapsed or refractory acute myeloid leukemia. *Blood.* 2017;.

29. Tateishi K, Wakimoto H, Iafrate AJ, et al. Extreme vulnerability of IDH1 mutant cancers to NAD+ depletion. *Cancer Cell.* 2015;28(6):773–784.

30. Wakimoto H, Tanaka S, Curry WT, et al. Targetable signaling pathway mutations are associated with malignant phenotype in IDH-mutant gliomas. *Clin Cancer Res.* 2014;20(11):2898–2909.

31. Luchman HA, Stechishin OD, Dang NH, et al. An in vivo patient-derived model of endogenous IDH1-mutant glioma. *Neuro Oncol.* 2012;14 (2):184–191.

32. Lenting K, Verhaak R, Ter Laan M, Wesseling P, Leenders W. Glioma: experimental models and reality. *Acta Neuropathol.* 2017;133(2):263–282.

33. Johannessen TA, Mukherjee J, Viswanath P, et al. Rapid conversion of mutant IDH1 from driver to passenger in a model of human Gliomagenesis. *Mol Cancer Res.* 2016;14(10):976–983.

34. Brat DJ, Verhaak RG, Aldape KD, et al. Comprehensive, integrative genomic analysis of diffuse lower-grade gliomas. *N Engl J Med.* 2015;372 (26):2481–2498.

35. Wahl M, Chang SM, Phillips JJ, et al. Probing the phosphatidylinositol 3-kinase/mammalian target of rapamycin pathway in gliomas: a phase 2 study of everolimus for recurrent adult low-grade gliomas. *Cancer.* 2017;123(23):4631–4639.

36. Sulkowski PL, Corso CD, Robinson ND, et al. 2-Hydroxyglutarate produced by neomorphic IDH mutations suppresses homologous recombination and induces PARP inhibitor sensitivity. *Sci Transl Med.* 2017;9(375).

37. van den Bent MJ, Wefel JS, Schiff D, et al. Response assessment in neuro-oncology (a report of the RANO group): assessment of outcome in trials of diffuse low-grade gliomas. *Lancet Oncol.* 2011;12(6):583–593.

38. Seltzer MJ, Bennett BD, Joshi AD, et al. Inhibition of glutaminase preferentially slows growth of glioma cells with mutant IDH1. *Cancer Res.* 2010;70 (22):8981–8987.

39. Emadi A, Jun SA, Tsukamoto T, Fathi AT, Minden MD, Dang CV. Inhibition of glutaminase selectively suppresses the growth of primary acute myeloid leukemia cells with IDH mutations. *Exp Hematol.* 2014;42(4):247–251.

40. Sappington DR, Siegel ER, Hiatt G, et al. Glutamine drives glutathione synthesis and contributes to radiation sensitivity of A549 and H460 lung cancer cell lines. *Biochim Biophys Acta.* 2016;1860(4):836–843.

41. Tran TQ, Ishak Gabra MB, Lowman XH, et al. Glutamine deficiency induces DNA alkylation damage and sensitizes cancer cells to alkylating agents through inhibition of ALKBH enzymes. *PLoS Biol.* 2017;15(11).

42. Meric-Bernstam F, DeMichele A, Telli ML, et al. Abstract C49: Phase 1 study of CB-839, a first-in-class, orally administered small molecule inhibitor of glutaminase in patients with refractory solid tumors. *Molecular Cancer Therapeutics.* 2015;14(12 Suppl 2):C49.

43. Pellegatta S, Valletta L, Corbetta C, et al. Effective immuno-targeting of the IDH1 mutation R132H in a murine model of intracranial glioma. *Acta Neuropathol Commun.* 2015;3:4.

44. Schumacher T, Bunse L, Pusch S, et al. A vaccine targeting mutant IDH1 induces antitumour immunity. *Nature.* 2014;512(7514):324–327.

45. Kohanbash G, Carrera DA, Shrivastav S, et al. Isocitrate dehydrogenase mutations suppress STAT1 and CD8 + T cell accumulation in gliomas. *J Clin Invest.* 2017;127(4):1425–1437.

46. Berghoff AS, Kiesel B, Widhalm G, et al. Correlation of immune phenotype with IDH mutation in diffuse glioma. *Neuro Oncol.* 2017;19 (11):1460–1468.

47. Flavahan WA, Drier Y, Liau BB, et al. Insulator dysfunction and oncogene activation in IDH mutant gliomas. *Nature.* 2016;529(7584):110–114.

48. Heldin CH. Targeting the PDGF signaling pathway in tumor treatment. *Cell Commun Signal.* 2013;11:97.

49. Borodovsky A, Salmasi V, Turcan S, et al. 5-azacytidine reduces methylation, promotes differentiation and induces tumor regression in a patient-derived IDH1 mutant glioma xenograft. *Oncotarget.* 2013;4(10):1737–1747.

50. Turcan S, Fabius AW, Borodovsky A, et al. Efficient induction of differentiation and growth inhibition in IDH1 mutant glioma cells by the DNMT inhibitor Decitabine. *Oncotarget.* 2013;4(10):1729–1736.

51. Fathi AT, Wander SA, Faramand R, Emadi A. Biochemical, epigenetic, and metabolic approaches to target IDH mutations in acute myeloid leukemia. *Semin Hematol.* 2015;52(3):165–171.

52. Bell RJ, Rube HT, Xavier-Magalhães A, et al. Understanding TERT promoter mutations: a common path to immortality. *Mol Cancer Res.* 2016;14 (4):315–323.

53. Bell RJ, Rube HT, Kreig A, et al. Cancer. The transcription factor GABP selectively binds and activates the mutant TERT promoter in cancer. *Science.* 2015;348(6238):1036–1039.

54. Misner DL, Kauss MA, Singh J, et al. Cardiotoxicity associated with nicotinamide phosphoribosyltransferase inhibitors in rodents and in rat and human-derived cells lines. *Cardiovasc Toxicol.* 2016;.

55. Zabka TS, Singh J, Dhawan P, et al. Retinal toxicity, in vivo and in vitro, associated with inhibition of nicotinamide phosphoribosyltransferase. *Toxicol Sci.* 2015;144(1):163–172.

56. Tateishi K, Higuchi F, Miller JJ, et al. The alkylating chemotherapeutic Temozolomide induces metabolic stress in IDH1-mutant cancers and potentiates NAD(+) depletion-mediated cytotoxicity. *Cancer Res.* 2017;77(15):4102–4115.

Chapter 32

Bevacizumab for recurrent anaplastic oligodendroglial tumors

Sophie Taillibert* and Marc C. Chamberlain[†]

*Pitié-Salpétrière University Hospital, Paris VI UPMC University, APHP, Paris, France; [†]Seattle Genetics, Seattle, WA, United States

Abbreviations

AE	adverse event
AG	anaplastic glioma
AO	anaplastic oligodendroglioma
AOA	anaplastic oligoastrocytoma
ATRX	alpha thalassemia/mental retardation syndrome X-linked
CIMP	CpG island hypermethylated phenotype
EGFR	epidermal growth factor receptor
FDA	Food and Drug Administration
FISH	fluorescence in situ hybridization
FUBP 1	far upstream binding element protein 1
GI	gastrointestinal
ICH	intracranial hemorrhage
IDH	isocitrate dehydrogenase
KPS	Karnofsky performance status
MGMT	O6-methylguanine-DNA methyltransferase
NCCN	National Comprehensive Cancer Network
NOS	not otherwise specified
O	oligodendroglioma
OA	oligoastrocytoma
ORR	overall response rate
OS	overall survival
PCV	procarbazine, lomustine, vincristine
PFS	progression free survival
RANO	Response Assessment in Neuro-Oncology
RR	response rates
RT	radiation
SAE	serious adverse event
TMZ	temozolomide
TE	thromboembolism
TERT	telomerase reverse transcriptase
TP53	tumor protein p53
US	United States
VEGF	vascular endothelial growth factor
WHO	World Health Organization

Oligodendroglioma. https://doi.org/10.1016/B978-0-12-813158-9.00032-3

Introduction

Oligodendroglioma (O: WHO grade 2), oligoastrocytoma (OA: WHO grade 2), anaplastic oligodendroglioma (AO: WHO grade 3), and anaplastic oligoastrocytoma (AOA: WHO grade 3) are glial tumors composed of neoplasticcellular features that morphologically appear like oligodendrocytes or oligodendrocytes and astrocytes (in the case of OA and AOA).

The treatment of recurrent, alkylator chemotherapy refractory oligodendroglial tumors is challenging given the scarcity of effective therapeutic options and lack of randomized controlled trials. This review discusses the use of bevacizumab and bevacizumab-based regimens for treatment of recurrent oligodendroglial tumors.

The review first provides an overview of the new WHO classification of oligodendroglial tumors and its implication in terms of decision-making regarding treatment. Secondly, bevacizumab, mechanisms of action, current use in glial tumors, side-effect profiles, and lessons learnt from its common use in patients with glioblastoma is discussed. Finally, the available data regarding bevacizumab-based regimens that have been evaluated in patients with recurrent oligodendroglial tumors are discussed.

Overview of oligodendroglial tumor classification integrating molecular characteristics

There are noteworthy distinctions differentiating oligodendrogliomas from other gliomas with respect to pathology, molecular pathogenesis, natural history, prognosis, and response to treatment. The new WHO classification, integrating molecular prognostic factors such as IDH-mutation status, 1p/19q co-deletion in O/AO is briefly discussed here. An update is also provided regarding the current treatment in first-line and recurrent glial tumors based on the molecular status.

As mentioned above, the 2016 WHO classification of central nervous system tumors now integrates histopathology and molecular characteristics.[1] According to this classification, the definition of oligodendroglial tumors is now genetically defined as diffuse gliomas harboring both a mutation in isocitrate dehydrogenase gene (*IDH1* or *IDH2*) and co-deletion of chromosomes 1p and 19q.[1] Although the classic histopathologic aspect is that of oligodendroglioma, microscopically some tumors display astrocytic features and were previously characterized as mixed OA. These mixed tumors no longer exist as separate entities except in the circumstance when referring to previously diagnosed glioma and for which molecular testing is not available [i.e., OA or AOA not otherwise specified (NOS)].[1] The WHO grading system differentiates two histopathologic grades of oligodendroglioma: grade II (low-grade) and grade III AO. Grade III tumors are characterized by the presence of anaplastic features (high cell density, mitosis, nuclear atypia, microvascular proliferation, and necrosis) and historically were associated with a worse prognosis compared with grade II tumors.[1,2] Based on recent comparative studies, this separation of oligodendrogliomas based on grade appears not to exist and consequently these two entities are considered to have a similar biology and response to treatment.[3,4]

Mutations in IDH1 and less frequently, IDH2, are very early events in the pathogenesis and a defining feature of the majority of WHO grade II and III diffuse gliomas, including oligodendrogliomas. These mutations can occur with and without the 1p/19q co-deletion. Oligodendrogliomas are defined by the combination of these two genetic features, while IDH1/2 mutations without 1p/19q co-deletion now characterize diffuse astrocytomas.[5]

Immunohistochemical staining for the most common mutant form of IDH1 (R132H) identifies 90% of all IDH mutations. When negative by immunohistochemistry, sequencing of IDH1 (codon 132) and IDH2 (codon 172) for hotspot mutations is currently recommended.

1p/19q status is most often defined using fluorescence in situ hybridization (FISH), but only loss of heterozygosity (LOH) studies can assess multiple regions of loss on 1p and 19q. The combined whole-arm loss of 1p and 19q is consistently correlated with IDH mutation.[2,6,7]

Detection of 1p/19q co-deletion by FISH without an IDH mutation (confirmed by IDH1/2 sequencing) suggests partial loss of 1p and/or 19q and should trigger supplementary analysis for genetic modifications related to glioblastoma such as gain of chromosome 7, loss of 10 or 10q, and amplification of the epidermal growth factor receptor (EGFR) gene.

In oligodendroglial tumors the deletion of 1p and 19q is retained, the tumor life cycle suggesting an early event in tumorigenesis.[8] Additional chromosomal abnormalities become more frequent in AO tumors.[8,9]

All 1p/19q co-deleted gliomas display IDH mutations as well as the CpG island hypermethylated phenotype (CIMP) in which the O^6-methylguanine-DNA methyltransferase (*MGMT*) promoter region is also methylated. Testing for *MGMT* promotor methylation is systematic in all glioblastomas, but not in oligodendrogliomas in which its clinical significance remains unclear.[2]

The IDH-mutant, 1p/19q co-deleted oligodendrogliomas also display telomerase reverse transcriptase (*TERT*) mutations, mutations in the capicua transcriptional repressor (*CIC*), and the far upstream binding element protein 1 (*FUBP1*) genes.[10,11] Currently, the analysis of *TERT* promoter mutations is not considered standard or essential in the diagnostic process in the WHO 2016 classification.[1,2,6] Historically, OA or AOA were defined as tumors containing histopathologic features of both O and OA. It has been shown that when molecular criteria are applied, mixed gliomas present either an O genotype (IDH mutated, 1p/19q co-deleted) or an astrocytic genotype (IDH, alpha thalassemia/mental retardation syndrome X-linked [*ATRX*], and tumor protein p53 [*TP53*] mutated).[12] As a result, diffuse gliomas are categorized as either O or OA.[1] Limited data indicate that among IDH-mutant, 1p/19q co-deleted AO, microvascular proliferation, necrosis, number of mitoses, and Ki-67 labeling index have prognostic significance.[13]

Overview of oligodendroglial tumor treatment in the adjuvant and recurrent settings

The presence of 1p/19q co-deletion is correlated with better survival in patients with IDH-mutant diffuse gliomas.[14–17] This may reflect both the natural evolution of the tumor as well as an improved response to therapy.[10] Compared with other diffuse gliomas, IDH-mutant, 1p/19q co-deleted oligodendrogliomas are highly chemosensitive. This appears related to metabolically triggered epigenetic modifications, including CIMP and O^6-MGMT promotor methylation.

The general therapeutic approach in patients with newly diagnosed AO or AOA initially includes a maximal safe surgical resection consistent with preservation of neurologic function. Subsequently all patients should receive an adjuvant chemoradiation combining radiation (RT) and chemotherapy.[6] The selection of either PCV (procarbazine, lomustine, vincristine) or single agent temozolomide (TMZ) should be individualized based on molecular features and patient preferences.

In patients with a 1p/19q co-deleted glioma, PCV and TMZ are two acceptable options. Two phase III randomized trials have confirmed the significant improvement in survival in patients with O/AO when treated with PCV, while TMZ has a proven impact on survival in other anaplastic glioma (AG) patients, displays a better tolerance profile, and is more convenient to administer.[14,17] For patients with a non-co-deleted 1p/19q AO, 12 cycles of adjuvant TMZ after completion of the radiation therapy are recommended based on results of the CATNON trial. The benefit of the concomitant administration of TMZ during RT is not established and is associated with more side effects.[18]

The optimal sequencing of RT and chemotherapy in the adjuvant setting currently is unknown. The consensus-based guidelines from the National Comprehensive Cancer Network (NCCN) recommend that PCV be administered after RT (as per EORTC 26951).[6,19] Likewise, TMZ is usually given after radiotherapy.

In the recurrent setting, both PCV and TMZ are active but with lower response rates (RR) and shorter tumor control in patients who have received a first-line neoadjuvant or adjuvant alkylator-based chemotherapy regimen.[20,21]

After progression on both TMZ and PCV, there is no consensus regarding next line of therapy. Paclitaxel, irinotecan, cyclophosphamide, carboplatin, and cisplatin combined with etoposide have shown modest activity as second-line agents, but RR is <15%, and most patients progress within 6-months, illustrating the unmet need for improved treatment options.[22–32]

The treatment of diffuse gliomas continues to be actively studied in ongoing trials, particularly given that molecular diagnostics have altered the classification of gliomas. The largest ongoing trial is a phase III Alliance for Clinical Trials in Oncology/EORTC intergroup trial ("CODEL") in which patients with 1p19q co-deleted O/AO in the adjuvant setting are randomly assigned to one of two treatment arms: RT followed by PCV; and RT with concurrent and adjuvant TMZ.[33] As mentioned above, in the recurrent setting, there remains a need for improved therapeutic alternatives, once patients have progressed following treatment with PCV and TMZ. These observations in recurrent O/AO led to the assessment of bevacizumab in the salvage setting, based on prior experience of this targeted agent in the treatment of glioblastoma.

Bevacizumab use in glioblastoma
Mechanism of action

Bevacizumab is a recombinant, humanized monoclonal antibody that prevents the binding of vascular endothelial growth factor (VEGF) to the endothelial receptors Flt-1 (VEGFR) and KDR, consequently inhibiting its pro-angiogenic activity.

Angiogenesis, which leads to the development of new blood vessels from preexisting ones, through endothelial cell migration and proliferation, is required for neoplastic growth beyond a tumor mass of 20 μm in diameter.[34] The VEGF pathway is fundamental to the neovascularization process of high-grade gliomas through the endothelial receptor VEGFR2.[35] The survival, proliferation, migration, and permeability of endothelial cells depend on this specific signaling pathway. Bevacizumab was approved in 2009 by the Food and Drug Administration (FDA) for recurrent or progressive glioblastoma. Bevacizumab was approved as monotherapy, based on two prospective phase II clinical trials showing an improvement in PFS and associated objective RR.[36,37]

Dosing schedule in glioblastoma

In glioblastoma, bevacizumab was approved using a dosing of 10 mg/kg every 2 weeks as a single agent until disease progression or unacceptable toxicity.

Notably a dose–response effect was never established in comparative trials, though many other dosing schedules are prescribed in other cancers.

In other observational studies, in which recurrent high-grade glioma patients were treated with doses varying from 5 to 15 mg/kg every two to three weeks, no difference in PFS or overall survival (OS) was observed.[38–40] Further, a trend toward improved OS, and a significant decrease of serious adverse events (SAEs) were observed in two retrospective studies dosing with 7.5 mg/kg every 3 weeks when compared to the standard schedule.[41,42] Based on these observations, and the lack of proven dose–response effect, less frequent administration (every three weeks) might be an option to achieve improved tolerability with fewer side effects and by treating less frequently, a better quality of life for patients.

Toxicities

A wide range of adverse events (AEs) has been reported with the use of bevacizumab. These are classically divided into two categories; one impacting the cardiovascular system mainly as hypertension and thromboembolism (TE), and the second global effects such as proteinuria, impaired wound healing, hemorrhage, and gastrointestinal (GI) perforation.[43] The overall incidence of mild proteinuria ranges from 21% to 63%, but grade 3 and 4 proteinuria (defined as 3+ on dipstick, >3.5 g/ 24 h, or nephrotic syndrome) occurs in only 2% of patients.[44] A dose dependency and an increase in risk when bevacizumab is combined with chemotherapy has been postulated.[45]

There is a relationship between treatment duration and proteinuria such that proteinuria often occurs later and may worsen with continued exposure. Whether the occurrence of proteinuria serves as a surrogate marker of antitumor efficacy is uncertain.[46] There are no evidence-based guidelines for management of proteinuria in patients receiving bevacizumab. Nevertheless, the prescribing information for bevacizumab recommends monitoring for the development of proteinuria, including a temporary withholding of the drug if protein excretion is above 2 g/24 h and a permanent discontinuation in case of the nephrotic syndrome.[47]

Hypertension is dose-dependent and one of the most frequent AEs of bevacizumab, with an overall incidence of 24%.[48] The blood pressure is significantly increased in 8% of patients, requiring at least two antihypertensive medications for control.[48] As a consequence, the Investigational Drug Steering Committee of the NCI formed a Cardiovascular Toxicities Panel and published guidelines for monitoring and management of elevated blood pressure in patients receiving bevacizumab.[48] There is an established increased risk of arterial TE complications with bevacizumab, as well as an increased risk for venous TE, which may lead to discontinuance of bevacizumab.[49]

GI perforations are observed in 0.3%–3% of patients treated with bevacizumab and may be fatal. The concomitant use of steroids appears to add to the risk of GI perforation, and may complicate the diagnosis by masking classic abdominal signs of peritonitis. Impaired wound healing can also be a complication of bevacizumab administration and is particularly challenging in patients requiring reoperation wherein a drug hold is recommended for 4 weeks before and after a major surgery. The frequency of severe or fatal bleeding is increased by up to 5 times in cancer patients treated with bevacizumab. Therefore, this drug is contraindicated in cases of recent bleeding such as hemoptysis or intratumoral hemorrhage.

Bevacizumab-related toxicities in glioblastoma

Glioblastoma patients display a similar safety profile as other cancer patients treated with bevacizumab and in most instances, there is excellent tolerability. Grade 3 or higher AEs occur in patients with glioblastoma treated with bevacizumab between 18% and 66% and at a lower rate when used as monotherapy (43% vs 66%).[37,50,51] Typical class-specific grade 1–2 AEs such as hypertension, delayed wound healing, proteinuria, and systemic or intracranial hemorrhage (ICH) are the most likely to occur.[37,51,52] The incidence of SAEs such as GI perforation, nephrotic syndrome, reversible posterior leukoencephalopathy, stroke, and myocardial infarction is seen in 2% or less. Life-threatening intracranial bleeding occurs in up to 3% of glioblastoma patients not treated with therapeutic doses of anticoagulants. This rate of ICH is in the estimated range for glioblastoma patients independent of bevacizumab treatment.[37,50–53] According to retrospective data, the overall risk of ICH was increased from 3% to 11% in patients treated with bevacizumab and serious ICH rate increased from 1% to 3% in patients anticoagulated at a therapeutic dose.[50,54,55] The majority of ICH (>75%) is intratumoral.[53] Thus, the risk–benefit ratio has to be carefully assessed individually, before prescribing bevacizumab in patients who are anticoagulated. A higher frequency of thrombocytopenia may occur when bevacizumab is coadministered with chemotherapy.[56] Similarly, there may be more venous TE complications in this population of patients in which there is a preexisiting high rate, though this has not been demonstrated in other cancers.[56] Patients with glioma often manifest with seizures and TE and the administration of bevacizumab with anticoagulants or antiepileptic drugs (enzyme-inducing or not) is neither contraindicated nor necessitates any dose adjustment.

Whether bevacizumab is more effective when administered as a single agent or combined with a cytotoxic or another targeted drug has been controversial. Unlike other bevacizumab responsive cancers in which chemo-synergy is evident, the value of combination therapy in gliomas is uncertain and currently not evidence-based. Two phase II randomized studies have compared bevacizumab monotherapy vs combination therapy in malignant gliomas. The phase II randomized study CABARET that compared bevacizumab alone vs bevacizumab with carboplatin in patients with recurrent glioblastoma demonstrated that adding carboplatin resulted in more toxicity and showed no additional clinical benefit (PFS, RR, OS).[57]

The contrasting data from a Dutch phase II study suggested that the addition of bevacizumab to lomustine might translate in a survival advantage in recurrent glioblastoma patients as compared to bevacizumab alone or lomustine alone.[58] However, the EORTC 26101 phase III randomized trial comparing bevacizumab with lomustine to lomustine as single agent in recurrent glioblastoma failed to show any survival advantage of the combination, despite a nonsignificant increase of 2.7 months in median PFS and showed that it was significantly more toxic with grade 3 to 5 AEs observed in 63.6% vs 38.1% of patients.[59]

At present there are no confirmed biomarkers that distinguish the patients who are most likely to benefit from bevacizumab.

Bevacizumab use in oligodendroglial tumors

Based on the bevacizumab experience in glioblastoma, the role of bevacizumab has been investigated in recurrent or progressive oligodendroglioma and AO. Only a few prospective trials have included a limited number of AO patients, and only

TABLE 1 Bevacizumab studies in patients with oligodendroglial tumors

Summary of studies assessing bevacizumab-based regimens in patients with a progressive or recurrent oligodendroglial tumors			
Study author (year)	Design, treatment	Tumor type/ number of patients	Results
Chamberlain and Johnston[61]	Retrospective; single agent bevacizumab	AO = 22	RR = 68% (PR = 100%) PFS6 = 68%, PFS12 = 23%
Taillibert et al.[62]	Retrospective; bevacizumab + irinotecan	O = 7, AO = 12, OA = 1, AOA = 5	RR = 72% PFS6 = 42%
Kreisl et al.[51]	Prospective: single agent bevacizumab	AA = 21, AO = 4 AOA = 6	*Results for all grade 3 AG:* RR = 43%, PFS6 = 20.9%, mPFS = 2.9 m, mOS = 12 m
Desjardins et al.[63]	Prospective; bevacizumab + irinotecan	AO = 8	RR = 87.5% (1 CR + 6 PR) PFS6 = 62%, mPFS = 50 wks, mOS = 61 wks
Reardon et al.[64]	Prospective; bevacizumab + etoposide	AA = 18, AO = 13, PXA = 1	*Results for all grade 3 AG:* RR = 22% PFS6 = 40.6%, mPFS 30 wks, mOS = 63.1 wks
Sathornsumetee et al.[60]	Prospective: bevacizumab + erlotinib	AA = 24, AO = 7, PXA = 1	*Results for all grade 3 AG:* RR = 31% (1 CR + 9 PR), PFS6 = 43.8%, mPFS = 23.4 wks, mOS = 71.3 wks

AA: Anaplastic astrocytoma; AG: Anaplastic glioma; AO: Anaplastic oligodendroglioma; AOA: Anaplastic oligoastrocytoma; Bev: Bevacizumab; CPT11: Irinotecan; CR: Complete response; m: Months; mOS: Median overall survival; mPFSs: Median progression-free survival; O: World Health Organization grade II oligodendroglioma; PFS6: 6-month progression-free survival; PFS12: 12-months progression-free survival; PR: Partial response; PXA: Pleomorphic xanthroastrocytoma; RR: Radiographic response; Wks: Weeks.
Adapted from Grimm SA, Chamberlain MC. Bevacizumab and other novel therapies for recurrent oligodendroglial tumors. *CNS Oncol.* 2015;4(5):333–339. https://doi.org/10.2217/cns.15.27.

two retrospective studies have specifically addressed this population.[51,60–64] A summary of these studies and their results is shown in Table 1.

In one study, 25 heavily pretreated patients with AO were treated with bevacizumab and irinotecan (CPT-11).[62] The objective RR was 72% (20% complete response, 52% partial response) and the clinical benefit rate (complete and partial response plus stable disease for >6-months) was 88%. After a median follow-up of 6.7 months, the median PFS was 4.6 months (95% CI: 3.9–∞), and OS had not been reached. The 6-month progression-free survival rate was 42% (95% CI 26%–67%). The molecular status was known in 17 patients, but no correlation was found between the RR and a type of genomic alteration (including 1p19q co-deletion). In all patients, the response was detected at the first on study time point, 1 month from the start of the treatment. Corticosteroids could be rapidly tapered (within 2 months) or discontinued in all responders; a clinically meaningful result due to complications of long-term exposure to steroids is decreased.

A striking clinical improvement was associated with radiologic response in patients whose initial KPS was 70% or less. Three responders were observed among the five patients who started treatment with a KPS of 50%. Three patients achieved a KPS of 80% or better, indicating that this regimen may be clinically relevant in severely disabled patients.

Among all high-grade gliomas, AO are known to demonstrate relatively robust chemosensitivity. At first recurrence after surgery and radiation therapy, RR in oligodendroglial tumors vary from 42% to 73% when treated with PCV or TMZ regimens.[20,21,65–70] After the failure of alkylator chemotherapy, the rate of response to second- or third-line chemotherapy in oligodendroglial tumors decreases sharply to 8%–17%.[16,21,31,67,68,70–72] Topoisomerase I inhibitors such as irinotecan alone also produce a relatively modest 13%–23% RR, and 33% PFS-6.[22,25,26,73,74] Considering these data, the reported 72% rate of ORR with bevacizumab in the above mentioned study is notable, considering that all patients had progressed on TMZ, and half were previously treated with nitrosourea-based chemotherapy. The results in terms of PFS were less marked, but they compare favorably with previously published data regarding AO patients in second or third line of chemotherapy.[26,65,67,70,73] Tolerability of the combined bevacizumab regimen was acceptable with discontinuation rate of only 20% of patients. A notable side effect was the occurrence of intratumoral bleeding in 24% of patients, which is higher than that in previous reports.[50,52]

However hemorrhage was asymptomatic in nearly all patients, with only a single patient discontinuing treatment due to intratumoral hemorrhage. Because none of these patients were receiving an anticoagulant, the authors concluded that bevacizumab might increase the well-established natural trend of oligodendroglial tumors to bleed, either by a direct effect, or by rapid tumor shrinkage.

In the other retrospective series, 22 patients with a recurrent 1p19q co-deleted AO received bevacizumab as monotherapy.[61] All had received prior treatment including surgery, radiation, adjuvant chemotherapy (PCV or TMZ) followed by a first-line of chemotherapy (cross over to either TMZ or PCV) at first relapse. All patients were in second-line of treatment at the time of bevacizumab administration and were resistant to alkylating chemotherapy (both temozolomide or nitrosoureas). A total of 391 cycles of bevacizumab (median, 14.5 cycles; range, 2–39 cycles) were administered. A total of 68% of patients displayed a partial radiographic response, 5% stable disease, and 27% progressive disease after two cycles of bevacizumab. In responders, the median time to tumor progression was 6.75 months (1–18 months). The median survival was 8.5 months (3–19 months). At 6-month and 12-month PFS were 68% and 23%, respectively. In all, 68% percent of patients were able to reduce the dexamethasone dose, and in 45% of patients, dexamethasone was discontinued. In this study again, a steroid-sparing effect of bevacizumab was shown. A total of nine grade, three AEs were seen in 41% of patients and no grade 4 or 5 AEs were observed. The toxicity profile was safe and comparable to the published data with bevacizumab and comprised hypertension, proteinuria, delayed wound healing, and low-grade bleeding. Two patients developed asymptomatic small intratumoral bleeding and in both bevacizumab was continued without evidence of hemorrhagic progression by serial magnetic resonance images.

As mentioned earlier, a limited number of patients with a pure AO or AOA have been included in prospective trials. In a phase II trial, bevacizumab was assessed as monotherapy in 31 patients with recurrent AG including 4 AO and 6 AOA (32% of all patients).[51] In the total population of AG, the reported radiologic ORR, 6-month PFS, median PFS, and median OS were 43%, 20.9% (95% CI: 10.3%–42.5%), 2.9 months (95% CI: 2.01–4.93), and 12 months (CI 95%:6.08–22.8), respectively.

In another phase II trial evaluating the efficacy of bevacizumab combined with irinotecan in recurrent grade 3 gliomas, eight patients were diagnosed with an AO (Desjardin 2008).[63] Although outcome in terms of survival did not differ between treatment cohorts, AO patients trended toward a better outcome in terms of PFS than patients with an anaplastic astrocytoma (AA). The statistical power was limited by the small number of patients, making extrapolation to a larger population problematic. In eight patients with AO, the ORR was 87.5%, with six partial and one complete response. One patient with AO had radiologic stable disease. The 6-month PFS was 62% (95% CI: 23%–86%), and the median PFS was 50 weeks (95% CI: 21–undetermined). The 6-month and median OS were 88% and 61 weeks, respectively.

Another phase II study assessed bevacizumab with metronomic etoposide in 32 patients with a recurrent grade 3 glioma, including 13 AO (41%).[64] The authors reported the results overall for grade 3 anaplastic gliomas without differentiating between AO and AA. The radiologic RR, the 6-month PFS, median PFS, and the median OS were 22%, 40.6% (CI 95%: 24–57), 24 weeks (CI 95%: 16–33), and 63.1 weeks (CI 95%: 36–∞), respectively.

A phase II trial evaluated bevacizumab in combination with the EGFR tyrosine kinase inhibitor erlotinib in 32 patients with AG, including 7 with AO (22%), though the authors do not describe the results separately for AO.[60] In the total population of AG, the ORR was 31%, the 6 month-PFS, median PFS, and median OS were 43.8% (95% CI: 26.5–59.8), 23.4 weeks (95% CI: 44.7–123.6), and 71.3 weeks (95% CI: 44.7–123.6), respectively.

In summary, prospective randomized trials in recurrent oligodendroglial tumors are lacking such that there is no evidence-based trial data to affirm the role of bevacizumab in this indication. Moreover, the recent reclassification of oligodendroglial tumors according to their molecular profile makes any interpretation of past studies challenging. From the published retrospective data in AOs, bevacizumab appears to have efficacy with an improvement seen in ORR and PFS. But the lack of impact on OS similar to that seen in the treatment of recurrent glioblastoma indicates treatment with bevacizumab is palliative only and of relatively brief duration.[58,59,75,76]

In patients with recurrent glioblastoma, notwithstanding the absence of survival advantage, bevacizumab does decrease peritumoral edema, shows a steroid-sparing effect, is well tolerated, improves performance status, and quality of life, and comprises a real benefit to patients. Based on this aggregate effect, the FDA has granted bevacizumab full approval in recurrent glioblastoma in December 2017.[77]

Similarly, based on the existing data, bevacizumab can currently be considered as a reasonable option in recurrent O/AO, pending further supportive studies.

Practical issues regarding bevacizumab administration

Controversy exists with bevacizumab therapy for recurrent high-grade gliomas in regard to timing, duration, and response assessment of treatment. Which patients benefit the most from bevacizumab is also unclear.

Which patients?

Patients with large contrast enhancing tumors and surrounding peritumoral edema may be the best candidates for bevacizumab therapy, as well as patients with a low performance status and those dependent on large doses of corticosteroids. Peritumoral edema is reduced rapidly as a result of the treatment-induced decrease in vascular permeability and stabilization of the blood–brain barrier caused by the VEGF neutralization. Further, patients in whom steroid reduction or discontinuation is achieved are at decreased risk of significant and chronic morbidity (diabetes, cushingoïd weight gain, osteoporosis leading to fractures and vertebral compression, proximal myopathy, TE, skin fragility, delay in wound healing, and opportunistic infections related to chronic lymphopenia) that attends chronic steroid use.

As mentioned above, not all AG patients respond to bevacizumab, and there are currently no predictive biomarkers that have been validated. Attempts to predict survival outcome in patients with glioblastoma patients treated with bevacizumab have demonstrated limited success with imaging, such as the relative cerebral blood volume changes on perfusion MRI, apparent diffusion coefficients modifications on diffusion MRI, and radiomic signatures; however, these remain hypothetical and prospective validation is needed.[78–81]

Timing of bevacizumab administration and discontinuation

Some preclinical and clinical data suggest that bevacizumab promotes invasion and at least in animal models may result in a gliomatosis-like pattern of recurrence that is highly resistant to subsequent treatments.[82,83] It is believed that the inhibition of VEGF induces an invasive phenotype that co-opts available cerebral vasculature rather than utilizing angiogenesis. These data, though not confirmed clinically, suggest use of bevacizumab be deferred until a later stage of disease evolution. Further, data from retrospective and prospective studies suggest that diffuse brain invasion reflects the natural history of glioblastoma in late stages and that a diffuse pattern of recurrence is common regardless of the nature of the treatment.[83–86]

Similarly, duration of the treatment and when to discontinue bevacizumab in both responding patients and after tumor progression are still debated. The outcome of stopping therapy in patients responding to bevacizumab has been addressed in the literature, in the randomized trial CABARET. At the time of progression, patients either continued bevacizumab or changed to an alternative therapy. No difference in post-bevacizumab progression was seen nor was there any evidence of rebound edema or more rapid clinical deterioration. Consequently, most clinicians use bevacizumab until toxicity or clear evidence of clinical and radiographic disease progression and then discontinue bevacizumab.[87] Nonetheless, preclinical studies have shown that when the inhibition of VEGF is suspended, a rebound effect leading to enhanced tumor growth and local invasion and distant metastases occurs.[88] In a retrospective analysis of five placebo-controlled clinical trials in non-neural systemic tumors, stopping bevacizumab before progression was not correlated with a decrease in time to disease progression, a decrease in survival, or a different mode of progression.[88]

Response assessment

Response assessment in patients treated with anti-VEGF agents has been formally addressed by the Response Assessment in Neuro-Oncology (RANO) working group.[89] It remains controversial whether the radiologic response reported with bevacizumab treatment is secondary to a true antitumor effect, or rather a restoration of the blood–brain barrier with subsequent improvement in contrast enhancement, a so called pseudo-response. As a consequence of decreased enhancement, tumor progression on bevacizumab may manifest as non-enhancing enlargement of the tumor mass.

The response assessment recommendation from RANO have included parameters that define tumor progression with the use of T2-weighted or FLAIR MRI imaging to assess for worsening of non-enhancing disease (Wen 2010).[89] These sequences in combination with clinical symptoms appear to be the best indicator of progressive disease, and change in T2-weighted/FLAIR often precedes the recurrence of contrast-enhancing tumor.[85,88]

Conclusion

Based on limited data and mostly retrospective data that evaluated bevacizumab in patients with recurrent O/AO, there appears to be a benefit as determined by RR and PFS-6, with acceptable toxicity. More problematic, however, is the absence of benefit in OS, since this issue has been constantly observed in randomized studies performed in glioblastoma. Furthermore, due to lack of therapeutic options in alkylator-resistant recurrent O/AOs combined with the recent changes in the pathologic classification, prospective randomized studies targeting these tumors are needed to evaluate the exact role of bevacizumab in this indication. The recent results of the Phase II TARVEC trial in 155 patients with first recurrent Grade 2 or 3 non-co-deleted gliomas comparing temozolomide to temozolomide and bevacizumab further confound these conclusions in that the study failed to indicate a survival benefit in the bevacizumab arm.[90]

References

1. Louis DN, Perry A, Reifenberger G, et al. The 2016 World Health Organization classification of tumors of the central nervous system: a summary. *Acta Neuropathol (Berl).* 2016;131(6):803–820. https://doi.org/10.1007/s00401-016-1545-1.

2. van den Bent MJ, Smits M, Kros JM, Chang SM. Diffuse infiltrating oligodendroglioma and astrocytoma. *J Clin Oncol.* 2017;35(21):2394–2401. https://doi.org/10.1200/JCO.2017.72.6737.

3. Olar A, Wani KM, Alfaro-Munoz KD, et al. IDH mutation status and role of WHO grade and mitotic index in overall survival in grade II-III diffuse gliomas. *Acta Neuropathol (Berl).* 2015;129(4):585–596. https://doi.org/10.1007/s00401-015-1398-z.

4. Suzuki H, Aoki K, Chiba K, et al. Mutational landscape and clonal architecture in grade II and III gliomas. *Nat Genet.* 2015;47(5):458–468. https://doi.org/10.1038/ng.3273.

5. Watanabe T, Nobusawa S, Kleihues P, Ohgaki H. IDH1 mutations are early events in the development of astrocytomas and oligodendrogliomas. *Am J Pathol.* 2009;174(4):1149–1153. https://doi.org/10.2353/ajpath.2009.080958.

6. Weller M, van den Bent M, Tonn JC, et al. European Association for Neuro-Oncology (EANO) guideline on the diagnosis and treatment of adult astrocytic and oligodendroglial gliomas. *Lancet Oncol.* 2017;18(6):e315–e329. https://doi.org/10.1016/S1470-2045(17)30194-8.

7. Van Den Bent MJ, Bromberg JEC, Buckner J. Low-grade and anaplastic oligodendroglioma. *Handb Clin Neurol.* 2016;134:361–380. https://doi.org/10.1016/B978-0-12-802997-8.00022-0.

8. Fallon KB, Palmer CA, Roth KA, et al. Prognostic value of 1p, 19q, 9p, 10q, and EGFR-FISH analyses in recurrent oligodendrogliomas. *J Neuropathol Exp Neurol.* 2004;63(4):314–322.

9. Bigner SH, Matthews MR, Rasheed BK, et al. Molecular genetic aspects of oligodendrogliomas including analysis by comparative genomic hybridization. *Am J Pathol.* 1999;155(2):375–386. https://doi.org/10.1016/S0002-9440(10)65134-6.

10. Dubbink HJ, Atmodimedjo PN, Kros JM, et al. Molecular classification of anaplastic oligodendroglioma using next-generation sequencing: A report of the prospective randomized EORTC brain tumor group 26951 phase III trial. *Neuro-Oncol.* 2016;18(3):388–400. https://doi.org/10.1093/neuonc/nov182.

11. Reuss DE, Sahm F, Schrimpf D, et al. ATRX and IDH1-R132H immunohistochemistry with subsequent copy number analysis and IDH sequencing as a basis for an "integrated" diagnostic approach for adult astrocytoma, oligodendroglioma and glioblastoma. *Acta Neuropathol (Berl).* 2015;129(1):133–146. https://doi.org/10.1007/s00401-014-1370-3.

12. Sahm F, Reuss D, Koelsche C, et al. Farewell to oligoastrocytoma: In situ molecular genetics favor classification as either oligodendroglioma or astrocytoma. *Acta Neuropathol (Berl)*. 2014;128(4):551–559. https://doi.org/10.1007/s00401-014-1326-7.

13. Figarella-Branger D, Mokhtari K, Dehais C, et al. Mitotic index, microvascular proliferation, and necrosis define 3 pathological subgroups of prognostic relevance among 1p/19q co-deleted anaplastic oligodendrogliomas. *Neuro-Oncol*. 2016;18(6):888–890. https://doi.org/10.1093/neuonc/now085.

14. Cairncross G, Wang M, Shaw E, et al. Phase III trial of chemoradiotherapy for anaplastic oligodendroglioma: long-term results of RTOG 9402. *J Clin Oncol*. 2013;31(3):337–343. https://doi.org/10.1200/JCO.2012.43.2674.

15. Intergroup Radiation Therapy Oncology Group Trial 9402, Cairncross G, Berkey B, et al. Phase III trial of chemotherapy plus radiotherapy compared with radiotherapy alone for pure and mixed anaplastic oligodendroglioma: Intergroup Radiation Therapy Oncology Group Trial 9402. *J Clin Oncol*. 2006;24(18):2707–2714. https://doi.org/10.1200/JCO.2005.04.3414.

16. van den Bent MJ, Carpentier AF, Brandes AA, et al. Adjuvant procarbazine, lomustine, and vincristine improves progression-free survival but not overall survival in newly diagnosed anaplastic oligodendrogliomas and oligoastrocytomas: A randomized European Organisation for Research and Treatment of Cancer phase III trial. *J Clin Oncol Off J Am Soc Clin Oncol*. 2006;24(18):2715–2722. https://doi.org/10.1200/JCO.2005.04.6078.

17. van den Bent MJ, Brandes AA, Taphoorn MJB, et al. Adjuvant procarbazine, lomustine, and vincristine chemotherapy in newly diagnosed anaplastic oligodendroglioma: Long-term follow-up of EORTC brain tumor group study 26951. *J Clin Oncol Off J Am Soc Clin Oncol*. 2013;31(3):344–350. https://doi.org/10.1200/JCO.2012.43.2229.

18. van den Bent MJ, Baumert B, Erridge SC, et al. Interim results from the CATNON trial (EORTC study 26053-22054) of treatment with concurrent and adjuvant temozolomide for 1p/19q non-co-deleted anaplastic glioma: A phase 3, randomised, open-label intergroup study. *Lancet Lond Engl*. 2017;390(10103):1645–1653. https://doi.org/10.1016/S0140-6736(17)31442-3.

19. Nabors LB, Portnow J, Ammirati M, et al. Central nervous system cancers, version 1.2015. *J Natl Compr Canc Netw*. 2015;13(10):1191–1202. https://doi.org/10.6004/jnccn.2015.0148.

20. Chinot OL, Honore S, Dufour H, et al. Safety and efficacy of temozolomide in patients with recurrent anaplastic oligodendrogliomas after standard radiotherapy and chemotherapy. *J Clin Oncol Off J Am Soc Clin Oncol*. 2001;19(9):2449–2455. https://doi.org/10.1200/JCO.2001.19.9.2449.

21. Triebels VHJM, Taphoorn MJB, Brandes AA, et al. Salvage PCV chemotherapy for temozolomide-resistant oligodendrogliomas. *Neurology*. 2004;63(5):904–906.

22. Chamberlain MC. Salvage chemotherapy with CPT-11 for recurrent oligodendrogliomas. *J Neurooncol*. 2002;59(2):157–163.

23. Chamberlain MC, Kormanik P. Salvage chemotherapy with paclitaxel for recurrent primary brain tumors. *J Clin Oncol Off J Am Soc Clin Oncol*. 1995;13(8):2066–2071. https://doi.org/10.1200/JCO.1995.13.8.2066.

24. Chang SM, Kuhn JG, Robins HI, et al. A phase II study of paclitaxel in patients with recurrent malignant glioma using different doses depending upon the concomitant use of anticonvulsants: A north American brain tumor consortium report. *Cancer*. 2001;91(2):417–422.

25. Cloughesy TF, Filka E, Kuhn J, et al. Two studies evaluating irinotecan treatment for recurrent malignant glioma using an every-3-week regimen. *Cancer*. 2003;97(9 Suppl):2381–2386. https://doi.org/10.1002/cncr.11306.

26. Batchelor TT, Gilbert MR, Supko JG, et al. Phase 2 study of weekly irinotecan in adults with recurrent malignant glioma: final report of NABTT 97-11. *Neuro-Oncol*. 2004;6(1):21–27. https://doi.org/10.1215/S1152-8517-03-00021-8.

27. Friedman HS, Petros WP, Friedman AH, et al. Irinotecan therapy in adults with recurrent or progressive malignant glioma. *J Clin Oncol*. 1999;17(5):1516–1525. https://doi.org/10.1200/JCO.1999.17.5.1516.

28. Macdonald D, Cairncross G, Stewart D, et al. Phase II study of topotecan in patients with recurrent malignant glioma. National Clinical Institute of Canada clinical trials group. *Ann Oncol Off J Eur Soc Med Oncol*. 1996;7(2):205–207.

29. Warnick RE, Prados MD, Mack EE, et al. A phase II study of intravenous carboplatin for the treatment of recurrent gliomas. *J Neurooncol*. 1994;19(1):69–74.

30. Poisson M, Péréon Y, Chiras J, Delattre JY. Treatment of recurrent malignant supratentorial gliomas with carboplatin (CBDCA). *J Neurooncol*. 1991;10(2):139–144.

31. Yung WK, Mechtler L, Gleason MJ. Intravenous carboplatin for recurrent malignant glioma: a phase II study. *J Clin Oncol*. 1991;9(5):860–864. https://doi.org/10.1200/JCO.1991.9.5.860.

32. Fulton D, Urtasun R, Forsyth P. Phase II study of prolonged oral therapy with etoposide (VP16) for patients with recurrent malignant glioma. *J Neurooncol*. 1996;27(2):149–155.

33. Jaeckle K, Vogelbaum M, Ballman K, et al. CODEL (Alliance-N0577; EORTC-26081/22086; NRG-1071; NCIC-CEC-2): Phase III Randomized Study of RT vs. RT+TMZ vs. TMZ for Newly Diagnosed 1p/19q-Codeleted Anaplastic Oligodendroglial Tumors. Analysis of Patients Treated on the Original Protocol Design (PL02.005). *Neurology*. 2016;86 [16 Supplement]. http://n.neurology.org/content/86/16_Supplement/PL02.005. abstract.

34. Folkman J. Tumor angiogenesis: Therapeutic implications. *N Engl J Med*. 1971;285(21):1182–1186. https://doi.org/10.1056/NEJM197111182852108.

35. Jain RK, di Tomaso E, Duda DG, Loeffler JS, Sorensen AG, Batchelor TT. Angiogenesis in brain tumours. *Nat Rev Neurosci*. 2007;8(8):610–622. https://doi.org/10.1038/nrn2175.

36. Kreisl TN, Kim L, Moore K, et al. Phase II trial of single-agent bevacizumab followed by bevacizumab plus irinotecan at tumor progression in recurrent glioblastoma. *J Clin Oncol*. 2009;27(5):740–745. https://doi.org/10.1200/JCO.2008.16.3055.

37. Friedman HS, Prados MD, Wen PY, et al. Bevacizumab alone and in combination with irinotecan in recurrent glioblastoma. *J Clin Oncol*. 2009;27(28):4733–4740. https://doi.org/10.1200/JCO.2008.19.8721.

38. Blumenthal DT, Mendel L, Bokstein F. The optimal regimen of bevacizumab for recurrent glioblastoma: Does dose matter? *J Neurooncol*. 2016; 127(3):493–502. https://doi.org/10.1007/s11060-015-2025-5.

39. Wong ET, Gautam S, Malchow C, Lun M, Pan E, Brem S. Bevacizumab for recurrent glioblastoma multiforme: a meta-analysis. *J Natl Compr Cancer Netw*. 2011;9(4):403–407.

40. Raizer JJ, Grimm S, Chamberlain MC, et al. A phase 2 trial of single-agent bevacizumab given in an every-3-week schedule for patients with recurrent high-grade gliomas. *Cancer*. 2010;116(22):5297–5305. https://doi.org/10.1002/cncr.25462.

41. Ajlan A, Thomas P, Albakr A, Nagpal S, Recht L. Optimizing bevacizumab dosing in glioblastoma: Less is more. *J Neurooncol*. 2017;135(1):99–105. https://doi.org/10.1007/s11060-017-2553-2.

42. Levin VA, Mendelssohn ND, Chan J, et al. Impact of bevacizumab administered dose on overall survival of patients with progressive glioblastoma. *J Neurooncol*. 2015;122(1):145–150. https://doi.org/10.1007/s11060-014-1693-x.

43. Brandes AA, Bartolotti M, Tosoni A, Poggi R, Franceschi E. Practical management of bevacizumab-related toxicities in glioblastoma. *Oncologist*. 2015;20(2):166–175. https://doi.org/10.1634/theoncologist.2014-0330.

44. Wu S, Kim C, Baer L, Zhu X. Bevacizumab increases risk for severe proteinuria in cancer patients. *J Am Soc Nephrol JASN*. 2010;21(8):1381–1389. https://doi.org/10.1681/ASN.2010020167.

45. Zhu X, Wu S, Dahut WL, Parikh CR. Risks of proteinuria and hypertension with bevacizumab, an antibody against vascular endothelial growth factor: systematic review and meta-analysis. *Am J Kidney Dis*. 2007;49(2):186–193. https://doi.org/10.1053/j.ajkd.2006.11.039.

46. Izzedine H, Massard C, Spano JP, Goldwasser F, Khayat D, Soria JC. VEGF signalling inhibition-induced proteinuria: mechanisms, significance and management. *Eur J Cancer Oxf Engl*. 2010;46(2):439–448. https://doi.org/10.1016/j.ejca.2009.11.001.

47. Yeh J, Frieze D, Martins R, Carr L. Clinical utility of routine proteinuria evaluation in treatment decisions of patients receiving bevacizumab for metastatic solid tumors. PubMed, NCBI. https://www.ncbi.nlm.nih.gov/pubmed/?term=Clinical+utility+of+routine+proteinuria+evaluation+in+treatment+decisions+of+patients+receiving+bevacizumab+for+metastatic+solid+tumors. Accessed 2 April 2018.

48. Maitland ML, Bakris GL, Black HR, et al. Initial assessment, surveillance, and management of blood pressure in patients receiving vascular endothelial growth factor signaling pathway inhibitors. *J Natl Cancer Inst*. 2010;102(9):596–604. https://doi.org/10.1093/jnci/djq091.

49. Zangari M, Fink LM, Elice F, Zhan F, Adcock DM, Tricot GJ. Thrombotic events in patients with cancer receiving antiangiogenesis agents. *J Clin Oncol*. 2009;27(29):4865–4873. https://doi.org/10.1200/JCO.2009.22.3875.

50. Norden AD, Bartolomeo J, Tanaka S, et al. Safety of concurrent bevacizumab therapy and anticoagulation in glioma patients. *J Neurooncol*. 2012; 106(1):121–125. https://doi.org/10.1007/s11060-011-0642-1.

51. Kreisl TN, Zhang W, Odia Y, et al. A phase II trial of single-agent bevacizumab in patients with recurrent anaplastic glioma. *Neuro-Oncol*. 2011; 13(10):1143–1150. https://doi.org/10.1093/neuonc/nor091.

52. Vredenburgh JJ, Desjardins A, Herndon JE, et al. Phase II trial of bevacizumab and irinotecan in recurrent malignant glioma. *Clin Cancer Res*. 2007; 13(4):1253–1259. https://doi.org/10.1158/1078-0432.CCR-06-2309.

53. Fraum TJ, Kreisl TN, Sul J, Fine HA, Iwamoto FM. Ischemic stroke and intracranial hemorrhage in glioma patients on antiangiogenic therapy. *J Neurooncol*. 2011;105(2):281–289. https://doi.org/10.1007/s11060-011-0579-4.

54. Simonetti G, Trevisan E, Silvani A, et al. Safety of bevacizumab in patients with malignant gliomas: a systematic review. *Neurol Sci*. 2014;35(1): 83–89. https://doi.org/10.1007/s10072-013-1583-6.

55. Nghiemphu PL, Green RM, Pope WB, Lai A, Cloughesy TF. Safety of anticoagulation use and bevacizumab in patients with glioma. *Neuro-Oncol*. 2008;10(3):355–360. https://doi.org/10.1215/15228517-2008-009.

56. Grimm SA, Chamberlain MC. Bevacizumab and other novel therapies for recurrent oligodendroglial tumors. *CNS Oncol*. 2015;4(5):333–339. https://doi.org/10.2217/cns.15.27.

57. Field KM, King MT, Simes J, et al. Health-related quality of life outcomes from CABARET: A randomized phase 2 trial of carboplatin and bevacizumab in recurrent glioblastoma. *J Neurooncol*. 2017;133(3):623–631. https://doi.org/10.1007/s11060-017-2479-8.

58. Taal W, Oosterkamp HM, Walenkamp AME, et al. Single-agent bevacizumab or lomustine versus a combination of bevacizumab plus lomustine in patients with recurrent glioblastoma (BELOB trial): A randomised controlled phase 2 trial. *Lancet Oncol*. 2014;15(9):943–953. https://doi.org/10.1016/S1470-2045(14)70314-6.

59. Wick W, Gorlia T, Bendszus M, et al. Lomustine and Bevacizumab in progressive Glioblastoma. *N Engl J Med*. 2017;377(20):1954–1963. https://doi.org/10.1056/NEJMoa1707358.

60. Sathornsumetee S, Desjardins A, Vredenburgh JJ, et al. Phase II trial of bevacizumab and erlotinib in patients with recurrent malignant glioma. *Neuro-Oncol*. 2010;12(12):1300–1310. https://doi.org/10.1093/neuonc/noq099.

61. Chamberlain MC, Johnston S. Bevacizumab for recurrent alkylator-refractory anaplastic oligodendroglioma. *Cancer*. 2009;115(8):1734–1743. https://doi.org/10.1002/cncr.24179.

62. Taillibert S, Vincent LA, Granger B, et al. Bevacizumab and irinotecan for recurrent oligodendroglial tumors. *Neurology*. 2009;72(18):1601–1606. https://doi.org/10.1212/WNL.0b013e3181a413be.

63. Desjardins A, Reardon DA, Herndon JE, et al. Bevacizumab plus irinotecan in recurrent WHO grade 3 malignant gliomas. *Clin Cancer Res Off J Am Assoc Cancer Res*. 2008;14(21):7068–7073. https://doi.org/10.1158/1078-0432.CCR-08-0260.

64. Reardon DA, Desjardins A, Vredenburgh JJ, et al. Metronomic chemotherapy with daily, oral etoposide plus bevacizumab for recurrent malignant glioma: a phase II study. *Br J Cancer*. 2009;101(12):1986–1994. https://doi.org/10.1038/sj.bjc.6605412.

65. Glass J, Hochberg FH, Gruber ML, Louis DN, Smith D, Rattner B. The treatment of oligodendrogliomas and mixed oligodendroglioma-astrocytomas with PCV chemotherapy. *J Neurosurg*. 1992;76(5):741–745. https://doi.org/10.3171/jns.1992.76.5.0741.

66. van den Bent MJ, Kros JM, Heimans JJ, et al. Response rate and prognostic factors of recurrent oligodendroglioma treated with procarbazine, CCNU, and vincristine chemotherapy. Dutch Neuro-oncology Group. *Neurology*. 1998;51(4):1140–1145.

67. van den Bent MJ, Chinot O, Boogerd W, et al. Second-line chemotherapy with temozolomide in recurrent oligodendroglioma after PCV (procarbazine, lomustine and vincristine) chemotherapy: EORTC brain tumor group phase II study 26972. *Ann Oncol Off J Eur Soc Med Oncol*. 2003;14(4):599–602.

68. Cairncross JG, Wang M, Jenkins RB, et al. Benefit from procarbazine, lomustine, and vincristine in oligodendroglial tumors is associated with mutation of IDH. *J Clin Oncol*. 2014;32(8):783–790. https://doi.org/10.1200/JCO.2013.49.3726.

69. Brandes AA, Tosoni A, Vastola F, et al. Efficacy and feasibility of standard procarbazine, lomustine, and vincristine chemotherapy in anaplastic oligodendroglioma and oligoastrocytoma recurrent after radiotherapy. A Phase II study. *Cancer*. 2004;101(9):2079–2085. https://doi.org/10.1002/cncr.20611.

70. Lassman AB, Iwamoto FM, Cloughesy TF, et al. International retrospective study of over 1000 adults with anaplastic oligodendroglial tumors. *Neuro-Oncol*. 2011;13(6):649–659. https://doi.org/10.1093/neuonc/nor040.

71. Soffietti R, Nobile M, Rudà R, et al. Second-line treatment with carboplatin for recurrent or progressive oligodendroglial tumors after PCV (procarbazine, lomustine, and vincristine) chemotherapy: a phase II study. *Cancer*. 2004;100(4):807–813. https://doi.org/10.1002/cncr.20042.

72. Brandes AA, Basso U, Vastola F, et al. Carboplatin and teniposide as third-line chemotherapy in patients with recurrent oligodendroglioma or oligoastrocytoma: a phase II study. *Ann Oncol Off J Eur Soc Med Oncol*. 2003;14(12):1727–1731.

73. Chamberlain MC, Glantz MJ. CPT-11 for recurrent temozolomide-refractory 1p19q co-deleted anaplastic oligodendroglioma. *J Neurooncol*. 2008;89(2):231–238. https://doi.org/10.1007/s11060-008-9613-6.

74. Buckner JC, Reid JM, Wright K, et al. Irinotecan in the treatment of glioma patients: current and future studies of the north central Cancer treatment group. *Cancer*. 2003;97(9 Suppl):2352–2358. https://doi.org/10.1002/cncr.11304.

75. Gilbert MR, Dignam JJ, Armstrong TS, et al. A randomized trial of bevacizumab for newly diagnosed glioblastoma. *N Engl J Med*. 2014;370 (8):699–708. https://doi.org/10.1056/NEJMoa1308573.

76. Chinot OL, Wick W, Mason W, et al. Bevacizumab plus radiotherapy-temozolomide for newly diagnosed glioblastoma. *N Engl J Med*. 2014;370 (8):709–722. https://doi.org/10.1056/NEJMoa1308345.

77. Schiff D, Wen PY. The siren song of bevacizumab: Swan song or clarion call? *Neuro-Oncol*. 2018;20(2):147–148. https://doi.org/10.1093/neuonc/nox244.

78. Ellingson BM, Gerstner ER, Smits M, et al. Diffusion MRI phenotypes predict overall survival benefit from anti-VEGF Monotherapy in recurrent Glioblastoma: Converging evidence from phase II trials. *Clin Cancer Res Off J Am Assoc Cancer Res*. 2017;23(19):5745–5756. https://doi.org/10.1158/1078-0432.CCR-16-2844.

79. Bennett IE, Field KM, Hovens CM, et al. Early perfusion MRI predicts survival outcome in patients with recurrent glioblastoma treated with bevacizumab and carboplatin. *J Neurooncol*. 2017;131(2):321–329. https://doi.org/10.1007/s11060-016-2300-0.

80. Kickingereder P, Götz M, Muschelli J, et al. Large-scale Radiomic profiling of recurrent Glioblastoma identifies an imaging predictor for stratifying anti-angiogenic treatment response. *Clin Cancer Res Off J Am Assoc Cancer Res*. 2016;22(23):5765–5771. https://doi.org/10.1158/1078-0432.CCR-16-0702.

81. Schmainda KM, Zhang Z, Prah M, et al. Dynamic susceptibility contrast MRI measures of relative cerebral blood volume as a prognostic marker for overall survival in recurrent glioblastoma: Results from the ACRIN 6677/RTOG 0625 multicenter trial. *Neuro-Oncol*. 2015;17(8):1148–1156. https://doi.org/10.1093/neuonc/nou364.

82. de Groot JF, Fuller G, Kumar AJ, et al. Tumor invasion after treatment of glioblastoma with bevacizumab: Radiographic and pathologic correlation in humans and mice. *Neuro-Oncol*. 2010;12(3):233–242. https://doi.org/10.1093/neuonc/nop027.

83. Pope WB, Xia Q, Paton VE, et al. Patterns of progression in patients with recurrent glioblastoma treated with bevacizumab. *Neurology*. 2011;76(5):432–437. https://doi.org/10.1212/WNL.0b013e31820a0a8a.

84. Wick A, Dörner N, Schäfer N, et al. Bevacizumab does not increase the risk of remote relapse in malignant glioma. *Ann Neurol*. 2011;69(3):586–592. https://doi.org/10.1002/ana.22336.

85. Chamberlain MC. Radiographic patterns of relapse in glioblastoma. *J Neurooncol*. 2011;101(2):319–323. https://doi.org/10.1007/s11060-010-0251-4.

86. Wick W, Wick A, Weiler M, Weller M. Patterns of progression in malignant glioma following anti-VEGF therapy: Perceptions and evidence. *Curr Neurol Neurosci Rep*. 2011;11(3):305–312. https://doi.org/10.1007/s11910-011-0184-0.

87. Field KM, Simes J, Nowak AK, et al. Randomized phase 2 study of carboplatin and bevacizumab in recurrent glioblastoma. *Neuro-Oncol*. 2015;17(11):1504–1513. https://doi.org/10.1093/neuonc/nov104.

88. Miles D, Harbeck N, Escudier B, et al. Disease course patterns after discontinuation of bevacizumab: Pooled analysis of randomized phase III trials. *J Clin Oncol*. 2011;29(1):83–88. https://doi.org/10.1200/JCO.2010.30.2794.

89. Wen PY, Macdonald DR, Reardon DA, et al. Updated response assessment criteria for high-grade gliomas: Response assessment in neuro-oncology working group. *J Clin Oncol*. 2010;28(11):1963–1972. https://doi.org/10.1200/JCO.2009.26.3541.

90. van de Bent M, Klein M, Smits M, et al. Bevacizumab and temozolomide in patients with first recurrence of WHO grade II and III glioma, without 1p/19q co-deletion (TAVAREC): a randomised controlled phase 2 EORTC trial. *Lancet Oncol*. 2018;19:1170–1179.

Chapter 33

Pediatric oligodendroglioma

Lennox Byer*, Cassie Kline-Nunnally[†], Tarik Tihan[‡] and Sabine Mueller[§]

*University of California, San Francisco (UCSF), School of Medicine, San Francisco, CA, United States; [†]UCSF, Departments of Pediatrics and Neurology, Division of Hematology/Oncology, San Francisco, CA, United States; [‡]UCSF, Department of Pathology, San Francisco, CA, United States [§]UCSF, Departments of Neurology, Neurosurgery, Pediatrics, San Francisco, CA, United States

Introduction

Oligodendroglioma (OG) is a clinicopathological entity among diffuse gliomas that is typically encountered in the adult population, where its definition includes mutations in *IDH* genes and co-deletion of chromosomes 1p and 19q.[1] The current classification scheme recognizes tumors within this category without these molecular alterations and considers these tumors "pediatric-type oligodendrogliomas." Therefore, even though the name implies a similar category, the definition implies pediatric oligodendrogliomas are a separate entity within glial neoplasms. Currently, it is not clear whether these neoplasms constitute a subset of classical oligodendrogliomas or a distinct diagnosis.

There is a paucity of studies for pediatric OG, likely due to the rarity of the tumors in children, with most literature based on retrospective reviews of individual institutional experience. Additionally, previous studies must be viewed with caution, especially if central review was not performed. In a recent analysis by Rodriguez et al., the authors could only confirm in 50% (50/100) of the samples the institutional diagnosis of OG following a histopathological review by two neuropathologists.[2] Another study that performed a central rereview found that only 25% (1/4) of anaplastic OG and 35% (9/26) of mixed oligoastrocytoma were actually confirmed with those respective diagnoses.[3] The 2016 WHO classification for tumors of the central nervous system more clearly defined molecular subtypes of adult OG.

Epidemiology

In adults, both low-grade and anaplastic OGs have been estimated to make up roughly 5.5% of brain tumor histologies. In pediatric and adolescent patients <19 years old, this number is estimated to be even lower at only 1%. The overall average incidence rate of low-grade OG appear to be about five times that of anaplastic OG. These figures though are likely to be significantly different as we have moved into the advent of molecular diagnoses. Out of all children, patients between the ages of 15 and 19 have the highest rates of both low- and high-grade OG. There is no gender predilection, but, as with most CNS tumors, OGs are most frequently reported in the Caucasian population.[4]

Clinical presentation

Children with OG frequently present with seizures at diagnosis. Other symptoms can include headache, focal neurologic deficits, hydrocephalus, and signs of increased intracranial pressure, depending on the anatomic location of the tumor. A retrospective review of 15 cases of pediatric oligodendroglioma found the presence of symptoms such as seizures and headaches was associated with grade II OG, while grade III anaplastic OG present more likely with hydrocephalus and increased intracranial pressure.[5]

Anatomy

Pediatric OG can arise in nearly all parts of the central nervous system. It most commonly arises in the cerebral hemispheres but can also present in thalamus and basal ganglia,[6–9] brainstem,[2,10–13] cerebellum,[2,6,11–16] meninges,[17,18] and spinal cord.[13,19] The rarest site for OG to occur is within the brainstem. A literature review of 1593 cases of both pediatric and adult OG found only six cases of intrinsic brainstem OG. Interestingly, five of the six occurred in children.[20] Dissemination of the tumor is also rare, though it has been reported.[21]

Histology

Histologically, tumors that qualify as pediatric OG are diffusely infiltrative, monomorphous tumors composed of a uniform population of cells demonstrating a typical "fried-egg" appearance (Fig. 1) vested in a delicate vasculature, often referred as a "chicken-wire" pattern. On morphological grounds, they are indistinguishable from an adult oligodendroglioma. The tumor cells often have optically clear cytoplasm with a centrally located nucleus that has a speckled "salt-and-pepper" chromatin pattern (Fig. 2). Most tumors have small or indistinct nucleoli. In some foci, more spindle-shaped tumor cells with processes, known as gliofibrillary oligodendroglia, can be observed. Tumors often demonstrate secondary structures (also known as secondary structures of Sherer) such as satellitosis, chain-like linear arrangement, and subpial or perivascular accumulation of tumor cells. Calcifications are also common, and are often associated with cortical involvement by

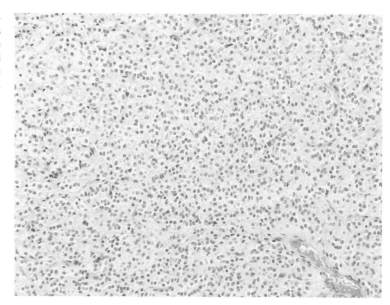

FIG. 1 Histological appearance of pediatric OG, indistinguishable from its adult counterpart. The section displays cells with a prominent "fried-egg" appearance due to the presence of cytoplasmic clearing and centrally located nuclei. The delicate vascularity imparting a chicken-wire meshwork appearance is recognizable. Original magnification X200.

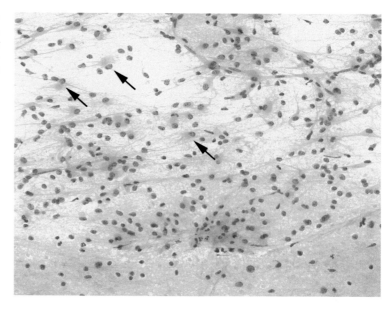

FIG. 2 Intraoperative smear from a pediatric OG demonstrating tumor cells in a markedly fibrillary background. The tumor cells have the typical delicate, salt-and-pepper type chromatin with indistinct nucleoli. Numerous "minigemistocytes" are also noted with conspicuous eosinophilic cytoplasm (arrows). Original magnification ×200.

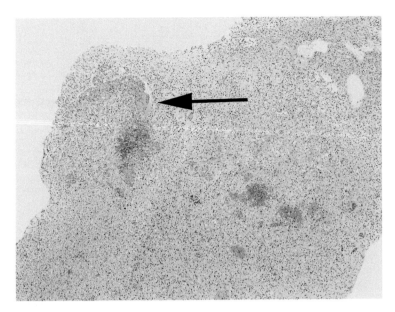

FIG. 3 Anaplastic OG with focal microvascular proliferation (arrow). The term microvascular proliferation refers to proliferation of vascular endothelial cells to outgrow and obliterate the vascular lumina. Along with increased mitoses and pseudopalisading necrosis, this feature is one of the histological criteria for anaplasia. Original magnification ×40.

the tumor. Microcysts and focal collections of minigemistocytes are often seen. Some of the tumors have focal hypercellular regions that appear as nodules. Most pediatric OGs demonstrate strong and diffuse positivity with the Olig-2 antibody, and since some tumors are only focally positive with the GFAP (glial fibrillary acidic protein), the former is more helpful in diagnosis.

Tumors that are reminiscent of oligodendrogliomas, but belong to other diagnostic categories are more common in the pediatric population, and these tumors should be excluded before recognizing a pediatric neoplasm as OG. Specifically, pilocytic astrocytomas in the posterior fossa often have oligodendroglioma-like patterns, and tumors in the posterior fossa should be investigated for *MAPK* pathway alterations (such as *BRAF* duplication or *FGFR1* mutation) before they are diagnosed as OG. Similarly, supratentorial tumors resembling oligodendrogliomas more often belong to categories such as dysembryoplastic neuroepithelial tumor, ganglioglioma, or other glioneuronal tumor types, and in such cases the diagnosis of oligodendroglioma rests upon reliably excluding these other possibilities.

Designation of anaplastic OG has been controversial in the pediatric population, though some studies demonstrate worse outcomes in tumors classified as anaplastic pediatric oligodendroglioma.[2] It remains the general tendency to still categorize these neoplasms as WHO grade II or III, similar to their adult counterparts. In many studies involving adult OG, the typical histological parameters of malignancy (i.e., high mitotic rate, vascular endothelial or microvascular proliferation, and palisading necrosis) imply survival differences between low-grade (WHO grade II) vs anaplastic (WHO grade III) tumors. Typically, anaplastic OG is a highly cellular tumor with numerous mitoses and apoptotic bodies. There is often unequivocal vascular endothelial proliferation that is recognized as a tuft of endothelial cells with scant or no vascular lumen (Fig. 3). Necrosis, especially pseudopalisading necrosis, can be seen in a subset of anaplastic OG.

Molecular drivers

Molecular drivers can be used to aid in the diagnosis and prognostication of adult OG, but are not yet readily applied in the pediatric setting. As described above, the 2016 WHO Classification of Tumors of the Central Nervous System diagnosis of either OG or anaplastic OG in adults requires the presence of both an *IDH* gene mutation and whole-arm 1p/19q co-deletion. Pediatric OG though seems to share little molecular similarity with its adult counterpart as the molecular changes seen in adults are often missing in children.

The paucity of 1p/19q co-deletions in pediatric OG is supported by the literature. When these alterations are seen, they are often restricted to older children, perhaps suggesting the presence of two different subtypes—an adolescent/young adult variant that is more similar to adult OG and another arising in younger patients. Retrospective reviews have estimated the presence of 1p/19q co-deletions to occur in about 10%–35% of pediatric OG[2,22–24] and most predominantly in patients older than 15 years of age.[2,23,24] A summary of the prevalence of 1p/19q co-deletions occurring in pediatric OG is presented in Table 1.

TABLE 1 Summary of 1p/19q co-deletion in pediatric OG

First author	Number of pediatric cases	Presence of 1p/19q co-deletion, % (n)	Presence of co-deletions in patients >15 years, % (n)
F. Rodriguez[2]	50	25 (10)	80 (8)
V. Suri[23]	14 (patients <25 years)	35 (5)	100 (5)
D. Nauen[22]	28	10 (3)	66.7 (2)
T. Lundar[11]	10	0	n/a
P. Kreiger[12]	16	0	n/a
R. Raghavan[24]	25	12 (3)	100 (3)

Much like 1p/19q co-deletions, *IDH* mutations also occur at a much lower frequency in pediatric OG compared to their adult counterparts. One review found *IDH1* mutations, as identified by immunohistochemistry, in only 4 of 22 (18%) cases of pediatric OG.[2] Another retrospective review of 14 cases of OG in people under 25 years of age found zero cases with *IDH1* mutation.[23] These findings support that the molecularly based diagnostic criteria used in adult OG cannot be applied in the pediatric setting, and support ongoing efforts to molecularly characterize pediatric OG.

One molecular characteristic that pediatric OG seems to hold similar to their adult counterparts is the demonstration of methylguanine-methyltransferase (*MGMT*) promoter methylation. Suri et al. evaluated 16 cases of pediatric OG and found *MGMT* methylation in 71% of these cases.[23]

In adults, molecular alterations play an important role in assessing disease prognosis and therapy response; however, it remains unclear how to apply these characterizations in children. Several studies have shown significant association between increased progression-free survival (PFS) and overall survival (OS) and chemosensitivity in adult patients with 1p/19q co-deletion and/or *IDH*-mutant OG.[25–27] Given the few studies that reported molecular alterations in pediatric OG, it is not surprising that there are no similar findings of the prognostic role of 1p/19q co-deletion or *IDH* mutations in children.

There is limited data available on the molecular differences among grade II and III pediatric OG. One retrospective review reported that 53% of grade II OG compared to 100% of grade III anaplastic OG carried cytogenetic alterations.[22] Additionally, they found that grade III anaplastic cases had statistically significantly more somatic alterations, with a mean of 11.6 mutations per anaplastic tumor vs 1.4 per non-anaplastic tumor.

Clearly, there is still much more need for research investigating the role of molecular alterations in pediatric OG, especially 1p/19q co-deletions, *IDH* family mutations, and *MGMT* promoter methylation status. The need is particularly true given the correlation of molecular findings with histology and response to therapy in adult disease.[28,29]

Therapy

Given the rarity of pediatric OG and lack of structured studies on dedicated therapy regimens, treatment plans typically follow the approach of adult OG or their pediatric low-grade glioma counterparts. Most frequently, grade II OG is treated with surgery, while chemotherapy and radiation are reserved for patients with anaplastic or recurrent disease. In particular, radiation is often avoided as long as possible in children given its association with long-term sequalae related to endocrine function, neurocognition, vascular injury, and hearing.[30–32]

Surgery

Some studies report better prognosis in terms of improved PFS and OS with gross total resection (GTR) compared to sub-total resection (STR) of pediatric OG.[19,33] Others did not find a significant relationship between outcome and extent of resection.[2,6,7,34] In comparison, larger cohorts of pediatric gliomas as a whole have reported improved outcomes with more complete resection.[35,36] Nonetheless, in the setting of low-grade OG and given the relatively high survival for patients undergoing even partial resection and the toxicities of chemotherapy and radiation, a "watchful waiting" period with regular surveillance imaging is frequently warranted before the initiation of chemotherapy or radiation.

Chemotherapy

As mentioned previously, chemotherapy regimens are not well standardized or extensively studied in pediatric OG. Outcomes related to specific regimens are limited to retrospective reviews or case reports. There are currently no published trials dedicated solely for the treatment of pediatric OG. Notably, nearly all studies reporting the use of chemotherapy for pediatric OG are in the setting of anaplastic OG or recurrent grade II disease. Table 2 lists chemotherapy regimens used for pediatric OG that have been reported in the literature.

Many cases of pediatric OG have been treated with a combination of procarbazine, lomustine, and vincristine (termed 'PCV'), likely due to adult literature showing sensitivity of adult OG to this chemotherapy regimen, particularly in patients with a 1p/19q co-deletion.[25,26,39] The translation of outcomes between adult and pediatric patients though remains unknown. For instance, Rodriguez et al. reported on two children with anaplastic OG who received PCV and progressed early in therapy. Of note, both were lacking 1p/19q co-deletion.[2] They also reported on two patients with anaplastic OG carrying 1p/19q co-deletions, who received alkylator-based therapy [e.g., procarbazine, lomustine, temozolomide (TMZ), cyclophosphamide]. These patients experienced disease progression early in therapy as well. In another retrospective review, Raghavan et al. reported the use of PCV in two patients with 1p/19q co-deleted, anaplastic disease.[24] Unfortunately, both patients experienced OS of <1 year. In general, across all reported cases of pediatric OG, there appears to be a trend toward worse survival in patients receiving chemotherapy than those without, which is most likely due to the correlation that patients receiving chemotherapy are also more likely to carry the most aggressive disease.

Citing support in the adult literature,[40,41] Sorge et al. used TMZ alone following resection in one pediatric patient with anaplastic OG.[37] At the time of publication, this patient was still alive and had no progression at 42 months post-therapy. Kumar et al. also reported on a patient with anaplastic OG treated with adjuvant radiation and TMZ. This patient was still alive and without progression at approximately 30 months post-resection.[38] The efficacy of TMZ has not been consistent across all patients though and some have shown continued progression, despite this intervention.[15] Another case report applied radiation and TMZ in combination with an alternating regimen of carboplatin, etoposide, and cyclophosphamide in the setting of a grade II, 1p/19q co-deleted OG that evolved to a grade III anaplastic lesion with metastatic disease.[21] This patient went through multiple rounds of varying treatments, including combination therapy with irinotecan and TMZ and then thalidomide, celecobrix, etoposide, and cyclophosphamide. With these combinations, the patient experienced OS of about 5 years from initial diagnosis.

TABLE 2 Chemotherapy regimens in the literature

	Number of pediatric cases (histology)	Chemotherapy	Reported survival, months	PFS, months
F. Rodriguez[2]	4 (anaplastic)	PCV or alkylator-based therapy	Not reported	Not reported
R. Raghavan[24]	2 (anaplastic)	PCV	<12 months (average)	<12
C. Sorge[37]	1 (anaplastic)	TMZ	>42	>42
A. Kumar[38]	1 (anaplastic)	TMZ	>30	>30
S. Furtado[15]	1 (anaplastic)	TMZ	Not reported	Not reported
C. Bruggers[21]	1 (low-grade progression to anaplastic)	TMZ + carboplatin, etoposide, cyclophosphamide	60	Not reported
T. Rizk[5]	4 (anaplastic)	8-in-1 + etoposide, thiotepa, BMT[a]	17 (median)	Not reported
D. Hyder[3]	1 (anaplastic)	8-in-1	9	8

[a]Bone marrow transplantation.

Pediatric low-grade gliomas have also been treated with a multidrug chemotherapy regimen using "eight-drugs-in-1-day" (8-in-1). This regimen includes high-dose methylprednisolone, vincristine, lomustine, hydroxyurea, cisplatin, cytarabine, procarbazine, and dacarbazine and has been used for the treatment of both low- and high-grade pediatric tumors.[42-44] Rizk et al. reported on four children with anaplastic OG receiving the 8-in-1 regimen. These patients experienced a median survival of 17 months. Three of these four patients also received etoposide and thiotepa followed by a bone marrow transplant.[5] Hyder et al. reviewed oligodendrogliomas that were treated on the Children's Cancer Group CCG-945 study. CCG-945 applied vincristine and radiotherapy followed by prednisone, lomustine, and vincristine vs the 8-in-1 regimen for pediatric patients with malignant gliomas.[3] There was one confirmed case of anaplastic oligodendroglioma in this cohort. This patient received the 8-in-1 regimen and experienced a PFS of 8 months and OS of 9 months.

Prognosis and outcome

Analysis of SEER data in patients up to 24 years of age has shown a 5-year OS rate of about 90% for grade II OG and 53% for anaplastic OG.[45] Retrospective reviews, which did not differentiate between tumor grade, have reported 5-year PFS varying between 49% and 75% for all pediatric OGs.[2,18,19,33] Peters et al. looked solely at 32 grade II pediatric OGs and found a 5-year PFS of 81.3%.[6] Overall, very few of these tumors seem to progress histologically to become more aggressive.[2,11,21,] When these tumors are located within midline structures of the CNS, PFS and OS have been reported to be significantly worse.[6]

Findings regarding age at diagnosis of pediatric OG and correlation with outcomes have been mixed. Two retrospective reviews noted improved prognosis in patients who were first diagnosed younger than 12 years of age.[33,34] One cohort though was ultimately found to be composed of 25% oligoastrocytoma cases vs pure OG.[33] The large review by Rodriguez et al. did not find a significant association between age at diagnosis and PFS or OS.[2]

In adults, there is significant difference in outcome based on the presence of 1p/19q co-deletion. Only one study in pediatric OG has investigated the role of 1p/19q co-deletion. This study reported no significant difference in survival based on 1p/19q co-deletion.[2] Additional studies either had no patients with 1p/19q co-deletions or did not analyze co-deletion relationship with survival.[11,22-24] These findings though may be supported by the lack of correlation with 1p/19q co-deletion with survival in pediatric high-grade gliomas.[46]

Conclusion

Pediatric OG remains a very rare disease without definitive molecular characteristics to aid in diagnosis. 1p/19q co-deletion is less common in pediatric OG and even more so in younger children. Prognosis appears to be determined by WHO grade II vs III and thus, therapy decision-making should be made taking into consideration tumor grade. Given the lack of treatment data specific to pediatric OG, there remains a need for further trial investigation into this cohort and additional information regarding molecular subgrouping of these tumors.

References

1. Louis DN, Perry A, Reifenberger G, et al. The 2016 World Health Organization classification of tumors of the central nervous system: a summary. *Acta Neuropathol.* 2016;131(6):803–820.
2. Rodriguez FJ, Tihan T, Lin D, et al. Clinicopathologic features of pediatric oligodendrogliomas: a series of 50 patients. *Am J Surg Pathol.* 2014;38(8):1058.
3. Hyder DJ, Sung L, Pollack IF, et al. Anaplastic mixed gliomas and anaplastic oligodendroglioma in children: results from the CCG 945 experience. *J Neurooncol.* 2007;83(1):1–8.
4. Ostrom QT, Gittleman H, Xu J, et al. CBTRUS statistical report: primary brain and other central nervous system tumors diagnosed in the United States in 2009–2013. *Neuro Oncol.* 2016;18(suppl_5):v1–v75.
5. Rizk T, Mottolese C, Bouffet E, et al. Cerebral oligodendrogliomas in children: an analysis of 15 cases. *Childs Nerv Syst.* 1996;12(9):527–529.
6. Peters O, Gnekow A, Rating D, Wolff J. Impact of location on outcome in children with low-grade oligodendroglioma. *Pediatr Blood Cancer.* 2004;43(3):250–256.
7. Bowers DC, Mulne AF, Weprin B, Bruce DA, Shapiro K, Margraf LR. Prognostic factors in children and adolescents with low-grade oligodendrogliomas. *Pediatr Neurosurg.* 2002;37(2):57–63.
8. Hirsch J-F, Sainte Rose C, Pierre-Kahn A, Pfister A, Hoppe-Hirsch E. Benign astrocytic and oligodendrocytic tumors of the cerebral hemispheres in children. *J Neurosurg.* 1989;70(4):568–572.
9. Favier J, Pizzolato G, Berney J. Oligodendroglial tumors in childhood. *Childs Nerv Syst.* 1985;1(1):33–38.
10. Mohindra S, Savardekar A, Bal A. Pediatric brainstem oligodendroglioma. *J Neurosci Rural Pract.* 2012;3(1):52.

11. Lundar T, Due-Tønnessen BJ, Egge A, Scheie D, Stensvold E, Brandal P. Neurosurgical treatment of oligodendroglial tumors in children and adolescents: a single-institution series of 35 consecutive patients. *J Neurosurg Pediatr.* 2013;12(3):241–246.
12. Kreiger PA, Okada Y, Simon S, Rorke LB, Louis DN, Golden JA. Losses of chromosomes 1p and 19q are rare in pediatric oligodendrogliomas. *Acta Neuropathol.* 2005;109(4):387–392.
13. Wang KC, Chi JG, Cho BK. Oligodendroglioma in childhood. *J Korean Med Sci.* 1993;8(2):110–116.
14. Geçirilmesi LG. Infratentorial oligodendroglioma in a child: a case report and review of the literature. *Turk Neurosurg.* 2012;22(4):461–464.
15. Furtado SV, Venkatesh PK, Ghosal N, Murthy GK, Hegde AS. Clinical and radiological features of pediatric cerebellar anaplastic oligodendrogliomas. *Ind J Pediatr.* 2011;78(7):880–883.
16. Packer RJ, Sutton LN, Rorke LB, et al. Oligodendroglioma of the posterior fossa in childhood. *Cancer.* 1985;56(1):195–199.
17. Chen R, Macdonald DR, Ramsay DA. Primary diffuse leptomeningeal oligodendroglioma: Case report. *J Neurosurg.* 1995;83(4):724–728.
18. Tice H, Barnes PD, Goumnerova L, Scott RM, Tarbell NJ. Pediatric and adolescent oligodendrogliomas. *Am J Neuroradiol.* 1993;14(6):1293–1300.
19. Wu C-T, Tsay P-K, Jaing T-H, Chen S-H, Tseng C-K, Jung S-M. Oligodendrogliomas in children: clinical experiences with 20 patients. *J Pediatr Hematol Oncol.* 2016;38(7):555–558.
20. Alvarez JA, Cohen ML, Hlavin ML. Primary intrinsic brainstem oligodendroglioma in an adult: Case report and review of the literature. *J Neurosurg.* 1996;85(6):1165–1169.
21. Bruggers C, White K, Zhou H, Chen Z. Extracranial relapse of an anaplastic oligodendroglioma in an adolescent: case report and review of the literature. *J Pediatr Hematol Oncol.* 2007;29(5):319–322.
22. Nauen D, Haley L, Lin MT, et al. Molecular analysis of pediatric oligodendrogliomas highlights genetic differences with adult counterparts and other pediatric gliomas. *Brain Pathol.* 2016;26(2):206–214.
23. Suri V, Jha P, Agarwal S, et al. Molecular profile of oligodendrogliomas in young patients. *Neuro Oncol.* 2011;13(10):1099–1106.
24. Raghavan R, Balani J, Perry A, et al. Pediatric oligodendrogliomas: a study of molecular alterations on 1p and 19q using fluorescence in situ hybridization. *J Neuropathol Exp Neurol.* 2003;62(5):530–537.
25. Cairncross JG, Wang M, Jenkins RB, et al. Benefit from procarbazine, lomustine, and vincristine in oligodendroglial tumors is associated with mutation of IDH. *J Clin Oncol.* 2014;32(8):783–790.
26. Cairncross G, Wang M, Shaw E, et al. Phase III trial of chemoradiotherapy for anaplastic oligodendroglioma: long-term results of RTOG 9402. *J Clin Oncol.* 2012;31(3):337–343.
27. van den Bent MJ, Carpentier AF, Brandes AA, et al. Adjuvant procarbazine, lomustine, and vincristine improves progression-free survival but not overall survival in newly diagnosed anaplastic oligodendrogliomas and oligoastrocytomas: a randomized European Organisation for Research and Treatment of Cancer phase III trial. *J Clin Oncol.* 2006;24(18):2715–2722.
28. Sasaki H, Zlatescu MC, Betensky RA, et al. Histopathological-molecular genetic correlations in referral pathologist-diagnosed low-grade "oligodendroglioma". *J Neuropathol Exp Neurol.* 2002;61(1):58–63.
29. Jenkins RB, Blair H, Ballman KV, et al. A t(1; 19)(q10; p10) mediates the combined deletions of 1p and 19q and predicts a better prognosis of patients with oligodendroglioma. *Cancer Res.* 2006;66(20):9852–9861.
30. Merchant TE, Conklin HM, Wu S, Lustig RH, Xiong X. Late effects of conformal radiation therapy for pediatric patients with low-grade glioma: Prospective evaluation of cognitive, endocrine, and hearing deficits. *J Clin Oncol.* 2009;27(22):3691–3697.
31. Mueller S, Sear K, Hills NK, et al. Risk of first and recurrent stroke in childhood cancer survivors treated with cranial and cervical radiation therapy. International Journal of Radiation Oncology Biology Physics. 2013;86(4):643–648.
32. Mueller S, Fullerton HJ, Stratton K, et al. Radiation, atherosclerotic risk factors, and stroke risk in survivors of pediatric cancer: a report from the Childhood Cancer Survivor Study. *International Journal of Radiation Oncology, Biology, Physics.* 2013;86(4):649–655.
33. Creach KM, Rubin JB, Leonard JR, et al. Oligodendrogliomas in children. *J Neurooncol.* 2012;106(2):377–382.
34. Razack N, Baumgartner J, Bruner J. Pediatric oligodendrogliomas. *Pediatr Neurosurg.* 1998;28(3):121–129.
35. Qaddoumi I, Sultan I, Gajjar A. Outcome and prognostic features in pediatric gliomas. *Cancer.* 2009;115(24):5761–5770.
36. Kramm CM, Wagner S, Van Gool S, et al. Improved survival after gross total resection of malignant gliomas in pediatric patients from the HIT-GBM studies. *Anticancer Res.* 2006;26(5B):3773–3779.
37. Sorge C, Li R, Singh S, et al. Complete durable response of a pediatric anaplastic oligodendroglioma to temozolomide alone: case report and review of literature. *Pediatr Blood Cancer.* 2017;.
38. Kumar A, Pathak P, Purkait S, et al. Oncogenic KIAA1549-BRAF fusion with activation of the MAPK/ERK pathway in pediatric oligodendrogliomas. *Cancer Genet.* 2015;208(3):91–95.
39. van den Bent MJ, Brandes AA, Taphoorn MJ, et al. Adjuvant procarbazine, lomustine, and vincristine chemotherapy in newly diagnosed anaplastic oligodendroglioma: long-term follow-up of EORTC brain tumor group study 26951. *J Clin Oncol.* 2012;31(3):344–350.
40. Brandes AA, Tosoni A, Cavallo G, et al. Correlations between O6-methylguanine DNA methyltransferase promoter methylation status, 1p and 19q deletions, and response to temozolomide in anaplastic and recurrent oligodendroglioma: a prospective GICNO study. *J Clin Oncol.* 2006;24(29):4746–4753.
41. Gan HK, Rosenthal MA, Dowling A, et al. A phase II trial of primary temozolomide in patients with grade III oligodendroglial brain tumors. *Neuro Oncol.* 2010;12(5):500–507.
42. Pendergrass TW, Milstein J, Geyer J, et al. Eight drugs in one day chemotherapy for brain tumors: experience in 107 children and rationale for preradiation chemotherapy. *J Clin Oncol.* 1987;5(8):1221–1231.
43. Geyer JR, Zeltzer PM, Boyett JM, et al. Survival of infants with primitive neuroectodermal tumors or malignant ependymomas of the CNS treated with eight drugs in 1 day: a report from the Childrens Cancer group. *J Clin Oncol.* 1994;12(8):1607–1615.

44. Finlay JL, Boyett JM, Yates AJ, et al. Randomized phase III trial in childhood high-grade astrocytoma comparing vincristine, lomustine, and prednisone with the eight-drugs-in-1-day regimen. Childrens Cancer Group. *J Clin Oncol.* 1995;13(1):112–123.

45. Achey RL, Khanna V, Ostrom QT, Kruchko C, Barnholtz-Sloan JS. Incidence and survival trends in oligodendrogliomas and anaplastic oligodendrogliomas in the United States from 2000 to 2013: a CBTRUS Report. *J Neurooncol.* 2017;1–9.

46. Pollack IF, Finkelstein SD, Burnham J, et al. Association between chromosome 1p and 19q loss and outcome in pediatric malignant gliomas: results from the CCG-945 cohort. *Pediatr Neurosurg.* 2003;39(3):114–121.

Immune checkpoint blockade in glioma

Sherise D. Ferguson*, Shiao-Pei Weathers[†] and Amy B. Heimberger*

*Department of Neurosurgery, The University of Texas MD Anderson Cancer Center, Houston, TX, United States; [†]Department of Neuro-Oncology, The University of Texas MD Anderson Cancer Center, Houston, TX, United States

Introduction

Immune system evasion is an established hallmark of cancer. As such, immunotherapy has become an embedded treatment algorithm of multiple solid malignancies, particularly melanoma.[1,2] and lung cancer,[3,4] and is now under intense investigation regarding central nervous system (CNS) malignancies. Among immunotherapeutic strategies, checkpoint modulation, especially, has had significant clinical success. Aberration in immune checkpoint signaling is a critical mechanism that allows tumors to evade immune detection. Therapies targeting immune checkpoints are unique in that instead of targeting tumor cells, they target molecules essential for T-cell regulation and remove inhibitory pathways that block an effective antitumor immune response. In this chapter, we review the rationale and principles behind the development of checkpoint inhibition for the treatment of glioma. We focus on the preclinical data supporting this treatment approach in glioma and the ongoing clinical trials dedicated to it. Additionally, we address the challenges in implementing this immunotherapeutic strategy. Most of the data regarding checkpoint modulation are dedicated to high-grade astrocytoma, but with further research, these principles may potentially be applied to oligodendroglioma.

What are immune checkpoints?

In the setting of solid malignancies, including glioma, a local immune response is elicited. Released tumor antigens are taken up by antigen-presenting cells (APCs), including dendritic cells (DCs), microglia, and tumor-associated macrophages. The APCs subsequently present these antigens to T-cell receptors. T cells surveying the tumor microenvironment recognize these tumor antigens and can become activated, differentiate, and proliferate in an attempt to mount an effective immune response. In addition to this APC/T-cell interaction, co-stimulatory molecules are also required for T-cell activation. Specifically, the co-stimulatory molecule CD28 expressed on T cells binds to B-7 ligands (B-7.1 and B-7.2) on APCs, which results in T-cell activation and proliferation.[5,6] T cells that lack appropriate co-stimulation become anergic, whereas the upregulation of the immune checkpoints induces T-cell exhaustion. Regulation of an effective and appropriate T-cell response is complex, requiring both co-stimulatory and inhibitory signaling in order to maintain immune homeostasis and prevent autoimmunity. Immune checkpoints are critical in the maintenance of this homeostasis/balance to optimize an appropriate immune response. The expression of immune checkpoints is induced after the activation of T cells to limit autoreactivity. Tumor cells can exploit these checkpoints in order to evade the immune antitumor response; hence, targeting of these immune modulators has substantially changed cancer therapy over the last decade.

Programmed cell death protein 1 (PD-1) and its associated ligand, PD-L1, represent a key checkpoint signaling axis that has been extensively studied. PD-1 is expressed on activated T cells and multiple other immune cells (B-cells, natural killer cells, and myeloid cells),[7] and its ligand, PD-L1, is a B7 homolog (B7·H1).[8] PD-1/PD-L1 is an inhibitory signaling axis that negatively regulates the immune response. The initiation of the immune signaling cascade, including the release of pro-inflammatory cytokines such as interferon-γ, causes the upregulation of PD-L1. Subsequent PD-1/PD-L1 binding attenuates the immune response by inducing T-cell exhaustion or apoptosis.[9] Data also suggest that this interaction can promote proliferation of immunosuppressive T-regulatory cells (Tregs) and that it reduces the respective immune responses of natural killer (NK) cells and B cells.[9–11] Targeting of the PD-1/PD-L1 axis was initially clinically achieved via the use of two FDA-approved antibodies, nivolumab and pembrolizumab, and both demonstrated success in multiple clinical trials, particularly among melanoma and lung cancer patient populations.[12,13] With the success of these antibodies in a variety of malignancies, there are many clinical trials now recruiting glioma patients to study the effects of immune checkpoint inhibitors, including the use of a wide variety of combinatorial strategies.

Cytotoxic T lymphocyte antigen-4 (CTLA-4), also referred to as CD152, is another well-described immune checkpoint expressed on activated T cells and Tregs. As mentioned above, CD28-mediated signaling is required for T-cell activation. This signaling is also a mechanism of immune regulation, as CD28 binding to B7 results in the expression of CTLA-4 on the T-cell surface. CTLA-4 has a 10–20-fold higher affinity for CD28 than B7, and thus it blocks T-cell co-stimulatory signaling by competitively binding to B7, thereby abrogating the T-cell response.[14,15] Furthermore, CTLA-4 can also cause T-cell inactivation through B7-independent mechanisms.[16–18] CTLA-4 may also increase the presence of immunosuppressive myeloid-derived suppressor cells (MDSCs) while downregulating T-helper cell activity.[19] Preclinical support for CTLA-4 blockade in glioma was demonstrated by Fecci et al., who reported 80% long-term survival in mice treated with a monoclonal anti-CTLA-4 antibody.[20] Ipilimumab was the first FDA-approved monoclonal antibody to human CTLA-4, and has since become the standard of care in melanoma treatment after showing significant durable tumor control in patients with advanced disease in phase III clinical trials.[21,22] Regarding glioma, there is currently a phase I trial to assess the safety of ipilimumab and/or nivolumab together with temozolomide administration in newly diagnosed glioblastoma (NCT02311920) patients. The results of this trial are pending.

Lymphocyte-activation gene 3 (LAG-3, also known as CD223) is another recently described checkpoint that is being tested in preclinical models. It is a CD4-related transmembrane protein that competitively binds MHC II and acts as a co-inhibitory checkpoint for T-cell activation.[23] Under inflammatory conditions, LAG-3 is expressed on the surface of activated T cells (CD4+, CD8+) and NK cells and plays a role in T-cell expansion and Treg functionality.[23,24] Workman et al. demonstrated that compared with LAG3+ T cells, LAG3 knockout T cells showed increased proliferation after stimulation in vivo and delayed cycle cell arrest.[23,24] An in vivo study demonstrated that anti-LAG-3 antibodies resulted in a reversal of Treg-mediated immune suppression.[24] Moreover, a study using an LAG-3 knockout in mice demonstrated that relative to wild-type Tregs, more than double the number of LAG-3(−/−) Tregs were required to control CD4+ helper-T-cell proliferation.[24] LAG-3 may also play a role in regulating DC function.[25]

T-cell immunoglobulin mucin 3 (TIM-3) is another negative regulator of T-cell function. It is a type I glycoprotein receptor that binds to S-type lectin galectin-9 (Gal-9), and is expressed on lymphocytes. Binding of Gal-9 to the TIM-3 receptor triggers downstream signaling to negatively regulate T-cell survival and function via calcium influx.[26] TIM-3 expression has been shown to be elevated in the tumor-infiltrating lymphocytes of glioma patients and is correlated with tumor grade.[27,28] A large study analyzing 325 glioma patients with RNA sequencing data from the Chinese Glioma Genome Atlas (CGGA project) confirmed that TIM-3 was enriched in glioblastoma (GBM), IDH wild-type tumors, and that high expression of TIM-3 was an independent indicator of poor prognosis.[29] These data indicated that TIM-3 may be a targetable checkpoint in glioma, but its protein expression has not yet been validated in GBM patients.

Predictors of checkpoint modulation efficacy

Since the clinical impact of checkpoint inhibition has been established in certain solid tumors, significant research effort has been dedicated to determining which patients will benefit from checkpoint modulation. Exploration of biomarkers can help create rational clinical trials and target specific populations to maximize therapeutic benefit.

Tumor mutational burden

Tumor mutational burden has become a marker of susceptibility to checkpoint inhibition. High tumor mutational burden results in an increased number of potential neoantigens that may be presented to the host immune system and increase the effector T-cell (CD8+) population.[30–32] Carcinogen-induced tumors have a relatively high mutational burden relative to other cancers. As such, multiple landmark studies have described the impact of mutational burden on immunotherapeutic efficacy. In non-small cell lung cancer, Rizvi et al. sequenced the exosomes and matched normal DNA of patients in two independent cohorts treated with checkpoint inhibition. These authors reported that a higher somatic mutation burden was associated with a clinical response to pembrolizumab (anti-PD-1). Specifically, 73% of patients with a high mutational burden were afforded a durable clinical response (partial or stable, lasting >6 months), unlike the 13% of patients with a low mutational status.[30–32] Additionally, patients with an increased mutational load had a longer progression-free survival than their low-burden counterparts. Studies of patients with advanced melanoma yielded similar results.[33,34] Snyder et al. analyzed whole-exosome sequencing of tumor and matched blood samples in melanoma patients and reported that mutational burden impacted the clinical benefit of CTLA-4 inhibition.[33]

In light of these findings, the role of mutational burden on checkpoint inhibition efficacy has been examined in glioma. A notable study using the cancer genome atlas (TCGA) demonstrated that mutational load correlated with tumor grade.[35] However, compared with carcinogen-induced cancer such as melanoma (UV light) and lung cancer (smoking), gliomas have a comparatively low mutational burden. A recent study employing a detailed biomarker analysis of 327 gliomas found that high tumor mutational load was only detected in 3.5% of GBMs and reported that mutational load was associated with loss of mismatch-repair protein expression. This indicates that only a small subset of glioma patients is likely to benefit from checkpoint monotherapy.[36] Even though gliomas harbor limited exome-wide mutations,[37,38] there is a subset of hypermethylated gliomas in which the mutational load is presumed to be notably higher.[39,40] This phenotype is present in 20%–30% of recurrent gliomas posttreatment with temozolomide, which has been postulated to result from an acquired somatic mismatch-repair deficiency.[41,42] Interestingly, significant clinical responses have been reported in patients with mismatch-repair-deficiency (MMRD) tumors, as these tumors have a high mutational load.[30,43,44] Bouffet et al. reported the case of two siblings with a bi-allelic mismatch-repair defect-induced GBM treated with PD-1 blockade. Treatment resulted in a significant clinical and radiographic response, highlighting the potential benefit of checkpoint blockade in hypermutated glioma.[45]

In addition to MMRD, tumors carrying a mutation in the catalytic subunit of the DNA polymerase epsilon (POLE) also have a highly mutated genome and are thus also likely to benefit from immunotherapy.[46–48] A hyper-mutated genotype was also recently reported in a newly diagnosed GBM associated with somatic POLE mutations, each carrying >100 times the average number of mutations.[49] A recent case report described a GBM patient with a hypermutated genotype in the setting of a POLE germline mutation, who was treated with checkpoint blockade (anti-PD-1) and had a robust clinical response.[38] The patient's intracranial disease showed a radiographic response, and the resected spinal metastasis had prominent infiltration of tumor-infiltrating lymphocytes. Notably, POLE mutations are found in <1% of all gliomas.[36] But with such encouraging results, early phase clinical trials are underway to further investigate the utility of checkpoint blockade in recurrent glioma with high mutational burden (NCT02658279).

PD-1/PDL-1 expression

PD-1/PD-L1 expression is the most extensively studied biomarker for determining which patients may benefit from immune checkpoint modulation. A phase I trial examining the safety/efficacy of nivolumab in multiple solid malignancies reported a significant relationship between intratumoral PD-L1 expression and clinical response to checkpoint inhibition.[50] Multiple subsequent studies have confirmed this association.[3,11,51] The expression of PD-L1 in glioma tumor cells and PD-1 in tumor-infiltrating lymphocytes remains controversial. Nduom et al. evaluated the incidence of PD-1 expression in 92 patients with high-grade glioma and found that 2.7% of GBM cells expressed PD-L1. In the majority of cases (61%), at least 1% of tumor cells expressed PD-L1. This result was confirmed with ex vivo staining of GBMs by flow cytometry, which demonstrated a median of 3.5% of PD-L1 surface-expressing cells. Additionally based on TCGA, they reported that PD-L1 and PD-1 expression impacted patient survival. Specifically, patients with high PD-L1 mRNA expression survived for 11.4 months relative to 14.9 months in the lower expression group ($p = 0.02$). This finding was validated at the protein level with immunohistochemistry using the previously determined median as the cut point (2.7%). When this cutoff was adjusted to 5%, they found that decreased survival was significantly associated with PD-L1 expression.[52] Garber et al. evaluated PD-L1 expression across glioma lineages and grades using CLIA-compliant testing. This study included 347 cases, and the authors found a correlation between PD-L1 expression and glioma grade. All PD-L1-positive cases were GBM. Garber et al. reported PD-L1 expression in 7.8% of GBM specimens using cut points established for other cancers.[53] Notably, it is still unclear how much PD-1/PD-L1 expression is needed in order to correlate with therapeutic responses, as clinical responses to checkpoint blockade have been described in patients without PD-L1 expression.[50]

T-cell infiltration

The density of effector T-cell infiltration has also been associated with the efficacy of checkpoint blockade. This link was first described in a landmark study of advanced melanoma in which the authors observed that patients with a significant clinical response to checkpoint blockade had higher numbers of CD8[+,] PD-1[+], and PD-L1[+] cells at the tumor margin and intratumorally. Specifically, preexisting CD8[+] cell density at the invading tumor margin was found to be the most predictive parameter of clinical response to PD-1 blockade.[54] Unlike melanoma, gliomas are inherently less immunogenic.[55] Berhoff et al. reported that CD8[+] tumor-infiltrating lymphocytes are sparse in 50% of GBMs and moderate in 7% of cases.[55] Furthermore, gliomas are well known to have multiple mechanisms that promote an immunosuppressive tumor microenvironment, including but not limited to secretion of immunosuppressive cytokines, decreased activity of effector T cells, skewing of macrophages to an immunosuppressive M2 phenotype, increased presence of immunosuppressive Tregs, immune-suppressive cytokines, inducers of T-cell apoptosis, etc. All of these factors contribute to making gliomas immunologically "cold" and are a barrier to effective implementation of immune checkpoint blockade. In light of the impact of T-cell density, methods to increase T-cell infiltration are an area of active investigation. Furthermore, strategies to increase intratumoral T-cell density are the rationale behind multiple combinatory immunotherapeutic strategies.

Checkpoint combination strategies

Dual-checkpoint blockade

Individual checkpoint blockade of CTLA-4, PD-1, or PD-L1 has achieved noteworthy benefit for patients with advanced solid tumors. Dual-checkpoint inhibition is a clear, rational combination therapy, particularly considering that the most clinically utilized checkpoints, PD-1 and CTLA-4, have non-redundant signaling pathways. As the mechanisms of CTLA-4 and PD-1 do not overlap, combination therapies have also been evaluated in clinical trials with great success and limited treatment-induced toxicity.[1,56,57] In a critical study, Larkin et al. evaluated treatment with nivolumab plus ipilimumab relative to ipilimumab and nivolumab as monotherapies in patients with unresectable melanoma. Combination therapy resulted in a median progression-free survival time of 11.5 months compared with 2.9 and 6.9 months for ipilimumab and nivolumab monotherapies, respectively.[56]

Preclinical studies in glioma have also evaluated dual-checkpoint modulation and reported encouraging results. Reardon et al. performed a detailed evaluation of checkpoint inhibition in a murine glioma model (GL261). Use of single agent treatments of anti-PD-L1, PD-1, and CTLA-4 resulted in long-term survival rates of 20%, 50%, and 15% of mice, respectively. However, anti-CTLA-4 in combination with anti-PD-1 resulted in a long-term survival of 75% of treated mice.[58] Another recent study showed that adding anti-TIM3 therapy to anti-PD-1 in glioma-bearing mice improved the median survival time from 33 to 100 days and improved the overall survival rate from 28% to 58%.[59]

Indoleamine, 2, 3 dioxygenase (IDO) is an intracellular catabolic enzyme involved in immunosuppression and could also be considered for immune-modulatory combinations. The IDO is upregulated in the setting of inflammation and is associated with the conversion of tryptophan to kynurenines. Tryptophan depletion in the vicinity of T cells results in T-cell apoptosis and deactivation. Moreover, it can result in the recruitment of immunosuppressive Tregs. The IDO activity is induced in the majority of GBMs, and its expression is associated with shorter patient survival times.[60,61] In a preclinical investigation of triple therapy using inhibition of IDO, CTLA-4, and PD-1, there was a marked reduction in the Treg population in glioma-bearing mice, and there were marked increases in survival. Specifically, relative to control animals (median survival 32 days; 100% mortality), IDO monotherapy resulted in a 45-day survival time and a 38% long-term survival rate (>90 days). Triple therapy resulted in long-term survival in 78% of treated animals.[62] Such results have led to clinical investigation of this immunotherapeutic strategy (NCT02327078, NCT03058289). One does need to bear in mind that the immune competent murine model typically used in these studies express high levels of PD-L1, are clonotypic, and their tumors lack glioma heterogeneity and are small in size.[52] Thus far, results in these models have not correlated with therapeutic activity in human subjects.

Radiation therapy

Radiation therapy remains a cornerstone of cancer therapy, and several studies have reported a synergistic effect between radiation and immunotherapy.[63–66] Radiation has several immunomodulatory effects. It causes the release of tumor antigens after cell lysis and enhances the presentation of these antigens to circulating APCs, resulting in increased infiltration of effector T cells to the tumor itself and the non-irradiated tumor tissue (referred to as the abscopal effect).[67,68] Sharabi et al. used established murine model systems of melanoma and breast cancer (with the respective tumors transplanted to the flanks of mice) to demonstrate that combination treatment with radiotherapy and an anti-PD-1 antibody resulted in an increased percentage of tumor-specific intratumoral T cells relative to anti-PD-1 monotherapy. Furthermore, this combination therapy produced a significant decrease in the tumor volume of each respective tumor type relative to control animals.[68] In an in vivo murine breast cancer model, Deng et al. showed that local inflammation induced by radiation therapy caused upregulation of PD-L1 in tumor cells, macrophages, and DCs.[69] In addition to delayed growth with combination therapy, this study also demonstrated that irradiation in combination with anti-PD-1 therapy resulted in a decreased accumulation of immunosuppressive MDSCs.

Preclinical studies in experimental glioma models have also examined the efficacy of this therapeutic approach. A notable preclinical study tested the combination of anti-PD-1 treatment with stereotactic radiosurgery in an animal model of high-grade glioma (GL261). Compared with single modality treatment, combination treatment resulted in increased infiltration of CD8[+] T cells and decreased Tregs. Furthermore, combined therapy resulted in twice the survival time seen for single modality-treated counterparts.[70] Belcaid et al. examined the efficacy of combination therapy with focal radiation and CTLA-4 blockade on established intracranial GL261 tumors. All mice treated with monotherapy (anti-CTLA-4 or focal radiation alone) died before day 30, whereas 50% of animals receiving combination therapy were still alive after 30 days.[71] Blockade of novel checkpoints such as TIM-3 combined with radiation therapy also showed improved survival in an animal model. Dual TIM-3 and PD-1 inhibition plus radiotherapy resulted in 100% overall survival in a murine glioma model system (GL261).[59] These supportive data spurred multiple clinical trials investigating this combination strategy (NCT02648633, NCT02313272, NCT02866747, NCT0282993).

Tumor vaccines

Checkpoint blockade has also been combined with other immunotherapeutic strategies such as DC vaccination. For vaccine preparation, DCs are extracted from the patient, cultured ex vivo and pulsed with tumor-associated antigens, tumor peptides, whole tumor lysate, glioma stem cells, or transfected tumor-specific RNA.[72] These primed DCs are subsequently reintroduced to the patient to facilitate an antigen-specific T-cell activation. The DC vaccination has been extensively studied in the setting of glioma in both preclinical and clinical studies.[73–75] The DC vaccines are also known to generate an antitumor response through the priming of host DCs. These powerful APCs in turn induce T-cell infiltration and a T cell-mediated immune response.[74,76] This impact on T-cell infiltration makes DC vaccines a rational combination with checkpoint blockade. A study showed that the addition of PD-1 inhibition to DC vaccination (tumor lysate loaded) enhanced the migration and activation of T cells in a murine glioma model. Furthermore, the addition of PD-1 inhibition to vaccination resulted in a significantly longer survival time in mice with established gliomas relative to those treated with either individual monotherapy.[77] Agarwalla et al. also demonstrated that although high-dose anti-CTLA-4 alone was ineffective against well-established tumors, the addition of a whole tumor-cell vaccination (Gvax) significantly improved long-term survival in GL261 tumor-bearing mice.[78] Thus far, there are no open clinical trials evaluating the combination of DCs with immune checkpoint inhibitors.

Oncolytic virotherapy

Similar to DC vaccination, oncolytic virotherapy can also result in a robust immune-stimulatory response, albeit by a different mechanism. Oncolytic viruses selectively replicate in and kill tumor cells, spreading in the tumor while sparing normal tissue. They promote an antitumor response by direct tumor lysis, resulting in the spreading of tumor-associated antigens that can be presented by APCs to tumor-infiltrating lymphocytes.[79,80] Several studies in a variety of solid malignancies have demonstrated that oncolytic virotherapy plus checkpoint modulation can be an effective combined strategy.[81–83] For example, intratumoral injection of an oncolytic Newcastle disease virus into murine B16 melanomas implanted in mouse flanks was shown to induce a robust inflammatory response and delayed tumor growth. Combination of this virus with systemic CTLA-1 blockade led to the regression of the established tumors and long-term survival in the animals. Moreover, mice treated with dual therapy demonstrated protection against tumor rechallenge in 80% of the cases compared with only 40% of animals treated with anti-CTLA-4 monotherapy.[83]

In gliomas, specifically, preclinical evidence is also accumulating supporting this treatment combination.[84–87] A notable study demonstrated in vivo that measles-based virotherapy combined with PD-1 blockade enhanced the survival of glioma-bearing mice.[84] Saha et al. reported that triple therapy, consisting of anti-PD-1, anti-CTLA-4, plus an oncolytic herpes virus resulted in a significantly increased survival in two murine glioma models. Interestingly, in addition to an increased T-cell influx with combination treatment, macrophage influx was also postulated to contribute to this impressive synergistic effect.[86] In a recent study by Jiang et al., the authors constructed an oncolytic adenovirus expressing the immune co-stimulator OX40 ligand, which induced potent lymphocyte recruitment to the tumor site in vivo. Intratumoral injection of this modified virus (Delta-24-RGDOX) plus anti-PD-L1 therapy showed synergistic inhibition of gliomas and significantly increased animal median survival time.[88] Clinically, a phase II trial is currently evaluating the efficacy of a modified oncolytic adenovirus (DNX-2401; delivered directly) combined with systemic pembrolizumab (NCT02798406).

Challenges to checkpoint modulation in Glioma
Toxicity

The potential toxicity associated with the immune-related adverse events linked to checkpoint blockade in the treatment of glioma has not been fully elucidated. In particular, the integration of checkpoint modulation with standard-of-care therapy is yet to be standardized. Moreover, the optimal timing, dosage, and implementation of checkpoint modulation in the treatment algorithm are not yet well defined, indicating that clinical guidelines are clearly needed.[89] In addition to the innate immunosuppression induced by gliomas, standard chemotherapy and radiation therapy treatments can result in T-cell depletion and inhibit the efficacy of checkpoint blockade.[90,91] Furthermore, the routine administration of steroids to brain tumor patients is a serious issue because it confounds the activity of immunotherapeutic strategies. Glioma patients frequently require steroid treatment preoperatively or postoperatively to combat peritumoral edema, which can cause neurological symptoms. Dexamethasone, the most commonly used agent, has been shown to decrease T-cell infiltration in a dose-dependent manner in vivo.[92,93]

Accurate evaluation of treatment response

A unique challenge in immunology and its implementation is accurate evaluation of a radiographically observed response to treatment, which is critical for accurate therapeutic monitoring. Problems here mostly stem from therapy-induced inflammation, which can mimic the radiographic features of tumor progression, making it difficult to distinguish true progression from pseudo-progression. As such, the standard RANO (Response Assessment in Neuro-Oncology) criteria,[94] used to assess glioma treatment response have recently been modified to account for potentially confounding immune-related changes.[95]

Preclinical models

One of the greatest challenges in immunotherapy research is lack of an ideal animal model system. Immunocompetent models of glioma are in limited supply. The majority of preclinical studies rely on orthotopic implantation of GL261, a high-grade glioma cell line. This is the model most commonly used to investigate the glioma-immune interaction. Direct implantation of this cell line is straightforward and reliably results in tumor formation, but there are several limitations to its use. As with all cell lines, it is dependent on culture for production of cells to implant. It has been demonstrated that gliomas cells that are cultured and removed from their native environment are immunologically different from glioma cells immediately assessed ex vivo [e.g., with respect to MHC (major histocompatibility complex) expression and cytokine production], and these changes start as early as in the first passage of cells.[96,97] Furthermore, gene expression profiles from patient tumors relative to their corresponding cultured cells demonstrate alterations in gene expression after growth in vitro.[98] Additionally, mouse tumors are typically less necrotic and smaller, thus making the delivery of therapeutics easier. Transgenic models that have been engineered to spontaneously develop gliomas in an immunocompetent host have been increasingly utilized to study immunology. Even though such models have several advantages over a clonotypic model, latency of tumor development and cost may be prohibitive factors to their use.[99] Additionally, the genetic alterations needed to produce tumors may impact immune function.[97]

Oligodendrogliomas and immunotherapy: Future directions

Data regarding the potential of immunotherapy in the treatment of oligodendrogliomas are very limited and markedly understudied. Garber et al.[53] performed one of the few studies addressing PD-1/PD-L1 expression in oligodendrogliomas. Of 19 grade II oligodendrogliomas, none was positive for PD-L1 or PD-1+ tumor-infiltrating lymphocytes. Furthermore, among grade III oligodendrogliomas, only one of nine tumors displayed positive PD-1 tumor-infiltrating lymphocytes.[53] These data indicate that exclusively targeting PD-L1 is likely to be an unsuccessful strategy in oligodendroglioma. In regard to mutational load, even though oligodendroglioma is not specifically addressed, Hodges et al. assessed the correlation between mutational load and tumor grade and reported that overall, 91% of grade II lesions carried a low mutational burden (10 or less mutations per 1.4 Mb). Moreover, only 11.8% of grade III tumors carried a high mutational load (20 mutations per 1.4 Mb).[100] In reference to newer immune checkpoints such as LAG3 and TIM3, there is almost no knowledge about their expression, activity, or activating ligands in oligodendroglioma. A notable study by Kohanbash et al. reported that IDH mutations in glioma resulted in decreased T-cell accumulation via decreased production of STAT1.[101] This is another avenue that may warrant additional investigation in oligodendroglioma. It is clear that future studies should focus on a comprehensive analysis of the immune microenvironment of oligodendroglioma to identify potential targets for immunotherapeutic intervention. Additional immunocompetent models of oligodendroglioma are also needed for preclinical testing. Currently, the transgenic mouse system (RCAS/Ntv-A) has been used with success. Specifically, when the PDGFB transgene is introduced, there is reproducible and specific induction of oligodendrogliomas.[102–104] The RCAS/Ntv-A system is particularly valuable for immune investigation, as animals typically survive for 3 months, which gives them adequate time to mount an immune response. This model has been previously used to study glioma immunity,[105] but further studies focusing on oligodendroglioma are warranted.

In summary, oligodendroglioma remains an incurable disease. Immunotherapeutic strategies, such as checkpoint inhibition, have been under heavy preclinical and clinical investigation in high-grade gliomas, particularly astrocytomas. Yet there is limited knowledge regarding the innate immune infiltration of oligodendrogliomas and subsequently, the activity of immune checkpoints in this disease. A detailed examination of immune-modulatory mechanisms present in oligodendrogliomas will be a key step in understanding which immunotherapeutic agents would be efficacious for treatment.

References

1. Hodi FS, Chesney J, Pavlick AC, et al. Combined nivolumab and ipilimumab versus ipilimumab alone in patients with advanced melanoma: 2-year overall survival outcomes in a multicentre, randomised, controlled, phase 2 trial. *Lancet Oncol.* 2016;17(11):1558–1568.
2. Wolchok JD, Rollin L, Larkin J. Nivolumab and Ipilimumab in advanced melanoma. *N Engl J Med.* 2017;377(25):2503–2504.
3. Garon EB, Rizvi NA, Hui R, et al. Pembrolizumab for the treatment of non-small-cell lung cancer. *N Engl J Med.* 2015;372(21):2018–2028.
4. Reck M, Rodriguez-Abreu D, Robinson AG, et al. Pembrolizumab versus chemotherapy for PD-L1-positive non-small-cell lung Cancer. *N Engl J Med.* 2016;375(19):1823–1833.
5. Greenwald RJ, Freeman GJ, Sharpe AH. The B7 family revisited. *Annu Rev Immunol.* 2005;23:515–548.
6. Sharma P, Allison JP. The future of immune checkpoint therapy. *Science.* 2015;348(6230):56–61.
7. Topalian SL, Drake CG, Pardoll DM. Immune checkpoint blockade: A common denominator approach to cancer therapy. *Cancer Cell.* 2015;27(4):450–461.
8. Dong H, Zhu G, Tamada K, Chen L. B7-H1, a third member of the B7 family, co-stimulates T-cell proliferation and interleukin-10 secretion. *Nat Med.* 1999;5(12):1365–1369.
9. Huang J, Liu F, Liu Z, et al. Immune checkpoint in Glioblastoma: Promising and challenging. *Front Pharmacol.* 2017;8:242.
10. Jackson CM, Lim M, Drake CG. Immunotherapy for brain cancer: Recent progress and future promise. *Clin Cancer Res.* 2014;20(14):3651–3659.
11. Taube JM, Klein A, Brahmer JR, et al. Association of PD-1, PD-1 ligands, and other features of the tumor immune microenvironment with response to anti-PD-1 therapy. *Clin Cancer Res.* 2014;20(19):5064–5074.
12. Garon EB. Current perspectives in immunotherapy for non-small cell lung Cancer. *Semin Oncol.* 2015;42(Suppl 2):S11–S18.
13. Robert C, Schachter J, Long GV, et al. Pembrolizumab versus Ipilimumab in advanced melanoma. *N Engl J Med.* 2015;372(26):2521–2532.
14. Alegre ML, Frauwirth KA, Thompson CB. T-cell regulation by CD28 and CTLA-4. *Nat Rev Immunol.* 2001;1(3):220–228.
15. Collins AV, Brodie DW, Gilbert RJ, et al. The interaction properties of costimulatory molecules revisited. *Immunity.* 2002;17(2):201–210.
16. Bour-Jordan H, Esensten JH, Martinez-Llordella M, Penaranda C, Stumpf M, Bluestone JA. Intrinsic and extrinsic control of peripheral T-cell tolerance by costimulatory molecules of the CD28/ B7 family. *Immunol Rev.* 2011;241(1):180–205.
17. Chikuma S, Imboden JB, Bluestone JA. Negative regulation of T cell receptor-lipid raft interaction by cytotoxic T lymphocyte-associated antigen 4. *J Exp Med.* 2003;197(1):129–135.
18. Rudd CE, Taylor A, Schneider H. CD28 and CTLA-4 coreceptor expression and signal transduction. *Immunol Rev.* 2009;229(1):12–26.
19. Topalian SL, Sharpe AH. Balance and imbalance in the immune system: Life on the edge. *Immunity.* 2014;41(5):682–684.
20. Fecci PE, Ochiai H, Mitchell DA, et al. Systemic CTLA-4 blockade ameliorates glioma-induced changes to the CD4 + T cell compartment without affecting regulatory T-cell function. *Clin Cancer Res.* 2007;13(7):2158–2167.
21. Hodi FS, O'Day SJ, McDermott DF, et al. Improved survival with ipilimumab in patients with metastatic melanoma. *N Engl J Med.* 2010;363(8):711–723.
22. Robert C, Thomas L, Bondarenko I, et al. Ipilimumab plus dacarbazine for previously untreated metastatic melanoma. *N Engl J Med.* 2011;364(26):2517–2526.
23. Workman CJ, Vignali DA. The CD4-related molecule, LAG-3 (CD223), regulates the expansion of activated T cells. *Eur J Immunol.* 2003;33(4):970–979.
24. Huang CT, Workman CJ, Flies D, et al. Role of LAG-3 in regulatory T cells. *Immunity.* 2004;21(4):503–513.
25. Triebel F. LAG-3: A regulator of T-cell and DC responses and its use in therapeutic vaccination. *Trends Immunol.* 2003;24(12):619–622.
26. Zhu C, Anderson AC, Kuchroo VK. TIM-3 and its regulatory role in immune responses. *Curr Top Microbiol Immunol.* 2011;350:1–15.
27. Han S, Feng S, Xu L, et al. Tim-3 on peripheral CD4(+) and CD8(+) T cells is involved in the development of glioma. *DNA Cell Biol.* 2014;33(4):245–250.
28. Liu Z, Han H, He X, et al. Expression of the galectin-9-Tim-3 pathway in glioma tissues is associated with the clinical manifestations of glioma. *Oncol Lett.* 2016;11(3):1829–1834.

29. Li X, Wang B, Gu L, et al. Tim-3 expression predicts the abnormal innate immune status and poor prognosis of glioma patients. *Clin Chim Acta.* 2018;476:178–184.

30. Le DT, Uram JN, Wang H, et al. PD-1 blockade in tumors with mismatch-repair deficiency. *N Engl J Med.* 2015;372(26):2509–2520.

31. McGranahan N, Furness AJ, Rosenthal R, et al. Clonal neoantigens elicit T cell immunoreactivity and sensitivity to immune checkpoint blockade. *Science.* 2016;351(6280):1463–1469.

32. Rizvi NA, Hellmann MD, Snyder A, et al. Cancer immunology. Mutational landscape determines sensitivity to PD-1 blockade in non-small cell lung cancer. *Science.* 2015;348(6230):124–128.

33. Snyder A, Makarov V, Merghoub T, et al. Genetic basis for clinical response to CTLA-4 blockade in melanoma. *N Engl J Med.* 2014; 371(23):2189–2199.

34. Van Allen EM, Miao D, Schilling B, et al. Genomic correlates of response to CTLA-4 blockade in metastatic melanoma. *Science.* 2015; 350(6257):207–211.

35. Draaisma K, Wijnenga MM, Weenink B, et al. PI3 kinase mutations and mutational load as poor prognostic markers in diffuse glioma patients. *Acta Neuropathol Commun.* 2015;3:88.

36. Hodges TR, Ott M, Xiu J, et al. Mutational burden, immune checkpoint expression, and mismatch repair in glioma: Implications for immune checkpoint immunotherapy. *Neuro Oncol.* 2017;19(8):1047–1057.

37. genes CGARNCgcdhg, core p. *Nature.* 2008;455(7216):1061–1068.

38. Johanns TM, Miller CA, Dorward IG, et al. Immunogenomics of Hypermutated Glioblastoma: A patient with Germline POLE deficiency treated with checkpoint blockade immunotherapy. *Cancer Discov.* 2016;6(11):1230–1236.

39. Johnson BE, Mazor T, Hong C, et al. Mutational analysis reveals the origin and therapy-driven evolution of recurrent glioma. *Science.* 2014; 343(6167):189–193.

40. Kim H, Zheng S, Amini SS, et al. Whole-genome and multisector exome sequencing of primary and post-treatment glioblastoma reveals patterns of tumor evolution. *Genome Res.* 2015;25(3):316–327.

41. Cahill DP, Levine KK, Betensky RA, et al. Loss of the mismatch repair protein MSH6 in human glioblastomas is associated with tumor progression during temozolomide treatment. *Clin Cancer Res.* 2007;13(7):2038–2045.

42. van Thuijl HF, Mazor T, Johnson BE, et al. Evolution of DNA repair defects during malignant progression of low-grade gliomas after temozolomide treatment. *Acta Neuropathol.* 2015;129(4):597–607.

43. Le DT, Durham JN, Smith KN, et al. Mismatch repair deficiency predicts response of solid tumors to PD-1 blockade. *Science.* 2017;357(6349): 409–413.

44. Westdorp H, Kolders S, Hoogerbrugge N, de Vries IJM, Jongmans MCJ, Schreibelt G. Immunotherapy holds the key to cancer treatment and prevention in constitutional mismatch repair deficiency (CMMRD) syndrome. *Cancer Lett.* 2017;403:159–164.

45. Bouffet E, Larouche V, Campbell BB, et al. Immune checkpoint inhibition for Hypermutant Glioblastoma Multiforme resulting from Germline Biallelic mismatch repair deficiency. *J Clin Oncol.* 2016;34(19):2206–2211.

46. Bourdais R, Rousseau B, Pujals A, et al. Polymerase proofreading domain mutations: New opportunities for immunotherapy in hypermutated colorectal cancer beyond MMR deficiency. *Crit Rev Oncol Hematol.* 2017;113:242–248.

47. Gong J, Wang C, Lee PP, Chu P, Fakih M. Response to PD-1 blockade in microsatellite stable metastatic colorectal Cancer harboring a POLE mutation. *J Natl Compr Canc Netw.* 2017;15(2):142–147.

48. Howitt BE, Shukla SA, Sholl LM, et al. Association of Polymerase e-mutated and microsatellite-instable endometrial cancers with Neoantigen load, number of tumor-infiltrating lymphocytes, and expression of PD-1 and PD-L1. *JAMA Oncol.* 2015;1(9):1319–1323.

49. Erson-Omay EZ, Caglayan AO, Schultz N, et al. Somatic POLE mutations cause an ultramutated giant cell high-grade glioma subtype with better prognosis. *Neuro Oncol.* 2015;17(10):1356–1364.

50. Topalian SL, Hodi FS, Brahmer JR, et al. Safety, activity, and immune correlates of anti-PD-1 antibody in cancer. *N Engl J Med.* 2012;366(26): 2443–2454.

51. Daud AI, Loo K, Pauli ML, et al. Tumor immune profiling predicts response to anti-PD-1 therapy in human melanoma. *J Clin Invest.* 2016;126(9): 3447–3452.

52. Nduom EK, Wei J, Yaghi NK, et al. PD-L1 expression and prognostic impact in glioblastoma. *Neuro Oncol.* 2016;18(2):195–205.

53. Garber ST, Hashimoto Y, Weathers SP, et al. Immune checkpoint blockade as a potential therapeutic target: Surveying CNS malignancies. *Neuro Oncol.* 2016;18(10):1357–1366.

54. Tumeh PC, Harview CL, Yearley JH, et al. PD-1 blockade induces responses by inhibiting adaptive immune resistance. *Nature.* 2014; 515(7528):568–571.

55. Berghoff AS, Kiesel B, Widhalm G, et al. Programmed death ligand 1 expression and tumor-infiltrating lymphocytes in glioblastoma. *Neuro Oncol.* 2015;17(8):1064–1075.

56. Larkin J, Hodi FS, Wolchok JD. Combined Nivolumab and Ipilimumab or Monotherapy in untreated melanoma. *N Engl J Med.* 2015;373(13): 1270–1271.

57. Wolchok JD, Chiarion-Sileni V, Gonzalez R, et al. Overall survival with combined Nivolumab and Ipilimumab in advanced melanoma. *N Engl J Med.* 2017;377(14):1345–1356.

58. Reardon DA, Gokhale PC, Klein SR, et al. Glioblastoma eradication following immune checkpoint blockade in an Orthotopic, Immunocompetent model. *Cancer Immunol Res.* 2016;4(2):124–135.

59. Kim JE, Patel MA, Mangraviti A, et al. Combination therapy with anti-PD-1, anti-TIM-3, and focal radiation results in regression of murine Gliomas. *Clin Cancer Res.* 2017;23(1):124–136.

60. Mitsuka K, Kawataki T, Satoh E, Asahara T, Horikoshi T, Kinouchi H. Expression of indoleamine 2,3-dioxygenase and correlation with pathological malignancy in gliomas. *Neurosurgery*. 2013;72(6):1031–1038 [discussion 1038-1039].

61. Uyttenhove C, Pilotte L, Theate I, et al. Evidence for a tumoral immune resistance mechanism based on tryptophan degradation by indoleamine 2,3-dioxygenase. *Nat Med*. 2003;9(10):1269–1274.

62. Wainwright DA, Chang AL, Dey M, et al. Durable therapeutic efficacy utilizing combinatorial blockade against IDO, CTLA-4, and PD-L1 in mice with brain tumors. *Clin Cancer Res*. 2014;20(20):5290–5301.

63. Derer A, Frey B, Fietkau R, Gaipl US. Immune-modulating properties of ionizing radiation: Rationale for the treatment of cancer by combination radiotherapy and immune checkpoint inhibitors. *Cancer Immunol Immunother*. 2016;65(7):779–786.

64. Esposito A, Criscitiello C, Curigliano G. Immune checkpoint inhibitors with radiotherapy and locoregional treatment: Synergism and potential clinical implications. *Curr Opin Oncol*. 2015;27(6):445–451.

65. Ngiow SF, McArthur GA, Smyth MJ. Radiotherapy complements immune checkpoint blockade. *Cancer Cell*. 2015;27(4):437–438.

66. Twyman-Saint Victor C, Rech AJ, Maity A, et al. Radiation and dual checkpoint blockade activate non-redundant immune mechanisms in cancer. *Nature*. 2015;520(7547):373–377.

67. Demaria S, Ng B, Devitt ML, et al. Ionizing radiation inhibition of distant untreated tumors (abscopal effect) is immune mediated. *Int J Radiat Oncol Biol Phys*. 2004;58(3):862–870.

68. Sharabi AB, Nirschl CJ, Kochel CM, et al. Stereotactic radiation therapy augments antigen-specific PD-1-mediated antitumor immune responses via cross-presentation of tumor antigen. *Cancer Immunol Res*. 2015;3(4):345–355.

69. Deng L, Liang H, Burnette B, et al. Irradiation and anti-PD-L1 treatment synergistically promote antitumor immunity in mice. *J Clin Invest*. 2014; 124(2):687–695.

70. Zeng J, See AP, Phallen J, et al. Anti-PD-1 blockade and stereotactic radiation produce long-term survival in mice with intracranial gliomas. *Int J Radiat Oncol Biol Phys*. 2013;86(2):343–349.

71. Belcaid Z, Phallen JA, Zeng J, et al. Focal radiation therapy combined with 4-1BB activation and CTLA-4 blockade yields long-term survival and a protective antigen-specific memory response in a murine glioma model. *PLoS One*. 2014;9(7):e101764.

72. Srinivasan VM, Ferguson SD, Lee S, Weathers SP, Kerrigan BCP, Heimberger AB. Tumor vaccines for malignant Gliomas. *Neurotherapeutics*. 2017;14(2):345–357.

73. Yamanaka R, Yajima N, Abe T, et al. Dendritic cell-based glioma immunotherapy (review). *Int J Oncol*. 2003;23(1):5–15.

74. Yu JS, Liu G, Ying H, Yong WH, Black KL, Wheeler CJ. Vaccination with tumor lysate-pulsed dendritic cells elicits antigen-specific, cytotoxic T-cells in patients with malignant glioma. *Cancer Res*. 2004;64(14):4973–4979.

75. Yu JS, Wheeler CJ, Zeltzer PM, et al. Vaccination of malignant glioma patients with peptide-pulsed dendritic cells elicits systemic cytotoxicity and intracranial T-cell infiltration. *Cancer Res*. 2001;61(3):842–847.

76. Antonios JP, Everson RG, Liau LM. Dendritic cell immunotherapy for brain tumors. *J Neurooncol*. 2015;123(3):425–432.

77. Antonios JP, Soto H, Everson RG, et al. PD-1 blockade enhances the vaccination-induced immune response in glioma. *JCI Insight*. 2016;1(10).

78. Agarwalla P, Barnard Z, Fecci P, Dranoff G, Curry Jr WT. Sequential immunotherapy by vaccination with GM-CSF-expressing glioma cells and CTLA-4 blockade effectively treats established murine intracranial tumors. *J Immunother*. 2012;35(5):385–389.

79. Aurelian L. Oncolytic viruses as immunotherapy: progress and remaining challenges. *Onco Targets Ther*. 2016;9:2627–2637.

80. Kaufman HL, Kohlhapp FJ, Zloza A. Oncolytic viruses: A new class of immunotherapy drugs. *Nat Rev Drug Discov*. 2015;14(9):642–662.

81. Hou W, Sampath P, Rojas JJ, Thorne SH. Oncolytic virus-mediated targeting of PGE2 in the tumor alters the immune status and sensitizes established and resistant tumors to immunotherapy. *Cancer Cell*. 2016;30(1):108–119.

82. Rojas JJ, Sampath P, Hou W, Thorne SH. Defining effective combinations of immune checkpoint blockade and Oncolytic Virotherapy. *Clin Cancer Res*. 2015;21(24):5543–5551.

83. Zamarin D, Holmgaard RB, Subudhi SK, et al. Localized oncolytic virotherapy overcomes systemic tumor resistance to immune checkpoint blockade immunotherapy. *Sci Transl Med*. 2014;6(226):226ra232.

84. Hardcastle J, Mills L, Malo CS, et al. Immunovirotherapy with measles virus strains in combination with anti-PD-1 antibody blockade enhances antitumor activity in glioblastoma treatment. *Neuro Oncol*. 2017;19(4):493–502.

85. Panek WK, Kane JR, Young JS, et al. Hitting the nail on the head: Combining oncolytic adenovirus-mediated virotherapy and immunomodulation for the treatment of glioma. *Oncotarget*. 2017;8(51):89391–89405.

86. Saha D, Martuza RL, Rabkin SD. Macrophage polarization contributes to Glioblastoma eradication by combination Immunovirotherapy and immune checkpoint blockade. *Cancer Cell*. 2017;32(2):253–267 [e255].

87. Saha D, Martuza RL, Rabkin SD. Curing glioblastoma: Oncolytic HSV-IL12 and checkpoint blockade. *Oncoscience*. 2017;4(7–8):67–69.

88. Jiang H, Rivera-Molina Y, Gomez-Manzano C, et al. Oncolytic adenovirus and tumor-targeting immune modulatory therapy improve autologous Cancer vaccination. *Cancer Res*. 2017;77(14):3894–3907.

89. Tan AC, Heimberger AB, Khasraw M. Immune checkpoint inhibitors in Gliomas. *Curr Oncol Rep*. 2017;19(4):23.

90. Grossman SA, Ye X, Lesser G, et al. Immunosuppression in patients with high-grade gliomas treated with radiation and temozolomide. *Clin Cancer Res*. 2011;17(16):5473–5480.

91. Yovino S, Kleinberg L, Grossman SA, Narayanan M, Ford E. The etiology of treatment-related lymphopenia in patients with malignant gliomas: Modeling radiation dose to circulating lymphocytes explains clinical observations and suggests methods of modifying the impact of radiation on immune cells. *Cancer Invest*. 2013;31(2):140–144.

92. Badie B, Schartner JM, Paul J, Bartley BA, Vorpahl J, Preston JK. Dexamethasone-induced abolition of the inflammatory response in an experimental glioma model: A flow cytometry study. *J Neurosurg*. 2000;93(4):634–639.

93. Benedetti S, Pirola B, Poliani PL, et al. Dexamethasone inhibits the anti-tumor effect of interleukin 4 on rat experimental gliomas. *Gene Ther.* 2003; 10(2):188–192.

94. Wen PY, Macdonald DR, Reardon DA, et al. Updated response assessment criteria for high-grade gliomas: Response assessment in neuro-oncology working group. *J Clin Oncol.* 2010;28(11):1963–1972.

95. Okada H, Weller M, Huang R, et al. Immunotherapy response assessment in neuro-oncology: A report of the RANO working group. *Lancet Oncol.* 2015;16(15):e534–e542.

96. Anderson RC, Elder JB, Brown MD, et al. Changes in the immunologic phenotype of human malignant glioma cells after passaging in vitro. *Clin Immunol.* 2002;102(1):84–95.

97. Sughrue ME, Yang I, Kane AJ, et al. Immunological considerations of modern animal models of malignant primary brain tumors. *J Transl Med.* 2009;7:84.

98. Li A, Walling J, Kotliarov Y, et al. Genomic changes and gene expression profiles reveal that established glioma cell lines are poorly representative of primary human gliomas. *Mol Cancer Res.* 2008;6(1):21–30.

99. Rankin SL, Zhu G, Baker SJ. Review: Insights gained from modelling high-grade glioma in the mouse. *Neuropathol Appl Neurobiol.* 2012; 38(3):254–270.

100. Hodges TR, Ott M, Xiu J, et al. Mutational burden, immune checkpoint expression, and mismatch repair in glioma: implications for immune check-point immunotherapy. Neuro Oncol. (in press).

101. Kohanbash G, Carrera DA, Shrivastav S, et al. Isocitrate dehydrogenase mutations suppress STAT1 and CD8 + T cell accumulation in gliomas. *J Clin Invest.* 2017;127(4):1425–1437.

102. Appolloni I, Calzolari F, Tutucci E, et al. PDGF-B induces a homogeneous class of oligodendrogliomas from embryonic neural progenitors. *Int J Cancer.* 2009;124(10):2251–2259.

103. Doucette T, Yang Y, Zhang W, et al. Bcl-2 promotes malignant progression in a PDGF-B-dependent murine model of oligodendroglioma. *Int J Cancer.* 2011;129(9):2093–2103.

104. Moore LM, Holmes KM, Smith SM, et al. IGFBP2 is a candidate biomarker for Ink4a-Arf status and a therapeutic target for high-grade gliomas. *Proc Natl Acad Sci U S A.* 2009;106(39):16675–16679.

105. Xu S, Wei J, Wang F, et al. Effect of miR-142-3p on the M2 macrophage and therapeutic efficacy against murine glioblastoma. *J Natl Cancer Inst.* 2014;106(8).

Index

Note: Page numbers followed by *f* indicate figures and *t* indicate tables.

CPI Antony Rowe
Eastbourne, UK
December 10, 2019